MW00974203

METHODS IN MOLECULAR BIOLOGY™

For further volumes:
http://www.springer.com/series/7651

Computational Toxicology

Volume I

Edited by

Brad Reisfeld

Chemical and Biological Engineering & School of Biomedical Engineering
Colorado State University, Colorado, USA

Arthur N. Mayeno

Chemical and Biological Engineering,
Colorado State University, Colorado, USA

Editors
Brad Reisfeld
Chemical and Biological Engineering
& School of Biomedical Engineering
Colorado State University
Colorado, USA

Arthur N. Mayeno
Chemical and Biological Engineering
Colorado State University
Colorado, USA

ISSN 1064-3745 ISSN 1940-6029 (electronic)
ISBN 978-1-62703-049-6 ISBN 978-1-62703-050-2 (eBook)
DOI 10.1007/978-1-62703-050-2
Springer New York Heidelberg Dordrecht London

Library of Congress Control Number: 2012947026

Printed on acid-free paper

Humana Press is a brand of Springer
Springer is part of Springer Science+Business Media (www.springer.com)

Preface

Rapid advances in computer science, biology, chemistry, and other disciplines are enabling powerful new computational tools and models for toxicology and pharmacology. These computational tools hold tremendous promise for advancing applied and basic science, from streamlining drug efficacy and safety testing to increasing the efficiency and effectiveness of risk assessment for environmental chemicals. These approaches also offer the potential to improve experimental design, reduce the overall number of experimental trials needed, and decrease the number of animals used in experimentation.

Computational approaches are ideally suited to organize, process, and analyze the vast libraries and databases of scientific information and to simulate complex biological phenomena. For instance, they allow researchers to (1) investigate toxicological and pharmacological phenomena across a wide range of scales of biological organization (molecular ↔ cellular ↔ organism), (2) incorporate and analyze multiple biochemical and biological interactions, (3) simulate biological processes and generate hypotheses based on model predictions, which can be tested via targeted experimentation in vitro or in vivo, (4) explore the consequences of inter- and intra-species differences and population variability on the toxicology and pharmacology, and (5) extrapolate biological responses across individuals, species, and a range of dose levels.

Despite the exceptional promise of computational approaches, there are presently very few resources that focus on providing guidance on the development and practice of these tools to solve problems and perform analyses in this area. This volume was conceived as part of the Methods in Molecular Biology series to meet this need and to provide both biomedical and quantitative scientists with essential background, context, examples, useful tips, and an overview of current developments in the field. To this end, we present a collection of practical techniques and software in computational toxicology, illustrated with relevant examples drawn principally from the fields of environmental and pharmaceutical sciences. These computational techniques can be used to analyze and simulate a myriad of multi-scale biochemical and biological phenomena occurring in humans and other animals following exposure to environmental toxicants or dosing with drugs.

This book (the first in a two-volume set) is organized into four parts each covering a methodology or topic, subdivided into chapters that provide background, theory, and illustrative examples. Each part is generally self-contained, allowing the reader to start with any part, although some knowledge of concepts from other parts may be assumed. Part I introduces the field of computational toxicology and its current or potential applications. Part II outlines the principal elements of mathematical and computational modeling, and accepted best practices and useful guidelines. Part III discusses the use of computational techniques and databases to predict chemical properties and toxicity, as well as the use of molecular dynamics. Part IV delineates the elements and approaches to pharmacokinetic and pharmacodynamic modeling, including non-compartmental and compartmental modeling, modeling of absorption, prediction of pharmacokinetic parameters, physiologically based pharmacokinetic modeling, and mechanism-based pharmacodynamic modeling; chemical mixture and population effects, as well as interspecies extrapolation, are also described and illustrated.

Although a complete picture of toxicological risk often involves an analysis of environmental transport, we believe that this expansive topic is beyond the scope of this volume, and it will not be covered here; overviews of computational techniques in this area are contained in a variety of excellent references [1–4].

Computational techniques are increasingly allowing scientists to gain new insights into toxicological phenomena, integrate (and interpret) the results from a wide variety of experiments, and develop more rigorous and quantitative means of assessing chemical safety and toxicity. Moreover, these techniques can provide valuable insights before initiating expensive laboratory experiments and into phenomena not easily amenable to experimental analysis, e.g., detection of highly reactive, transient, or trace-level species in biological milieu. We believe that the unique collection of explanatory material, software, and illustrative examples in Computational Toxicology will allow motivated readers to participate in this exciting field and undertake a diversity of realistic problems of interest.

We would like to express our sincere thanks to our authors whose enthusiasm and diverse contributions have made this project possible.

Colorado, USA

Brad Reisfeld
Arthur N. Mayeno

References

1. Clark, M.M., Transport modeling for environmental engineers and scientists. 2nd ed. 2009, Hoboken, N.J.: Wiley.
2. Hemond, H.F. and E.J. Fechner-Levy, Chemical fate and transport in the environment. 2nd ed. 2000, San Diego: Academic Press. xi, 433 p.
3. Logan, B.E., Environmental transport processes. 1999, New York: Wiley. xiii, 654 p.
4. Nirmalakhandan, N., Modeling tools for environmental engineers and scientists. 2002, Boca Raton, Fla.: CRC Press. xi, 312 p.

Contents

Part IV Pharmacokinetic and Pharmacodynamic Modeling

List of Contributors

HERVÉ ABDI • *School of Behavioral and Brain Sciences, The University of Texas at Dallas, Richardson, TX, USA*
BILLY AMZAL • *LA-SER Europe Ltd, London, UK*
MELVIN E. ANDERSEN • *The Hamner Institutes for Health Sciences, Research Triangle Park, NC, USA*
JAMES B. BASSINGTHWAIGHTE • *Department of Bioengineering, University of Washington, Seattle, WA, USA*
FRÉDÉRIC Y. BOIS • *Royallieu Research Center, Technological University of Compiegne, Compiegne, France; INERIS, DRC/VIVA/METO, Verneuil en Halatte, France*
MICHAEL B. BOLGER • *Simulations Plus, Inc., Lancaster, CA, USA*
ERIK BUTTERWORTH • *Department of Bioengineering, University of Washington, Seattle, WA, USA*
JERRY L. CAMPBELL JR. • *The Hamner Institutes for Health Sciences, Research Triangle Park, NC, USA*
DANIEL T. CHANG • *National Exposure Research Laboratory, US Environmental Protection Agency, Research Triangle Park, NC, USA*
XIAOLIN CHENG • *Oak Ridge National Laboratory, UT/ORNL Center for Molecular Biophysics, Oak Ridge, TN, USA; Department of Biochemistry and Cellular and Molecular Biology, University of Tennessee, Knoxville, TN, USA*
HARVEY J. CLEWELL III • *The Hamner Institutes for Health Sciences, Research Triangle Park, NC, USA*
REBECCA A. CLEWELL • *The Hamner Institutes for Health Sciences, Research Triangle Park, NC, USA*
JEAN PAUL COMET • *I3S laboratory, UMR 6070 CNRS, University of Nice-Sophia Antipolis, Sophia Antipolis, France*
RORY CONOLLY • *National Health and Environmental Effects Research Laboratory, U.S. Environmental Protection Agency, Research Triangle Park, NC, USA*
AMÉLIE CRÉPET • *French Agency for Food, Environment and Occupational Health Safety (ANSES), Maisons-Alfort, France*
CURTIS C. DARY • *National Exposure Research laboratory, US Environmental Protection Agency, Research Triangle Park, NC, USA*
LISETTE G. DE PILLIS • *Department of Mathematics, Harvey Mudd College, Claremont, CA, USA*
JEAN LOU DORNE • *Emerging Risks Unit, European Food Safety Authority, Parma, Italy*
STEPHEN B. DUFFULL • *School of Pharmacy, University of Otago, Otago, New Zealand*
HARISH DUREJA • *M. D. University, Rohtak, India*

SEAN EKINS • *Collaborations in Chemistry, Fuquay Varina, NC, USA; Department of Pharmaceutical Sciences, University of Maryland, Baltimore, MD, USA; Department of Pharmacology, University of Medicine & Dentistry of New Jersey (UMDNJ)-Robert Wood Johnson Medical School, Piscataway, NJ, USA*
MELANIE A. FELMLEE • *Department of Pharmaceutical Sciences, University at Buffalo, State University of New York, Buffalo, NY, USA*
JOHAN GABRIELSSON • *Division of Pharmacology and Toxicology, Department of Biomedical Sciences and Veterinary Public Health, Swedish University of Agricultural Sciences, Uppsala, Sweden*
P. ROBINAN GENTRY • *Environ International Corporation, Monroe, LA, USA*
MICHAEL R. GOLDSMITH • *National Exposure Research Laboratory, US Environmental Protection Agency, Research Triangle Park, NC, USA*
JANNA HASTINGS • *European Bioinformatics Institute, Hinxton, UK*
GEOFFREY K. ISBISTER • *Department of Clinical Toxicology and Pharmacology, Calvary Mater Newcastle, University of Newcastle, Newcastle, NSW, Australia; Discipline of Clinical Pharmacology, University of Newcastle, Newcastle, NSW, Australia*
IVAYLO IVANOV • *Department of Chemistry, Georgia State University, Atlanta, GA, USA*
BARTHOLOMEW JARDINE • *Department of Bioengineering, University of Washington, Seattle, WA, USA*
ZARA JOSEPHS • *European Bioinformatics Institute, Hinxton, UK*
MUTHUKUMARASAMY KARTHIKEYAN • *National Chemical Laboratory, Digital Information Resource Centre & Centre of Excellence in Scientific Computing, Pune, India*
ELAINA KENYON • *Pharmacokinetics Branch, Integrated Systems Toxicology Division, MD B105-03, National Health and Environmental Effects Research Laboratory, Office of Research and Development, U.S. Environmental Protection Agency, Research Triangle Park, NC, USA*
JOHN KERRIGAN • *Cancer Institute of New Jersey, Robert Wood Johnson Medical School, New Brunswick, NJ, USA*
SANDHYA KORTAGERE • *Department of Microbiology and Immunology, Drexel University College of Medicine, Philadelphia, PA, USA*
MATTHEW D. KRASOWSKI • *Department of Pathology, University of Iowa Hospitals and Clinics, Iowa City, IA, USA*
MARKUS LUKACOVA • *Department of Medicinal Chemistry and Molecular Pharmacology, Purdue University, West Lafaytte, IN, USA*
VIERA LUKACOVA • *Simulations Plus, Inc., Lancaster, CA, USA*
A.K. MADAN • *Pt. B.D. Sharma University of Health Sciences, Rohtak, India*
DONALD E. MAGER • *Department of Pharmaceutical Sciences, University at Buffalo, State University of New York, Buffalo, NY, USA*
ARTHUR N. MAYENO • *Department of Chemical and Biological Engineering, Colorado State University, Fort Collins, CO, USA*
MARILYN E. MORRIS • *Department of Pharmaceutical Sciences, University at Buffalo, State University of New York, Buffalo, NY, USA*

SHANE D. PETERSON • *National Exposure Research Laboratory, US Environmental Protection Agency, Research Triangle Park, NC, USA*
AMI E. RADUNSKAYA • *Department of Mathematics, Pomona College, Claremont, CA, USA*
GARY M. RAYMOND • *Department of Bioengineering, University of Washington, Seattle, WA, USA*
BRAD REISFELD • *Department of Chemical and Biological Engineering, Colorado State University, Fort Collins, CO, USA*
OLA SPJUTH • *Department of Pharmaceutical Biosciences, Uppsala University, Uppsala, Sweden; Swedish e-Science Research Center, Royal Institute of Technology, Stockholm, Sweden*
CHRISTOPH STEINBECK • *European Bioinformatics Institute, Hinxton, UK*
YU-MEI TAN • *National Exposure Research Laboratory, U.S. Environmental Protection Agency, Research Triangle Park, NC, USA*
ROGELIO TORNERO-VELEZ • *National Exposure Research Laboratory, U.S. Environmental Protection Agency, Research Triangle Park, NC, USA*
THOMAS R. TRANSUE • *Lockheed Martin Information Technology, Research Triangle Park, NC, USA*
JESSICA TRESSOU • *National Institute for Agronomic Research (INRA), Paris, France*
PAVAN VAJJAH • *School of Pharmacy, University of Otago, Otago, New Zealand; Systems Pharmacology Group, Simcyp Ltd, Sheffield, UK*
PHILIPPE VERGER • *Department of Food Safety and Zoonoses, World Health Organization, Geneva, Switzerland*
RENU VYAS • *Department of Bioinformatics and Computer Science, Dr. D.Y. Patil Biotechnology and Bioinformatics Institute, Pune, India*
DANIEL WEINER • *Division of Certara, Pharsight Corporation, Cary, NC, USA*
ANTONY J. WILLIAMS • *Royal Society of Chemistry, Wake Forest, NC, USA*
EGON L. WILLIGHAGEN • *Department of Pharmaceutical Biosciences, Uppsala University, Uppsala, Sweden; Division of Molecular Toxicology, Institute of Environmental Medicine, Karolinska Institutet, Stockholm, Sweden; Department of Bioinformatics - BiGCaT, Maastricht University Universiteitssingel 50, Maastricht, The Netherlands*
WALTER S. WOLTOSZ • *Simulations Plus, Inc., Lancaster, CA, USA*

Part I

Introduction

Chapter 1

What is Computational Toxicology?

Brad Reisfeld and Arthur N. Mayeno

Abstract

Computational toxicology is a vibrant and rapidly developing discipline that integrates information and data from a variety of sources to develop mathematical and computer-based models to better understand and predict adverse health effects caused by chemicals, such as environmental pollutants and pharmaceuticals. Encompassing medicine, biology, biochemistry, chemistry, mathematics, computer science, engineering, and other fields, computational toxicology investigates the interactions of chemical agents and biological organisms across many scales (e.g., population, individual, cellular, and molecular). This multidisciplinary field has applications ranging from hazard and risk prioritization of chemicals to safety screening of drug metabolites, and has active participation and growth from many organizations, including government agencies, not-for-profit organizations, private industry, and universities.

Key words: Computational toxicology, Computational chemistry, Computational biology, Systems biology, Risk assessment, Safety assessment

1. Introduction

There are over 80,000 chemicals in common use worldwide, and hundreds of new chemicals and chemical mixtures are introduced into commerce each year. Because chemical safety has traditionally been assessed using expensive and time-consuming animal-based toxicity tests, only a small fraction of these chemicals have been adequately assessed for potential risk.

Aside from these environmentally relevant chemicals, drugs are another class of chemicals for which toxicity testing is crucial. For new drugs, toxicity is still the cause of a significant number of candidate failures during later development stages. In the pharmaceutical industry, toxicity assessment is hampered by the large amounts of compound required for the in vivo studies, lack of reliable high-throughput in vitro assays, and inability of in vitro and animal models to correctly predict some human toxicities.

Brad Reisfeld and Arthur N. Mayeno (eds.), *Computational Toxicology: Volume I*, Methods in Molecular Biology, vol. 929, DOI 10.1007/978-1-62703-050-2_1, © Springer Science+Business Media, LLC 2012

For both environmental pollutants and drugs, there is a clear need for alternative approaches to traditional toxicity testing to help predict the potential for toxicity and prioritize testing in light of the limited resources. One such approach is computational toxicology.

2. What is Computational Toxicology?

The US Environmental Protection Agency (EPA) defines Computational Toxicology as "the application of mathematical and computer models to predict adverse effects and to better understand the single or multiple mechanisms through which a given chemical induces harm."

In a larger context, computational toxicology is an emerging multidisciplinary field that combines knowledge of toxicity pathways with relevant chemical and biological data to inform the development, verification, and testing of multi-scale computer-based models that are used to gain insights into the mechanisms through which a given chemical induces harm. Computational toxicology also seeks to manage and detect patterns and interactions in large biological and chemical data sets by taking advantage of high-information-content data, novel biostatistical methods, and computational power to analyze these data.

3. What Are Some Application Areas for Computational Toxicology?

Some of the principal application areas for computational toxicology are (1) hazard and risk prioritization of chemicals, (2) uncovering mechanistic information that is valuable in tailoring testing programs for each chemical, (3) safety screenings of food additives and food contact substances, (4) supporting more sophisticated approaches to aggregate and cumulative risk assessment, (5) estimating the extent of variability in response in the human population, (6) pharmaceutical lead selection in drug development, (7) safety screening and qualification of pharmaceutical contaminants and degradation products, and (8) safety screening of drug metabolites.

4. What Are the Major Fields Comprising Computational Toxicology?

Computational toxicology is highly interdisciplinary. Researchers in the field have backgrounds and training in toxicology, biochemistry, chemistry, environmental sciences, mathematics, statistics, medicine, engineering, biology, computer science, and many other disciplines.

In addition, the development of models in computational toxicology has been supported by the development of numerous "-omics" technologies, which have evolved into a number of scientific disciplines, including genomics, proteomics, metabolomics, transcriptomics, glycomics, and lipomics.

5. Who Uses Computational Toxicology?

A broad spectrum of international organizations are involved in the development, application, and dissemination of knowledge, tools, and data in computational toxicology. These include

- Government agencies in
  - The USA (EPA, Centers for Disease Control, Food and Drug Administration, National Institutes of Health, Agency for Toxic Substances and Disease Registry).
  - Europe (European Chemicals Agency, Institute for Health and Consumer Protection).
  - Canada (Health Canada, National Centre for Occupational Health and Safety Information).
  - Japan (National Institute of Health Sciences of Japan).
- The USA state agencies.
- Not-for-profit organizations.
- National laboratories.
- Nongovernment organizations.
- Military laboratories; private industry.
- Universities.

6. What Are Some Current Areas of Research in Computational Toxicology?

This volume covers a diverse range of applications for computational toxicology. This trend is also reflected by recent publications in the scientific literature. Some of the topics of papers published recently in the area of computational toxicology include

- Computational methods for evaluating genetic toxicology.
- Structure-based predictive toxicology.
- Informatics and machine learning in computational toxicology.
- Estimating toxicity-related biological pathways.
- Computational approaches to assessing human genetic susceptibility.

- Assessing activity profiles of chemicals evaluated across biochemical targets.
- Pharmacokinetic modeling for nanoparticles.
- Quantitative structure–activity relationships in toxicity prediction.
- In silico prediction of carcinogenic potency.
- Virtual tissues in toxicology.
- Public databases supporting computational toxicology.
- Regulatory use of computational toxicology tools and databases.
- Computational approaches to assess the impact of environmental chemicals on key transcription regulators.
- Molecular modeling for screening environmental chemicals for estrogenicity.
- Predicting inhibitors of acetylcholinesterase by machine learning approaches.
- Predicting activation enthalpies of cytochrome-P450-mediated reactions.

7. What Are Likely Future Directions in Computational Toxicology?

Progress in computational toxicology is expected to facilitate the transformative shift in toxicology called for by the US National Research Council in the recent report entitled "Toxicity Testing in the 21st Century: A Vision and a Strategy." Specifically, the following directions are among those that will be critical in enabling this shift:

- Broadening the usage of high-throughput screening methods to evaluate the toxicity for the backlog of thousands of industrial chemicals in the environment.
- Informing computational models through the continued and expanded use of "-omics" technologies.
- Acquiring new biological data at therapeutic or physiologically relevant exposure levels for more realistic computational endpoints.
- Predicting adverse outcomes of environmental chemical and drug exposure in specific human populations.
- Establishing curated and widely accessible databases that include both chemical and biological information.
- Creating models for characterizing gene–environment interactions.

- Developing approaches to predict cellular responses and biologically based dose–response.
- Utilizing toxicogenomics to illuminate mechanisms and bridge genotoxicity and carcinogenicity.
- Incorporating rigorous uncertainty estimates in models and simulations.
- Implementing strategies for utilizing animals more efficiently and effectively in bioassays designed to answer specific questions.
- Generating models to assess the effects of mixtures of chemicals by employing system-level approaches that encompass the underlying biological pathways.
- Developing virtual tissues for toxicological investigations.

Chapter 2

Computational Toxicology: Application in Environmental Chemicals

Yu-Mei Tan, Rory Conolly, Daniel T. Chang, Rogelio Tornero-Velez, Michael R. Goldsmith, Shane D. Peterson, and Curtis C. Dary

Abstract

This chapter provides an overview of computational models that describe various aspects of the source-to-health effect continuum. Fate and transport models describe the release, transportation, and transformation of chemicals from sources of emission throughout the general environment. Exposure models integrate the microenvironmental concentrations with the amount of time an individual spends in these microenvironments to estimate the intensity, frequency, and duration of contact with environmental chemicals. Physiologically based pharmacokinetic (PBPK) models incorporate mechanistic biological information to predict chemical-specific absorption, distribution, metabolism, and excretion. Values of parameters in PBPK models can be measured in vitro, in vivo, or estimated using computational molecular modeling. Computational modeling is also used to predict the respiratory tract dosimetry of inhaled gases and particulates [computational fluid dynamics (CFD) models], to describe the normal and xenobiotic-perturbed behaviors of signaling pathways, and to analyze the growth kinetics of preneoplastic lesions and predict tumor incidence (clonal growth models).

Key words: Computational toxicology, Source-to-effect continuum, Fate and transport, Dosimetry, Signaling pathway, Physiologically based pharmacokinetic model, Biologically based dose response model, Clonal growth model, Virtual tissue

1. Overview

Computational toxicology involves a variety of computational tools including databases, statistical analysis packages, and predictive models. In this chapter, we focus on computational models that describe various aspects of the source-to-health effect continuum (Fig. 1). Literature on the application of computational models across the continuum has been expanding rapidly in recent years. Using the Web of Science portal, we conducted a

Brad Reisfeld and Arthur N. Mayeno (eds.), *Computational Toxicology: Volume I*, Methods in Molecular Biology, vol. 929, DOI 10.1007/978-1-62703-050-2_2, © Springer Science+Business Media, LLC 2012

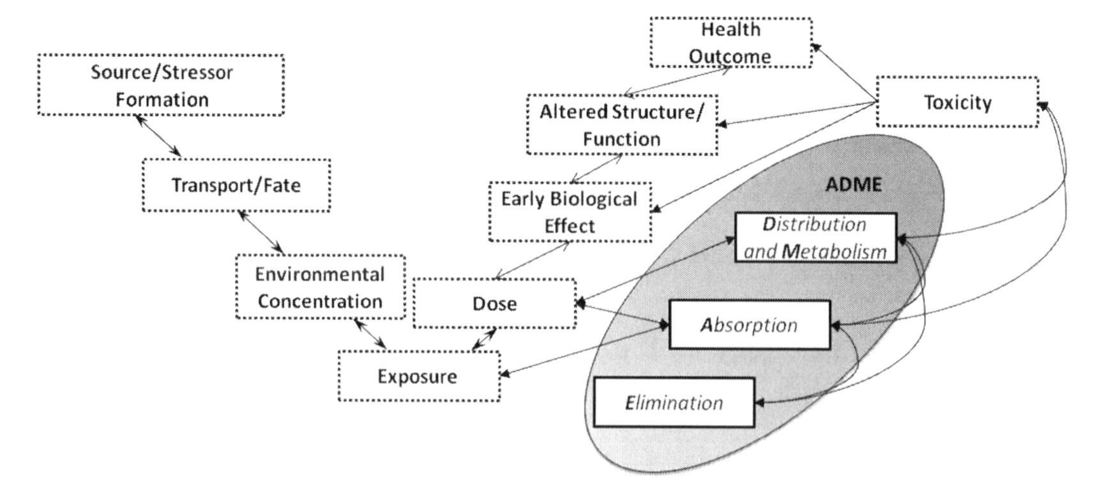

Fig. 1. Major components of the source-to-effect continuum.

bibliometric analysis of publications that appeared between 1970 and 2009. Using the search structure [TS = (computational OR "in silico" OR predictive OR model* OR virtual) AND TS = (toxicology) AND TS = (environment*)] (TS: Topic), a total of 397 articles were found. Adding "NOT pharmaceutic*" to the search structure above, found 371 articles, indicating only a small fraction of the 397 deal with aspects of drug development. A PubMed search (Feb 17, 2011) on "physiologically based pharmacokinetic (PBPK) modeling" found 769 articles, indicating that our search, which focused on computational modeling specifically in environmental toxicology, was quite restrictive.

Literature searches using specific terminology were performed to understand the publication frequency of some of the most common types of modeling used in computational toxicology, including fate and transport, exposure, PBPK, computational fluid dynamic (CFD), signaling pathway, biologically based dose–response (BBDR), and clonal growth modeling. Searches were restricted to original scientific publications only (i.e., reviews were excluded) and fields of science were restricted (e.g., "NOT eco*") in order to focus on applications relevant to human health effects. A yearly breakdown showing publication frequency over time is presented in Fig. 2. The data show a rapid increase in publication frequency for many of the modeling types beginning in the early 1990s and that PBPK, fate and transport and signaling pathways are the most common. BBDR and clonal growth modeling have received considerably less attention.

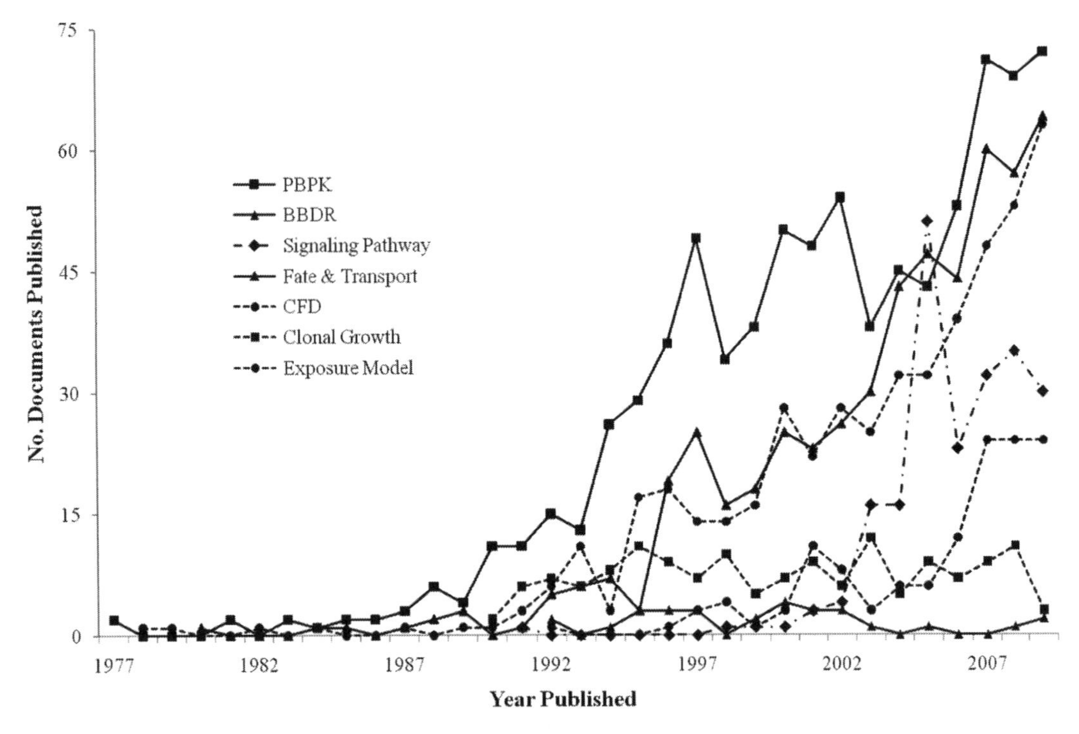

Fig. 2. Literature searches performed to understand publication frequency of common modeling types used in environmental computational toxicology.

2. Computational Models Along the Source-to-Health Effect Continuum

2.1. Fate and Transport

Fate and transport models describe the release, transportation, and transformation of chemicals from sources of emission throughout the general environment. Fate addresses persistence, dissipation, and loss of chemical mass along the migration pathway; and transport addresses mobility of a chemical along the migration pathway (1). Based on their complexity, models of fate and transport can be used for either "screening-level" or "higher-tiered" applications (2). Screening-level models often use default input parameters that tend to over-predict exposures (the preferred default approach used in the absence of data). These models are suitable for obtaining a first approximation or to screen out exposures that are not likely to be of concern (3). Screening-level models have limited spatial and temporal scope. Higher-tiered models are needed when analyses require greater temporal and spatial resolution, but much more information is required, such as site-specific data.

The processes that can be described in fate and transport models include advection, dispersion, diffusion, equilibrium partitioning between solid and fluid, biodegradation, and phase separation of immiscible liquids (1). In general, fate and transport models require information on physicochemical properties; mechanisms of release of chemicals to environmental media; physical, chemical, and biological properties of the media though which migration occurs; and interactions between the chemical and medium (1). For example, typical inputs to an air quality and dispersion model are source data (e.g., emission rates), meteorological data (e.g., temperature), and physicochemical properties of the chemical. Inputs to a surface water model, in addition to source data and physicochemical properties, may include water flows, soil properties and topography, and advective/dispersive movement (2).

2.2. Exposure

The outputs of a fate and transport model are concentrations to which humans may be exposed. These predicted concentrations are then used, in some cases, as surrogates for actual exposure (2). Since these provisional estimates do not provide sufficient resolution about variation of exposure among individuals and by time and location, they can also be used as inputs to exposure models. Exposure models integrate the microenvironmental concentrations with the amount of time an individual spends in these microenvironments to provide qualitative and quantitative evaluations of the intensity, frequency, and duration of contact with chemicals, and sometimes, the resulting amount of chemicals that is actually absorbed into the exposed organism. Exposure models vary considerably in their complexity. Some models are deterministic and generate site of contact-specific point estimates (e.g., dermal concentration × contact time). Others are probabilistic, describing spatial and temporal profiles of chemical concentrations in microenvironments. Both deterministic and probabilistic models may aggregate some or all of the major exposure pathways.

Probabilistic models can also be used to describe variability in human behavior. Human activities contribute to exposure variability, and at first glance appear to be arbitrary, yet patterns of behavior are known to be representative of different age groups (e.g., hand-to-mouth behavior among 3–5 year olds) and this information can be used to better inform stochastic exposure models (4). A major challenge in characterizing human activity is overcoming the cost of collecting information. For example, food consumption questionnaires are important in dietary modeling (e.g., estimating chronic arsenic exposure by shellfish consumption); however the accuracy in assessing chronic exposure is limited by the lack of longitudinal survey information in the surveys such as Continuing Survey of Food Intake by Individuals (CSFII) and National Health and Nutrition Examination Survey (NHANES) (5, 6). The recent study of Song et al. (7) examined how much information is needed in order to predict human behavior. The authors examined the predictability of *macro-scale* human mobility over a span of 3 months based on cell phone use—comparing a continuous record (e.g., hourly) of a user's momentary location with a less expensive measure of mobility. The authors found that there is a potential

93% average predictability in user mobility. This predictability reflects the inherent regularity of human behavior (7) and exemplifies an approach that holds promise for examining aspects of human mobility, thereby reducing the cost of exposure modeling.

The degree of complexity needed in an exposure model depends on (1) the nature of the chemical (e.g., volatility) and (2) the number and complexity of the most common exposure scenarios that the model is required to describe. The number of parameters in the model and their corresponding data needs are functions of model complexity. The first choice for obtaining input parameter data is direct measurement of the environment concentrations and observations of human activity patterns. When these specific data are not available, inputs may be obtained from population-based surveys, such as NHANES or the Exposure Factors Handbook (8). The outputs of fate, transport, and exposure models can serve as inputs to pharmacokinetic models for estimating internal tissue dosimetry.

2.3. Dosimetry

Pharmacokinetic processes translate the exposure or applied dose into a delivered dose at an internal site. Internal doses often correlate better with apical effects than do the external doses due to nonlinear pharmacokinetics (9). Pharmacokinetic data can be obtained from studies using laboratory animals (10) or from controlled human exposures (11, 12). Controlled human exposures are largely reserved for evaluating the safety and efficacy of drugs or therapies not for environmental chemicals. Human observational studies (13–15) may provide some insight into the disposition of environmental chemicals, but the relationship between exposure and systemic levels is obscured because of the complexity of exposure. Furthermore, the lack of control with regards to chemical co-occurrence may confound interpretation of the exposure–dose relationship.

The relationship between exposure to a chemical and its dose at an internal target site is determined by a set of chemical structure-dependent properties (e.g., solubility in water, blood, and tissues, volatility, susceptibility to biotransformation) and corresponding properties of the biological system (e.g., tissue volumes, blood flows, metabolic capabilities). Computational models that describe the minimum set of these characteristics needed to predict chemical-specific absorption, distribution, metabolism, excretion (ADME) are commonly referred to as PBPK models though PBTK, were the T stands for toxicokinetic, is also used. Because models of this type describe the relevant biology that determines ADME, they are useful not only for predicting pharmacokinetic behavior within the dose range and time course of available data but also for extrapolation outside these ranges. These characteristics make these models particularly useful in risk assessments, where extrapolation to doses well below those for which data are available is often necessary (16).

Many of the parameters used in PBPK models can be measured in vitro (17). Obach and colleagues (18, 19) observed that scaling in vitro metabolism data from human liver microsomes to in vivo clearance values yielded predictions that were within 70–80% of actual values. They also found that the clearance predictions were improved by accounting for plasma and microsomal protein binding. Tornero-Velez and colleagues (20) applied the same approach to account for deltamethrin's age-dependent pharmacokinetics in the maturing Sprague-Dawley rats using in vitro parameters for hepatic and plasma metabolic clearance of deltamethrin. Finding agreement between in vitro parameter values and in vivo parameter estimates is one way to explore pharmacokinetic mechanisms and reduce pharmacokinetic data gaps. In the absence of data, however, which may often be the case for new chemicals, the exposure–dose modeler may turn to the emerging field of molecular modeling and chemoinformatics to obtain provisional pharmacokinetic values.

Molecular modeling makes use of a wide variety of techniques to predict or understand chemical behavior at the atomic level. Modeling chemical interactions is an important step in understanding the molecular events encountered in both biological and environmental systems (21–23). These methods have the potential to explain the underlying molecular processes of chemical interactions and transformations in the source–exposure–dose–response continuum. Here, the primary use of such tools is to provide in silico predictions of relevant data where little or no actual data exist. Provisional estimates derived from structure–activity relationships may then be tested using focused methods to validate or augment parameter values.

The field of molecular modeling comprises a wide variety of tools from chemoinformatics-based disciplines [e.g., quantitative structure–activity relationships (QSAR)] and graph network theory (e.g., two-dimensional topological molecular descriptors) to detailed atomistic simulations (e.g., molecular dynamics) and quantum mechanical simulations of the electron distributions of a molecule. Chemoinformatic techniques have a long history in promoting simple concepts such as lipophilicity and partitioning (24) as indicators of persistence and toxicity within the environment (i.e., fate and transport) (25). These techniques are also used to obtain indicators of chemical disposition (26) and pharmacodynamics (27) within biological organisms (28). Many software packages exist whereby one can develop, augment, and utilize new or existing QSARs for parameters such as blood–brain barrier transfer coefficients, dermal permeation rates, cell line permeability, and octanol–water partition coefficient [e.g., molecular operating environment (MOE), QiKProP (https://www.schrodinger.com/products/14/17/], and OpenEye (http://www.eyesopen.com)]. These QSAR packages are generally confined within biological system analysis as seen on the right side of the source–exposure–dose–response continuum (Fig. 1).

For environmental fate and transport models, QSAR can be used to estimate the values of the physicochemical parameters describing the partitioning and transfer processes among air, water, and soil. For example, the US EPAs SPARC predictive modeling system is able to calculate large numbers of physical/chemical parameters from molecular structure and basic information about the environment (e.g., media, temperature, pressure, PH). These parameters are used in fate and transport modeling of organic pollutants, nutrients, and other stressors.

Techniques such as QSAR are ideally suited for rapid evaluation of parameters for pharmacokinetic and fate and transport models. However, development of these techniques is data intensive, requiring training sets with well-defined endpoints to develop the relationship between chemical structure and observed activity. In addition, QSAR models are fitted to specific molecular subsets (training set) and it is difficult to apply them to compounds outside the chemical space represented in the training set.

While QSAR is the more known molecular modeling technique within computational toxicology, there are other tools, such as classical force-field docking techniques, that can aid in understanding the biological processes which involve chemical interactions with biomolecular targets. Inter-molecular interactions between ligands and biomolecular targets determine binding mechanics that ultimately lead to altered physiological responses and potential toxicological effects. Thus, an understanding of the relevant binding interactions can lead to a better understanding of chemical function and provide a visual representation of chemical binding and mechanisms of toxicity. For example, estimating the relative binding affinities of 281 chemicals to a surrogate rat estrogen receptor, Rabinowitz et al. (29) utilized docking techniques to screen out 16 actives ("true competitive inhibitors") from nonactive substrates with no false negatives and eight misclassified false positives. Molecular dynamics (17, 30) or ab initio molecular dynamics (31) can be used to simulate time-evolving processes such as diffusion through environmental media, solvation effects, and "classical" kinetic rate constants (e.g., solvent-mediated hydrolysis, oxidation, and hydrogen abstraction rates). This information can be used as chemical-specific inputs to pharmacokinetic and environmental fate and transport models (32–37).

Computational models are also used to predict the respiratory tract dosimetry of inhaled gases and particulates. These models are needed because the complex shapes of the nasal airways and the branching pattern of the airways leading from the trachea to the alveoli often result in nonuniform deposition of inhaled materials. Models of the respiratory tract incorporate varying degrees of anatomical realism. CFD models of the nasal airways use accurate, three-dimensional reconstructions of the airways (38), while one-dimensional reconstructions have been more commonly used for the pulmonary airways (39).

2.4. Signaling Pathways

Signaling pathways such as the mitogen-activated protein kinase (MAPK) pathway (40) consist of one or more receptors at the cell surface that, when activated by their cognate ligands, transmit signals to cytosolic effectors and also to the genome. The cytosolic effects are rapid, occurring within seconds or minutes of receptor activation, while the effects on gene expression take longer, with changes in the associated protein levels typically occurring after one or more hours. A number of computational models of signaling pathways have been described (e.g., (41, 42).

The National Research Council (NRC) report, Toxicity Testing in the twenty-first century (18) introduced the concept of "toxicity pathways." Toxicity pathways were defined by the NAS as "interconnected pathways composed of complex biochemical interactions of genes, proteins, and small molecules that maintain normal cellular function, control communication between cells, and allow cells to adapt to changes in their environment" and which, "when sufficiently perturbed, are expected to result in adverse health effects are termed toxicity pathways" (43). The adverse effect is the clinically evident effect on health and is often referred to as the apical effect, denoting its placement at the terminal end of the toxicity pathways. Although not much work has been done to date, computational models of signaling pathways are expected to be integral components of toxicity pathway models.

2.5. BBDR/Clonal Growth

Cancer is a disease of cell division. In healthy tissue, the respective rates of cellular division and death are tightly regulated, allowing for either controlled growth or the maintenance of tissue size in adulthood. When regulation of division and death rates is disrupted, tumors can develop. (It should also be noted that embryonic development depends on tight regulation of division and death rates, where dysregulation can result in malformations). A number of computational models have been developed to describe tumor incidence and the growth kinetics of preneoplastic lesions. These vary from purely statistical models fit to incidence data (44) to models that track time-dependent division and death rates of cells in the various stages of multi-stage carcinogenesis (45). These latter kinds of models provide insights into how different kinds of toxic effects—e.g., direct reactivity with DNA versus cytolethality—can differentially affect tumor development.

BBDR models represent the entire exposure—target site dose—apical response continuum. These kinds of models require large amounts of supporting data but have the capability to predict both dose–response and time course for development of apical effects as well as for some intermediate effects (e.g., (46). This latter capability is important as it provides the opportunity to use data on biomarkers in support of model development. The resources needed to develop such models are, unfortunately, seldom available. In some cases, however, where the economic importance or the degree of human exposure is sufficient, development of BBDR models can be justified.

3. Virtual Tissues

The computational models described above incorporate varying degrees of biological detail. Over time, these models will be refined as new data and new degrees of understanding of the relevant biological processes become available. Taking a long-term view, the iterative refinement of such models will lead asymptotically to the development of virtual tissues, where multiple scales of biology—molecular, macromolecular, organelle, tissue—are described in a spatially and temporally realistic manner. Numerous efforts that are self-described as virtual tissues are underway (47–49). While important and useful, these are, however, preliminary steps toward actual development of virtual tissues, and we do not expect that this goal will be realized for some time. However, while a long-term goal of computational toxicology, virtual tissues and, by extension, virtual organisms, have the potential to eventually reduce and perhaps even eliminate the use of laboratory animals, thereby revolutionizing toxicity testing.

Disclaimer

The United States Environmental Protection Agency through its Office of Research and Development funded and managed the research described here. It has been subjected to Agency's administrative review and approved for publication.

References

1. ASTM (1998) 1998 Annual book of ASTM standards: standard guide for remediation of ground water by natural attenuation at petroleum release sites (Designation: E 1943-98), vol 11.04. American Society for Testing and Materials, West Conshohocken, pp 875–917
2. Williams PRD, Hubbell WBJ, Weber E et al (2010) An overview of exposure assessment models used by the U.S. Environmental Protection Agency. In: Hanrahan G (ed) Modeling of pollutants in complex environmental systems, vol 2. ILM Publications, St Albans
3. US EPA (1992) Guidelines for exposure assessment. EPA/600/Z-92/001. US Environmental Protection Agency, Washington, DC
4. Zartarian VG, Xue J, Ozkaynak H, Dang W, Glen G, Smith L, Stallings C (2006) A probabilistic arsenic exposure assessment for children who contact CCA-treated playsets and decks, Part 1: model methodology, variability results, and model evaluation. Risk Anal 26 (2):515–531
5. Glen G, Smith L, Isaacs K, Mccurdy T, Langstaff J (2008) A new method of longitudinal

diary assembly for human exposure modeling. J Expo Sci Environ Epidemiol 18(3):299–311

6. Tran NL, Barraj L, Smith K, Javier A, Burke T (2004) Combining food frequency and survey data to quantify long-term dietary exposure: a methyl mercury case study. Risk Anal 24 (1):19–30
7. Song C, Qu Z, Blumm N, Barabasi AL (2010) Limits of predictability in human mobility. Science 327(5968):1018–1021
8. US EPA (1997) Exposure factors handbook. US Environmental Protection Agency, Washington, DC. http://www.epa.gov/NCEA/pdfs/efh/front.pdf
9. Watanabe PG, Gehring PJ (1976) Dose-dependent fate of vinyl chloride and its possible relationship to oncogenicity in rats. Environ Health Perspect 17:145–152
10. Reddy MB, Yang RSH, Clewell HJ, Andersen ME (2005) Physiologically based pharmacokinetic modeling: science and applications. Wiley, Hoboken
11. Emmen HH, Hoogendijk EM, Klopping-Ketelaars WA, Muijser H, Duisterrnaat E, Ravensberg JC, Alexander DJ, Borkhataria D, Rusch GM, Schmit B (2000) Human safety and pharmacokinetics of the CFC alternative propellants HFC 134a (1,1,1,2-tetrafluoroethane) and HFC 227 (1,1,1,2,3,3,3-heptafluoropropane) following whole-body exposure. Regul Toxicol Pharmacol 32(1):22–35
12. Ernstgard L, Andersen M, Dekant W, Sjogren B, Johanson G (2010) Experimental exposure to 1,1,1,3,3-pentafluoropropane (HFC-245fa): uptake and disposition in humans. Toxicol Sci 113(2):326–336
13. Sexton K, Kleffman DE, Cailahan MA (1995) An introduction to the National Human Exposure Assessment Survey (NHEXAS) and related phase I field studies. J Expo Anal Environ Epidemiol 5(3):229–232
14. Shin BS, Hwang SW, Bulitta JB, Lee JB, Yang SD, Park JS, Kwon MC, do Kim J, Yoon HS, Yoo SD (2010) Assessment of bisphenol A exposure in Korean pregnant women by physiologically based pharmacokinetic modeling. J Toxicol Environ Health A 73 (21–22):1586–1598
15. Wilson NK, Chuang JC, Morgan MK, Lordo RA, Sheldon LS (2007) An observational study of the potential exposures of preschool to pentachlorophenol, bisphenol-A, and nonylphenol at home and daycare. Environ Res 103(1):9–20
16. Andersen ME (2003) Toxicokinetic modeling and its applications in chemical risk assessment. Toxicol Lett 138(1–2):9–27
17. Rapaport DC (2004) The art of molecular dynamics simulation, 2nd edn. Cambridge University, New York
18. Obach RS (1999) Prediction of human clearance of twenty-nine drugs from hepatic microsomal intrinsic clearance data: an examination of in vitro half-life approach and non-specific binding to microsomes. Drug Metab Dispos 27(11):1350–1359
19. Obach RS, Baxter JG, Liston TE, Silber BM, Jones BC, MacIntyre F, Rance DJ, Wastall P (1997) The prediction of human pharmacokinetic parameters from preclinical and in vitro metabolism data. J Pharmacol Exp Ther 283 (1):46–58
20. Tornero-Velez R, Mirfazaelina A, Kim KB, Anand SS, Kim HJ, Haines WT, Bruckner JV, Fisher JW (2010) Evaluation of deltamethrin kinetics and dosimetry in the maturing rats using a PBPK model. Toxicol Appl Pharmacol 244(2):208–217
21. Böhm G (1996) New approaches in molecular structure prediction. Biophys Chem 59 (1–2):1–32
22. Fielden MR, Matthews JB, Fertuck KC et al (2002) In silico approaches to mechanistic and predictive toxicology: an introduction to bioinformatics for toxicologists. Crit Rev Toxicol 32(2):67–112
23. Marrone TJ, Briggs JM, McCammon JA (1997) Structure-based drug design: computational advances. Annu Rev Pharmacol Toxicol 37:71–90
24. Leo A, Handsch C, Elkins D (1971) Partition coefficients and their uses. Chem Rev 71 (6):525–616
25. Valko K (2002) Measurements and predictions of physicochemical properties. In: Darvas F, Dorman G (eds) High-throughput ADMETox estimation. Eaton Publishing, Westborough
26. Topliss JG (ed) (1983) Quantitative structure-activity relationships of drugs. Academic, New York
27. Cronin MTD, Dearden JC, Duffy JC, Edwards R, Manga N, Worth AP, Worgan ADP (2002) The importance of hydrophobicity and electrophilicity descriptors in mechanistically-based QSARs for toxicological endpoints. SAR QSAR Environ Res 13:167–176
28. Pratt WB, Taylor P (eds) (1990) Principles of drug action. Churchill-Livingstone, Inc, New York
29. Rabinowitz JR, Little S, Laws SC, Goldsmith R (2009) Molecular modeling for screening environmental chemicals for estrogenicity: use of the toxicant-target approach. Chem Res Toxicol 22(9):1594–1602

30. Allen MP, Tildesley DJ (2002) Computer simulations of liquids. Oxford University, New York
31. Car R, Parrinello M (1985) Unified approach for molecular dynamics and density-functional theory. Phys Rev Lett 55(22):2471–2474
32. Colombo MC, Guidoni L, Laio A, Magistrato A, Maurer P, Piana S, Röhrig U, Spiegel K, Sulpizi M, VandeVondele J, Zumstein M, Röthlisberger U (2002) Hybrid QM/MM Carr-Parrinello simulations of catalytic and enzymatic reactions. CHIMIA 56(1–2):13–19
33. Geva E, Shi Q, Voth GA (2001) Quantum-mechanical reaction rate constants from centroid molecular dynamics simulations. J Chem Phys 115:9209–9222
34. Prezhdo OV, Rossky PJ (1997) Evaluation of quantum transition rates from quantum classical molecular dynamics simulation. J Chem Phys 107:5863
35. Truhlar DG, Garrett BC (1980) Variational transition-state theory. Acc Chem Res 13:440–448
36. Tuckerman M, Laasonen K, Sprik M, Parrinello M (1995) Ab initio molecular dynamics simulation of the solvation and transport of hydronium and hydroxyl ions in water. J Chem Phys 103:150–161
37. Wang H, Sun X, Miller WH (1998) Semiclassical approximations for trhe calculation of thermal rate constants for chemical reactions in complex molecular systems. J Phys Chem 108 (23):9726–9736
38. Kimbell JS, Gross EA, Joyner DR, Godo MN, Morgan KT (1993) Application of computational fluid dynamics to regional dosimetry of inhaled chemicals in the upper respiratory tract of the rat. Toxicol Appl Pharmacol 121:253–263
39. Overton JH, Kimbell JS, Miller FJ (2001) Dosimetry modeling of inhaled formaldehyde: the human respiratory tract. Toxicol Sci 64:122–134
40. Roux PP, Blenis J (2004) ERK and p38 MAPK-activated protein kinases: a family of protein kinases with diverse biological functions. Microbiol Mol Biol Rev 68:320–344
41. Bhalla US, Prahlad RT, Iyengar R (2002) MAP kinase phosphatase as a locus of flexibility in a mitogen-activated protein kinase signaling network. Science 297:1018–1023
42. Hoffman A, Levchenko A, Scott ML, Baltimore D (2005) The IκB-NF-κB signaling module: temporal control and selective gene activation. Science 298:1241–1245
43. National Research Council (NRC) Committee on Toxicity and Assessment of Environmental Agents (2007) Toxicity testing in the twenty-first century: a vision and a strategy. National Academies, Washington, DC. ISBN 0-309-10989-2
44. Crump KS, Hoel DG, Langley CH, Peto R (1976) Fundamental carcinogenic processes and their implications for low dose risk assessment. Cancer Res 36:2937–2979
45. Moolgavkar SH, Dewanji A, Venzon DJ (1988) A stochastic two-stage model for cancer risk assessment. The hazard function and probability of tumor. Risk Anal 8:383–392
46. Conolly RB, Kimbell JS, Janszen D, Schlosser PM, Kalisak D, Preston J, Miller FJ (2004) Human respiratory tract cancer risks of inhaled formaldehyde: dose-response predictions derived from biologically-motivated computational modeling of a combined rodent and human dataset. Toxicol Sci 82:279–296
47. Adra S, Sun T, MacNeil S, Holcombe M, Smallwood R (2010) Development of a three-dimensional multiscale computational model of the human epidermis. PLoS One 5(1): e8511. doi:10.1371/journal.pone.0008511
48. Shah I, Wambaugh J (2010) Virtual tissues in toxicology. J Toxicol Environ Health 13 (2–4):314–328
49. Wambaugh J, Shah I (2010) Simulating microdosimetry in a virtual hepatic lobule. PLoS Comput Biol 6(4):e1000756. doi:10.1371/journal.pcbi. 1000756

Chapter 3

Role of Computational Methods in Pharmaceutical Sciences

Sandhya Kortagere, Markus Lill, and John Kerrigan

Abstract

Over the past two decades computational methods have eased up the financial and experimental burden of early drug discovery process. The in silico methods have provided support in terms of databases, data mining of large genomes, network analysis, systems biology on the bioinformatics front and structure–activity relationship, similarity analysis, docking, and pharmacophore methods for lead design and optimization. This review highlights some of the applications of bioinformatics and chemoinformatics methods that have enriched the field of drug discovery. In addition, the review also provided insights into the use of free energy perturbation methods for efficiently computing binding energy. These in silico methods are complementary and can be easily integrated into the traditional in vitro and in vivo methods to test pharmacological hypothesis.

Key words: ADME/Tox, Bioinformatics, Chemoinformatics, Protein structure prediction, Homology models, Virtual screening, Ligand-based methods, Structure-based methods, Drug design, Docking and scoring, Protein–protein interactions, Protein networks, Systems biology, PK–PD modeling

1. Introduction

A living cell is composed of a number of processes that are well networked and compartmentalized in both space and time. These complex networks hold clues to the normal and diseased physiology of the living organism. Thus understanding the molecular interaction network to delineate the normal from the disease phenotype could help diagnose and treat the symptoms of disease states. While this may sound simplistic, building the complete network of molecular interactions is highly challenging and could at best be described through model systems by incorporating evidence of interactions obtained from biochemical experiments. Such evidence is now obtained in a large scale with the advent of the various "omics and omes" such as genomics, proteomics, and interactome to name a few (1–3). The field of "omics" has revolutionized the way drug

Brad Reisfeld and Arthur N. Mayeno (eds.), *Computational Toxicology: Volume I*, Methods in Molecular Biology, vol. 929, DOI 10.1007/978-1-62703-050-2_3, © Springer Science+Business Media, LLC 2012

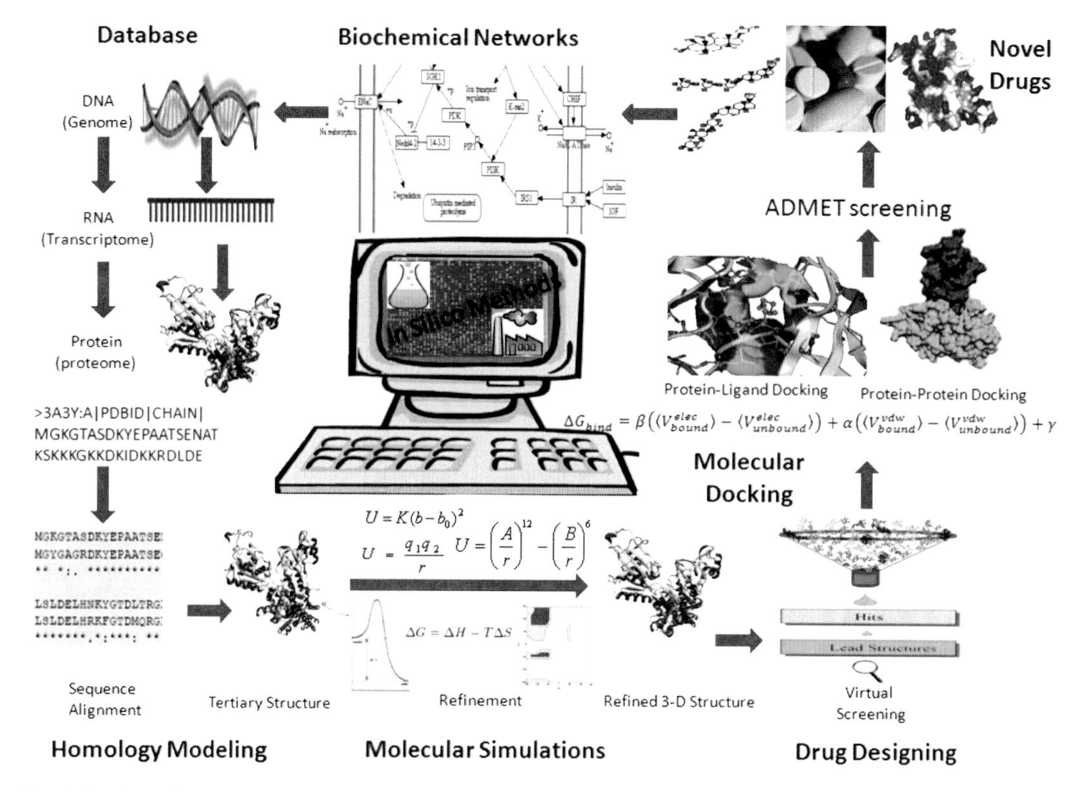

Fig. 1. A schematic representation of the role of in silico methods in pharmaceutical sciences.

discovery and development has been done in the past decade. This transformation has been achieved mainly due to the increase in computational power and development of novel bioinformatics algorithms that have helped mine the "omics" data and present it in usable formats for target and lead discovery (4–6). An associated wing of bioinformatics that has taken a lead role in drug discovery is the field of chemoinformatics. The term was aptly coined by Brown (7, 8) as a way to integrate informatics and chemistry in 1998. However, prior to being integrated under the roof of chemoinformatics, the algorithms were functional under the broad title of ligand-based methods (9). Thus in a broad sense the role of in silico methods in drug discovery can be discussed under two categories namely bioinformatics (as a tool for finding new targets) and chemoinformatics (for lead identification and optimization) methods as summarized in Fig. 1.

2. Bioinformatics

The precise definition of the term bioinformatics is not known. It varies from being as simple as using computers to study biological problems to as complex as methods for generating, storing, retrieving, analyzing, and modeling biological data such as genomic, proteomic, and biological pathways. The goal of bioinformatics from a pharmaceutical sciences perspective is to mine the vast data available and correlate the data with disease phenotypes to discover new target proteins for further use in developing new drugs. However, in recent years, there seems to be a paradigm shift in single target hypothesis and moving towards protein–proteininteraction (PPI) inhibitors, pathway inhibitors and giving way to the concept of polypharmacology (10–12). The field of polypharmacology is synonymous to the early medicines developed by ayurvedic and traditional Chinese medicines, where in the goal was to treat with a concoction of compounds that could hit multiple targets to provide a systemic relief from the symptoms. Thus systems biology is trending towards becoming a reality and in silico models to delineate these pathways are being developed. Further, with improvements in robotic technologies and their utility in protein crystallography, characterizing the newly identified targets has become a possibility and this field is called structural bioinformatics.

As the name suggests, structural bioinformatics deals with protein structure and algorithms that can be used to predict, analyze, and model three-dimensional structure of protein. This is one of the most popular fields of computational biology with algorithms designed as early as 1960. Protein structure can be explained at four levels of complexity: the primary, secondary, tertiary, and quaternary structure. Hence algorithms were developed to deal with each of these levels. The role of bioinformatics starts with the sequencing of the genomes during mapping of the loci, fragment assembly and annotations of the sequenced genomes (13, 14). Gene annotations are complex with algorithms needed to identify coding and noncoding regions and those for deriving information from homologous and orthologous genomes using evolutionary dendrograms (15, 16) and sequence alignment programs (17–19).

3. Protein Structure and Prediction

Protein structure prediction algorithms can be classified into three categories namely, secondary structure prediction, ab initio structure modeling, and homology modeling. Given that there are 20 amino acids, the number of possibilities for a given sequence of amino acids that constitute the primary structure of the protein to fold into a tertiary structure is astronomical. However, protein folding in physiological conditions probably follows the path of least complexity and is therefore highly efficient. To mimic this process, it is prudent to first predict the secondary structural elements namely alpha helix, beta sheet, and gamma turns which are guided by evolutionary and functional relationships among homologous proteins (20, 21). The algorithms for searching databases and predicting secondary structural elements are based on heuristic statistical methods. These algorithms include sequence database search methods such as BLAST (22) and FASTA (23); sequence alignment methods such as ClustalW (24), KALLIGN (25), muscle (26), T-Coffee (27); secondary structure prediction methods such as HMM (28), Predict Protein (29), JPred (30); and other methods listed in repositories such as ExPASy tools (http://expasy.org/tools/). These methods assign the most preferred secondary structural element to a stretch of amino acids based on the statistical propensity of a given amino acid to be a part of alpha helix or beta sheet or a turn (31, 32). Ab initio methods on the other hand use only sequence information to predict the tertiary structure and rely on conformational analysis methods such as Monte-Carlo or molecular dynamics simulations to assess whether the predicted sequence can correctly fold into the assigned structure (33–35). Homology modeling on the other hand uses the structure of a known protein called the template to model the three-dimensional structure of the unknown (36). The resulting model is then refined using energy minimization and molecular dynamics simulations and further validated using evidence from biochemical experiments such as site-directed mutagenesis and fluorescence measurements (37, 38). All these methods are highly resourceful in the context of structure-based drug design wherein the models built can be directly used for screening compounds or to understand the mode of binding of an inhibitor and hence designing mutants or in rational lead optimization (39). Several examples of structure-based drug design are available in the literature including the early success of HIV protease inhibitors (40).

4. Protein–Protein Interactions

It can be envisioned that the next paradigm in drug discovery will be to design inhibitors to key protein–protein interactions (PPIs). These PPIs could be present between host and pathogen or entirely belonging to a host or a pathogen. The feasibility of such drug design has been shown recently by our group in designing small molecule interaction inhibitors of key PPIs of the malaria parasite (41) and other infectious agents such as *Toxoplasma gondii* and HIV (42, 43). However, the bottle neck in this design process lies in identifying key PPIs given only a handful of crystal structures of such complexes are currently available in the protein databank. Understanding PPIs are also important from other pharmacological and biochemical perspectives such as in signaling, cellular adhesion, enzyme kinetics, pathways, etc. Thus understanding, predicting, and cataloging these PPIs using bioinformatics and structural biology methods is crucial. A significant step towards achieving this goal is the availability of the genomic information of species of interest. A general hypothesis about PPIs is that if the proteins coevolve then they have a higher probability of being interaction partners (44). Several experimental techniques such as yeast two hybrid, mammalian hybrid methods, protein fragment complementation assays, fluorescence resonance energy transfer can detect or validate if the two proteins form interaction partners (45–48). Given the complexities of establishing these in vitro methods, a number of computational methods have been developed to predict PPIs. Based on the protein coevolution hypothesis, phylogenetic methods that can make inferences about interactions between pairs of proteins based on their phylogenetic distances have been designed (49, 50). Others include use of PSI-BLAST (51) or BLAST algorithms to query for templates either from a set of nonredundant sequences or from a library of known protein–protein interfaces (52, 53). These libraries can be built from databases that maintain information about PPIs such as BIND (54), BioGRID (55), DIP (56), HPRD (57), IntAct (58), and MINT (59). Some methods of prediction use homology modeling techniques to build the complexes and to score them, using statistical residue interaction energetic (60–62). Machine learning methods such as SVM and Bayesian network models have also been used to predict the interacting partners of a given protein (63).

5. Systems Biology and Protein Networks

PPIs can be called as the minimal subunit of protein networks. The concept of systems biology deals with systemically understanding the biological processes by incrementally building up the networks of interactions that underlie the biological process (64, 65). These networks then provide the molecular basis for the etiology of diseases and to rationally develop therapeutics that can work at one or more components of the protein network (66, 67). The systems approach also help in identifying key targets, biomarkers, and to quantify potential side effects of drugs due to off-target interactions (68, 69). In delineating these new networks, experimental methods such as microarrays work in close association with statistical methods such as Bayesian network models to uncover new protein networks (70, 71). In addition, the networks of lower organisms such as *S. cerevisiae* (72, 73), *Drosophila melanogaster*, have been identified (74, 75) and stored in databases such as KEGG (http://www.genome.jp/kegg/pathway.html), UniPathway (http://www.grenoble.prabi.fr/obiwarehouse/unipathway) and clearly serve as models to understand and derive the networks of higher organisms.

Mathematical modeling of the pathways is another tool that has added to the understanding of these biochemical networks often called "reaction networks." Given a few parameters, mathematical modeling can help derive the unknowns in the equations for flux modeling and hence help in modeling the networks (76, 77). A major utility of such networks in the pharmaceutical industry is the pharmacokinetics and pharmacodynamics models (PK–PD). Pharmacokinetics (PK) characterizes the absorption, distribution, metabolism, and elimination properties of a drug. Pharmacodynamics (PD) defines the physiological and biological response to the administered drug. PK–PD modeling establishes a mathematical and theoretical link between these two processes and helps better predict drug action (78, 79). Recent models for PK–PD include mechanism-based models which aim to link the drug administration and effect such as target-site distribution, receptor binding kinetics, and the dynamics of receptor activation and transduction (80). Algorithms for PK–PD modeling include GASTRO-PLUS (Simulations, Inc.), WinNonLin (Pharsight, Inc.) which are based on a knowledgebase of known drugs experimentally measured PK–PD parameters. These modeling programs use data such as the presence of drug in plasma at particular points in time and allows for calculation and estimation of critical PK parameters such as maximum concentration, total exposure (i.e., area under the curve), half-life, clearance rate, and volume of distribution. Other utilities of mathematical modeling involve deriving flux-based networks in modeling phenotypes in response to a toxic agent such as receptor overexpression induced by environmental agents (81). In general, most mathematical models are deterministic in nature and hence limited in ability for use in small scales and so extending their utility to derive large networks is not feasible. However, other newer methods such as stochastic networks, Bayesian models can conquer such limitations (82, 83).

6. Chemoinformatics

Analogous to bioinformatics, chemoinformatics is a field of science that is involved in management of chemical data using computational methods. This area of science has gained tremendous significance in the past few decades due to the availability of chemical data in the form of combinatorial chemistry and high throughput screening (84). These two aspects have set new trends in drug discovery in which attrition rates are now being seriously taken into account at early stages of drug discovery. Chemoinformatics plays a major role in designing models for virtual screening, lead design, lead optimization, preclinical filtering schemes for drug like properties, ADMET and PK–PD modeling (85).

Chemoinformatics models were earlier referred to as ligand-based methods. As the name suggests ligand-based methods are derived solely with information from a molecule(s). A variety of techniques such as CoMFA, CoMSIA, QSAR (1D or multidimensional), Bayesian and other numerical and statistical methods and pharmacophore methods can be classified under ligand-based methods.

7. Classical QSAR

Quantitative structure–activity relationships (QSAR) aim to quantitatively relate the biological activity (e.g, inhibition constants, rate constants, bioavailability, or toxicity) of a series of ligands with the similarities between the chemical structures of those molecules. It requires, first, the consistent measurement of biological activity, second, the quantitative encoding of the chemical structure of the studied molecules, also known as molecular descriptors, and third, the derivation of mathematical equations quantitatively relating molecular descriptors with biological activity.

Many different means of encoding the chemical structure of ligands have been devised since the pioneering work of (86) who used hydrophobicity and Hammett constants to describe the varying substituent of a common scaffold of a series of studied molecules. Classical Hansch-type QSAR models utilize atomic or group properties describing the physicochemical properties of the substitutions such as hydrophobic (e.g., log P, π = partial log P), electronic (e.g., Hammett σ, quantum-mechanical indices such as electronegativity or hardness index), or steric and polarizability properties (e.g., molecular volume, molecular refractivity). Also, the spatial distribution of physicochemical properties of a ligand in 3D can be mapped onto molecular surfaces and used as molecular descriptors (e.g., polar surface area). Several molecular descriptors d_{ik} for each ligand k in the dataset are usually combined in a multi-linear regression model representing a correlation with the biological activity A_k of compound k:

$$A_k = c_0 + \sum_i c_i \times d_{ik} \quad \text{for all ligands } k. \tag{1}$$

The prefactors c_0 and c_i can be derived using multi-linear regression analysis to fit the experimental activity data of a preselected set of compounds that are used to train the regression model, named the training set. Whereas parameters such as the regression coefficient r^2, which measures the ratio of explained variance to total variance in the training set's activity data, and the Fisher value F, which measures the statistical significance of the model, describe how well the model fits the experimental data of the training set, they do not provide information about the predictive quality of the model for new compounds that are not included in the training set. Leave-one-out cross-validation, a technique used in the past to provide this information, proved to be insufficient to measure predictive power (87–89), but leaving out larger groups throughout cross-validation or scrambling tests (the activity data of the compounds is randomly reordered among the dataset and no QSAR model with comparable regression quality should be obtained for any reordering) have been shown to more reliably estimate the predictive quality of the QSAR model. As an ultimate test, however, the QSAR model should always be validated by its potential to predict compounds, called the external test set, not included at any stage in the training process.

In parallel to Hansch and Fujita, Free and Wilson (90) derived QSAR models using indicator variables. Indicator variables $[a_{im}]_k$ describe the absence or presence of a chemical substituent i_m (e.g., Cl, Br, I, Me) at position m of a common ligand scaffold with values of 0 (absence) and 1 (presence)

$$A_k = c_0 + \sum_{i_0} c_{i0} \times [a_{i0}]_k + \cdots + \sum_{i_N} c_{iN} \times [a_{iN}]_k \quad \text{for all ligands } k. \tag{2}$$

N is the number of substitutions. Only one indicator variable in each sum of equation (2) can have a value of 1 for each ligand. Although original Free–Wilson type QSAR analysis displayed some shortcomings over Hansch-type QSAR models (e.g., activity predictions are only possible for new combinations of substituents already included in the training set; more degrees of freedom are necessary to describe every substitution), this QSAR scheme has become popular again with the onset of structural fingerprints or hashed keys (91, 92) (Fig. 2) describing the topology of the molecules in the data set.

8. 3D-QSAR and Extensions

With the introduction of comparative molecular field analysis (CoMFA) (93), for the first time structure–activity relationships were based on the three-dimensional structure of the ligand molecules (3D-QSAR). In 3D-QSAR, the ligands' interaction with chemical probes or the ligands' property fields (such as electrostatic fields) are mapped onto a surface or grid surrounding a series of compounds. The values on the grid or surface points are utilized as individual descriptors, which are usually grouped into a smaller number of descriptors, for use in a regression. The quality of the 3D-QSAR model critically depends on the correct superposition of

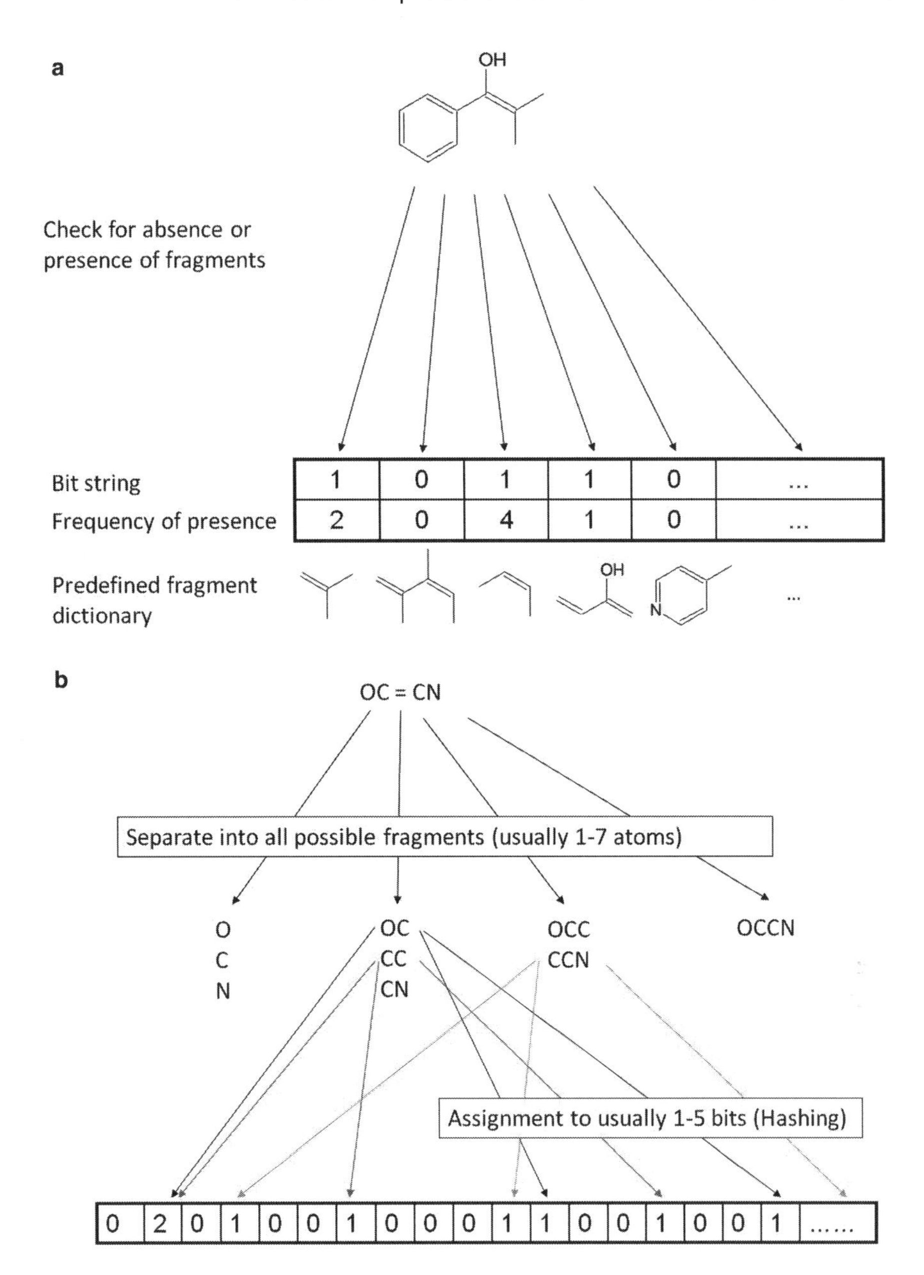

Fig. 2. (**a**) A structural fingerprint for a chemical is generated by determining the frequency a specific fragment of a predefined library is present in the ligand. The frequencies of all fragments are stored as individual bits in a bit string. The individual bits are used as individual indicator variables. (**b**) In a hashed key the fragments are generated on-the-fly for all ligands in the training set and the frequency of presence of a fragment is distributed to a hash key with fixed length.

the ligands representing the native conformations and orientations of the ligands, a very difficult task, particularly, in the absence of the X-ray structure of the target protein.

QSAR based on "alignment-independent descriptors" (AIDs) (94–96) was proposed to surpass the necessity of the correct ligand alignment. In these methods, properties of compounds, such as hydrophobicity or hydrogen bonds, are projected on its molecular surface. The surface properties are then transformed into position-independent characteristics, such as the terms of a moment expansion of the associated physicochemical fields of a molecule. However, the selection of the native ligand conformations is likewise critical to the quality of an QSAR model based on such descriptors. Alternatively, 4D-QSAR concepts (97–100) approach the alignment issue by representing each molecule in different conformations, orientations, tautomers, stereoisomers, or protonation states. The true binding mode (or the bioactive conformation) is then identified throughout the training procedure of the QSAR model, e.g., by the energetically most favorable ligand pose with respect to its interaction with the surrounding molecular probes. 4D-QSAR not only addresses the uncertainties of the alignment process, but can also model multimode binding targets, such as cytochrome P450 enzymes. Those enzymes, critical for drug metabolism, are known to accommodate a ligand in different binding poses, each yielding different metabolic products of a given compound (101). Standard 3D-QSAR methods do not incorporate protein flexibility upon ligand binding. To model this important factor in protein–ligand association, 5D-QSAR techniques that simulate a topological adaptation of the QSAR model to the individual ligand have been devised (100, 102, 103).

9. Applications of QSAR

QSAR has become an integral component in pharmaceutical research to optimize lead compounds. Whereas QSAR is widely used to identify ligands with high affinity for a given target protein, more recently QSAR methodology has been extended to predict pharmacokinetic properties, such as adsorption, distribution, metabolism, elimination (ADME) properties (104) or the oral bioavailability of compounds (105, 106), as well as the toxicity of drug candidates. Furthermore, in the context of the Registration, Evaluation, and Authorization of Chemicals (REACH) legislation of the European Union, the prediction of the toxic potential of environmental chemicals using QSAR has created public interest (107).

10. Pharmacophore-Based Modeling

Another ligand-based method that has found utility in pharmaceutical industry is pharmacophore-based modeling. A pharmacophore can be defined as a molecular framework required for the biological activity of a compound or a set of compounds (108). The concept of pharmacophores have found wide spread utility in virtual screening, similarity analysis, and lead optimization. Three popular pharmacophore modeling tools are DIStance COmparisons (DISCO) (109), Genetic Algorithm Similarity Program (GASP) (110), and Catalyst (111) and these have been thoroughly described and compared by Patel and colleagues (112). The Catalyst program has been used widely by researchers (113–120) and has two methods for generating pharmacophores namely HIPHOP (121) and HYPOGEN (122). HIPHOP uses few active molecules to derive the common chemical features, while HYPOGEN derives models based on a series of molecules with varying structure activity and function. Both these methods have found utility in deriving models for enzymes, nuclear receptors, ion channels, and transporters. The types of transporter pharmacophores that have been published to date have been recently reviewed (123) and along with 3D-QSAR these methods have become widely accepted methods for assessing the drug–transporter interactions (124). The pharmacophore methods have been used to discover new inhibitors or substrates for transporters by first searching a database then generating in vitro data (114, 125–127). The pharmacophore methods can also have a good hit rate which may be used alongside other QSAR methods for database screening and ADMET modeling (128, 129). In addition, 3D pharmacophore methods are being used for lead design (41, 130).

11. ADMET Modeling

One of the major applications of ligand-based method is to predict absorption, distribution, metabolism, excretion, and toxicity (ADME/Tox) applications. A number of studies have utilized different ligand-based models (131–139). In addition there are many studies that have provided an extensive comparison of these programs that have been designed to predict the ADMET properties (140–142). ADMET properties can be described using a set of physicochemical properties such as solubility, log P, log D, pK_a, polar surface area that describe permeability, intestinal absorption, blood brain barrier penetration, and excretion. Solubility is modeled as logarithm of solubility (log S) using molecular descriptors that govern shape, size, interatomic forces, and polarity (143–147). Permeability is a measure of the compound's bioavailability and is modeled using molecular descriptors that code for hydrophobicity, steric and electronic properties of molecules (148–150). In addition to passive diffusion across cellular membranes, permeability could also be through active transport by membrane bound transporters and pumps (151, 152). In silico models of such active transport have been modeled using ligand-based methods (153, 154). Permeability across the blood brain barrier is an associated parameter that is computed exclusively for CNS drugs and for other compounds as an off-target filter. It is computed as logarithm of BB (log BB) which is a measure of the ratio of concentration of the drug in the brain to that in the blood (155, 156). Several in silico models have been proposed that utilize molecular descriptors such as log P, pK_a, TPSA, and molecular weight with a variety of methods such as ANN, multiple regression models (MLR), QSAR, Support vector machines (SVM), and other statistical techniques (157–159). Similarly several in silico models have been proposed to model drug metabolism and toxicity predictions which is reviewed in a recent report (160).

12. Structure-Based Methods

Structure-based methods as the name suggests relies on the three-dimensional structure of the target and small molecule. Three-dimensional (3D) structure of the target can be obtained by experimental methods such as X-ray crystallography or NMR methods or by homology modeling methods. Several review articles provide additional details about the methods and utility of homology models (160–162). Here we discuss the utility of structure-based methods in virtual screening applications.

13. Virtual Screening

High-throughput screening (HTS) has become a common tool for drug discovery used to identify new hits interacting with a certain biological target. Virtual screening technologies are applied to either substitute or aid HTS. Both ligand-based methods that use similarity to previously identified hits and structure-based methods that use existing protein structure information can be used to perform virtual high-throughput screening (VHTS) to identify potentially active compounds. In ligand-based VHTS, one or several hits must be identified first, for example, from previous HTS experiments using a smaller subset of a ligand library or from previously published hits. Factors such as the set of molecular descriptors, the measurement of similarity, size, and diversity of the virtual ligand library, and the similarity threshold value separating potential active from inactive compounds are critical to the success of VHTS and must be carefully tuned. Techniques used in ligand-based VHTS include methods based on 2D and 3D similarity. Examples of such methods are substructure comparison (163), shape matching (164), or pharmacophore methods (165–167).

In structure-based VHTS, automated docking is commonly used to identify potential active compounds by ranking the ligand library based on the strength of protein–ligand interactions evaluated by a scoring function. Throughout the docking process, many different ligand orientations and conformations (binding poses) are generated in the binding site of the protein using a search algorithm. Docking methods can be classified by the level of flexibility that will be allowed for the protein and ligands (168). However, with the increase in computational resources, most recently developed algorithms allow complete flexibility for the ligand molecules and varying levels of flexibility to the amino acid side chains that are involved in binding the ligands. Conformational analysis of ligands can be performed using several algorithms such as systemic search, stochastic or random search and using molecular simulation techniques (169, 170). Systemic search includes performing conformational search along each dihedral in the small molecule. This could lead to an exponential number of conformations that needs to be docked and scored and may not be practical in many cases. To avoid this, several algorithms utilize stored conformations of fragments to limit the number of dihedrals that can be sampled (e.g., FLOG (171)). Other alternatives include splitting the ligand into the core and side chain regions and docking the core first and incrementally sampling and adding the side chains (172). This method has been adopted by several programs including DOCK (173), FlexX (174), GLIDE (175), and Hammerhead (176). In stochastic search, flexibility is computed by introducing random changes to the chosen dihedrals and sampled using a Monte Carlo method or genetic algorithm method. In each case, the newly formed conformation is accepted or rejected using a probabilistic function that uses a library of previously computed conformations (177). Autodock (178), GOLD (179), MOE-DOCK (http://www.chemcomp.com/) are some of the well-known programs that utilize the random search method. Molecular simulations help derive conformations that may be compatible in a dynamic model. Molecular dynamics methods are very efficient in deriving such conformations but are expensive in terms of time and resources. However simulated annealing, and accelerated and high-temperature MD studies have helped conquer some of the issues associated with MD studies, which also addresses the local minima problem (180, 181). Many programs such as DOCK, GLIDE, MOE-DOCK utilize MD simulations to refine conformations obtained from other methods. Protein pockets are generally represented as grid points, surface, or atoms (180). Thus protein ensemble grids can also be used to provide information on protein flexibility. The atomistic representation of proteins are generally used for computing scoring functions, while surface representations are more useful in the case of protein–protein docking methods.

14. Scoring Functions

Scoring functions are used to estimate the protein–ligand interaction energy of each docked pose. The pose with the most favorable interaction score represents the predicted bioactive pose and in principle can be used as a starting point for subsequent rational structure-based drug design. Scoring functions can be classified into three types, namely force field, knowledge based, and empirical based on the type of parameters used for scoring protein–ligand interactions. The force field method uses the molecular mechanics energy terms to compute the internal energy of the ligand and the binding energy. However, entropic terms are generally omitted in the calculations as they are computationally expensive to be computed. Various scoring schemes are built on different force fields such as Amber (182), MMFF (183), and Tripos (184). In general a force field-based scoring function consists of the van der Waals term approximated by a Lennard Jones potential function and an electrostatics term in the form of a Coulombic potential with distance-dependent dielectric function to reduce the effect of charge–charge interactions (180). Additional terms can be incorporated in certain cases where in the contributions of water molecules or metal ions are distinctly known that can increase the accuracy of the scoring function. Empirical scoring functions are devised to fit known experimental data derived from a number of protein–ligand complexes (185). Regression equations are derived using these known protein–ligand complexes and regression coefficients are computed and these coefficients are used to derive information about energetic of other protein–ligand complexes. Scoring schemes employing empirical methods include LUDI (185), Chemscore (186), and *F*-score (174). Knowledge-based scoring functions are used to score simple pairwise atom interactions based on their environment. A set of known protein–ligand complexes are used to build the knowledge database about the type of interactions that can exist. Because of their simplistic approach they are advantageous to be used in scoring large databases in relatively short time scales. Scoring functions that use knowledge-based methods include Drugscore (187), PMF (188), and SMOG (189).

Each of the scoring schemes mentioned have their advantages and disadvantages. Hence the concept of consensus scoring schemes was introduced to limit the dependency on any of the schemes (190). A number of publications in the literature describe comparative studies employing different docking and scoring schemes (168, 180). There is no set rule to combine scoring schemes, deriving a consensus score should be customized to every application to limit amplifying errors and balancing the right set of parameters that can be useful to identify the correctly docked pose. Although there are several scoring schemes, it should be noted that existing scoring functions are not accurate enough to reliably predict the native binding mode and associated free energy of binding. This limitation originates from the necessity to find a balance between accuracy and efficiency in order to screen large ligand libraries. Consequently, scoring functions quantify a simplified representation of the full protein–ligand interaction by only including critical elements such as hydrogen bonds and hydrophobic contacts and neglecting effects such as polarization and entropy. In addition to using a simplified scoring function to reduce the computational time required for VHTS, only critical degrees of freedom, such as translation, rotation and torsional rotations of the ligand, are considered during the search algorithm to limit the conformational space that must be sampled. To compensate for the tradeoff between efficiency and accuracy, more accurate post-processing techniques such as free-energy methods based on molecular dynamics simulations are required to confirm the predicted bioactive binding pose or to more reliably rank the ligand library according to their free energy of binding.

15. Binding Energy Estimation

Estimation of binding free energy for the case of protein–protein and protein–drug complexes has remained a daunting challenge; however, significant progress in free energy calculations has been made over the past 20 years. This section will touch on the rigors of binding free energy calculations with a focus on the linear interaction energy method covering the past few years. The free energy, G, is represented as follows

$$G = -k_{\mathrm{B}}T\ \ln(Z), \tag{3}$$

where k_{B} is the Boltzmann constant, T is the temperature, and Z is the partition function described as

$$Z = \sum_{i} \mathrm{e}^{-\beta E_i}, \tag{4}$$

where $\beta = 1/k_{\mathrm{B}}T$. The equation describes samples of configurations i (also referred to as a "microstate") following a Boltzmann distribution. These samples of configurations i can be thought of as the snapshots generated by a molecular dynamics (MD) or Monte Carlo (MC) simulation. The quantity E_i is the potential energy of configuration i. When it comes to binding free energy, we are most interested in the differences between two states (e.g., the bound Z_{B} versus the unbound state Z_{A}). Equation 3 below is known as Zwanzig's formula (191).

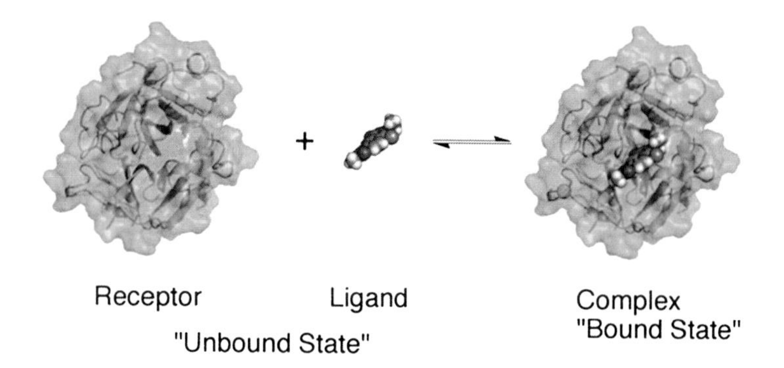

Fig. 3. Illustration of the unbound receptor plus unbound ligand in equilibrium with the receptor–ligand complex (the bound state).

$$\Delta G = G_B - G_A = -k_B T \ln\left(\frac{Z_B}{Z_A}\right). \tag{5}$$

The equations below describe the relation to chemical equilibrium (Fig. 3). K_A is the equilibrium constant for association of receptor (R) and ligand (L) to the complex (RL). The dissociation of the complex (RL) back to receptor (R) and ligand (L) is described by the dissociation constant K_D which is also the inhibition constant K_i.

$R + L \rightleftharpoons RL$	$K_A = \frac{[RL]}{[R][L]}$
$RL \rightleftharpoons R + L$	$K_D = K_i = \frac{[R][L]}{[RL]}$

These constants are related to the binding free energy ΔG_{bind}, enthalpy ΔH, and entropy ΔS via the following relationship.

$$\Delta G_{bind} = -RT \ln K_A = RT \ln K_D = \Delta H - T\Delta S. \tag{6}$$

All of the components of the molecular mechanics force field contribute to the enthalpy (ΔH) of the system. The change in entropy (ΔS) associated with the motions (conformational changes) of the ligand and the protein upon binding is a key contributor to the overall system free energy. For example, the release of highly ordered water molecules in a hydrophobic pocket of the protein upon binding of the ligand gives a positive change in entropy resulting in more favorable ΔG_{bind}. If the bound conformation of the ligand is not a stable conformation when the ligand is in the free or unbound state the binding to protein will be less favored entropically. However, if the bound state conformation of the ligand is also the most stable conformation when the ligand is free in solution (unbound) the binding will be more favored entropically and the ligand is said to be *preorganized* for binding (192). The estimation of system entropy is an extensive area of research and an ongoing challenge in computational chemistry today. A brief, yet elegant discussion of the pitfalls of methods used to calculate the entropy term can be found in Singh and Warshel's excellent review paper on binding free energy calculations (193).

16. Free Energy Perturbation (FEP)

The principal goal of all free energy simulations is the computation of the ratio of the two states (Z_B/Z_A). Rigorous methods require exact sampling of the configurations between the two states. A well-known technique is the free energy perturbation method based on Zwanzig's formula in Eq. 5. The potential energy difference between the two states can be computed using MD or MC simulations. Intermediate states represented by lambda λ (also referred to as windows) are introduced to cover space between the two states.

The energy difference between the two states (e.g., ligand free in solution versus the ligand bound to protein) is often too large to be computed directly using FEP and that the two states are too distinct in their conformations. Hence a relative ΔG ($\Delta\Delta G$) can be computed using a thermodynamic cycle.

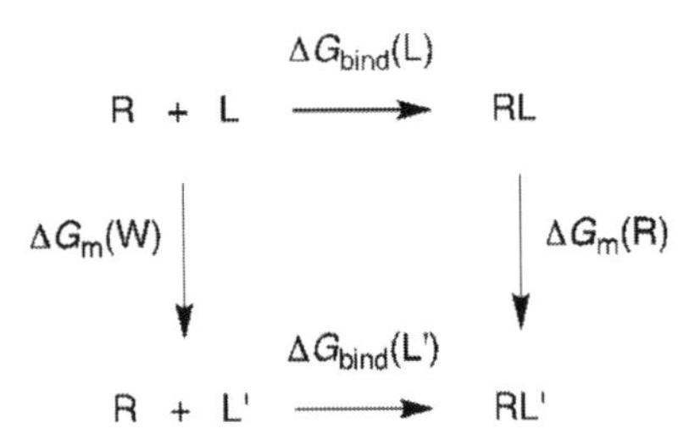

Two sets of FEP calculations need to be performed. The more simple of the two is the calculation of $\Delta G_m(W)$, where state A is ligand L in water and state B is ligand L′ in water. The second more expensive calculation is that for $\Delta G_m(R)$, where we mutate ligand L to L′ in the presence of the receptor (protein). The relative free energy can be calculated using Eq. 7.

$$\Delta\Delta G = \Delta G_{\text{bind}}(\text{L'}) - \Delta G_{\text{bind}}(\text{L}) = \Delta G_m(\text{R}) - \Delta G_m(\text{W}). \qquad (7)$$

There are several inaccuracies, which can plague free energy calculations. One is error in the force field used. Another issue is adequate sampling of phase space. This can be addressed by running the simulation for a longer time period for molecular dynamics calculations or use more iterations in the case of Monte Carlo calculations. The FEP calculation should be run in both forward and reverse directions. The difference in the free energies of the forward and reverse direction calculations provides a lower bound estimate of the error in the calculation (194). Changing one atom to another atom or group in the perturbation near the end of the intermediate states can result in well-known endpoint problems like numerical instability and singularities. These "endpoint" issues can be addressed using "soft-core" potentials for the van der Waals component of the force field (195, 196).

17. Linear Interaction Energy (LIE)

Two methods, LIE and Molecular Mechanics-Poisson Boltzmann-Surface Area (MM-PBSA), are often referred to as "endpoint" methods because they neglect the intermediate states in the transition (192). The LIE method developed by Åqvist uses a scaling factor β, based on the linear response approximation for the electrostatic component (197, 198), while estimating the van der Waals term using a scaling factor, α (199). This approach only considers the endpoints: the bound ligand and the unbound or "free" ligand.

$$\Delta G_{\mathrm{bind}} = \beta(\langle V^{\mathrm{elec}}_{\mathrm{bound}}\rangle - \langle V^{\mathrm{elec}}_{\mathrm{unbound}}\rangle) + \alpha(\langle V^{\mathrm{vdw}}_{\mathrm{bound}}\rangle - \langle V^{\mathrm{vdw}}_{\mathrm{unbound}}\rangle) + \gamma. \tag{8}$$

The sometimes-used parameter γ is a scaling factor used to account for the medium in computing absolute free energies. The values for the scaling factors α, β, γ are dependent on the nature of binding pocket, functional groups of ligands (Table 1), force fields, solvent models (200–205). Several studies in the literature have proposed the use of LIE methods to efficiently design small molecule inhibitors such as antimalarials (206), antibiofilm agent Dispersin B (207), HIV-1 reverse transcriptase inhibitors (208), glucose binding to insulin (209), BACE-1 inhibitors (210), tubulin (211), and CDK-2 inhibitors (212). The LIE method continues to evolve and grow as

Table 1
Optimal β parameters based on compound type (203)

Compound	β
Alcohols	0.37
1° Amides	0.41
1°, 2° Amines	0.39
Carboxylic acids	0.40
Cations	0.52
Anions	0.45
Other compounds	0.43

a mainstay tool of the computational chemist or biologist. The technique was born from free energy perturbation methods, is faster than the FEP or thermodynamic integration (TI) techniques, and with careful fit to training sets of experimental data is highly accurate in its prediction. The implicit solvent methods used in LIE calculations enjoy a one order of magnitude increase in speed over the older explicit solvent model LIE calculations (208). Even in its functional form, the LIE equation is useful in a qualitative sense for early investigative work. The LIE method can be used in the study of protein–protein and small molecule complexes with nucleic acids.

In general, docking methods are significantly more time demanding than similarity-based methods and thus are often only applied to a subset of the full ligand library that is pre-filtered by a similarity-based method. As well as being applied consecutively in VHTS applications, structure-based and ligand-based methods can also be applied in parallel or as a single integrated method which occurs in structure-derived pharmacophore models (213, 214) and hybrid structure-based methods that integrate both ligand-based and structure-based methods (41, 130).

To estimate the expected success of VHTS or to optimize the procedure, retrospective screening is often performed; known actives are mixed with a large number of proposed nonactive molecules (decoys) and the percentage of identified actives as a function of the number of tested molecules is plotted in so-called enrichment plots. While this may show enrichment, it should be noted that successful retrospective screening does not always imply successful prospective screening of a novel ligand library. For example, similarity-based screening methods may be biased towards the original set of active compounds and does not identify a novel class of molecules as potential hits.

In many instances, virtual screening is not used to replace HTS but rather to pre-filter the ligand library to a smaller subset of compounds (focused library) that are more likely have the required properties to be active compounds for the biological target of interest. As part of this library design process, computational methods are also used to estimate lead-like, drug-like, or pharmacokinetic properties that are utilized to preselect compounds with reasonable properties for the drug-discovery project.

18. Conclusions

In this review we have introduced a number of in silico methods that have a myriad of applications in pharmacology. We have classified these approaches into bioinformatics based that deal with protein targets and chemoinformatics based that deal with small molecules. In addition, we have also discussed in a slightly detailed manner the role of free energy methods in estimating binding energies. Although the FEP methods are computationally intensive, their applications to designing new inhibitor molecules and to understand the structure–activity relationship of a series of compounds are unmatched. Taken together, the in silico methods clearly complement the in vitro and in vivo methods in pharmacology and have become an integral part of the drug discovery process.

Acknowledgments

We would like to thank our collaborators for their views and comments on the manuscript and Bharat Kumar Stanam for his help in designing figures. SK is funded by American Heart Association, scientist development grant.

References

1. Figeys D (2004) Combining different 'omics' technologies to map and validate protein–-protein interactions in humans. Brief Funct Genomic Proteomic 2:357–365
2. Cusick ME, Klitgord N, Vidal M, Hill DE (2005) Interactome: gateway into systems biology. Hum Mol Genet 14(Spec No. 2): R171–R181
3. Chakravarti B, Mallik B, Chakravarti DN (2010) Proteomics and systems biology: application in drug discovery and development. Methods Mol Biol 662:3–28
4. Butcher EC, Berg EL, Kunkel EJ (2004) Systems biology in drug discovery. Nat Biotechnol 22:1253–1259
5. Cho CR, Labow M, Reinhardt M, van Oostrum J, Peitsch MC (2006) The application of systems biology to drug discovery. Curr Opin Chem Biol 10:294–302
6. Chen C, McGarvey PB, Huang H, Wu CH (2010) Protein bioinformatics infrastructure for the integration and analysis of multiple high-throughput "omics" data. Adv Bioinform 423589:19
7. Gund P, Maliski E, Brown F (2005) Editorial overview: whither the pharmaceutical industry? Curr Opin Drug Discov Dev 8:296–297
8. Brown FK (1998) Chemoinformatics: what is it and how does it impact drug discovery. Annu Rep Med Chem 33:9
9. Gasteiger J, Engel T (2004) Chemoinformatics: a textbook. Wiley, Weinheim
10. Hopkins AL (2008) Network pharmacology: the next paradigm in drug discovery. Nat Chem Biol 4:682–690
11. Metz JT, Hajduk PJ (2010) Rational approaches to targeted polypharmacology: creating and navigating protein-ligand interaction networks. Curr Opin Chem Biol 14:498–504
12. Morrow JK, Tian L, Zhang S (2010) Molecular networks in drug discovery. Crit Rev Biomed Eng 38:143–156
13. Scheibye-Alsing K, Hoffmann S, Frankel A, Jensen P, Stadler PF, Mang Y, Tommerup N, Gilchrist MJ, Nygard AB, Cirera S, Jorgensen CB, Fredholm M, Gorodkin J (2009) Sequence assembly. Comput Biol Chem 33:121–136
14. Huang X (2002) Bioinformatics support for genome sequencing projects. In: Lengauer T (ed) *Bioinformatics—from genomes to drugs.* Wiley-VCH Verlag GmbH, Weinheim
15. Mihara M, Itoh T, Izawa T (2010) SALAD database: a motif-based database of protein annotations for plant comparative genomics. Nucleic Acids Res 38:D835–D842
16. Katayama S, Kanamori M, Hayashizaki Y (2004) Integrated analysis of the genome and the transcriptome by FANTOM. Brief Bioinform 5:249–258
17. Blanchette M (2007) Computation and analysis of genomic multi-sequence alignments. Annu Rev Genomics Hum Genet 8:193–213
18. Mungall CJ, Misra S, Berman BP, Carlson J, Frise E, Harris N, Marshall B, Shu S,

Kaminker JS, Prochnik SE, Smith CD, Smith E, Tupy JL, Wiel C, Rubin GM, Lewis SE (2002) An integrated computational pipeline and database to support whole-genome sequence annotation. Genome Biol 3: RESEARCH0081

19. Lewis SE, Searle SM, Harris N, Gibson M, Lyer V, Richter J, Wiel C, Bayraktaroglir L, Birney E, Crosby MA, Kaminker JS, Matthews BB, Prochnik SE, Smithy CD, Tupy JL, Rubin GM, Misra S, Mungall CJ, Clamp ME (2002) Apollo: a sequence annotation editor. Genome Biol 3:RESEARCH0082
20. Pirovano W, Heringa J (2010) Protein secondary structure prediction. Methods Mol Biol 609:327–348
21. Cozzetto D, Tramontano A (2008) Advances and pitfalls in protein structure prediction. Curr Protein Pept Sci 9:567–577
22. Altschul SF, Gish W, Miller W, Myers EW, Lipman DJ (1990) Basic local alignment search tool. J Mol Biol 215:403–410
23. Pearson WR (1990) Rapid and sensitive sequence comparison with FASTP and FASTA. Methods Enzymol 183:63–98
24. Thompson JD, Higgins DG, Gibson TJ (1994) CLUSTAL W: improving the sensitivity of progressive multiple sequence alignment through sequence weighting, position-specific gap penalties and weight matrix choice. Nucleic Acids Res 22:4673–4680
25. Lassmann T, Sonnhammer EL (2005) Kalign—an accurate and fast multiple sequence alignment algorithm. BMC Bioinform 6:298
26. Edgar RC (2004) MUSCLE: multiple sequence alignment with high accuracy and high throughput. Nucleic Acids Res 32:1792–1797
27. Notredame C, Higgins DG, Heringa J (2000) T-Coffee: a novel method for fast and accurate multiple sequence alignment. J Mol Biol 302:205–217
28. Tusnady GE, Simon I (1998) Principles governing amino acid composition of integral membrane proteins: application to topology prediction. J Mol Biol 283:489–506
29. Rost B, Liu J (2003) The PredictProtein server. Nucleic Acids Res 31:3300–3304
30. Cole C, Barber JD, Barton GJ (2008) The Jpred 3 secondary structure prediction server. Nucleic Acids Res 36:W197–W201
31. Guzzo AV (1965) The influence of amino-acid sequence on protein structure. Biophys J 5:809–822
32. Chou PY, Fasman GD (1974) Prediction of protein conformation. Biochemistry 13:222–245
33. Bonneau R, Baker D (2001) Ab initio protein structure prediction: progress and prospects. Annu Rev Biophys Biomol Struct 30:173–189
34. Simons KT, Kooperberg C, Huang E, Baker D (1997) Assembly of protein tertiary structures from fragments with similar local sequences using simulated annealing and Bayesian scoring functions. J Mol Biol 268:209–225
35. Hardin C, Pogorelov TV, Luthey-Schulten Z (2002) Ab initio protein structure prediction. Curr Opin Struct Biol 12:176–181
36. Sali A, Blundell TL (1993) Comparative protein modelling by satisfaction of spatial restraints. J Mol Biol 234:779–815
37. Kriwacki RW, Wu J, Tennant L, Wright PE, Siuzdak G (1997) Probing protein structure using biochemical and biophysical methods. Proteolysis, matrix-assisted laser desorption/ionization mass spectrometry, high-performance liquid chromatography and size-exclusion chromatography of p21Waf1/Cip1/Sdi1. J Chromatogr A 777:23–30
38. Kasprzak AA (2007) The use of FRET in the analysis of motor protein structure. Methods Mol Biol 392:183–197
39. Takeda-Shitaka M, Takaya D, Chiba C, Tanaka H, Umeyama H (2004) Protein structure prediction in structure based drug design. Curr Med Chem 11:551–558
40. Wlodawer A, Erickson JW (1993) Structure-based inhibitors of HIV-1 protease. Annu Rev Biochem 62:543–585
41. Kortagere S, Welsh WJ, Morrisey JM, Daly T, Ejigiri I, Sinnis P, Vaidya AB, Bergman LW (2010) Structure-based design of novel small-molecule inhibitors of *Plasmodium falciparum*. J Chem Inf Model 50:840–849
42. Kortagere S, Mui E, McLeod R, Welsh WJ. Rapid discovery of inhibitors of *Toxoplasma gondii* using hybrid structure-based computational approach. J Comput Aided Mol Des. 2011 May;25(5):403–11
43. Kortagere S, Madani N, Mankowski MK, Schön A, Zentner I, Swaminathan G, Princiotto A, Anthony K, Oza A, Sierra LJ, Passic SR, Wang X, Jones DM, Stavale E, Krebs FC, Martín-García J, Freire E, Ptak RG, Sodroski J, Cocklin S, Smith AB 3rd. Inhibiting Early-Stage Events in HIV-1 Replication by Small-Molecule Targeting of the HIV-1 Capsid. J Virol. 2012 Aug;86 (16):8472–81
44. Pazos F, Valencia A (2008) Protein co-evolution, co-adaptation and interactions. EMBO J 27:2648–2655

45. Hu CD, Chinenov Y, Kerppola TK (2002) Visualization of interactions among bZIP and Rel family proteins in living cells using bimolecular fluorescence complementation. Mol Cell 9:789–798
46. Chien CT, Bartel PL, Sternglanz R, Fields S (1991) The two-hybrid system: a method to identify and clone genes for proteins that interact with a protein of interest. Proc Natl Acad Sci U S A 88:9578–9582
47. Selbach M, Mann M (2006) Protein interaction screening by quantitative immunoprecipitation combined with knockdown (QUICK). Nat Methods 3:981–983
48. Gavin AC, Aloy P, Grandi P, Krause R, Boesche M, Marzioch M, Rau C, Jensen LJ, Bastuck S, Dumpelfeld B, Edelmann A, Heurtier MA, Hoffman V, Hoefert C, Klein K, Hudak M, Michon AM, Schelder M, Schirle M, Remor M, Rudi T, Hooper S, Bauer A, Bouwmeester T, Casari G, Drewes G, Neubauer G, Rick JM, Kuster B, Bork P, Russell RB, Superti-Furga G (2006) Proteome survey reveals modularity of the yeast cell machinery. Nature 440:631–636
49. Pellegrini M, Marcotte EM, Thompson MJ, Eisenberg D, Yeates TO (1999) Assigning protein functions by comparative genome analysis: protein phylogenetic profiles. Proc Natl Acad Sci U S A 96:4285–4288
50. Dandekar T, Snel B, Huynen M, Bork P (1998) Conservation of gene order: a fingerprint of proteins that physically interact. Trends Biochem Sci 23:324–328
51. Tan SH, Zhang Z, Ng SK (2004) ADVICE: automated detection and validation of interaction by co-evolution. Nucleic Acids Res 32: W69–W72
52. Aloy P, Russell RB (2003) InterPreTS: protein interaction prediction through tertiary structure. Bioinformatics 19:161–162
53. Aytuna AS, Gursoy A, Keskin O (2005) Prediction of protein-protein interactions by combining structure and sequence conservation in protein interfaces. Bioinformatics 21:2850–2855
54. Bader GD, Donaldson I, Wolting C, Ouellette BF, Pawson T, Hogue CW (2001) BIND—the biomolecular interaction network database. Nucleic Acids Res 29:242–245
55. Stark C, Breitkreutz BJ, Reguly T, Boucher L, Breitkreutz A, Tyers M (2006) BioGRID: a general repository for interaction datasets. Nucleic Acids Res 34:D535–D539
56. Xenarios I, Salwinski L, Duan XJ, Higney P, Kim SM, Eisenberg D (2002) DIP, the database of interacting proteins: a research tool for studying cellular networks of protein interactions. Nucleic Acids Res 30:303–305
57. Peri S, Navarro JD, Amanchy R, Kristiansen TZ, Jonnalagadda CK, Surendranath V, Niranjan V, Muthusamy B, Gandhi TK, Gronborg M, Ibarrola N, Deshpande N, Shanker K, Shivashankar HN, Rashmi BP, Ramya MA, Zhao Z, Chandrika KN, Padma N, Harsha HC, Yatish AJ, Kavitha MP, Menezes M, Choudhury DR, Suresh S, Ghosh N, Saravana R, Chandran S, Krishna S, Joy M, Anand SK, Madavan V, Joseph A, Wong GW, Schiemann WP, Constantinescu SN, Huang L, Khosravi-Far R, Steen H, Tewari M, Ghaffari S, Blobe GC, Dang CV, Garcia JG, Pevsner J, Jensen ON, Roepstorff P, Deshpande KS, Chinnaiyan AM, Hamosh A, Chakravarti A, Pandey A (2003) Development of human protein reference database as an initial platform for approaching systems biology in humans. Genome Res 13:2363–2371
58. Kerrien S, Alam-Faruque Y, Aranda B, Bancarz I, Bridge A, Derow C, Dimmer E, Feuermann M, Friedrichsen A, Huntley R, Kohler C, Khadake J, Leroy C, Liban A, Lieftink C, Montecchi-Palazzi L, Orchard S, Risse J, Robbe K, Roechert B, Thorneycroft D, Zhang Y, Apweiler R, Hermjakob H (2007) IntAct—open source resource for molecular interaction data. Nucleic Acids Res 35:D561–D565
59. Zanzoni A, Montecchi-Palazzi L, Quondam M, Ausiello G, Helmer-Citterich M, Cesareni G (2002) MINT: a molecular INTeraction database. FEBS Lett 513:135–140
60. Ogmen U, Keskin O, Aytuna AS, Nussinov R, Gursoy A (2005) PRISM: protein interactions by structural matching. Nucleic Acids Res 33: W331–W336
61. Keskin O, Ma B, Nussinov R (2005) Hot regions in protein–protein interactions: the organization and contribution of structurally conserved hot spot residues. J Mol Biol 345:1281–1294
62. Chen YC, Lo YS, Hsu WC, Yang JM (2007) 3D-partner: a web server to infer interacting partners and binding models. Nucleic Acids Res 35:W561–W567
63. Jansen R, Yu H, Greenbaum D, Kluger Y, Krogan NJ, Chung S, Emili A, Snyder M, Greenblatt JF, Gerstein M (2003) A Bayesian networks approach for predicting protein-protein interactions from genomic data. Science 302:449–453
64. Monk NA (2003) Unravelling nature's networks. Biochem Soc Trans 31:1457–1461
65. Uetz P, Finley RL Jr (2005) From protein networks to biological systems. FEBS Lett 579:1821–1827

66. Schrattenholz A, Groebe K, Soskic V (2010) Systems biology approaches and tools for analysis of interactomes and multi-target drugs. Methods Mol Biol 662:29–58
67. Lowe JA, Jones P, Wilson DM (2010) Network biology as a new approach to drug discovery. Curr Opin Drug Discov Dev 13:524–526
68. Kell DB (2006) Systems biology, metabolic modelling and metabolomics in drug discovery and development. Drug Discov Today 11:1085–1092
69. Xie L, Bourne PE (2011) Structure-based systems biology for analyzing off-target binding. Curr Opin Struct Biol 21(2):189–199
70. Imoto S, Higuchi T, Goto T, Tashiro K, Kuhara S, Miyano S (2003) Combining microarrays and biological knowledge for estimating gene networks via Bayesian networks. Proc IEEE Comput Soc Bioinform Conf 2:104–113
71. Needham CJ, Manfield IW, Bulpitt AJ, Gilmartin PM, Westhead DR (2009) From gene expression to gene regulatory networks in *Arabidopsis thaliana*. BMC Syst Biol 3:85
72. Otero JM, Papadakis MA, Udatha DB, Nielsen J, Panagiotou G (2010) Yeast biological networks unfold the interplay of antioxidants, genome and phenotype, and reveal a novel regulator of the oxidative stress response. PLoS One 5:e13606
73. Teusink B, Westerhoff HV, Bruggeman FJ (2010) Comparative systems biology: from bacteria to man. Wiley Interdiscip Rev Syst Biol Med 2:518–532
74. Neumuller RA, Perrimon N (2010) Where gene discovery turns into systems biology: genome-scale RNAi screens in Drosophila. Wiley Interdiscip Rev Syst Biol Med 3:471–478
75. Bier E, Bodmer R (2004) Drosophila, an emerging model for cardiac disease. Gene 342:1–11
76. Gianchandani EP, Chavali AK, Papin JA (2010) The application of flux balance analysis in systems biology. Wiley Interdiscip Rev Syst Biol Med 2:372–382
77. Neves SR, Iyengar R (2009) Models of spatially restricted biochemical reaction systems. J Biol Chem 284:5445–5449
78. Czock D, Markert C, Hartman B, Keller F (2009) Pharmacokinetics and pharmacodynamics of antimicrobial drugs. Exp Opin Drug Metab Toxicol 5:475–487
79. Chien JY, Friedrich S, Heathman MA, de Alwis DP, Sinha V (2005) Pharmacokinetics/pharmacodynamics and the stages of drug development: role of modeling and simulation. AAPS J 7:E544–E559
80. Danhof M, de Jongh J, De Lange EC, Della Pasqua O, Ploeger BA, Voskuyl RA (2007) Mechanism-based pharmacokinetic-pharmacodynamic modeling: biophase distribution, receptor theory, and dynamical systems analysis. Annu Rev Pharmacol Toxicol 47:357–400
81. Paul Lee WN, Wahjudi PN, Xu J, Go VL (2010) Tracer-based metabolomics: concepts and practices. Clin Biochem 43:1269–1277
82. Chipman KC, Singh AK (2011) Using stochastic causal trees to augment Bayesian networks for modeling eQTL datasets. BMC Bioinform 12:7
83. Hou L, Wang L, Qian M, Li D, Tang C, Zhu Y, Deng M, Li F (2011) Modular analysis of the probabilistic genetic interaction network. Bioinformatics 27:853
84. Villar HO, Hansen MR (2009) Mining and visualizing the chemical content of large databases. Curr Opin Drug Discov Dev 12:367–375
85. Langer T, Hoffmann R, Bryant S, Lesur B (2009) Hit finding: towards 'smarter' approaches. Curr Opin Pharmacol 9:589–593
86. Fujita T, Hansch C. Analysis of the structure-activity relationship of the sulfonamide drugs using substituent constants. J Med Chem. 1967 Nov;10(6):991–1000
87. Golbraikh A, Tropsha A (2002) Beware of q2! J Mol Graph Model 20:269–276
88. Kubinyi H (2002) High throughput in drug discovery. Drug Discov Today 7:707–709
89. Kubinyi H, Hamprecht FA, Mietzner T (1998) Three-dimensional quantitative similarity-activity relationships (3D QSiAR) from SEAL similarity matrices. J Med Chem 41:2553–2564
90. Free SM Jr, Wilson JW (1964) A mathematical contribution to structure-activity studies. J Med Chem 7:395–399
91. Brown RD, Martin YC (1996) Use of structure activity data to compare structure-based clustering methods and descriptors for use in compound selection. J Chem Inform Comput Sci 36:12
92. Brown RD, Martin YC (1997) The information content of 2D and 3D structural descriptors relevant to ligand-receptor binding. J Chem Inform Comput Sci 37:9
93. Cramer RD, Patterson DE, Bunce JD (1988) Comparative molecular-field analysis (Comfa). 1. Effect of shape on binding of steroids to carrier proteins. J Am Chem Soc 110:8

94. Bravi G, Gancia E, Mascagni P, Pegna M, Todeschini R, Zaliani A (1997) MS-WHIM, new 3D theoretical descriptors derived from molecular surface properties: a comparative 3D QSAR study in a series of steroids. J Comput Aided Mol Des 11:79–92
95. Belvisi L, Bravi G, Scolastico C, Vulpetti A, Salimbeni A, Todeschini R (1994) A 3D QSAR approach to the search for geometrical similarity in a series of nonpeptide angiotensin II receptor antagonists. J Comput Aided Mol Des 8:211–220
96. Silverman BD, Platt DE (1996) Comparative molecular moment analysis (CoMMA): 3D-QSAR without molecular superposition. J Med Chem 39:2129–2140
97. Hopfinger AJ, Wang S, Tokarski JS, Jin B, Albuquerque M, Madhav PJ, Duraiswami C (1997) Construction of 3D-QSAR models using the 4D-QSAR analysis formalism. J Am Chem Soc 119:15
98. Vedani A, Briem H, Dobler M, Dollinger H, McMasters DR (2000) Multiple-conformation and protonation-state representation in 4D-QSAR: the neurokinin-1 receptor system. J Med Chem 43:4416–4427
99. Lukacova V, Balaz S (2003) Multimode ligand binding in receptor site modeling: implementation in CoMFA. J Chem Inf Comput Sci 43:2093–2105
100. Lill MA, Vedani A, Dobler M (2004) Raptor: combining dual-shell representation, induced-fit simulation, and hydrophobicity scoring in receptor modeling: application toward the simulation of structurally diverse ligand sets. J Med Chem 47:6174–6186
101. Lill MA, Dobler M, Vedani A (2006) Prediction of small-molecule binding to cytochrome P450 3A4: flexible docking combined with multidimensional QSAR. ChemMedChem 1:73–81
102. Vedani A, Dobler M, Lill MA (2005) Combining protein modeling and 6D-QSAR. Simulating the binding of structurally diverse ligands to the estrogen receptor. J Med Chem 48:3700–3703
103. Vedani A, Dobler M (2002) 5D-QSAR: the key for simulating induced fit? J Med Chem 45:2139–2149
104. Norinder U (2005) In silico modelling of ADMET-a minireview of work from 2000 to 2004. SAR QSAR Environ Res 16:1–11
105. Yoshida F, Topliss JG (2000) QSAR model for drug human oral bioavailability. J Med Chem 43:2575–2585
106. Martin YC (2005) A bioavailability score. J Med Chem 48:3164–3170
107. Worth AP, Bassan A, De Bruijn J, Gallegos Saliner A, Netzeva T, Patlewicz G, Pavan M, Tsakovska I, Eisenreich S (2007) The role of the European Chemicals Bureau in promoting the regulatory use of (Q)SAR methods. SAR QSAR Environ Res 18:111–125
108. Leach AR, Gillet VJ, Lewis RA, Taylor R (2010) Three-dimensional pharmacophore methods in drug discovery. J Med Chem 53:539–558
109. Martin YC, Bures MG, Danaher EA, DeLazzer J, Lico I, Pavlik PA (1993) A fast new approach to pharmacophore mapping and its application to dopaminergic and benzodiazepine agonists. J Comput Aided Mol Des 7:83–102
110. Jones G, Willett P, Glen RC (1995) A genetic algorithm for flexible molecular overlay and pharmacophore elucidation. J Comput Aided Mol Des 9:532–549
111. Chang C, Swaan PW (2006) Computational approaches to modeling drug transporters. Eur J Pharm Sci 27:411–424
112. Patel Y, Gillet VJ, Bravi G, Leach AR (2002) A comparison of the pharmacophore identification programs: Catalyst, DISCO and GASP. J Comput Aided Mol Des 16:653–681
113. Ekins S, Johnston JS, Bahadduri P, D'Souza VM, Ray A, Chang C, Swaan PW (2005) In vitro and pharmacophore-based discovery of novel hPEPT1 inhibitors. Pharm Res 22:512–517
114. Chang C, Bahadduri PM, Polli JE, Swaan PW, Ekins S (2006) Rapid identification of P-glycoprotein substrates and inhibitors. Drug Metab Dispos 34:1976–1984
115. Ekins S, Kim RB, Leake BF, Dantzig AH, Schuetz EG, Lan LB, Yasuda K, Shepard RL, Winter MA, Schuetz JD, Wikel JH, Wrighton SA (2002) Application of three-dimensional quantitative structure-activity relationships of P-glycoprotein inhibitors and substrates. Mol Pharmacol 61:974–981
116. Ekins S, Kim RB, Leake BF, Dantzig AH, Schuetz EG, Lan LB, Yasuda K, Shepard RL, Winter MA, Schuetz JD, Wikel JH, Wrighton SA (2002) Three-dimensional quantitative structure-activity relationships of inhibitors of P-glycoprotein. Mol Pharmacol 61:964–973
117. Bednarczyk D, Ekins S, Wikel JH, Wright SH (2003) Influence of molecular structure on substrate binding to the human organic cation transporter, hOCT1. Mol Pharmacol 63:489–498
118. Chang C, Pang KS, Swaan PW, Ekins S (2005) Comparative pharmacophore

modeling of organic anion transporting polypeptides: a meta-analysis of rat Oatp1a1 and human OATP1B1. J Pharmacol Exp Ther 314:533–541
119. Suhre WM, Ekins S, Chang C, Swaan PW, Wright SH (2005) Molecular determinants of substrate/inhibitor binding to the human and rabbit renal organic cation transporters hOCT2 and rbOCT2. Mol Pharmacol 67:1067–1077
120. Ekins S, Swaan PW (2004) Computational models for enzymes, transporters, channels and receptors relevant to ADME/TOX. Rev Comp Chem 20:333–415
121. Clement OO, Mehl AT (2000) HipHop: pharmacophore based on multiple common-feature alignments. IUL, San Diego, CA
122. Evans DA, Doman TN, Thorner DA, Bodkin MJ (2007) 3D QSAR methods: phase and catalyst compared. J Chem Inf Model 47:1248–1257
123. Bahadduri PM, Polli JE, Swaan PW, Ekins S (2010) Targeting drug transporters—combining in silico and in vitro approaches to predict in vivo. Methods Mol Biol 637:65–103
124. Ekins S, Ecker GF, Chiba P, Swaan PW (2007) Future directions for drug transporter modeling. Xenobiotica 37:1152–1170
125. Diao L, Ekins S, Polli JE (2010) Quantitative structure activity relationship for inhibition of human organic cation/carnitine transporter. Mol Pharm 7(6):2120–2131
126. Zheng X, Ekins S, Rauffman J-P, Polli JE (2009) Computational models for drug inhibition of the human apical sodium-dependent bile acid transporter. Mol Pharm 6:1591–1603
127. Diao L, Ekins S, Polli JE (2009) Novel inhibitors of human organic cation/carnitine transporter (hOCTN2) via computational modeling and in vitro testing. Pharm Res 26:1890–1900
128. Gao Q, Yang L, Zhu Y (2010) Pharmacophore based drug design approach as a practical process in drug discovery. Curr Comput Aided Drug Des 6:37–49
129. Keri G, Szekelyhidi Z, Banhegyi P, Varga Z, Hegymegi-Barakonyi B, Szantai-Kis C, Hafenbradl D, Klebl B, Muller G, Ullrich A, Eros D, Horvath Z, Greff Z, Marosfalvi J, Pato J, Szabadkai I, Szilagyi I, Szegedi Z, Varga I, Waczek F, Orfi L (2005) Drug discovery in the kinase inhibitory field using the Nested Chemical Library technology. Assay Drug Dev Technol 3:543–551
130. Kortagere S, Welsh WJ (2006) Development and application of hybrid structure based method for efficient screening of ligands binding to G-protein coupled receptors. J Comput Aided Mol Des 20:789–802
131. Ekins S, Waller CL, Swaan PW, Cruciani G, Wrighton SA, Wikel JH (2000) Progress in predicting human ADME parameters in silico. J Pharmacol Toxicol Methods 44:251–272
132. Ekins S, Ring BJ, Grace J, McRobie-Belle DJ, Wrighton SA (2000) Present and future in vitro approaches for drug metabolism. J Pharm Tox Methods 44:313–324
133. Ekins S, Ring BJ, Bravi G, Wikel JH, Wrighton SA (2000) Predicting drug-drug interactions in silico using pharmacophores: a paradigm for the next millennium. In: Guner OF (ed) Pharmacophore perception, development, and use in drug design. IUL, San Diego, pp 269–299
134. Ekins S, Obach RS (2000) Three dimensional-quantitative structure activity relationship computational approaches of prediction of human in vitro intrinsic clearance. J Pharmacol Exp Ther 295:463–473
135. Ekins S, Bravi G, Binkley S, Gillespie JS, Ring BJ, Wikel JH, Wrighton SA (2000) Three and four dimensional-quantitative structure activity relationship (3D/4D-QSAR) analyses of CYP2C9 inhibitors. Drug Metab Dispos 28:994–1002
136. Ekins S, Bravi G, Wikel JH, Wrighton SA (1999) Three dimensional quantitative structure activity relationship (3D-QSAR) analysis of CYP3A4 substrates. J Pharmacol Exp Ther 291:424–433
137. Ekins S, Bravi G, Ring BJ, Gillespie TA, Gillespie JS, VandenBranden M, Wrighton SA, Wikel JH (1999) Three dimensional-quantitative structure activity relationship analyses of substrates for CYP2B6. J Pharm Exp Ther 288:21–29
138. Ekins S, Bravi G, Binkley S, Gillespie JS, Ring BJ, Wikel JH, Wrighton SA (1999) Three and four dimensional-quantitative structure activity relationship (3D/4D-QSAR) analyses of CYP2D6 inhibitors. Pharmacogenetics 9:477–489
139. Ekins S, Bravi G, Binkley S, Gillespie JS, Ring BJ, Wikel JH, Wrighton SA (1999) Three and four dimensional-quantitative structure activity relationship analyses of CYP3A4 inhibitors. J Pharm Exp Ther 290:429–438
140. Lagorce D, Sperandio O, Galons H, Miteva MA, Villoutreix BO (2008) FAF-Drugs2: free ADME/tox filtering tool to assist drug discovery and chemical biology projects. BMC Bioinform 9:396

141. Villoutreix BO, Renault N, Lagorce D, Sperandio O, Montes M, Miteva MA (2007) Free resources to assist structure-based virtual ligand screening experiments. Curr Protein Pept Sci 8:381–411
142. Ekins S (2007) Computational toxicology: risk assessment for pharmaceutical and environmental chemicals. Wiley, Hoboken, NJ
143. Wang J, Hou T (2009) Recent advances on in silico ADME modeling. Annu Rep Comput Chem 5:101–127
144. Jorgensen WL, Duffy EM (2002) Prediction of drug solubility from structure. Adv Drug Deliv Rev 54:355–366
145. Wang J, Hou T, Xu X (2009) Aqueous solubility prediction based on weighted atom type counts and solvent accessible surface areas. J Chem Inf Model 49:571–581
146. Delaney JS (2005) Predicting aqueous solubility from structure. Drug Discov Today 10:289–295
147. Votano JR, Parham M, Hall LH, Kier LB, Hall LM (2004) Prediction of aqueous solubility based on large datasets using several QSPR models utilizing topological structure representation. Chem Biodivers 1:1829–1841
148. Hou TJ, Zhang W, Xia K, Qiao XB, Xu XJ (2004) ADME evaluation in drug discovery. 5. Correlation of Caco-2 permeation with simple molecular properties. J Chem Inf Comput Sci 44:1585–1600
149. Jung E, Kim J, Kim M, Jung DH, Rhee H, Shin JM, Choi K, Kang SK, Kim MK, Yun CH, Choi YJ, Choi SH (2007) Artificial neural network models for prediction of intestinal permeability of oligopeptides. BMC Bioinform 8:245
150. Thomas VH, Bhattachar S, Hitchingham L, Zocharski P, Naath M, Surendran N, Stoner CL, El-Kattan A (2006) The road map to oral bioavailability: an industrial perspective. Exp Opin Drug Metab Toxicol 2:591–608
151. Zheng X, Ekins S, Raufman JP, Polli JE (2009) Computational models for drug inhibition of the human apical sodium-dependent bile acid transporter. Mol Pharm 6:1591–1603
152. Varma MV, Ambler CM, Ullah M, Rotter CJ, Sun H, Litchfield J, Fenner KS, El-Kattan AF (2010) Targeting intestinal transporters for optimizing oral drug absorption. Curr Drug Metab 11:730–742
153. Chang C, Swaan PW (2006) Computer optimization of biopharmaceutical properties. In: Ekins S (ed) Computer applications in pharmaceutical research and development. Wiley, Hoboken, NJ, pp 495–512
154. Chang C, Ekins S, Bahadduri P, Swaan PW (2006) Pharmacophore-based discovery of ligands for drug transporters. Adv Drug Del Rev 58:1431–1450
155. Hamilton RD, Foss AJ, Leach L (2007) Establishment of a human in vitro model of the outer blood-retinal barrier. J Anat 211:707–716
156. Loscher W, Potschka H (2005) Drug resistance in brain diseases and the role of drug efflux transporters. Nat Rev Neurosci 6:591–602
157. Abraham MH, Ibrahim A, Zhao Y, Acree WE Jr (2006) A data base for partition of volatile organic compounds and drugs from blood/plasma/serum to brain, and an LFER analysis of the data. J Pharm Sci 95:2091–2100
158. Kortagere S, Chekmarev D, Welsh WJ, Ekins S (2008) New predictive models for blood-brain barrier permeability of drug-like molecules. Pharm Res 25:1836–1845
159. Zhang L, Zhu H, Oprea TI, Golbraikh A, Tropsha A (2008) QSAR modeling of the blood-brain barrier permeability for diverse organic compounds. Pharm Res 25:1902–1914
160. Kortagere S, Ekins S (2010) Troubleshooting computational methods in drug discovery. J Pharmacol Toxicol Methods 61:67–75
161. Grant MA (2009) Protein structure prediction in structure-based ligand design and virtual screening. Comb Chem High Throughput Screen 12:940–960
162. Sjogren B, Blazer LL, Neubig RR (2010) Regulators of G protein signaling proteins as targets for drug discovery. Prog Mol Biol Transl Sci 91:81–119
163. Willett P (2003) Similarity-based approaches to virtual screening. Biochem Soc Trans 31:603–606
164. Ebalunode JO, Zheng W (2010) Molecular shape technologies in drug discovery: methods and applications. Curr Top Med Chem 10:669–679
165. Horvath D (2011) Pharmacophore-based virtual screening. Methods Mol Biol (Clifton, NJ) 672:261–298
166. Yang SY (2010) Pharmacophore modeling and applications in drug discovery: challenges and recent advances. Drug Discov Today 15:444–450
167. Ebalunode JO, Zheng W, Tropsha A (2011) Application of QSAR and shape pharmacophore modeling approaches for targeted chemical library design. Methods Mol Biol (Clifton, NJ) 685:111–133

168. Halperin I, Ma B, Wolfson H, Nussinov R (2002) Principles of docking: an overview of search algorithms and a guide to scoring functions. Proteins 47:409–443
169. Lorber DM, Shoichet BK (2005) Hierarchical docking of databases of multiple ligand conformations. Curr Top Med Chem 5:739–749
170. Koca J (1998) Travelling through conformational space: an approach for analyzing the conformational behaviour of flexible molecules. Prog Biophys Mol Biol 70:137–173
171. Miller MD, Kearsley SK, Underwood DJ, Sheridan RP (1994) FLOG: a system to select 'quasi-flexible' ligands complementary to a receptor of known three-dimensional structure. J Comput Aided Mol Des 8:153–174
172. Sousa SF, Fernandes PA, Ramos MJ (2006) Protein-ligand docking: current status and future challenges. Proteins 65:15–26
173. Kuntz ID, Blaney JM, Oatley SJ, Langridge R, Ferrin TE (1982) A geometric approach to macromolecule-ligand interactions. J Mol Biol 161:269–288
174. Rarey M, Kramer B, Lengauer T, Klebe G (1996) A fast flexible docking method using an incremental construction algorithm. J Mol Biol 261:470–489
175. Halgren TA, Murphy RB, Friesner RA, Beard HS, Frye LL, Pollard WT, Banks JL (2004) Glide: a new approach for rapid, accurate docking and scoring. 2. Enrichment factors in database screening. J Med Chem 47:1750–1759
176. Welch W, Ruppert J, Jain AN (1996) Hammerhead: fast, fully automated docking of flexible ligands to protein binding sites. Chem Biol 3:449–462
177. Junmei Wang TH, Chen L, Xiaojie Xu (1999) Conformational analysis of peptides using Monte Carlo simulations combined with the genetic algorithm. Chemom Intell Lab Syst 45:5
178. Goodsell DS, Olson AJ (1990) Automated docking of substrates to proteins by simulated annealing. Proteins 8:195–202
179. Jones G, Willett P, Glen RC, Leach AR, Taylor R (1997) Development and validation of a genetic algorithm for flexible docking. J Mol Biol 267:727–748
180. Kitchen DB, Decornez H, Furr JR, Bajorath J (2004) Docking and scoring in virtual screening for drug discovery: methods and applications. Nat Rev Drug Discov 3:935–949
181. Verkhivker GM, Bouzida D, Gehlhaar DK, Rejto PA, Arthurs S, Colson AB, Freer ST, Larson V, Luty BA, Marrone T, Rose PW (2000) Deciphering common failures in molecular docking of ligand-protein complexes. J Comput Aided Mol Des 14:731–751
182. Cornell WD, Cieplak P, Bayly CI, Gould IR, Merz KM Jr, Ferguson DM, Spellmeyer DC, Fox T, Caldwell JW, Kollman PA (1995) A second generation force field for the simulation of proteins, nucleic acids, and organic molecules. J Am Chem Soc 117:19
183. Halgren T (1996) Merck molecular force field. I. Basis, form, scope, parameterization, and performance of MMFF94. J Comput Chem 17:490–519
184. Clark M, Crammer RD, Van Opdenbosch N (1989) Validation of the general purpose tripos 5.2 force field. J Comput Chem 10:30
185. Bohm HJ (1992) LUDI: rule-based automatic design of new substituents for enzyme inhibitor leads. J Comput Aided Mol Des 6:593–606
186. Eldridge MD, Murray CW, Auton TR, Paolini GV, Mee RP (1997) Empirical scoring functions: I. The development of a fast empirical scoring function to estimate the binding affinity of ligands in receptor complexes. J Comput Aided Mol Des 11:425–445
187. Gohlke H, Hendlich M, Klebe G (2000) Knowledge-based scoring function to predict protein-ligand interactions. J Mol Biol 295:337–356
188. Muegge I, Martin YC (1999) A general and fast scoring function for protein-ligand interactions: a simplified potential approach. J Med Chem 42:791–804
189. DeWitte RS, Shakhnovich EI (1996) SMoG: de novo design method based on simple, fast, and accurate free energy estimates. 1. Methodology and supporting evidence. J Am Chem Soc 118:11
190. Charifson PS, Corkery JJ, Murcko MA, Walters WP (1999) Consensus scoring: a method for obtaining improved hit rates from docking databases of three-dimensional structures into proteins. J Med Chem 42:5100–5109
191. Zwanzig R (1954) High-temperature equation of state by a perturbation method. J Chem Phys 22:1420–1426
192. Gilson MK, Zhou HX (2007) Calculation of protein-ligand binding affinities. Annu Rev Biophys Biomol Struct 36:21–42
193. Singh N, Warshel A (2010) Absolute binding free energy calculations: on the accuracy of computational scoring of protein-ligand interactions. Proteins 78:1705–1723
194. Leach AR (2001) Molecular modelling principles and applications, 2nd edn. Pearson Education Ltd, New York, NY

195. Beutler TC, Mark AE, Vanschaik RC, Gerber PR, van Gunsteren WF (1994) Avoiding singularities and numerical instabilities in free-energy calculations based on molecular simulations. Chem Phys Lett 222:529–539
196. Zacharias M, Straatsma TP, McCammon JA (1994) Separation-shifted scaling, a new scaling method for Lenard-Jones interactions in thermodynamic integration. J Chem Phys 100:9025–9031
197. Jorgensen W, Chandrasekhar J, Madura J, Klein M (1983) Comparison of simple potential functions for simulating liquid water. J Chem Phys 79:926–935
198. Berendsen HJ, Postma JP, van Gunsteren WF, Hermans J (1981) Interaction models for water in relation to protein hydration. In: Pullman B (ed) Intermolecular forces. D. Reidel Publishing Co., Dordrecht, pp 331–342
199. Åqvist J, Medina C, Samuelsson JE (1994) A new method for predicting binding affinity in computer-aided drug design. Protein Eng 7:385–391
200. Åqvist J, Hansson T (1996) On the validity of electrostatic linear response in polar solvents. J Phys Chem 100:9512–9521
201. Hansson T, Marelius J, Åqvist J (1998) Ligand binding affinity prediction by linear interaction energy methods. J Comput Aided Mol Des 12:27–35
202. Almlöf M, Carlsson J, Åqvist J (2007) Improving the accuracy of the linear interaction energy method for solvation free energies. J Chem Theory Comput 3:2162–2175
203. Almlöf, M. (2007) Computational Methods for Calculation of Ligand-Receptor Binding Affinities Involving Protein and Nucleic Acid Complexes, In Cell and Molecular Biology, p 53, Uppsala University, Uppsala, Sweden
204. Almlof M, Brandsdal BO, Aqvist J (2004) Binding affinity prediction with different force fields: examination of the linear interaction energy method. J Comput Chem 25:1242–1254
205. Jorgensen WL, Maxwell DS, Tirado-Rives J (1996) Development and testing of the OPLS all-atom force field on conformational energetics and properties of organic liquids. J Am Chem Soc 118:11225–11236
206. Orrling KM, Marzahn MR, Gutierrez-de-Teran H, Aqvist J, Dunn BM, Larhed M (2009) α-Substituted norstatines as the transition-state mimic in inhibitors of multiple digestive vacuole malaria aspartic proteases. Bioorg Med Chem 17:5933–5949
207. Kerrigan JE, Ragunath C, Kandra L, Gyemant G, Liptak A, Janossy L, Kaplan JB, Ramasubbu N (2008) Modeling and biochemical analysis of the activity of antibiofilm agent Dispersin B. Acta Biol Hung 59:439–451
208. Zhou R, Friesner RA, Ghosh A, Rizzo RC, Jorgensen WL, Levy RM (2001) New linear interaction method for binding affinity calculations using a continuum solvent model. J Phys Chem B 105:10388–10397
209. Zoete V, Meuwly M, Karplus M (2004) Investigation of glucose binding sites on insulin. Proteins 55:568–581
210. Liu S, Zhou LH, Wang HQ, Yao ZB (2010) Superimposing the 27 crystal protein/inhibitor complexes of beta-secretase to calculate the binding affinities by the linear interaction energy method. Bioorg Med Chem Lett 20:6533–6537
211. Alam MA, Naik PK (2009) Applying linear interaction energy method for binding affinity calculations of podophyllotoxin analogues with tubulin using continuum solvent model and prediction of cytotoxic activity. J Mol Graph Model 27:930–943
212. Alzate-Morales JH, Contreras R, Soriano A, Tunon I, Silla E (2007) A computational study of the protein-ligand interactions in CDK2 inhibitors: using quantum mechanics/molecular mechanics interaction energy as a predictor of the biological activity. Biophys J 92:430–439
213. Wolber G, Langer T (2005) LigandScout: 3-D pharmacophores derived from protein-bound ligands and their use as virtual screening filters. J Chem Inf Model 45:160–169
214. Tan L, Batista J, Bajorath J (2010) Computational methodologies for compound database searching that utilize experimental protein-ligand interaction information. Chem Biol Drug Des 76:191–200

Part II

Mathematical and Computational Modeling

Chapter 4

Best Practices in Mathematical Modeling

Lisette G. de Pillis and Ami E. Radunskaya

Abstract

Mathematical modeling is a vehicle that allows for explanation and prediction of natural phenomena. In this chapter we present guidelines and best practices for developing and implementing mathematical models, using cancer growth, chemotherapy, and immunotherapy modeling as examples.

Key words: Mathematical modeling, Modeling tutorial, Cancer, Immunology, Chemotherapy, Immunotherapy

1. Introduction and Overview

Mathematics is a concise language that encourages clarity of communication. Mathematical modeling is a process that makes use of the power of mathematics as a language and a tool to develop helpful descriptions of natural phenomena. Mathematical models of biological and medical processes are useful for a number of reasons. These include the following: clarity in communication; safe hypothesis testing; predictions; treatment personalization; new treatment protocol testing.

1.1. Clarity in Communication

Describing a phenomenon using mathematics forces clarity of communication. The process of choosing mathematical terms requires one to be precise, and implicit assumptions are less likely to slip by. This formulation in mathematical terms is sometimes called a "formal model" to distinguish it from, for example, an experimental model, such as a "mouse model" (1).

Brad Reisfeld and Arthur N. Mayeno (eds.), *Computational Toxicology: Volume I*, Methods in Molecular Biology, vol. 929, DOI 10.1007/978-1-62703-050-2_4, © Springer Science+Business Media, LLC 2012

For example, suppose a clinician wants to model treating a cancer with chemotherapy. Describing the process using mathematics forces us to clarify certain assumptions such as the following:

1. Is the tumor heterogeneous? Is it liquid or solid?
2. Does the immune system have any effect on the tumor; does it slow growth or stimulate growth? Do we need to include the immune system in the model? If so, what effect does the chemotherapy have on the immune system?
3. Does the treatment depend on tumor vasculature and, therefore, does the vasculature need to be included in the mathematical model?
4. Does the tumor develop resistance to the drug? Is the phase of the cell cycle an important consideration in treatment?

A mathematical, or "formal," model makes clear which features are most important when considering chemotherapy treatment in a particular case. In the remainder of this chapter we illustrate the process of building up the formal model from the simplest level to a more complicated model, as required to answer a specific question. We use the treatment of cancer as our running example.

1.2. Safe Hypothesis Testing

A useful mathematical model may allow one to test the possible mechanisms behind certain observed phenomena. For example, the reasons some patients go into remission from cancer and never relapse, while others do relapse, are not fully understood. A mathematical model, however, can allow us to test the hypothesis that the strength of a patient's immune system plays a significant role in whether or not a patient will experience relapse.

1.3. Predictions

Mathematical models can be used to predict system performance under otherwise un-testable conditions. One cannot and should not experiment on patients the way one can experiment with a mathematical model. For example, with a mathematical model, one can make predictions about disease progression if a patient does not receive any treatment, and one can also test new combination therapies and alternate protocols without endangering a patient's health or safety.

1.4. Treatment Personalization

A calibrated mathematical model can be used to test personalized treatments. In practice, essentially identical treatments are given to a broad array of patients who are not identical. A mathematical model, however, allows us to take into consideration patient-specific features such as the strength of their immune response and their response to treatments. A variety of scenarios and patient profiles can therefore be efficiently and safely addressed using the mathematical model.

1.5. New Treatment Protocol Testing

A useful mathematical model may allow one to test new medical interventions: a variety of hypothetical interventions can be analyzed through the formal model much more quickly, inexpensively and safely than can be done using clinical trials.

2. Modeling Philosophy

There are two main approaches to developing a mathematical model from the ground up. One is to start with the most complicated model that includes everything and pare it down. Alternatively, in the Occam's Razor approach, one starts with the simplest model possible and then builds it up as necessary. We recommend using the Occam's Razor approach, only adding complexity when the simple model is not sufficient to achieve the desired results.

Paul Sanchez (1) suggests keeping in mind the following guidelines when developing a mathematical model.

- Start with the simplest model possible that captures the essential features of the system of interest.
- Build in small steps—Once your simple model is working, add features incrementally to make it more realistic. Be sure to only add *one thing at a time*, and *test* the model after each addition.
- Keep only those additional features that actually improve the model: is the more complicated model more useful in answering the questions that you need addressed?
- Always compare incremental improvements with previous, simpler versions of the model. Do not hesitate to go back to earlier models, or to start over, considering a different approach.

We can summarize these guidelines in the *Goldilocks Principle*: a mathematical model should be not too complicated, but not too simple either. There is always a trade-off between complexity and tractability. On the one hand, a highly complicated model can have many variables and many parameters, making it appear more realistic. It could be difficult, however, if not impossible, to estimate the large number of parameters. Each parameter estimate introduces some new degree of uncertainty, thus potentially diminishing the usefulness of the model. In addition, it is typically very difficult to mathematically analyze a system with a large number of variables. On the other hand, a simpler model might be analytically tractable but too unrealistic: the simple model may not be able to answer the question of interest.

There is an art to modeling: there is not necessarily one correct model. Part of this art is to determine which elements are important to have in the model, and which elements we can ignore.

Keeping these philosophical guidelines in mind, we view the modeling process in terms of the following five step approach.

We illustrate each step in the context of modeling chemotherapy of cancer. A more detailed implementation of the five-step process follows in the next section.

STEP 1: Ask the question.

Which chemotherapy protocol will most effectively control a patient's cancer?

STEP 2: Select the modeling approach.

There are many individuals, with many types of cancer. We first pick a type of cancer, for example a cancer of the blood. This allows us to use a modeling approach with no spatial component. Since the cell populations are large, a continuous and deterministic modeling approach is appropriate. Therefore, we develop an *ordinary differential equations model.*

STEP 3: Formulate the model.

We describe the interactions between the model elements as functional expressions. This is the step in which we write mathematical formulas that describe the system behavior.

STEP 4: Solve and validate.

Find a solution to the mathematical model, either numerically or analytically. This typically involves the estimation of model parameters, and *calibration* of the model to the specific situation being studied. In this example we would need tumor growth data from patients, and data on responses to a particular chemotherapy against which we could compare model outcomes. Does the model adequately describe observed behavior? If necessary, go back to step 2 and revise the model.

STEP 5: Answer the question.

Interpret the results from step 4. Have we determined a reasonable chemotherapy protocol? Does this give rise to new questions? If so, return to step 1.

Figure 1 gives a visual representation of our five-step modeling process.

3. Example: Implementation of the Modeling Process

In this section we illustrate the five-step modeling process by using it to develop a mathematical model of tumor response to the immune system. We then extend this model to study the effects of chemotherapy on the system.

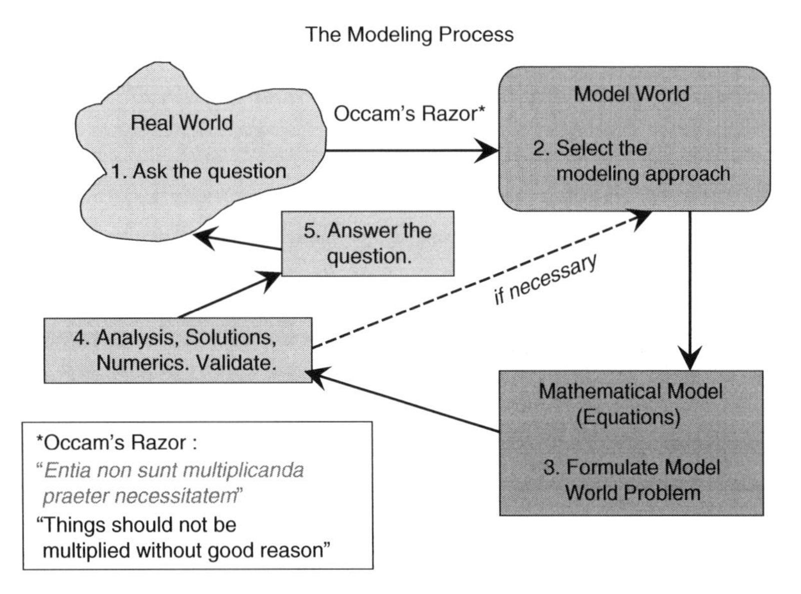

Fig. 1. The Modeling Process. This diagram shows the five steps of the modeling process, with possible loops to illustrate successive model refinement.

Step 1: Ask the Question.

How does the immune system affect tumor cell growth? Could it be responsible for tumor "dormancy"—when the tumor apparently regresses for a significant amount of time, followed by aggressive recurrence?

Step 2: Select the Modeling Approach.

We need the model to track tumor and immune populations over time. Following the philosophy of Occam's Razor, we assume that the populations are homogeneous, i.e., individual cells are identical, and we assume that populations are well mixed, which means that each individual in one population is equally likely to interact with any individual in the other population. This simplification allows us to neglect spatial changes, and so we only track population size changes. The next step is to consider which intrinsic population model we will use.

Population models can be roughly divided into several different types. One type of model considers the evolution of the populations by keeping track of the population sizes at discrete points in time, for example every year, day, or hour. These are (appropriately) called *discrete time* models. *Continuous time* models represent time as a continuum, and describe the evolution of the population sizes as continuous functions of time. These two types of models are illustrated in Fig. 2.

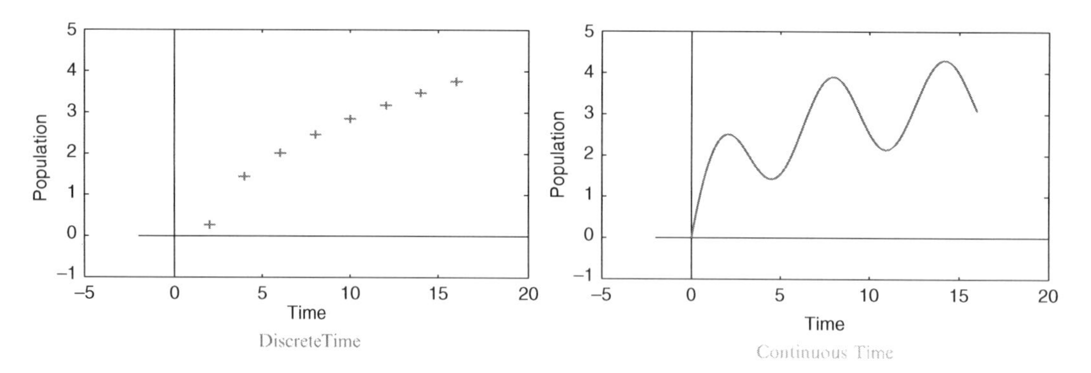

Fig. 2. Discrete versus continuous models. In a *discrete* model, the system is defined at a finite set of points (*left panel*), while in a *continuous* model, the system is defined on a continuous interval of time points (*right panel*).

Another broad distinction between mathematical models is whether the law of evolution is *deterministic* or *stochastic*.

In deterministic models, populations evolve according to a fixed law that can be formulated as a function. Formally, this function describes a *difference equation* in the discrete case:

$$P(t_{i+1}) = P(t_i) + F[P(t_i)]$$

and a *differential equation* in the continuous case:

$$\frac{\mathrm{d}P}{\mathrm{d}t} = F[P(t)].$$

In these equations, P is the population size, t_i is the ith time point in the discrete model, t is the time in the continuous model, and F is the law of evolution. In general, there could be many populations interacting, so F could be a multivariable function. It is possible that the evolution of the population is affected by time or other variables, so that these might be independent variables of the function, F as well.

A stochastic, or probabilistic model, assumes that the evolution of the populations cannot be described by a function. Rather, it assumes that the population's size over time is determined by random events that are described by some probability distribution. Since some aspects of the environment might be random, while others might not, it is also reasonable to consider models that combine both deterministic and stochastic elements.

In our example, we want to describe the evolution of large cell populations in the body. Physiological processes can be described at any point in time, and, at least on an observational scale, time appears to be continuous. If populations are relatively large, as in the case of a population of tumor cells forming a tumor, or the immune cells in an individual's body, they appear to evolve continuously over time. We therefore choose to use a continuous model to describe these populations. Furthermore, inter-cellular reactions are described by empirically determined rates, so that we are able to formulate deterministic laws that describe how the two populations change over time. Hence, for this example, we choose a model that is *continuous* and *deterministic*. We remark that there are analytical advantages to using continuous functions as well: we can integrate them and, if they are smooth enough, we can differentiate them. We would like to take advantage of the analytical tools applicable to systems of differential equations, and so at this time we choose to represent our system of two populations as a *system of differential equations*:

$$\frac{\mathrm{d}E}{\mathrm{d}t} = F_1(E, T)$$
$$\frac{\mathrm{d}T}{\mathrm{d}t} = F_2(E, T),$$

where E denotes the immune cell populations, and T denotes the tumor cell population.

Remark: We point out that, while focusing here on differential equations models, these are appropriate only when a continuous description of the variables is appropriate (many cells, many time points). Also, ordinary differential equations (ODE's) often assume that the populations are *well-mixed*. For an example of a model of tumor—immune interactions that has discrete and stochastic elements, see ref. (2). In the more complicated model described in the referenced paper, spatial variation and a variety of tumor—immune interactions can be explored, at the cost of analytical tractability. The general modeling philosophy is independent of the choice of model type (continuous versus discrete), but we emphasize that the choice of model must be informed by the question being asked, as well as the analytical tools at hand.

Step 3: Formulate the Model.

Applying our philosophy of starting with the simplest possible model, we choose to track two cell populations: effector cells and tumor cells. Thus, our model contains two dependent variables:

$T(t)$ = Tumor cells (number of cells or density),
$E(t)$ = Immune Cells that kill tumor cells (Effectors) (number of cells or density),
and one independent variable: t (time).

The growth of each cell population can be divided into two components: population growth in isolation, and competitive interactions between populations. We initially focus on the first component, and begin by examining how each cell population might grow if it were isolated from the other.

3.1. Modeling Tumor Growth

The simplest growth process involves the cells dividing at a *constant rate*, giving the differential equation:

$$\frac{\mathrm{d}T}{\mathrm{d}t} = kT, k > 0.$$

This equation has the solution:

$$T(t) = T_0 \mathrm{e}^{kt},$$

where $T_0 = T(0)$ is the initial tumor population, the number of tumor cells, at time zero. A tumor large enough to be detected clinically is approximately 7 mm in diameter. This corresponds to a population of approximately 10^8 cells. Exponential growth implies that, with a doubling rate of 2 days, the tumor population would grow to 10^{11} cells and *weigh 1 kg, in only 20 days*! Thus, exponential growth seems physically unrealistic, even if the growth rate were much smaller, say, a doubling every 2 weeks.

Experiments show that tumor cells grow exponentially when the population is small, but growth slows down when the population is large. Figure 3 compares exponential growth and self-limiting growth on the same graph.

Figure 4 shows a graph plotting tumor growth rate ($\mathrm{d}T/\mathrm{d}t$) as a function of tumor size (T).

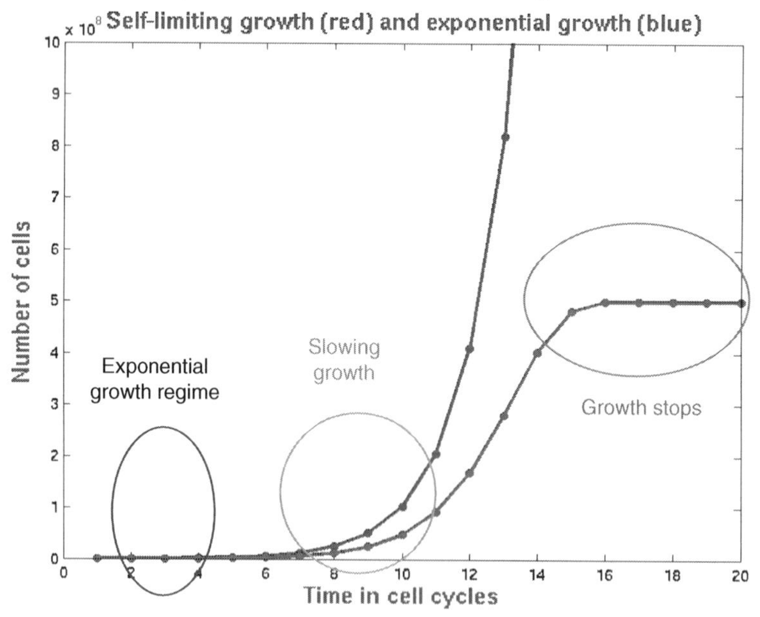

Fig. 3. Exponential versus self-limiting growth. A plot of the time-evolution of two models of population growth on the same axes. In the initial time period, both graphs look the same: this is the initial *exponential growth phase* of the self-limiting growth model. By time $t = 6$, the graphs of the two models begin to separate, and by time $t = 15$ the graph of the self-limiting growth model has leveled off, indicating that growth has stopped.

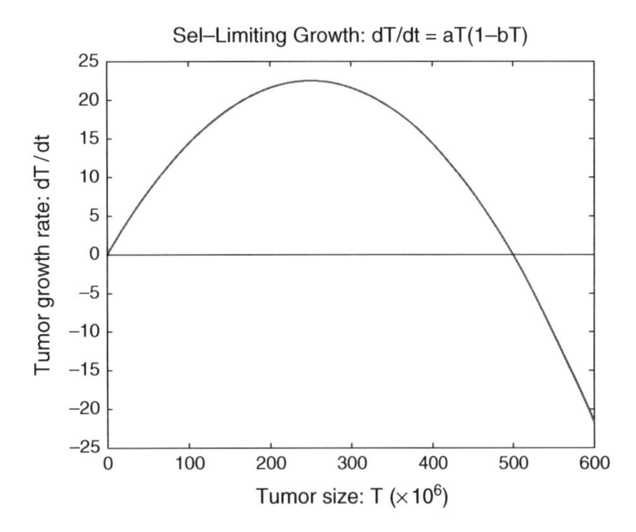

Fig. 4. Self-limiting growth. This graph of growth rate (dT/dt) against population (T, in units of 10^6 cells) shows one self-limiting growth model, the *logistic model*, (see Table 1: population growth models). Note that, for small population values, growth rate increases as the population increases, but as the population increases beyond $T = 250$, growth rate begins to decrease until it reaches zero at the *carrying capacity*, $T = 500$. This corresponds to the value at which the population levels off in Fig. 3. For values of the population, T, larger than 500, dT/dt is negative, indicating that a population larger than the carrying capacity will decrease until it reaches 500. Models of growth may be developed starting with an empirically derived graph, similar to this one, of the rate of growth against the population.

Note that the curve in Fig. 4 is closely approximated by a line with positive slope for small values of T, the slope decreases as T increases, and the growth *rate*, dT/dt, reaches a maximum when T is approximately 250. For larger population values, the growth rate decreases, reaching zero when $T = 500$. For values of T larger than 500, the growth rate is negative, indicating that the tumor cell population will *decrease* if the population exceeds 500. This curve could be described by the quadratic equation:

$$\frac{dT}{dt} = aT(1 - bT).$$

This is also known as the *logistic growth equation*. In this particular example, $b = 0.002$, and the model predicts that the population will grow if $0 < T < 500$, it will decrease if $T > 500$ and, if $T = 500$, the population size will remain constant. The quantity, $1/b$ gives the limiting tumor size (in this case 500), and it is called the carrying *capacity*. The model also predicts that, if $T = 0$, the growth rate is also zero. In the equation, a is the slope of the graph when $T = 0$, so we can interpret a as the *intrinsic growth rate* of the population: the rate of growth that occurs when the population is small enough to grow almost exponentially. The terms a and b are the two model *parameters*. They could be

Table 1
Growth laws

Growth law	Equation	Number of parameters
Logistic	$\frac{dN}{dt} = aN(1 - bN)$	Two parameters: a and b
Power	$\frac{dN}{dt} = aN^b$	Two parameters: a and b
Gompertz	$\frac{dN}{dt} = aN\ln(1/bN)$	Two parameters: a and b
von Bertalanffy	$\frac{dN}{dt} = aN((bN)^c - 1)$	Three parameters: a, b and c

This Table gives the equations and the number of parameters for four commonly used laws of population growth. A growth law should be chosen to fit to experimental data and according to the principle of "parameter parsimony." see Fig. 5

measured from data, or hypothesized from basic knowledge of the system.

There are many other possible formulas that could describe self-limiting growth, which can be found in the literature. Growth functions, in addition to logistic growth, that are frequently used to model tumor growth include the power growth law, von Bertalanffy growth, and Gompertz growth. The equations for these growth laws are given in Table 1.

Which intrinsic growth model is best? The answer depends on several factors, including the type of cancer being modeled, the location of the cancer in the body, etc. However, if we have tumor growth data against which to calibrate potential models, we recommend choosing an intrinsic growth model based on a best fit to the data.

In Fig. 5 we show an example of calibrating the four different intrinsic growth models against the same tumor growth data sets. Diefenbach et al. (3) provided tumor growth data from immunocompromised mice in three different sets of experiments, where the mice were challenged with 10^3, 10^4, and 10^5 tumor cells.

We used the Matlab® routine *fminsearch* to find the parameter values for each growth model that minimized the least squares distance to all three data sets simultaneously. Figure 5 shows the resulting model curves using the best-fit parameter values, with residual error bars (the model value minus the data value) shown under each of the four panels. We then compared the residual error for each of the four models. The two models shown in the right panels, the logistic (top) and von Bertalanffy (bottom), have smaller residuals than the other two models. We chose the logistic over von Bertalanffy using the principle of *parameter parsimony*, which prefers models with the fewest parameters. Note that the logistic model has two parameters: a and b, while the von Bertalanffy model has three parameters: a, b and c.

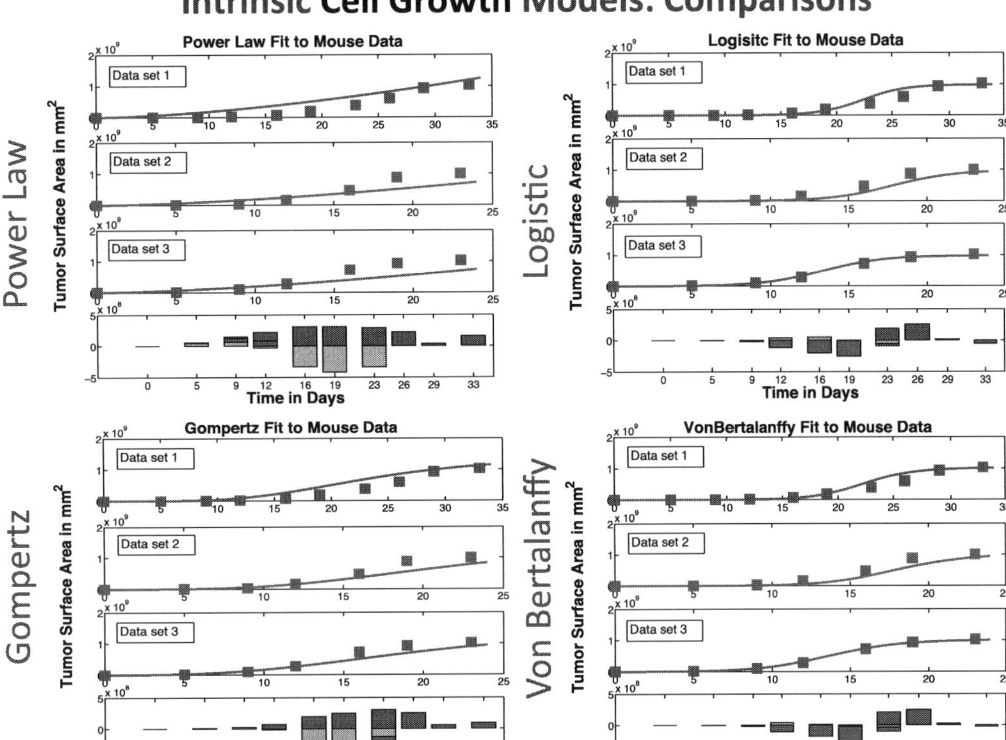

Fig. 5. A comparison of four growth laws. Data from ref. (3), which describes three different mouse experiments (marked as "Data set 1," "Data set 2," and "Data set 3," respectively), is used to fit four different growth laws. The model result is shown in *solid curves*, while the data points are shown by *filled squares*. In each case, the parameters of the models are chosen to minimize the least squares distance from the model's predicted values to the data. Residuals showing the difference between the predicted values and the data are shown as *bars* below the graphs in each case. Note that the first data set has more time points than the other two, so that the last three residuals are due to differences coming only from the first data set. The two models shown in the *left column*, the power law and the Gompertz models, show larger residuals than the two models depicted on the *right*, the logistic and the von Bertalanffy models. In this sense, the two models on the *right* are "better" than the two on the *left*. Using the principle of "parameter parsimony," which indicates that the model with the fewest parameters is preferable, the logistic model is "better" than the von Bertalanffy in predicting the outcome of the experiments represented by these data.

3.2. Modeling the Growth of Effector Cells

Cytotoxic effector cells include those immune cells that are capable of killing tumor cells, for example *Natural Killer (NK)* cells or *Killer T Cells (CTL)*; we combine all cytotoxic effector cells into one population, referred to as *Effector Cells*, denoted by E.

If we assume that there is a constant source of effector cells providing cells at a fixed rate, s, cells per unit time, and that the fraction of these cells that die off per unit time is another constant, d, we get the differential equation:

$$\frac{\mathrm{d}E}{\mathrm{d}t} = s - dE \quad \text{with} \quad s, d > 0.$$

3.3. Modeling Tumor and Immune Interactions

We now need to add to the immune cell equation a term that represents how the production of tumor-specific effector cells responds to the presence of the tumor. This function should incorporate the recognition of antigen by the immune system, and could take several different forms. The most common function describing the interaction of two populations describes the *Law of Mass Action*, which assumes that the effect of the interaction is directly proportional to the product of the two populations. For example, if the effector cells were stimulated by the tumor cells according to the Law of Mass Action, the equation would take on the following form:

$$\frac{\mathrm{d}E}{\mathrm{d}t} = s - dE + rET \quad \text{with} \quad s, d, r > 0.$$

Note, however, that the term rET implies that the larger the tumor, the greater the response of the immune system, and that this response could get infinitely large. The rate of production of tumor-specific effector cells is difficult to measure experimentally, however, it is known that the effector response saturates, and does not grow indefinitely. Therefore, the response function should be an *increasing* function of the number of tumor cells, but should be bounded above by some constant. One such function has a *Michaelis–Menten* form, where we replace rET by $rET/(\sigma + T)$ and the equation becomes:

$$\frac{\mathrm{d}E}{\mathrm{d}t} = s - dE + \frac{rET}{\sigma + T} \quad \text{with} \quad s, d, r, \sigma > 0.$$

We now must include the destructive effect that interactions between tumor cells and immune cells have on the cell populations. Biologically, we assume that tumor cells can be killed by effector cells while, in the process, the effector cells are deactivated following interactions with tumor cells. In this case, a mass action law makes sense, since the maximum number of cells that can be killed under this law will not exceed the total population. Putting these terms into the differential equations for effector cell and tumor growth gives the system of equations:

$$\frac{\mathrm{d}E}{\mathrm{d}t} = s - dE + \frac{rET}{\sigma + T} - c_1 ET \tag{1}$$

$$\frac{\mathrm{d}T}{\mathrm{d}t} = aT(1 - bT) - c_2 ET, \tag{2}$$

where c_1 and c_2 are also positive constants.

Note that the following assumptions are now included in the two-dimensional model:

- Effectors have a constant source.
- Effectors are recruited by tumor cells.

- Tumor cells can deactivate effectors (assume mass action law).
- Effectors have a natural death rate.
- Tumor cell population grows logistically (includes death already).
- Effector cells kill tumor cells (assume mass action law).

3.4. Dimensional Analysis

Once the model has been developed, it is very important to make sure that the units for the parameters and populations ("state variables") are balanced. Inappropriately mixing units is a common mistake that beginning modelers make. The first step in checking whether the equations are balanced is to specify the units of each model parameter. For example, suppose we want to examine the units in the Michaelis–Menten term. On the left, the units of dE/dt are cells per time, or #cells/day, since E represents the number of effectors cells, and t represents time. Each term of the equation on the right must then work out to #cells/day. Looking at the Michaelis–Menten term in Eq. 1, with the units in parentheses, and only the units of r and σ as yet to be determined, we have:

$$\frac{dE(\#\text{cells})}{dt(\text{day})} = \ldots \frac{r(?)E(\#\text{cells})T(\#\text{cells})}{\sigma(?) + T(\#\text{cells})} \ldots \tag{3}$$

We can now determine the proper units for the parameters r and σ in the Michaelis–Menten term. First, it is clear that the units of σ must be in units of cells, that is $\sigma(?)=\sigma(\#\text{ cells})$, for the denominator to make sense (we add number of cells to number of cells). Secondly, the parameter r must be in units of 1/day, since the left hand side of Eqs. 1 and 3 are in units of (# cells/day), each additive term on the right hand side of those equations must also be in units of (# cells/day). Thus, with $r \times (1/\text{day})$, the Michaelis–-Menten term units work out to be $r \times (1/\text{day}) \times (\#\text{cells})^2/(\#\text{cells}) = (\#\text{cells/day})$. The rest of the parameter units can be worked out in a similar fashion, giving the values for all of the parameters, which are shown in Table 2.

In summary, the differential equations model has one equation per variable. According to the mass balance principle, each equation gives the *total change* in the variable, which is equal to the *amount in* minus the *amount out*. Positive terms represent the amount going in, while negative terms represent the amount going in, while negative terms represent the amount going out. The form of each term is derived through biological considerations such as knowing there is a constant source of immune cells, and empirical observations, such as knowing that tumor growth is self-limiting. Figure 6 summarizes the elements in the formal mathematical model.

Step 4: Solve the System.

In general we cannot find an explicit function or formula for the solution (also known as a *closed form* solution) for a mathematical

Table 2
Description of parameters and their units

Parameter name	Description	Units
s	Constant immune cells source rate	# cells/day
σ	Steepness coefficient	# cells
r	Tumor recruitment rate of effector cells	1/day
c_1	Tumor deactivation rate of effectors	1/(cell–day)
d	Effector death rate	1/day
a	Intrinsic tumor growth rate	1/day
$1/b$	Tumor population carrying capacity	# cells
c_2	Effector kill rate of tumor cells	1/(cell–day)

The parameters in a model give insight into the quantities that affect the long-term behavior of the solutions. A dimensional analysis is useful as a check that the model equations are balanced, and this analysis should always be performed before solving the model and interpreting the results

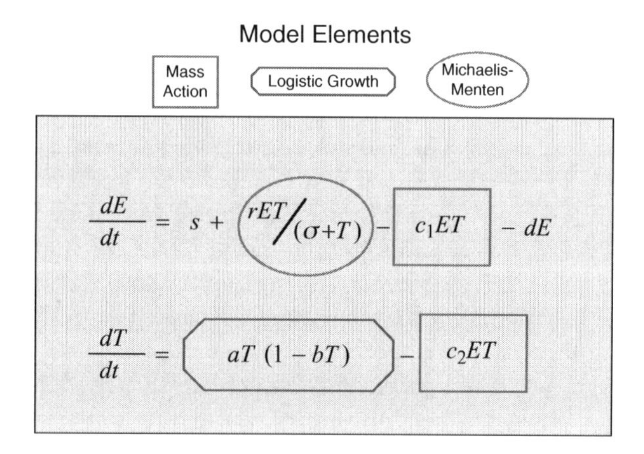

Fig. 6. Elements of the tumor-immune model. This figure shows the equations of the tumor immune model, with key functional elements *highlighted*. The negative mass action terms are inside a *rectangular border*, indicating that we assume that any cell is equally likely to interact with any other cell. The logistic growth term in the second differential equation describing tumor growth is shown inside a *hexagonal border*, this term indicates our assumption that tumor growth is self-limiting, even in the absence of immune cells. The recruitment term in the differential equation describing immune cell growth is shown *circled by an oval*. The Michaelis–Menten form of this term indicates our assumption that immune cell production in response to the presence of a tumor saturates at the level given by the parameter, *r*. Each term in a model is developed using knowledge of the system being modeled, for example the dynamics of cell growth and the kinetics of the immune response, as well as available experimental data showing the shape of growth and response curves.

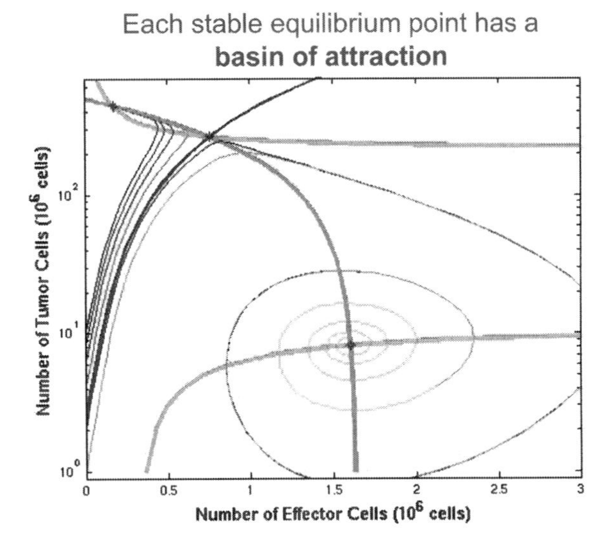

Fig. 7. A phase portrait of the tumor-immune model. This figure shows *orbits*, the result of numerical solutions to the tumor-immune model given in Eqs. 1 and 2. For this set of parameter values (given in Table 3), there are three equilibria, indicated by *asterisks*. Two equilibria are *stable*, so that nearby solutions are attracted to them. One of these represents a relatively large tumor size near the carrying capacity of 5×10^8 cells, while the other represents a relatively small tumor size of approximately 8.5×10^5 cells. The third equilibrium is *unstable*, but the orbits that converge to it, indicated by *dark lines*, serve to separate the two *basins of attraction* of the stable equilibria. This figure also depicts several representative orbits, or solutions of the model. Some of these orbits move up towards the large-tumor stable equilibrium, while the others spiral towards the small-tumor stable equilibrium. This phase portrait captures *all possible* behaviors of the model.

model, so we must use numerical solvers (computational solutions). However, we can use mathematical analysis to determine some of the *qualitative features* of the solution. The details are beyond the scope of this chapter, but see ref. (4) for a good reference.

3.5. Interpreting the Solution: The Phase Portrait

Figures 7 and 8 show a numerical solution of the model given in Eqs. 1 and 2. Figure 7 shows a *Phase Portrait*, which is a graph showing the evolution of the *state variables* of the system, in this case the effector cell population, $E(t)$, on the horizontal axis, and the tumor cell population, $T(t)$, plotted on the vertical axis. The phase portrait does not show time explicitly, but rather it shows how the two state variables of the system change over time with respect to each other. A phase portrait is a compact way of displaying all possible behaviors of the system at once. An *orbit* of the differential equation is a solution plotted as a curve in the phase plane. The graph shows six orbits (thin lines), going to one of three *equilibria*, or *steady states*, marked by asterisks in Fig. 7. The equilibria are found by setting the differential equations, Eqs. 1 and 2,

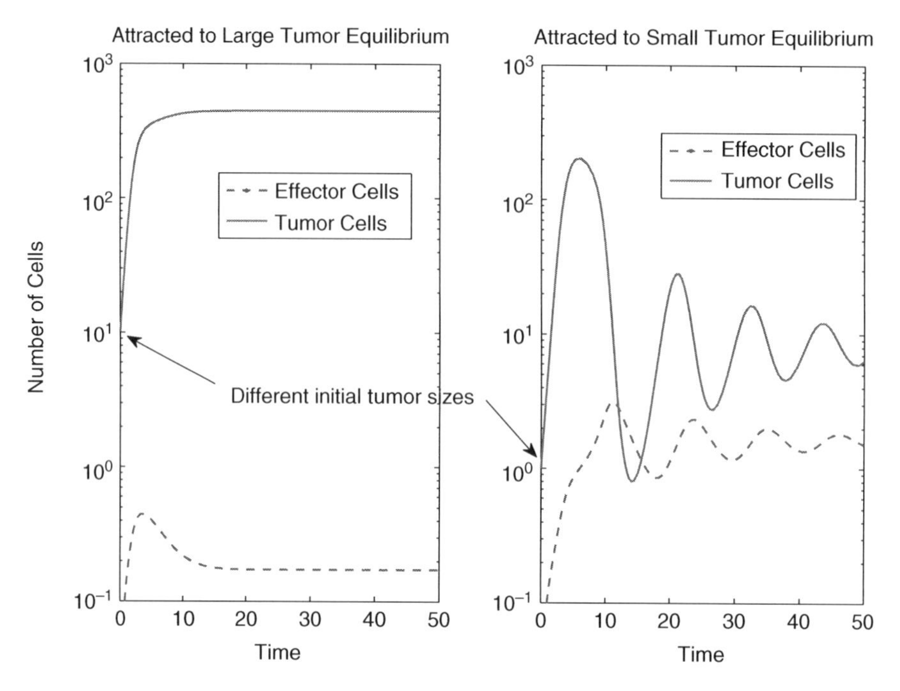

Fig. 8. Tumor-immune solution curves. This figure shows two distinct solutions of the tumor immune model using the same parameter values as in Fig. 7, (see Table 3). On the *left*, the solution converges to the large tumor equilibrium, while the graphs on the *right* show a solution converging to the smaller tumor equilibrium. The initial tumor population values for the two solutions are the only thing that differs: on the *left*, $T(0) = 10$, while on the *right* $T(0) = 1$ (in units of 10^6 cells).

equal to zero, and they describe possible long-term behaviors of the system. Equilibria are determined graphically by plotting *nullclines*, the set of points where each differential equation is zero (the thicker lines) and finding their points of intersection (the three asterisks). In this case, there are two *stable equilibria*, one at the upper left of the graph, representing a high tumor burden and a relatively low effector cell population, and one in the middle of the graph, representing a smaller tumor population and a higher effector cell population (notice that the tumor population is plotted using a *logarithmic scale* on the vertical axis.) Most orbits will approach one or the other of these equilibria, depending on where they initiate. The *Basin of Attraction* of a stable equilibrium is the set of orbits that converge to it: from a modeling perspective, determining the basins of attraction is equivalent to being able to predict the long-term outcome of a particular scenario. For example, if a patient's condition is represented by an orbit that starts in the basin of attraction of the small-tumor equilibrium, we predict that, even without treatment, the tumor will soon reach a relatively small steady state size. The third equilibrium, in between the other two, is *unstable*, and has only one orbit that approaches it from each side, one from below and one from above. This special orbit separates the two basins of attraction.

Figure 8 shows two sets of solution curves corresponding to two different scenarios. The set of solution curves on the left corresponds to an orbit in the basin of attraction of the high tumor equilibrium, while the set of curves on the right corresponds to an orbit in the basin of attraction of the smaller tumor equilibrium. The initial tumor size is $T(0) = 10$ in the first case, and $T(0) = 1$ in the second case. The initial effector cell populations and all parameter values are identical in both cases. Note that in the set of solutions on the right, the populations oscillate with decreasing amplitude before settling at the equilibrium. This is observed as a *spiraling* orbit in the phase portrait of Fig. 7.

Step 5: Interpret the Results—Did We Answer the Question?

Our question was: does the immune system explain the observed clinical phenomena known as "tumor dormancy" and "creeping through." In both of these scenarios, the tumor shrinks to an undetectable size, but after some time begins to grow again, sometimes quite aggressively. In the phase portrait analysis above, we do see solutions in the basin of attraction of the lower tumor equilibrium in which the tumor size oscillates around the equilibrium value. However, we do not see dormancy followed by aggressive regrowth, or "creeping through."

3.6. Reexamine Model Parameters and/or Assumptions

At this stage, before abandoning the model as inadequate, we choose to test other parameter ranges. In fact, with one change in a parameter value, we can get the "creeping through" phenomenon. The new phase portrait is shown in Fig. 9. Note that in this simulation, the value of s, the constant source rate of immune cells, has been decreased by a factor of 10, making the immune response particularly susceptible to the strength of the adaptive immune response. The parameter values are given in Table 3. Two solutions are shown in Fig. 9. The initial values of the immune and tumor cell populations are close to each other (the points are labeled on the graph), but one initial value results in a tumor that is small for a time, i.e., ***dormant***, then increases in size, gradually reaching the lower tumor equilibrium in an oscillatory manner. The other initial value results in a dormant phase followed by aggressive growth towards the large tumor equilibrium. Solution curves showing the tumor populations over time are shown in Fig. 10. In this current model, we now see both tumor dormancy and aggressive growth, so we have answered the question in the affirmative: the immune response ***can*** explain both tumor dormancy and creeping through.

Step 5, continued: Can We Use the Mathematical Model to Answer Other Questions? Now that we have a mathematical model of tumor growth with the immune response, we can return to our original question: "What is the best way to administer chemotherapy?" In particular, how can we determine the optimal doses and timings for treatments? It is probably impossible to find the "best" treatment, so

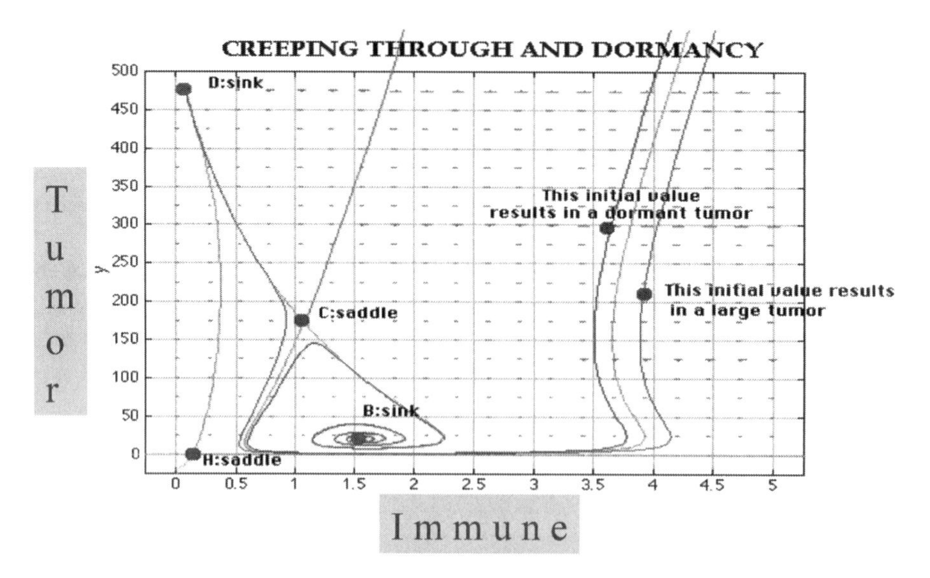

Fig. 9. Creeping-through and dormancy. This figure shows a phase portrait of the tumor-immune system, similar to Fig. 7. However, the parameter representing the constant immune source rate is smaller in this numerical solution of the model, resulting in a phase portrait with a different structure. There is an fourth, unstable equilibrium (marked "H"), in addition to the three equilibria shown in Fig. 7. In addition, the shape of the basins of attraction of the two stable equilibria ("D" and "B") are different, giving rise to the possibility of an observed phenomenon known as *creeping through.* In this scenario, a tumor decreases in size and remains small for some time (*dormancy*), but then begins to regrow until it reaches a dangerously large size. This phase portrait shows that, depending on the initial conditions, a tumor can exhibit dormancy followed by oscillating growth and shrinking until it approaches the smaller-tumor equilibrium (labeled "B"), or it can experience shrinkage, followed by aggressive growth, or creeping-through, to the larger-tumor equilibrium (labeled "D"). These two solutions are graphed as a function of time in Fig. 10. Parameter values are given in Table 3.

Table 3
Parameter values used in the simulations

Parameter name	Figures 7 and 8	Figures 9 and 10	Units: $c = 10^6$ cells, $d = 10^2$ days
s	0.1181	0.01	c/d
σ	20.19	20.19	c
r	1.131	1.131	$1/d$
c_1	0.00311	0.00311	$1/(c{-}d)$
d	0.3743	0.3743	$1/d$
a	1.636	1.636	$1/d$
b	0.002	0.002	$1/c$
c_2	1	1	$1/(c{-}d)$
$E(0)$	0.00001	3.5 and 3.6	c
$T(0)$	10 and 1	300	c

This Table lists all of the parameter values used to numerically solve the system given by Eqs. 1 and 2. These solutions are shown in Figs. 7 through 10

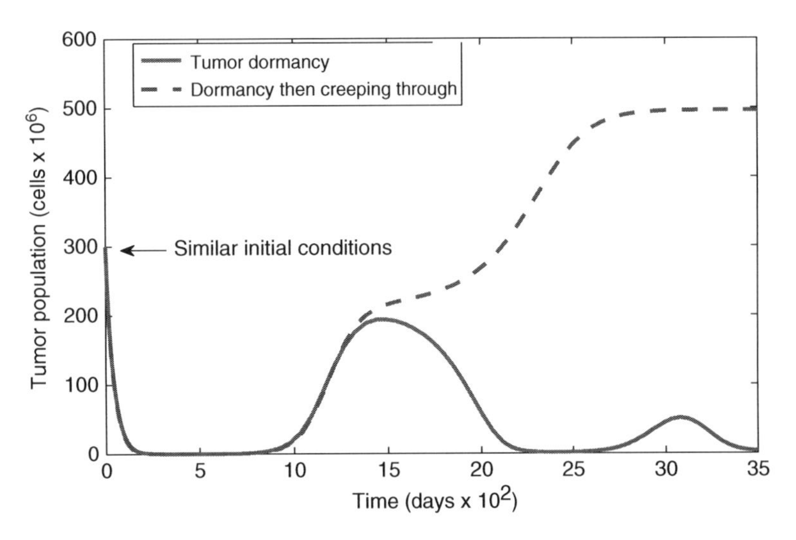

Fig. 10. Creeping-through solutions. This figure shows two solutions that are also shown in the phase portrait in Fig. 9. Only the values of the tumor populations over time are shown in this figure, so that the two initial conditions look identical, with $T(0) = 300$ ($\times\ 10^6$ cells). However, the initial effector cell populations differed, with $E(0) = 3.5 \times 10^6$ cells (*solid line*) for the solution that converges to the small tumor equilibrium ("dormancy"), and $E(0) = 3.6 \times 10^6$ cells (*dashed line*) for the solution that converges to the large tumor equilibrium. The fact that a larger initial immune cell population results in a worse outcome is a nonintuitive result of the structure of the phase portrait. To explain this phenomenon: if the initial immune population is large, the immune system is able to control the tumor for a while (for approximately 900 days, or 2½ years), but during this time the immune *response* declines to a point at which the tumor is able to escape the immune surveillance entirely. A similar scenario ensues with a slightly *smaller* initial effector cell population, but the tumor population does not get quite as small, and hence the immune response is just a bit larger, preventing the tumor from escaping to the large tumor equilibrium. Parameter values are given in Table 3.

instead we might be satisfied with answering whether there is a way to improve existing protocols. When we compare treatments, we must be able to make quantitative comparisons between outcomes. In this context, we could decide that a treatment is better if it minimizes the amount of tumor after a specific amount of time, while keeping toxicity levels low. In fact, other quantitative comparisons could be made: for example, we might want to minimize the total amount of drug given, or we might want to minimize the amount of time until the tumor is reduced to a specific size. In these cases the steps in the modeling process would be the same, and we select one set of criteria here for illustrative purposes. We refine our example model following the steps outlined above. In the process, we illustrate how mathematical techniques, in this case optimal control theory, can be applied to a mathematical model to answer a specific question.

Step 2: Select the Modeling Approach.

Since we have an existing ODE (ordinary differential equation) model, we add state variable, $u(t)$, to model the drug, and modify

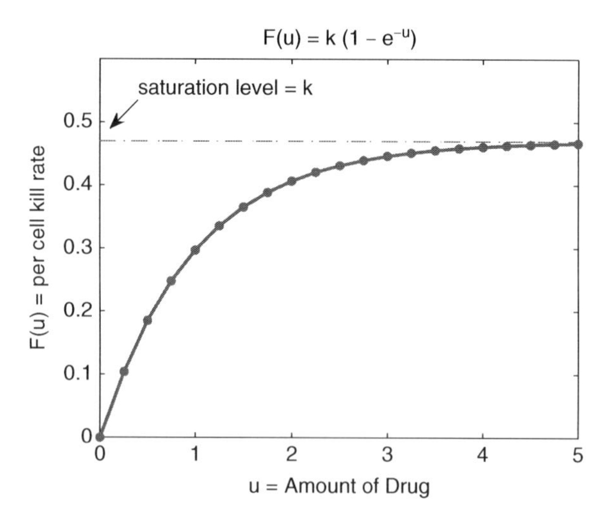

Fig. 11. Drug kill rate. This figure shows the graph of a function describing the fractional cell kill as a function of the amount of drug at the tumor site. The function was chosen to conform with the assumptions that (1) the kill rate increases with the amount of drug present, and (2) the kill rate saturates at a fixed level, in this case given by the parameter $k = 0.47$, indicated by the *dashed line*.

the equations to describe the interaction between the drug and the cells. We note, however, that our choice of modeling approach implies certain assumptions. For example, since we only describe total populations, all interactions are assumed to be homogeneous. In particular, we assume that the drug at the tumor site reaches all tumor cells with equal likelihood, and that all cells are affected in the same way. Since we want to monitor toxicity, we introduce another cell population that we denote by $N(t)$ for "normal" cells. This population will be adversely affected by the drug, and we consider a treatment to be "not too toxic" if this normal cell population is maintained above a certain prespecified amount.

Step 3: Formulate the Model.

As before, we use the mass conservation principle to develop the formal model for the extended model:

The change in population over time is equal to "amount in"−"amount out."

Applying this to the amount of drug, we assume that the drug goes in at a rate that we can control. Incorporating the time it takes for the drug to be transported from the injection site to the tumor site we denote the rate of influx of the drug to the tumor site by a function of time, $v(t)$. We assume that the drug decays exponentially, at a rate proportional to the existing amount:

$$\frac{du}{dt} = v(t) - d_2 u.$$

To model the interaction between the drug and the tumor cells, we use a mass action term of the form $F(u)T$, where $F(u)$ is a saturating function of the amount of drug, $u(t)$:

$$F(u) = k(1 - e^{-u}).$$

Figure 11 shows a graph of the function $F(u)$.

We assume that the drug is toxic to all cells, with the toxicity varying according to cell type. The parameter k is varied for each cell type to indicate the different levels of toxicity. To model the normal cell population, we use a formal model similar to the differential equation describing tumor cell growth, using a logistic growth law to describe the "amount in." To model the "amount out," in addition to the drug kill rate, a mass action term is used to describe competition with tumor cells. Finally, since we assume that normal cells and tumor cells compete, we add another competition term to reflect this in the equation for the tumor cells. This gives the following set of four differential equations for the extended model:

$$\frac{dE}{dt} = s - dE + \frac{rET}{\sigma + T} - c_1 ET - k_1(1 - e^{-u}) \tag{4}$$

$$\frac{dT}{dt} = a_1 T(1 - b_1 T) - c_2 ET - c_3 NT - k_2(1 - e^{-u}) \tag{5}$$

$$\frac{dN}{dt} = a_2 N(1 - b_2 N) - c_4 NT - k_3(1 - e^{-u}) \tag{6}$$

$$\frac{du}{dt} = v(t) - d_2 u. \tag{7}$$

Step 4: Solve the System.

Our question is now turned into the following mathematical problem that we must solve. We must find the function, $v(t)$, that will minimize the value of the tumor variable, T, while maintaining the normal cell variable, N, above a prescribed level. A mathematical solution to this problem can be obtained using optimization techniques based on Pontryagin's Maximum Principle. For more details see ref. (5). The results are shown in Fig. 12, which compares treatments on two hypothetical individuals, one whose initial immune population is 0.15 (in units of 10^5 cells), and the other with a slightly smaller immune cell population initially: $E(0) = 0.1 = 10^4$ cells. The top row of Fig. 12 shows simulations of the model when no chemotherapy treatment is given: the tumor population in both "patients" increases to a dangerous level. The middle row of Fig. 12 shows the results from the model if $v(t)$, the function describing the drug dosing, mimics that of a (hypothetical) "traditional" chemotherapy protocol, where drug is administered in bolus injections regularly every 3 days for 90 days. We note that this regular, pulsed therapy does

No Chemotherapy

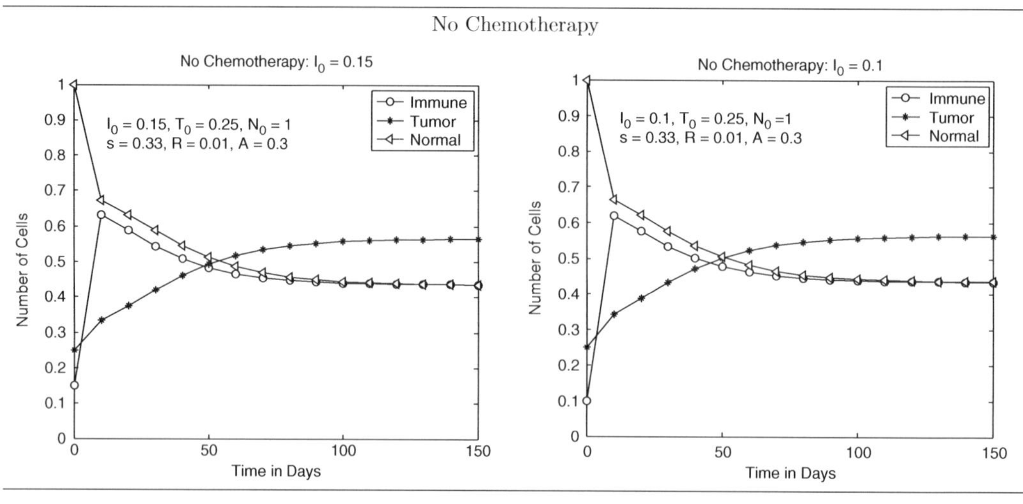

Traditional Pulsed Chemotherapy

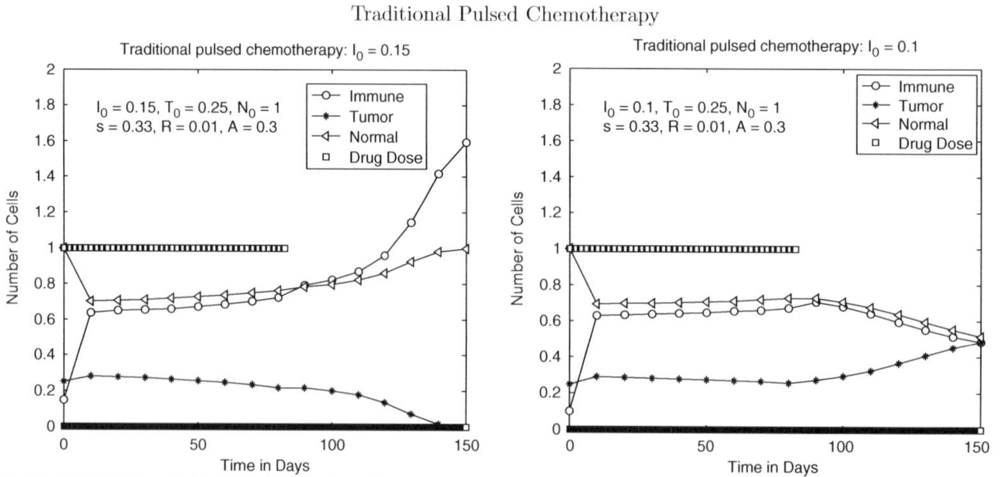

Solution of Optimization Problem: Irregularly-space, Extended Doses

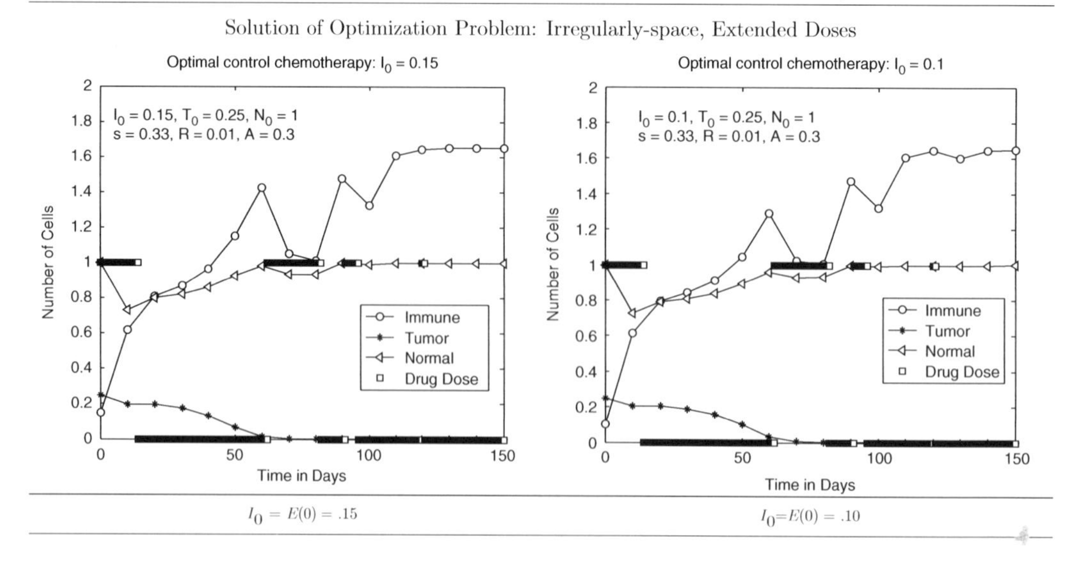

$I_0 = E(0) = .15$ $\qquad$ $I_0=E(0) = .10$

eliminate the tumor when the initial immune cell count is 0.15, but is unable to control tumor growth in the right-hand panel, when the initial immune cell count is only 0.1. The bottom row of Fig. 12 shows the numerical solution to the optimization problem, which indicates that the function $v(t)$ should be a step function that changes irregularly from the minimum (zero) to the maximum dosage. With this treatment schedule, tumor growth is controlled in *both* cases. We note that the total amount of drug given is identical in the treatment scenarios depicted in the first and second rows of Fig. 12. Furthermore, the value of the normal cells never falls below 0.75, which, in these scaled variables, is three-fourths of its usual level.

Step 5: Answer the Question.

We have seen that the mathematical model, which reproduced the clinically observed phenomena of tumor dormancy and "creeping through," can be used to study the effect of chemotherapy treatments. Using optimization theory, treatment protocols can be identified that, according to the model, improve the treatment outcomes, in the sense that the tumor volume is smaller at the end of a prescribed amount of time, while normal cells never get below three-fourths of their usual levels.

4. Conclusion

In this chapter we have given an overview of the modeling process, and a detailed example of the process applied to the development of a model of tumor growth, the immune response, and chemotherapy. We illustrated how new questions might arise as the model is developed, and how these new questions can lead to new model developments. In the case of the treatment of tumors, new therapies can

Fig. 12. A comparison of chemotherapy treatment protocols. This figure depicts three different solutions of the model system given by Eqs. 4 through 7. In the *top row*, the control function, $v(t)$, which describes the rate of drug delivery, is zero (no treatment). The *left graph* shows a solution with an initial value describing a patient with a strong immune system [$E(0) = 1.5 \times 10^5$ cells], while the *right graph* shows a patient with a weaker immune system [$E(0) = 1 \times 10^5$ cells]. Without treatment, both solutions show the tumor population (*asterisks*) converging to the larger tumor equilibrium. In the *second row*, the control function, $v(t)$ describes regularly spaced treatments: a bolus injection every 3 days for 90 days (the amount of drug at the tumor site is depicted with *squares*). The *left panel* shows that this treatment does control the tumor in the patient with the stronger immune system, but it is ineffective in the case of the patient with the weaker immune system, shown in the *right panel*. Note that in the *right panel*, the normal cells (*triangles*) and the immune cells (*circles*) are decreasing after treatment as the tumor grows. In the *third row* of the figure, the control function $v(t)$ is the solution to the optimal control problem. It is an irregularly varying function, and in both cases, the stronger and the weaker immune response shown in the *left* and *right panels*, the tumor is reduced to zero. Note that the total amount of drug delivered, the area under the graph of the function $v(t)$ is the same in the second and third rows.

also lead to new models. For example, the development of cancer vaccines has meant the introduction of new compartments into the model, such as the spleen, in order to include the proliferation of different types of immune cells due to the vaccine. Delays, representing the time required for the immune response to initiate, must also be included in the equations. Further model refinements include spatial variables to include tumor heterogeneity and the inclusion of stochastic components to model uncertainties and variation in the response of individual cells. see refs. (2), (6) and (7) for further model refinements. General articles about the modeling process that might be useful are refs. (1) and (8).

References

1. Sanchez PJ (2006) As simple as possible, but no simpler: a gentle introduction to simulation modeling. Proceedings of the 2006 winter simulation conference, pp 2–10
2. de Pillis LG, Mallett D, Radunskaya AE (2006) Spatial tumor—immune modeling. Comput Math Model Med 7(2):159–176
3. Diefenbach A, Jensen ER, Jamieson A, Raulet DH (2001) Rae1 and H60 ligands of the NKG2D receptor stimulate tumour immunity. Nature 413:165–171
4. Borrelli R, Coleman C (2004) Differential equations—a modeling perspective. Wiley, New York, NY
5. de Pillis LG, Radunskaya AE (2001) A mathematical tumor model with immune resistance and drug therapy: an optimal control approach. J Theor Med 3:79–100
6. de Pillis LG, Radunskaya AE, Wiseman CL (2005) A validated mathematical model of cell-mediated immune response to tumor growth. Cancer Res 65:7950–7958
7. de Pillis LG, Radunskaya AE (2006) Some promising approaches to tumor—immune modeling. In: Gumel A, Castillo-Chavez C, Mickens R and Clemence DP(eds) Mathematical studies on human disease dynamics: emerging paradigms and challenges. AMS Contemporary Mathematics Series
8. Cooke K (1981) On the construction and evaluation of mathematical models. In: Sabloff JA (ed) Simulations in Archaeology. University of New Mexico Press, Albuquerque, NM

Chapter 5

Tools and Techniques

Arthur N. Mayeno and Brad Reisfeld

Abstract

This chapter lists some of the software and tools that are used in computational toxicology, as presented in this volume.

Key words: Software, Mathematical modeling, Algorithms, Computational toxicology, Computational chemistry, Computational biology

1. Introduction

As the name "Computational Toxicology" implies, this field involves computations, often in the form of mathematical equations and computer algorithms, implemented either as part of a dedicated software package or through any variety of software languages. By algorithm, we mean a list of instructions that calculates a quantity of interest or performs a task. The need for computations arises due to the vast quantity of complex information that must be organized, filtered, and analyzed, and the desire to perform elaborate calculations that are impossible without the use of a computer, such as solving complex systems of coupled differential equations or performing Monte Carlo simulations, where a large number of repeated random samplings are required.

Over the course of these past few decades, as the field has matured, an increasing number of software packages have become available. Before acquiring a software application, it is crucial for prospective users to evaluate a number of factors, in addition to technical considerations. Some software packages are available at no cost while others are relatively expensive and may require the payment of annual, upgrade, and per-seat licensing fees. The documentation and level of support also vary widely across

Brad Reisfeld and Arthur N. Mayeno (eds.), *Computational Toxicology: Volume I*, Methods in Molecular Biology, vol. 929, DOI 10.1007/978-1-62703-050-2_5,

the range of available software. Many packages come with some level of technical support and, sometimes, documentation of validation of certain aspects of its performance, while others are completely unsupported, with no guarantee of performance or accuracy. Aside from evaluating the current level of support, it is also important to consider the long-term availability, vitality, and maintenance of the software. For instance, a software product with a strong user base and active maintainers and support staff may be updated with new features and bug fixes on an ongoing basis, while a niche software application produced in a research lab may not be available, or be maintained, past its period of use for a particular project. Finally, there is the issue of the license under which the software is distributed. Commercial software is often distributed under a proprietary software license where ownership of the software remains with the software publisher and the user is restricted from an extensive number of activities, such as reverse engineering, simultaneous use of the software by multiple users, and publication of benchmarks or performance tests. On the other end of the spectrum are a variety of free and open source licenses, some of which essentially grant the end users permission to do anything they wish with the source code in question, including the right to take the code and use it as part of closed-source software or software released under a proprietary software license.

Performing the calculations requires hardware, namely, computers. The type of computer and the computing "power" (the speed at which a computer can perform calculations) required depends on the nature of the calculations performed. For example, linear regression calculations (e.g., often used in quantitative structure–activity relationships) can be accomplished instantly on a spreadsheet, while other types of calculations (e.g., molecular dynamics of enzyme catalysis, which involve simulating the movement of atoms and molecules) may require weeks or months (or longer), even on "supercomputers." Thus, in addition to the software, computing resources are required, which can dramatically increase the cost of performing research in computational toxicology. Today, most software packages can run on one or more of the three most common computing platforms, i.e., Windows, Macintosh, and Linux/Unix, although some functionality (such as parallelization) may not be available on all platforms. Thus, it is essential to carefully investigate the software and hardware before purchase.

Another cautionary note worth mentioning has arisen due to the ease of performing calculations. As software packages have allowed calculations to be performed more easily, for example through the use of simplified user interfaces such as a "wizard," the user many not understand what is actually being performed. On the other hand, new users who attempt to use a complex analysis program without this simplified interface may be overwhelmed with the number of program options and settings that must be made to correctly solve their problem of interest. As such, researchers new to the area of computational sciences are wise to remember the phrase, "Garbage in, garbage out": the output results will be worthless if the input quantities are flawed. These input quantities comprise essentially everything used to perform the calculations, including the quality of the data itself, parameter values, assumptions made, structure of the model used, and so forth. A common mistake that new modelers make is the belief that the result the software package gives is "correct" and that the computed number is exact (i.e., precise to the number of digits output). Neither is true: the results are only as good as the input used and the appropriateness and correctness of the computational algorithm and implementation, and the precision of values produced is typically a result of a set threshold value and algorithm dependent. When using unvalidated software, the danger of erroneous calculations is even more likely. Even with the theoretical advances in the understanding of biology and the physical universe, experimental results are still the "gold standard" in nearly all cases (the exception being when experiments cannot be performed).

These caveats are not intended, by any means, to dissuade scientists and researchers from using and applying computational methodologies. Rather, they are mentioned here to help those who are new to the field to avoid these pitfalls. We believe that the "Golden Age of Computational Toxicology" is underway, and we hope that others will take advantage of and contribute to this burgeoning and practical field.

Table 1 outlines many of the software packages and tools presented in this volume. This list is not a comprehensive list of software and tools available, just a summary of some of those mentioned in the chapters. Depending on the purpose of their work, users—before acquiring software—should take the time to perform a thorough investigation of the software packages available and to understand which packages are acceptable for their needs, either as a purely research tool or as one used to support a regulatory filing.

Table 1
List of software packages, tools, and companies presented in this volume

Name/Developer	Description/URL
Chapter 1.2	
MOE (Molecular Operating Environment)	Software package contains Structure-Based Design; Pharmacophore Discovery; Protein & Antibody Modeling; Molecular Modeling & Simulations; Cheminformatics & (HTS) Quantitative Structure/Activity Relationship (QSAR); and Medicinal Chemistry Applications.
Chemical Computing Group, Inc.	http://www.chemcomp.com/software.htm

(continued)

Table 1 (continued)

Name/Developer	Description/URL
QiKProP	Package to predict parameters such as octanol/water and water/gas logP, logs, logBB, overall CNS activity, Caco-2 and MDCK cell permeabilities, human oral absorption, log K_{hsa} for human serum albumin binding, and log IC_{50} for HERG K^+-channel blockage.
Schrődinger, LLC	http://www.schrodinger.com/products/14/17/
SPARC[a]	Predictive modeling system that calculates large number of physical/chemical parameters from molecular structure and basic information about the environment.
US EPA, University of Georgia	http://www.epa.gov/extrmurl/research/projects/sparc/index.html; http://archemcalc.com/index.php
OpenEye	Company that offers various software packages, libraries, and toolkits.
OpenEye Scientific Software, Inc.	www.eyesopen.com
Chapter 1.3	
BLAST[a]	Sequence database search methods: finds regions of local similarity between sequences. The program compares nucleotide or protein sequences to sequence databases and calculates the statistical significance of matches. BLAST can be used to infer functional and evolutionary relationships between sequences as well as help identify members of gene families.
National Center for Biotechnology Information	http://blast.ncbi.nlm.nih.gov
FASTA[a]	Sequence database search methods: The FASTA programs find regions of local or global (new) similarity between Protein or DNA sequences, either by searching Protein or DNA databases or by identifying local duplications within a sequence. Other programs provide information on the statistical significance of an alignment. Like BLAST, FASTA can be used to infer functional and evolutionary relationships between sequences as well as help identify members of gene families.
William R. Pearson and the University of Virginia	http://fasta.bioch.virginia.edu/fasta_www2/fasta_list2.shtml
ClustalW[a]	Multiple sequence alignment methods.
Ch.EMBnet.org	http://www.ch.embnet.org/index.html
Kalign[a]	Multiple sequence alignment methods: a method employing the Wu-Manber string-matching algorithm.
European Bioinformatics Institute	http://www.ebi.ac.uk/Tools/msa/kalign/
MUSCLE[a]	Multiple sequence alignment method.
Robert C. Edgar	http://www.drive5.com/muscle/
T-Coffee[a]	Multiple sequence alignment methods: A collection of tools for Computing, Evaluating, and Manipulating Multiple Alignments of DNA, RNA, Protein Sequences, and Structures.
Center for Genomic Regulation (CRG)	http://www.tcoffee.org/

(continued)

Table 1 (continued)

Name/Developer	Description/URL
Predict Protein[b]	Secondary structure prediction methods: PredictProtein integrates feature prediction for secondary structure, solvent accessibility, transmembrane helices, globular regions, coiled-coil regions, structural switch regions, B-values, disorder regions, intra-residue contacts, protein–protein and protein–DNA binding sites, subcellular localization, domain boundaries, beta-barrels, cysteine bonds, metal binding sites, and disulfide bridges.
ROSTLAB.ORG	http://www.predictprotein.org/
JPred[a]	Secondary structure prediction method.
Geoff Barton, Bioinformatics and Computational Biology Research, University of Dundee, Scotland, UK	http://www.compbio.dundee.ac.uk/Software/JPred/jpred.html
ExPASy-tools[a]	ExPASy is the new SIB Bioinformatics Resource Portal which provides access to scientific databases and software tools in different areas of life sciences including proteomics, genomics, phylogeny, systems biology, evolution, population genetics, transcriptomics, etc.
Swiss Institute of Bioinformatics	http://expasy.org/tools/
PSI-BLAST[a]	PSI-BLAST is similar to NCBI BLAST except that it uses position-specific scoring matrices derived during the search; this tool is used to detect distant evolutionary relationships.
European Bioinformatics Institute	http://www.ebi.ac.uk/Tools/sss/psiblast/
BLAST[a]	The Basic Local Alignment Search Tool (BLAST) finds regions of local similarity between sequences. The program compares nucleotide or protein sequences to sequence databases and calculates the statistical significance of matches. BLAST can be used to infer functional and evolutionary relationships between sequences as well as help identify members of gene families.
National Center for Biotechnology Information	http://blast.ncbi.nlm.nih.gov/Blast.cgi
BIND[a]	Biomolecular Interaction Network Database (BIND) as a Web database. BIND offers methods common to related biology databases and specializations for its protein interaction data.
Christopher Hogue's Research Lab	http://www.blueprint.org/
BioGRID[a]	Biological General Repository for Interaction Datasets (BioGRID) database was developed to house and distribute collections of protein and genetic interactions from major model organism species.
TyersLab.com	http://www.thebiogrid.org
DIP[a]	The DIP™ database catalogs experimentally determined interactions between proteins. It combines information from a variety of sources to create a single, consistent set of protein–protein interactions.

(continued)

Table 1 (continued)

Name/Developer	Description/URL
Regents of the University of California and David Eisenberg	http://dip.doe-mbi.ucla.edu/dip/Main.cgi
HPRD[b]	Human Protein Reference Database (HPRD) is a protein database accessible through the Internet.
Institute of Bioinformatics in Bangalore, India, and the Pandey lab at Johns Hopkins University	http://www.hprd.org/
IntAct[a]	IntAct provides a freely available, open source database system and analysis tools for protein interaction data.
European Bioinformatics Institute (EBI)	http://www.ebi.ac.uk/intact/
MINT[a]	MINT, the Molecular INTeraction database. MINT focuses on experimentally verified protein–protein interactions mined from the scientific literature by expert curators.
University of Rome Tor Vergataand IRCCS Fondazione Santa Lucia	http://mint.bio.uniroma2.it/mint/Welcome.do
KEGG	KEGG (Kyoto Encyclopedia of Genes and Genomes) is a collection of online databases dealing with genomes, enzymatic pathways, and biological chemicals. The PATHWAY database records networks of molecular interactions in the cells, and variants of them specific to particular organisms.
Kanehisa Laboratories at Kyoto University and the University of Tokyo	http://www.genome.jp/kegg/pathway.html
UniPathway[a]	UniPathway is a curated resource of metabolic pathways for UniProtKB/Swiss-Prot knowledge base.
Swiss Institute of Bioinformatics (SIB), French National Institute for Research in Computer Science and Control (INRIA Rhone-Alpes)—HELIX/BAMBOO group, Laboratoire d'Ecologie Alpine, Pôle Rhône-Alpin de Bioinformatique	http://www.grenoble.prabi.fr/obiwarehouse/unipathway

(continued)

Table 1 (continued)

Name/Developer	Description/URL
GastroPlus	Advanced software program that simulates the absorption, pharmacokinetics, and pharmacodynamics for drugs in human and preclinical species. The underlying model is the Advanced Compartmental Absorption and Transit (ACAT) model.
Simulations-Plus, Inc.	http://www.simulations-plus.com/
WinNonLin	WinNonlin is the industry standard for pharmacokinetic, pharmacodynamic, and noncompartmental analysis. In addition to its extensive library of built-in PK, PD, and PK/PD models, WinNonlin supports custom, user-defined models to address any kind of data.
Pharsight Inc.	http://www.pharsight.com
DOCK[b]	DOCK addresses the problem of "docking" molecules to each other. In general, "docking" is the identification of the low-energy binding modes of a small molecule, or ligand, within the active site of a macromolecule, or receptor, whose structure is known.
Kuntz Lab program	http://dock.compbio.ucsf.edu/
FlexX	Docking software: Binding mode prediction (predicts the geometry of the protein–ligand complex) and Virtual high-throughput screening (vHTS).
BioSolveIT	http://www.biosolveit.de/FlexX/
Glide	Docking software: full spectrum of speed and accuracy from high-throughput virtual screening of millions of compounds to extremely accurate binding mode predictions.
Schrödinger, LLC	http://www.schrodinger.com/products/14/5/
Autodock[a]	A suite of automated docking tools. It is designed to predict how small molecules, such as substrates or drug candidates, bind to a receptor of known 3D structure.
The Scripps Research Institute	http://autodock.scripps.edu/
GOLD	Protein–Ligand Docking: a program for calculating the docking modes of small molecules in protein binding sites and is provided as part of the GOLD Suite.
University of Sheffield, GlaxoSmithKline plc and CCDC	http://www.ccdc.cam.ac.uk/products/life_sciences/gold/
Amber	A suite of programs that allow users to carry out molecular dynamics simulations, particularly on biomolecules.
Multiple collaborator: http://ambermd.org/contributors	http://ambermd.org/
Tripos	Software company offering a variety of packages, including SYBYL-X Suite, Muse, and D360.
Tripos	http://tripos.com/
Chapter 3.1	
SMILES	A compact machine and human-readable chemical nomenclature
Daylight Chemical Information Systems Inc.	www.daylight.com/dayhtml_tutorials/languages/smiles/index.html

(continued)

Table 1 (continued)

Name/Developer	Description/URL
QSAR Toolbox[a]	Software to fill gaps in (eco-)toxicity data needed for assessing the hazards of chemicals.
Organisation for Economic Co-operation and Development & European Chemicals Agency	http://www.qsartoolbox.org/index.html
CODESSA	An advanced, full featured QSAR program that ties information from AMPAC™ to experimental data.
SemiChem Inc.	http://www.semichem.com/
MoKa	A novel approach for in silico computation of p*Ka* values; trained using a very diverse set of more than 25,000 p*Ka* values, it provides accurate and fast calculations using an algorithm based on descriptors derived from GRID molecular interaction fields.
Molecular Discovery	http://www.moldiscovery.com/index.php
COSMO-RS	A program for the quantitative calculation of solvation mixture thermodynamics based on quantum chemistry.
COSMOlogic	http://www.cosmologic.de/index.php
SPARC[a]	Predictive modeling system that calculates large number of physical/chemical parameters from molecular structure and basic information about the environment.
US EPA, University of Georgia	http://www.epa.gov/extrmurl/research/projects/sparc/index.html; http://archemcalc.com/index.php
PREDICTPlus	Prediction of Thermodynamic and Transport Properties.
Dragon Technology Inc.	http://www.mwsoftware.com/dragon/
PhysProps	Thermodynamic/chemical physical property database and property estimation tool.
G&P Engineering Software	http://www.gpengineeringsoft.com/
ChemDBsoft	Searches databases by structure and substructure.
Chemistry Database Software	http://www.chemdbsoft.com/
EPI Suite[a]	A Windows®-based suite of physical/chemical property and environmental fate estimation programs developed by the EPA's Office of Pollution Prevention Toxics and Syracuse Research Corporation (SRC).
EPA	http://www.epa.gov/opptintr/exposure/pubs/episuite.htm
CSLog D	Prediction gives the log of the water/octanol partition coefficient for charged molecules.
ChemSilico	http://www.chemsilico.com/index.html
CSLog P	Prediction gives the log of the water/octanol partition coefficient for neutral molecules.
ChemSilico	http://www.chemsilico.com/index.html
Marvin Suite	A collection of tools for drawing, displaying, and characterizing chemical structures, substructures, and reactions.
ChemAxon	http://www.chemaxon.com/

(continued)

Table 1 (continued)

Name/Developer	Description/URL
Physicochemical & ADMET Prediction Software	A complete array of tools for the prediction of molecular physical properties from structure. The ability to train allows for the inclusion of novel chemical space in many modules. The value of predictions has also been extended to include a tool for property-based structure design.
ACD/Labs	http://www.acdlabs.com/home/
Molconn-Z	The standard program for generation of Molecular Connectivity, Shape, and Information Indices for QSAR Analyses.
eduSoft	http://www.edusoft-lc.com/molconn/
Jaguar	A ab initio quantum chemistry package for the calculation of molecular properties, including NMR, IR, p*K*a, partial charges, multipole moments, polarizabilities, molecular orbitals, electron density, electrostatic potential, Mulliken population, and NBO analysis.
Schrödinger	http://www.schrodinger.com/
Statistica	A comprehensive package for data analysis, data management, data visualization, and data mining procedures.
StatSoft	http://www.statsoft.com/#
SPSS Software	Advanced mathematical and statistical packages to extract predictive knowledge that when deployed into existing processes makes them adaptive to improve outcomes.
IBM	http://www.spss.com/
Scigress Explorer	A computational chemistry package to calculate molecular properties and energy values. The software provides insight into chemical structure, properties, and reactivity.
SCUBE	http://www.scubeindia.com/scigress_bio.html
ProChemist	A molecular modeling package to simulate on screen the physical-chemical properties of organic molecules, design graphically new structures, and predict their behavior.
Cadcom	http://pro.chemist.online.fr/
SAS	A widely used, comprehensive statistical analysis software package.
SAS Institute Inc.	http://www.sas.com/
PSPP[a]	A program for statistical analysis of sampled data; similar to SPSS; free.
http://savannah.gnu.org/projects/pspp	http://www.gnu.org/software/pspp/
NCSS	A statistical and power analysis software.
NCSS	http://www.ncss.com/
Minitab	A leading statistical software.
Minitab	http://www.minitab.com/
MATLAB	A high-level language and interactive environment that enables you to perform computationally intensive tasks faster than with traditional programming languages such as C, C++, and Fortran.
MathWorks	http://www.mathworks.com/
ADMET Predictor	A software for advanced predictive modeling of ADMET properties.
Simulations Plus	http://www.simulations-plus.com/Default.aspx
PLSR[a]	Performs model construction and prediction of activity/property using the Partial Least Squares (PLS) regression technique.
VCCLab	http://www.vcclab.org/lab/pls/

(continued)

Table 1 (continued)

Name/Developer	Description/URL
Chapter 3.2	
Coot[a]	For macromolecular model building, model completion, and validation, particularly suitable for protein modeling using X-ray data.
Paul Emsley et al.	www.biop.ox.ac.uk/coot/
Chapter 3.6	
ProSa 2003[b]	Software tool for the analysis of 3D structures of proteins.
CAME	http://www.came.sbg.ac.at/prosa_details.php
Chapter 4.2	
GastroPlus	An advanced software program that simulates the absorption, pharmacokinetics, and pharmacodynamics for drugs in human and preclinical species.
Simulations Plus	www.simlations-plus.com
PK-Sim	A software tool for summary and evaluation of preclinical and clinical experimental results on absorption, distribution, metabolism, and excretion (ADME) by means of whole-body physiologically based pharmacokinetic (PBPK) modeling and simulation.
Bayer Technology	http://www.systems-biology.com/products/pk-sim.html
SimCYP	Population-based Simulator for drug development through the modeling and simulation of pharmacokinetics and pharmacodynamics in virtual populations.
Simcyp Limited	http://www.simcyp.com/
Chapter 4.3	
ADME Suite	A collection of software modules that provide predictions relating to the pharmacokinetic profiling of compounds, specifically their ADME properties.
ACD Labs	www.acdlabs.com
ADME Descriptors	ADMET Descriptors in Discovery Studio® include models for intestinal absorption, aqueous solubility, blood–brain barrier penetration, plasma protein binding, cytochrome P450 2D6 inhibition, and hepatotoxicity.
Accelrys	www.accelrys.com
Cloe PK	A software system that uses PBPK modeling for pharmacokinetic prediction.
Cyprotex	www.cyprotex.com
Gastroplus	An advanced software program that simulates the absorption, pharmacokinetics, and pharmacodynamics for drugs in human and preclinical species.
Simulations Plus	www.simlations-plus.com
META	Program for metabolic pathways prediction of molecules, using provided dictionaries.
Multicase	www.multicase.com
MetabolExpert	Program for prediction of the metabolic fate of a compound in the drug discovery process or during the dispositional research phase.
Compudrug	www.compudrug.com
QikProp	Software for rapid ADME predictions of drug candidates.
Schrodinger	www.schrodinger.com
Volsurf+	Software that creates 128 molecular descriptors from 3D Molecular Interaction Fields (MIFs) produced by Molecular Discovery's software GRID, which are particularly relevant to ADME prediction.

(continued)

Table 1 (continued)

Name/Developer	Description/URL
Molecular Discovery	www.moldiscovery.com
MetaCore	An integrated knowledge database and software suite for pathway analysis of experimental data and gene lists.
GeneGo	www.genego.com
MetaDrug	A unique systems pharmacology platform designed for evaluation of biological effects of small molecule compounds on the human body, with pathway analysis and other bioinformatics applications from toxicogenomics to translational medicine.
GeneGo	www.genego.com
Metasite	A computational procedure that predicts metabolic transformations related to cytochrome-mediated reactions in phase I metabolism.
Molecular Discovery	www.moldiscovery.com
MEXAlert	A quick and sensitive tool for indicating possibilities of first-pass metabolism.
Compudrug	www.compudrug.com
RetroMEX	RetroMex predicts the structure of the retro-metabolites, collects them in compound databases, and presents the results in tree format. Retro-metabolic drug design encompasses a series of concepts, e.g., prodrugs, soft drugs, and chemical drug-delivery systems.
Compudrug	www.compudrug.com
MoKa	A novel approach for in silico computation of p*Ka* values; trained using a very diverse set of more than 25,000 p*Ka* values, it provides accurate and fast calculations using an algorithm based on descriptors derived from GRID molecular interaction fields.
Molecular Discovery	www.moldiscovery.com
MCASE/MC4PC	Windows-based Structure–Activity Relationship (SAR) automated expert system: the program will automatically evaluate the dataset and try to identify the structural features responsible for activity (biophores). It then creates organized dictionaries of these biophores and develops ad hoc local QSAR correlations that can be used to predict the activity of unknown molecules.
Multicase	www.multicase.com
METAPC	Windows-based Metabolism and Biodegradation Expert System: uses provided dictionaries to create metabolic paths of molecules submitted to it. Each product is tested in silico for carcinogenicity.
Multicase	www.multicase.com
ADMET Predictor	A software for advanced predictive modeling of ADMET properties.
Simulations Plus	http://www.simulations-plus.com/Default.aspx
Meteor	The program uses expert knowledge rules in metabolism to predict the metabolic fate of chemicals and the predictions are presented in metabolic trees.
LHASA	www.lhasalimited.org
PK Solutions	An automated Excel-based program that does single- and multiple-dose pharmacokinetic data analysis of concentration–time data from biological samples (blood, serum, plasma, lymph, etc.) following intravenous or extravascular routes of administration.
Summit PK	www.summitpk.com

(continued)

Table 1 (continued)

Name/Developer	Description/URL
PK-Sim	A software tool for summary and evaluation of preclinical and clinical experimental results on ADME by means of whole-body PBPK modeling and simulation.
Bayer Technology	http://www.systems-biology.com/products/pk-sim.html
Chapter 4.4	
Discovery Studio	A software suite of life science molecular design solutions for computational chemists and computational biologists.
Accelrys	http://accelrys.com/products/discovery-studio/index.html
MOE	Software package contains Structure-Based Design; Pharmacophore Discovery; Protein & Antibody Modeling; Molecular Modeling & Simulations; Cheminformatics & (HTS) QSAR; and Medicinal Chemistry Applications.
Chemical Computing Group, Inc.	http://www.chemcomp.com/software.htm
CORINA	A fast and powerful 3D structure generator for small and medium-sized, typically drug-like molecules.
Molecular Networks	http://www.molecular-networks.com/products/corina
FlexX	Docking software: Binding mode prediction (predicts the geometry of the protein–ligand complex) and vHTS.
BioSolveIT	http://www.biosolveit.de/FlexX/
Chapter 4.6	
Jsim[b]	A Java-based simulation system for building quantitative numeric models and analyzing them with respect to experimental reference data.
Physiome Project	http://www.physiome.org/jsim
Chapter 4.7	
acslX	A modeling, execution, and analysis environment for continuous dynamic systems and processes.
The AEgis Technologies Group, Inc.	http://www.acslx.com/
Berkeley Madonna	General-purpose differential equation solver.
Robert I. Macey & George F. Oster	http://www.berkeleymadonna.com
MATLAB	A high-level language and interactive environment that enables you to perform computationally intensive tasks faster than with traditional programming languages such as C, C++, and Fortran.
MathWorks	http://www.mathworks.com/
ModelMaker	Two-way class tree oriented productivity, refactoring and UML-style CASE tool. Native Refactoring and UML 2.0 modeling for Delphi and C#.
ModelMaker Tools	http://www.modelmakertools.com
SCOP	Database and tool to provide a detailed and comprehensive description of the structural and evolutionary relationships between all proteins whose structure is known.
SCOP	http://scop.mrc-lmb.cam.ac.uk/scop

(continued)

Table 1 (continued)

Name/Developer	Description/URL
Chapter 5.1	
QSAR Toolbox[a]	Software for grouping chemicals into categories and filling gaps in (eco) toxicity data needed for assessing the hazards of chemicals.
OECD (Organisation for Economic Co-operation and Development) & European Chemicals Agency	http://www.qsartoolbox.org/index.html
Statistica	A comprehensive package for data analysis, data management, data visualization, and data mining procedures.
StatSoft	http://www.statsoft.com/#
SIMCA-P	A tool for scientists, researchers, product developers, engineers, and others who have huge datasets.
Umetrics	http://www.umetrics.com/simca
Chapter 5.4	
Toxtree[a]	A full-featured and flexible user-friendly open source application, which is able to estimate toxic hazard by applying a decision tree approach.
Ideaconsult Ltd.	http://toxtree.sourceforge.net/
OncoLogic[a]	A desktop computer program that evaluates the likelihood that a chemical may cause cancer.
EPA	http://www.epa.gov/oppt/sf/pubs/oncologic.htm
QSAR Toolbox[a]	Software for grouping chemicals into categories and filling gaps in (eco) toxicity data needed for assessing the hazards of chemicals.
OECD (Organisation for Economic Co-operation and Development) & European Chemicals Agency	http://www.qsartoolbox.org/index.html
OpenTox[a]	An interoperable predictive toxicology framework which may be used as an enabling platform for the creation of predictive toxicology applications.
OpenTox	http://www.opentox.org
Chapter 5.6	
Caesar[a]	An EC-funded project (Project no. 022674—SSPI), which was specifically dedicated to develop QSAR models for the REACH legislation.
Istituto di Ricerche Farmacologiche Mario Negri	http://www.caesar-project.eu
Derek Nexus	Expert knowledge base system that predicts whether a chemical is toxic in humans, other mammals, and bacteria.
Lhasa Limited	https://www.lhasalimited.org
HazardExpert	A software tool for initial estimation of toxic symptoms of organic compounds in humans and in animals.
Compudrug	http://www.compudrug.com
Lazar[a]	An open source software program that makes predictions of toxicological endpoints (e.g., mutagenicity, rodent and hamster carcinogenicity, maximum recommended daily dose) by analyzing structural fragments in a training set.

(continued)

Table 1 (continued)

Name/Developer	Description/URL
In Silico Toxicology gmbh	http://lazar.in-silico.de
Meteor	The program uses expert knowledge rules in metabolism to predict the metabolic fate of chemicals and the predictions are presented in metabolic trees.
LHASA	www.lhasalimited.org
QSAR Toolbox[a]	Software for grouping chemicals into categories and filling gaps in (eco) toxicity data needed for assessing the hazards of chemicals.
OECD (Organisation for Economic Co-operation and Development) & European Chemicals Agency	http://www.qsartoolbox.org/index.html
OncoLogic[a]	A desktop computer program that evaluates the likelihood that a chemical may cause cancer.
EPA	http://www.epa.gov/oppt/sf/pubs/oncologic.htm
TOPKAT	Discovery Studio tool that makes predictions of a range of toxicological endpoints, including mutagenicity, developmental toxicity, rodent carcinogenicity, rat chronic Lowest Observed Adverse Effect Level (LOAEL), rat Maximum Tolerated Dose, and rat oral LD50.
Accelrys	http://accelrys.com
ACD/Tox Suite	A collection of software modules that predict probabilities for basic toxicity endpoints.
ACD/Labs	http://www.acdlabs.com/home
Toxtree[a]	A full-featured and flexible user-friendly open source application, which is able to estimate toxic hazard by applying a decision tree approach.
Ideaconsult Ltd.	http://toxtree.sourceforge.net/
Chapter 6.2	
CellNetOptimizer[a]	A MATLAB toolbox for creating logic-based models of signal transduction networks, and training them against high-throughput biochemical data.
Saez-Rodriguez Group Software	http://www.ebi.ac.uk/saezrodriguez/software.html#CellNetOptimizer
MATLAB	A high-level language and interactive environment that enables you to perform computationally intensive tasks faster than with traditional programming languages such as C, C++, and Fortran.
MathWorks	http://www.mathworks.com/
Chapter 6.3	
SMBioNet	A tool for modeling genetic regulatory systems, based on the multivalued logical formalism of René Thomas and the Computational Tree Logic (CTL).
Laboratoire I3S—CNRS & Université de Nice	http://www.i3s.unice.fr/~richard/smbionet

(continued)

Table 1 (continued)

Name/Developer	Description/URL
Chapter 6.4	
Pajek[a]	A program, for Windows, for analysis and visualization of large networks having some thousands or even millions of vertices.
Vladimir Batagelj and Andrej Mrvar	http://vlado.fmf.uni-lj.si/pub/networks/pajek/
Cytoscape[a]	An open source software platform for visualizing complex networks and integrating these with any type of attribute data.
Cytoscape Consortium	http://www.cytoscape.org
Chapter 7.1	
MetaCore	An integrated knowledge database and software suite for pathway analysis of experimental data and gene lists.
GeneGo	www.genego.com
GoMiner[a]	A tool for biological interpretation of "omic" data—including data from gene expression microarrays.
NCI/LMP Genomics and Bioinformatics group	http://discover.nci.nih.gov/gominer/index.jsp
Chapter 9.3	
PASSI Toolkit[a]	The PTK tool is Rational Rose plug-in that offers a support for PASSI (a Process for Agent Societies Specification and Implementation), a step-by-step requirement-to-code methodology for designing and developing multi-agent societies.
mcossentino, mr_lombardo, sirtoy	http://sourceforge.net/projects/ptk
MASON[a]	A fast discrete-event multi-agent simulation library core in Java, designed to be the foundation for large custom-purpose Java simulations.
George Mason University's Evolutionary Computation Laboratory and the GMU Center for Social Complexity	http://cs.gmu.edu/~eclab/projects/mason
NetLogo[a]	A multi-agent programmable modeling environment.
Uri Wilensky	http://ccl.northwestern.edu/netlogo
SeSAm[a]	A generic environment for modeling and experimenting with agent-based simulation.
University of Würzburg	http://www.simsesam.de

[a]Open source or free access
[b]Free for noncommercial and/or academic use
Packages and tools are listed based on their occurrence in a chapter; as such, some listings are repeated. Descriptions are generally excerpts from information obtained at the listed URLs. This list is not comprehensive; if a package, tool, or company mentioned in a chapter is not listed, it is due to an inability to locate the appropriate URL or an oversight by the authors

Part III

Cheminformatics and Chemical Property Prediction

Chapter 6

Prediction of Physicochemical Properties

John C. Dearden

Abstract

Physicochemical properties are key factors in controlling the interactions of xenobiotics with living organisms. Computational approaches to toxicity prediction therefore generally rely to a very large extent on the physicochemical properties of the query compounds. Consequently it is important that reliable *in silico* methods are available for the rapid calculation of physicochemical properties. The key properties are partition coefficient, aqueous solubility, and p*K*a and, to a lesser extent, melting point, boiling point, vapor pressure, and Henry's law constant (air–water partition coefficient). The calculation of each of these properties from quantitative structure–property relationships (QSPRs) and from available software is discussed in detail, and recommendations made. Finally, detailed consideration is given of guidelines for the development of QSPRs and QSARs.

Key words: Physicochemical properties, Prediction, Quantitative structure–property relationships, QSPR, Prediction software, Partition coefficient, Aqueous solubility, p*K*a, Melting point, Boiling point, Vapor pressure, Henry's law constant, QSPR guidelines

1. Introduction

Why, in a book devoted to the prediction of chemical toxicity, is it necessary to consider the prediction of physicochemical properties of chemicals? The answer is quite simple: it is largely their physicochemical properties that determine and control the absorption, distribution, metabolism, excretion, and toxicity (ADMET) of chemicals (1). For example, a chemical needs to have some aqueous solubility in order to be absorbed by an organism. It also needs to possess some hydrophobicity (lipophilicity) if it is to be absorbed through lipid membranes. Both solubility and hydrophobicity are affected by p*K*a, which is also a function of chemical structure.

There is a vast range of physicochemical properties that can affect the toxicity of chemicals, and clearly it would be impossible in this chapter to consider more than a fraction of these. I have

Brad Reisfeld and Arthur N. Mayeno (eds.), *Computational Toxicology: Volume I*, Methods in Molecular Biology, vol. 929, DOI 10.1007/978-1-62703-050-2_6, © Springer Science+Business Media, LLC 2012

therefore chosen to deal with the main physicochemical properties that have been found from experience to be important in modeling toxicity. There have been several reviews of the calculation of these properties (2–4). Other properties concerned with environmental toxicity are bioconcentration factor (BCF) and soil sorption of chemicals, and these have been reviewed by Dearden (5, 6).

Physicochemical properties are predicted largely through the use of quantitative structure–property relationships (QSPRs), which in effect relate such properties to other, more fundamental properties and/or structural features such as the presence or absence of molecular substructures (7). An example is the modeling of octanol–water partition coefficient (P, K_{ow}) (representing hydrophobicity) (8):

$$\log P = 0.088 + 0.562R - 1.054\pi^{H} + 0.034\Sigma\alpha^{H} - 3.460\Sigma\beta^{H} + 3.814V_X, \quad (1)$$

$n = 613$, $R^2 = 0.995$, and $s = 0.116$,

where R = excess molar refractivity (a measure of polarizability), π^{H} = a polarity term, $\Sigma\alpha^{H}$ and $\Sigma\beta^{H}$ = hydrogen bond donor and acceptor abilities, respectively, V_X = the McGowan characteristic molecular volume, n = number of chemicals used to develop the model, R = correlation coefficient, and s = standard error of the estimate (log units). Since the descriptors in this QSPR are approximately auto-scaled, the magnitudes of the coefficients give an indication of the relative contribution of each descriptor to log P. Thus it can be seen that hydrogen bond acceptor ability and molecular size make the most important contributions to log P; on the other hand the contribution of hydrogen bond donor ability is negligible, and this is attributed to the hydrogen bond acceptor abilities of both water and octanol being very similar, while in contrast the hydrogen bond donor ability of water being very strong, accounting for the high negative coefficient on the $\Sigma\beta^{H}$ term. The standard error of prediction is very low compared with the typical experimental error on log P of about 0.35 log unit (9), and may indicate some over-fitting of the data (see Subheading 4.12).

Many QSPRs are available in the literature for the prediction of the most important physicochemical properties. For example, Dearden (10) reported 93 published QSPRs for prediction of aqueous solubility in diverse data sets between 1990 and 2005. How then does one select the best, or most appropriate, QSPR for one's own purposes? Firstly, choose one with good statistics that has been validated by the use of data not used in the development of the QSPR (see Subheadings 4.17 and 4.20). Then check that the descriptors used in the QSPR are accessible, and preferably have some clear physicochemical significance. Finally consider whether

Table 1
Software predictions of aqueous solubility[a]

	Atropine	Caffeine	Butylparaben
Measured	−2.18	−1.02	−2.96
Software no. 1	−1.87	−1.87	−3.09
Software no. 2	−2.06	−0.65	−3.05
Software no. 3	−1.01	−0.27	−3.07
Software no. 4	−2.03	−0.56	−2.58
Consensus prediction	−1.74	−0.84	−2.94

[a]Solubility values given as log *S*, with *S* in mol/L

the use of more than one such QSPR, to give a consensus of predictions, is appropriate.

It may be noted that perhaps the majority of QSPR modeling is carried out using multiple linear regression (MLR), in which it is assumed that the property being modeled correlates in a rectilinear (straight line) manner with each descriptor used. That assumption is not always valid, and the artificial neural networks (ANN) approach is sometimes used to overcome this problem. The ANN approach brings its own problems, however, such as overfitting and lack of an interpretable QSPR. There are numerous other QSPR approaches (11), the discussion of which is outside the scope of this chapter.

Clearly, for someone not familiar with QSPR, such a task could be daunting. It is therefore not surprising that numerous software programs are now available for the prediction of physicochemical properties (as well as, of course, for the prediction of a range of toxicity endpoints). Some of these software programs are freely available, although most have to be purchased. They almost invariably use QSPR methodology for their predictions, but all that the user has to do is to input a chemical structure, for example as a SMILES string (12) or a molfile. Again, the use of more than one prediction program can allow a consensus prediction to be made. The example in Table 1 shows predicted aqueous solubilities of three compounds, atropine, caffeine, and butylparaben, using four different software programs.

For atropine, it is clear that three programs give similar predictions, well within the experimental error of ±0.6 log unit, whereas software no. 3 gives a poor prediction. The consensus (mean) of the three good predictions is −1.99, which is only 0.2 log unit different from the experimental value of −2.18. However, it may be noted that even if the poor prediction from software no. 3 is included, the consensus predicted value is −1.74, which is still within the experimental error of the measured value.

For caffeine, there is a considerable divergence of predicted values, indicating that the solubility of this compound is difficult to predict. Only two of the four predictions are within the experimental error of ±0.6 log unit, but the consensus of all four predictions is −0.84, which is well within the experimental error. This example really emphasizes the value of consensus modeling.

Butylparaben has a simpler chemical structure than atropine and caffeine, and this is reflected in the more accurate predictions of aqueous solubility, with all four being within the experimental error of ±0.6 log unit, and the consensus of all four predictions being −2.94.

It is recommended that, wherever possible, predictions be obtained from more than one software program and/or QSPR, and that the consensus of all the predictions be used, unless one of the predicted values is clearly very different from the others, in which case that prediction should be rejected.

2. Methods

2.1. QSPR Development

If one wishes to develop a QSPR, there are three main steps to be followed. Firstly, it is necessary to obtain experimental values of the property of interest, for a sufficiently large number of chemicals. "How large?" is an open question; QSPRs have been published based on training sets of a few to several thousands of chemicals. The more chemicals that are in the training set, the more robust will be the QSPR—provided that the experimental data are accurate, and that the chemicals selected are consistent with the purpose for which the QSPR is to be developed (13). For example, if one wished to develop a QSPR for the prediction of aqueous solubility of aliphatic amines, it would be wrong to include data for aromatic amines. A number of published papers (14, 15) have discussed the selection of chemicals for QSPR development. Table 2 lists some sources of physicochemical property data that could be useful for QSPR modeling, and more are given by Wagner (27).

If a QSPR applicable to a diverse range of chemicals is required, then of course the training set data should also be diverse. The actual diversity is often limited by the availability of data, and care should be taken in the application of the QSPR. Oyarzabal et al. (28) have recently described a novel approach to minimize the problems of using data from various sources and of data sets with restricted coverage of chemical space.

Secondly, one has to obtain values for descriptors that will model the specified property well (29, 30). One of the two approaches can be taken here. If one believes that one or two descriptors will serve, then those are all that may be required. For example, Yalkowsky and coworkers (31) have found that

Table 2
Some sources of data for modeling of physicochemical properties

Database	Reference
Aquasol	(16)
Benchware Discovery 360	(17)
Chem. & Physical Properties Database	(18)
Chemical Database Service	(19)
ChemSpider	(20)
Crossfire	(21)
OCHEM	(22)
OECD eChemPortal	(23)
OECD Toolbox	(24)
OSHA	(25)
PhysProp	(26)

aqueous solubility can often be modeled quite well with log *P* and melting point. However, generally one does not know a priori which descriptor(s) will best model a given property, in which case the approach is usually to assemble a large pool of descriptors (from the thousands that are now available from which to select a pool). It should be noted also that whilst experimentally measured descriptor values are generally more accurate than are calculated values, the use of the latter means that one can use a QSPR to predict the requisite properties of chemicals not yet synthesized or available (Table 3).

The third step is to use a statistical method that will select from the pool the "best" descriptor(s) based on appropriate statistical criteria, and will generate a QSPR based on those descriptors (42, 43). Typical techniques include stepwise regression and genetic algorithms, and most commercially available statistical packages include these (Table 4).

Tetko et al. (60) have discussed the accuracy of ADMET predictions, and emphasized the importance of accuracy in order to avoid filtering out promising series of compounds because of, for example, wrongly predicted log *P* or aqueous solubility.

Even if a good QSPR is obtained, it may not be a good predictor of the property in question for compounds not in the training set. Hence some measure of predictive ability is required. The best way for predictivity to be assessed is to use the QSPR to predict the property in question for a number of compounds that

Table 3
Some sources of descriptors for modeling of physico-chemical properties

Database	Reference
ADAPT	(32)
Almond	(33)
CODESSA	(34)
C-QSAR	(35)
Discovery Studio	(36)
Dragon	(37)
eDragon	(38)
MOE	(39)
MOLCONN-Z	(40)
OCHEM	(22)
QSARpro	(41)
Volsurf	(33)

were not used in the training set, but for which the measured value of the property is known; such a set of compounds is called a test set. The test set compounds must be reasonably similar to those of the training set; that is, they must lie within the applicability domain of the QSPR. This is often achieved by dividing the total number of compounds into two groups; the larger group forms the training set, and the smaller group (typically 5–50% of the total) forms the test set. If the standard error for the test set is much larger than that for the training set, then the QSPR does not have good predictivity, and it should not be used for predictive purposes.

If the total number of compounds is small, then it may not be practicable to split it into training and test sets. In that case a procedure called internal cross-validation can be used, whereby each compound in turn is deleted from the training set, the QSPR is developed with the remaining compounds, and is used to predict the property value of the omitted compound. That compound is then returned to the training set and a second compound is deleted, and so on until every compound has been left out in turn. A cross-validated R^2 value, called Q^2, is then calculated, which is an indicator of the internal predictivity of the QSPR. It is, however, not considered to be as good an indicator as is obtained using an external test set. Walker et al. (61) have proposed that an indicator of good predictivity is that Q^2 should not be more than

Table 4
Statistical packages for QSPR modeling of physicochemical properties

Software	Reference
ADMET Modeler	(44)
ADMEWORKS ModelBuilder	(45)
ASNN	(46)
Cerius2	(47)
CODESSA	(34)
C-QSAR	(35)
GENSTAT[a]	(48)
MATLAB[a]	(49)
Minitab[a]	(50)
MOE	(39)
NCSS[a]	(51)
OCHEM	(22)
Pentacle	(33)
Pipeline Pilot	(36)
PNN	(46)
PredictionBase	(52)
ProChemist	(53)
PSPP[a]	(54)
QSARpro	(41)
SAS[a]	(55)
Scigress Explorer	(56)
SPSS[a]	(57)
Statistica[a]	(58)
Strike	(59)
SYBYL-X	(17)

[a]General-purpose statistical software

0.3 lower than R^2, whilst Eriksson et al. (62) have proposed a minimal acceptable value of 0.5 for Q^2.

2.2. Published QSPRs

Many thousands of QSPRs are available in the open literature. A large number of these have been referenced in review papers

Table 5
Some QSAR/QSPR databases

Database	Reference
C-QSAR	(35)
JRC QMRF Database	(65)
Danish QSAR Database	(66)
OCHEM	(22)

(for example (10, 63, 64)), and many more can be found with the use of a search engine. Thus, inputting "QSPR p*K*a" into Scholar Google produced 830 hits.

Compilations of QSARs and QSPRs are also available in a number of databases, some of which are listed in Table 5.

2.3. QSPR Software

There are now numerous software packages available for the prediction of physicochemical properties. These vary in their performance and in the range of properties that they predict. They almost all use QSPR modeling approaches for their predictions. Some are available free of charge, and some are very expensive. Most give some indication of their performance, but what is lacking in general is independent comparative assessment of performance. Many of these software packages are listed in Table 6. The ChemProp software (70) is unique in that it selects the best QSPR or software program, from those it holds, for the prediction of a given property of a given compound.

3. Prediction of Selected Physicochemical Properties

As pointed out above, it is beyond the scope of this chapter to consider all physicochemical properties that might play a part in modeling toxicity. Those considered most important are octanol–water partition coefficient, aqueous solubility, p*K*a, melting point, boiling point, vapor pressure, and Henry's law constant (air–water partition coefficient), and they are discussed below.

3.1. 1-Octanol–Water Partition Coefficient (Log *P*, Log K_{ow})

Partition coefficient is defined (84) as the ratio of concentrations at equilibrium of a solute distributed between two immiscible liquid phases; the concentration in the more lipophilic phase is, by convention, the numerator. The term "immiscible" does not preclude the two phases' having partial miscibility. For ionizable solutes, the partition coefficient *P* relates to the undissociated species only, and is thus (approximately) independent of pH.

Table 6
Physicochemical properties estimated by some commercially and freely available software

Software	Log *P*	Log *D*	Aqueous solubility	p*Ka*	Melting point	Boiling point	Vapor pressure	Henry's law constant	Availability	References
Absolv	✓				✓	✓	✓	✓	Purchase	(67)
ACD/PhysChem Suite	✓	✓	✓	✓		✓	✓		Purchase	(67)
ADMET Predictor	✓	✓	✓	✓					Purchase	(44)
ADMEWORKS Predictor	✓								Purchase	(45)
ChemAxon	✓	✓		✓					Purchase	(68)
ChemOffice	✓				✓	✓		✓	Purchase	(69)
ChemProp	✓		✓		✓	✓	✓	✓	Free online	(70)
ChemSilico	✓	✓	✓						Purchase	(71)
C log *P*	✓								Purchase	(35, 72)
Episuite	✓		✓		✓	✓	✓	✓	Free download	(73)
MOE	✓								Purchase	(39)
Molecular Modeling Pro	✓		✓		✓	✓	✓		Purchase	(74)
MoKa				✓					Purchase	(33)
Molinspiration	✓								Free online	(75)

(continued)

Table 6 (continued)

Software	Log *P*	Log *D*	Aqueous solubility	p*K*a	Melting point	Boiling point	Vapor pressure	Henry's law constant	Availability	References
MOLPRO	✓	✓	✓	✓					Purchase	(76)
OECD Toolbox[a]	✓		✓		✓	✓	✓	✓	Free download	(24)
Pallas	✓	✓		✓					Purchase	(77)
PhysProps					✓	✓	✓		Purchase	(78)
Pipeline Pilot	✓	✓	✓	✓					Purchase	(36)
PredictionBase			✓						Purchase	(52)
PREDICTPlus					✓	✓	✓		Purchase	(79)
ProChemist	✓	✓	✓	✓					Purchase	(53)
ProPred	✓		✓	✓	✓	✓	✓	✓	Consortium	(80)
Schrödinger	✓		✓	✓				✓	Purchase	(59)
SPARC	✓	✓	✓	✓		✓	✓	✓	Free online	(81)
TerraQSAR-LOGP	✓								Purchase	(82)
StarDrop[b]	✓	✓	✓						Purchase	(83)
VCCLAB	✓	✓	✓	✓					Free online	(46)

[a]The OECD Toolbox uses the Episuite software
[b]StarDrop offers log *D* at pH 7.4 only

The 1-octanol/water solvent pair was first used by Hansch et al. (85) as a surrogate for lipid–water partitioning, and on the whole has worked well. About 70% of published QSARs and QSPRs incorporate a log *P* term, indicating the importance of this property for modeling biological activities. It is essentially a transport term, reflecting the rate of movement of a chemical through biological membranes.

Many publications have dealt with the estimation of log *P* values from molecular structure, and there have been a number of reviews of the subject (9, 86–91). Mannhold et al. (9) in particular give a detailed critical analysis of available methods. The main prediction methodologies are based on physicochemical, structural, and/or topological descriptors, or on atomic or group contributions.

The earliest work on log *P* prediction was that of Hansch and coworkers, who developed (92) a hydrophobic substituent constant π, which was, to a first approximation, additive, although it required numerous correction factors. Rekker and coworkers (93, 94) developed a fragmental approach which proved easier to use. Extension of the fragmental approach by Leo et al. (95) led to the development of the *C* log *P* software (35) for log *P* prediction.

Bodor et al. (96) developed a QSPR with 14 physicochemical and quantum chemical descriptors to model log *P* of a diverse set of 118 organic chemicals, with $R^2 = 0.882$ and a standard error of prediction of 0.296 log unit. Klopman and Wang (97) used their MCASE group contribution approach to predict the log *P* values of 935 organic compounds with a standard error of 0.39 log unit. These errors are close to the typical experimental error of 0.35 log unit on log *P* (89). The method of Ghose et al. (98) used atomic contributions, and on a set of 893 compounds the standard error was 0.496 log unit.

Liu and Zhou (99) used molecular fingerprints to model log *P* values of a 9,769-chemical training set, with $R^2 = 0.926$ and root mean square error (RMSE) = 0.511 log unit. Chen (100) compared the ability of multiple linear regression, radial basis neural networks, and support vector machines to model log *P* of about 3,500 diverse chemicals, and found R^2 values of 0.88, 0.90, and 0.92, respectively. Tetko et al. (101) used E-state indices (102) and artificial neural networks to model log *P* values of 12,777 chemicals, with $R^2 = 0.95$ and RMSE = 0.39 log unit.

There are numerous software programs available for the estimation of log *P* of organic chemicals, and some of these give good predictions. A recent comparison of 14 such programs (103) found that, using a 138-chemical test set, the percentage of chemicals with log *P* predicted within ±0.5 log unit of the measured log *P* value ranged from 94 to 50%; the performances of the top ten programs are shown in Table 7.

Table 7
Log *P* predictions from ten software packages for a 138-chemical test set (103)

Software	% of chemicals with log *P* prediction error ≤0.5 log unit	r^2	*s*
QMPRPlus[a] (44)	94.2	0.965	0.272
ACD/Labs (67)	93.5	0.965	0.271
ChemSilico (71)	93.5	0.958	0.297
ProPred (80)	89.9	0.945	0.342
A log *P* (75)	89.1	0.948	0.332
KOWWIN (Episuite) (73)	89.1	0.947	0.335
SPARC (81)	88.5	0.941	0.330
C log *P* (35)	88.4	0.961	0.287
Prolog *P*[b] (77)	86.2	0.949	0.329
MOLPRO (76)	81.1	0.847	0.568

[a]Now ADMET Predictor
[b]Now in PALLAS

Sakuratani et al. (104) tested six software programs, using a test set of 134 simple organic compounds. None of the programs predicted log *P* values of all the compounds. Their results were as follows: Episuite (KOWWIN) (73), $n = 130$, $s = 0.94$; *C* log *P* (35), $n = 131$, $s = 0.95$; ACD/Labs (67), $n = 127$, $s = 1.09$; VLOGP (36), $n = 122$, $s = 1.11$; SLOGP (39), $n = 132$, $s = 1.34$; and COSMO (105), $n = 129$, $s = 1.35$. These standard prediction errors are all very high.

Mannhold et al. (9) tested 35 log *P* software programs, and found wide variations in performance, with RMSE values ranging from 0.41 to 1.98 log unit.

It is recommended that at least two of the better software programs be used for the prediction of log *P*. If possible, the average of several predictions should be taken. It should be noted that the VCCLAB Web site (46), as well as giving its own log *P* prediction, gives predictions from nine other software packages, together with the mean of all ten.

The distribution coefficient (*D*) is the ratio of total concentrations, both dissociated and undissociated, of the solute at a given pH in two liquid phases. It is related to *P* by Eq. 2 for acids and by Eq. 3 for bases:

$$\log D_{(\mathrm{pH})} = \log P - \log(1 + 10^{\mathrm{pH}-\mathrm{pKa}}). \quad (2)$$

$$\log D_{(\text{pH})} = \log P - \log(1 + 10^{\text{pKa}-\text{pH}}). \tag{3}$$

There are very few published studies on QSPR modeling of log *D*. One such study derived a QSPR for uranyl extracted by podands from water to 1,2-dichloroethane (106). However, log *D* values can generally readily be calculated from Eqs. 2 and 3, using either measured or calculated log *P* and p*Ka* values (see above and Subheading 3.3).

When a toxicity data set includes ionizable chemicals, it is worth considering whether or not using log *D* instead of log *P* would result in an improved QSPR correlation. This was the case for a study of the toxicity of substituted nitrobenzenes to *Tetrahymena pyriformis* (107).

3.2. Aqueous Solubility

Aqueous solubility depends not only on the affinity of a solute for water, but also on its affinity for its own crystal structure. Molecules that are strongly bound in their crystal lattice require considerable energy to remove them. This also means that such compounds have high melting points, and high-melting compounds generally have poor solubility in any solvent. Note that solubility can vary considerably with temperature, and it is important that solubility data are reported at a given temperature.

Removal of a molecule from its crystal lattice also means an increase in entropy, and this can be difficult to model accurately. For this reason, as well as the fact that the experimental error on solubility measurements is estimated to be about 0.6 log unit (108), the prediction of aqueous solubility is not as accurate as is the prediction of partition coefficient. Nevertheless, many papers (10) and a book (109) have been published on the prediction of aqueous solubility, as well as a number of reviews (10, 87, 110–112).

There are also a number of commercial software programs available for that purpose (10, 113). Livingstone (90) has discussed the reliability of aqueous solubility predictions from both QSPRs and commercial software. It should be noted that there are various ways that aqueous solubilities can be reported: in pure water, at a specified pH, at a specified ionic strength, as the undissociated species (intrinsic solubility), or in the presence of other solvents or solutes. Solubilities are also reported in different units, for example g/100 mL, mol/L, and mole fraction. The use of mol/L is recommended, as this provides a good basis for comparison.

The aqueous solubility (*S*) of a chemical could be described as its hydrophilicity, and so one could perhaps expect an inverse relationship between aqueous solubility and hydrophobicity (as measured by partition coefficient). This is in fact so for organic liquids (114), but does not hold well for solids, because the thermodynamics of melting means that there are significant enthalpy and entropy changes when a molecule is removed from its crystal lattice (10). The melting point (MP) of a chemical is a reasonable measure of these changes, and Yalkowsky and Valvani (115) found that aqueous solubility could be modeled by log *P* and MP:

$$\log S = 0.87 - 1.05 \log P - 0.012\,\mathrm{MP}. \quad (4)$$

$n = 155$, $R^2 = 0.978$, and $s = 0.308$.

Note that Eq. 4 was developed using experimental log *P* and MP (°C) values. Had predicted values been used, the statistics would probably have been a little worse; melting point in particular cannot yet be predicted very well (60, 116); see also Subheading 3.4 below.

Equation 4 has now been modified (117), and is called the General Solubility Equation (GSE):

$$\log S = 0.5 - \log P - 0.01(\mathrm{MP} - 25). \quad (5)$$

The melting point term is taken as zero for compounds melting at or below 25°C. Aqueous solubilities of 1,026 nonelectrolytes, with a log *S* range of −13 to +1 (*S* in mol/L), calculated with the GSE had a standard error of 0.38 log unit.

Good predictions for a large diverse data set have been obtained by the use of linear solvation energy descriptors (118):

$$\log S = 0.518 - 1.004R + 0.771\pi^{\mathrm{H}} + 2.168\Sigma\alpha^{\mathrm{H}} + 4.238\Sigma\beta^{\mathrm{H}} - 3.362\Sigma\alpha^{\mathrm{H}}\Sigma\beta^{\mathrm{H}} - 3.987V_X. \quad (6)$$

$n = 659$, $R^2 = 0.920$, and $s = 0.557$.

It can be seen from Eq. 6 that the main factors controlling aqueous solubility are hydrogen bond acceptor ability and molecular size. The $\Sigma\alpha^{\mathrm{H}}\cdot\Sigma\beta^{\mathrm{H}}$ term models intramolecular hydrogen bonding, which lowers aqueous solubility. Increased molecular size also lowers solubility, since a larger cavity has to be created in water to accommodate a larger solute molecule.

Katritzky et al. (108) used their CODESSA descriptors to model the aqueous solubilities of a large diverse set of organic chemicals:

$$\log Saq = -16.1\,Q\min - 0.113\,N\mathrm{el} + 2.55\,\mathrm{FHDSA}(2) + 0.781\,\mathrm{ABO}(N) + 0.328\,0\mathrm{SIC} - 0.0143\,\mathrm{RNCS} - 0.882, \quad (7)$$

$n = 411$, $R^2 = 0.879$, and $s = 0.573$,

where Qmin = most negative partial charge, *N*el = number of electrons, FHDSA(2) = fractional hydrogen bond donor area, ABO(N) = average bond order of nitrogen atoms, 0SIC = an information content topological descriptor, and RNCS = relative negatively charged surface area. The CODESSA software is available from SemiChem Inc.(34).

Electrotopological state descriptors (119), hydrogen bonding and nearest-neighbor similarities (120), and group contributions (121) have also been used to model the aqueous solubilities of large diverse data sets of organic chemicals. Palmer et al. (122) used random forest models for solubility prediction of a diverse data set, with good results; for an external test set of 330 chemicals, they obtained $R^2 = 0.89$ and $s = 0.69$ log unit. Similar results were obtained with support vector machines (123): $R^2 = 0.88$ and RMSE = 0.62.

Several recent papers (124–126) have reported QSPRs specifically for the aqueous solubility of drugs and drug-like chemicals.

There are relatively few studies of solubility prediction within specific chemical classes. Yang et al. (127) found a good correlation of log *S* with mean molecular polarizability for a small set of dioxins ($n = 12$, $r^2 = 0.978$, $s = 0.30$). Wei et al. (128) obtained a good correlation of log *S* of all 209 polychlorinated biphenyls with quantum chemical descriptors and positions of chlorine substitution.

Dearden et al. (129) compared 11 commercial software programs for aqueous solubility prediction (as log *S*), and found considerable variation in performance against a 113-chemical test set of organic chemicals that included 17 drugs and pesticides. The performances of the top ten programs are shown in Table 8.

Table 8
Aqueous solubility (*S*) predictions from ten software packages for a 113-chemical test set (129)

Software	% of chemicals with log *S* prediction error ≤ 0.5 log unit	r^2	*s*
ChemSilico (71)	79.7	0.951	0.451
WATERNT (Episuite) (73)	79.6	0.954	0.437
VCCLAB (46)	77.0	0.943	0.487
QMPRPlus[a] (44)	74.3	0.939	0.501
ACD/Labs (67)	72.6	0.940	0.498
WSKOWWIN (Episuite) (73)	69.9	0.923	0.562
SPARC (81)	68.1	0.853	0.779
ABSOLV (36)	61.9	0.888	0.680
QikProp (59)	55.7	0.867	0.742
MOLPRO (76)	50.4	0.766	0.984

[a]Now ADMET Predictor

Table 9
Aqueous solubility (*S*) predictions from ten software packages for a 122-chemical test set of drugs (10)

Software	% of chemicals with log *S* prediction error ≤0.5 log unit	r^2	*s*
Admensa[a] (83)	72.1	0.76	0.65
ADMET Predictor (44)	64.8	0.82	0.47
MOLPRO (76)	62.3	0.44	1.22
ChemSilico (71)	59.8	0.67	0.73
ACD/Labs (67)	59.0	0.73	0.66
VCCLAB (46)	51.6	0.67	0.73
QikProp (59)	47.6	0.57	0.97
PredictionBase (52)	46.7	0.48	1.07
SPARC (81)	42.9	0.73	0.96
WSKOWWIN (Episuite) (73)	41.0	0.51	1.17

[a]Now known as StarDrop

The Episuite predictions were made without the input of measured melting point values.

Dearden (10) tested 16 commercially available software programs for their ability to predict the aqueous solubility of a 122-compound test set of drugs with accurately measured solubilities in pure water. Again there was considerable variation in performance. The performances of the top ten programs are shown in Table 9.

Dearden (130) also tested the relatively new fragment-based WATERNT module in the Episuite software on both the 113-chemical test set (comprising mostly simple organic chemicals) and the 122-drug test set. He found it to be the best for the former (79.6% within ±0.5 log unit of measured value; standard error = 0.44 log unit) and among the worst for the latter (38.5% within ±0.5 log unit of measured value; standard error = 0.93 log unit). Investigation indicated that this was caused by the program's not including all fragments and/or correction factors in its calculations.

3.3. pKa

Within a congeneric series of chemicals, p*K*a is often closely correlated with the Hammett substituent constant, and this is the basis for a number of attempts at p*K*a prediction. Harris and Hayes (131) and Livingstone (90) have reviewed the published literature in this area.

The Hammett substituent constant σ was derived from a consideration of acid dissociation constants Ka, and most noncomputerized methods of calculating Ka and p*K*a values are based on σ values:

$$\mathrm{p}K\mathrm{a(derivative)} = \mathrm{p}K\mathrm{a(parent)} - \rho\sigma, \quad (8)$$

where ρ is the series constant, which is 1.0 for benzoic acids. Harris and Hayes (131) list ρ values for other series.

Harris and Hayes (131) give several examples of p*K*a calculation, for example for 4-*t*-butylbenzoic acid. The p*K*a value of benzoic acid is 4.205, the ρ value for benzoic acids is 1.0, and the ρ value for 4-*t*-butyl is −0.197. Hence the p*K*a value of 4-*t*-butylbenzoic acid is calculated as 4.205−(−0.197) = 4.402. This value is virtually identical to the measured value for this compound.

A number of publications have dealt with estimation of p*K*a values from chemical structure, but these relate mostly to specific chemical classes, e.g., benzimidazoles (132), 4-aminoquinolines (133), and imidazol-1-ylalkanoic acids (134).

There have also been a number of attempts to model p*K*a values of diverse sets of chemicals. Klopman and Fercu (135) used their MCASE methodology to model the p*K*a values of a set of 2,464 organic acids, and obtained good predictions; a test set of about 600 organic acids yielded a standard error of 0.5 p*K*a unit. The COSMO-RS methodology was used to predict p*K*a values of 64 organic and inorganic acids, with a standard error of 0.49 p*K*a unit (136) and of 43 organic bases, with an RMSE of 0.66 p*K*a unit (137). Lee et al. (138) used a decision tree approach and SMARTS strings to model a diverse set of 1,693 monoprotic chemicals, with an RMSE of 0.80 p*K*a unit. Milletti et al. (139) used GRID interaction fields to model p*K*a on a very large training set of 24,617 p*K*a values; for a class of 947 6-membered *N*-heterocyclic bases they found an RMSE of 0.60 p*K*a unit. The MoKa program developed by Cruciani et al. (140) was reported as giving predictions within 0.5 p*K*a units. A model based on group philicity (141) yielded a standard error of 0.57 p*K*a unit for 63 chemicals of various chemical classes.

There are a number of software programs that predict multiple p*K*a values of organic chemicals. ACD/p*K*a has a claimed standard error of 0.39 p*K*a unit for 22 compounds, and one of 0.36 p*K*a unit for 26 drugs. pKalc (part of the PALLAS suite) is claimed to be accurate to within 0.25 p*K*a unit (142), Schrödinger's p*K*a calculator (59) is claimed to have a mean absolute error (MAE) of 0.19 p*K*a unit, and SPARC is claimed to have an RMSE of 0.37 p*K*a unit when evaluated on 3,685 compounds (143), although Lee et al. (144) found an RMSE of 1.05 p*K*a unit when they tested SPARC on 537 drug-like chemicals. ADMET Predictor is claimed to have

Table 10
Some software predictions of p*Ka* by Dearden et al. (146) and Liao and Nicklaus (147)

Software	Dearden et al.		Liao and Nicklaus	
	r^2	MAE	r^2	MAE
ACD/Labs (67)	0.922	0.54	0.908	0.478
ADME Boxes[a]	0.959	0.32	0.944	0.389
ADMET Predictor (44)	0.899	0.67	0.837	0.659
ChemSilico[a] (71)	0.565	1.48	–	–
Epik (59)	0.768	0.93	0.802	0.893
Jaguar (59)	–	–	0.579	1.283
ChemAxon/Marvin (68)	0.778	0.90	0.763	0.872
PALLAS (77)	0.656	1.17	0.803	0.787
Pipeline Pilot (36)	0.852	0.43	0.757	0.769
SPARC (81)	0.846	0.78	0.894	0.651
VCCLAB (46)	0.931	0.40	–	–

[a]No longer available

an MAE of 0.56 p*Ka* unit for a test set of 2,143 diverse chemicals. ChemSilico's p*Ka* predictor was reported to have an MAE of 0.99 p*Ka* unit for a test set of 665 diverse chemicals, many of them multiprotic. However, this module does not appear currently to be available, although it is stated to be used in ChemSilico's log *D* predictor (71). The ChemProp software (70) uses a novel approach whereby, for a given compound, the best prediction method is selected based on prediction errors for structurally similar compounds (145).

Dearden et al. (146) tested the performance of ten available software programs that calculate p*Ka* values. Some of these programs will calculate p*Ka* values of all ionizable sites. However, the test set of 665 chemicals that they used, which was kindly supplied by ChemSilico Inc. and used by them as their test set, had measured p*Ka* values only for the prime ionization site in each molecule. There were doubts about the correct structures of 11 of the test set chemicals, and so the programs were tested on 654 chemicals. Some of the software companies kindly ran our compounds through their software in-house. The results are given in Table 10. It should be noted that the ACD/p*Ka* predictions were incorrectly reported by Dearden et al. (146), for which the author apologizes. The ACD/p*Ka* predictions given in Table 10 are correct.

Liao and Nicklaus (147) carried out a comprehensive comparison of nine p*K*a prediction software programs on a test set of 261 drugs and drug-like chemicals. Their results are also given in Table 10, and on the whole are comparable with those of Dearden et al. (146).

Meloun and Bordovská (148) compared the p*K*a predictions from ACD/p*K*a, Marvin, PALLAS, and SPARC for 64 drug-like chemicals, and found MAE values of 0.12, 0.23, 0.55, and 1.64 p*K*a unit, respectively. Balogh et al. (149) compared the performance of five available p*K*a software programs (ACD/p*K*a (67), Epik (59), Marvin (68), PALLAS (77), and VCCLAB (46)), using a 248-chemical Gold Standard test set, and found standard errors of 0.808, 2.089, 0.957, 1.229, and 0.615 p*K*a unit, respectively. Manchester et al. (150), using a test set of 211 drug-like chemicals, found the following RMSE values (p*K*a unit): ACD/p*K*a (67) v. 12, 0.8; Epik (59), 3.0; Marvin (68), 0.9, MoKa (33), 1.0; and Pipeline Pilot (36), 2.6.

It should be noted that the ADME Boxes software tested by Liao and Nicklaus (147) is no longer available. It was offered by Pharma Algorithms, which has now been taken over by ACD/Labs. Similarly the VCCLAB software tested by Balogh et al. (149) was also that of Pharma Algorithms. The current VCCLAB p*K*a prediction tool uses ACD/p*K*a.

It thus appears that ACD/p*K*a is generally the best overall p*K*a prediction software currently available, with ADMET Predictor also performing well.

3.4. Melting Point

Melting point is an important property for two main reasons. Firstly, it indicates whether a chemical will be solid or liquid at particular temperatures, which will dictate how it is handled. Secondly, it is used in the GSE (117) to predict aqueous solubility.

The melting point of a crystalline compound is controlled largely by two factors—intermolecular interactions and molecular symmetry. For example, 3-nitrophenol, which can hydrogen-bond via its −OH group, melts at 97°C, whereas its methyl derivative, 3-nitroanisole, which cannot hydrogen-bond with itself, melts at 39°C. The symmetrical 1,4-dichlorobenzene melts at 53°C, whilst its less-symmetrical 1,3-isomer melts at −25°C. These and other effects have been discussed in detail by Dearden (151).

There have been many attempts to predict the melting point of organic chemicals, and these have been reviewed by Horvath (152), Reinhard and Drefahl (87), Dearden (63, 153), and Tesconi and Yalkowsky (154). It may be noted that in 1884 Mills (155) developed a QSPR based on carbon chain length for melting points of homologous series of compounds that was accurate to ±2°:

$$\mathrm{MP} = (\beta(x - c))/(1 + \gamma(x - c)), \tag{9}$$

where x = number of CH_2 groups in the chain, and β, γ, and c are constants depending on the series.

Essentially two approaches have been used in the prediction of melting point—the physicochemical/structural descriptor approach and the group contribution approach. The former is exemplified by the work of Katritzky et al. (156), who used nine of their CODESSA descriptors to model a diverse set of 443 aromatic chemicals with $R^2 = 0.837$ and $s = 30.2°$. The CODESSA software is available from SemiChem Inc. (34). This is a complex QSPR, with descriptors that are not easy to comprehend, and reflects the difficulty of modeling the melting points of diverse data sets. Even for a set of 58 PCB congeners with 1–10 chlorine atoms, a 5-term QSPR was required (157), with $R^2 = 0.83$ and $s = 22.1°$. Yalkowsky and coworkers have published extensively on the prediction of melting point. They incorporated terms to account for conformational flexibility and rotational symmetry (158) and molecular eccentricity (159) to try to account for the entropic contributions to melting. They were able (160) to model the melting points of 1,040 aliphatic chemicals, using a combination of molecular geometry and group contributions, with a standard error of 34.4°.

Todeschini et al. (161) used their WHIM descriptors to model the melting points of 94 European Union environmental priority chemicals, with a standard error of 32.8°. Bergström et al. (162) used principal components analysis and partial least squares to model the melting points of 227 diverse drugs. They used 2-D, 3-D, and a combination of 2-D and 3-D descriptors to give three separate models. A consensus of all three models gave the best results, with $R^2 = 0.63$ and RMSE = 35.1°. Modarresi et al. (163) used eight descriptors from the no-longer-available Tsar program (36), CODESSA (34) and Dragon (37), to model the melting points of 323 drugs, with $R^2 = 0.660$ and RMSE = 41.1 °. Godavarthy et al. (164) used genetic algorithms and neural networks to model the melting points of over 1,250 chemicals, with $R^2 = 0.95$ and RMSE = 12.6°, although it must be said that those results are so good that over-fitting could have occurred.

Karthikeyan et al. (165) used a very large diverse training set of 4,173 chemicals to develop a QSPR based on a neural network approach using principal components. They found 2-D descriptors to be better than 3-D descriptors; their results were as follows:

Training set	Internal validation	Test set	Test set (drugs)
$n = 2{,}089$	$n = 1{,}042$	$n = 1{,}042$	$n = 277$
$R^2 = 0.661$	$Q^2 = 0.645$	$Q^2 = 0.658$	$Q^2 = 0.662$
MAE = 37.6°	MAE = 39.8°	MAE = 38.2°	MAE = 32.6°

Table 11
Some software predictions of melting points of a 96-compound test set (63)

Software	Mean absolute error
Episuite (73)	26.3°
ChemOffice (69)	27.0°
ProPred (80)	25.8°

Considering the size and diversity of the data sets, the statistics are quite good. However, the methodology used was complex, and could not readily be applied.

The group contribution approach to melting point prediction was first used by Joback and Reid (166). Simamora and Yalkowsky (167) modeled the melting points of a diverse set of 1,690 aromatic compounds using a total of 41 group contributions and four intramolecular hydrogen bonding terms, and found a standard error of 37.5°. Constantinou and Gani (168) used two levels of group contributions to model the melting points of 312 diverse chemicals, and obtained a mean absolute error of prediction of 14.0°, compared with an MAE of 22.6° for the Joback and Reid method. Marrero and Gani (169) extended this approach to predict the melting points of 1,103 diverse chemicals with a standard error of 25.3°. Tu and Wu (170) used group contributions to predict melting points of 1,310 diverse chemicals with an MAE of 8.2%.

There are several software programs that predict melting point (see Table 6); they all use one or more group contribution approaches. Dearden (63) used a 96-compound test set to compare the performances of three of these programs. Episuite (73) calculates melting point by two methods, that of Joback and Reid (166) and that of Gold and Ogle (171), and takes their mean. ChemOffice (69) uses the method of Joback and Reid (166), and ProPred (80) uses the Gani approach (168, 169). The results are given in Table 11.

An ECETOC report (113) mentions a 1999 US Environmental Protection Agency (EPA) test of the performance of the Episuite MPBPVP module; for two large, diverse test sets the performance was as follows: (1) $n = 666$, $r^2 = 0.73$, MAE = 45°; (2) $n = 1{,}379$, $r^2 = 0.71$, MAE = 44°.

Molecular Modeling Pro uses the Joback and Reid (166) method, so its performance should be the same as that of ChemOffice. Four other programs, Absolv (67), ChemProp (70), OECD Toolbox (24), and PREDICTPlus (79), also predict melting point.

It can be seen that there is little to choose between the programs in terms of accuracy of prediction. They can all operate in batch mode. It is therefore recommended that the Episuite software, which is freely downloadable, and at least one other method be used to calculate melting point.

It should be noted that currently both QSPR methods and software programs have prediction errors well in excess of the error on experimental measurement of melting point, which is usually <2°. Therefore it is preferable to use measured melting points if at all possible.

3.5. Boiling Point

Boiling point (Tb) is an important property since it is an indicator of volatility, and can be used to predict vapor pressure. From the Clausius–Clapeyron equation, boiling point is inversely proportional to the logarithm of vapor pressure. Boiling point also indicates whether a chemical is gaseous or liquid at a given temperature.

Lyman (172) has discussed seven recommended methods for the prediction of boiling point. The methods are based on physicochemical and structural properties and group contributions. Perhaps the simplest of those methods is that of Banks (173), who developed the following QSPR:

$$\log Tb(\mathrm{K}) = 2.98 - 4/\sqrt{\mathrm{MW}}, \quad (10)$$

where MW = molecular weight. No statistics were given for this QSPR.

Rechsteiner (174), Reinhard and Drefahl (87), and Dearden (63) have reviewed the QSPR prediction of boiling point.

Many studies of boiling point prediction have dealt with specific chemical classes, and very good correlations have generally been obtained. In 1884 Mills (155) modeled the boiling points of a number of homologous series with QSPRs based on carbon chain length, and claimed accuracy to within about 2° (see Eq. 9). Ivanciuc et al. (175) used four topological descriptors to model the boiling points of 134 alkanes with a standard error of 2.7°, whilst Gironés et al. (176) used only one quantum chemical descriptor (electron–electron repulsion energy) to model the boiling points of 15 alcohols with a standard error of 5.6°.

Models based on diverse training sets are, however, more widely applicable. Katritzky et al. (177) used four CODESSA descriptors to model the boiling points of 298 diverse organic compounds:

$$Tb(\mathrm{K}) = 67.4\,\mathrm{GI}^{1/3} + 21{,}540\,\mathrm{HDSA(2)} + 140.4\delta^{-}{}_{\max} + 17.5\,N_{\mathrm{Cl}} - 151.3, \quad (11)$$

$n = 298$, $R^2 = 0.973$, and $s = 12.4°$,
where GI = gravitational index, HDSA(2) = area-weighted surface charge of hydrogen-bond donor atoms, $\delta_{\max}{}^{-}$ = most negative atomic partial charge, and N_{Cl} = number of chlorine atoms. The CODESSA software is available from SemiChem Inc. (34).

Sola et al. (178) used eight CODESSA descriptors to model the boiling points of 135 diverse chemicals, with an RMSE of 9.1°.

Wessel and Jurs (179) used their ADAPT descriptors to develop two QSPRs for the prediction of boiling point—one for compounds containing O, S, and halogens, and the other for compounds containing N. The QSPR for O, S, and halogens is

$$\begin{aligned} Tb(\mathrm{K}) = {} & 0.3009\,\mathrm{PPSA} - 3.690\,\mathrm{PNSA} - 51.78\,\mathrm{RPCG} \\ & + 9.515\,N_{\mathrm{RA}} + 19.21\,\mathrm{SQMW} + 554.7\,\mathrm{SADH} \\ & - 25.52\,N_{\mathrm{F}} + 19.52\,\mathrm{KETO} + 50.84\,N_{\mathrm{sulf}} \\ & - 135.0\,\mathrm{S/NA} + 59.86, \end{aligned} \qquad (12)$$

$n = 248$, $R^2 = 0.991$, and RMSE $= 11.6°$,

where PPSA = partial positive surface area, PNSA = partial negative surface area, RPCG = relative positive charge, N_{RA} = number of ring atoms, SQMW = square root of molecular weight, SADH = surface area of donatable hydrogen atoms, N_{F} = number of fluorine atoms, KETO = indicator variable for ketone, N_{sulf} = number of sulfide groups, and S/NA = (number of sulfur atoms)/(total number of atoms).

Basak and Mills (180) used eight topochemical, topological, and hydrogen bonding descriptors to model the boiling points of 1,015 diverse organic compounds, with a standard error of 15.7°. Probably the best QSPR developed to date is that of Hall and Story (181), who used atom-type electrotopological descriptors (182) and a neural network to obtain an MAE of 3.9° for a set of 298 diverse chemicals with a boiling point range of about 430°.

The group contribution approach was used first by Joback and Reid (166), who obtained an MAE of 12.9° for a set of 438 diverse chemicals. Stein and Brown (183) devised a simple group contribution method to model boiling points of a very large set of 4,426 diverse chemicals, with an MAE of 15.5°. A group contribution approach was also used by Marrero and Gani (169) to model the boiling points of 1,794 organic compounds with a standard error of 8.1°, whilst Labute (184) used 18 atomic contributions on a set of 298 diverse organics, to give a standard error of 15.5°. Simamora and Yalkowsky (167) used 36 group contributions and four intramolecular hydrogen bonding terms to model the boiling points of a diverse set of 44 aromatic compounds, with a standard error of 17.6°. Ericksen et al. (185) modeled the boiling points of 1,141 chemicals with second-order group contributions, and found an MAE of 7.8°.

There are a number of software programs available for the prediction of boiling point, and Dearden (63) compared the performance of six of these using a 100-compound test set. The results are shown in Table 12.

The ACD/Labs result is based on the 54 chemicals in the test set that were not included in the ACD/Labs training set. Clearly

Table 12
Some software predictions of boiling points of a 100-compound test set (63)

Software	Mean absolute error
ACD/Labs (67)	1.0°
SPARC (81)	6.3°
Episuite (73)	13.8°
ChemOffice (69)	13.8°
ProPred (80)	16.1°
Molecular Modeling Pro (74)	21.7°

the ACD/Labs software gives by far the best predictions, but has to be purchased. SPARC is freely accessible, but operates only in manual mode, with SMILES input. Episuite can be freely downloaded, but its standard error of prediction was more than twice that of SPARC. ECETOC (113) quotes the US EPA testing of the MPBPVP module of the Episuite software; two very large diverse test sets yielded the following: $n = 4{,}426$, MAE = 15.5°; $n = 6{,}584$, MAE = 20.4°. PREDICTPlus is claimed to have an MAE of 12.9°. These results are comparable with those of Dearden given above. Five other software programs, Absolv (67), ChemProp (70), PhysProps (78), OECD Toolbox (24), and PREDICTPlus (79), also predict boiling point.

It is recommended that at least two predictions be obtained, and their average used.

3.6. Vapor Pressure

The vapor pressure (VP) of a chemical controls its release into the atmosphere, and thus is an important factor in the environmental distribution of chemicals. Vapor pressure is highly temperature dependent. Most literature values are at ambient temperature, but some QSPRs allow predictions over a range of temperatures. The variation of vapor pressure with temperature is given by the Clausius–Clapeyron equation:

$$\ln\left(\frac{\mathrm{VP}_2}{\mathrm{VP}_1}\right) = -\left(\frac{L}{R}\right)\left(\left(\frac{1}{T_2}\right) - \left(\frac{1}{T_1}\right)\right), \tag{13}$$

where L = latent heat of vaporization, and R = universal gas constant.

If the latent heat of vaporization is high, vapor pressure changes markedly with temperature, which is why some chemicals (e.g., PCBs) deposit out in polar regions.

Numerous methods are available for the estimation of vapor pressure, and Grain (186), Schwarzenbach et al. (111), Delle Site (187), Sage and Sage (188), and Dearden (63) have reviewed many of these. The descriptors used in vapor pressure QSPRs include physicochemical, structural, and topological descriptors, and group contributions. Katritzky et al. (189) used their CODESSA descriptors to model the vapor pressure of a large set of diverse organic chemicals:

$$\log \mathrm{VP} = -0.00559\, G_{\mathrm{I}} - 0.708\, \mathrm{HDCA(1)} + 0.767\, N_{\mathrm{F}} - 0.00757\, \mathrm{WNSA} - 1 + 7.01, \quad (14)$$

$n = 645$, $R^2 = 0.937$, and $s = 0.366$,

where G_{I} = gravitational index, HDCA(1) = hydrogen-bond donor solvent-accessible surface area, N_{F} = number of fluorine atoms, and WNSA-1 = weighted partial negative surface area. The CODESSA software is available from SemiChem Inc. (34).

Liang and Gallagher (190) used polarizability and seven structural descriptors to model the vapor pressure of 479 diverse organic chemicals, using both multiple linear regression and an artificial neural network. There was little difference between the two methods with MLR giving a standard error of 0.534 log unit and ANN yielding 0.522 log unit.

Tu (191) used a group contribution method to model the vapor pressure of 1,410 diverse organic chemicals. Using 81 group contributions, 2 hydrogen bonding terms, and melting point he obtained a standard error of 0.36 log unit. Öberg and Liu (192) used a partial least squares regression (PLSR) method to model the vapor pressures of a set of 1,340 diverse organic chemicals, with RMSE = 0.410 log unit. Basak and Mills (193) used topological descriptors to model vapor pressures of 121 chlorinated organic chemicals, and obtained an RMSE of 0.130 log unit.

The vapor pressures of 352 hydrocarbons and halohydrocarbons were modeled by Goll and Jurs (194), using seven of their ADAPT descriptors. Vapor pressure was recorded in pascals, and the data covered the log VP range −1.016 to +6.65:

$$\log \mathrm{VP} = -0.670\,{}^{\circ}\chi + 0.204\, N_{\mathrm{F}} + 5.47 \times 10^{-2}\, N_{\mathrm{SB}} - 0.121\, N_{\mathrm{RA}} - 6.35 \times 10^{-2}\, \mathrm{DPSA} + 0.117\, N_{3\mathrm{C}} + 0.518\, \mathrm{RPCG} + 8.15, \quad (15)$$

$n = 352$, $R^2 = 0.983$, and RMSE = 0.186 log unit,

where ${}^{\circ}\chi$ = zero order molecular connectivity, N_{F} = number of fluorine atoms, N_{SB} = number of single bonds, N_{RA} = number of atoms in ring systems, DPSA = difference between partial positive surface area and partial negative surface area, $N_{3\mathrm{C}}$ = number of 3rd order clusters, and RPCG = relative positive charge.

Table 13
Some software predictions of vapor pressures at 25 °C of a 100-compound test set (63)

Software	Mean absolute error (log unit)
SPARC (81)	0.105
ACD/Labs (67)	0.107
Episuite (73)	0.285
Molecular Modeling Pro (74)	0.573

Some of the ADAPT descriptors are difficult to interpret, but have been found to give good correlations of a number of physicochemical properties. The very low standard RMSE reflects the fact that there was little chemical diversity within the compounds used.

An interesting approach was used by Staikova et al. (195), who modeled the vapor pressures of nonpolar chemicals with a single descriptor, average polarizability, and found a standard error of prediction of 0.313 log unit. It is unlikely, however, that this approach would work with polar chemicals.

A number of studies (196–199) allow the estimation of vapor pressures over a range of temperatures.

There are several commercially available software programs that will calculate vapor pressure; one of them (ACD/Labs) will allow the calculation of vapor pressure over a temperature range. Using a 100-compound test set of organic chemicals with vapor pressures measured at 25°C, Dearden (63) compared the performance of four software programs that calculate log (vapor pressure). The test results are given in Table 13.

The programs can operate in batch mode, except for SPARC. The ACD/Labs result was determined on only 42 compounds; 46 test set compounds that were used in the ACD/Labs training set were deleted, and in addition the ACD/Labs software did not give a vapor pressure at 25°C for 18 very volatile compounds. ECETOC (113) quotes the US EPA testing of the MPBPVP module of the Episuite software in 1999: $n = 805$, $r^2 = 0.941$, and MAE $= 0.476$ log unit. This MAE probably reflects either the greater diversity of the US EPA test set or improvements made in the software since 1999.

The prediction errors of the PREDICTPlus software (79) are reported to be 2–5%, depending on the method of calculation. Five other software programs, Absolv (67), ChemProp (70), PhysProps (78), OECD Toolbox (24), and ProPred (80), also predict vapor pressure.

It is recommended that SPARC, ACD/Labs, or Episuite software be used for the calculation of vapor pressure. Predictions from at least two different sources should be obtained if possible.

3.7. Henry's Law Constant (Air–Water Partition Coefficient)

The air–water partition coefficient is important in the distribution of chemicals between the atmosphere and water in the environment. The prediction of Henry's law constant (H) has been reviewed by Schüürmann and Rothenbacher (200), Schwarzenbach et al. (111), Reinhard and Drefahl (87), Mackay et al. (201), and Dearden and Schüürmann (64).

One simple way of calculating H is to use the ratio of vapor pressure and aqueous solubility (VP/C_w). It is not a highly accurate method, but neither is the measurement of H, especially for chemicals with very high or very low H values. VP/C_w can be converted to the dimensionless form of H (ratio of concentrations in air and water, $C_\mathrm{a}/C_\mathrm{w}$, or K_aw) by the following equation, which is valid for 25°C:

$$\frac{C_\mathrm{a}}{C_\mathrm{w}} = 40.874\,\frac{\mathrm{VP}}{C_\mathrm{w}}. \tag{16}$$

Most prediction methods for H use a group or bond contribution approach, although some have used physicochemical properties (202). The group and bond contribution methods were first used by Hine and Mookerjee (203), who obtained, for a set of 263 diverse simple organic chemicals, a standard deviation of 0.41 log unit for the group contribution method and one of 0.42 for the bond contribution method. Cabani et al. (204) claimed an improvement in the group contribution method over that of Hine and Mookerjee, whilst Meylan and Howard (205) extended the bond contribution method and obtained, for a set of 345 diverse chemicals, a standard error of 0.34 log unit. Their method, together with a group contribution method, is incorporated in the HENRYWIN module of the Episuite software (73).

Several workers have used physicochemical and/or structural descriptors to model H.

Nirmalakhandan and Speece (206) developed a QSPR using a polarizability descriptor, a molecular connectivity term, and an indicator variable for hydrogen bonding. However, Schüürmann and Rothenbacher (200) found it to have poor predictive power.

Russell et al. (207) used their ADAPT software to develop a 5-descriptor model of $\log K_\mathrm{aw}$ for a relatively small but diverse data set:

$$\begin{aligned}\log K_\mathrm{aw} = &-0.547\,N_\mathrm{HEAVY} + 0.0402\,\mathrm{WPSA} + 0.0360\,\mathrm{RNCS}\\ &+ 10.1\,\mathrm{QHET} - 215\,\mathrm{QRELSQ} + 0.73,\end{aligned} \tag{17}$$

$n = 63$, $R^2 = 0.956$, and $s = 0.375$,

where N_{HEAVY} = number of heavy atoms, WPSA = (total solvent-accessible surface area) × (sum of surface areas of positively charged atoms), RNCS = (charge on most negative atom) × (surface area of most negative atom)/(sum of charges on negatively charged atoms), QHET = (total charge on heteroatoms)/(number of heteroatoms), and QRELSQ = square of (total charge on heteroatoms)/(number of atoms).

Recently QSPRs have been developed by Modarresi et al. (208) using a very large (940-compound) diverse data set. Using genetic algorithm selection of descriptors, they obtained a 10-descriptor QSPR with an RMSE of 0.571 log unit.

The Ostwald solubility coefficient L (the reciprocal of K_{aw}) of a very diverse data set of chemicals was modeled by Abraham et al. (209):

$$\log L = 0.577R + 2.549\pi + 3.813\Sigma\alpha + 4.841\Sigma\beta - 0.869V_X + 0.994, \quad (18)$$

$n = 408$, $R^2 = 0.996$, and $s = 0.151$,

where R = excess molar refractivity (a measure of polarizability), π = a polarity/polarizability term, $\Sigma\alpha$ and $\Sigma\beta$ = sum of hydrogen bond donor and acceptor abilities, respectively, and V_X = the McGowan characteristic volume. The Abraham descriptors are approximately auto-scaled, so that the magnitudes of the coefficients in Eq. 18 indicate the relative contributions of each term. It is clear that hydrogen bonding is the most important factor controlling water–air distribution; the greater magnitude of the $\Sigma\beta$ term probably reflects the strong hydrogen bond donor ability of water. Molecular size, represented by V_X, appears to play only a minor role in determining air–water partitioning. It may be noted that the very high correlation coefficient and low standard error of Eq. 6.22 suggest possible over-fitting; no external validation of Eq. 18 was provided. The Abraham descriptors are available in the Absolv software (67).

Over-fitting of data seems likely in two other papers also (210, 211), where prediction errors of 0.03 and 0.1 log unit, respectively, were reported.

Katritzky et al. (212) used their CODESSA software to model the data set of Abraham et al. [1994]:

$$\log L = 42.37\,\mathrm{HDCA(2)} + 0.65[N_{\mathrm{O}} + N_{\mathrm{N}}] - 0.16\,\mathrm{DE} + 0.12\,\mathrm{PCWT} + 0.82\,N_{\mathrm{R}} + 2.65, \quad (19)$$

$n = 406$, $R^2 = 0.942$, and $s = 0.52$,

where HDCA(2) = hydrogen bond donor ability, ($N_{\mathrm{O}} + N_{\mathrm{N}}$) = a linear combination of the number of oxygen and nitrogen atoms, DE = HOMO–LUMO energy difference, PCWT = most negative partial charge-weighted topological electronic index, and N_{R} = number of rings. It may be noted that the standard error of 0.52 log unit is more realistic than that of 0.151 reported by Abraham et al. (209).

Katritzky et al. (108) used predicted vapor pressure and aqueous solubility to calculate Henry's law constant according to Eq. 20 for 411 diverse chemicals. The table giving their results was inadvertently omitted in their paper, but they reported a standard error of 0.63 log unit, which is not very much greater than that found (0.52 log unit) in their correlation shown in Eq. 23 above.

There are eight software programs that calculate Henry's law constant, namely, Absolv (67), ChemOffice (69), ChemProp (70), Episuite (73), OECD Toolbox (24), Schrödinger (59), ProPred (80), Schrödinger (59) and SPARC (81). The performances of most of them are not known.

Dearden and Schüürmann (64) tested a number of methods for prediction of log H, using a large, diverse test set of 700 chemicals. Only one of the methods, the bond contribution method in the HENRYWIN module of the Episuite software, allowed prediction of log H for all 700 chemicals, with an MAE of prediction of 0.63 log unit.

It is recommended that the HENRYWIN module of the Episuite software be used for the prediction of Henry's law constant.

4. Guidelines for Developing QSARs and QSPRs

A number of publications have offered guidelines on how to develop QSARs and QSPRs (11, 61, 213–215). In March 2002 a meeting of QSAR/QSPR experts was held in Setúbal, Portugal, to formulate a set of guidelines for the validation of QSARs/QSPRs, in particular for regulatory purposes. Six guidelines were drawn up, which were later adopted by the OECD (216) and modified to five. The guidelines are the following:

A valid QSAR/QSPR should have:

1. A defined endpoint.
2. An unambiguous algorithm.
3. A defined domain of applicability.
4. Appropriate measures of goodness of fit, robustness, and predictivity.
5. A mechanistic interpretation, if possible.

The guidelines are now known as the OECD Principles for the Validation of (Q)SARs, although they are intended to apply to QSPRs also. The OECD has also provided a checklist to provide guidance on the interpretation of the principles (217).

Table 14
Types of error in QSAR/QSPR development and use (from Dearden et al. (218), by kind permission of Taylor & Francis Ltd., publishers (www.informaworld.com))

No.	Type of error	Relevant OECD principle(s)
1	Failure to take account of data heterogeneity	1
2	Use of inappropriate endpoint data	1
3	Use of collinear descriptors	2, 4, 5
4	Use of incomprehensible descriptors	2, 5
5	Error in descriptor values	2
6	Poor transferability of QSAR/QSPR	2
7	Inadequate/undefined applicability domain	3
8	Unacknowledged omission of data points	3
9	Use of inadequate data	3
10	Replication of compounds in data set	3
11	Too narrow a range of endpoint values	3
12	Over-fitting of data	4
13	Use of excessive number of descriptors in a QSAR/QSPR	4
14	Lack of/inadequate statistics	4
15	Incorrect calculation	4
16	Lack of descriptor auto-scaling	4
17	Misuse/misinterpretation of statistics	4
18	No consideration of distribution of residuals	4
19	Inadequate training/test set selection	4
20	Failure to validate a QSAR/QSPR correctly	4
21	Lack of mechanistic interpretation	5

Recently Dearden et al. (218) published an analysis of 21 types of error made in the development of QSARs and QSPRs, with examples taken from the literature, including some of their own. The different types of error are shown in Table 14, and are discussed briefly below. However, the reader is directed to Dearden et al. (218) for further details.

4.1. Heterogeneous Descriptors

There is a temptation, especially if data are scarce, to use values that are not strictly comparable. For example, aqueous solubilities can be measured in pure water, as undissociated species, at a given pH, or at different temperatures. It is important, in the development of a QSPR, to use data that were obtained under the same conditions, and if possible using the same protocol. Failure to do so will result in a less than satisfactory QSPR.

4.2. Inappropriate Endpoint Data

The values of the property of interest must be in molar units, and not weight units. This is an important matter, but one that is frequently not recognized. Again using aqueous solubility as an example, values should be in units of mol/L, and not g/L. The reason is that the effect of a chemical (be it physicochemical or biological) is determined by the number of molecules present, and not by how much they weigh. Consider two chemicals, A with a molecular weight of 100 and B with a molecular weight of 200. Both are found to have an aqueous solubility of 100 mg/L. However, the molar solubility of A is 1 (i.e., 100/100) mmol/L, whilst that of B is 0.5 (i.e., 100/200) mmol/L, so B is really only half as soluble as A.

4.3. Descriptor Collinearity

If two descriptors in a QSPR are themselves highly correlated (collinear), they contribute essentially the same information. In addition, one could be misled into misinterpretation of the QSPR. A more serious problem is that highly collinear descriptors in a QSPR or a QSAR can give rise to a spurious model. Dearden et al. (218) gave an example where two collinear ($r^2 = 0.959$) descriptors separately yielded two good QSARs, both with positive descriptor coefficients. However, if both were included in the same QSAR, one of the descriptors appeared with a negative coefficient and was statistically insignificant.

4.4. Incomprehensible Descriptors

There are now thousands of molecular descriptors available for use in QSPR and QSAR model development, and many of them have no clear physicochemical meaning. Whilst it is not essential for descriptors to be clearly understood, it is helpful and satisfying if they are, as well as aiding in the interpretation of the model. It must nevertheless be recognized that the existence of a correlation, however good, is not a guarantee of causality.

4.5. Errors in Descriptor Values

Most descriptor values, be they measured or calculated, contain error, and so it behoves the QSAR/QSPR practitioner to try to ensure that descriptor values are as accurate as possible (60). For example, measured log P values have a typical error of about 0.35 log unit (9), and hence calculated values must not contain errors significantly lower than that.

Ghafourian and Dearden (219) examined the use of three different quantum chemical approaches to the calculation of atomic charges and orbital energies as descriptors of hydrogen bonding, and found MNDO and AM1 methods to be better than PM3.

It is the opinion of this author that much more work needs to be done on comparison and accuracy of descriptor values, as a means of improving the accuracy of QSAR and QSPR property predictions.

4.6. Poor Transferability of QSARs and QSPRs

One of the main values of a QSAR or a QSPR is that it can be used by others for predictive purposes. Hence it has to be transferable and reproducible. Unfortunately, for various reasons (e.g., lack of availability of software) this is often not the case. Hartung et al. (220) have suggested the following criteria for transferability of a QSAR or a QSPR to a different operator:

(a) Descriptor values can be reproduced.

(b) Model definition can be confirmed.

(c) Goodness of fit and statistical robustness can be confirmed.

(d) Reproducibility of predictions can be confirmed.

(e) An assessment is given of the adequacy of documentation on the development and application of the model.

4.7. Inadequate/ Undefined Applicability Domain

The applicability domain (AD) of a QSAR or a QSPR has been defined as: "the response and chemical structure space in which the model makes predictions with a given reliability" (221, 222). It is permissible to use a QSAR/QSPR to make predictions a little way outside its AD, but one should have less confidence in the accuracy of such predictions. For example, if a given toxicity endpoint correlated well with log *P*, and the log *P* range of the training set chemicals was 0–6, one could not expect an accurate toxicity prediction for a chemical with a log *P* value of 9.

At present, very few published QSAR/QSPR papers give an indication of AD, although if descriptor values are given one can see what ranges of descriptor values were used in the training set. In addition, of course, one should not use a QSAR or a QSPR to make predictions for a chemical that is not structurally similar to at least some of the chemicals in the training set. Again, this guideline is not always adhered to.

4.8. Unacknowledged Omission of Data Points

Data used for the development of a QSAR/QSPR are often taken from published literature, and one may, for a number of reasons, wish to use only a selection of those available. If some data are omitted, that must be stated, with reasons (e.g., to keep the training set to a reasonable number of chemicals, or to examine a particular class of chemicals). However, it is not uncommon to find that data have been pruned without a reason being given, or even without a mention that data have been omitted. Probably the main reason for omission of data is that the omitted chemicals were found to be outliers; that is, their property of interest was not well predicted by the QSAR/QSPR. If this is the case, it must be stated clearly, and preferably a reason should be given (e.g., the omitted chemicals were the only ones that were strongly dissociated).

4.9. Use of Inadequate Data

Inadequacy of data can occur in a number of ways. It can include heterogeneity (Subheading 4.1), inappropriate data (Subheading 4.2), an undefined applicability domain (Subheading 4.7), and omission of data (Subheading 4.8). Another common problem is the accuracy of data, which is often very difficult to determine.

Sometimes one finds incorrect or inadequately defined chemical names. For example, a QSAR study of skin absorption (223) listed 4-chlorocresol and chloroxylenol in the training set used. The former has two isomers (4-chloro-2-cresol and 4-chloro-3-cresol), whilst chloroxylenol has 18 isomers.

A recent study (224) found that incorrect chemical structures in a number of public and private databases ranged from 0.1 to 3.4%, and observed that even slight structural errors could cause pronounced changes in the accuracy of QSAR/QSPR predictions.

It should also be noted that it is unacceptable to use predicted values of the property of interest, when developing a QSAR or a QSPR, as one is then making predictions about predictions. An example is a study of skin permeability of 114 chemicals, 63 of which had calculated permeability values (225).

4.10. Replication of Chemicals in a Data Set

Replication of chemicals in a QSAR/QSPR data set clearly distorts the development and predictivity of the model. Replicate structures can occur for a number of reasons. The same chemical can sometimes have more than one CAS number, more than one chemical name (226), different numbering systems, and different values of the property of interest (227).

How, then, can one avoid replication in QSAR/QSPR data sets? Probably the best way is by the use of unique structural codes such as the InChI code (228), and using a computer program to check for replicates. An alternative method is to sort all data by each of the available parameters, particularly the chemical formula, for possible replicates. Failure to eliminate replicates will result in invalid correlations and poor predictivity.

4.11. Too Narrow a Range of Endpoint Values

The greater the range of endpoint values used in the development of a QSPR model, the better is its predictivity (229). It is recommended that a range of endpoint values of at least 1.0 log unit be used, if possible, for a good QSAR/QSPR model to be developed (229). Sometimes, of course, that cannot be achieved, perhaps through lack of availability of sufficient data, or through sheer impossibility. For example, a QSPR for the melting points of substituted anilines used chemicals with a melting range of 244.5–461.5 K, or 0.276 log unit (151).

4.12. Over-fitting of Data

In the development of a QSAR or a QSPR, one aims to achieve as good a model as possible. Sometimes this is done by the use of many descriptors, by the removal of outliers, or by the use of a certain statistical technique. In order to reduce the risk of chance correlations, it is recommended (230) that the ratio of number of training set chemicals to number of descriptors in the model is at least 5:1. This "rule" has often been broken, with probably the worst example being the use of nine descriptors to model the aquatic toxicities of 12 alcohols (231). Some statistical techniques, such as fuzzy ARTMAP, appear (232) to produce over-fitted models in which the standard error of prediction is much lower than the error on the experimental data used to develop the model.

It is recommended that *y*-scrambling be carried out to check for over-fitting and chance correlations. This procedure involves randomizing the values of the property being modeled, and then developing a new model. This procedure is repeated for, say, 100 times, and the R^2 values of the correlations compared with the R^2 value of the true correlation. If the true R^2 value is well above all of the R^2 values from the randomized models, then one can have confidence that the original model is valid.

4.13. Use of Excessive Number of Descriptors in a QSAR/QSPR

QSPRs with a large number of descriptors are difficult to interpret (233). Dearden et al. (218) recommended a maximum of five or six descriptors as a general rule, largely on the grounds of understanding. The principle of Occam's razor is apposite here: "Entia non sunt multiplicanda praeter necessitatem" ("One should not increase beyond what is necessary the number of entities required to explain anything"). However, occasionally QSARs/QSPRs are developed with a large number of descriptors, as for example the use of 55 descriptors to model the aqueous solubility of 1,050 chemicals (234). This practice should be avoided because of the difficulty of comprehension and the risk of over-fitting (see Subheading 4.12).

4.14. Lack of/ Inadequate Statistics

The statistics provided with a QSAR/QSPR are an indication of how well the model fits the training set data, and how predictive the model is. Many QSAR and QSPR models are still published without full statistics, which means that it is difficult, if not impossible, to judge the validity of a model. Dearden et al. (218) have recommended that the following statistical indicators are included with each published QSAR/QSPR: *n* (number of chemicals in the training set); r^2 or R^2 (coefficient of determination or squared correlation coefficient, lower case for a 1-descriptor model and upper case for a multi-descriptor model); q^2 or Q^2 (squared cross-validated correlation coefficient, an internal indicator of predictivity); ${R_{adj}}^2$ (squared correlation coefficient adjusted for degrees of freedom, which allows comparison between QSARs and QSPRs containing different number of descriptors); *s* (standard error of the estimate) or RMSE (very similar to standard error of the estimate, especially when n is large); and full F statistics (F is the Fisher statistic or variance ratio, which indicates the confidence level of the model).

4.15. Incorrect Calculation

It is the duty of authors to ensure that calculations that are made to obtain their results are as accurate as possible. Editors and manuscript reviewers expect that to be the case, and it is often impossible to check accuracy because insufficient data are supplied. There is no doubt that incorrectly calculated QSARs and QSPRs have been published, but no one has investigated this problem. A few instances have come to light. Dearden et al. (218) reported a case in which a 5-descriptor QSAR was published with $R^2 = 0.958$, $s = 0.14$, and $F = 79$. Because all descriptor values were given, it was possible to recalculate the QSAR, and the statistics were found to be $R^2 = 0.298$, $s = 0.56$, and $F = 1.4$.

It is recommended that all calculations be double-checked, preferably by two different people, before a QSAR or a QSPR is published.

4.16. Lack of Descriptor Auto-Scaling

Auto-scaling is the modification of descriptor values by subtracting the mean from the value of each descriptor, and then dividing by the standard deviation. This yields descriptor values with a mean of zero and a variance of one, which means that modeling is less susceptible to the influence of chemicals with extreme values (42), and avoids the risk of descriptors with large numerical values dominating those with small values (62). Another important result of auto-scaling is that the relative contributions of each descriptor to the model can readily be seen (see Eq. 1).

Regrettably, very few QSAR/QSPR publications use auto-scaling, and it is recommended that it should be standard practice.

4.17. Misuse or Misinterpretation of Statistics

Not many QSAR/QSPR practitioners are statisticians, and few statisticians have experience of QSAR/QSPR modeling. Nevertheless, statistics is an essential QSAR/QSPR tool. It is therefore not surprising to find that it is sometimes misused, or its results misinterpreted. Livingstone's book (42) is very helpful in this respect, and a survey of QSAR/QSPR statistics available on the Internet (235) gives useful guidance. Useful Web sites are those of the Scripps Institute (236) and QSAR World (237).

Two cases involving the misuse or misinterpretation of statistics (231, 232) have already been mentioned in Subheading 4.12.

It behoves QSAR/QSPR workers to familiarize themselves sufficiently with the requisite statistics, or to enlist the help of a knowledgeable statistician, in order to ensure that the statistical techniques and results that they use are valid.

4.18. No Consideration of Distribution of Residuals

Residuals (differences between measured and predicted endpoint values) can arise from both random error and systematic error. Random error arises from lack of data and/or descriptor value reproducibility, whilst systematic error generally results from biases in descriptor values, perhaps from poor choice of descriptors.

If residuals are plotted against measured endpoint values, random distribution around the zero residual line indicates random error. If, however, all or most of the residuals lie on one side of the zero residual line, or show a consistent variation of residuals with increasing measured values, systematic error is indicated. This suggests that the model needs to be reexamined in order to eliminate such error.

Currently very few QSAR/QSPR publications include a consideration of the distribution of residuals. It is recommended that residual plots be included, as a guide to model improvement.

4.19. Inadequate Training/Test Set Selection

Data for QSAR/QSPR modeling are usually divided into training and test sets either randomly or by ordering the chemicals according to endpoint values and then selecting every *n*th chemical for the test set. These approaches have, however, been shown to be suboptimal (14, 238). The training set should cover a good range of endpoint values and (for diverse data sets) have an adequate coverage of the requisite chemical space. The test set chemicals should also cover a good range of endpoint values, be sufficiently diverse in nature, and be similar (but not too similar) to training set chemicals. Various techniques (14, 15, 239, 240) have been proposed for the rational selection of training and test sets.

It is essential that proper attention is paid to training and test set selection when preparing to develop a QSAR/QSPR model. It is recognized, however, that one often serious drawback to this is lack of availability of satisfactory and appropriate data.

4.20. Failure to Validate a QSAR/QSPR Correctly

The main use of a QSAR or a QSPR is to make predictions of properties of chemicals that were not in the training set used to develop the model. So the prediction ability (the predictivity) of the model needs to be assessed (241). Tropsha et al. (242) discussed a number of ways of doing this, and they recommended that both internal and external validation be carried out; these procedures were described briefly in Subheading 2.1. It is now generally accepted that external validation is the better way to validate a QSAR/QSPR model (233, 243).

A problem arises if there are only a relatively few chemicals available from which to develop the model. Suppose that one has only 15 chemicals in one's data set. Removal of 20% of them to use as a test set would leave only 12 chemicals in the training set, which would allow, from the Topliss and Costello rule (230), the inclusion of only two descriptors in the model. An alternative acceptable procedure in such a case is to remove one chemical, and develop the model on the remaining chemicals, from the whole pool of descriptors. That model is then used to predict the endpoint value of the omitted chemical. That chemical is then returned to the training set, a second chemical is removed, and the model is redeveloped, again from the whole pool of descriptors. This procedure is repeated until every chemical has been removed in turn, and one then has external predictions for every chemical. The technique is not as satisfactory as using separate training and test sets, but is better than internal cross-validation (244).

It should be noted that at least four journals (SAR & QSAR in Environmental Research, Molecular Informatics, Journal of Medicinal Chemistry and Journal of Chemical Information and Modeling) require external validation of published QSARs/QSPRs. It is thus essential that all QSARs and QSPRs are fully validated before being used predictively.

4.21. Lack of Mechanistic Interpretation

The documentation for the OECD Principles for the Validation of (Q)SARs (216) states: *It is recognised that it is not always possible, from a scientific viewpoint, to provide a mechanistic interpretation of a given (Q)SAR (Principle 5), or that there even be multiple mechanistic interpretations of a given model. The absence of a mechanistic interpretation for a model does not mean that a model is not potentially useful in the regulatory context. The intent of Principle 5 is not to reject models that have no apparent mechanistic basis, but to ensure that some consideration is given to the possibility of a mechanistic association between the descriptors used in a model and the endpoint being predicted, and to ensure that this association is documented.*

The OECD guidance on the Principles for the Validation of (Q)SARs/(Q)SPRs (217) recommends that the following questions be asked regarding the mechanistic basis of a QSAR/QSPR:

1. Do the descriptors have a physicochemical interpretation that is consistent with a known mechanism?
2. Can any literature references be cited in support of the purported mechanistic basis of the QSAR/QSPR?

If the answers to both questions are positive, one may have some confidence in the proposed mechanism of action. If the answer to one or both questions is negative, then the level of confidence will be lower. In all cases, it must be remembered that the existence of a correlation does not imply causality.

Johnson (245) recently commented that QSAR has devolved into a kind of logical fallacy: *cum hoc ergo propter hoc* (with this, therefore because of this). He also stated: “rarely, if ever, are any designed experiments presented to test or challenge the interpretation of the descriptors . . . Statistical methodologies should be a tool of QSAR but instead have often replaced the craftsman tools of our trade—rational thought, controlled experiments, and personal observation.” The QSAR/QSPR practitioner would do well to take those strictures to heart, as well as the recommendations made in a recent extensive review of QSPR prediction of physicochemical properties (246).

References

1. van de Waterbeemd H (2009) Improving compound quality through *in vitro* and *in silico* profiling. Chem Biodivers 6:1760–1766
2. Cronin MTD, Livingstone DJ (2004) Calculation of physicochemical properties. In: Cronin MTD, Livingstone DJ (eds) Predicting chemical toxicity and fate. CRC, Boca Raton, FL, pp 31–40
3. Fisk PR, McLaughlin L, Wildey RJ (2004) Good practice in physicochemical property prediction. In: Cronin MTD, Livingstone DJ (eds) Predicting chemical toxicity and fate. CRC, Boca Raton, FL, pp 41–59
4. Webb TH, Morlacci LA (2010) Calculation of physic-chemical and environmental fate properties. In: Cronin MTD, Madden JC (eds) *In silico* toxicology: principles and applications. RSC Publishing, Cambridge, pp 118–147
5. Dearden JC (2004) QSAR modeling of bioaccumulation. In: Cronin MTD, Livingstone DJ (eds) Predicting chemical toxicity and fate. CRC, Boca Raton, FL, pp 333–355
6. Dearden JC (2004) QSAR modeling of soil sorption. In: Cronin MTD, Livingstone DJ (eds) Predicting chemical toxicity and fate. CRC, Boca Raton, FL, pp 357–371
7. Schüürmann G, Ebert R-U, Nendza M et al (2007) Predicting fate-related physicochemical properties. In: van Leeuwen CJ, Vermeire TG (eds) Risk assessment of chemicals: an introduction, 2nd edn. Springer, Dordrecht, pp 375–426
8. Abraham MH, Chadha HS, Mitchell RC (1994) Hydrogen bonding. 32. An analysis of water–octanol and water–cyclohexane partitioning and the Δlog P parameter of Seiler. J Pharm Sci 83:1085–1100
9. Mannhold R, Poda GI, Ostermann C et al (2009) Calculation of molecular lipophilicity: state-of-the-art and comparison of log P methods on more than 96,000 compounds. J Pharm Sci 98:861–893
10. Dearden JC (2006) *In silico* prediction of aqueous solubility. Exp Opin Drug Discov 1:31–52
11. Livingstone DJ (2004) Building QSAR models: a practical guide. In: Cronin MTD, Livingstone DJ (eds) Predicting chemical toxicity and fate. CRC, Boca Raton, FL, pp 151–170
12. SMILES: www.daylight.com/dayhtml_tutorials/languages/smiles/index.html
13. Nendza M, Aldenberg T, Benfenati E et al (2010) Data quality assessment for *in silico* methods: a survey of approaches and needs. In: Cronin MTD, Madden JC (eds) *In silico* toxicology: principles and applications. RSC Publishing, Cambridge, pp 59–117
14. Leonard JT, Roy K (2006) On selection of training and test sets for the development of predictive QSAR models. QSAR Comb Sci 25:235–251
15. Golbraikh A, Shen M, Xiao Z et al (2003) Rational selection of training and test sets for the development of validated QSAR models. J Comput Aided Mol Des 17:241–253
16. Aquasol: www.pharmacy.arizona.edu/outreach/aquasol/
17. Tripos: www.tripos.com
18. Chemical & Physical Properties Database: www.dep.state.pa.us/physicalproperties/CPP_search.htm
19. Chemical Database Service: cds.dl.ac.uk
20. ChemSpider: www.chemspider.com
21. Crossfire: info.crossfiredatabases.com
22. OCHEM: www.ochem.eu
23. OECD eChemPortal: www.echemportal.org
24. OECD QSAR Toolbox: www.qsartoolbox.org
25. OSHA: www.osha.gov/web/dep/chemicaldata/
26. PhysProp: www.syrres.com/what-we-d0/product.aspx?id=133
27. Wagner AB (2001) Finding physical properties of chemicals: a practical guide for scientists, engineers, and librarians. Sci Technol Lib 21(3/4):27–45
28. Oyarzabal J, Pastor J, Howe TJ (2009) Optimizing the performance of in silico ADMET general models according to local requirements: MARS approach. Solubility estimations as case study. J Chem Inf Model 49:2837–2850
29. Dearden JC (1990) Physico-chemical descriptors. In: Karcher W, Devillers J (eds) Practical applications of quantitative structure–activity relationships (QSARs) in environmental chemistry and toxicology. Kluwer Academic, Dordrecht, pp 25–59
30. Maran U, Sild S, Tulp I et al (2010) Molecular descriptors from two-dimensional chemical structure. In: Cronin MTD, Madden JC (eds) *In silico* toxicology: principles and applications. RSC Publishing, Cambridge, pp 148–192
31. Ran YQ, Jain N, Yalkowsky SH (2001) Prediction of aqueous solubility of organic compounds by the general solubility equation (GSE). J Chem Inf Comput Sci 41: 1208–1217

32. ADAPT: research.chem.psu.edu/pcjgroup/adapt.html
33. Molecular Discovery: www.moldiscovery.com
34. SemiChem: www.semichem.com
35. Biobyte: www.biobyte.com
36. Accelrys: www.accelrys.com
37. Dragon: www.talete.mi.it/products/dragon_description.htm
38. eDragon: www.vcclab.org/lab/edragon/
39. ChemComp: www.chemcomp.com
40. EduSoft: www.edusoft-lc.com/molconn/
41. vLifeSciences: www.vlifesciences.com
42. Livingstone D (1995) Data analysis for chemists. Oxford University Press, Oxford
43. Rowe PH (2010) Statistical methods for continuous measured endpoints in *in silico* toxicology. In: Cronin MTD, Madden JC (eds) *In silico* toxicology: principles and applications. RSC Publishing, Cambridge, pp 228–251
44. SimulationsPlus: www.simulations-plus.com
45. FQS Poland: www.fqs.pl
46. VCCLAB: www.vcclab.org
47. MSI: www.msi.umn.edu/sw/cerius2
48. VSN International: www.vsni.co.uk/software/genstat/
49. MathWorks: www.mathworks.com
50. Minitab: www.minitab.com
51. NCSS: www.ncss.com
52. IDBS: www.idbs.com
53. ProChemist: pro.chemist.online.fr
54. GNU: www.gnu.org/software/pspp/
55. SAS: www.sas.com
56. Scigress Explorer: www.scigress-explorer.software.informer.com
57. SPSS: www.spss.com
58. StatSoft: www.statsoft.com
59. Schrödinger: www.schrodinger.com
60. Tetko IV, Bruneau P, Mewes H-W et al (2006) Can we estimate the accuracy of ADME-Tox predictions? Drug Disc Today 11:700–707
61. Walker JD, Dearden JC, Schultz TW et al (2003) QSARs for new practitioners. In: Walker JD (ed) Quantitative structure–activity relationships for pollution prevention, toxicity screening, risk assessment, and web applications. SETAC, Pensacola, FL, pp 3–18
62. Eriksson L, Jaworska J, Worth AP et al (2003) Methods for reliability and uncertainty assessment and for applicability evaluations of classification- and regression-based QSARs. Environ Health Perspect 111:1361–1375
63. Dearden JC (2003) Quantitative structure--property relationships for prediction of boiling point, vapor pressure, and melting point. Environ Toxicol Chem 22:1696–1709
64. Dearden JC, Schüürmann G (2003) Quantitative structure–property relationships for predicting Henry's law constant from molecular structure. Environ Toxicol Chem 22:1755–1770
65. QMRF Database: qsardb.jrc.it/qmrf/
66. Danish QSAR Database: www.130.226.165.14/index.html
67. ACD/Labs: www.acdlabs.com
68. ChemAxon: www.chemaxon.com
69. CambridgeSoft: www.cambridgesoft.com
70. UFZ: www.ufz.de/index.php?en=6738
71. ChemSilico: www.chemsilico.com
72. Daylight: www.daylight.com
73. Episuite: www.epa.gov/opptintr/exposure/pubs/episuite.htm
74. ChemSW: www.chemsw.com
75. Molinspiration: www.molinspiration.com
76. Chemistry Database Software: www.chemdbsoft.com
77. CompuDrug: www.compudrug.com
78. G & P Engineering Software: www.gpengineeringsoft.com
79. MW Software: www.mwsoftware.com/dragon
80. ProPred: www.capec.kt.dtu.dk
81. SPARC: ibmlc2.chem.uga.edu/sparc
82. TerraBase: www.terrabase-inc.com
83. Optibrium: www.optibrium.com
84. Dearden JC (1985) Partitioning and lipophilicity in quantitative structure–activity relationships. Environ Health Perspect 61:203–228
85. Hansch C, Maloney PP, Fujita T et al (1962) Correlation of biological activity of phenoxyacetic acids with Hammett substituent constants and partition coefficients. Nature 194:178–180
86. Nendza M (1998) Structure–activity relationships in environmental sciences. Chapman & Hall, London
87. Reinhard M, Drefahl A (1999) Estimating physicochemical properties of organic compounds. Wiley, New York, NY
88. Leo A (2000) Octanol/water partition coefficients. In: Boethling RS, Mackay D (eds) Handbook of property estimation methods for chemicals. Lewis, Boca Raton, FL, pp 89–114

89. Mannhold R, van de Waterbeemd H (2001) Substructure and whole molecule approaches for calculating log P. Comput Aided Mol Des 15:337–354
90. Livingstone DJ (2003) Theoretical property predictions. Curr Top Med Chem 3: 1171–1192
91. Klopman G, Zhu H (2005) Recent methodologies for the estimation of n-octanol/ water partition coefficients and their use in the prediction of membrane transport properties of drugs. Mini Rev Med Chem 5:127–133
92. Fujita T, Iwasa J, Hansch C (1964) A new substituent constant, π, derived from partition coefficients. J Am Chem Soc 86:5175–5180
93. Nys GG, Rekker RF (1973) Statistical analysis of a series of partition coefficients with special reference to the predictability of folding of drug molecules. Introduction of hydrophobic fragmental constants (*f* values). Chim Ther 8:521–535
94. Rekker RF (1977) The hydrophobic fragmental constant. Elsevier, Amsterdam
95. Leo A, Jow PYC, Silipo C et al (1975) Calculation of hydrophobic constant (log P) from π and *f* values. J Med Chem 18:865–868
96. Bodor N, Gabanyi NZ, Wong C-K (1989) A new method for the estimation of partition coefficient. J Am Chem Soc 111:3783–3786
97. Klopman G, Wang S (1991) A computer automated structure evaluation (CASE) approach to calculation of partition coefficient. J Comput Chem 12:1025–1032
98. Ghose AK, Pritchett A, Crippen GM (1988) Atomic physicochemical parameters for three dimensional structure directed quantitative structure–activity relationships: III. Modeling hydrophobic interactions. J Comput Chem 9:80–90
99. Liu R, Zhou D (2008) Using molecular fingerprint as descriptors in the QSPR study of lipophilicity. J Chem Inf Model 48:542–549
100. Chen H-F (2009) *In silico* log P prediction for a large data set with support vector machines, radial basis neural networks and multiple linear regression. Chem Biol Drug Des 74:142–147
101. Tetko IV, Tanchuk VYu, Villa AEP (2001) Prediction of *n*-octanol/water partition coefficients from PHYSPROP database using artificial neural networks and E-state indices. J Chem Inf Comput Sci 41:1407–1421
102. Hall LH, Kier LB (1999) Molecular structure description: the electrotopological state. Academic, New York, NY
103. Dearden JC, Netzeva TI, Bibby R (2003) A comparison of commercially available software for the prediction of partition coefficient. In: Ford M, Livingstone D, Dearden J et al (eds) Designing drugs and crop protectants: processes, problems and solutions. Blackwell, Oxford, pp 168–169
104. Sakuratani Y, Kasai K, Noguchi Y et al (2007) Comparison of predictivities of log P calculation models based on experimental data for 134 simple organic compounds. QSAR Comb Sci 26:109–116
105. COSMO*logic*: www.cosmologic.de
106. Varnek A, Fourches D, Solov'ev VP et al (2004) "In silico" design of new uranyl extractants based on phosphoryl-containing podands: QSPR studies, generation and screening of virtual combinatorial library, and experimental tests. J Chem Inf Comput Sci 44:1365–1382
107. Dearden JC, Cronin MTD, Schultz TW et al (1995) QSAR study of the toxicity of nitrobenzenes to *Tetrahymena pyriformis*. Quant Struct Act Relat 14:427–432
108. Katritzky AR, Wang Y, Sild S et al (1998) QSPR studies on vapor pressure, aqueous solubility, and the prediction of air–water partition coefficients. J Chem Inf Comput Sci 38:720–725
109. Yalkowsky SH, Banerjee S (1992) Aqueous solubility: methods of estimation for organic compounds. Dekker, New York, NY
110. Mackay D (2000) Solubility in water. In: Boethling RS, Mackay D (eds) Handbook of property estimation methods for chemicals: environmental and health sciences. Lewis, Boca Raton, FL, pp 125–139
111. Schwarzenbach RP, Gschwend PM, Imboden DM (1993) Environmental organic chemistry. Wiley, New York, NY
112. Johnson SR, Zheng W (2006) Recent progress in the computational prediction of aqueous solubility and absorption. AAPS J 8: E27–E40
113. ECETOC Technical Report No. 89 (2003) (Q)SARs: evaluation of the commercially available software for human health and environmental endpoints with respect to chemical management applications. ECETOC, Brussels
114. Hansch C, Quinlan JE, Lawrence GL (1968) The linear free energy relationship between partition coefficients and aqueous solubility of organic liquids. J Org Chem 33:347–350
115. Yalkowsky SH, Valvani SC (1980) Solubility and partitioning I: solubility of nonelectrolytes in water. J Pharm Sci 69:912–922

116. Hughes LD, Palmer DS, Nigsch F et al (2008) Why are some properties more difficult to predict than others? A study of QSPR models of solubility, melting point, and log P. J Chem Inf Model 48:220–232
117. Sanghvi T, Jain N, Yang G et al (2003) Estimation of aqueous solubility by the general solubility equation (GSE) the easy way. QSAR Comb Sci 22:258–262
118. Abraham MH, Le J (1999) The correlation and prediction of the solubility of compounds in water using an amended solvation energy relationship. J Pharm Sci 88:868–880
119. Votano JR, Parham M, Hall LH et al (2004) Prediction of aqueous solubility based on large datasets using several QSPR models utilizing topological structure representation. Chem Biodivers 11:1829–1841
120. Raevsky OA, Raevskaja OE, Schaper K-J (2004) Analysis of water solubility data on the basis of HYBOT descriptors. Part 3. Solubility of solid neutral chemicals and drugs. QSAR Comb Sci 23:327–343
121. Klopman G, Zhu H (2001) Estimation of the aqueous solubility of organic molecules by the group contribution approach. J Chem Inf Comput Sci 41:439–445
122. Palmer DS, O'Boyle NM, Glen RC et al (2007) Random forest models to predict aqueous solubility. J Chem Inf Model 47: 150–158
123. Lind P, Maltseva T (2003) Support vector machines for the estimation of aqueous solubility. J Chem Inf Comput Sci 43:1855–1859
124. Duchowicz PR, Talevi A, Bruno-Blanch LE et al (2008) New QSPR study for the prediction of aqueous solubility of drug-like compounds. Bioorg Med Chem 16:7944–7955
125. Duchowicz PR, Castro EA (2009) QSPR studies on aqueous solubilities of drug-like compounds. Int J Mol Sci 10:2558–2577
126. Huuskonen J, Livingstone DJ, Manallack DT (2008) Prediction of drug solubility from molecular structure using a drug-like training set. SAR QSAR Environ Res 19:191–212
127. Yang G-Y, Yu J, Wang Z-Y et al (2007) QSPR study on the aqueous solubility (−lgS(w)) and *n*-octanol/water partition coefficients (lgK(ow)) of polychlorinated dibenzo-*p*-dioxins (PCDDs). QSAR Comb Sci 26: 352–357
128. Wei X-Y, Ge Z-G, Wang Z-Y et al (2007) Estimation of aqueous solubility (−lgS(w)) of all polychlorinated biphenyl (PCB) congeners by density function theory and position of Cl substitution (N-PCS) method. Chinese J Struct Chem 26:519–528
129. Dearden JC, Netzeva TI, Bibby R (2003) A comparison of commercially available software for the prediction of aqueous solubility. In: Ford M, Livingstone D, Dearden J et al (eds) Designing drugs and crop protectants: processes, problems and solutions. Blackwell, Oxford, pp 169–171
130. Dearden JC. Unpublished information
131. Harris JC, Hayes MJ (1990) Acid dissociation constant. In: Lyman WJ, Reehl WF, Rosenblatt DH (eds) Handbook of chemical property estimation methods. American Chemical Society, Washington, DC, pp 6.1–6.28
132. Brown TN, Mora-Diez N (2006) Computational determination of aqueous pKa values of protonated benzimidazoles (Part 2). J Phys Chem B 110:20546–20554
133. Kaschula CH, Egan TJ, Hunter R et al (2002) Structure–activity relationships in 4-aminoquinoline antiplasmodials. The role of the group at the 7-position. J Med Chem 45:3531–3539
134. Soriano E, Cerdan S, Ballesteros P (2004) Computational determination of pK(a) values. A comparison of different theoretical approaches and a novel procedure. J Mol Struct Theochem 684:121–128
135. Klopman G, Fercu D (1994) Application of the multiple computer automated structure evaluation methodology to a quantitative structure–activity relationship study of acidity. J Comput Chem 15:1041–1050
136. Klamt A, Eckert F, Diedenhofen M et al (2003) First principles calculations of aqueous pK(a) values for organic and inorganic acids using COSMO-RS reveal an inconsistency in the slope of the pK(a) scale. J Phys Chem A 107:9380–9386
137. Eckert F, Klamt A (2006) Accurate prediction of basicity in aqueous solution with COSMO-RS. J Comput Chem 27:11–19
138. Lee AC, Yu J-Y, Crippen GM (2008) pKa prediction of monoprotic small molecules the SMARTS way. J Chem Inf Model 48:2042–2053
139. Milletti F, Storchi L, Sforna G et al (2007) New and original pKa prediction method using GRID molecular interaction fields. J Chem Inf Model 47:2172–2181
140. Cruciani G, Milletti F, Storchi L et al (2009) *In silico* prediction and ADME profiling. Chem Biodivers 6:1812–1821
141. Parthasarathi R, Padmanabhan J, Elango M et al (2006) pKa prediction using group philicity. J Phys Chem A 110:6540–6544
142. Tsantili-Kakoulidou A, Panderi I, Csizmadia F et al (1997) Prediction of distribution

coefficient from structure 2. Validation of PrologD, an expert system. J Pharm Sci 86: 1173–1179
143. Hilal SH, Karickhoff SW, Carreira LA (1995) A rigorous test for SPARC's chemical reactivity models: estimation of more than 4300 ionisation pKa's. Quant Struct Act Relat 14: 348–355
144. Lee PH, Ayyampalayam SN, Carreira LA et al (2007) In silico prediction of ionization constants of drugs. Mol Pharm 4:498–512
145. Kühne R, Ebert R-U, Schüürmann G (2006) Model selection based on structural similarity—method description and application to water solubility prediction. J Chem Inf Model 46:636–641
146. Dearden JC, Cronin MTD, Lappin DC (2007) A comparison of commercially available software for the prediction of pKa. J Pharm Pharmacol 59(suppl 1):A-7
147. Liao C, Nicklaus MC (2009) Comparison of nine programs predicting pKa values of pharmaceutical substances. J Chem Inf Model 49:2801–2812
148. Meloun M, Bordovská S (2007) Benchmarking and validating algorithms that estimate pKa values of drugs based on their molecular structure. Anal Bioanal Chem 389:1267–1281
149. Balogh GT, Gyarmati B, Nagy B et al (2009) Comparative evaluation of in silico pKa prediction tools on the Gold Standard dataset. QSAR Comb Sci 28:1148–1155
150. Manchester J, Walkup G, Rivin O et al (2010) Evaluation of pKa estimation methods on 211 druglike compounds. J Chem Inf Model 50:565–571
151. Dearden JC (1991) The QSAR prediction of melting point, a property of environmental relevance. Sci Total Environ 109(110):59–68
152. Horvath AL (1992) Molecular design: chemical structure generation from the properties of pure organic compounds. Elsevier, Amsterdam
153. Dearden JC (1999) The prediction of melting point. In: Charton M, Charton B (eds) Advances in quantitative structure–property relationships, vol 2. JAI Press, Stamford, CT, pp 127–175
154. Tesconi M, Yalkowsky SH (2000) Melting point. In: Boethling RS, Mackay D (eds) Handbook of property estimation methods for chemicals. Lewis, Boca Raton, FL, pp 3–27
155. Mills EJ (1884) On melting point and boiling point as related to composition. Phil Mag 17:173–187
156. Katritzky AR, Maran U, Karelson M et al (1997) Prediction of melting points for the substituted benzenes. J Chem Inf Comput Sci 37:913–919
157. Abramowitz R, Yalkowsky SH (1990) Estimation of aqueous solubility and melting point of PCB congeners. Chemosphere 21:1221–1229
158. Tsakanikas PD, Yalkowsky SH (1988) Estimation of melting point of flexible molecules: aliphatic hydrocarbons. Toxicol Environ Chem 17:19–33
159. Abramowitz R, Yalkowsky SH (1990) Melting point, boiling point and symmetry. Pharm Res 7:942–947
160. Zhao L, Yalkowsky SH (1999) A combined group contribution and molecular geometry approach for predicting melting points of aliphatic compounds. Ind Eng Chem Res 38:3581–3584
161. Todeschini R, Vighi M, Finizio A et al (1997) 3-D modelling and prediction by WHIM descriptors. Part 8. Toxicity and physicochemical properties of environmental priority chemicals by 2D-TI and 3D-WHIM descriptors. SAR QSAR Environ Res 7:173–193
162. Bergström CAS, Norinder U, Luthman K et al (2003) Molecular descriptors influencing melting point and their role in classification of solid drugs. J Chem Inf Comput Sci 43: 1177–1185
163. Modarresi H, Dearden JC, Modarress H (2006) QSPR correlation of melting point for drug compounds based on different sources of molecular descriptors. J Chem Inf Model 46:930–936
164. Godavarthy SS, Robinson RL, Gasem KAM (2006) An improved structure–property model for predicting melting-point temperatures. Ind Eng Chem Res 45:5117–5126
165. Karthikeyan M, Glen RC, Bender A (2005) General melting point prediction based on a diverse compound data set and artificial neural networks. J Chem Inf Model 45:581–590
166. Joback KG, Reid RC (1987) Estimation of pure-component properties from group contributions. Chem Eng Commun 57:233–243
167. Simamora P, Yalkowsky SH (1994) Group contribution methods for predicting the melting points and boiling points of aromatic compounds. Ind Eng Chem Res 33: 1405–1409
168. Constantinou L, Gani R (1994) New group contribution method for estimating properties of pure compounds. Am Inst Chem Eng J 40:1697–1710

169. Marrero J, Gani R (2001) Group-contribution based estimation of pure component properties. Fluid Phase Equil 183–184:183–208
170. Tu C-H, Wu Y-S (1996) Group-contribution estimation of normal freezing points of organic compounds. J Chin Inst Chem Eng 27:323–328
171. Gold PI, Ogle GJ (1969) Estimating thermophysical properties of liquids. Part 4—Boiling, freezing and triple-point temperatures. Chem Eng 76:119–122
172. Lyman WJ (2000) Boiling point. In: Boethling RS, Mackay D (eds) Handbook of property estimation methods for chemicals: environmental and health sciences. Lewis, Boca Raton, FL, pp 29–51
173. Banks WH (1939) Considerations of a vapour pressure-temperature equation, and their relation to Burnop's boiling point function. J Chem Soc 292–295
174. Rechsteiner CE (1990) Boiling point. In: Lyman WJ, Reehl WF, Rosenblatt DH (eds) Handbook of chemical property estimations methods. American Chemical Society, Washington, DC, pp 12.1–12.55
175. Ivanciuc O, Ivanciuc T, Cabrol-Bass D et al (2000) Evaluation in quantitative structure–property relationship models of structural descriptors derived from information-theory operators. J Chem Inf Comput Sci 40: 631–643
176. Gironés X, Amat L, Robert D et al (2000) Use of electron–electron repulsion energy as a molecular descriptor in QSAR and QSPR studies. J Comput Aided Mol Des 14: 477–485
177. Katritzky AR, Mu L, Lobanov VS et al (1996) Correlation of boiling points with molecular structure. 1. A training of 298 diverse organics and a test set of 9 simple inorganics. J Phys Chem 100:10400–10407
178. Sola D, Ferri A, Banchero M et al (2008) QSPR prediction of N-boiling point and critical properties of organic compounds and comparison with a group-contribution method. Fluid Phase Equil 263:33–42
179. Wessel MD, Jurs PC (1995) Prediction of normal boiling points for a diverse set of industrially important organic compounds from molecular structure. J Chem Inf Comput Sci 35:841–850
180. Basak SC, Mills D (2001) Use of mathematical structural invariants in the development of QSPR models. Commun Math Comput Chem 44:15–30
181. Hall LH, Story CT (1996) Boiling point and critical temperature of a heterogeneous data set. QSAR with atom type electrotopological state indices using artificial neural networks. J Chem Inf Comput Sci 36:1004–1014
182. Kier LB, Hall LH (1999) Molecular structure description: the electrotopological state. Academic, San Diego, CA
183. Stein SE, Brown RL (1994) Estimation of normal boiling points from group contributions. J Chem Inf Comput Sci 34:581–587
184. Labute P (2000) A widely applicable set of descriptors. J Mol Graph Model 18:464–477
185. Ericksen D, Wilding WV, Oscarson JL et al (2002) Use of the DIPPR database for development of QSPR correlations: normal boiling point. J Chem Eng Data 47:1293–1302
186. Grain CF (1990) Vapor pressure. In: Lyman WJ, Reehl WF, Rosenblatt DH (eds) Handbook of chemical property estimation methods. American Chemical Society, Washington, DC, pp 14.1–14.20
187. Delle Site A (1996) The vapor pressure of environmentally significant organic chemicals: a review of methods and data at ambient temperature. J Phys Chem Ref Data 26: 157–193
188. Sage ML, Sage GW (2000) Vapor pressure. In: Boethling RS, Mackay D (eds) Handbook of property estimation methods for chemicals: environmental and health sciences. Lewis, Boca Raton, FL, pp 53–65
189. Katritzky AR, Slavov SH, Dobchev DA et al (2007) Rapid QSPR model development technique for prediction of vapor pressure of organic compounds. Comput Chem Eng 31:1123–1130
190. Liang CK, Gallagher DA (1998) QSPR prediction of vapor pressure from solely theoretically-derived descriptors. J Chem Inf Comput Sci 38:321–324
191. Tu C-H (1994) Group-contribution method for the estimation of vapor pressures. Fluid Phase Equil 99:105–120
192. Öberg T, Liu T (2008) Global and local PLS regression models to predict vapor pressure. QSAR Comb Sci 27:273–279
193. Basak SC, Mills D (2009) Predicting the vapour pressure of chemicals from structure: a comparison of graph theoretic versus quantum chemical descriptors. SAR QSAR Environ Res 20:119–132
194. Goll ES, Jurs PC (1999) Prediction of vapor pressures of hydrocarbons and halohydrocarbons from molecular structure with a

computational neural network model. J Chem Inf Comput Sci 39:1081–1089

195. Staikova M, Wania F, Donaldson DJ (2004) Molecular polarizability as single-parameter predictor of vapor pressures and octanol-air partitioning coefficients of nonpolar compounds: a priori approach and results. Atmos Environ 38:213–225
196. Andreev NN, Kuznetsov SE, Storozhenko SY (1994) Prediction of vapour pressure and boiling points of aliphatic compounds. Mendeleev Commun 173–174
197. Kühne R, Ebert R-U, Schüürmann G (1997) Estimation of vapour pressures for hydrocarbons and halogenated hydrocarbons from chemical structure by a neural network. Chemosphere 34:671–686
198. Yaffe D, Cohen Y (2001) Neural network based temperature-dependent quantitative structure property relationships (QSPRs) for predicting vapor pressure of hydrocarbons. J Chem Inf Comput Sci 41:463–477
199. Godavarthy SS, Robinson RL, Gasem KAM (2006) SVRC-QSPR model for predicting saturated vapor pressure of pure fluids. Fluid Phase Equil 246:39–51
200. Schüürmann G, Rothenbacher C (1992) Evaluation of estimation methods for the air–water partition coefficient. Fresenius Environ Bull 1:10–15
201. Mackay D, Shiu WY, Ma KC (2000) Henry's law constant. In: Boethling RS, Mackay D (eds) Handbook of property estimation methods for chemicals: environmental and health sciences. Lewis, Boca Raton, FL, pp 69–87
202. Dearden JC, Cronin MTD, Ahmed SA et al (2000) QSPR prediction of Henry's law constant: improved correlation with new parameters. In: Gundertofte K, Jørgensen FS (eds) Molecular modeling and prediction of bioactivity. Kluwer Academic/Plenum, New York, NY, pp 273–274
203. Hine J, Mookerjee PK (1974) The intrinsic hydrophilic character of organic compounds. Correlations in terms of structural contributions. J Org Chem 40:292–298
204. Cabani S, Gianni P, Mollica V et al (1981) Group contributions to the thermodynamic properties of non-ionic organic solutes in dilute aqueous solution. J Solut Chem 10:563–595
205. Meylan WM, Howard PH (1991) Bond contribution method for estimating Henry's law constants. Environ Toxicol Chem 10: 1283–1293
206. Nirmalakhandan NN, Speece RE (1988) QSAR model for predicting Henry's constant. Environ Sci Technol 22:1349–1357
207. Russell CJ, Dixon SL, Jurs PC (1992) Computer-assisted study of the relationship between molecular structure and Henry's law constant. Anal Chem 64:1350–1355
208. Modarresi H, Modarress H, Dearden JC (2007) QSPR model of Henry's law constant for a diverse set of organic chemicals based on genetic algorithm-radial basis function network approach. Chemosphere 66:2067–2076
209. Abraham MH, Andonian-Haftvan J, Whiting GS et al (1994) Hydrogen bonding. Part 34. The factors that influence the solubility of gases and vapours in water at 298 K, and a new method for its determination. J Chem Soc Perkin Trans 2:1777–1791
210. Yaffe D, Cohen Y, Espinosa G et al (2003) A fuzzy ARTMAP-based quantitative structure–property relationship (QSPR) for the Henry's law constant of organic compounds. J Chem Inf Comput Sci 43:85–112
211. Gharagheizi F, Abbasi R, Tirandazi B (2010) Prediction of Henry's law constant of organic compounds in water from a new group-contribution-based model. Ind Eng Chem Res 49:10149–10152
212. Katritzky AR, Mu L, Karelson M (1996) A QSPR study of the solubility of gases and vapors in water. J Chem Inf Comput Sci 36: 1162–1168
213. Walker JD, Jaworska J, Comber MHI et al (2003) Guidelines for developing and using quantitative structure–activity relationships. Environ Toxicol Chem 22:1653–1665
214. Dearden JC, Cronin MTD (2006) Quantitative structure–activity relationships (QSAR) in drug design. In: Smith HJ (ed) Introduction to the principles of drug design and action, 4th edn. Taylor & Francis, Boca Raton, FL, pp 185–209
215. Madden JC (2010) Introduction to QSAR and other *in silico* methods to predict toxicity. In: Cronin MTD, Madden JC (eds) *In silico* toxicology: principles and applications. RSC Publishing, Cambridge, pp 11–30
216. OECD Principles: www.oecd.org/dataoecd/33/37/37849783.pdf
217. OECD Guidelines: www.olis.oecd.org/olis/2004doc.nsf/LinkTo/NT00009192/$FILE/JT00176183.PDF
218. Dearden JC, Cronin MTD, Kaiser KLE (2009) How not to develop a quantitative

structure–activity or structure–property relationship (QSAR/QSPR). SAR QSAR Environ Res 20:241–266

219. Ghafourian T, Dearden JC (2000) The use of atomic charges and orbital energies as hydrogen-bonding-donor parameters for QSAR studies: comparison of MNDO, AM1 and PM3 methods. J Pharm Pharmacol 52: 603–610
220. Hartung T, Bremer S, Casati S et al (2004) A modular approach to the ECVAM principles on test validity. ATLA 32:467–472
221. Netzeva TI, Worth A, Aldenberg T et al (2005) Current status of methods for defining the applicability domain of (quantitative) structure–activity relationships. The report and recommendations of ECVAM Workshop 52. ATLA 33:155–173
222. Hewitt M, Ellison CM (2010) Developing the applicability domain of *in silico* models: relevance, importance and methods. In: Cronin MTD, Madden JC (eds) *In silico* toxicology: principles and applications. RSC Publishing, Cambridge, pp 301–333
223. Flynn GL (1990) Physicochemical determinants of skin absorption. In: Gerrity TR, Henry CJ (eds) Principles of route-to-route extrapolation for risk assessment. Elsevier, Amsterdam, pp 93–127
224. Young D, Martin T, Venkatapathy R et al (2008) Are the chemical structures in your QSAR correct? QSAR Comb Sci 27: 1337–1345
225. Cronin MTD, Dearden JC, Moss GP et al (1999) Investigation of the mechanism of flux across human skin *in vitro* by quantitative structure–permeability relationships. Eur J Pharm Sci 7:3250330
226. Hewitt M, Madden JC, Rowe PH, Cronin MTD (2007) Structure-based modelling in reproductive toxicology: (Q)SARs for the placental barrier. SAR QSAR Environ Res 18: 57–76
227. Doniger S, Hofmann T, Yeh J (2002) Predicting CNS permeability of drug molecules: comparison of neural network and support vector machine algorithms. J Comput Biol 9:849–864
228. IUPAC InChI code: www.iupac.orgt.inchi
229. Gedeck P, Rohde B, Bartels C (2006) QSAR—how good is it in practice? Comparison of descriptor sets on an unbiased cross section of corporate data sets. J Chem Inf Model 46:1924–1936
230. Topliss JG, Costello RJ (1972) Chance correlations in structure–activity studies using multiple regression analysis. J Med Chem 15:1066–1068
231. Romanelli GP, Cafferata LFR, Castro EA (2000) An improved QSAR study of toxicity of saturated alcohols. J Mol Struct Theochem 504:261–265
232. Yaffe D, Cohen Y, Espinosa G et al (2001) A fuzzy ARTMAP based on quantitative structure–property relationships (QSPRs) for predicting aqueous solubility of organic compounds. J Chem Inf Comput Sci 41:1177–1207
233. Aptula AO, Jeliazkova NG, Schultz TW et al (2005) The better predictive model: high q^2 for the training set or low root mean square error of prediction for the test set? QSAR Comb Sci 24:385–396
234. Erös D, Kéri G, Kövesdi I et al (2004) Comparison of predictive ability of water solubility QSPR models generated by MLR, PLS and ANN methods. Mini Rev Med Chem 4:167–177
235. Devillers J, Doré JC (2002) e-Statistics for deriving QSAR models. SAR QSAR Environ Res 13:409–416
236. Scripps Institute: www.scripps.edu/rc/softwaredocs/msi/cerius45/qsar/working_with_stats.html
237. QSAR World: www.qsarworld.com/statistics.php
238. Golbraikh A, Tropsha A (2002) Predictive QSAR modeling based on diversity sampling of experimental datasets for the training and test set selection. J Comput Aided Mol Des 16:357–369
239. Eriksson L, Johansson E, Müller M et al (2000) On the selection of the training set in environmental QSAR analysis when compounds are clustered. J Chemom 14: 599–616
240. Hemmateenajad B, Javadnia K, Elyasi M (2007) Quantitative structure-retention relationship for the Kovats retention indices of a large set of terpenes: a combined data splitting-feature selection strategy. Anal Chim Acta 592:72–81
241. Cronin MTD (2010) Characterisation, evaluation and possible validation of *in silico* models for toxicity: determining if a prediction is valid. In: Cronin MTD, Madden JC (eds) *In silico* toxicology: principles and applications. RSC Publishing, Cambridge, pp 275–300
242. Tropsha A, Gramatica P, Gombar VK (2003) The importance of being earnest: validation is the absolute essential for successful

application and interpretation of QSPR models. QSAR Comb Sci 22:69–77

243. Benigni R, Bossa C (2008) Predictivity of QSAR. J Chem Inf Model 48:971–980
244. Dearden JC, Hewitt M, Geronikaki AA et al (2009) QSAR investigation of new cognition enhancers. QSAR Comb Sci 28: 1123–1129
245. Johnson SR (2008) The trouble with QSAR (or how I learned to stop worrying and embrace fallacy). J Chem Inf Model 48:25–26
246. Katritzky AR, Kuanar M, Slavov S et al (2010) Quantitative correlation of physical and chemical properties with chemical structure: utility for prediction. Chem Rev 110:5714–5789

Chapter 7

Informing Mechanistic Toxicology with Computational Molecular Models

Michael R. Goldsmith, Shane D. Peterson, Daniel T. Chang, Thomas R. Transue, Rogelio Tornero-Velez, Yu-Mei Tan, and Curtis C. Dary

Abstract

Computational molecular models of chemicals interacting with biomolecular targets provides toxicologists a valuable, affordable, and sustainable source of *in silico* molecular level information that augments, enriches, and complements in vitro and in vivo efforts. From a molecular biophysical ansatz, we describe how 3D molecular modeling methods used to numerically evaluate the classical pair-wise potential at the chemical/biological interface can inform mechanism of action and the dose–response paradigm of modern toxicology. With an emphasis on molecular docking, 3D-QSAR and pharmacophore/toxicophore approaches, we demonstrate how these methods can be integrated with chemoinformatic and toxicogenomic efforts into a tiered computational toxicology workflow. We describe generalized protocols in which 3D computational molecular modeling is used to enhance our ability to predict and model the most relevant toxicokinetic, metabolic, and molecular toxicological endpoints, thereby accelerating the computational toxicology-driven basis of modern risk assessment while providing a starting point for rational sustainable molecular design.

Key words: Docking, Molecular model, Virtual ligand screening, Virtual screening, Enrichment, Toxicity, Toxicoinformatics, Discovery, Prediction, 3D-QSAR, Toxicophore, Toxicant, *In silico*, Pharmacophore

Brad Reisfeld and Arthur N. Mayeno (eds.), *Computational Toxicology: Volume I*, Methods in Molecular Biology, vol. 929, DOI 10.1007/978-1-62703-050-2_7,

1. Introduction

1.1. Overview of Molecular Modeling and Its Role in Computational Toxicology: Filling the Data Gaps in Mechanistic Toxicology

Modern computational molecular modeling methods are some of the most well-established, versatile, and vital computational chemistry methods that are at the very core of the emerging field of both mechanistic (1) and computational toxicology and sustainable molecular design.[1] The use of molecular modeling coupled to mathematical and chemical–biological inquiry is crucial "to better understand the mechanisms through which a given chemical induces harm and, ultimately, to be able to predict adverse effects of the toxicants on human health and/or the environment" (2).

A first step in considering the use of three-dimensional (3D) computer-assisted molecular modeling (CAMM) methods is the awareness of molecular level questions one can address with the various techniques. Molecular modeling can be used in the context of toxicological inquiry to address three molecular level aspects of both individual small-molecule (ligand) or biological macromolecules (targets) and the resultant interactions of the ligand/target complex, namely (1) *Structure*, (2) *properties, and* (3) *(re)activity.* In the context of toxicological and chemical genomic research (or toxicogenomics), one is interested in or requires downstream information that makes use of "optimized" structures or geometries of ligands or biological targets. Of the properties one may be interested in, molecular complementarity is a key objective along with catalytic competence of a chemical and possibly molecular susceptibility (or reactivity of a molecule). Similarly, there are two main research efforts one wishes to inform in mechanistic toxicology, namely

1. Toxico*kinetics* or ADME (rate of fate within the body).
2. Toxico*dynamics* or molecular toxicological interactions that result in a cellular response.

By considering two principal research paths and the biological macromolecular target space to which these coupled processes are related (Table 1) it becomes evident that the subset of molecular modeling tools that will be used by a toxicologist is not much different than the *in silico* drug discovery workflows (3), with the exception that there is less of an emphasis on lead optimization and more of an incentive on modeling approaches that possess an ability to both accurately and efficiently prioritize and categorize chemicals to their respective macromolecular targets; *in silico* methods

[1]Historically molecular modeling methods, stemming from roots in theoretical and computational chemistry, are composed of an ensemble of developed and thoroughly vetted computational approaches used to investigate molecular-level processes and phenomena including but not limited to molecular structure, chemical catalysis, geochemistry, interfacial chemistry, nanotechnology, conformational analysis, stereoselectivity, enzyme biochemistry, chemical reaction dynamics, solvation, molecular aggregation, and molecular design.

Table 1
This table shows the overlap between macroscopic and mechanistic toxicology, examples of targets for which pair-wise ligand/target interactions are most often sought after, and the molecular modeling methods used to inform the toxicological questions. Toxicology research streams (toxicokinetics/dynamics), specific toxicology-related processes (ADME/T), examples of toxicologically related biological macromolecules implicated in specific processes and the three-dimensional Computational Molecular Modeling methods (3D-CAMM) elaborated in this text

"Toxico-"	Coupled processes in toxicology	Examples of process relevant macromolecular protein targets	Molecular modeling methods
(I) -KINETICS (biological fate models of chemicals or disposition models)	**(A)**bsorption (i.e., dermal, oral, inhalation)	Ion channels (PgP) Molecular transporters Cell membranes (lipid bilayer considered for passive properties)	(A) Geometry **optimization** (B) Partial charge **calculation/ assignment**
	(D)istribution (i.e., Target Tissue of target organ)	Extracellular protein binding (e.g., human serum albumin or alpha-fetoprotein or immunoglobulin binding)	Target-specific endpoint data available (C) **Pharmacophore** modeling and (D) **3D-QSAR**
	(M)etabolism (enzyme-mediated chemical transformations typically associated with hepatic clearance mechanisms)	Intracellular solute carrier proteins (e.g., SHBG or FABP) (inhibitors or substrate binding related properties) Phase I/II enzymes (i.e., CYP450s, oxidoreductases, carboxylesterases) *For kinetic properties, such as rate constants consider QM formalism (see * in methods as well)	Target Structure available (E) **Target geometry Optimization and/or homology modeling** (F) A priori Small-molecule/target interaction evaluation by **molecular docking** (G) Molecular mechanics or empirical **pose scoring** Structure-based **Virtual Ligand Screening** (SB-VLS)
	(E)limination/ excretion (Renal or Biliary elimination processes)	Ion channels, organic molecular ionic transporters, globulins, active transporters	...+SB-VLS + SB-VLS + SB-VLS ... =***in silico* chemical genomics**
(II) - DYNAMICS (response or effect models)	Molecular **(T)** oxicology (ligand/ receptor pathways)	Nuclear receptors, G-protein-coupled receptors, ion channels (e.g., for neuronal impulse propagation or cardiac charge regulation, examples are Sodium-gated Ion channels, or HERG2 channel)	

that are complementary to modern experimental toxicogenomic inquiry.

It is estimated that there exist in the order of 7,000,000 chemical leads for small-molecule drug discovery and ~80,000–100,000 chemicals under the auspice of environmental chemicals for which the data matrix for risk is sparsely populated (i.e., environmental chemicals), and so there is a need for large-scale screening efforts for prioritization and categorization of these large inventories (1, 3–6). Due to the scale of chemical inventories of interest and variety of toxicologically implicated targets of interest, the most appropriate starting point for 3D-CAMM most frequently applied to toxicology (pharmacophores, 3D-QSAR and molecular docking) is the use of molecular mechanics force fields to describe or determine the 3D structure of a chemical/biological molecular system of interest (7–13). In this approach both ligands and biological macromolecular targets are mathematically described and modeled by applying classical Newtonian mechanics to atomic (not electronic) systems which in turn are numerically evaluated using modern computational implementations of the underlying biophysical models. We stress the importance of delineating the fundamental choice of molecular mechanics as opposed to quantum mechanical approaches for answering questions typical of chemical/biological perturbations due to the size domain, and information criteria of the part of mechanistic toxicology one most often wants to inform in the computational toxicology framework; the pair-wise interaction potential between ligand and macromolecular target. To better understand the difference between 2D/3D molecular modeling methods as applied to computational toxicology, we present a symbolic graphic (Fig. 1) outlining the three main 3D molecular modeling techniques used in this chapter.

Although intrinsic property or functional group chemical filters (e.g., Leadscope) in addition to classical QSAR approaches (14) and decision tree classifiers (15) are both pragmatic and parsimonious components of the chemoinformatic toolkit of computational toxicologists, they lack the intimate molecular level detail of the biomolecular interaction that could only be resolved by 3D-CAMM. Often chemoinformatics methods alone are unsuitable to address structurally related questions that require target-specific insight. For instance, for cases that fail to be able to resolve stereoisomerism and its implications in biomolecular interactions, species-related differences in sequences, polymorphism-related extrapolations in susceptible populations, and structural bases for mechanistic variability (inhibition versus substrate, agonist versus antagonist), there is little question that primarily 3D modeling methods such as (I) pharmacophore mapping, (II) 3D-QSAR, and (III) molecular docking methods that necessitate detailed structural information (i.e., Cartesian coordinates of atoms and

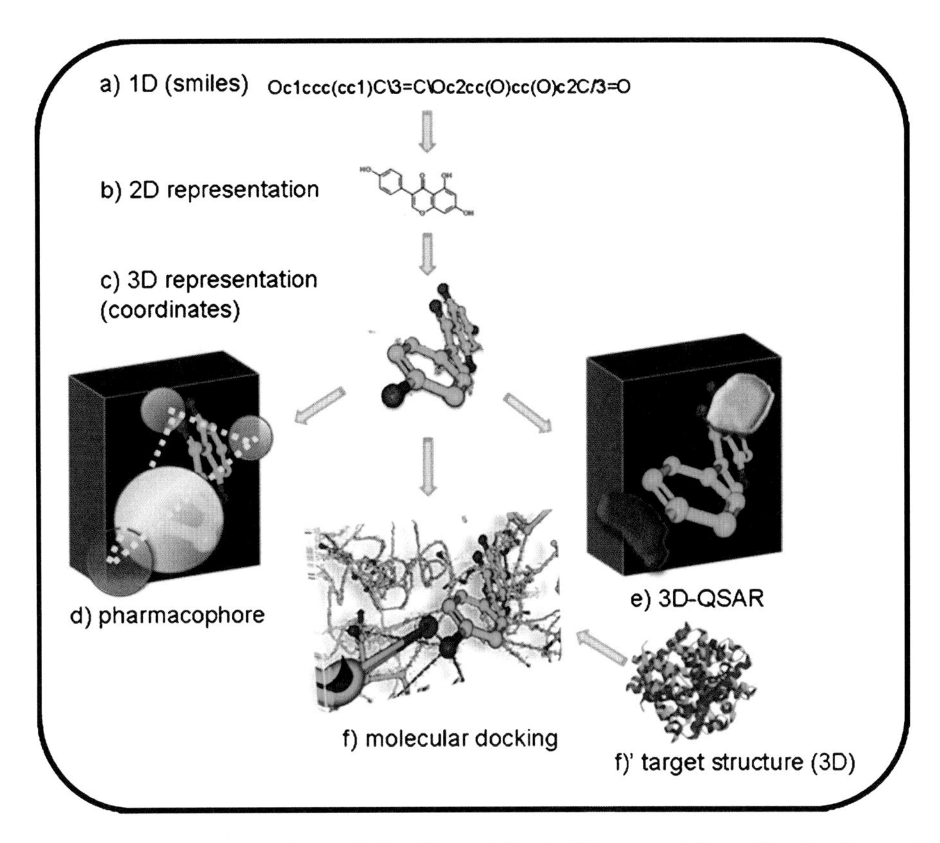

Fig. 1. Point of departure from (**a**) 1D chemical smiles notation to 3D representation, with atom type and specific coordinates spatially defined. (**d–f**) The three major classes of molecule modeling methods used to evaluate ligand/target interactions.

their specific connectivity) and are the only viable alternatives for reliable a priori estimates for risk assessment.

1.2. Exploring Ligand: Target Interactions Implicitly: 3D-QSAR and Pharmacophores

Unlike specific models of both biological macromolecules and small-molecule ligands, both 3D-QSAR and pharmacophore methods address the fundamental chemical/biological aspects of pair-wise interactions implicitly. Although both deal with the explicit (i.e., full 3D) structure of a chemical of interest and both require either a training/test set of chemicals with known activities for a given target for a given mode of action (i.e., agonist or antagonist, substrate or inhibitor), neither pharmacophore approaches nor 3D-QSAR approaches can provide specific molecular level detail between atoms on both macromolecule coupled to those of the ligands that give rise to said activity. Pair-wise interactions between the ligand and the target molecule must be spatially defined. Nonetheless, both methods are a step in the right direction from traditional 2D-QSAR since inherently both 3D-QSAR and pharmacophore models have the ability to discriminate activity based on three-dimensional topology (i.e., inform stereochemical interactions or regiospecific interactions) without providing residue-specific interactions that could give rise to the specific interaction.

According to IUPAC, a pharmacophore (or in the case of toxicology, a toxicophore) is "an ensemble of steric and electronic features that is necessary to ensure the optimal supramolecular interactions with a specific biological target and to trigger (or block) its biological response" (16). In this sense a pharmacophore model's objective is to characterize a molecule's atomic constituents in terms of the primary interaction types that give rise to pairwise interactions; from multiple atoms to a subset of binding "features." The features most common for a set of known chemical actors on a known biological target of undefined tertiary structure are hydrophobic, aromatic, hydrogen bond acceptor, hydrogen bond donor, and cationic, anionic, or metal interactions. Furthermore there may be exclusion volumes and feature directionality included in the pharmacophore. The pharmacophore features are elucidated by comparing multiple known actors in terms of common overlaid structural features (or alignments). Next, if one is to investigate a series of chemicals and test the pharmacophore one would sample conformational space and any structure that contained a conformation that satisfied the spatial and feature requirements of the pharmacophore model would be considered a complete or partial "hit."

On the other hand, being able to assess which functional groups or specific spatial features have the ability to modulate the chemical/biological interactions in a quantitative sense is the area of 3D-QSAR. Although there are cases of simple QSAR models dating back to the late 1800s (17), 3D-QSAR is a much more recent approach. While classical QSAR models are useful for rapidly predicting chemically induced effects based on physicochemical properties, its main weakness is that it does not account for three-dimensional molecular shape, a critical aspect of intermolecular interaction. Instead of relying on physicochemical properties as molecular descriptors, 3D-QSAR interprets molecular shape using interaction energies from force field calculations. The huge number of individual interaction energies was historically difficult to correlate with biological activity and it was not until the advent of computationally implemented Partial Least Squares (PLS) analysis (18) that 3D-QSAR became technically feasible.

The first, and still most widely used 3D-QSAR method, is known as Comparative Molecular Field Analysis (CoMFA) (19). Other methods have since emerged, including Comparative Molecular Similarity Indices Analysis (CoMSIA) (20), ALMOND (21), three-dimensional QSAR (TDQ) (22), Catalyst (23), and Phase (23), generally to either improve predictive performance or simplify the model development process. The main drawback of 3D-QSAR is the time requirements and difficulty in preparing the data set for model development.

Further details and steps required for both pharmacophore elucidation/mapping and 3D-QSAR as applied to toxicology are elaborated in Subheading 3.

1.3. Modeling Explicit Pair-Wise Interaction Potential of Ligand–Target: Molecular Mechanics, Empirical Scoring, and the Need for Structurally Informed Molecular Models

In the specific case of modeling ligand/target interactions for virtual ligand screening as applied to toxicology, certain methods for evaluating pair-wise interaction energy are too computationally expensive/intensive and scale poorly with system size; Quantum Mechanical (QM or sometimes referred to as quantum chemical or electronic structure theory methods) are highly accurate but not ideal (hence not pertinent) due to their computational demand for almost all of the said interaction partners and processes listed in Table 1, with perhaps the exception of bond breaking/making processes inherent in metabolic reactions or irreversible binding. Although the principal focus of *in silico* methods to estimate metabolic rate constants have been quantum mechanical (24, 25), the majority of pair-wise interactions a toxicologist will require are related to ligand/macromolecular target pair-wise interactions, composed of both bonded (ligand and target "self-energy") and nonbonded interactions *between* a small molecule and target biological macromolecule (i.e., receptor or enzyme) for which all structural optimization routines are adequate within a classical physics formalism, or more specifically, within a molecular mechanics (MM) framework in which the smallest unit of relevance are atoms (not electrons as in the case of QM approaches).

The classical physics approach to modeling molecules requires the assumptions of molecular mechanics which makes use of atom-specific functions, or force fields parameters, that have been developed by a variety of experimental or high-level theoretical calculations (i.e., ab initio or semi-empirical QM). These are related to atom-specific terms that describe all bonded and nonbonded interactions (conformational energy, as a function of dihedral angles, bond angles and bond lengths intramolecular electrostatic interactions and van der Waals, or dispersion forces) in Cartesian space that are ultimately integrated over all space of the individual molecule or ligand/biomolecule complex to estimate "intermolecular" interaction energy. The "pair-wise interaction potential" between a ligand and a macromolecular target is provided in a simplified form in Fig. 2.

It is well known that there is a relationship between free energy of a reaction or complex formation and reaction rate (i.e., chemical kinetic) variables. As provided in Eq. 2 of Fig. 2, if one has a method for capturing interaction free energy of complex formation (or association) of a ligand/target complex this thermodynamic

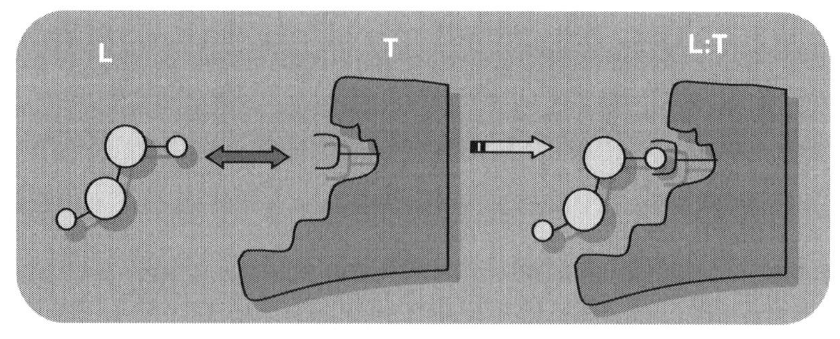

$$1.\ \Delta E_{L:T} = \sum_{L} \sum_{T} (E_{bonded} + E_{non-bonded} + E_{solvation})_{L:T} - \sum_{L,T} (E_{bonded} + E_{non-bonded} + E_{solvation})_{L,T}$$

$$2.\ \Delta E \approx \Delta G_{association} = \Delta H_{L:T} - T\Delta S_{L:T} = -RT\ln(K_{A_{L:T}}) = -RT\ln(1/K_{D_{L:T}})$$

$$3.\ K_{D_{L:T}} \approx K_i$$

Fig. 2. Fundamental classical expressions evaluated in pair-wise interaction modeling between ligand (L) and macromolecular target (T), and the resulting affinity of the complex (L:T). The first expression relates the energy of the molecular components as a difference of the complex's bonded/nonbonded atomic potential from the energy of the individual partners (L,T). (2) The approximation that the energy function is related to free energy of a system, the thermodynamic representation in terms of enthalpy (*H*) and entropy (*S*), and the thermodynamic interpretation of transition-state theory and molecular driving forces for association (K_a). Finally the relationship between the complex affinity or dissociation and a toxicologists metric of the inhibition constant (Cheng-Prusoff).

variable, dG can be cast in terms of an equilibrium process via the expression in lines 2 and 3, where the association constant of an L:T complex, $K_a = [LT]/[L][T]$ $K_d = 1/K_a$ and K_i, the inhibition constant from competitive inhibition assays that in vitro assays often quantify is directly proportional to K_d (dissociation constant) of the ligand with respect to a reference probe (26).

In theory, it is tempting to believe that the free energy from scoring or force-field functions should directly correlate with the experimentally determined biological activity (K_d or K_i) of complex formation as evaluated by pair-wise interaction schemes, the problem is significantly more involved. The complexity of the problem and inherent simplifications in molecular docking often result in an ability to enrich a data set in question in such a way that "actives" (i.e., biologically active molecules or "hits" for a target) considered above some threshold expectation value for binding are guessed several orders of magnitude better than a random guess. For screening this is a reasonable expectation. Details of the various steps for 3D molecular modeling are addressed briefly in Subheading 3, with focus on how to use these optimized structures for 3D pharmacophore elucidation, 3D-QSAR, and molecular docking. For more extensive methodological resources for any of the methods provided, we refer the reader to Table 3 which contains expansions of the topics covered in this chapter. We strongly encourage familiarization with these tools through practice. If one wants to apply these techniques to individual toxicological research efforts.

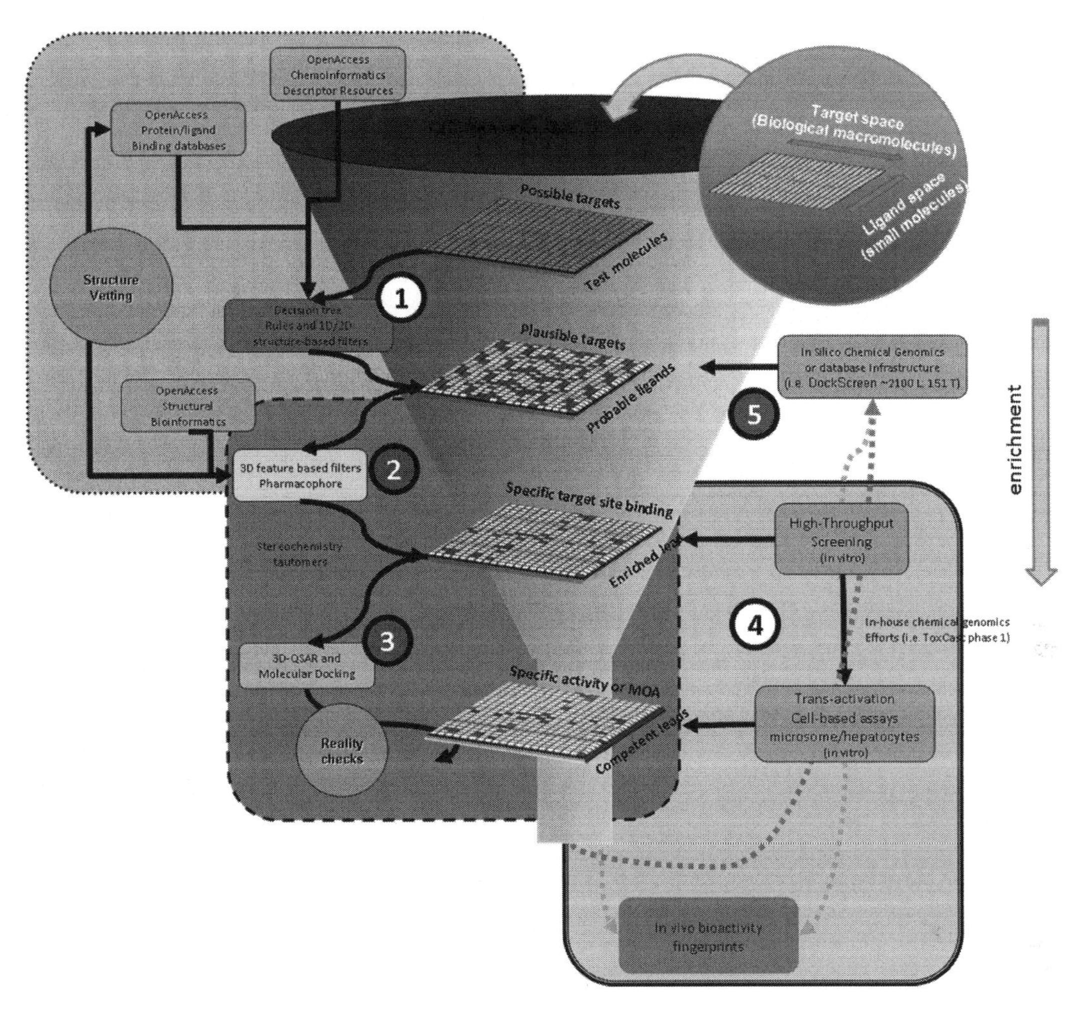

Fig. 3. Computational toxicology modeling workflow showcasing the *in silico*, in vitro, and in vivo integration of data and models within an informatics framework.

1.4. The Use of Molecular Modeling in Computational Toxicology: The integrated Modeling Workflow to In Silico Chemical Genomics

Although we have provided an overview of the most popular and useful aspects of 3D-CAMM that could be used to inform mechanistic toxicology, we need to understand how they fall into the computational toxicology framework. To know how and when these methods are applied in practice, and by whom, we have devised a workflow (Fig. 3) that highlights some of these components and how they may complement experimental High Throughput Screening protocols. The objective is to enrich the understanding of chemical/biological interactions through toxicogenomic inquiry. This is achieved by an *in silico* (filters → 2D QSAR > pharmacophore → docking/3D-QSAR) tiered approach that is tightly coupled to experimental in vitro screening efforts (i.e., protein ligand binding assays, transient activation assays, gene expression profiling, cytotoxicity assays, etc.) to encode a chemical-specific biological activity fingerprint or signature. This conceivably can also be performed *in silico* using multiple target screening and used as a metric for chemical/biological activity comparisons (i.e., similarity based on multi-target virtual affinity fingerprint as opposed to structure alone).

All of the data from a tiered approach to virtual ligand screening could and should ultimately be encoded or captured within a database framework so that easy recall could be performed to inform molecular level resolution data gaps as they arise. We add that the development of a database infrastructure that can subsequently capture the resultant poses and pair-wise interaction energy (surrogates for affinity) holds value in being able to query molecular-level insight for an experimental chemical genomics screen.

We provide, in brief, a workflow that demonstrates how to pair or couple experimental, *in silico*, and 3D *in silico* methods and the various pipes of data that allows one to build a virtual ligand–target complex structural database. This type of strategy had been adopted to build our own in-house resource to support *in silico* toxicogenomic inquiry (DockScreen) which is explained in Subheading 7.4.

2. Materials

There are well over 350 independent packages (computational codes) available for various aspects of the molecular modeling or Virtual Ligand Screening (or 3D VLS) paradigm that capture the various components required for 3D modeling of ligand/biomolecule interactions: all chemoinformatics and QSAR development, docking, homology modeling, pharmacophore elucidation, chemical structure manipulation, structure building, refinement, optimization, and finally bioinformatics applications.

For the case of computational toxicology, the lead optimization procedure/process typically associated with *in silico* drug discovery or rational drug design and associated methods and coded implementations are essentially dropped (although they may persist for sustainable molecular design). These packages run on many different platforms including but not limited to Windows/PCs, SGI, Mac, Linux (UNIX workstations), and some limited functionality molecular modeling utilities are even available for handheld devices and smartphones. For practical purposes, we have typically chosen one of several commercial suites that with the following features.

1. Platform independence (works on heterogeneous network architecture).
2. Token key license structure (check out by user when required).

Table 2
A list of several comprehensive software/tool/data resources lists available on the WWW that provide access to various commercial and open-source software packages, in addition to open-access database resources

Individual 3D CAMM lists	Uniform resource locator (URL)
Directory of *In Silico* Drug Design Tools (Swiss Institute of Bioinformatics)	http://www.click2drug.org
Universal Molecular Modeling List (NIH)	http://cmm.info.nih.gov/modeling/universal_software.html
Free computer tools in Structural Bioinformatics and Chemoinformatics	http://www.vls3d.com/links.html
Computational Chemistry List, ltd. (CCL.NET) Software-Related Sites (Note, these include "ALL" chemistry-related sites above and beyond the scope of this paper)	http://www.ccl.net/chemistry/links/software/index.shtml
Virtual Library: Science: Chemistry: Chemistry Software Houses from the University of Liverpool (UK)	http://www.liv.ac.uk/Chemistry/Links/softwarecomp.html

3. Many independent molecular modeling methods, bioinformatics, chemoinformatics and data mining methods combined.
4. Built-in functionality for scripting, piping data, and automated/macro-workflows.
5. Is well documented and has good active and passive support networks (technical service and FAQ/scripting forums).

Publically available resources for the "nonexpert" or experts are included in Table 2 and provide numerous links for a variety of software packages, both commercial and open-source, in addition to visualization tools and databases relevant for informing ligand/target pair-wise interaction modeling.

From an application standpoint, the authors have required both bioinformatics and chemoinformatics tools, structural database capabilities, and the ability to perform geometry optimization of structures, molecular docking, homology modeling of target structures, conformational searches, pharmacophore elucidation, and QSAR development. However, we have primarily used Chemical Computing Group's Molecular Operating Environment (MOE) (27) for all database manipulation, QSAR development, library development, structural optimization, and descriptor calculation. Similarly, for ADME-related parameter estimation via QSAR we use Schrodinger's QikProp (28) which has been vetted against various animal and human drug targets or ADME-related endpoints (i.e., $\mathrm{Log}P_{(\mathrm{Brain:Blood})}$, $\mathrm{Log}K_{(\mathrm{HSA})}$, #metabolites, $CACO_2$, or MDCK permeability, etc.).

3. Methods

As mentioned in the introduction, the classical physics approach to modeling molecules requires the assumptions of molecular mechanics which makes use of atom-specific functions, or force field parameters, that have been developed often for specific classes of molecules. Force field calculations have been successfully performed on larger polypeptides and protein structures. Park and Harris (24) utilized AMBER force fields to develop an all-atom model for CYP2E1 which was subsequently used for docking studies. A comprehensive review of AMBER protein force field development can be found elsewhere (7, 29). Several studies have also assessed the relative performance of CHARMM, MMx, OPLS, and AMBER force fields (30, 31). Gundertofte et al. (32) have assessed the relative accuracies of MMx and AMBER force fields. Jorgenson et al. (33, 34) have also examined the performance of their OPLS force field in the context of proteins and organic liquids. Regardless of the framework details, a molecular mechanics force field is always chosen for structural optimization, and the specific force field selected is usually chosen that best captures the atom-type diversity in the data set (i.e., chemical space of the training fragments or atoms). A broad overview of all the stepwise modeling procedures is provided in Fig. 4.

- For our molecular modeling needs (i.e., in the case of small-molecule ligands with environmental chemicals), we have almost exclusively used the MMFFx force field (MMFFx, (35)) for 3D geometry optimizations of entire libraries of chemicals.
- Subsequently, partial atomic charges are assigned either using empirical (i.e., Gasteiger) or semiempirical-based charge model representations of the electrostatics of the system (i.e., AM1-BCC) and are stored in a 3D chemical structural database.
- These structures could be used directly with target specific activity information as the seed for a conformational search (spanning all rotatable bonds to predict other relevant geometries) and aligned to other known biologically active chemicals to generate 3D-QSAR or pharmacophore models.
- However, if the structure is known and the target protein sequence is known and a crystal structure or near-neighbor homolog exists, it is conceivably simple to optimize hydrogens

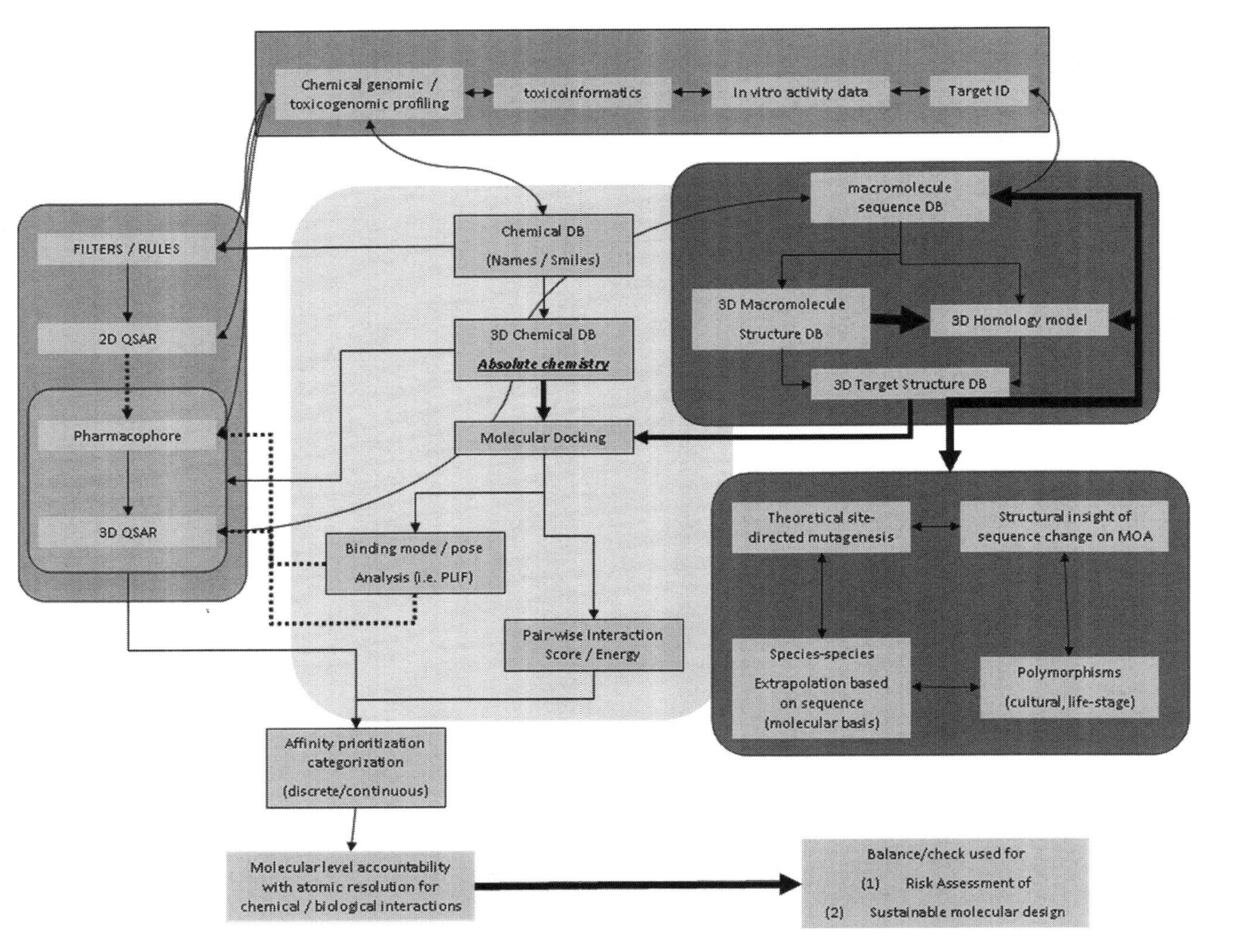

Fig. 4. Specific steps in 3D-CAMM (computer-assisted molecular modeling) workflows, chemical/biological knowledge-based boundary conditions (*top box*), ligand-based approaches (*left box*), structure-guided methods (*central box*), structural biological target inventories, and types of questions related to toxicologically relevant target–target extrapolations one can inform from structure-based approaches (*dark gray boxes*).

on the crystal structure obtained from the literature, or perform theoretical site-directed mutagenesis or threading, the basis for homology modeling based off of a known structural template.

- Finally, with an optimized target structure database and an optimized ligand database, one could perform molecular docking experiments where the pair-wise ligand:target interactions (bonded and nonbonded terms) are systematically evaluated. The resulting poses from such a docking "run" can each be individually scored based on known binding affinity. There are numerous online resources that provide ligand/target binding affinity data (i.e., www.bindingdb.org).
- Using these rank-ordered lists of chemicals based on scored docking poses between a small molecule and a macromolecular target is the starting point for a prioritization or rank-order scheme for screening a specific target: virtual structure (macromolecule)-based virtual ligand screening or structure-guided chemical prioritization methods.
- A library of structural targets of interest that may have been selected based on their role in a major toxicity pathway a researcher may be studying, has value in being able to fish for targets of any chemical (36). The next section elaborates on the capabilities of a large-scale ligand/target screening initiative.

For detailed description of external methods we encourage the readers to consult Table 3, which contain more detail for each aspect of the various steps of molecular modeling. Next we provide a stepwise breakdown of various modeling steps required for evaluating ligand/macromolecular target interactions.

3.1. Ligand Preparation

1. Collect a list of all the chemicals of interest (optional)—Data set augmentation (i.e., adding "simulated metabolites"—Enumerate metabolites using a heuristic (or knowledge base) metabolite enumeration algorithm (37, 38)).
2. Curate this list with the smiles representation of the structure of interest.
3. Convert this 2D chemical structure data set to a 3D representation by assigning the chemicals absolute configuration which includes the atom types but their 3D connectivity (bond type and orientation) and selecting an appropriate molecular mechanics force field and charge model of interest (depending on the chemical space of the chemicals of interest in addition to the magnitude of the screening initiative (i.e., hundreds to thousands of chemicals one would be better of going to no more than a classical physics approximation of the molecular geometry)).
4. Assign charges to the geometry optimized structures.

Table 3
A comprehensive set of combined reviews and/or methods papers for various aspects of the 3D molecular modeling methods discussed in this chapter. We urge the reader to familiarize themselves with each of the steps associated with their modeling method chosen and the particular toxicological data gaps they may wish to address

Step	Systematic methods reference
Molecular docking	Morris G, Lim-Wilby M (2008) Molecular docking. Methods in molecular biology, vol 443. Clifton, NJ, pp 365
A general introduction to molecular modeling techniques in the area of protein–ligand interactions	(a) Kroemer R (2003) Molecular modelling probes: docking and scoring. Biochem Soc Trans 31:980–984; (b) Van Dijk A, Boelens R Bonvin A (2005) Data driven docking for the study of biomolecular complexes. FEBS J 272(2):293–312
Docking scoring functions	Pick D (2004) Novel scoring methods in virtual ligand screening. Methods in molecular biology, vol 275, pp 439–448
Chemical database preparation	Bologa C, Olah M, Oprea T (2005) Chemical database preparation for compound acquisition or virtual screening. Methods in Molecular biology, vol 316. Clifton, NJ, p 375
Target selection criteria	Wishart D (2008) Identifying putative drug targets and potential drug leads: starting points for virtual screening and docking. Methods in molecular biology, vol 443. Clifton, NJ, p 333
Virtual or *in silico* affinity fingerprints	Briem H, Lessel U (2000) In vitro and in silico affinity fingerprints: Finding similarities beyond structural classes. Perspect Drug Discov Des 20(1):231–244
3D structure-based virtual ligand screening resources and brief overview	Villoutreix B et al (2007) Free resources to assist structure-based virtual ligand screening experiments. Curr Protein Pept Sci 8 (4):381–411
In silico chemical genomics—target and ligand preparation	Jongejan A et al (2005) The role and application of in silico docking in chemical genomics research. Methods in molecular biology, vol 310. Clifton, NJ, p 63
Analysis of chemical space in the context of domain of applicability	Jaworska J, Nikolova-Jeliazkova N, Aldenberg T (2005) QSAR applicability domain estimation by projection of the training set descriptor space: a review. Altern Lab Anim 33(5):445
Analysis of docking data	Bender A et al (2007) Chemogenomic data analysis: prediction of small-molecule targets and the advent of biological fingerprints. Comb Chem High Throughput Screen 10(8):719–731

5. Refine the data set (see Table of Methods)—(39).
   (a) Consider charge state and charge model, force field, and domain of applicability (DOA) dependent on the nature of your chemical.

6. Capture all 3D geometries into a database.
   (a) Almost all major molecular modeling suites (e.g., Chemical Computing group's MOE (27), Accelrys Discovery Suite (40), Schrodinger (28), and Tripos (41)) provide database representation of the chemicals of interest, so converting from smiles code to 3D optimized, cleaned, and charge-model applied 3D representation is relatively seamless.
   (b) STOP.

3.2. Target Preparation

1. Coupled to experimental knowledge, searching through chemical genomics databases such as http://stitch.embl.de or the Comparative Toxicogenomics Database (http://ctd.mdibl.org/) often identifies relevant targets for a chemical or analog of interest.
2. Finding a suitable target model (typically an X-ray crystal structure from http://www.pdb.org) is the next step. Before assuming that a given target structure will serve as a sound basis for molecular modeling studies, it is critical to understand that "protein structures" are models. Although they are based on experimental data, they are nonetheless prone to bias or ambiguity from several sources. While numeric metrics such as resolution, R-factor, free-R, redundancy, and average I/sigma (signal to noise ratio) are important considerations for the overall reliability of a crystal structure model, at least some local errors or ambiguities are found in nearly all structures. Active sites are often somewhat rigid, especially when bound to ligands, so one can hope that the structure of interest is a sound choice. However, there is usually no substitute for examining electron density (see Subheading 7.5 and (42)).
3. How does one select the "appropriate structure" if confronted with several? Perform an RMSD evaluation on a structural superposition. If the geometries are similar they may cluster into most probable conformation states. Select a representative from each cluster.
4. If (a) the target structure is known, (b) the sequence is known, and (c) a crystal structure or near-neighbor homolog exists, it is conceivably simple to optimize hydrogen atoms on the crystal structure obtained from the literature or perform theoretical site-directed mutagenesis or threading, the basis for homology modeling based off of a known structural template.
5. If the target structure is not known and one wishes to perform structure-based virtual screening or molecular docking, one must build a homology model of the structure of interest using the sequence of the desired target (from www.uniprot.org) and a crystal structure template of the nearest-neighbor homolog (template or crystal structures from www.pdb.org) and homology or sequence identify search using BLAST.

A protein homology model server is available for integrated web modeling at www.proteinmodelportal.org.

3.3. Molecular Docking

1. With an optimized target structure database, and an optimized ligand database one could perform molecular docking experiments where the pair-wise interactions (bonded and non-bonded terms) between the ligand and the macromolecular target are systematically evaluated. The resulting poses from such a docking "run" can each be individually scored based on known binding affinity data training set of chemicals for a given target.
2. A binding site is identified (co-crystallized ligand site or rationally selected site) and each ligand is subject to interact with the macromolecular target, where sampling and docking trajectories are subject to the force field approximations. Each individual "pose" is scored or captured for subsequent analysis.
3. Each of the poses are systematically *scored* using pair-wise interaction potentials that are either derived from classical physics approaches (i.e., force field approximation) or empirical scoring functions that have been optimized to reproduce either experimental in vitro binding affinities (trained scoring function).
4. The results are subsequently validated for their ability to enrich MOA data or rank-order chemical binding for a known target. Another common validation protocol that has less to do with the binding affinity and more with pose analysis is the ability for the docking algorithm to reproduce the original co-crystallized ligand in the same geometry. Methods that minimize the RMSD between known pose and docked pose are considered optimal. This approach of being able to reproduce experimental crystal structures is termed "pose fidelity" (43, 44) and references within.
5. Docking "experiments" can form the basis of a numerically continuous complementarity evaluation of ligand/target complexes (unlike experimental, that rely on binding stronger than a probe chemical threshold or limit of detection, or else result in an "NA" or blank result.). Since one can take the top and bottom rank-ordered chemicals for a target and deduce chemoinformatic filters (i.e., intrinsic functional property or functional group profiling) one could conceivably perform what is known as "progressive docking," where filters from molecular docking simulations are used to create "structure-guided" filters for subsequent chemicals (Progressive Docking: A Hybrid QSAR/Docking Approach for Accelerating *In silico* High Throughput Screening) (45).
6. Details about assumptions and expectations from structure-based virtual ligand screening models, and the very nature of

the target structure used are enumerated in Subheading 7.5. It is very important to be aware of the various issues that lead to mismatched expectations (i.e., surprise) when attempting to apply these 3D-CAMM approaches.

3.4. 3D-QSAR

This section is intended to provide a brief overview of 3D-QSAR, outlining its advantages and disadvantages and describing the basic steps required to derive and validate a model. For greater detail on the topic, the reader is referred to external references (46–48). Figure 5 outlines the basic steps involved with developing, validating, and using a 3D-QSAR model. Although not listed on the figure, the first step in deriving a 3D-QSAR model is really defining the applicability domain or the chemical space comprising the set of compounds for which the model is to be used. A good way of rapidly assessing the chemical space of two chemical lists is using ChemGPS-NP (49, 50). When that has been defined, a representative sample of that compound list with known biological activity must be chosen for further development. 3D-QSAR models have been developed using data sets ranging from as little as 20–30 compounds or as much as 200–300. A large data set of compounds

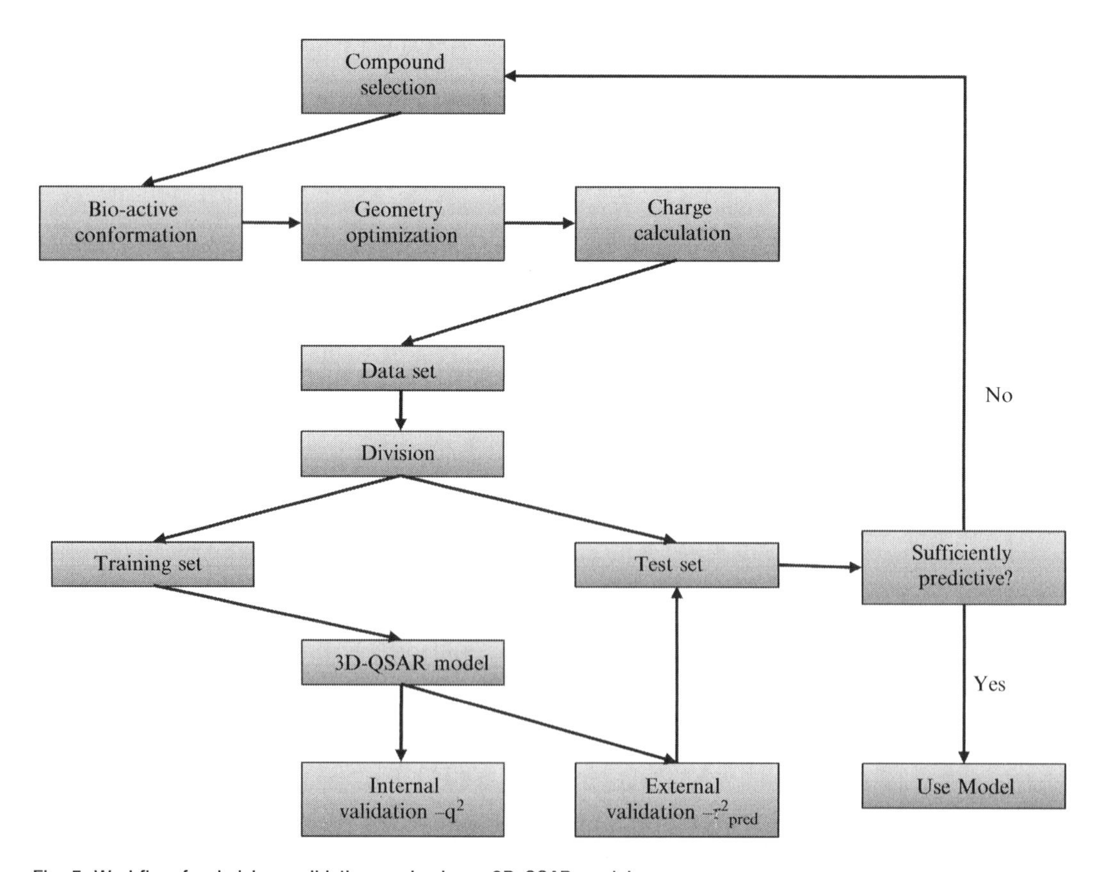

Fig. 5. Workflow for deriving, validating, and using a 3D-QSAR model.

will likely cover a larger chemical space, but considerations of the cost of biological testing usually limit this size. Another consideration in the initial stages of data set design is the overall span in biological activity values: a span in activity values of 5 log units is generally considered to be the minimum requirement.

1. After data set compounds have been carefully chosen, several steps of molecular modeling are necessary to ensure
   (a) They are in their biologically active conformation.
   (b) Geometry optimization has been performed.
   (c) Accurate partial atomic charges have been assigned.
2. Of these steps, identification of the biologically active conformation is the most important and most difficult. Geometry optimization and charge calculation methods generally have a much lesser effect on model predictive performance.
   (a) Data set is carefully divided into training and test sets. The training set is a subset of the data set used for deriving the 3D-QSAR model and usually comprises roughly two-thirds of the original data set. The test set is then composed of the remaining one-third of compounds and is used for evaluation of the predictive performance of the model. Care must be taken to ensure the training and test sets have similar coverage of chemical space as well as a similar span in activity.
   (b) Following data set division, the final software-dependent steps in 3D-QSAR model derivation may be taken. Since 3D-QSAR correlates biological activity with differences in structural features, these final steps generally involve aligning important pharmacophore groups or features of the chemical scaffold to bring to light those structural features that impact biological activity.
3. After the 3D-QSAR model has been derived it must be validated. The first step of the validation process normally includes internal validation by cross-validation, which gives an indication of the strength of correlation within the training set compounds. Although this is a useful metric, it gives little indication of how well the model can predict activity data for compounds not included in the training set. To understand this, the model is used to predict the activities of test set compounds. These predicted values are compared with the previously known activity values to calculate r_{pred}^2, a measure of external predictive performance.
4. If the model is found to be sufficiently predictive following these tests, it may be used to perform predictions on similar compounds for which biological activity is unknown. If it is

not, the user must repeat the previous steps leading up to model derivation until the model is sufficiently predictive.

3.5. Pharmacophore/ Toxicophore Elucidation

1. Taking the optimized ligand geometries and knowledge of the specific target for which these chemicals interact with it is possible to superimpose the various ligands, or conformations of the various ligands, to recreate or infer the optimal features for a specific experimental mode of action.
2. Negative data is especially useful in pharmacophore models as it allows one to rule out "impossible" superpositions (hence better molecular level boundary conditions) and increase the predictive accuracy of models to predict "hits" or "non-hits."
3. Flexible alignment or ligand superposition is preceded by geometry optimization and conformation enumeration of each of the 3D geometries that can be made by performing either stochastic or deterministic molecular simulations within a molecular mechanics framework to systematically alter the dihedral angles of the chemicals of interest and localize multiple low energy conformers or rotamers.
4. Once alignment procedures between chemicals have been completed, one often finds common molecular features that one can reduce the ligand structure into (i.e., hydrogen bond donor, hydrogen bond acceptor, hydrophobic contact, aromatic contacts, metal interactions, cationic interactions, and anionic interactions). These features can include exclusion volumes or cavities that "wrap" the outer volume of a set of known ligands. For any given chemical, if a conformation falls within the cavity, and the spatial relationship between features are either completely or partially satisfied one would identify "potential hits."

4. Examples

In order to familiarize toxicologists with pertinent examples for further exploration, we briefly outline some of the key papers in table format (Table 4) as they address mechanistic toxicology questions, the modeling approaches taken, the software used, and the literature reference of the research effort. Finally, we provide a brief description of an in-house *in silico* chemical genomics program and the key actions taken to bring an *in silico* molecular modeling results database to fruition to complement screening and toxicogenomic efforts.

Example(s) I: 3D CAMM to inform toxicology, single target research.

Table 4
Selected research papers that exemplify 3D CAMM to inform mechanistic toxicology

Tox data gap	Modeling method	Chemical/target space	Citation
1. Absorption	Catalyst (Accelrys) 3D-QSAR Pharmacophore Homology modeling,	TARGET: P-gp (P-glycoprotein efflux transporter) LIGANDS: 27 digoxin inhibitors, 21 + 17 vinblastine inhibitors	(51)
2. Distribution	Homology modeling, 3D-QSAR, molecular dynamics, and molecular docking (various—Glide—Schrodinger, and Chemical Computing and Tripos suites) Molecular docking (Autodock) molecular modeling (Macromodel/Schrodinger), theoretical site-directed mutagenesis (Sybyl, Tripos) 3D-QSAR and molecular docking	TARGET: sex hormone binding globulin LIGAND: 80,000 ligands TARGET: Human serum albumin binding (plasma binding) LIGAND: <10 structurally related chemicals to naturally occurring ochratoxin TARGET: Human Serum Albumin binding (plasma binding) LIGAND: 37 structurally related putative interleukin 8 inhibitors	(52–54)
3. Metabolism	Pharmacophore and 3D-QSAR	(a) TARGET: CYPs 1A2, 2B6, 2C9, 2D6, 3A4 LIGANDS: various drug-like/leads (b) TARGET: PXR LIGANDS: various drug-like/leads	(55–57)
4. Elimination	Pharmacophore and 3D-QSAR [various in-house packages such as CAMDA, as well as commercial suites such as Sybyl (TRIPOS)]	TARGET: rat multidrug resistance-associated protein 2 and 1 LIGAND: >4,000 conformers of >18 metabolism-like leads	(58, 59)
5. Molecular toxicology (receptors)	Homology modeling and molecular docking (OpenEye's FRED and SimBioSys Ltd. eHITS)	TARGET: Estrogen receptor alpha (i.e., nuclear receptor) or 5-HT6 receptor based off the rhodopsin crystal structure (GPCR modeling)	(60, 61)

Example II: Building an *in silico* chemical genomics framework using 3D CAMM to inform toxicology on multiple targets

This next example focuses primarily on the components required to build a multiple target large-scale docking inventory to support chemical genomic research. In an effort to dig deeper into toxicogenomics data we had built an in-house *in silico* chemical genomics infrastructure. The idea was to complement HTS and chemical genomics programs and support a fully integrated *in silico,* in vitro, and in vivo computational toxicology framework.

Particular examples of building a multiple target, multiple chemical docking database are relatively rare in the literature, and for the most part support drug discovery research. We wanted to demonstrate to the reader how this was done at an overview level of detail to show key considerations, infrastructure and coding requirements needed to be assessed. In this case, we were building a database to inform the data matrix required for risk assessment of many environmental chemicals. One of the efforts from the US-Environmental Protection Agency's National Center for Computational Toxicology is the ToxCast program (5). In this case, for phase I a total of 320 chemicals (DSSTox data set) were selected for thorough in vitro work-up. What was not provided by any of these assays, however, was the type of information at molecular resolution as obtained by structure-based virtual ligand screening. In this case an *in silico* chemical genomic initiative, DockScreen, was started (62–64).

The Dockscreen data is the result of nearly 2,500 ligands (environmental, therapeutic, metabolites and industrial chemicals) docked into about 151 binding sites of ~100 protein targets using the eHiTS (Simbiosys Ltd, Canada) software package. The result was over 350,000 docking runs resulting in over nine million ligand poses. These calculations were performed over a period of 2 months on 20 servers and collected a total of ~250 Gb of coordinate-specific pose data for each of the 2,100 chemicals on each of the 151 targets. To store and manage queries to access this data, a MySQL relational database schema was designed with separate tables for ligands, proteins, docking runs, and poses as well as some computational statistics. A custom interface to this database was built in a Linux OS environment with a PHP enabled Apache web server. The acronym "LAMP" is often used to refer to such a combination of Linux, Apache, MySQL, and PHP which have been used in combination to provide web access to many databases. For Dockscreen, only a dozen or so PHP scripts were needed to let users to view, query, and select groups of ligands, groups of targets, and statistical calculations on the distribution of docking scores for the runs including such ligands and targets. In addition to numeric statistics, histogram graphics were produced on the fly. An applet allowing users to draw chemicals and search against ligands is also built in. We believe in-house tools such as these that provide scientists relatively fast access to "molecular-level" target binding properties and poses is critical for those wishing to focus on chemical risk assessment at a molecular level of accountability.

5. Notes

- The good modeling practices as discussed more broadly throughout this book in other chapters still apply to molecular modeling, that is, in order to keep an audit trail of the steps and methods applied to virtual screening one must keep track of details used to obtain the numerical results.

- Keep track of crystal structure *PDB accession number*
- *Comment* on species type and co-crystallized ligands
- Consider any information with regards to MOA to be pertinent and capture (e.g., *IS inhibitor, IS substrate, IS agonist, or IS antagonist*)
- Consider *keeping the crystal structure of the co-crystallized ligands* as methods to test pose fidelity and "accuracy" of your modeling experiment
- When selecting a crystal structure it is good practice to inspect the atoms in the vicinity of the putative binding site for which you will perform docking. If the B or thermal factor is relatively low, then this is a good sign that the active site is relatively rigid and not an "ensemble of conformations." This information is explicitly found in the PDB file (and can be downloaded from http://www.pdb.org). Other derived information about the model geometry can be analyzed with free tools such as MolProbity (65). This free software helps identify model inconsistencies which may suggest not using a given structure. Similarly, it is vital to consider only targets for which the original X-ray data have been deposited. Using this data, electron density maps can be calculated or downloaded from places such as the Uppsala Electron Density server (66). The maps and models can be viewed using free programs such as Coot (67), Python Molecular Viewer (PMV) (68, 69), or SwissPDB's DeepView (70). Even with help from an experience X-ray crystallographer, one can confirm that (a) density clearly follows the shape of the model and (b) there is not substantial "difference electron density" to indicate that model atoms are incorrectly placed.
- It is good practice to *use crystal structures with relevant co-crystallized ligands* as opposed to only resolution criteria.

- *Every detail counts*: Knowledge of the pH, solvent medium, ionic strength, and buffers used can have implications on the model, the charge state of the model, and the type of charge model you would select to estimate atomic charges.
  - *Considering the subcellular localization* can often help in determining charge state for a chemical. For instance, the pH of the cytosol is ~7, the mitochondrial and ER pH is ~5, whereas the pH of the nucleus is ~7.5–8. This may affect the charge models you wish to capture.
- For *modeling protein binding* consider solvation/desolvation description as implemented in molecular mechanics frameworks or molecular docking algorithms as being inadequately captured or addressed.

- For *modeling more complex cellular activity* phenomena (such as receptor-mediated transient activation assays) consider cellular transport processes as surrogates to modeling a molecular MOA. For instance, estimating cellular membrane permeability, nonspecific target binding and specific target binding may assist in these efforts.
- When validation of pose fidelity is not optimal, *consider the reasons for failure*. "Reasons for Pose fidelity failure—Many of these pose fidelity failures could be attributed to one of four common causes: (a) insufficient conformational sampling of the ligand, particularly of aliphatic ring puckering, (b) symmetric or pseudo symmetric molecule, (c) critical missing water molecules, and (d) ligand pose dominated by electronic (orbital) effects. These issues are common to all docking methods and protocols" (71).
- *Putting the pieces together*: data and models and different software packages: One may want to consider either the purchase of a workflow manager such as Pipeline Pilot (40) or use of public domain versions such as KNIME (74), Bioclipse (72), or Taverna (73). Many of these packages contain the necessary elements to address step 1 and 2 of Fig. 2 in the *in silico*, in vitro workflow. Then, the selection of a docking package is the final step.

Acknowledgments

Michael-Rock Goldsmith would like to thank James Rabinowitz and Stephen Little (from the US-EPA's National Center for Computational Toxicology) for providing mentorship and assistance during his postdoctoral research, and providing the environment to explore molecular docking in the context of toxicology while providing insight and valuable discussion in the development of the in-house *in silico* chemical genomics initiative at the US-EPA.

References

1. Voutchkova A, Osimitz T, Anastas P (2010) Toward a comprehensive molecular design framework for reduced hazard. Chem Rev 110:5845–5882
2. Rusyn I, Daston G (2010) Computational toxicology: realizing the promise of the toxicity testing in the 21st century. Environ Health Perspect 118:1047–1050
3. Rabinowitz J, Goldsmith M, Little S, Pasquinelli M (2008) Computational molecular modeling for evaluating the toxicity of environmental chemicals: prioritizing bioassay requirements. Environ Health Perspect 116:573–577
4. Allinger N, Burkert U (1982) Molecular mechanics. American Chemical Society, Washington, DC

5. Dix D, Houck K (2007) The ToxCast program for prioritizing toxicity testing of environmental chemicals. Toxicol Sci 95:5–12
6. Villoutreix B, Renault N, Lagorce D, Sperandio O, Montes M, Miteva M (2007) Free resources to assist structure-based virtual ligand screening experiments. Curr Protein Pept Sci 8:381–411
7. Ponder J, Case D (2003) Force fields for protein simulations. Adv Protein Chem 66:27–85
8. Pearlman D, Case D, Caldwell J, Ross W, Cheathham T, DeBolt S, Ferguson D, Seibel G, Kollman P (1995) AMBER, a package of computer programs for applying molecular mechanics, normal mode analysis, molecular dynamics and free energy calculations to simulate the structural and energetic properties of molecules. Comput Phys Commun 91:1–41
9. MacKerell A, Brooks B, Brooks C, Nilsson L, Roux B, Won Y, Kaplus M (1998) CHARMM: the energy function and its parameterization with an overview of the program. In: Scheyer PVR et al (eds) The encyclopedia of computational chemistry. Wiley, Chichester
10. Case D, Cheatham T, Darden T, Gohlke H, Luo R, Merz K, Onufriev A, Simmerling C, Wang B, Woods R (2005) The AMBER biomolecular simulation programs. J Comput Chem 26:1668–1688
11. Brooks B, Brooks C, Mackerell A, Nilsson L, Petrella R, Roux B, Won Y, Archontis C, Bartels S, Caflish B, Caves L, Cui Q, Dinner A, Feig M, Fischer S, Gao J, Hodoscek M, Im W, Kuczera K, Lazaridi T, Ma J, Ovchinnikov V, Paci E, Pastor R, Post C, Pu J, Schaefer M, Tidor B, Venable T, Woodcock H, Wu X, Yah W, York D, Karplus M (2009) CHARMM: the biomolecular simulation program. J Comput Chem 30:1545–1615
12. Brooks B, Bruccoleri R, Olafson B, States D, Swaminathan S, Karplus M (1983) CHARMM: a program for macromolecular energy, minimization, and dynamics calculations. J Comput Chem 4:187–217
13. Allinger N, Yuh Y, Lii J (1989) Molecular mechanics: the MM3 force field for hydrocarbons. J Am Chem Soc 111:8551–8566
14. Leo A, Hansch C, Elkins D (1971) Partition coefficients and their uses. Chem Rev 71:525–616
15. Lipinski C, Lombardo F, Dominy B, Feeney P (1997) Experimental and computational approaches to estimate solubility and permeability in drug discovery and development settings. Adv Drug Deliv Rev 23:3–25
16. Wermuth C, Ganellin C, Lindberg P, Mitscher L (1998) Glossary of terms used in medicinal chemistry. Pure Appl Chem 70:1129–1143
17. Kubinyi H (2002) From narcosis to hyperspace: the history of QSAR. Quant Struct Act Relat 21:348–356
18. Wold S, Ruhe A, Wold H, Dunn W (1984) The collinearity problem in linear regression—the partial least squares (PLS) approach to generalized inverses. SIAM J Sci Stat Comput 5:735–743
19. Cramer R, Patterson D, Bunce J (1988) Comparative Molecular Field Analysis (CoMFA). 1. Effect of shape on binding of steroids to carrier proteins. J Am Chem Soc 110:5959–5967
20. Klebe G, Abraham U, Mietzner T (1994) Molecular similarity indices in a comparative analysis (CoMSIA) of drug molecules to correlate and predict their biological activity. J Med Chem 37:4130–4146
21. Pastor M, Cruciani G, McLay I, Pickett S, Clementi S (2000) GRid-INdependent descriptors (GRIND): a novel class of alignment-independent three-dimensional molecular descriptors. J Med Chem 43:3233–3243
22. Norinder U (1996) 3D-QSAR investigation of the Tripos benchmark steroids and some protein-tyrosine kinase inhibitors of styrene type using the TDQ approach. J Chemom 10:533–545
23. Kurogi Y, Guner O (2001) Pharmacophore modelling and three-dimensional database searching for drug design using catalyst. Curr Med Chem 8:1035–1055
24. Park J, Harris D (2003) Construction and assessment of models of CYP2E1: predictions of metabolism from docking, molecular dynamics and density functional theoretical calculations. J Med Chem 46:1645–1660
25. Jones J, Mysinger M, Korzekwa K (2002) Computational models for cytochrome P450: a predictive electronic model for aromatic oxidation and hydrogen atom abstraction. Drug Metab Dispos 30:7–12
26. Cheng Y, Prusoff W (1973) Relationship between the inhibition constant (Ki) and the concentration of inhibitor which causes 50 per cent inhibition (I50) of an enzymatic reaction. Biochem Pharmacol 22:3099–3108
27. MOE. Chemical Computing Group. Montreal, Quebec, Canada
28. Schrodinger, Inc. New York, NY
29. Cheatham T, Young M (2001) Molecular dynamics simulation of nucleic acids: successes, limitations and promise. Biopolymers 56: 232–256
30. Roterman I, Lambert M, Gibson K, Scheraga H (1989) A comparison of the CHARMM, AMBER and ECEPP potentials for peptides. 2. Phi-Psi maps for *n*-acetyl alanine N′-methyl amide—comparisons, contrasts and simple

experimental tests. J Biomol Struct Dyn 7:421–453
31. Roterman I, Gibson K, Scheraga H (1989) A comparison of the CHARMM, AMBER and ECEPP potential for peptides. 1. Conformational predictions for the tandemly repeated peptide (Asn-Ala-Asn-Pro)9. J Biomol Struct Dyn 7:391–419
32. Gundertofte K, Liljefors T, Norrby P, Petterson I (1996) A comparison of conformational energies calculated by several molecular mechanics methods. J Comput Chem 17:429–449
33. Jorgensen W, Maxwell D, Tirado-Rives J (1996) Development and testing of the OPLS all-atom force field on conformational energetics and properties of organic liquids. J Am Chem Soc 118:11225–11236
34. Jorgensen W, Tirado-Rives J (1988) The OPLS potential functions for proteins—energy minimizations for crystals of cyclic-peptides and crambin. J Am Chem Soc 110:1657–1666
35. Halgren T (1996) Merck molecular force field. I. Basis, form, scope parameterization and performance of MMFF94. J Comput Chem 17:490–519
36. Chen Y, Zhi D (2001) Ligand–protein inverse docking and its potential use in the computer search of protein targets of a small molecule. Proteins 43:217–226
37. Ellis L, Hou B, Kang W, Wackett L (2003) The University of Minnesota Biocatalysis/Biodegradation Database: post-genomic data mining. Nucleic Acids Res 31:262–265
38. MetaPrint2d http://www-metaprint2d.ch.cam.ac.uk/metaprint2d
39. Bologa C, Olah M, Oprea T (2005) Chemical database preparation for compound acquisition or virtual screening. Methods Mol Biol 316: 375
40. Accelrys Discovery Suite, Accelrys, Inc. San Diego, CA
41. Sybyl. Tripos, Inc. St. Louis, MO
42. Schwede T, Sali A, Honig B, Levitt M, Berman H, Jones D, Brenner S, Burley S, Das R, Dokholyan N, Dunbrack R, Fidelis K, Fiser A, Godzik A, Huang Y, Humblet C, Jacobsen M, Joachimiak A, Krystek S, Kortemme T, Kryshtafovych A, Montelione G, Moult J, Murray D, Sanchez R, Sosinick T, Standley D, Stouch T, Vajda S, Vasquez M, Westbrook J, Wilson I (2009) Outcome of a workshop on applications of protein models in biomedical research. Structure 17:151–159
43. Irwin J (2008) Community benchmarks for virtual screening. J Comput Aided Mol Des 22:193–199
44. Cross J, Thompson D, Rai B, Baber J, Fan K, Hu Y, Humblet C (2009) Comparison of several molecucular docking programs: pose prediction and virtual screening accuracy. J Chem Inf Model 49:1455–1474
45. Cherkasov A, Fuqiang B, Li Y, Fallahi M, Hammond G (2006) Progressive docking: a hybrid QSAR/Docking approach for accelerating in silico high throughput screening. J Med Chem 49:7466–7478
46. Peterson S (2007) Improved CoMFA modeling by optimization of settings: toward the design of inhibitors of the HCV NS3 protease. Uppsala University, Uppsala
47. Norinder U (1998) Recent progress in CoMFA methodology and related techniques. Perspect Drug Discov Des 12/13/14:25–39
48. Kim K, Grecco G, Novellino E (1998) A critical review of recent CoMFA applications. Perspect Drug Discov Des 12/13/14:257–315
49. Rosen J, Lovgren A, Kogej T, Muresan S, Gottfries J, Backlund A (2009) ChemGPS-NPWeb: chemical space navigation tool. J Comput Aided Mol Des 23:253–259
50. Larsson J, Gottfries J, Muresan S, Backlund A (2007) ChemGPS-NP: tuned for navigation in biologically relevant chemical space. J Nat Prod 70:789–794
51. Ekins S et al (2002) Three-dimensional quantitative structure-activity relationships of inhibitors of P-glycoprotein. Mol Pharmacol 61:964
52. Thorsteinson N, Ban F, Santos-Filho O, Tabaei S, Miguel-Queralt S, Underhill C, Cherkasov A, Hammond G (2009) In silico identification of anthropogenic chemicals as ligands of zebrafish sex hormone binding globulin. Toxicol Appl Pharmacol 234:47–57
53. Perry J, Goldsmith M, Peterson M, Beratan D, Wozniak G, Ruker F, Simon J (2004) Structure of the ochratoxin A binding site within human serum albumin. J Phys Chem B 108: 16960–16964
54. Aureli L, Cruciani G, Cesta M, Anacardio R, De Simone L, Moriconi A (2005) Predicting human serum albumin affinity of interleukin-8 (CXCL8) inhibitors by 3D-QSPR approach. J Med Chem 48:2469–2479
55. Ekins S, de Groot M, Jones J (2001) Pharmacophore and three-dimensional quantitative structure activity relationship methods for modeling cytochrome P450 active sites. Drug Metab Dispos 29:936–944
56. Ekins S, Erickson J (2002) A pharmacophore for human pregnane X receptor ligands. Drug Metab Dispos 30:96–99

57. Lewis D (2002) Molecular modeling of human cytochrome P450-substrate interactions. Drug Metab Rev 34:55–67
58. Hirono S, Nakagome L, Imai R, Maeda K, Kusuhara H, Sugiyama Y (2005) Estimation of the three-dimensional pharmacophore of ligands for rat multidrug-resistance-associated protein 2 using ligand-based drug design techniques. Pharm Res 22:260–269
59. DeGorter M, Conseil G, Deeley R, Campbell R, Cole S (2008) Molecular modeling of the human multidrug resistance protein 1 (MRP1/ABCC1). Biochem Biophys Res Commun 365:29–34
60. Rabinowitz J, Little S, Laws S, Goldsmith M (2009) Molecular modeling for screening environmental chemicals for estrogenicity: use of the toxicant-target approach. Chem Res Toxicol 22:1594–1602
61. Hirst W, Abrahamsen B, Blaney F, Calver A, Aloj L, Price G, Medhurst A (2003) Differences in the central nervous system distribution and pharmacology of the mouse 5-hydroxytryptamine-6 receptor compared with rat and human receptors investigated by radioligand binding, site-directed mutagenesis, and molecular modeling. Mol Pharmacol 64:1295–1308
62. http://oaspub.epa.gov/eims/eimscomm.getfile?p_download_id=466705
63. http://www.epa.gov/ncct/bosc_review/2009/posters/2-06_Rabinowitz_CompTox_BOSC09.pdf
64. Goldsmith M, Little S, Reif D, Rabinowitz J Digging deeper into deep data: molecular docking as a hypothesis-driven biophysical interrogation system in computational toxicology
65. http://molprobity.biochem.duke.edu
66. http://xray.bmc.uu.se/valid/density/form1.html
67. http://www.biop.ox.ac.uk/coot
68. http://pmvbase.blogspot.com/2009/04/electron-density-map.html
69. http://mgltools.scrips.edu/documentation/tutorial/python-molecular-viewer
70. http://spdbv.vital.it.ch
71. Irwin J, Shoichet B, Mysinger M, Huang N, Colizzi F, Wassam P, Cao Y (2009) Automated docking screens: a feasibility study. J Med Chem 52:5712–5720
72. Bioclipse. Proteometric Group, Department of Pharmaceutical Biosciences, Uppsala University, Sweden & Cheminformatics and Metabolism Team, European Bioinformatics Institute (EMBI)
73. Taverna. School of Computer Science, University of Manchester, UK
74. www.knime.org

Chapter 8

Chemical Structure Representations and Applications in Computational Toxicity

Muthukumarasamy Karthikeyan and Renu Vyas

Abstract

Efficient storage and retrieval of chemical structures is one of the most important prerequisite for solving any computational-based problem in life sciences. Several resources including research publications, text books, and articles are available on chemical structure representation. Chemical substances that have same molecular formula but several structural formulae, conformations, and skeleton framework/scaffold/functional groups of the molecule convey various characteristics of the molecule. Today with the aid of sophisticated mathematical models and informatics tools, it is possible to design a molecule of interest with specified characteristics based on their applications in pharmaceuticals, agrochemicals, biotechnology, nanomaterials, petrochemicals, and polymers. This chapter discusses both traditional and current state of art representation of chemical structures and their applications in chemical information management, bioactivity- and toxicity-based predictive studies.

Key words: Linear molecular representations, 2D and 3D representation, Chemical fragments and fingerprints, Chemical scaffolds, Toxicophores, Pharmacophores, Molecular similarity and diversity, Molecular descriptors, Structure activity relationship studies

1. Introduction

Chemists believe that everything is made of chemicals and work towards understanding their basic properties from their constituents elements and spatial arrangement to extend this knowledge for designing better chemicals. Chemical information is a knowledge based domain, where the primary chemical data is transcribed mainly in the form of chemical structures and commonly used as the international language of chemistry (1). Systematic naming of even a moderately complex chemical structure is a challenging task for a graduate student as the same chemical can have many

Brad Reisfeld and Arthur N. Mayeno (eds.), *Computational Toxicology: Volume I*, Methods in Molecular Biology, vol. 929, DOI 10.1007/978-1-62703-050-2_8, © Springer Science+Business Media, LLC 2012

different synonyms. The most commonly known chemical like formaldehyde is documented with more than 180 synonyms in publicly available databases (2). For instance a simple keyword search for "formaldehye" in PubChem as substance would return over 5181 hits (3). Here the PubChem compound ID for "formaldehyde" is assigned a value of 712 (4) with several user submitted synonyms. However searching the molecule formaldehyde with C(=)O as Simplified Molecular Input Line Entry System (SMILES) resulted in 11 entries (5). So the readers are encouraged to use structure based search for retrieving focused hits. Chemical structure representation requires therefore an unambiguous notation for the simplest atom to the most complicated molecule so that the end user can identify the relevant record in shortest possible time and need not browse through the voluminous literature that has minimum or no relevance to molecule of interest. There are several publicly available chemically relevant databases (6–12) and professional chemical information service providers like Chemical Abstract Service (13) and BEILSTEIN (14), from where one can quickly obtain relevant chemical structure information for nominal charges. Chemical Abstract Service (CAS) is a pioneer in chemical information management involved in collecting all the critical chemical data along with metadata from research publications, patents, research reports, theses, etc., and it is easy to search these resources accurately with the help of Chemical abstract registry numbers using user friendly web-based tools like Scifinder (15).

A brief look at history reveals that *in silico* representation of chemical structures proceeded almost parallely with advances in computer technology. Beginning with punch cards, light pen, keyboards, alphanumeric strings, graph theory methods to store few chemicals, today it is practically possible to store millions of chemical entities in an easy computer readable format (16). Traditionally hand-drawn chemical structures were used for chemical communications and eventually replaced by computer-generated images over a period of time. The preferred way of chemical structure generation today is via computers due to their flexibility in manipulation and reusability in simulations and modeling studies. These modern techniques of computer representation of structures effectively meet the following diverse needs of chemists.

- Information Storage, Search and Efficient Retrieval
- The Prediction of physicochemical, biological, and toxicological properties of compounds
- Spectra Simulation, Structure Elucidation
- Reaction Modeling
- Synthesis Planning
- Drug Design and Environmental Toxicity Assessment

2. Materials

In the early days of informatics attempts were made to digitize hand-drawn artistic chemical structures for facilitating structure-based searches. However, for the past two decades, the availability of advanced computer softwares and development of chemical structure representation standards has changed the scenario more towards the use of computer-generated chemical structures for chemical registration and communications. Both linear representations (WLN, SLN, and SMILES) and connection table-based matrix representation (MOL, MOL2, and SDF) of chemical structures were developed for compact storage and searching chemical structures efficiently (17, 18). The methods presently in vogue can be categorized mainly into compact line notation-based systems such as SMILES string which were developed when storing data was expensive and the more recent connection tables format in structure databases. An overview of the various methods of chemical structure representation methods so far reported in literature is depicted in Fig. 1. Here the readers are encouraged to refer a recent review article by Warr et al. (19) for a comprehensive coverage of general representation of chemical structures and references cited therein.

2.1. Linear Representation

Linear line notation is easily accessible to chemists and flexible enough to allow interpretation and generation of chemical notation similar to natural language. Alphanumeric string-based linear chemical structure encoding rules were developed by the pioneering contributions of Wiswesser, Morgan, Weininger, and Dyson, and eventually applied in machine description. In 1949, William Wiswesser introduced a line notation system based on the Berzelian symbols with the structural and connectivity features. This system named as Wiswesser Line Notation (WLN) was used online for structure and substructure searches (20). Unfortunately, the adoption of WLN was ignored due to the complexity of specification and inflexible rules associated with it. The development of SMILES notation made a significant effect on compact storage in chemical information systems and it has led to the development of modern form of representing chemical structures. This line notation system has several advantages over the older systems in terms of its compactness, simplicity, uniqueness, and human readability. David Weininger developed the DEPICT program to interpret SMILES into a molecular structure (21). A detailed description of many advanced versions of SMILES such as USMILES, SMARTS, STRAPS, and CHUCKLES are available at Daylight website (22). SMiles ARbitrary Target Specification (SMARTS) is basically an extension of SMILES used for describing molecular patterns as

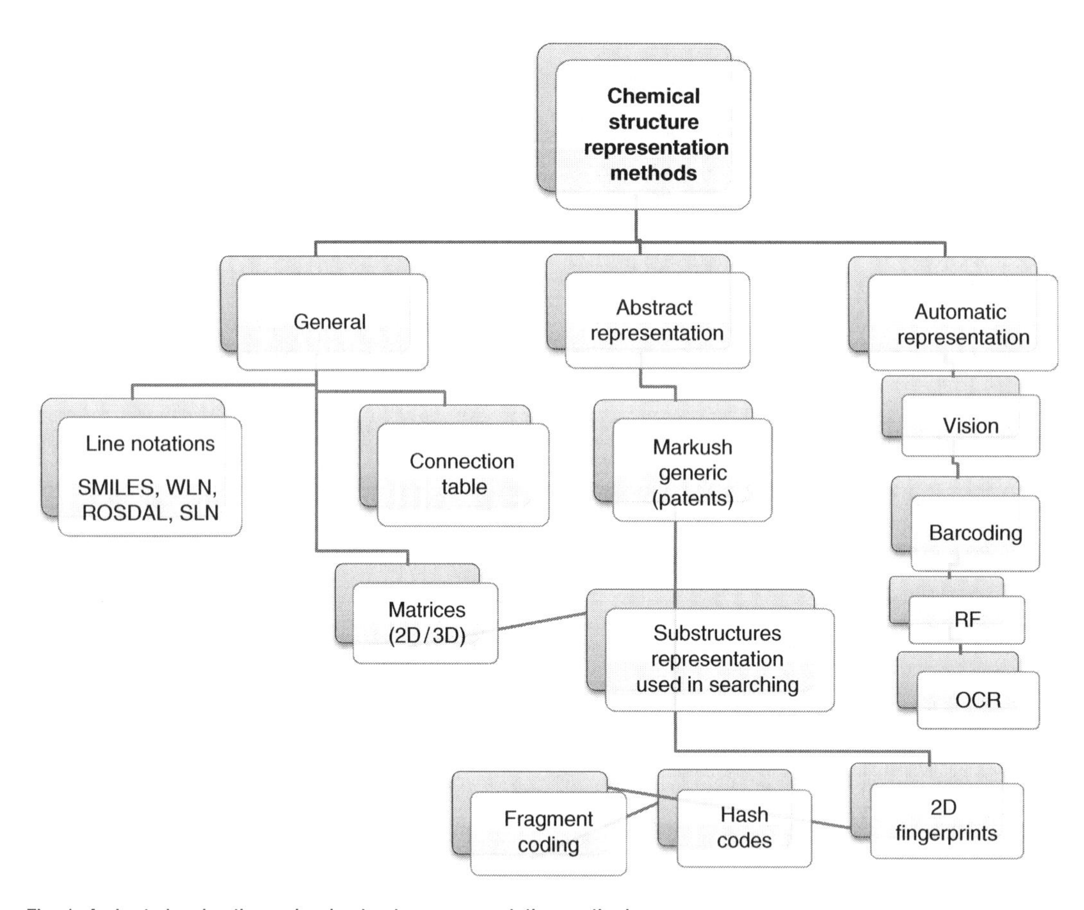

Fig. 1. A chart showing the molecule structure representation methods.

well as for substructure searching interpret (22). Sybyl Line Notation (SLN) is an another ASCII language which is almost similar to SMILES the difference being mainly in the representation of explicit hydrogen atoms (23). It can be used for substructure searching, Markush representation, database storage, and network communication but does not support reactions. An International Chemical Identifier (InChI) notation is a string of characters capable of uniquely representing a chemical substance (24). It is derived from a structural representation of that substance in a way designed to be independent of the way that the structure was drawn (thus a single compound will always produce the same identifier). It provides a precise, robust, IUPAC approved tag for representing a chemical substance. InChI is the latest and most modern of the line notations. It resolves many of the chemical ambiguities not addressed by SMILES, particularly with respect to stereocenters, tautomers, and other valence model problem. Table 1 shows the various line notations used for representing a chemical compound.

Table 1
Various notations of 3- (p-CHLOROPHENYL) -1, 1-DIMETHYLUREA TRICHLOROACETATE

- CAS No: *140-41-0*
- *Other Names:*
- EPA Pesticide Chemical Code 035502
- GC-2996
- LS-12938
- Caswell No. 583A
- PubChem : 8799
- EPA Pesticide Chemical Code: *035502*
- Urox
- Monuron trichloroacetate
- 3-(p-CHLOROPHENYL)-1,1-DIMETHYLUREA TRICHLOROACETATE
- Acetic acid, trichloro-, compd. with 3-(*p*-chlorophenyl)-1,1-dimethylurea (1:1)
- Urea, 3-(*p*-chlorophenyl)-1,1-dimethyl-, cmpd. with trichloroacetic acid (1:1)
- Acetic acid, trichloro-, compd. with *N'*-(4-chlorophenyl)-*N,N*-dimethylurea (1:1) (9CI)
- Trichloroacetic acid compound with 3-(*p*-chlorophenyl)-1,1-dimethylurea (1:1) (8CI)
- Wiswesser Line Notation (WLN): *GR DMVR N1&1 &GXGGVO*
- Canonical SMILES: *CN(C)C(=O)NC1 = CC = C(C = C1)Cl.C(=O)(C(Cl)(Cl)Cl)O*
- InChI:
- InChI = 1 S/C9H11ClN2O.C2HCl3O2/c1-12(2)9(13)11-8-5-3-7(10)4-6-8;3-2(4,5)1(6)7/h3-6 H,1-2 H3,(H,11,13);(H,6,7)
- InChIKey: *DUQGREMIROGTTD-UHFFFAOYSA-N*

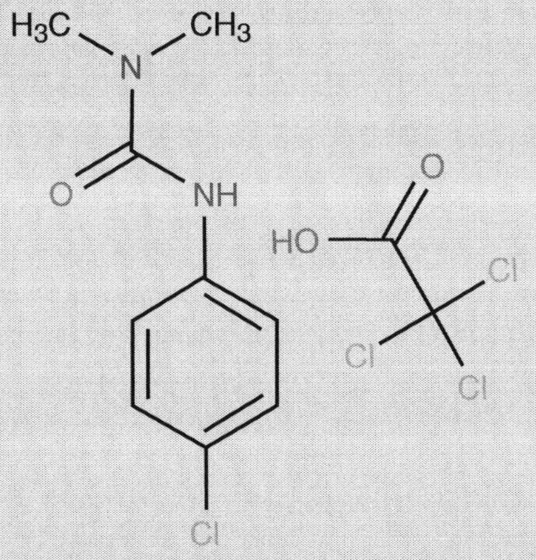

2.2. Graph/Matrix Representation of Molecules

According to graph theory a chemical structure is an undirectional, unweighted, and labeled graph with atoms as nodes and bonds as edges. Grave and Costa augmented molecular graphs with rings and functional groups by inserting additional vertices with corresponding edges (25). Matrix representation of graph was also used to denote chemical structure with *n* atoms as an array of $n \times n$ entries. There are several types of matrix representation such as adjacency matrix, distance matrix, atom connectivity matrix, incidence matrix, bond matrix, bond electron matrix each with its own set of merits and demerits.

2.3. Connection Tables

Contemporarily a new system of representation of molecular information is the form of a connection table is used. Simply defined a connection table is a list of atoms and bonds in a molecule. The connection table is a table enumerating the atoms and the bonds connecting specific atoms (26). The molecule shown in the connection table below has three atoms and two bonds. The table provides the three-dimensional (x, y, z) coordinates and the information about the bonds connecting the atoms along with the type of bonds (1 = single, 2 = double, etc.) alongside. Despite the size and format constraints, the connection tables are easily handled by the computers. However, it lacks in the human interpretability of the structural information. Owing to the constraints, the connection tables were widely adopted by the storage media file (Fig. 2).

2.4. 3D Structure Generation

Representing stereochemistry in computers during early days was a challenging task. In early 1970s Wipke et al. developed a rapid storage and retrieval of chemical structure based on SEMA Algorithm (Stereo chemically Extended Morgan Algorithm) (27). James G Nourse provided a method CFSG (ConFormation Symmetry Group) for specifying molecular conformation using bond attributes producing a unique name for conformation which can be coded for searching databases (28). Helmut described a matrix-based simple stereocode for structures wherein the stereochemistry was defined by the sequence of substituent indexes (29). Peter Willet and coworkers developed a similarity-based sophisticated method for searching in files of three-dimensional chemical structures (30). Standard methods for 3D structure generation used today are Concord, Corina, Chem Model, and Chem Axon (31). Nicklaus et al. compared chemmodel against concord and found that the earlier is superior due to its ability to map entire conformational space while comparing with X-ray crystallographic database (32). Clark et al. in their study found that genetic algorithm and directed tweak methods are efficient for searching database with flexible 3D structures (33).

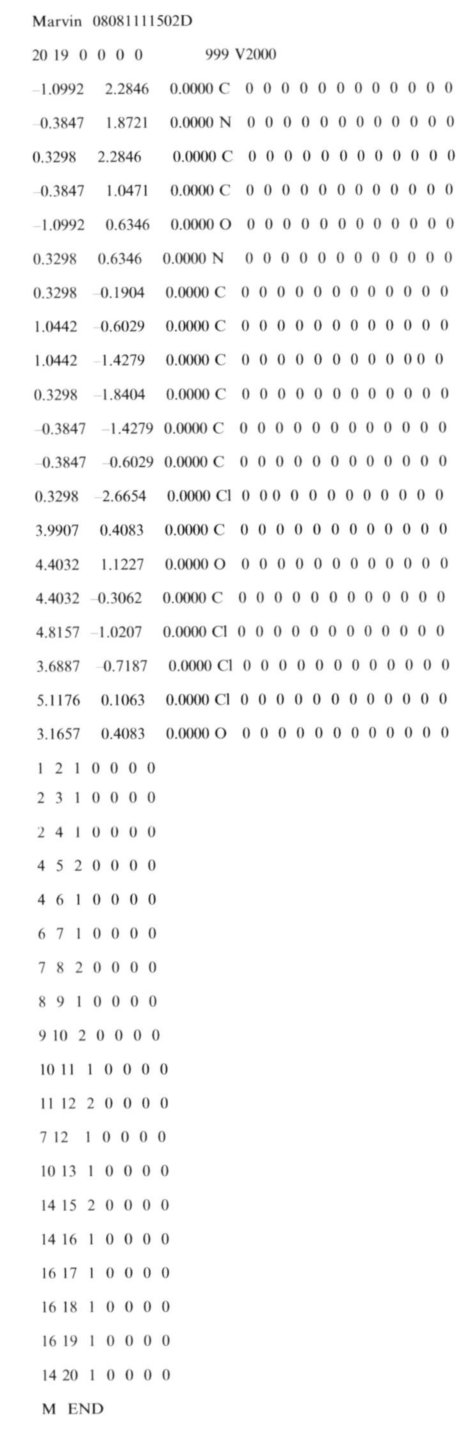

```
Marvin 08081111502D
20 19 0 0 0 0          999 V2000
-1.0992  2.2846  0.0000 C  0 0 0 0 0 0 0 0 0 0 0 0
-0.3847  1.8721  0.0000 N  0 0 0 0 0 0 0 0 0 0 0 0
0.3298  2.2846  0.0000 C  0 0 0 0 0 0 0 0 0 0 0 0
-0.3847  1.0471  0.0000 C  0 0 0 0 0 0 0 0 0 0 0 0
-1.0992  0.6346  0.0000 O  0 0 0 0 0 0 0 0 0 0 0 0
0.3298  0.6346  0.0000 N  0 0 0 0 0 0 0 0 0 0 0 0
0.3298  -0.1904  0.0000 C  0 0 0 0 0 0 0 0 0 0 0 0
1.0442  -0.6029  0.0000 C  0 0 0 0 0 0 0 0 0 0 0 0
1.0442  -1.4279  0.0000 C  0 0 0 0 0 0 0 0 0 00 0
0.3298  -1.8404  0.0000 C  0 0 0 0 0 0 0 0 0 0 0 0
-0.3847  -1.4279 0.0000 C  0 0 0 0 0 0 0 0 0 0 0 0
-0.3847  -0.6029 0.0000 C  0 0 0 0 0 0 0 0 0 0 0 0
0.3298  -2.6654  0.0000 Cl 0 0 0 0 0 0 0 0 0 0 0 0
3.9907  0.4083  0.0000 C  0 0 0 0 0 0 0 0 0 0 0 0
4.4032  1.1227  0.0000 O  0 0 0 0 0 0 0 0 0 0 0 0
4.4032  -0.3062  0.0000 C  0 0 0 0 0 0 0 0 0 0 0 0
4.8157  -1.0207  0.0000 Cl 0 0 0 0 0 0 0 0 0 0 0 0
3.6887  -0.7187  0.0000 Cl 0 0 0 0 0 0 0 0 0 0 0 0
5.1176  0.1063  0.0000 Cl 0 0 0 0 0 0 0 0 0 0 0 0
3.1657  0.4083  0.0000 O  0 0 0 0 0 0 0 0 0 0 0 0
1 2 1 0 0 0 0
2 3 1 0 0 0 0
2 4 1 0 0 0 0
4 5 2 0 0 0 0
4 6 1 0 0 0 0
6 7 1 0 0 0 0
7 8 2 0 0 0 0
8 9 1 0 0 0 0
9 10 2 0 0 0 0
10 11 1 0 0 0 0
11 12 2 0 0 0 0
7 12 1 0 0 0 0
10 13 1 0 0 0 0
14 15 2 0 0 0 0
14 16 1 0 0 0 0
16 17 1 0 0 0 0
16 18 1 0 0 0 0
16 19 1 0 0 0 0
14 20 1 0 0 0 0
M END
```

Fig. 2. Mol File format of 3- (p-CHLOROPHENYL) -1, 1-DIMETHYLUREA TRICHLOROACETATE.

3. Methods

3.1. Strategies Used in Searching Chemical Databases

To know which molecule is already published and studied in chemical or biological context, scientific literature search is conducted with the help of exact structure, similar structure, substructure and hyperstructures.

3.1.1. Fragment and Fingerprint Based Search Strategies

In the fragment-based approach in order to search a database to find similar structures, the query molecule (graph) is fragmented into various logical fragments (subgraphs such as functional groups, and rings) (34). The list of retrieved hit structures will include all those substructure (fragments) that were present in the query structure. Another approach is the fingerprint-based approach which is more of a partial description of the molecule as a set of fragments rather than a well-defined substructure (35). It essentially treats a molecule as a binary array of integers where each element (1 or 0) representing presence or absence of particular fragments as TRUE or FALSE. A given bit is set to 1 (True) if a particular structural feature is present and 0 (False). Hashed Fingerprints composed of bits of molecular information (fragments) such as types of rings, functional groups, and other types of molecular and atomic data. Comparing fingerprints will allow one to determine the similarity between two molecules A and B as shown in Fig. 3.

3.1.2. Similarity Search

The fingerprints of two structures can be compared and a distance score can be computed for their similarity. The similarity coefficient metrics such as Tanimoto, Cosine and Euclidean distance helps to filter relevant molecules rapidly from millions of entries (36). Takashi developed an approach for automatic identification of molecular

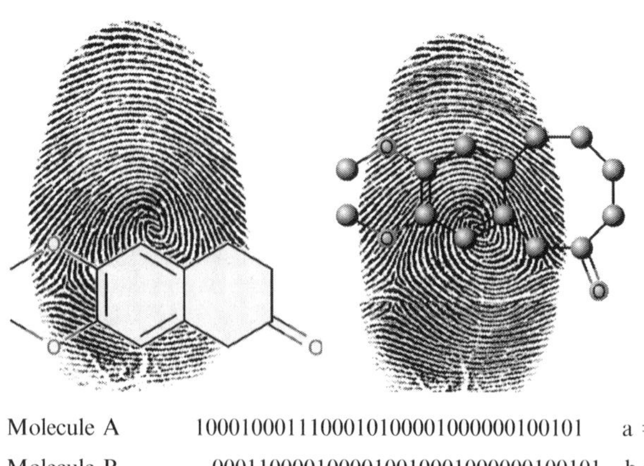

Molecule A 10001000111000101000010000000100101 a = 11
Molecule B 00011000010000100100010000000100101 b = 9
Similarity (A and B) 00001000010000100000010000000100101 c = 7

Fig. 3. Similarity and fingerprint analysis.

similarity using reduced graph representation and applied clique detection algorithm for finding similar structural features (37).

3.1.3. Hash Coding

Hash coding of molecules is an efficient way of substructure storing and searching where a molecule is assigned given a unique id key which is used for directly mapping the address of the compound in a computer system. There is, however, information loss in this approach as the source molecule cannot be reconstructed from the hash keys (38). In a report by Wipke et al. four different hash functions were used effectively for the rapid storage and retrieval of chemical structures (39). Hodes and Feldman developed a novel file organization for substructure searching via hash coding (40).

3.1.4. Hyperstructures

Hyperstructure has been defined by Robert Brown as a novel representation for a set of 2D chemical structures that has been suggested for increasing the speed of substructure searches, when compared with a sequential search. He reported two methods for construction of hyperstructures atom assignment method and maximal overlap-set method wherein the first one was found to be computationally less intensive (41).

3.1.5. Abstract Representation of Molecules

One of the major challenges confronting pharma industry is how to find a molecule or list of molecules with desired properties which are not yet published? One approach is to simulate all the possible chemicals for a given formula and enumerate them to full length. However it is impractical to do that and also searching each entry with millions of entries already known is another impossible task! Only solution is to map what is known from the literature and hence that would indirectly indicate what is "not yet reported." It does not necessarily imply that "those entries are not yet identified." It is known in business context that several thousand new molecules with desired properties are kept as trade secrets and superficially protected by patents as generic structures. There are occasions where the molecules are represented as markush structures in a generic context to cover a family of molecular structures and that sometimes go beyond millions. Markush structures are generic structures used in patent databases such as MARPAT maintained by Chemical abstracts service for protecting intellectual chemical information in patents (42) (Fig. 4).

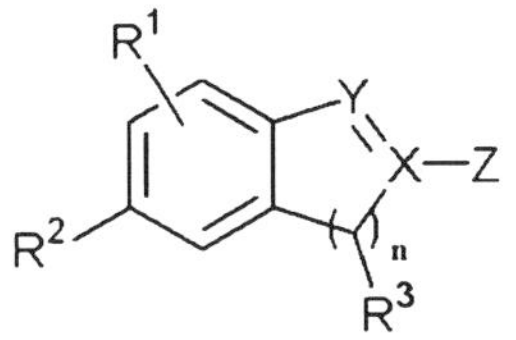

Fig. 4. A Markush structure.

A formal language GENSAL (generic structure description language) has been developed to describe generic structure from chemical patents (43). A seminal work by Stuart Kaback in 1980s highlights the structure searching parameters in Derwent world patent index (44).

3.2. Chemical Drawing Tools

There are several commonly available tools for generating chemical structures and storing them in standard file formats. Most popular ones among academia and industry are MarvinSketch from ChemAxon and ChemDraw from Cambridgesoft. Other tools exist for specific purposes like integrating with analytical data (^{1}H NMR), for example, ChemSketch and Inventory management using ISIS draw/ISIS Base(45). With the help of these drawing tools one can easily drag and drop necessary pre-built templates and build complex chemical structures. Interconversion of 2D to 3D structures is also possible through these tools. The quality of 3D structure however depends on the methods used within the system. The best 3D model is the one which is comparable with X-ray crystallographic data. Corina a software developed by Gasteiger et al. is very fast in computing 3D coordinates for 2D structures. Comparable or even better algorithms are implemented in ChemAxon tools. With the help of 3D structure it is possible to calculate energy of the molecule, volume, interatomic charge distribution, and other three-dimensional descriptors required for QSAR-based predictive studies.

A simple way to generate a 3D structure is to use Marvin View:

- Create and open a molecule in Marvin View tool
- Then go to edit and clean in 3D
- The output will be the 3D structure of the molecule as shown in Fig. 5

3.3. Current Trends in Chemical Structure Representation

The traditional methods of chemical structure representation and manipulation can be time consuming and tedious entailing long hours of searching. In this section we discuss the new emerging technologies being developed to handle chemical structures for automation and inventory management. Neural network-based chemical structure indexing is a technique where the chemical structure is presented as an image to the neural network using pulse-coupled neural networks (PCNN) to produce binary barcodes of chemical structures (46).

Karthikeyan et al. in their work on encoding of chemical structures reported that chemical structures can be encoded and read as 2D barcodes (PDF417 format) in a fully automated fashion (47). A typical linear barcode consists of a set of black bars in varying width separated by white spaces encoding alphanumeric characters. To reduce the amount of data that has to be encoded on the

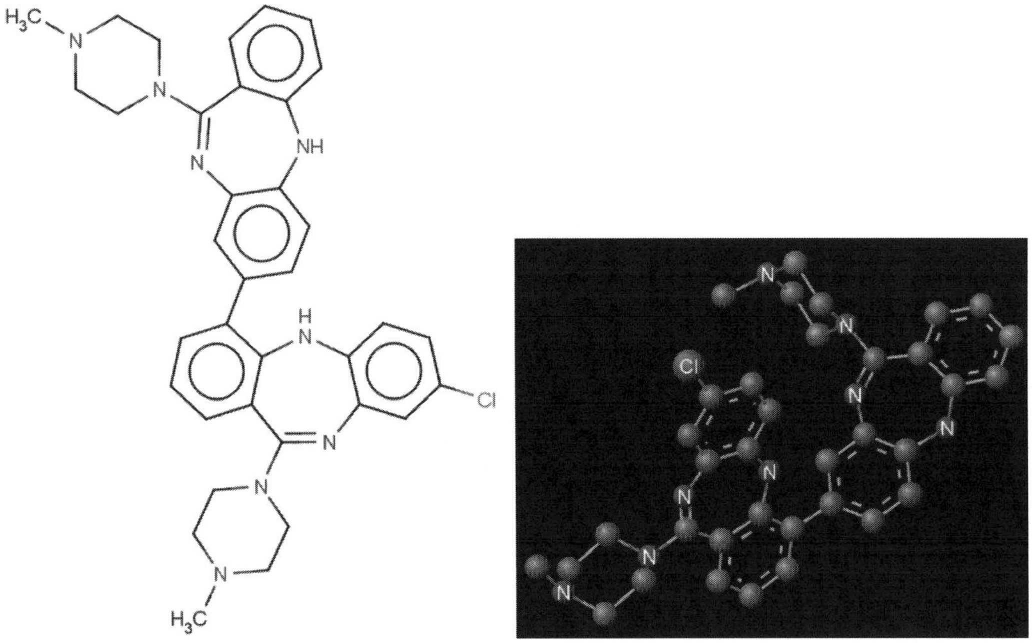

Fig. 5. Molecule 2D and 3D representation in Marvin View.

barcode, a template-based chemical structure encoding method was developed, the ACS file format (48). This method is based on the Computer-Generated Automatic Chemical Structure Database (CG-ACS-DB) originally developed to create a virtual library of molecules through enumeration from a selected set of scaffolds and functional groups (49). Scaffolds and groups are stored in Automatic Chemical Structure (ACS) format as a plain text file. In this ACS format, most commonly used chemical substructures are represented as templates (scaffolds or functional groups) through reduced graph algorithm along with their interconnectivity rather than atom-by-atom connectivity information. The barcoded chemical structures can be used for error-free chemical inventory management. One of the molecules containing over thousands of atoms can be easily represented as barcoded and can be decoded automatically and accurately in seconds without manual intervention.

Several file formats used for storing and retrieving chemical structures using the barcode method are shown in Fig. 6.

(a) *IUPAC NAME Representation*
6-chloro-10- (4-methylpiperazin-1-yl) -15-[10- (4-methylpiperazin-1-yl)-2,9-diazatricyclo[9.4.0.0^{3,8}]pentadeca-1 (11),3,5,7,9,12,14-heptaen-6-yl]-2,9-diazatricyclo[9.4.0.0^{3,8}]pentadeca-1 (11),3,5,7,9,12,14-heptaene

Fig. 6. Barcode representation of a molecule.

(b) *InChIKey = KEFNWQBDDAALRP-UHFFFAOYSA-N*
The InChI representation of chemical structure is most suitable for automatic and inventory management solutions using barcodes. For example, the InChI code of chemical structure is represented as barcode as shown in Fig. 6. The simple barcode code contains 1,309 characters representing chemical structures in InChI format describing the connectivity, atom information, etc. Earlier work on barcoding chemical structures used SMILES format.

Data (inside Barcode in InChI format)
InChI = 1/C36H37ClN8/c1-42-14-18-44 (19-15-42) 35-27-6-3-4-9-29 (27) 38-30-12-10-24 (22-32 (30) 40-35) 26-7-5-8-28-34 (26) 39-31-13-11-25 (37) 23-33 (31) 41-36 (28) 45-20-16-43 (2) 17-21-45/h3-13,22-23,38-39 H,14-21 H2,1-2H3AuxInfo = 1/0/N:45,22,37,36,14,38,13,15, 35,25,2,26,3,41,43,18,20,40,44,17,21............ (Lines removed for brevity) ....

.1175,6.7142,0;.5991,7.123,0;1.3114,6.7068,0;-.8298,7.1304,0;

(c) *SMILES Format*

CN1CCN (CC1) C1 = Nc2cc (ccc2Nc2ccccc12) -c1cccc2c1 Nc1ccc (Cl) cc1N = C2N1CCN (C) CC1

(d) *SMARTS:*

[#6]N1CCN (CC1) C1 = Nc2cc (ccc2Nc2ccccc12) -c1cccc 2c1Nc1ccc (Cl) cc1N = C2N1CCN ([#6]) CC1

(e) *MOL File format*
Structure #1
Marvin 07131117252D

45 52 0 0 0 0 999 V2000
6.1237 0.8510 0.0000 C 0 0 0 0 0 0 0 0 0 0 0 0
.... (Lines removed for brevity) ....
-0.8298 7.1304 0.0000 C 0 0 0 0 0 0 0 0 0 0 0 0
1 2 4 0 0 0 0
.... (Lines removed for brevity) ....
43 44 1 0 0 0 0
M END

(f) *Gaussian Cube file with atom data*
No surface data provided
45 0.000000 0.000000 0.000000
1 1.889726 0.000000 0.000000
.... (Lines removed for brevity) ....
6 6.000000 -2.927110 25.152404 0.000000
0.00000E00

(g) *PDB Format*
HEADER PROTEIN 13-JUL-11 NONE
TITLE NULL
COMPND MOLECULE: Structure #1
SOURCE NULL
KEYWDS NULL
EXPDTA NULL
AUTHOR Marvin
REVDAT 1 13-JUL-11 0
HETATM 1 C UNK 0 11.431 1.589 0.000 0.00 0.00 C + 0
.... (Lines removed for brevity) ....
HETATM 45 C UNK 0 -1.549 13.310 0.000 0.00 0.00 C + 0
CONECT 1 2 6 23
.... (Lines removed for brevity) ....
CONECT 45 42
MASTER 0 0 0 0 0 0 0 0 45 0 104 0
END

(h) *XYZ format*
45
Structure #1
C 11.43091 1.58853 0.00000
.... (Lines removed for brevity) ....
C -1.54896 13.31008 0.00000

Software programs like openBabel can interconvert molecules over 50 standard file formats required by several computational chemistry and chemoinformatics oriented programs (50). Dalby et al. in a classic paper have discussed all the file formats and their interrelations that are required for storing and managing chemical structures developed at Molecular Design Limited (MDL) (51).

RF tagging is a technology complementary to barcode representation of molecules which is commonly used technique in security and inventory management (52). Yet another upcoming technology based on OCR (Optical Character Recognition) can recognize molecular structures from scanned images of printed text, one can recognize structures, reactions, and text from scanned images of printed chemistry literature (53). This can save users valuable time of redrawing structures from printed material, as it directly transforms the "images" into a "real structures" that can then be saved into chemical databases. Programs such as CLiDE, OSRA, and ChemOCR are the known relevant softwares that recognize structures, reactions, and text from scanned images of printed chemistry literature (54–56).

One of the major breakthroughs due to the progress of WWW system is the evolution of content-based markup language based on XML syntax, the Chemical Markup Language (CML) developed by Peter Murray-Rust (57). Now the CML has become as a valuable tool with the functionalities to describe atomic, molecular, and crystallographic information. CML captures the structural information through a concise set of tags with the associated semantics. CML representation is well documented however due its size comparison with other existing file formats it is prohibitive for many applications. If a suitable tool is developed to store CML format in compressed mode without loss of information and freedom of use then it will encourage user community to apply CML for their applications. CCML is a methodology for encoding chemical structures as compressed CML generated by popular chemical structure generating programs like JME (58). This CCML format consists of both SMILES and/or equivalent data along with coordinate information about the atoms for generating chemical structures in plain text format. Each structure is either generated by JME in standalone mode or generated by virtual means that can be stored in this format for efficient retrieval, as it requires about one tenth or below of actual CML file format, since the SMILES describes the interconnectivity of the molecule. This CCML format is compatible for automated inventory application and is commonly used technique in security and inventory management.

An open source-based computer program called Chem Robot is developed which can use digital video devices to capture and analyze rapidly hand drawn or computer-generated molecular structures from plain papers (59). The computer program is capable of extracting molecular images from live streaming digital video signals and prerecorded chemistry-oriented educational videos. The images captured from these sources are further transformed into vector graphics for edge detection, node detection, Optical Character Recognition (OCR), and interpreted as bonds, atoms in molecular context. The molecular information generated is further transformed into reusable data formats (MOL, SMILES, InCHI, SDF) for modeling and simulation studies. The connection table and atomic coordinates (2D) generated through this automatic process can be further used for generation of IUPAC names of the molecules and also for searching the chemical data from public and commercial chemical databases. Applying this software the digital webcams, camcorders can be used for recognition of molecular structure from hand-drawn or computer-generated chemical images. The method and algorithms can be further used to harvest chemical structures from other digital documents or images such as PDF, JPEG formats. Effective implementation of this program can be further used for automatic translation of chemical images into common names or IUPAC names for chemical education and research. The performance and efficiency of this workflow can be extended to mobile devices (smart phones) with wi-fi and camera.

3.4. Handling Chemical Structure Information

Storage of chemical structure in proper file format is very important, for example, structures (html) stored as GIF, JPG, BMP, and PNG look alike but may not be compatible while database transfer or computer processing. In late 1970s, Blake et al. developed methods for processing chemical structure information using light pen to create high-quality structures which were used in chemical abstract volumes (60). A decade later all the features of structure formatting guidelines for chemical abstract publications appeared in a paper by Alan Goodson (61). Another interesting parallel development was the use of SNN (structure–Nomenclature Notation) wherein the molecule was split into fragments by structure determining vertices and linked by special signs for use in Beilstein system (62). Igor developed a compact code for storing structure formulas, performing substructure, and similarity searches and applied it on a set of 50,000 structures (63). A modular architecture for chemical structure elucidation called as mosaic or artel architecture was also developed (64). Hagadone and Schulz used a relational database system in conjunction with a chemical software component to create chemical databases with enhanced retrieval capabilities (65).

3.5. Cluster Analysis and Classification of Chemical Structures

Clustering is a process of finding the common features from a diverse class of compounds that requires multivariate analysis methods. One of the most suitable methods for this study is clustering where the consensus score and distance between set of compounds could be easily measured through mean/Euclidean distance measures. This score reflects the similarity or dissimilarity between classes of compounds and helps to identify potentially active or toxic substances through predictive studies. Peter Willet carried out a comparative study of various clustering algorithms for classifying chemical structures (66). An excellent review by Barnard and Downs illustrates the methods useful for clustering files of chemical structures (67). The jarvis patrick algorithm is useful for clustering chemical structures on the basis of 2D fragment descriptors (68). Lipinski rule of five is one such example where the similar

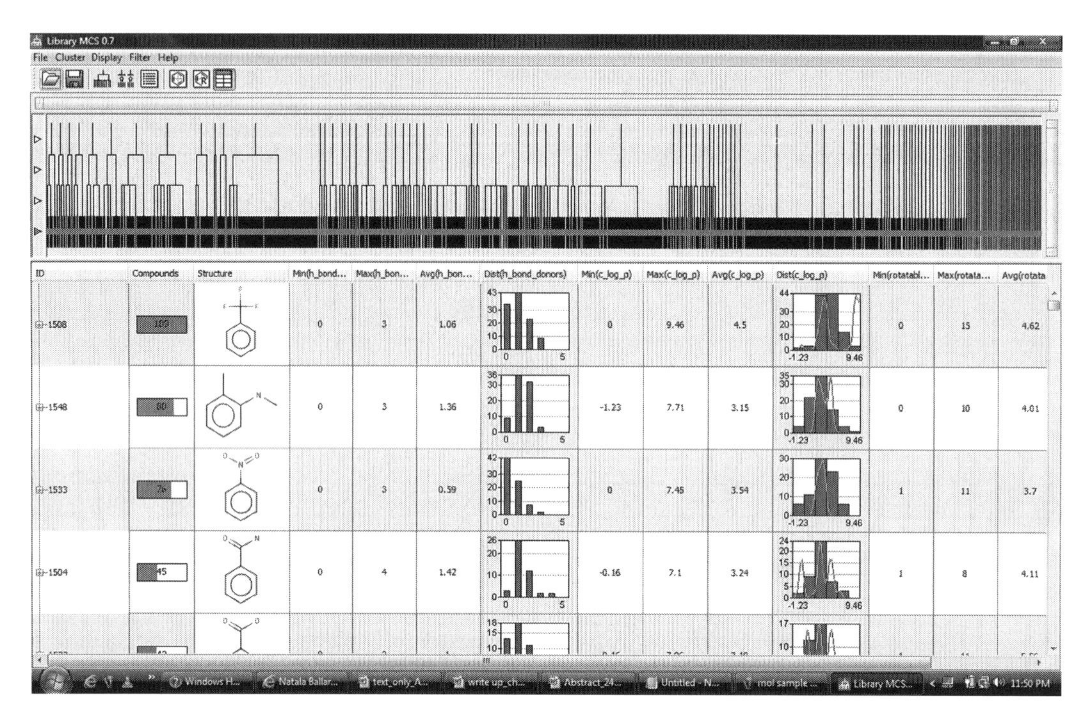

Fig. 7. Clustering of compounds in Library MCS module of chemaxon.

characteristics of drug molecules were derived by clustering large number of drugs and lead molecules. ChemAxon provides clustering tools to analyze hundreds and thousands of molecules (Library MCS) via maximum common substructures (69) (Fig. 7).

3.6. Internet as a Repository of Chemical Structures

In situations where there is no chemical structure information available, it is still possible to generate necessary chemical structure data through metadata harvesting method. Karthikeyan et al. developed ChemXtreme, a java-based computer program to harvest chemical information from Internet web pages using Google search engine and applying distributed computing environment (70). ChemXtreme employs the "search the search engine" strategy, where the URLs returned from the search engines are analyzed further via textual pattern analysis. This process resembles the manual analysis of the hit list, where relevant data are captured and, by means of human intervention, are mined into a format suitable for further analysis. ChemXtreme on the other hand transforms chemical information automatically into a structured format suitable for storage in databases and further analysis and also provides links to the original information source. The query data retrieved from the search engine by the server is encoded, encrypted, and compressed and then sent to all the participating active clients in the network for parsing. Relevant information identified by the clients on the retrieved web sites is sent back to the server, verified, and added to the database for data mining and further analysis. The chemical names including global identifiers like InChI or corporate identifiers like CAS registry numbers and Beilstein registry number could be mapped to corresponding structural information in a relational database systems.

3.7. Structure Property Correlation

Physicochemical properties, bioactivities, and toxicity-related data of chemicals available from scientific literature or from experimental results are used for building predictive models applying advanced mathematical methods or machine-learning techniques. The principle of "similar structure with similar property" is applied while building mathematical models. Chemical structure descriptors or the structural features with independent property of interest such as activity, property, and toxicity applying mathematical modeling or statistical techniques are always linked. The quality of predictive models basically depends on the type of relevant molecular descriptors and accuracy of experimental data. The applicability domain is one of the most important factors which should be taken into consideration while building mathematical models or while applying the prebuilt models for predictive studies. Explaining outliers in the training set, test set, and predicted set is one of the requirements in modern structure–property–activity relationship studies. There are several types of molecular descriptors and features used for structure property relationship studies. Most commonly used molecular descriptors are topological, electronic, and shape descriptors. These three molecular features are intimately related to each other. Most common programs used to compute descriptors are Dragon, Molconnz, MOE, JOELib, PaDEL, and Chem Axon (71). The predictive models could be either continuous model or binary model. In the continuous model, one can predict the property value in a range, where as in binary model the outcome would be either yes or no. Binary fingerprints, fragment keys (MACCS), predefined pharmacophore features including aromatic, hydrogen bond acceptors (HBA), or hydrogen bond donors (HBD) generated from chemical structures helped to design better and efficient lead molecules in drug discovery research. According to IUPAC "A pharmacophore is an ensemble of steric and electronic features that is necessary to ensure the optimal supramolecular interactions with a specific biological target and to trigger (or block) its biological response" (72). In simple terms, a pharmacophore represents the key features like hydrogen bond donors, acceptors, centroids, aromatic rings, hydrophobic regions present in the molecules which are likely to be responsible for bioactivity. In the pharmacophore-based modeling, the positions of consensus pharmacophore features from a set of aligned molecules with known bioactivity type is captured in three-dimensional space along with their interatomic distances. These features are used to identify the hits from a collection of molecules with unknown bioactivity. Commercially available programs like Molecular

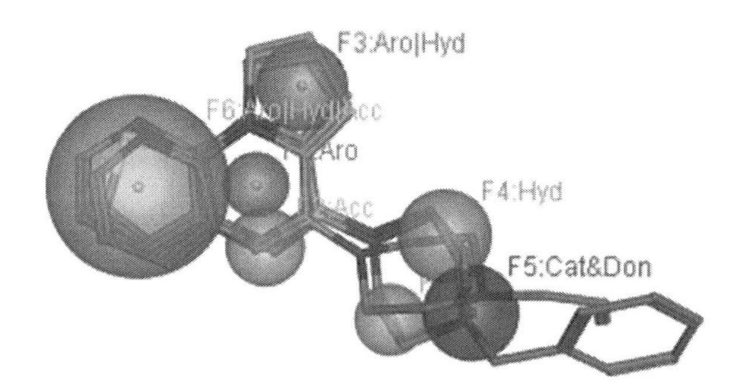

Fig. 8. Pharmacophore model generated by aligning molecules in MOE program.

Operating Environment (MOE), Catalyst (Accelerys), and Phase (Schrodinger) are equipped with these modules to design lead molecules (73) (Figs. 8 and 9).

Experimental chemists and biologists are actually interested in the properties of the chemicals and their response to biological systems both in beneficial and in adverse effects contexts. Several research groups across the world have compared chemical and drug databases to identify the molecular descriptors that can be used to classify molecules as drugs or nondrugs, toxins or non-toxins. In a classic paper by David Bawden, the applications of structure handling techniques for large-scale molecular property prediction along with relevant descriptors have been described in detail (74). Balaban et al. demonstrated correlation between chemical structures and boiling points of acyclic saturated compounds and haloalkanes using topological descriptors (75). Helge et al. used neural networks to predict the clearing temperatures of nematic liquid crystalline phases of several compounds from their chemical structures (76). Recently Karthikeyan et al. built a artificial neural network-based machine learning model for predicting melting point of diverse organic compounds (77). The quality of prediction depends on the primary experimental data used for mathematical modeling. A mathematical modeler is not expected to validate the experimental data cited in the scientific literature. Therefore it is the primary responsibility of the experimental chemist or biologist to publish high-quality data which is original, authentic and reproducible.

3.7.1. Structure Correlation with Toxicity

Here it is pertinent to discuss in detail the role of molecular structures/functional groups in toxicity context. Toxicity is another important parameter which needs to be assessed from molecular structures. Topkat is one such program which predict several toxicological-related data from given molecular structures based on the availability of selected structural patterns (78). The basic premise of the field is that molecular structure is related to the potential toxicity of a compound. This structure-based modeling

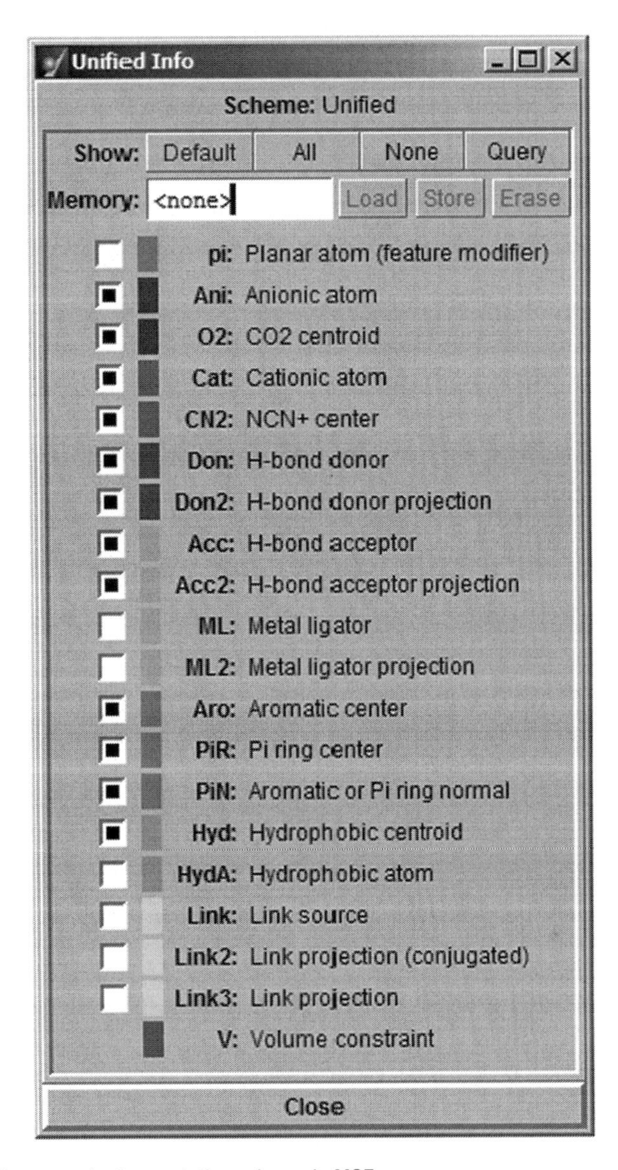

Fig. 9. Pharmacophoric annotation scheme in MOE.

finds application in QSAR, QSTR, environmental toxicity and high throughput screening.

The type of fragments present in a molecule which is responsible for a toxicity-eliciting interaction with the receptor or by metabolic activation is called as a toxicophore. In order to build models for predicting toxicology data, analysis of molecular structures in terms of presence or absence of these toxicophores is most important. Figure 10 shows common toxicophores which include Phosphoric groups, acetylenic, acetylide, acetylene halide, Diazo, Nitroso, Nitro, Nitrite, *N*-Nitroso, *N*-Nitro, Azoic, Peroxide, Hydroperoxide, Azide, Sulfur, carboxylates of diazonium, Halogenamine (79) (Fig. 10).

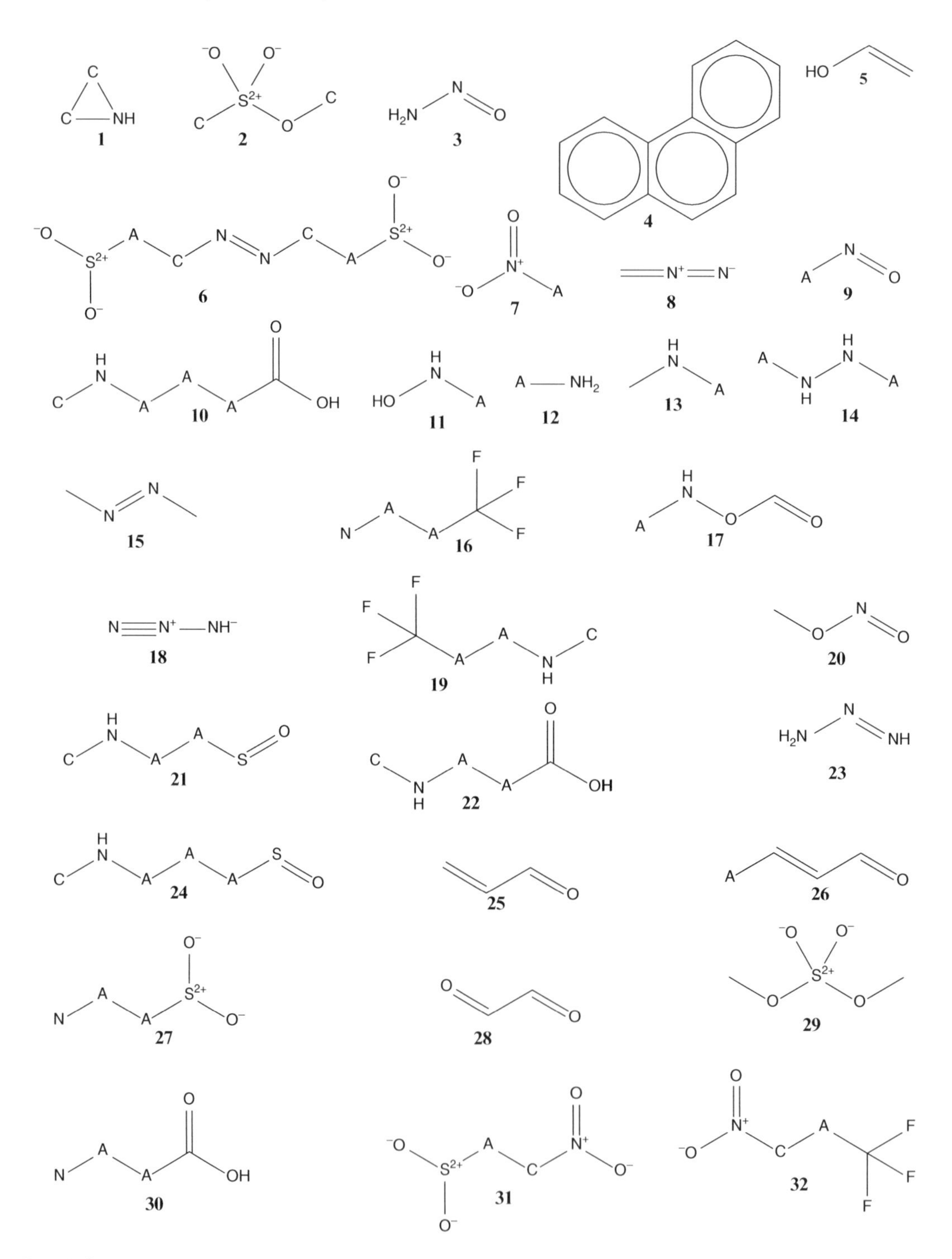

Fig. 10. Common toxicophores.

Among the toxicophores, epoxides, and aziridines are electrophilic, alkylating substructures with significant intrinsic reactivity and they are present in various compound data sets available publicly. It is necessary to identify these functional groups in early stage of drug design and eliminate them cautiously using virtual screening. Toxicophore models have been developed for mutagenicity prediction, HERG channel blockers, and hepatotoxicity (80). In an interesting study by Hughes et al. on evaluation of various physicochemical properties on in vivo toxicity TPSA (Topological Polar Surface Area) descriptor showed the maximum correlation to toxicity. A low TPSA allowed compounds to pass through lipid membranes and distribute into tissue compartments leading to high levels of toxicity. Other measures in this study included predicted blood brain barrier penetration and ClogP. It was revealed that high-ClogP compounds have a greater incidence of adverse outcomes relative to lower ClogP compounds (81).

3.7.2. Virtual Library Enumeration

In order to design a better lead molecule one has to perform sequence of several steps starting from collecting molecular data with known bioactivity, analysis of those chemical structures to extract significant features related to activity of interest and rebuild new molecules with promising and favorable bioactivity profiles.

Virtual library of diverse molecules which are not yet synthesized can be enumerated from a set of scaffolds and functional groups by combinatorial means. Here the scaffold represents a molecule containing at least one ring or several rings which are connected by linker atoms. Scaffolds can be generated from complex molecular structures by a systematic disconnection of functional groups connected by single bonds. Using this approach a scaffold translator (chemscreener and scaffoldpedia) was built and applied on large-scale data set of PubChem compounds containing over 12 million ring structures to generate a library of about 1 million scaffolds in a distributed computing environment (82). From this study, it was observed that over 300,000 molecules contain a common scaffold of "benzene" ring. The scaffolds and functional groups generated could be further enumerated to build virtual library of diverse organic molecules. An alternate approach namely "lead hopping" is also available to replace common scaffold by chemically and spatially equivalent core fragments (83).

Designing a better molecule is just one aspect of research, and getting it synthesized and biologically evaluated is the most critical component in drug discovery programs. Therefore it is necessary to rank molecules based on their synthetic accessibility score that represents the ease of synthesis of required molecule in a shortest possible way with high yield (84). Reactivity pattern fingerprints are being developed to characterize the molecules either as a reactant or product based on the nature of functional groups composition (85). The emerging trends in mapping reactivity pattern of molecule is illustrated with following example where molecule "A" from KEGG database with Reactant Like Score of (RLS) 52:22 contains more reactivity groups which is likely to be a reactant for 52 reactions whereas it is likely to be a product for 22 known reactions.

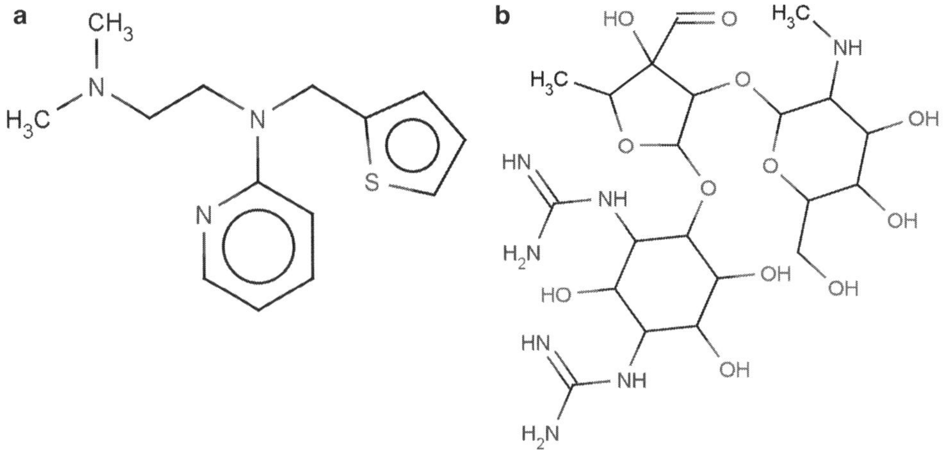

Fig. 11. (**a**) Reactant like (**b**) Product like molecule.

The converse is true for another molecule "B" with Product Like Score (PLS) of 13:5 where it is likely to be less reactive and could be synthesized through 13 different reactions but likely to undergo just 5 type of reactions based on their functional groups (Fig. 11). With the help of these PLS and RLS one can prioritize and characterize the compounds either as a reactant or as a product rapidly through *in silico* analysis and design efficient synthetic routes for the same.

3.7.3. Environmental Concerns

It is also important to discuss the characteristics of chemicals in the context of the environment. Many petrochemicals used in industrial, agricultural or domestic use reach the environment and become a source of grave concern in long run. There are certain chemicals such as polymers which require adverse conditions or long time to degrade into harmless chemicals in the environment. Some of the organic chemicals used as agrochemicals, for example pesticides and fungicides, become persistent in the environment and damages are significant. Now federal agencies are taking measures to control substances which are potential risk to the environment and human lives. Environmental Protection Agency (EPA) strives to develop models for predicting characteristics of controlled substances, chemicals of environmental concerns such as persistent organic substances (POS), carcinogens, and mutagens. There are software tools available to study the correlation between structure and biodegradability termed as QSBR (quantitative structure biodegradability relationship) (86). Environmental Protection Agency in USA and REACH initiatives by European Union are significant steps to control movement of large quantities of chemicals with potential risk (87). In the early days animals were used for toxicological testing of chemicals, however due to resistance by bioethics groups and law enforcement by government agencies the numbers of such tests are restricted to a bare minimum. One way to assess

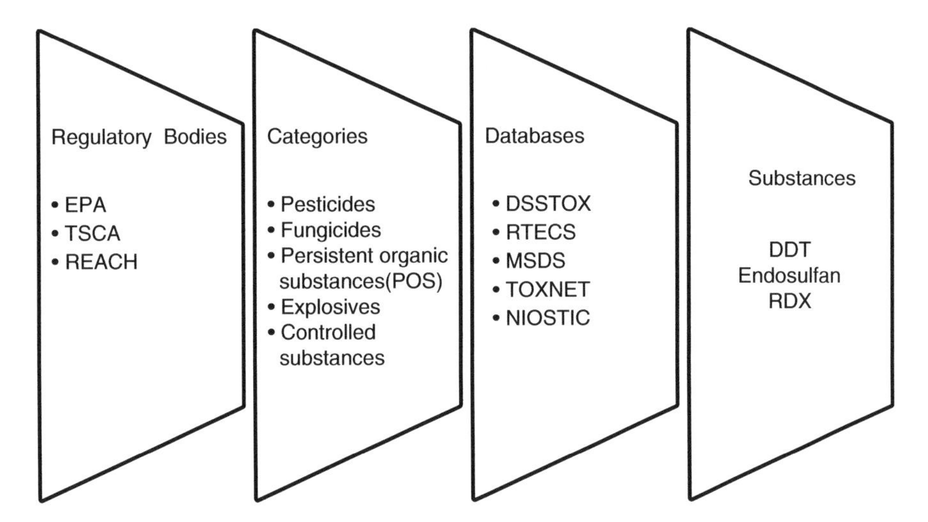

Fig. 12. Toxicological information system.

toxicity is to use data from past experiments and to build better predictive models as a substitute for animal toxicity testing and apply these models to develop risk assessment procedures. Over a period of time several toxicological models were evolved and frequently used for *in silico* prediction of environmental toxicity, mutagenicity, carcinogenicity, and ecotoxicity of chemicals. The EPI (Estimation Programs Interface) Suite developed by EPA is a physical/chemical property and environmental fate estimation program. EPI Suite uses a set of several estimation programs like KOWWIN, AOPWIN, HENRYWIN, MPBPWIN, BIOWIN, Bio HCwin, KOCWIN, WSKOWWIN, WATERNT, BCFBAF, HYD ROWIN, KOAWIN and AEROWIN, WVOLWIN, STPWIN, LEV 3EPI, and ECOSAR (88). Every module in this program and similar program has its own level of approximation and accuracy. It is also important to remember that computational or predictive studies have their own merits and demerits, and preference should be given to available experimental data over the computed or estimated data especially while decision making on processes or products of interest. DDT and Endosulfan are two compounds historically used as pesticides banned by most of the countries due to their adverse effect on human health and potential environmental toxicity (89) (Fig. 12).

Today predictive models are used extensively as guiding tools essentially as a replacement for animal experiments and environmental damage estimations. It is very important to learn from past experiments and incidents about the dangerous nature of chemicals and design better, safe, and green chemicals for the future with the help of these knowledge-based tools and methods.

4. Notes

Structural handling of chemicals efficiently is the basic need of computational toxicologist and any development in this area is sure to find many applications in organizations worldwide. The user has to choose his arsenal of methods and softwares in a cautious manner to make informed decisions. There is no substitute for human wisdom and care has to be taken while handling extremely sensitive data for building toxicity prediction models. In cases where wet lab validation is prohibitive comparison with data output of other modeling methods should be practiced. To conclude though the field of structure representation and manipulation appears to be mature, still there is enough scope for advancement in generic, complex, mixture or polymeric structure representations, and predictive studies with precision.

References

1. Ash JE, Warr WA, Willett P (1991) Chemical structure systems. Ellis Horwood, Chichester
2. http://pubchem.ncbi.nlm.nih.gov/. Accessed 11 July 2012
3. http://www.ncbi.nlm.nih.gov/sites/entrez?db=pccompound&term = formaldehyde. Accessed 11 July 2012
4. http://pubchem.ncbi.nlm.nih.gov/summary/summary.cgi?cid=712. Accessed 11 July 2012
5. http://pubchem.ncbi.nlm.nih.gov/search/search.cgi#. Accessed 10 July 2012
6. http://www.genome.jp/kegg/. Accessed 11 July 2012
7. https://www.ebi.ac.uk/chembldb/. Accessed 11 July 2012
8. http://www.drugbank.ca/. Accessed 11 July 2012
9. http://www.epa.gov/ncct/dsstox/index.html. Accessed 11 July 2012
10. http://www.cdc.gov/niosh/rtecs/. Accessed 11 July 2012
11. http://www.cdc.gov/niosh/database.html. Accessed 11 July 2012
12. http://www.msds.com/. Accessed 11 July 2012
13. http://www.cas.org. Accessed 11 July 2012
14. http://beilstein.com. Accessed 11 July 2012
15. https://scifinder.cas.org. Accessed 11 July 2012
16. Gasteiger J, Engel T (eds) (2003) Chemoinformatics: a textbook. Wiley-VCH, Weinheim
17. Quadrelli L, Bareggi V, Spiga S (1978) A new linear representation of chemical structures. J Chem Inf Comput Sci 18:37–40
18. www.ccl.net/cca/documents/molecular-modeling/node3.html. Accessed 11 July 2012
19. Wendy AW (2011) Representation of chemical structures. Vol 1. Wiley, New York. doi:10.1002/wcms.36
20. Fritts L, Schwind E, Margaret M (1982) Using the Wiswesser line notation (WLN) for online, interactive searching of chemical structures. J Chem Inf Comput Sci 22:106–9
21. Weininger D (1990) SMILES Graphical depiction of chemical structures. J Chem Inf Comput Sci 30:237–43
22. www.daylight.com/dayhtml/doc/theory/theory.smarts.html. Accessed 11 July 2012
23. Ash S, Malcolm AC, Homer RW, Hurst T, Smith GB (1997) SYBYL Line Notation (SLN): A versatile language for chemical structure representation. J Chem Inf Comput Sci 37:71–79
24. McNaught A (2006) The IUPAC International Chemical Identifier: InChI. Chemistry International (IUPAC) 28 (6). http://www.iupac.org/publications/ci/2006/2806/4_tools.html. Accessed 11 July 2012
25. Grave KD, Costa F (2010) Molecular graph augmentation with rings and functional groups. J Chem Inf Model 50:1660–8
26. www.lohninger.com/helpcsuite/connection_table.htm. Accessed 11 July 2012
27. Wipke WT, Dyott TM (1974) SEMA the stereochemically extended algorithm. J Am Chem Soc 96:4834
28. Christie BD, Leland BA, Nourse JG (1993) Structure searching in chemical databases by

direct lookup methods. J Chem Inf Comput Sci 33:545–7
29. Helmut B (1982) Stereochemical structure code for organic chemistry. J Chem Inf Comput Sci 22:215–22
30. Bath PA, Poirrette AR, Willett P, Allen FH (1994) Similarity searching in files of three-dimensional chemical structures: Comparison of fragment-based measures of shape similarity. J Chem Inf Comput Sci 34:141–7
31. http://www.molecular-networks.com. Accessed 11 July 2012
32. Nicklaus MC, Milne GW, Zaharevitz D (1993) ChemX and Cambridge Comparison of computer generated chemical structures with X ray crystallographic data. J Chem Inf Comput Sci 33:639–46
33. Clark DE, Jones G, Willet P et al (1994) Pharmacophoric pattern matching in files of three dimensional chemical structure: Comparison of conformational searching algorithms for flexible searching. J Chem Inf Comput Sci 34:197–206
34. Bond VL, Bowman CM, Davison LC et al (1979) On-line storage and retrieval of chemical information Substructure and biological activity screening. J Chem Inf Comput Sci 19:231–4
35. Wang Y, Bajorath J (2010) Advanced Fingerprint methods for similarity searching: balancing molecular complexity effects. Comb Chem High Throughput Screen 13:220–228
36. Andrew Leach (2007) An introduction to chemoinformatics Springer
37. Takahashi Y, Sukekawa M, Sasaki S (1992) Automatic identification of molecular similarity using reduced-graph representation of chemical structure. J Chem In Comput Sci 32:639–43
38. Zupan J (1989) Algorithms for chemists. Wiley, Chichester, UK
39. Wipke WT, Krishnan S, Ouchi GI (1978) Hash functions for rapid storage and retrieval of chemical structures. J Chem Inf Comput Sci 18:32–7
40. Hodes L, Feldman A (1978) An efficient design for chemical structure searching. II. The file organization. J Chem Inf Comput Sci 18:96–100
41. Brown R, Downs D, Geoffrey M et al (1992) Hyperstructure model for chemical structure handling: generation and atom-by-atom searching of hyperstructures. J Chem Inf Comput Sci 32:522–31
42. Barnard JM, Lynch MF, Welford SM (1982) Computer storage and retrieval of generic structures in chemical patents. 4. An extended connection table representation for generic structures. J Chem Inf Comput Sci 22:160–4
43. Barnard JM, Lynch MF, Welford SM (1981) Computer storage and retrieval of generic chemical structures in patents.; GENSAL, a formal language for the description of generic chemical structures. J Chem Inf Comput Sci 21:151–61
44. Kaback SM (1980) Chemical structure searching in Derwent's World Patents Index. J Chem Inf Comput Sci 20:1–6
45. www.chemaxon.com. Accessed 11 July 2012
46. Rughooputh SDDV, Rughooputh HCS (2001) Neural network based chemical structure indexing. J Chem Inf Comput Sci 41:713–717
47. Karthikeyan M, Bender A (2005) Encoding and decoding graphical chemical structures as two-dimensional (PDF417) barcodes. J Chem Info Model 45:572–580
48. Karthikeyan M, Krishnan S, Steinbeck C (2002) Text based chemical information locator from Internet (CILI) using commercial barcodes, 223rd American Chemical Society Meeting - Orlando, Florida, USA
49. Karthikeyan M, Uzagare D, Krishnan S. Compressed Chemical Markup Language for compact storage and inventory applications, 225th ACS Meeting - New Orleans, March 23–27, 2003. CG ACS
50. Guha R, Howard MT, Hutchinson GR et al (2006) The Blue Obelisk—interoperability in chemical informatics. J Chem Inf Model 46:991–998
51. Dalby A, Nourse JG, Hounshell WD et al (1992) Description of several chemical structure used by computer programs developed at molecular design limited. J Chem Inf Comput Sci 32:244–55
52. Xiai XY, Li RS (2000) Solid phase combinatorial synthesis using microkan reactors, Rf tagging and directed sorting. Biotech Bioeng 71:41–50
53. en.wikipedia.org/wiki/Optical_character_recognition
54. Valko AT, Peter JA (2009) CLiDE Pro: The Latest Generation of CLiDE, a Tool for Optical Chemical Structure Recognition. J Chem Info Model 49:780–787
55. cactus.nci.nih.gov/osra/
56. infochem.de/mining/chemocr.shtml
57. Murray-Rust P, Rzepa HS (2003) Chemical Markup, XML and the world wide web CML Schema. J Chem Inf Comput Sci 43:757–72
58. http://www.molinspiration.com/jme/index.html. Accessed 11 July 2012
59. http://moltable.ncl.res.in. Accessed 11 July 2012

60. Blake JE, Farmer NA, Haines RC (1977) An interactive computer graphics system for processing chemical structure diagrams. J Chem Inf Comput Sci 17:223–8
61. Goodson AL (1980) Graphical representation of chemical structures in Chemical Abstracts Service publication. J Chem Inf Comput Sci 20:212–17
62. Walentowski R (1980) Unique unambiguous representation of chemical structures by computerization of a simple notation. J Chem Inf Comput Sci 23:181–92
63. Strokov I (1995) Compact code for chemical structure storage and retrieval. J Chem Inf Comput Sci 35:939–44
64. Strokov I (1996) A new modular architecture for chemical structure elucidation systems. J Chem Inf Comput Sci 36:741–745
65. Hagadone TR, Schulz MW (1995) Capturing Chemical Structure Information in a Relational Database System: The Chemical Software Component Approach. J Chem Inf Comput Sci 35:879–84
66. Willett Peter J (1984) Evaluation of relocation clustering algorithms for the automatic classification of chemical structures. J Chem Inf Comput Sci 24:29–33
67. Barnard JM, Downs GM (1992) Clustering of chemical structures on the basis of two-dimensional similarity measures. J Chem Inf Comput Sci 32:644–9
68. Gu Q, Xu J, Gu L (2010) Selecting diversified compound to build a tangible library for biological and biochemical assay. Molecules 15:5031–44
69. www.chemaxon.com/jchem/doc/user/LibMCS.html. Accessed 11 July 2012
70. Karthikeyan M, Krishnan S, Pandey AK (2006) Harvesting chemical information from the internet using a distributed approach: Chem Extreme. J Chem Inf Model 46:452–461
71. www.moleculardescriptors.eu/softwares/softwares.htm. Accessed 11 July 2012
72. Horvath D (2011) Pharmacophore-based virtual screening In: Bajorath J (ed) Chemoinformatics and Computational Chemical Biology, Methods in Molecular Biology, Humana Press 672:261–298
73. Yang SY (2010) Pharmacophore modeling and applications in drug discovery challenges and recent advances. Drug Discov Today 15: 444–50
74. Adamson GW, Bawden D (1981) Comparison of hierarchical cluster analysis techniques for automatic classification of chemical structures. J Chem Inf Comput Sci 21:204–9
75. Balaban AT, Kier LB, Joshi N (1992) Correlations between chemical structure and physicochemical properties like boiling points. J Chem Inf Comput Sci 32:237–44
76. Helge V, Volkmar M (1996) Prediction of material properties from chemical structures. The clearing temperature of nematic liquid crystals derived from their chemical structures by artificial neural Networks. J Chem Inf Comput Sci 36:1173–1177
77. Karthikeyan M, Glen RC, Bender A (2005) General melting point prediction based on a diverse compound data set and artificial neural networks. J Chem Inf Model 45:581–90
78. http://accelrys.com/solutions/scientific-need/predictive-toxicology.html. Accessed 11 July 2012
79. Hakimelahi GH, Khodarahmi GA (2005) The identification of toxicophores for the prediction of mutagenicity, hepatotoxicity and cardiotoxicity. JICS 2:244–267
80. Garg D, Gandhi T, Gopi Mohan C (2008) Exploring QSTR and toxicophore of HERG K + channel blockers using GFA and HypoGen techniques. J Mol Graph Model 26:966–76
81. Hughes JD, Blagg J, Da P et al (2008) Physicochemical drug properties associated with in vivo toxicological outcomes. Bioorg Med Chem Lett 18:4872–4875
82. Karthikeyan M, Krishnan S, Pandey AK, Bender A, Tropsha A (2008) Distributed chemical computing using Chemstar: an open source Java Remote Method Invocation architecture applied to large scale molecular data from Pubchem. J Chem Info Comput Sci 48:691–703
83. Maass P (2007) Recore: A fast and versatile method for scaffold hopping based on small molecule crystal structure conformations. J Chem Inf Model 47:390–9
84. Ertl P, Schuffenhauer A (2009) Estimation of synthetic accessibility score of drug like molecules based on molecular complexity and fragment contributions. J Cheminform 1:8
85. Melvin JYu (2011) Natural product like virtual libraries: Recursive atom based enumeration. J Chem Inf Model 51:541–557
86. Yang H, Jiang Z, Shi S (2006) Aromatic compounds biodegradation under anaerobic conditions and their QSBR models. Sci Total Environ 358:265–76
87. http://www.epa.ie/whatwedo/monitoring/reach/. Accessed 11 July 2012
88. Zhang X, Brown TN, Wania F (2010) Assessment of chemical screening outcomes based on different partitioning property estimation methods. Environ Int 36:514–20
89. Qiu X, Zhu T, Wang FHJ (2008) Air water gas exchange of organochlorine pesticides in Taihu lake China. Environ Sci Technol 42: 1928–32

Chapter 9

Accessing and Using Chemical Property Databases

Janna Hastings, Zara Josephs, and Christoph Steinbeck

Abstract

Chemical compounds participate in all the processes of life. Understanding the complex interactions of small molecules such as metabolites and drugs and the biological macromolecules that consume and produce them is key to gaining a wider understanding in a systemic context. Chemical property databases collect information on the biological effects and physicochemical properties of chemical entities. Accessing and using such databases is key to understanding the chemistry of toxic molecules. In this chapter, we present methods to search, understand, download, and manipulate the wealth of information available in public chemical property databases, with particular focus on the database of Chemical Entities of Biological Interest (ChEBI).

Key words: Chemistry, Databases, Ontology, ChEBI, Chemical graph, Cheminformatics, Chemical properties, Structure search, Chemical nomenclature

1. Introduction

Small molecules are the chemical entities which are not directly encoded by the genome, but which nevertheless are essential participants in all the processes of life. They include many of the vitamins, minerals, and nutritional substances which we consume daily in the form of food, the bioactive content of most medicinal preparations we use for the treatment of diseases, the neurotransmitters which modulate our mood and experience, and the metabolites which are transformed and created in a multitude of complex pathways within our cells.

Understanding the complex interactions of small molecules such as metabolites and drugs with the biological macromolecules that consume and produce them is key to gaining a wider biological understanding in a systemic context, such as can enable the prediction of the therapeutic—and toxic—effects of novel chemical substances. Ultimately, all therapeutic and harmful effects of chemical substances depend on the constitution, shape, and chemical

Brad Reisfeld and Arthur N. Mayeno (eds.), *Computational Toxicology: Volume I*, Methods in Molecular Biology, vol. 929, DOI 10.1007/978-1-62703-050-2_9, © Springer Science+Business Media, LLC 2012

properties of the molecules which influence their interactions within the body. In order to study these effects, it is therefore crucial to access databases, which provide clear and accurate data on these aspects of chemical entities.

Chemical property databases collect information on the nature and properties of chemical entities. Accessing and using such databases is an essential tool in understanding the chemistry of toxic molecules. In this chapter, we present methods to search, understand, download, and manipulate the wealth of information available in public chemical property databases, with particular focus on the Chemical Entities of Biological Interest (ChEBI) database (1).

2. Materials

2.1. Describing Chemical Structures with Chemical Graphs

The shape and properties of chemical entities depends to a large extent on the molecular structure of the entity—that is, the arrangement of atoms and bonds. Chemical graphs are a way of representing these core elements in a concise formalism. The chemical (molecular) graph describes the atomic connectivity within a molecule by using labelled nodes for the atoms or groups within the molecule, and labelled edges for the (usually covalent) bonds between the atoms or groups (2).

The graph, strictly speaking, encodes only the constitution of molecules, i.e., their constituent atoms and bonds. However, the representational formalism is usually extended to include other information such as idealised 2D or 3D coordinates for the atoms, and bond order and type (single, double, triple, or aromatic). The chemical graph formalism is also accompanied by a standard for diagrammatic representation. Figure 1 illustrates an example of a chemical graph, 2D and 3D coordinates, and the corresponding 2D and 3D visualisations.

Hydrogen nodes, and the edges linking them to their nearest neighbouring atoms, are not explicitly displayed in Fig. 1, and carbon nodes, which form the edges of the illustration, are not explicitly labelled as such. These representational economies are due to the prevalence of carbon and hydrogen in organic chemistry, thus allowing these efficiencies, which have been introduced for clarity of depiction, and efficiency in storage. Hydrogen-suppressed graphs of this form are called skeleton graphs.

Chemical graphs simplify the complex nature of chemical entities and allow for powerful visualisations, which assist the work of both bench and computational chemists. They also enable many useful predictions to be made about those physical and chemical properties of molecules, which are based on connectivity. For these reasons, most chemical property databases make use of the chemical graph formalism for the storage of structural information about chemical entities.

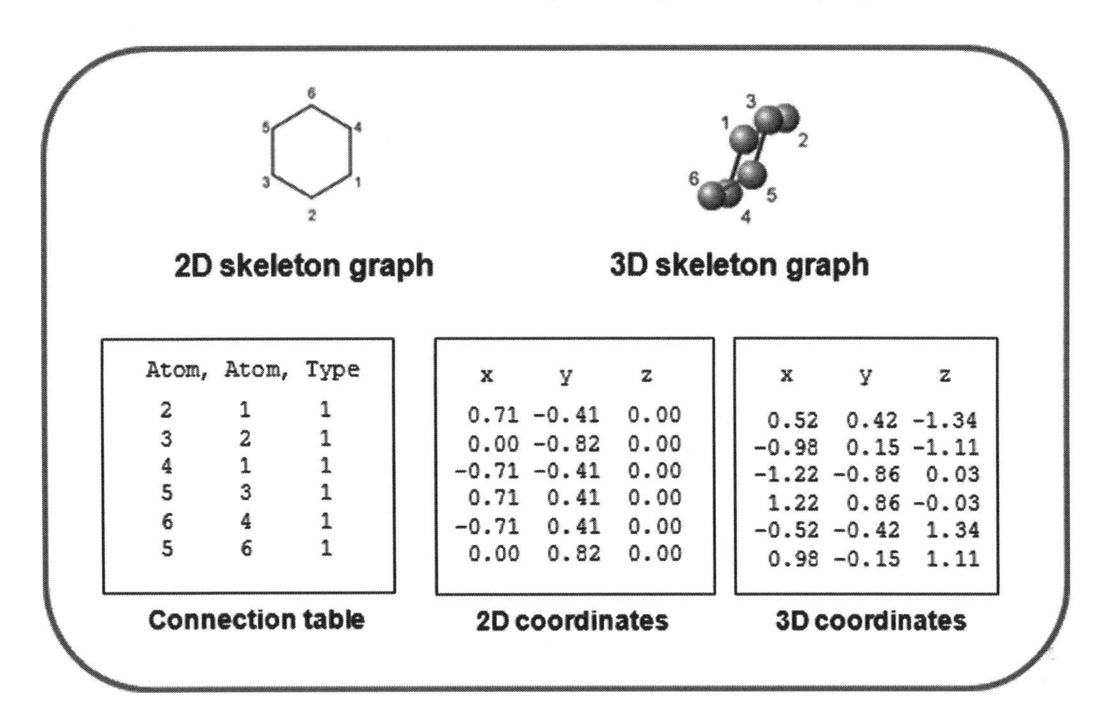

Fig. 1. The chemical graph (*connection table*) for a simple cyclohexane molecule, together with idealised 2D and 3D coordinates and visualisations (for the 3D coordinates and visualisation shows the chair conformer of cyclohexane).

2.2. Exchanging Data on Chemicals

The most common format used to encode and exchange chemical graphs is the MOLfile format owned by Elsevier's MDL (3). The MOLfile is a flat ASCII text file with a specific structured format, consisting of an atom table, describing the atoms contained in the chemical entity, and a bond table, describing the bonds between the atoms. Both the atom table and the bond table are extended with additional properties including the isotope and charge of the individual atoms, and the bond order and type of the bonds. The MOLfile representation of the cyclohexane molecule is illustrated in Fig. 2.

The content of a MOLfile depends on the way in which the chemical structure is drawn. For this reason, it is not possible to efficiently check whether two representations are of the same chemical based on the MOLfile, and therefore a canonical (i.e., the same regardless of which way the molecule is drawn) and unique representation of the molecule is needed for efficient identification. An international standard for identification of chemical entities is the IUPAC International Chemical Identifier (InChI) code. The InChI is a nonproprietary, structured textual identifier for chemical entities which is generated by an algorithm from a MOLfile representation of the chemical entity (4). The generated InChI identifier is not intended to be read and understood by humans, but is particularly useful for computational matching of chemical entities, such as when a computer has to align two different sets of data derived

```
  Marvin  10260612372D

  6  6  0  0  0  0            999 V2000
     0.7145   -0.4125    0.0000 C   0  0  0  0  0  0  0  0  0  0  0  0
     0.0000   -0.8250    0.0000 C   0  0  0  0  0  0  0  0  0  0  0  0
    -0.7145   -0.4125    0.0000 C   0  0  0  0  0  0  0  0  0  0  0  0
     0.7145    0.4125    0.0000 C   0  0  0  0  0  0  0  0  0  0  0  0
    -0.7145    0.4125    0.0000 C   0  0  0  0  0  0  0  0  0  0  0  0
     0.0000    0.8250    0.0000 C   0  0  0  0  0  0  0  0  0  0  0  0
  2  1  1  0  0  0  0
  3  2  1  0  0  0  0
  4  1  1  0  0  0  0
  5  3  1  0  0  0  0
  6  4  1  0  0  0  0
  5  6  1  0  0  0  0
M  END
```

Fig. 2. The MOLfile format for a cyclohexane molecule. The carbon atoms appear in the atom table and in the bond table the bonds between atoms are listed, with line numbers from the atom table representing the atoms participating in the bonds. Additional property columns allow for representation of many more features such as charge and stereochemistry.

from different sources. InChIs have variable lengths depending on the complexity of the chemical structure, and as a result, database lookups can be slow. The InChIKey is a hashed key for the InChI, which allows easier database lookups as it has fewer characters and is of a specified, invariant length of 14 characters. The InChI and InChIKey for paracetamol are as follows

$$InChI = 1/C8H9NO2/c1 - 6(10)9 - 7 - 2 - 4 - 8(11)5 - 3 \\ - 7/h2 - 5, 11H, 1H3, (H, 9, 10)/f/h9H$$

$$InChIKey = RZVAJINKPMORJF - BGGKNDAXCW$$

The InChI is an example of a line notation for chemical structures, expressing structural information on a single line. But the InChI is not intended to be read by humans. Another line notation for chemical structure representation which is somewhat easier for human parsing yet still providing a compact representation is the SMILES (Simplified Molecular Input Line Entry Specification) format representation. The original SMILES specification was developed by Arthur Weininger and David Weininger in the late 1980s. It has since been modified and extended by others, most notably by Daylight Chemical Information Systems, Inc. (5). The SMILES for paracetamol is

$$CC(=O)Nc1ccc(O)cc1$$

Note that Hydrogen atoms are implicit in SMILES (as they were in the example MOLfile representation for cyclohexane given above), and to further reduce the space used in the representation, single bonds are also implicit between neighbouring atoms. Atoms are numbered when necessary to illustrate where the same atom appears again (which is the case when the molecule contains a cycle). Brackets indicate branching, and "=" indicates a double rather than a single bond. Lowercase letters indicate aromaticity.

The SMILES representation retains a high degree of human readability while compressing the structural encoding of the molecule into as few characters as possible. However, different algorithms exist which produce different SMILES codes for the same molecule, so care must be taken to cite the implementation used when storing and comparing structures of molecules using SMILES.

2.3. Repositories of Chemical Property Data

Until fairly recently, the bulk of chemical data was only available through proprietary chemical databases such as Chemical Abstracts Service (CAS), provided by the American Chemical Society (6), and Beilstein which since 2007 has been provided by Elsevier (7). However, in recent years this has been changing, with more and more chemical data being brought into the public domain. This change has been brought about partly through the efforts of the bioinformatics community, which needed access to chemistry data to support systems-wide integrative research, and partly through the joint efforts of pharmaceutical companies to reduce the expense of pre-competitive research, since pharmaceutical companies had historically each maintained their own database of chemicals for pre-competitive research (8).

In 2004, two complementary open access databases were initiated by the bioinformatics community, ChEBI (1) and PubChem (9). PubChem serves as automated repository on the structures and biological activities of small molecules, containing 29 million compound structures. ChEBI is a manually annotated database of small molecules containing around 620,000 entities. Both resources provide chemical structures and additional useful information such as names and calculated chemical properties. ChEBI additionally provides a chemical ontology, in which a structure-based and role-based classification for the chemical entities is provided. Additional publicly available resources for chemistry information which are becoming more widely used are ChemSpider (10), which provides an integrated cheminformatics search platform across many publicly available databases, and the Wikipedia Chemistry pages (11). Many smaller databases also exist, often dedicated to particular topic areas or types of chemicals. For a full listing of publicly available chemistry data sources, see (12), and for a discussion see (13). Table 1 gives a brief listing of some of the publicly available databases for chemical structures and properties.

2.4. Software Libraries

Manipulating chemical data and generating chemical properties programmatically requires the use of a cheminformatics software library which is able to perform standard transformations and provides implementations for common algorithms. The Chemistry

Table 1
A listing of several chemical property databases which are publicly available (i.e., not commercial)

Name	Description	URL
ChEBI	A freely available database and ontology of chemical entities of biological interest	http://www.ebi.ac.uk/chebi
PubChem	A deposition-supplied database of publicly available chemical entities and bioactivity assays	http://pubchem.ncbi.nlm.nih.gov/
DrugBank	A database collecting information about drugs, drug targets, and drug active ingredients	http://www.drugbank.ca/
Spectral DB for organic compounds (SDBS)	A database for 1 H NMR and 13 C NMR spectra; also FT-IR, Raman, ESR, and MS data	http://riodb01.ibase.aist.go.jp/sdbs/cgi-bin/cre_index.cgi
Nmr spectra DB	A resource dedicated to biomolecules (proteins, nucleic acids, etc.), providing raw spectral data	http://www.bmrb.wisc.edu/
NmrShift DB	Organic structures and their NMR (H-1 and C-13) spectra	http://www.ebi.ac.uk/nmrshiftdb/
MassBank	Mass spectral data of metabolites	http://www.massbank.jp/index.html?lang = en
caNanolab	A portal for accessing data on nanoparticles important in cancer research	http://nano.cancer.gov/collaborate/data/cananolab.asp
Japan Chemical Substance Dictionary (JCSD, Nikkaji web)	Structure, synonyms, systematic names, and MOL files for wide variety of chemical entities	http://nikkajiweb.jst.go.jp/nikkaji_web/pages/top_e.jsp
GlycosuiteDB	A curated glycan database	http://glycosuitedb.expasy.org/glycosuite/glycodb
Cambridge Structural DB	From the CCDC—crystal structure data	http://www.ccdc.cam.ac.uk/products/csd/
Chemexper chemical directory	Gives structure and mol files for a large number of entities, as well as seller info	http://www.chemexper.com/
ChemicalBook	Structure, CAS no., mol file, synonym list, loads of physicochemical data (melting, boiling points, etc.)	http://www.chemicalbook.com/
IUPAC gold book	Useful reference work for terminology and definitions	http://goldbook.iupac.org/
ChemBlink	An online database of chemicals from around the world. Gives molecular structure and formula, CAS no., hyperlinks to lists of suppliers and market analysis reports	http://www.chemblink.com/index.htm

Development Kit (CDK) is a collaboratively developed open source software library providing implementations for many of the common cheminformatics problems (14, 15). These range from Quantitative Structure–Activity Relationship descriptor calculations to 2D and 3D model building, input and output in different formats, SMILES parsing and generation, ring searches, isomorphism checking, and structure diagram generation. It is written in Java and forms the backbone of a growing number of cheminformatics and bioinformatics applications.

When interacting with publicly available chemical property databases, it is often necessary to draw or upload a chemical structure for the purpose of searching. For this purpose, it will be necessary to make use of a chemical structure editor, usually embedded into the website of the chemical database. JChemPaint is one such editor, a freely available open source Java-based 2D editor for molecular structures (16). It provides most of the standard drawing features of commercial chemical structure editors, including bonds of different orders, stereodescriptors, structure templates with complex structures, atom type selection from the periodic table, rotation and zooming of the image, selection and deletion of parts of the structure, and input and output of chemical structure in various common formats including SMILES and InChI.

3. Methods

3.1. Searching for Entities

3.1.1. Searching Using Text and Property Values

The first entry point in accessing chemical data is the search interface provided by the database one is accessing. Generally, such databases provide a "simple" search as a first entry point, which takes the search string and searches across all data fields in the database to bring back candidate hits. For example, ChEBI provides a search box in the centre of the ChEBI front page and in the top right of every entry page. The search query may be any data associated with an entity, such as names, synonyms, formulae, CAS or Beilstein Registry numbers, or InChIs.

When searching any database with free text, it is important to know the wildcard character, which allows one to match part of a word or phrase. For the ChEBI database, the wildcard character is "*." A wildcard character allows one to find compounds by typing in a partial name. The search engine will then try to find names matching the pattern one has specified.

- To match words *starting with* a search term, add the wildcard character to the end of the search term. For example, searching for *aceto** will find compounds such as *acetochlor, acetophenazine,* and *acetophenazine maleate.*

- To match words *ending with* a search term, add the wildcard character to the start of the search term. For example, searching for **azine* will find compounds such as *2-(pentaprenyloxy)dihydrophenazine*, *acetophenazine*, and *4-(ethylamino)-2-hydroxy-6-(isopropylamino)-1,3,5-triazine*.
- To match words *containing* a search term, add the wildcard character to the start and the end of the search term. For example, searching for **propyl** will find compounds such as *(R)-2-hydroxypropyl-CoM*, *2-isopropylmaleic acid*, and *2-methyl-1-hydroxypropyl-TPP*.
- Any number of wildcard characters may be used *within* a search term, thus making the search facility very powerful.

Even using wildcards to extend the search capability, simple text searching in chemistry databases can be problematic. Many of the interesting properties for search purposes are structural or numeric and therefore not easily expressed in text. To this end, most chemistry databases provide a chemistry-specific advanced search interface which allows for searches based on chemical properties such as mass and formula and, for example, value ranges for properties such as charge. In ChEBI, the advanced text search provides for additional granularity by allowing one to specify which category to search in, as well as providing the option of using Boolean operations when searching. Figure 3 illustrates the ChEBI advanced search page with search options, using which the user may combine multiple search criteria.

The user has the option to search by mass range (i.e., to find compounds within a certain minimum and maximum mass), charge

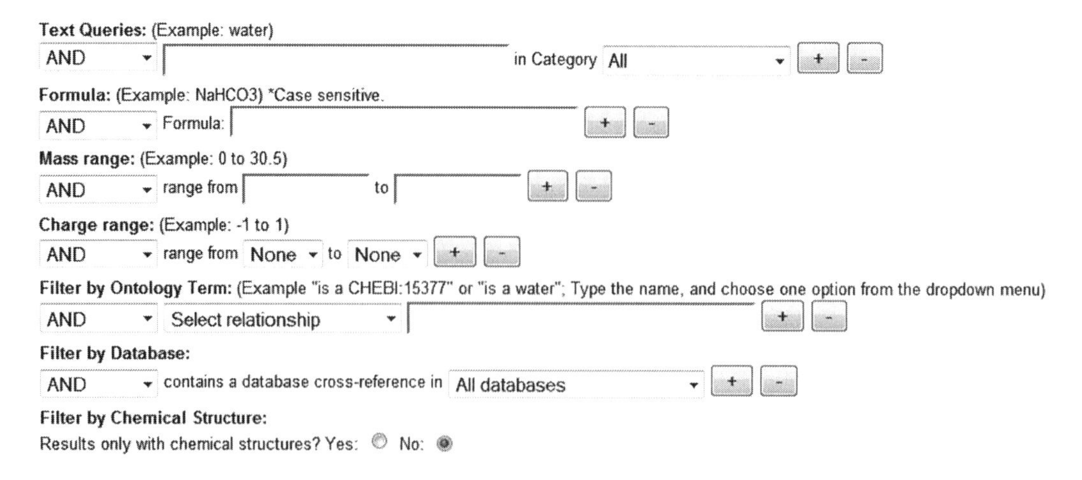

Fig. 3. The ChEBI advanced search showing textual and chemical property-based search options. Multiple search options may be specified and combined with the logical operators AND, OR, and BUT NOT. Text and property-based searches may be combined with chemical structure searches.

range, chemical formula, and to combine these searches based on Boolean operators. The standard Boolean operators are as follows.

- *AND*
  This operator allows one to find a compound which contains all of the specified search terms. For example, when searching for a pyruvic acid with formula C5H6O4, specifying **pyruvic acid C5H6O4* as the search term and selecting AND as the search option will retrieve *acetylpyruvic acid*.
- *OR*
  This operator allows one to type two or more words. It then tries to find a compound which contains at least any one of these words. For example, if one wanted to find all compounds containing iron in the database, one could type in the search string *iron fe Fe2 Fe3*.
- *BUT NOT*
  Sometimes, common words can be a problem when searching, as they can provide too many results. The "BUT NOT" operator can be used to limit the result set. For example, if one was looking for a compound related to chlorine but excluding acidic compounds, one could specify **chlor* as the search string but qualify the search by specifying *acid* in the BUT NOT operator.

3.1.2. Chemical Structure Searching

An important search method to master in chemical databases is searching based on chemical structures. Structure searching is the method by which a user is able to specify a chemical structure and thereafter search through the database of chemical structures for the search structure or similar structures. Different forms of structure searching include identity search, in which the exact entity drawn is searched for in the database (which is useful if the name of the entity is not known), substructure search, in which the entity drawn is searched for as a wholly contained part of the structures in the database, and similarity search, in which the entity drawn is matched for similarity of features with the structures in the database.

In order to perform a chemical structure search in ChEBI, the structure must be provided via the JChemPaint applet embedded in the structure search screen. The ChEBI structure search interface is illustrated in Fig. 4.

Once the search is executed, a results page is shown with matching hits. Clicking on the relevant ChEBI accession hyperlinked under the search result image navigates to the entry page for that entity. In addition, further structure-based searches may be performed by hovering the mouse pointer over the displayed image in the results grid and clicking on one of the resulting popup search options. This is a shortcut which passes the illustrated structure directly to the search facility.

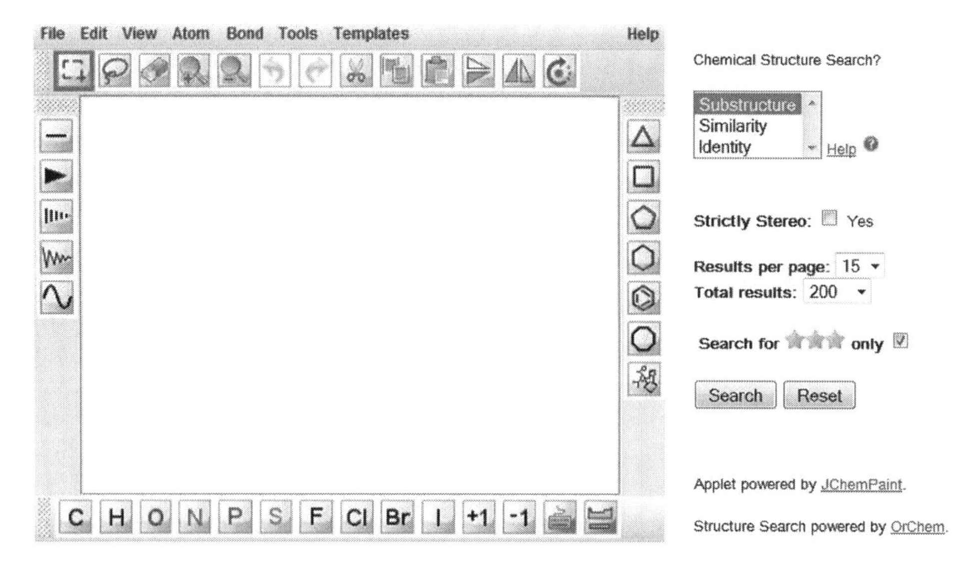

Fig. 4. The JChemPaint applet for drawing chemical structures inside of web pages for performing chemical structure-based searches. The bond selection utility is *on the left* and the atom selection *along the bottom. The top menu* includes file tools and utilities, and *along the right hand menu* are common structures and an extended structure template library. Structure search types are "Identity," "Substructure," and "Similarity".

In ChEBI, structure searching is executed on the database in the background using the ORCHEM library (17). When performing identity searching, the search is based on the InChI, which means that an InChI is generated from the drawn or uploaded structure, and the database is then searched for exact matches to that InChI, a straightforward string match. The substructure and similarity search options are both based on *chemical fingerprints.*

A fingerprint of a chemical structure is a way of representing special characteristics of that structure in an easily searchable form. Fingerprints are necessary because the problem of finding whether a given chemical structure is a substructure of or is similar to, another structure, is a computationally very expensive problem. In fact, in the worst case, the time taken will increase exponentially with the number of atoms. This makes running these searches across whole databases almost completely intractable. Fortunately, a general (and intractable) substructure or similarity search algorithm does not have to be used across the full database, since various different heuristics can be used to drastically narrow the number of candidates for the algorithm to be applied to. For example, the chemical formula could be used as one such heuristic. Consider a search for all structures which have paracetamol as a substructure. The chemical formula of paracetamol is $C8H9NO2$. This means that we could immediately eliminate from the search candidates any structures which don't contain *at least* those quantities of carbon, hydrogen, nitrogen, and oxygen. This simple heuristic acts as a screen which cuts down the number of structures required to perform the full substructure search against, by a large percentage.

Fingerprints are designed to operate similarly as a screening device, however, they are more generalised and abstract, allowing more information about the structure to be encoded, thus eliminating a far larger percentage of search candidates. A fingerprint is a boolean array (i.e. an array of 1s and 0s), or bitmap, in which the characteristic features of structural patterns are encoded. Fingerprints are created by an algorithm which generate a pattern for

- each atom in the structure,
- then each atom and its nearest neighbours, including the bonds between them,
- then each group of atoms connected by paths up to two bonds long,
- ... continuing, with paths of lengths 3, 4, 5, 6, 7, and 8 bonds long.

For example, the water molecule generates the following patterns.

water (HOH)

0-bond paths H O H

1-bond paths HO OH

2-bond paths HOH

Every pattern in the molecule up to 8 bonds length is generated. Each generated pattern is hashed to create a bit string output, then, to create the final bitmap representation of the fingerprint, the individual bit strings are added together (using the logical OR operation) to create the final 1,024 bit fingerprint. Table 2 shows how individual structural patterns might look for the example of water, assuming for purposes of illustration that we have only a 10-bit result to create the fingerprint.

Since the final result represents "infinitely" many structural possibilities (as there are infinitely many chemical structures, at least in theory) in a fixed length bitmap, it is inevitable that collisions will occur—bits may be set already when they appear in a subsequent pattern. Thus, fingerprints do not uniquely represent chemical structures. However, fingerprints do have the very useful property that every bit set in the fingerprint of a substructure of a given structure, will also be set in the fingerprint of the full structure. So, for example, the fingerprint pattern for the water substructure hydroxide (OH), which is a part of the full water molecule, would be computed as shown in Table 3. By comparing the fingerprint for hydroxide with the fingerprint for water given in Table 2, it is easy to see that every bit set in the hydroxide fingerprint is also set in the water fingerprint.

Table 2
Bitmaps for individual structural patterns and the resulting combined fingerprint for a water molecule in a simplified fingerprinting example which consists only of a bitmap of length 10

Pattern	Hashed bitmap
H	0000010000
O	0010000000
HO	1010000000
OH	0000100010
HOH	0000000101
Result:	1010110111

Table 3
Bitmaps for individual structural patterns and the resulting combined fingerprint for hydroxide, which is a substructure of the water molecule

Pattern	Hashed bitmap
H	0000010000
O	0010000000
OH	0000100010
Result:	0010110010

For substructure searching, fingerprints are used as effective screening devices to narrow the set of candidates for a full substructure search. If all bits in a query fingerprint are also present in the target fingerprint of a stored database structure, this structure is subjected to the computationally expensive subgraph matching algorithm. Those bit operations are very fast and independent of the number of atoms in a structure due to the fixed length of the fingerprint. For similarity searching, fingerprints are used as input to the calculation of the similarity of two molecules in the form of the Tanimoto coefficient, a measure of similarity. It is important to remember that fingerprints have limitations: they are good at indicating that a particular structural feature is *not* present, but they can only indicate a structural feature's presence with some probability.

Table 4
The calculation of a Tanimoto similarity coefficient between two chemical structures as represented by fingerprints

Object	Fingerprint
Object A	0010110010
Object B	1010110001
c (bits on in both)	3
a (bits on in Obj A)	4
b (bits on in Obj B)	5
Tanimoto (c/(a + b-c))	3/(4 + 5-3) = 0.5

The Tanimoto coefficient between two fingerprints is calculated as the ratio

$$T(a, b) = \frac{c}{a + b - c}$$

where c is the count of bits "on" (i.e., 1 not 0) in the same position in both the two fingerprints, a is the count of bits on in object A, and b is the count of bits on in object B. An example of a Tanimoto calculation of similarity between two fingerprints is given in Table 4.

The Tanimoto coefficient varies in the range 0.0–1.0, with a score of 1.0 indicating that the two structures are very similar (i.e., their fingerprints are the same). This is represented on the search page in terms of a percentage of similarity. The percentage cutoff for a similarity search may be specified, either at 70% (the default), higher at 80% or 90%, or lower at 50% and (the minimum) 25%.

For all of the structure-based searches in ChEBI, the user has the option to select "explicit stereo" to include explicit stereochemistry in the search term. Additionally, the user has the option to select more results than the default 200 and to determine the pagination size of the results page (default 15 results per page).

3.2. Viewing a Typical Database Entry

After a successful search for an entity in a chemical property database, the user will be presented with an entry page for that entity. The entry page usually displays a visual representation of the structure of the entity together with useful information such as names and synonyms, calculated properties, and classification information.

ChEBI database entries contain as follows

- A unique, unambiguous, recommended ChEBI name and an associated stable unique identifier

- An illustration of the chemical structure where appropriate (compounds and groups, but generally not classes), as well as secondary structures such as InChI and related chemical data such as formula
- A definition, giving an important textual description of the entity
- A collection of synonyms, including the IUPAC recommended name for the entity where appropriate, and brand names and INNs for drugs
- A collection of cross-references to other databases (where these are sourced from nonproprietary origins)
- Links to the ChEBI ontology
- Citation information where the chemical has been cited in publication

ChEBI entities are illustrated by means of a diagram of the chemical structure. Following best practices, the default illustration is an unambiguous, two-dimensional representation of the structure of the entity. Additional structures may be present; for example, a three-dimensional structure. Where available, this can be accessed by clicking on "More structures >>" on the main entity page. Structures are stored as MOLfiles within ChEBI, and other structural representations such as InChI are also provided.

The chemical structure can be interactively explored by activating the ChemAxon's MarvinView applet (http://www.chemaxon.com) by selecting the "applet" checkbox next to the structure image. MarvinView allows the user to quickly and easily control many aspects of the display, such as molecule display format, colour scheme, dimension, and point of view. It is possible to move the displayed molecule by translation, dragging, changing the zoom, or rotating in three dimensions. It is also possible to animate the display. The format of the display can be switched among common formats such as wireframe, ball and stick, and spacefill. The colour scheme can be changed. Also, the display of implicit and explicit hydrogens in the image can be altered.

The entry page collates names and synonyms. Chemical names may be trivial or systematic. In systematic names, the name encodes structural features, and indeed, a fully specified systematic name can be automatically transformed into a structural representation. Trivial names are often assigned in common use and supersede the systematic name for communication purposes, since trivial names are easier to pronounce and write. Table 5 shows some examples of trivial and systematic names for chemical entities.

The entry page also collects together database cross-references and citations. Cross-references to other databases allow navigation of the broader knowledge space pertaining to that entity. In particular, ChEBI provides many links of biological relevance, allowing

Table 5
Some examples of trivial and systematic names for ChEBI entities

ChEBI ID	Trivial name	Systematic name
CHEBI:16285	Phytanic acid	3,7,11,15-Tetramethylhexadecanoic acid
CHEBI:17992	Sucrose	β-D-Fructofuranosyl α-D-glucopyranoside
CHEBI:16494	Lipoic acid	5-(1,2-Dithiolan-3-yl)pentanoic acid
CHEBI:15756	Palmitic acid	Hexadecanoic acid

Trivial names are generally shorter and easier to remember and pronounce than their systematic counterparts. However, systematic names are more informative with respect to the chemical structural information

browsing to references to chemical entities appearing in diverse biological databases. Citations are to literature resources in which the entity, or significant properties of the entity, is described.

3.3. Understanding Chemical Classification

Annotation of data is essential for capturing and transmitting the knowledge associated with data in databases. Annotations are often captured in the form of free text, which is easy for a human audience to read and understand, but is difficult for computers to parse; can vary in quality from database to database; and can use different terminology to mean the same thing (even within the same database, if for example different human annotators used different terminology). A core structure for the organisation of terminologies used in annotation is into an *ontology*, which consists of a structured vocabulary with explicit semantics attached to the relationships between the terms.

ChEBI provides such an ontology in the domain of chemistry, in which are organised the terms describing structure-based chemical classes in a structure-based hierarchy, and function-based classes including terms for bioactivity in a function-based hierarchy.

The ChEBI ontology is an ontology for biologically interesting chemistry. It consists of three sub-ontologies, namely

Chemical entity, in which molecular entities or parts thereof and chemical substances are classified according to their structural features;

Role, in which entities are classified on the basis of their role within a biological context, e.g., as antibiotics, antiviral agents, coenzymes, enzyme inhibitors, or on the basis of their intended use by humans, e.g. as pesticides, detergents, healthcare products, and fuel;

Subatomic Particle, in which are classified particles which are smaller than atoms.

Chemical entities are linked to their structure-based classification with the ontology "is a" relationship and to the roles or bioactivities which they are known to exhibit under the relevant circumstances with the "has role" relationship.

Molecular entities with defined connectivity are classified under the *chemical entity* sub-ontology. These include the chemical compounds which themselves could exist in some form in the real world, such as in drug formulations, insecticides, or different alcoholic beverages. In addition, classes of molecular entities are classified under the *chemical entity* sub-ontology. Classes may be structurally defined, but do not represent a single structural definition; rather a generalisation of the structural features which all members of that class share. It is often useful to define the interesting parts of molecular entities as *groups*. Groups have a defined connectivity with one or more specified attachment points.

The *role* sub-ontology is further divided into three distinct types of role, namely biological role, chemical role, and application. Roles do not themselves have chemical structures, but rather it is the case that items in the *role* ontology are linked to the molecular entities which have those roles.

The ChEBI ontology uses two generic ontology relationships, namely

- *Is a*. Entity A is an instance of Entity B. For example, chloroform *is a* chloromethanes.
- *Has part*. Indicates relationship between part and whole, for example, potassium tetracyanonickelate(2−) *has part* tetracyanonickelate(2−).

In addition, the ChEBI ontology contains several chemistry-specific relationships which are used to convey additional semantic information about the entities in the ontology. These are as follows.

- *Is conjugate base of and is conjugate acid of*. Cyclic relationships used to connect acids with their conjugate bases, for example, pyruvic acid is the conjugate acid of the pyruvate anion, while pyruvate is the conjugate base of the acid.
- *Is tautomer of*. Cyclic relationship used to show relationship between two tautomers, for example, L-serine and its zwitterion are tautomers.
- *Is enantiomer of*. Cyclic relationship used in instances when two entities are mirror images and nonsuperimposable upon each other. For example, D-alanine is enantiomer of L-alanine and vice versa.
- *Has functional parent*. Denotes the relationship between two molecular entities or classes, one of which possesses one or more characteristic groups from which the other can be derived by functional modification. For example, 16α-hydroxyprogesterone

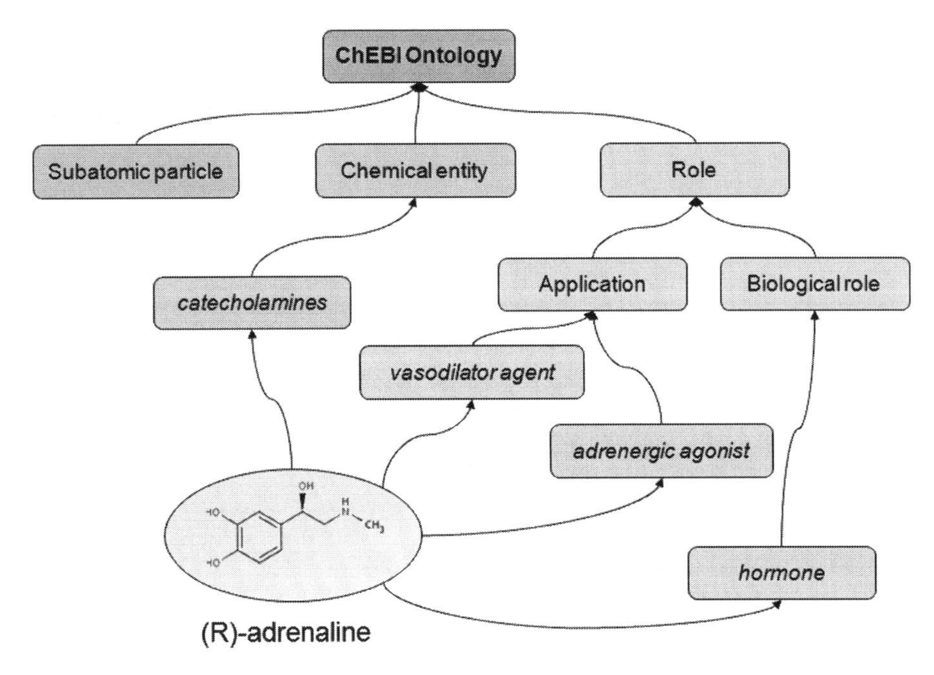

Fig. 5. A high-level illustration of the ChEBI ontology for the entity *R*-adrenaline. The entity is given a structure-based classification as "catecholamines" in the chemical entity ontology, and role classification both in terms of an application which the entity is used for, "vasodilator agent," and in terms of the biological role, "hormone".

can be derived by functional modification (i.e., 16α-hydroxylation) of progesterone.

- *Has parent hydride.* Denotes the relationship between an entity and its parent hydride, for example, 1,4-napthoquinone has parent hydride naphthalene.
- *Is substituent group from.* Indicates the relationship between a substituent group/atom and its parent molecular entity, for example, the L-valino group is derived by a proton loss from the N atom of L-valine.
- *Has role.* Denotes the relationship between a molecular entity and the particular behaviour which the entity may exhibit either by nature or by human application, for example, morphine *has role* opioid analgesic.

Figure 5 gives a high-level illustration of the ChEBI ontology for *R*-adrenaline.

The ChEBI ontology may be browsed online by navigating between entities using the links in the ChEBI ontology section of the main entry page. To view all the paths from the entry to the root, select the "Tree view" option on this screen. On the other hand, to retrieve all terms which hold a specified relationship to another term, the Advanced Search page provides an ontology filter. For example, to find all carboxylic acids in ChEBI, go to the

Advanced Search, scroll down to the "Ontology Filter" selection, and enter the ChEBI ID for carboxylic acid in the filter input box: CHEBI:33575, and then click "Search." The Advanced Search ontology filter can also be used to retrieve all chemical entities which participate in a particular role, even though such entities may have vastly different structures. For example, searching for all entities which have the *has role* relationship to the role "vasodilator agent" retrieves 47 chemical entities (in the current release) including convallatoxin and amlodipine.

3.4. Programmatically Accessing and Manipulating Chemical Data

While web interfaces provide an easy point of entry for explorative research surrounding the data in chemical property databases and allow access to powerful searches, ultimately in order to drive any large-scale computational research it is important to be able to download the data, in a computationally accessible format, to the local machine. The primary format for downloading chemical data is the SDF format, which consists of a collection of MOLfile resources, together with custom properties.

In ChEBI, targeted results of chemical searches can be downloaded directly from the search results page by clicking the icon entitled "Export your search results," SDF file format. On the other hand, the full database can be downloaded directly from the Downloads page available at http://www.ebi.ac.uk/chebi/downloadsForward.do. To download the chemical structures (and structure-related properties), an SDF file download is available. The data is provided in two flavours as follows.

- Chebi_lite.sdf file contains only the chemical structure, ChEBI identifier, and ChEBI Name.
- Chebi_complete.sdf file contains all the chemical structures and associated information. Note that it excludes any ontological information as, since they do not contain a structure, ontological classes are not able to be represented. To download the ontology classification, the popular ontology formats OBO (18) and OWL (19) are available.

Once one has an SDF file locally on the machine, several publicly available tools facilitate manipulation of the file data, for example, the Chemistry Development Kit. Figure 6 shows a brief snippet of code necessary to extract data from an SDF file using version 1.0.2 of the CDK.

An alternative to parsing and interpreting the full downloadable data file is to make use of the web service facility to programmatically retrieve exactly the entries needed in real time. ChEBI provides a web service implemented in SOAP (Simple Object Access Protocol) (http://www.w3.org/TR/soap/), which allows programmatic retrieval of all the data in the database based on a search facility which mimics the search interface available online. The WSDL

```
String filename = "input.sdf"; // a multiple molecule SDF file
InputStream ins = this.getClass().getClassLoader().getResourceAsStream(filename);
MDLV2000Reader reader = new MDLV2000Reader(ins);
ChemFile fileContents = (ChemFile)reader.read(new ChemFile());

IChemSequence sequence = fileContents.getChemSequence(0);

for (int i=0; i<sequence.getChemModelCount(); i++) {
      IChemModel model = sequence.getChemModel(i);
      org.openscience.cdk.interfaces.IMoleculeSet som = model.getMoleculeSet();

      for (int j=0; j<som.getMoleculeCount(); j++) {
            org.openscience.cdk.interfaces.IMolecule m = som.getMolecule(j);


      }
}
```

Fig. 6. A code snippet used to parse and extract the data from an SDF file using the Chemistry Development Kit.

(Web Services Description Language) XML document describing the ChEBI web service implementation is available at http://www.ebi.ac.uk/webservices/chebi/2.0/webservice?wsdl and contains the specification of the methods available for programmatic invocation by clients connected to the ChEBI web service, as well as details of where the service is running. The WSDL can be used to generate client applications for various different programming languages. ChEBI provides such clients in Java and Perl for ease of use. The clients are available for download from http://www.ebi.ac.uk/chebi/webServices.do.

In the ChEBI web service, there are seven methods provided with which to access data:

- *getLiteEntity*
- *getCompleteEntity* and *getCompleteEntityByList*
- *getOntologyParents*
- *getOntologyChildren* and *getAllOntologyChildrenInPath*
- *getStructureSearch*

The search method *getLiteEntity* retrieves a *LiteEntityList* and takes as parameters a search string and a search category (which may be null to search across all categories). The *getCompleteEntity* method retrieves the full data for a specified entity including synonyms, database links, and structures and takes as parameter a ChEBI identifier. The *getCompleteEntityByList* is similar but retrieves full data for a specified list of multiple ChEBI identifiers rather than a single one. To browse the ontology programmatically, there is a *getOntologyParents* method, which retrieves the parents of the given entity (specified by ChEBI identifier) in the ChEBI ontology, a *getOntologyChildren* method, which retrieves the children of the given entity (specified by ChEBI identifier), and a *getAllOntologyChildrenInPath* method, which returns all entities found when navigating down the hierarchy of the ontology to the leaf terms. Finally, the *getStructureSearch* method allows programmatic access to chemical structure searching.

The *getLiteEntity* search method allows the specification of a search string and a search category. Allowed search categories are (available as the *SearchCategory* enumeration in the domain model)

- ALL
- CHEBI ID
- CHEBI NAME
- SYNONYM
- IUPAC NAME
- DATABASE LINK
- FORMULA
- REGISTRY NUMBER
- COMMENT
- INCHI
- SMILES

The search method returns a *LiteEntityList*, which may contain many *LiteEntity*s. For each *LiteEntity* contained in the list, the ChEBI ID may then be used to retrieve the full dataset by passing it as a parameter to the *getCompleteEntity* method. The *Entity* object which is then returned contains the full ChEBI dataset linked to that identifier, including structures, database links and registry numbers, formulae, names, synonyms, and parent and children ontology relationships.

Navigating the ontology, without retrieving a complete Entity for each data item in the ontology, is accomplished by using the methods *getOntologyParents* (for navigating up towards the root) and *getOntologyChildren* (for navigating downwards towards the leaves). Note that an *OntologyDataItem* represents a relationship, specifying the ChEBI ID of the related term and the type of the relationship. The ChEBI ID can then be used to access the complete *Entity* if required. Some relationship types are cyclic, in this case the flag cyclicRelationship will be set to true. Cyclic relationships should be ignored for purposes of navigation or, if navigating them is required, care should be taken to access them only once.

By using the web service to access the ChEBI data, details of the underlying database model are not required to be known, and the user need not worry about updates to the underlying database model which do not affect the web service object model. This makes using web services a very user-friendly mechanism for programmatically accessing chemical databases.

4. Examples

4.1. Understanding Ranking in a Simple Search for Mercury (CHEBI:16170)

Heavy metals—trace metals at least five times denser than water—are generally systemic toxins with neurotoxic, nephrotoxic, fetotoxic, and teratogenic effects (http://tuberose.com/Heavy_Metal_Toxicity.html). Mercury, one such metal, is a well-known environmental toxin. Its most common organic form is methylmercury, to which exposure occurs through consumption of seafood (20). Exposure to elemental metallic mercury occurs from dental amalgam restorations (21). Also found in batteries, fluorescent and mercury lamps, vaccines and dental amalgams, its effects include pulmonary, cutaneous and neurological disorders as well as immunological glomerular disease (22).

When searching ChEBI for "mercury" using the simple search box, the search retrieves (at the current release) 35 search results. These results include all the entities in which the word "mercury" appears in any associated data field. The results are *ranked*, that is, sorted by relevance, depending on where the search term appeared in the searched entity. A search hit in the primary name, for example, is promoted (receives a higher search score) above a search hit in the cross-references or other associated data. The first five results include all the forms of pure mercury in ChEBI, while the remainder of the search results include molecules and salts containing mercury as part of a larger species. This ranking is designed to assist the user in discovering the most relevant information more easily.

4.2. Browsing the Cross-References Associated with Sulphur Dioxide (ChEBI:18422)

Sulphur dioxide (SO_2) is a poisonous gas produced by volcanoes, in the combustion of fossil fuels and in various industrial processes. It is used as a bleaching agent and, due to its germicidal and insecticidal properties, as a disinfectant and pesticide (http://en.wikipedia.org/wiki/Sulfur_dioxide). Its preservative and antioxidant properties make it useful in the food and drink industries (23). Exposure to high levels of SO_2 can lead to headache, acute pharyngitis, acute airway problems, and asthma (24).

In ChEBI, sulphur dioxide has a standard entry page with a structural representation, a collection of names and synonyms, an ontology classification, and various database links and cross-references. As we mentioned before, the cross-references, both manually annotated (on the main entry page) and automatically collected (on the separate "Automatic Xrefs" tab), provide a way to navigate from a chemical entity into the wider space of biological and chemical knowledge about that chemical. Sulphur dioxide is manually cross-referenced to the KEGG COMPOUND resource (25) and the PDBeChem resource (26). It is also automatically cross-referenced to several resources. Collection of automatic cross-references in ChEBI is often driven by the partner resource performing annotation of ChEBI IDs to their data records. These annotations are then collected together and passed back to ChEBI, together with the IDs and links to the partner resource. Such resources, linked to from sulphur dioxide, include the UniProt knowledge base of proteins (27), the NMRShiftDB database for organic structures and their nuclear magnetic resonance (NMR) spectra (28), and the SABIO-RK database for reaction kinetics (29).

4.3. Exploring the Ontology Structural Relationships with Phenol (ChEBI:15882)

Phenol is an aromatic alcohol with the chemical formula C_6H_5OH, used commonly in molecular biology as an organic solvent in bacterial plasmid extraction protocols. It is also used in medical treatments, petroleum refineries and the production of glue, fibre, and nylon (30). Although only mildly acidic, it can cause burns (31) and is also associated with cardiac dysfunction in certain contexts (32).

The ontology for phenol contains several structural relationships (conjugate base/acid; has functional parent).

4.4. Exploring the Role Ontology with Bafilomycin A (ChEBI:22689)

Bafilomycin A1 is the most commonly used of the bafilomycins, a family of toxic macrolide antibiotics derived from *Streptomyces griseus*. It has a wide range of biological actions, including antibacterial, antifungal, antineoplastic, immunosuppressive, and antimalarial activities, as well as a tendency to reduce multidrug resistance (33).

In ChEBI, several roles are annotated to this entity, including the biological role "toxin," due to its toxic properties, and the application "fungicide." By clicking the link to the target role in the ontology and selecting "tree view" in the ontology viewer, it is possible to view the role hierarchy for the role terms and to navigate to other chemicals which have been annotated with those roles.

4.5. Interacting with 2D and 3D Structural Representations of Fusicoccin (ChEBI:51015)

Fusicoccin is a polycyclic organic compound whose structure contains three fused carbon rings and another ring containing an oxygen atom and five carbons. A phytotoxin produced by the fungus *Fusicoccum amygdali*, it causes membrane hyperpolarisation and proton extrusion in plants (http://en.wikipedia.org/wiki/Fusicoccin). The toxin acts by stimulating the H + -ATPases of the guard cells, leading to plasma membrane hyperpolarisation and the irreversible opening of stomata. The ensuing wilting of the leaves is followed by cell death (34). Fusicoccin has also been shown to cause cell death in cultured cells of the sycamore (35).

When viewing the ChEBI entry page for Fusicoccin, the default structural illustration is the schematic 2D diagram, however, clicking "More structures" reveals a 3D structure associated with the entry page. Selecting the checkbox "applet" loads the MarvinView applet which allows for interactive exploration of the structure, including rotation in three dimensions.

4.6. Interacting with the ChEBI Web Service Using Ciguatoxin (ChEBI:36467)

Ciguatoxin is a cyclic polyether produced by the dinoflagellate *Gambierdiscus toxicus*. A neurotoxin, it is the agent of ciguatera shellfish poisoning (36, 37). Ciguatera, an important form of human poisoning caused by the consumption of seafood, is an important medical entity in tropical and subtropical Pacific and Indian Ocean regions and in the tropical Caribbean (38). It is characterised by gastrointestinal, neurological, and cardiovascular disturbances, which may lead to paralysis, coma, and death. Although it is the most widespread marine-borne disease affecting humans, there is no immunity, and the toxins are cumulative. Symptoms may persist for months or years or recur periodically.

One common task that a user may wish to perform when interacting with the ChEBI web service is performing a structure-based similarity search for structures similar to the structure of a given entity. Web services are accessed programmatically, but there is an interface provided for graphically invoking the web service methods at http://www.ebi.ac.uk/chebi/webServices.do. This interface allows the web service methods to be tested and the output examined, which can serve as input for designing and fine tuning the programmatic pipeline. For example, there is a table for executing the structure search method, allowing the input of a structure in MOLfile or SMILES format. By copying the SMILES from ciguatoxin into the search box and then executing a similarity search, the browser displays the XML corresponding to the SOAP of the web service response, which would be parsed automatically by the client library in a programmatic execution. A similarity search executed in this fashion with a 25% Tanimoto cutoff returns many results, of which okadaic acid (CHEBI:44658) is the most highly ranked at 84% similarity.

4.7. Linking to the Literature Involving PDC-E2 173–184 Peptide (ChEBI:60738)

The PDC-E2 173-184 peptide is the linear dodecapeptide sequence DKATIGFEVQEE corresponding to residues 173–184 of dihydrolipoyl transacetylase, the second (E2) of three enzymes in the pyruvate dehydrogenase enzyme complex (PDC) (http://en.wikipedia.org/wiki/Pyruvate_dehydrogenase_complex). In the autoimmune disease primary biliary cirrhosis (PBC), the major autoepitope recognised by both T and B cells lies in the inner lipoyl domain of PDC-E2, which contains the PDC-E2 173–184 peptide (39). Lipoylated in vivo at the lysine (K) residue, this dodecapeptide has been shown to act as an autoepitope, eliciting autoantibody formation in the sera of PBC patients (40).

When browsing the ChEBI entry for PDC-E2 173–184 peptide, notice that the entity is referenced to the literature via the citations feature. This provides a direct link from the database record to relevant publications in which properties of the chemical entity have been described. Sometimes, the only way to establish the complex properties of chemical entities is to glean them from the primary literature in this fashion.

5. Notes

5.1. Cross-Database Integration and Chemical Identity

When searching public chemical databases and, in particular, when searching across multiple databases with a view to collecting as complete a view as possible of the available information for a particular chemical entity, redundancy of chemical information is a large problem, as multiple records proliferate across public databases. PubChem is a database which accepts depositions of chemical data from a vast array of public databases and thereafter performs structure-based integration to provide a unified compound detail page for each separately identifiable chemical entity. Ideally, the user would like to see one record for each distinct chemical entity and only one such record.

However, the process of deciphering, which chemical entities in different data sources are the same and which are not, is a challenging one which is not yet fully solved by the wider community. The InChI code is intended to solve the problem by providing a standard identifier which can be used for data integration. It goes a long way towards achieving this goal, but falls short of a full solution for several reasons, each of which is active areas of discussion and research within the chemical database community.

One problem with the InChI as a tool for solving all the data integration issues facing the public compound databases is that, although it is intended to resolve to the same code regardless of the way in which a chemical entity is drawn, but unfortunately, the InChI cannot resolve differences in what is depicted where differences in drawing represent real chemical distinctions. For example, salts may be depicted in different ways in chemical drawing software. One possibility is to depict the molecular structure of the salt as a bonded whole with the ionised bond drawn as charge separated. Another possibility is to draw the two ions as distinct units, disconnected from a graph perspective, with explicit charges. Yet another possibility includes several copies of the charged ions in order to illustrate the final quantitative ratio of the different ions within the salt. Each of these depictions results—inevitably—in a different InChI code, rendering programmatic unification more difficult.

Another reason that the InChI is not able to fully resolve data integration issues across different resources is that the basic unit of identity for a chemical entity is regarded differently in different resources where there is a different application context. For example, a chemical entity database such as ChEBI regards different salts as different chemical entities, while a drug database such as DrugBank regards all different salts of the same active ingredient as the same drug entity. Integration between such resources is a matter of maintaining one-to-many, and in some cases even many-to-many, mappings between identifiers. Similar such complications arise because some resources unify all tautomers of a given chemical, while other resources (including ChEBI) distinguish separate tautomers. Again, from a biological perspective, conjugate bases and acids are mergeable entities, since in physiological conditions these forms of a chemical may readily interconvert, while from a chemist's perspective these are very different entities, with different names and properties.

5.2. Calculated Properties and Experimentally Derived Properties

Many properties provided by public chemical databases are calculated directly from the chemical structure. For example, the formula, molecular weight, and overall charge are generally calculated directly from the chemical structure. Unfortunately, some of the properties of interest to toxicology cannot be calculated directly from the chemical

Table 6
Lists some examples of chemical properties of interest

Chemical property	Importance of property
Hydrogen bond index	H-bonding (the formation of weak intermolecular bonds between a hydrogen carrying a partial positive charge on one molecule and a partially negatively charged N, O, or F on another molecule) plays pivotal role in biological systems, contributing to the structures and interactions of molecules such as carbohydrates, nucleotides, and amino acids. The transience of hydrogen bonds means that they can be switched on or off with energy values that lie within the range of thermal fluctuations of life temperatures (1)
Reactivity with water	The bodies of all living organisms are chiefly (70–80%) water, the unique, life-sustaining properties of which arise from H-bonding
pH tendencies	pH, a measure of hydrogen ion concentration, is a critical factor controlling life processes such as enzyme function. For example, a pH change of as little as 0.2 from the normal value of 7.4 in humans is life-threatening. pH levels similarly determine the functions of enzymes and other molecules in the wide variety of living systems
Toxicity	Toxins (substances capable of causing harm to living organisms) include heavy metals such as mercury, lead, cadmium, and aluminium; biotoxins, produced by living cells and organism; xenobiotics, substances that are not a normal component of the organism in question; or food preservatives and cosmetics

structure. Such properties are predicted using *models*, which correlate chemical structural features with properties based on a large set of training data which is then used for prediction. But the model is only as good as the training data, and for these purposes, to generate input training data, there is no substitute for experimental measurement of property values.

Table 6 lists some properties of interest in biology, including some properties which can only be measured experimentally.

References

1. de Matos P, Alcántara R, Dekker A, Ennis M, Hastings J, Haug K, Spiteri I, Turner S, Steinbeck C (2010) Nucl Acids Res 38:D249–D254
2. Trinajstic N (1992) Chemical graph theory. CRC Press, Florida, USA
3. MDL (2010) http://www.mdl.com/company/about/history.jsp. Last accessed December 2010
4. IUPAC The IUPAC International Chemical Identifier (InChI) (2010) \http://www.iupac.org/inchi/. Last accessed July 2012
5. Daylight, inc. (2010) http://www.daylight.com/dayhtml/doc/theory/theory.smiles.html. Last accessed July 2012
6. CAS Chemical Abstracts Service (2010) http://www.cas.org/. Last accessed July 2012
7. Beilstein Crossfire Beilstein Database (2010) \http://info.crossfiredatabases.com/beilstein acquisitionpressreleasemarch607.pdf. Last accessed December 2010
8. Marx V (2009) *GenomeWeb BioInform News* \http://www.genomeweb.com/informatics/tear-down-firewall-pharma-scientists-call-pre-competitive-approach-bioinformatic?page = show. Last accessed July 2012
9. Sayers E (2005) PubChem: An Entrez Database of Small Molecules. NLM Tech Bull (342):e2
10. Williams A (2008) Chemistry International 30(1):1
11. Wikipedia Chemistry (2010) http://en.wikipedia.org/wiki/Chemistry. Last accessed July 2012
12. CHEMBIOGRID CHEMBIOGRID: chemistry databases on the web (2010) http://www.chembiogrid.org/related/resources/about.html. Last accessed July 2012
13. Williams AJ (2009) Drug Discov Today 13:495–501
14. Steinbeck C, Han Y, Kuhn S, Horlacher O, Luttmann E, and Willighagen E (2003) J chem Inf Comput Sci 43(2):493–500
15. Steinbeck C, Hoppe C, Kuhn S, Floris M, Guha R, Willighagen EL (2006) Curr Pharm Des 12:2111 2120
16. Krause S, Willighagen EL, Steinbeck C (2000) Molecules 5:93–98
17. Rijnbeek M, Steinbeck C (2009) J Cheminformatics 1(1):17
18. The Gene Ontology Consortium The OBO language, version 1.2 (2010) http://www.geneontology.org/GO.format.obo-1_2.shtml. Last accessed July 2012
19. Smith MK, Welty C, McGuinness DL (2010) The web ontology language http://www.w3.org/TR/owl-guide/. Last accessed July 2012
20. Berry MJ, Ralston NV (2008) Ecohealth 5:456–459
21. Guzz G, Fogazzi GB, Cantù M, Minoia C, Ronchi A, Pigatto PD, Severi G (2008) J Environ Pathol Toxicol Oncol 27:147–155
22. Haley BE (2005) Medical Veritas 2:535–542
23. Freedman BJ (1980) Br J Dis Chest 74:128–34
24. Longo BM, Yang W, Green JB, Crosby FL, Crosby VL (2010) Toxicol Environ Health A 73:1370–8
25. Kanehisa M, Goto S, Hattori M, Aoki-Kinoshita K, Itoh M, Kawashima S, Katayama T, Araki M, Hirakawa M (2006) Nucleic Acids Res 34: D354–357
26. PDB The world wide protein data bank (2010) http://www.wwpdb.org/. Last accessed July 2012
27. The UniProt Consortium (2010) Nucleic Acids Res 38:D142–D148
28. Steinbeck C, Krause S, Kuhn S (2003) J Chem Inf Comput Sci 43(6):1733–1739
29. Wittig U, Golebiewski M, Kania R, Krebs O, Mir S, Weidemann A, Anstein S, Saric J, Rojas I (2006) In: Proceedings of the 3rd international workshop on data integration in the life sciences 2006 (DILS'06), Hinxton, UK, pp 94–103
30. AÅŸkin H, Uysal H, Altun D (2007) Toxicol Ind Health 23:591–8
31. Lin TM, Lee SS, Lai CS, Lin SD (2006) Burns 4:517–21

32. Warner MA, Harper JV (1985) Anesthesiology 62:366–7
33. van Schalkwyk DA, Chan XW, Misiano P, Gagliardi S, Farina C, Saliba KJ (2010) Biochem Pharmacol 79:1291–9
34. Lanfermeijer FC, Prins H (1994) Plant Physiol 104:1277–1285
35. Malerba M, Contran N, Tonelli M, Crosti P, Cerana R (2008) Physiol Plant 133:449–57
36. Mattei C, Wen PJ, Nguyen-Huu TD, Alvarez M, Benoit E, Bourdelais AJ, Lewis RJ, Baden DG, Molgó J, Meunier FA (2008) PLoS One 3:e3448
37. Nguyen-Huu TD, Mattei C, Wen PJ, Bourdelais AJ, Lewis RJ, Benoit E, Baden DG, Molgó J, Meunier FA (2010) Toxicon 56:792–6
38. Lehane L, Lewis RJ (2000) Int J Food Microbiol 61:91–125
39. Long SA, Quan C, Van de Water J, Nantz MH, Kurth MJ, Barsky D, Colvin ME, Lam KS, Coppel RL, Ansari A, Gershwin ME (2001) J Immunol 167:2956–63
40. Amano K, Leung PS, Xu Q, Marik J, Quan C, Kurth MJ, Nantz MH, Ansari AA, Lam KS, Zeniya M, Coppel RL, Gershwin ME (2004) J Immunol 172:6444–52

Chapter 10

Accessing, Using, and Creating Chemical Property Databases for Computational Toxicology Modeling

Antony J. Williams, Sean Ekins, Ola Spjuth, and Egon L. Willighagen

Abstract

Toxicity data is expensive to generate, is increasingly seen as precompetitive, and is frequently used for the generation of computational models in a discipline known as computational toxicology. Repositories of chemical property data are valuable for supporting computational toxicologists by providing access to data regarding potential toxicity issues with compounds as well as for the purpose of building structure–toxicity relationships and associated prediction models. These relationships use mathematical, statistical, and modeling computational approaches and can be used to understand the mechanisms by which chemicals cause harm and, ultimately, enable prediction of adverse effects of these chemicals to human health and/or the environment. Such approaches are of value as they offer an opportunity to prioritize chemicals for testing. An increasing amount of data used by computational toxicologists is being published into the public domain and, in parallel, there is a greater availability of Open Source software for the generation of computational models. This chapter provides an overview of the types of data and software available and how these may be used to produce predictive toxicology models for the community.

Key words: Bioinformatics, Cheminformatics, Computational toxicology, Public domain toxicology data, QSAR, Toxicology databases

1. Introduction

Since the inception of computational toxicology as a subdiscipline of toxicology there have likely been many hundreds of publications and several books that address approaches for modeling various toxicity endpoints. The reader is especially recommended to read the available chapters in several books (to which this volume will contribute) to see the broad diversity of computational approaches applied to date (1–3).

Computational toxicology approaches are used in industries that need to estimate risk early on. For example, the pharmaceutical, agrochemical, or consumer product industries may want to determine the effects of a molecule on cytotoxicity. This can be

Brad Reisfeld and Arthur N. Mayeno (eds.), *Computational Toxicology: Volume I*, Methods in Molecular Biology, vol. 929, DOI 10.1007/978-1-62703-050-2_10,

achieved using in-house data for the end point of interest or using freely available or commercial data sets for modeling the property. At a deeper level, the lack of appreciable genotoxicity or cardiotoxicity may be important determinants of whether a molecule is approved for human use or not by government regulatory authorities. As the assays become more specific it is likely that the sources of data may be restricted. Computational toxicology models could be used early in the R&D process so that compounds with lower risk may be progressed as part of a multidimensional process that also considers other properties of molecules (4). The application of an array of computational models to fields such as green chemistry has also recently been reviewed (5). The advantages of such methods could be that they save money or prevent the physical testing of compounds (and concomitant animal or other tissue usage) which may be undesirable or impractical at high volume.

In the past 5 years, 2D-ligand-based approaches have been increasingly used along with sophisticated algorithms and networks to form a systems-biology approach for toxicology (6–10). These studies commonly require compounds with toxicity data and molecular descriptors to be generated or retrieved from databases. Several of these recent approaches have described how machine learning methods can be used for modeling binary or continuous data relevant to toxicology. We have experience in this domain and examples of our own activities include studying drug-induced liver injury using data from a previously published study (11) and collaborating with Pfizer to generate models for the time-dependent inhibition of CYP3A4 (12). Machine learning models for predicting cytotoxicity using data from many different in-house assays that were combined has also been published separately by Pfizer (13).

As examples of how databases of toxicity information can be used or created to enable computational modeling we have provided a few recent examples. There are several examples of quantitative structure activity relationship (QSAR) or machine learning methods for predicting hepatotoxicity (14, 15) or drug–drug interactions (12, 16–18). Drug metabolism in the liver can convert some drugs into highly reactive intermediates (19–22) and hence cause drug-induced liver injury (DILI). DILI is the number one reason why drugs are not approved or are withdrawn from the market after approval (23). Idiosyncratic liver injury is much harder to predict from the preclinical in vitro or in vivo situation so we frequently become aware of such problems once a drug reaches large patient populations in the clinic and this is generally too late for the drug developer to identify an alternative and safer drug molecule. One study assembled a list of approximately 300 drugs and chemicals, with a classification scheme based on human clinical data for hepatotoxicity, for the purpose of evaluating an in vitro testing methodology based on cellular imaging of primary human hepatocyte cultures (24). It was found that the 100-fold C_{max} scaling factor represented a reasonable threshold to differentiate safe versus toxic drugs, for an orally dosed drug and with regard to hepatotoxicity (24). The concordance of the in vitro human hepatocyte imaging assay technology (HIAT) applied to about 300 drugs and chemicals, is about 75% with regard to clinical hepatotoxicity, with very few false positives (24). An alternative is to use the human clinical DILI data to create a computational model, and then validate it with enough compounds to provide confidence in its predictive ability so it can be used as a prescreen before in vitro testing.

In a recent study a training set of 295 compounds and a test set of 237 molecules were used with a Bayesian classification approach (25, 26). The Bayesian model generated was evaluated by leaving out either 10, 30, or 50% of the data. In each case, the leave-out testing was comparable to the leave-one-out approach and these values were very favorable indicating good model robustness. The mean concordance >57%, specificity >61%, and sensitivity >52% did not seem to differ depending on the amount of data left out. Molecular features such as long aliphatic chains, phenols, ketones, diols, α-methyl styrene (represents a polymer monomer), conjugated structures, cyclohexenones, and amides predominated in DILI active compounds (11). The Bayesian model was tested with 237 new compounds. The concordance ~60%, specificity 67%, and sensitivity 56% were comparable with the internal validation statistics. A subset of 37 compounds of most interest clinically showed similar testing values with a concordance greater than 63% (11). Compounds of most interest clinically are defined as well-known hepatotoxic drugs plus their less hepatotoxic comparators. These less hepatotoxic comparators are approved drugs that typically share a portion of the chemical core structure as the hepatotoxic ones. The purpose of this test set was to explore whether the Bayesian in silico method could differentiate differences in DILI potential between or among closely related compounds, as this is likely the most useful case in the real-world drug discovery setting.

In order to develop any computational model, there are numerous requirements for the data utilized as the foundation of the algorithm development. There are also independent requirements for the data, simply as content of a look-up database for medicinal chemists and toxicologists, for example, to enable search and retrieval of property data of interest. For the purpose of lookup and retrieval, the data need to be of high "quality." Text searches based on chemical identifiers such as chemical names, CAS numbers and international identifiers such as European Inventory of Existing Chemical Substances (EINECS) numbers (http://www.ovid.com/site/products/fieldguide/EINECS/eine.htm#abouthealth) should result in the retrieval of an accurate representation of the chemical compound with associated data, preferably including attribution to the source data. While any aggregated database is sure to contain errors this has been shown to be a considerable issue when accessing data from public domain databases (27) and will be examined in more detail later. For the purpose of sourcing data to be utilized for modeling purposes, it is preferable that the data span a diverse area of structure space (although this may be dependent on whether "local" or "global" models are being developed) that the assays captured in the database provide responses over many orders of magnitude and that the data have been acquired with repeatable measurements and, when possible, validated against other data. Clearly these criteria are rather challenging (and variable depending on the type of end point) and as a result the production of a high-quality data set for the purpose of modeling can be both tremendously time consuming and exacting in its assembly.

2. Public Domain Databases

Computational toxicologists have a choice when accessing data for the purpose of reference or to create models. They can source data in house or from collaborators or commercial sources [e.g., data and computational models from Leadscope (http://www.leadscope.com), Accelrys (http://accelrys.com), LHASA (https://www.lhasalimited.org), and Aureus (http://www.aureus-sciences.com/aureus/web/guest/adme-overview)] or, as is increasingly more common, from online resources (28). Online resources should, in general, be expected to be more favorable to scientists as the data are available at no cost and can contain a broader range of information. However, it should be noted that all structure-based databases can be prone to error and care needs to be taken when choosing data from such sources. This is especially true of public domain structure-based data sources as has been discussed elsewhere (29).

Online resources hosting toxicity data include the multiple databases integrated via the Toxicity Network, ToxNet (http://toxnet.nlm.nih.gov), the Distributed Structure Searchable Toxicity (DSSTox) database hosted by the EPA (30), the Hazardous Substances Databank (http://www.nlm.nih.gov/pubs/factsheets/hsdbfs.html), ACToR (31), ChEMBL (32), and ToxRefDB (http://www.epa.gov/ncct/toxrefdb/) to name just a few. These are discussed in more detail below.

TOXNET, the TOXicology data NETwork, is a cluster of databases covering toxicology, hazardous chemicals, environmental health, and related areas. It is managed by the Toxicology and Environmental Health Information Program (TEHIP) in the Division of Specialized Information Services (SIS) of the National Library of Medicine (NLM). It is a central portal integrating a series of databases related to toxicology. These are Integrated Risk Information System (IRIS), International Toxicity Estimates for

Risk (ITER), Chemical Carcinogenesis Research Information System (CCRIS), Genetic Toxicology (GENE-TOX), the Household Products Database, Carcinogenic Potency Database (CPDB), and a number of other related databases. It also integrates to HSDB, the Hazardous Substances Data Bank.

HSDB focuses on the toxicology of potentially hazardous chemicals and contains information on human exposure, industrial hygiene, emergency handling procedures, environmental fate, regulatory requirements, and other related areas. It is peer reviewed by a committee of experts and presently contains over 5,000 records.

The EPA recently released the ACToR database online (31). It is made up of 500 public data sources on over 500,000 chemicals and contains data regarding chemical exposure, hazard, and potential risks to human health and the environment. ACToR is integrated to a toxicity reference database called ToxRefDB which allows searching and downloading of animal toxicity testing results on hundreds of chemicals. The database captures 30 years and $2 billion worth of animal testing chemical toxicity data in a publicly accessible searchable format.

ACToR also links to DSSTox, the EPA DSSTox database project (30, 33), that provides a series of documented, standardized, and fully structure-annotated files of toxicity information. The initial intention for the project was to deliver a public central repository of toxicity information to allow for flexible analogue searching, structure–activity relationship (SAR) model development and the building of chemical relational databases. In order to ensure maximum uptake by the public and allow users to integrate the data into their own systems, the DSSTox project adopted the use of a common standard file format to include chemical structure, text, and property information. The DSSTox data sets are among the most highly curated public data sets available and, in the judgment of these authors, likely the reference standard in publicly available structure-based toxicity data. Through ACToR scientists can also access data from the ToxCast™ database (which in itself can be used for computational toxicology modeling (34)). This is a long-term, multi-million dollar effort that hopes to understand biological processes impacted by chemicals that may lead to adverse health effects, as well as generate predictive models that should enable cheaper predictions of toxicity (35).

ChEMBL is a database of drugs and other small molecules of biological interest. The database contains information for >500,000 compounds and >3 million records of their effects on biological systems including target binding, the effect of these compounds on cells and organisms (e.g. IC_{50}), and associated ADMET data (>60,000 records of which >7,000 relate to toxicity). The database contains manually curated SAR data from almost 40,000 papers from the medicinal chemistry and pharmacology literature and therefore provides data that may be used for

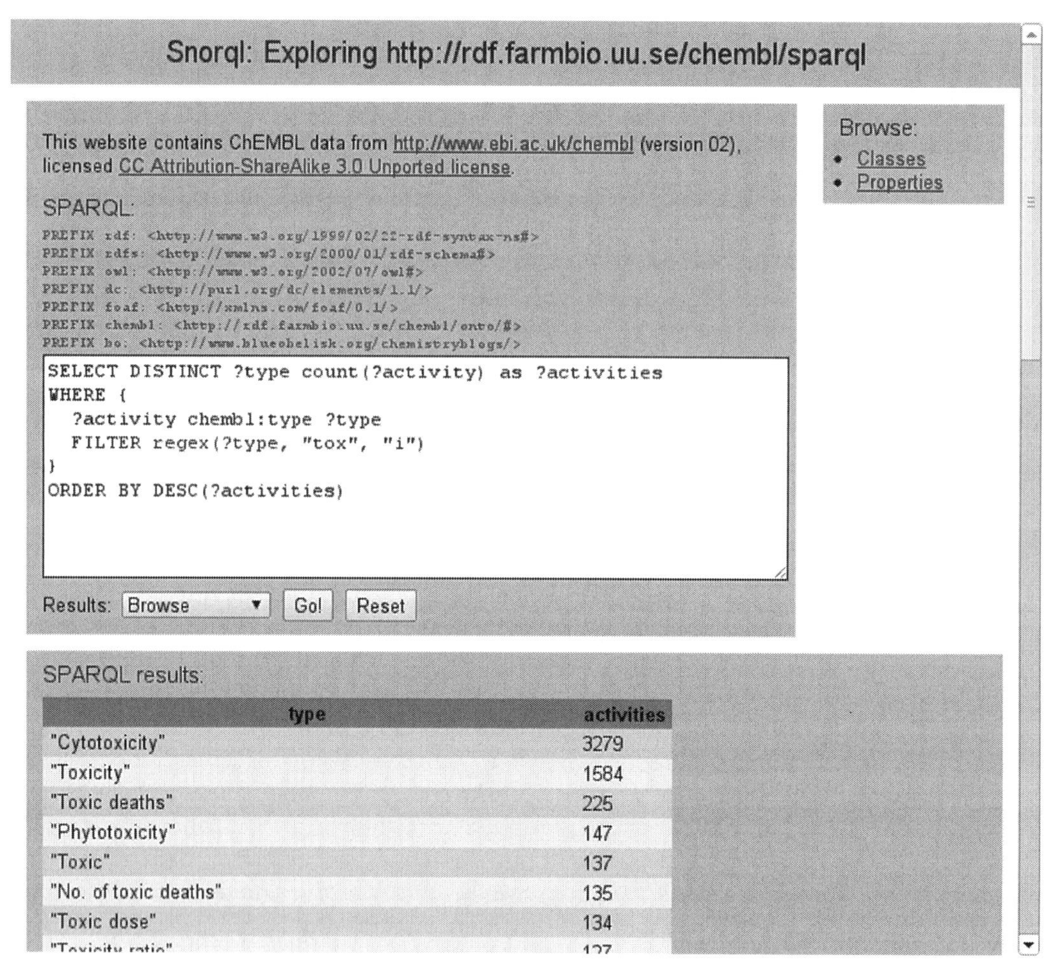

Fig. 1. The Snorql graphical user interface showing the SPARQL query and the seven most frequent toxicity measurements recorded in ChEMBL.

computational toxicology purposes. Modern application programming interfaces (APIs) make it easy to query such databases. Version 2 of the ChEMBL database is available as resource description framework (RDF) (36) via web services that allow searching of the database with the query language known as SPARQL (37). Using a SPARQL query we can summarize what experimental toxicology data is available in the database. Figure 1 shows a graphical user interface around the SPARQL end point (called Snorql), showing the SPARQL query and the seven most frequent toxicity measurements recorded in ChEMBL.

Using such RDF and SPARQL queries proteochemometrics studies have recently been performed, linking small molecules to their targets. In one study, metadata on the assessed assay quality provided by the ChEMBL curators was used in the modeling of IC_{50} values (38), hinting at the potential that the inclusion of complementary information has in understanding complex

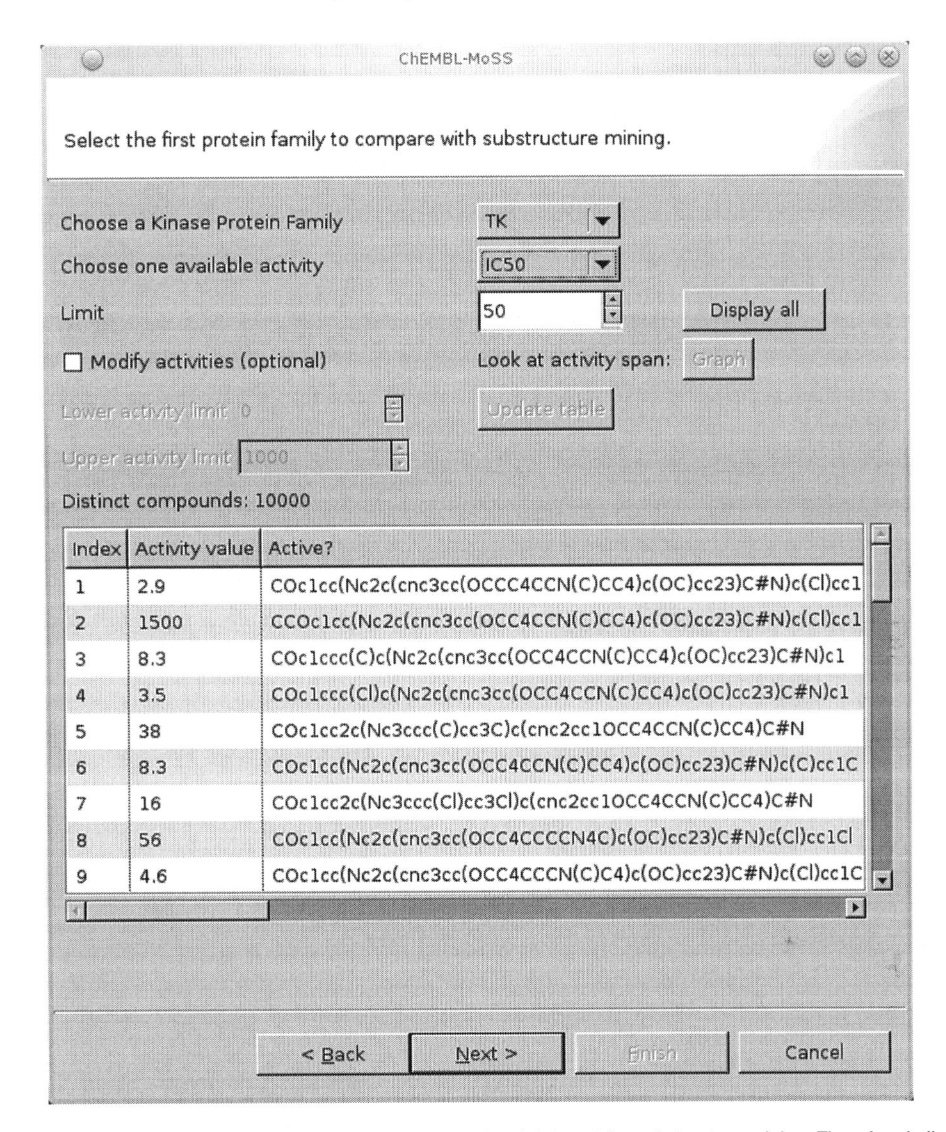

Fig. 2. A screenshot of the Bioclipse wizard used to create a local data set for substructure mining. The wizard allows the user to select two protein families that will be compared. The biological activity, IC_{50}, is selected in this example and is dynamically discovered in the online database.

biological patterns. Similarly, in an unpublished work, kinase families have been compared in a database-wide substructure mining study, where molecular fragments were identified in the small molecules that were more present in the database for one kinase family compared to others (Fig. 2).

It is important to note that both studies were acting directly on the remote ChEMBL database, and the use of the SPARQL standard is crucial, thereby allowing any tool to query the ChEMBL database using the same Open Standard as an increasing amount of other databases in life sciences. These include general life sciences databases exposed by Chem2Bio2RDF (39) and Bio2RDF (40, 41) as well as toxicology-specific databases.

The power of semantic web technologies lies primarily in the fact that they describe more explicitly what the content of the database is (often using standardized ontologies) and also provide a unified approach to access data. A nice example of the power of this approach is the use of a federated SPARQL query by Prud'hommeaux in 2007 (42). A federated SPARQL query hypothetically searches across multiple different databases simultaneously. The example in this use case links five online databases, searching for chemical compounds causing over-expression of a particular gene involved in apoptosis and which have a chemical similarity to compounds showing a low lethal dose in mice (35). The databases providing the data to answer this query are available as Open Data. The fact that these queries can easily be changed shows how the use of databases will likely impact computational toxicology in the next few years. Some examples of online databases that will likely contribute to this shift are given below.

PubChem is the highest profile online database serving our community and was launched by the NIH in 2004 to support the "New Pathways to Discovery" component of their roadmap initiative (43). The primary purpose for the database was to act as a *repository* for biological properties of chemical probes. PubChem archives and organizes information about the biological activities of chemical compounds into a comprehensive biomedical database and is intended to empower the scientific community to use small molecule chemical compounds in their research. It contains screening data that could be used for building models relevant to toxicity.

One example is the availability of data to enable modeling of the potassium channel human Ether-à-go-go-related gene (hERG). This channel is particularly important pharmaceutically as many drugs interact and cause hERG-related cardiotoxicity. Numerous blockbuster drugs have recently been removed from the market due to QT syndrome side effects, an abnormality associated with the hERG and associated channels (44). Several groups have used this hERG data for computational model testing (45–47). Despite the authoritative position granted to PubChem, as evidenced in a recent online questionnaire (Antony J. Williams, personal communication), PubChem data have been shown to be rather low quality in many cases, especially when it comes to providing data sets for the purpose of modeling. While the number of chemical probes screened by the National Screening Libraries to date is limited to 348,258, the PubChem database contains over 31 million unique chemicals with PubChem's content derived from the voluntary contributions of commercial, academic, and government organizations and, unfortunately, much of this has contributed to the pollution of the database. Chemical structures often do not accurately represent the expected chemical and the quality of the screening data has been questioned by Shoichet and others (42). No data curation is performed other than cheminformatics filters and standardization approaches at the time of deposition. We have found retrieval of available molecular structures for well-known FDA-approved drugs from PubChem is difficult and this can severely impact current and future efforts at drug repurposing using computational methods (48). For example, Symbicort is a well-known combination drug of budesonide and formoterol. A search on PubChem should return both drugs as one record but instead returns only budesonide. Searching on budesonide itself returns nine hits, one of which is Symbicort but eight of which are the single component, all of the same molecular mass but differing in stereochemistry. Searching for formoterol returns six hits all differing in their stereochemistry. These are not isolated examples (27).

The NIMH Psychoactive Drug Screening Program (PDSP, http://pdsp.med.unc.edu/indexR.html) (49–51) is a large data set of compounds associated with biology data (over 55,000 values) for many G-protein-coupled receptors. Several of these receptors are associated with toxicity of drugs and may also be off-targets (49, 50, 52–54). The website has links to PubChem, however a recent upload of this data in another database (Collaborative Drug Discovery, http://www.collaborativedrug.com/) has provided over 20,000 searchable structures, although the quality of these may be questionable (incorrect stereochemistry, etc.) as all the data came from PubChem originally. A subset of this data for the 5-HT_{2B} receptor was previously used for building machine learning models but required extensive sourcing of compounds without structures provided in the database (55).

These public domain databases contain potentially valuable data that can be used for the purposes of toxicology but, unfortunately, have a number of challenges. There are, of course, the issues regarding the quality of the experimental data (42) in terms of the quality of measurement, reproducibility, and appropriateness of the assays. Richard (33) has discussed the challenges of assembling high-quality data based on the experiences of creating the DSSTox data set, while Tropsha et al. (56) have examined the issues of data quality in terms of the development of QSAR models. Williams (27, 29) analyzed the quality of public domain databases in relation to the curation of the ChemSpider database and has identified common issues in regards to the relationships between chemical structures and associated chemical names, generally drug names and associated synonyms. As a result of the processes used to assemble the databases, especially for repositories such as PubChem (43), many of the public domain databases are contaminated. As discussed earlier for PubChem, querying based on chemical names can result in the retrieval of incorrect chemical structures that are then used in the development of computational toxicology models. The studies of Ekins and Williams (11, 29, 57) requiring the assembly of structure data files have shown numerous challenges with regards to the assembly of quality data from public domain data (27). Williams et al. have initiated a study to examine 200 of the top selling drugs and the consistency between a gold standard set of compound structures and their presence in a series of public compound databases. Early reports show that there are significant issues with data sourced from PubChem, Wikipedia, Drugbank (58), and others.

Williams and Ekins (27) have recently raised an alert in regards to data quality for internet-based chemistry resources if they and their content are to be used for drug repurposing or integrated into other cheminformatics or bioinformatics resources for drug discovery. It is clear that it is not yet appropriate to treat any of these chemistry databases as authoritative and users should be vigilant in their use and reliance on the chemical structures and data derived from these sources. They identified an urgent need for government funding of data curation for public domain databases to improve the overall quality of chemistry on the internet and stem the proliferation of errors.

3. Utilizing Databases for Computational Toxicity

Computational access to the databases discussed in this section has not been formalized and, at best, one can download the full data from a download page. Recently however the OpenTox project proposed a standardized API for accessing data sets (59). The resulting OpenTox Framework describes the use of a number of technologies to implement this API, specifically RESTful services to facilitate web access, and use of the RDF as an open standard for communication. This combination makes it easy for third party software projects to integrate databases exposed via the framework. Bioclipse (60) was recently extended with a number of scripting extensions to interact with OpenTox servers, and Table 1 shows two approaches of accessing toxicological data sets made available via the OpenTox framework (38). The first approach lists all data sets and takes the first (index = 0) to be saved in a standard SDF file format. The second approach uses an OpenTox ontology server and queries the underlying RDF data directly with the SPARQL query language. The SPARQL requests shown in the listing shows queries for data sets related to mutagenicity. At the time of writing only one data set was available but this will hopefully change as the community embraces and adopts the OpenTox Framework. Figure 3 shows how this data set was subsequently downloaded as

Table 1
JavaScript scripts that show two approaches to how Bioclipse can retrieve information from OpenTox servers

```
//Approach 1: list all data sets provided by a OpenTox server
datasets = opentox.listDataSets("http://apps.ideaconsult.net:8080/
   ambit2/")
opentox.downloadDataSetAsMDLSDfile(
"http://apps.ideaconsult.net:8080/ambit2/",
datasets.get(0), "/OpenTox/dataset1.sdf"
)
```

```
//Approach 2: query an OpenTox ontology server
var sparql = " \
PREFIX ot:<http://www.opentox.org/api/1.1#> \
PREFIX rdfs: <http://www.w3.org/2000/01/rdf-schema#> \
PREFIX rdf:<http://www.w3.org/1999/02/22-rdf-syntax-ns#> \
select ?dataset ?comment where {\
?dataset rdf:type ot:Dataset . \
OPTIONAL {?dataset ?p ?comment .} \
FILTER regex(?comment, \"mutagenicity\") . \
}\
";
rdf.sparqlRemote("http://apps.ideaconsult.net:8080/ontology/", sparql)
```

an SDF file and then displayed in a Bioclipse molecule table view. With these types of approaches and with adoption of the standards offered by the OpenTox Framework, it is expected that many of the public domain databases will become more accessible to the community via interfaces such as that offered by Bioclipse.

4. Utilizing Databases for Prediction of Metabolic Sites

The toxicity of compounds is not always directly caused by the compound itself, but may also be due to metabolism. Public databases related to drug metabolism are rare. The lack of such databases makes methods that predict metabolism, or just the likelihood that structures are metabolized, all the more relevant. Predicted metabolites can be screened against toxicity databases and existing predictive models, while the likelihood that a drug can undergo metabolism can be used in the overall analysis of a compound's toxicity. The cytochrome P450 (CYPs) family of heme-thiolate enzymes are the cause of the majority of drug–drug interactions and metabolism-dependent toxicity issues (61, 62). The identification of the site-of-metabolism (SOM) for CYPs is an important

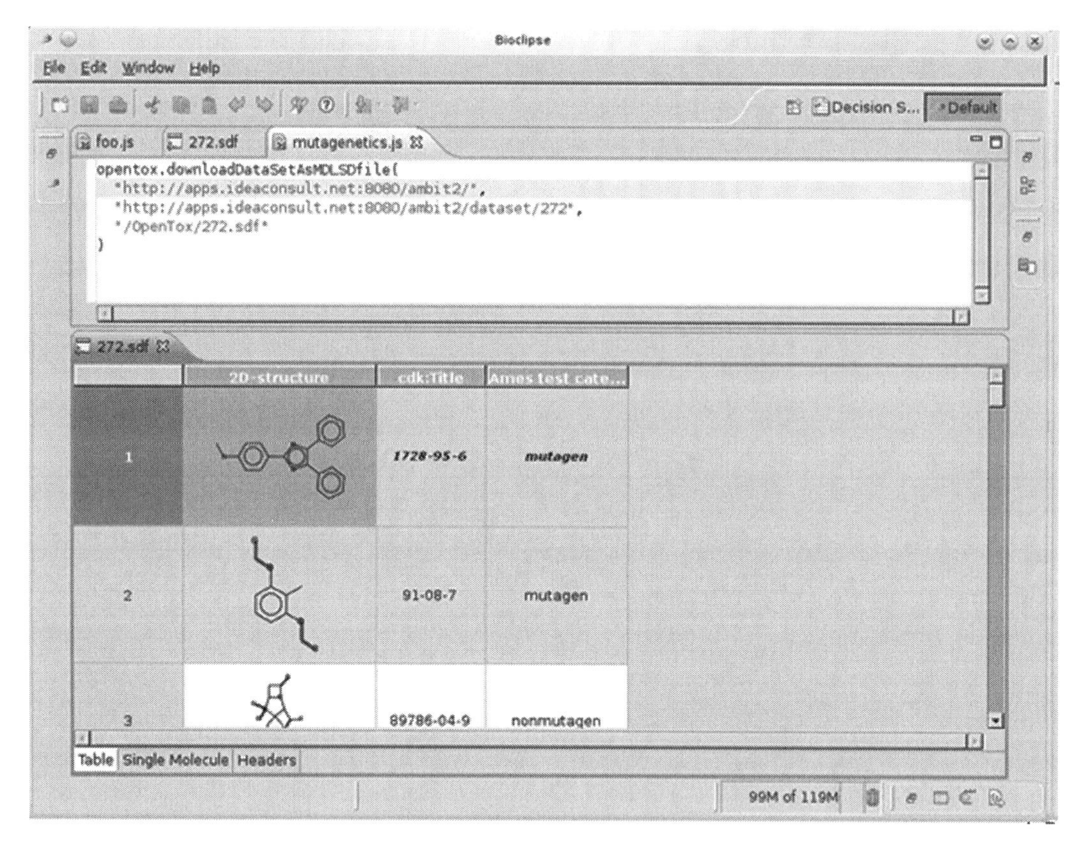

Fig. 3. A screenshot of a Bioclipse script that downloads a data set with mutagenicity information from an OpenTox server and shows the resulting set of data in a molecules table view.

problem. Several methods have been reported to predict CYP metabolism using QSAR, docking, pharmacophore modeling, and statistical methods (61–64).

In the absence of free CYP metabolism databases the MetaPrint2D (65, 66) method has been developed for the prediction of metabolic sites from input chemical structures (see Figs. 4 and 5). The method uses an internal database based on historical metabolite data derived from the proprietary Accelrys Metabolite database (http://accelrys.com/products/databases/bioactivity/metabolite.html), which is preprocessed using circular fingerprints capable of accurately describing individual atom environments. New molecules can then be processed with the same circular fingerprints, and the probability of metabolism can be calculated for each atom by querying the database. This method is very fast (ca. 50 m s per compound) (61) and performs well compared to other algorithms in the field such as SmartCyp (67) and MetaSite (68).

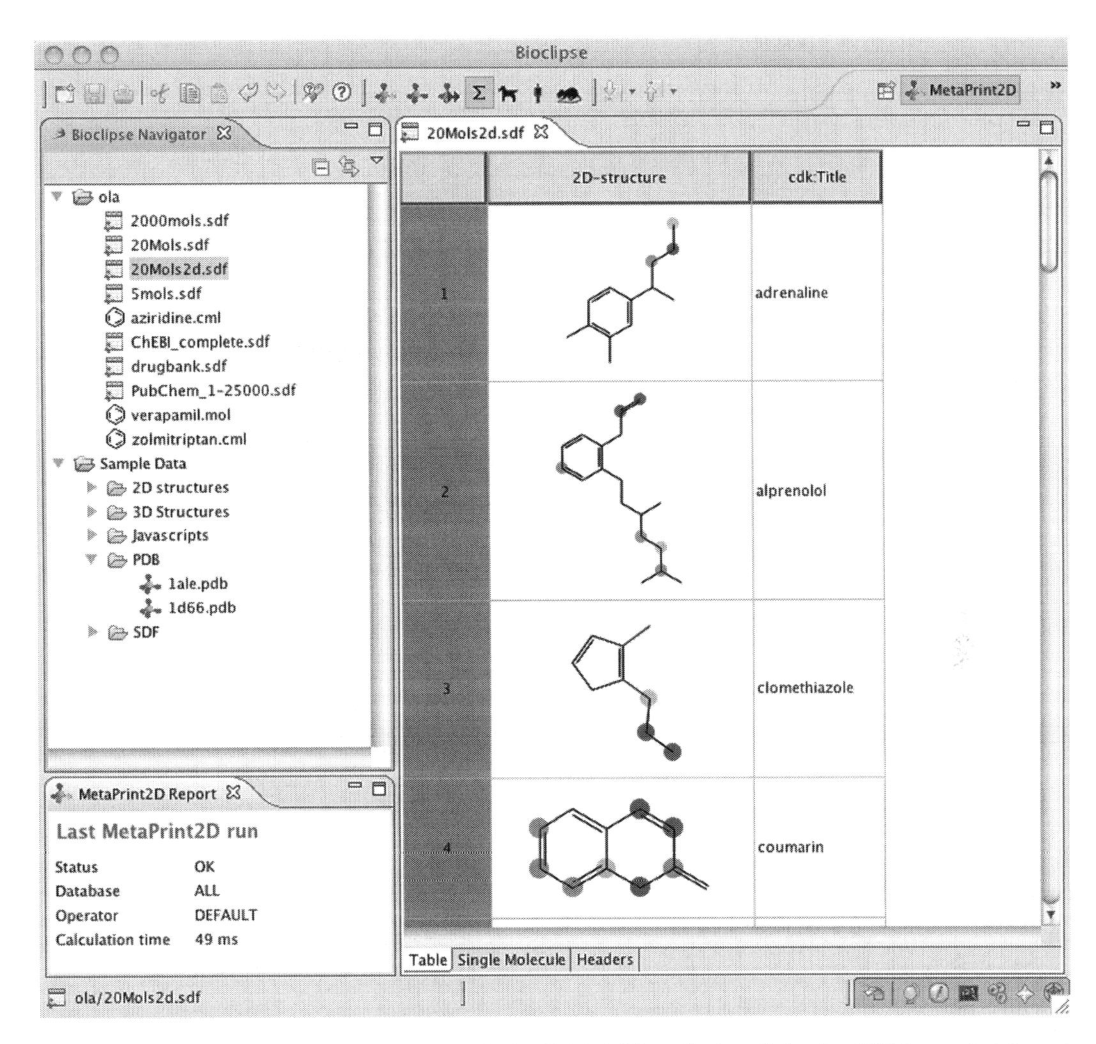

Fig. 4. A screenshot of the Bioclipse workbench with the MetaPrint2D method predicting the SOM for a set of drugs. In the interface individual atoms are colored according to their likelihood of metabolism (the color scheme is as follows in the interface *red* = high, *yellow* = medium, and *green* = low). Please see color figure online to see color coding in the interface.

5. Standardizing QSAR Experiments

Setting up data sets for QSAR analysis is not without complications. There are numerous software tools available for the calculation of descriptors and the conversion of chemical structures into a form which can be used in statistical and machine-learning methods. These software packages are generally incompatible, have proprietary file formats, or are available as standalone applications. A lack of standardization in terms of descriptors and the associated descriptor implementations has likely contributed to the poor quality of the supplemental data for published QSAR models and has made it an

Fig. 5. A screenshot from Bioclipse with a script calling the MetaPrint2D method and outputting the predicted likelihood for the SOM per atom.

often impossible task to reproduce both the formation of the data set and hence the entire analysis.

QSAR-ML (69) is a new file format for exchanging QSAR data sets, which consists of an open XML format (QSAR-ML) and builds on the Blue Obelisk descriptor ontology (70). The ontology provides an extensible manner of uniquely defining descriptors for use in QSAR experiments. The exchange format supports multiple versioned implementations of these descriptors. As a result a data set described by QSAR-ML makes setup by others completely reproducible. A Bioclipse plug-in is available for working with

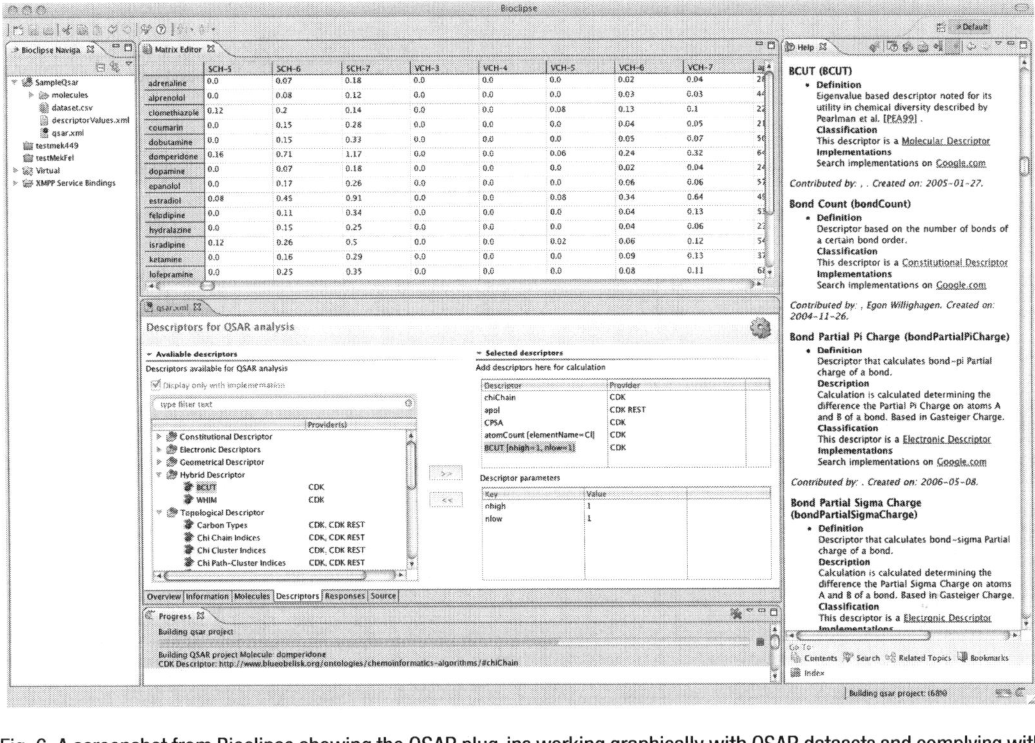

Fig. 6. A screenshot from Bioclipse showing the QSAR plug-ins working graphically with QSAR datasets and complying with the QSAR-ML standard for interoperable QSAR datasets.

QSAR data and provides graphical tools for importing molecules, selecting descriptors, performing calculations, and exporting data sets in QSAR-ML as well as in the more traditional CSV file formats (Fig. 6). Descriptors can take advantage of software installed on desktop computers, as well as calculations on networked computers via Web services (71).

Pfizer has recently evaluated open source descriptors and model building algorithms using a training set of approximately 50,000 molecules and a test set of approximately 25,000 molecules with human liver microsomal metabolic stability data (72). A C5.0 decision tree model demonstrated that the Chemistry Development Kit (73) descriptors together with a set of SMARTS keys had good statistics (Kappa = 0.43, sensitivity = 0.57, specificity 0.91, positive predicted value (PPV) = 0.64) equivalent to models built with commercial MOE2D software and the same set of SMARTS keys (Kappa = 0.43, sensitivity = 0.58, specificity 0.91, PPV = 0.63). This observation was also confirmed upon extension of the data set to ~193,000 molecules and generation of a continuous model using Cubist (http://www.rulequest.com/download.html). When the continuous predictions and actual values were binned to get a categorical score an almost identical Kappa statistic (0.42) was observed (72).

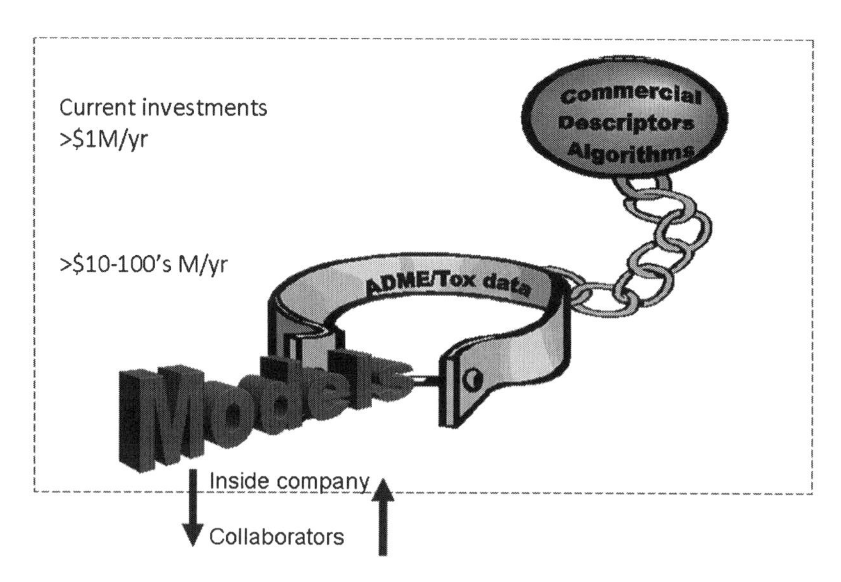

Fig. 7. A schematic indicating the release of ADME/Tox models from commercial descriptors and algorithms, and how it may facilitate data sharing.

The advantages of models and data built with open source technologies are the reduced costs and overheads as well as the ease of sharing with other researchers (Fig. 7). This can enable groups outside of pharmaceutical companies to have some of the capabilities that are taken for granted such as models for computational toxicology based on the abundance of data available to them. We have recently proposed that companies could assist in providing ADME/Tox data to a central repository as such data could be considered precompetitive (57) and such a database would be valuable for building and validating computational models. Making such data available to the OpenTox Framework could dramatically expand the availability of computational toxicology across the community.

If toxicology data is to be integrated on a large scale then it is necessary that it is available in a form which can be processed by computers. The topic of scientific data processing by computers has been mentioned as one of the primary bottlenecks for research (74). There are numerous standardization initiatives in the life sciences, with the Minimum Information standards (http://en.wikipedia.org/wiki/Minimum_Information_Standards) as examples of standards which specify the minimum amount of metadata and data required to meet a specific aim in a field. The MGED consortium pioneered this approach in bioinformatics with the definition for Minimum Information About Microarray Experiments (MIAME) (75), and it has now become a requirement that data must be deposited in MIAME-compliant public repositories using the MAGE-ML exchange format in order to publish microarray experiments in most journals. QSAR-ML is a step towards the same goal for QSAR studies but in order to meet this demand a database for sharing QSAR-ML-compliant experiments must first be developed.

6. Conclusions

Computational toxicology modeling depends on the availability of data to both generate and validate models and access to software to generate the appropriate descriptors and develop the resulting algorithms. At this stage in the development cycle for tools to support computational toxicology we should expect that high-quality data sets which can be used as the basis of model development and validation would be available. In recent years, especially with the development of the internet as a distribution network for data and as a result of investments in bioinformatics and public domain data sources there has been a dramatic increase in the availability of data. While data are available caution is warranted as in many cases data are not curated and scientists are encouraged to understand the nature of the data sources from which they are extracting data before they use it for modeling.

We have proposed the construction of a validated ADME/Tox database and suggest an updated strategy for how the scientific community could build such a resource (57).

1. Identify all available publications containing toxicity data relating to molecular structures tested in animal or human tissues in vitro or in vivo. Mine the data from these publications relating to toxicity properties.
2. Clean and organize the data from these publications, e.g., relate by species, tissue, cell types, and capture experimental conditions using manual curation and create an ontology.
3. Provide a means for other scientists to update and include new properties.
4. Encourage pharmaceutical companies to publish their previously "unpublished toxicology data" in exchange for access to a duplicate of the database of toxicity data for their own in-house efforts for internal deployment.
5. As an example of the development and value of such a toxicology database, toxicity computational models could be built, validated, and provided over the Web for free.

Assuming that such a database, or series of interconnected databases, could be developed, it will be necessary that the community agrees to and adopts standards for describing the data and, in parallel, develops similar approaches for the distribution of resulting computational models. The efforts of the OpenTox Framework or other organizations to develop standards are likely to catalyze significant shifts. Coupling high-quality data with new standards in interoperability, model interpretability, and usability will hopefully enable the development of improved models for computational toxicology. We can then envisage a deeper understanding of the toxicity mechanisms of molecules and this will ultimately enable the prediction of adverse effects of these chemicals to human health and/or the environment.

Acknowledgments

SE gratefully acknowledges the many collaborators involved in the cited work. Contributions by OS and ELW were supported by Uppsala University (KoF 07).

SE consults for Collaborative Drug Discovery, Inc. on a Bill and Melinda Gates Foundation Grant#49852 "Collaborative drug discovery for TB through a novel database of SAR data optimized to promote data archiving and sharing."

References

1. Helma C (ed) (2005) Predictive toxicology. Taylor and Francis, Boca Raton
2. Cronin MTD, Livingstone DJ (2004) Predicting chemical toxicity and fate. CRC, Boca Raton
3. Ekins S (2007) Computational toxicology: risk assessment for pharmaceutical and environmental chemicals. Wiley, Hoboken
4. Ekins S, Boulanger B, Swaan PW, Hupcey MAZ (2002) Towards a new age of virtual ADME/TOX and multidimensional drug discovery. J Comput Aided Mol Des 16:381–401
5. Voutchkova AM, Osimitz TG, Anastas PT (2010) Toward a comprehensive molecular design framework for reduced hazard. Chem Rev 110:5845–5882
6. Ekins S, Giroux C (2006) Computers and systems biology for pharmaceutical research and development. In: Ekins S (ed) Computer applications in pharmaceutical research and development. John Wiley, Hoboken, pp 139–165
7. Ekins S, Bugrim A, Brovold L, Kirillov E, Nikolsky Y, Rakhmatulin EA, Sorokina S, Ryabov A, Serebryiskaya T, Melnikov A, Metz J, Nikolskaya T (2006) Algorithms for network analysis in systems-ADME/Tox using the MetaCore and MetaDrug platforms. Xenobiotica 36(10–11):877–901
8. Ekins S (2006) Systems-ADME/Tox: resources and network approaches. J Pharmacol Toxicol Methods 53:38–66
9. Nikolsky Y, Ekins S, Nikolskaya T, Bugrim A (2005) A novel method for generation of signature networks as biomarkers from complex high throughput data. Toxicol Lett 158:20–29
10. Ekins S, Nikolsky Y, Nikolskaya T (2005) Techniques: application of systems biology to absorption, distribution, metabolism, excretion, and toxicity. Trends Pharmacol Sci 26:202–209
11. Ekins S, Williams AJ, Xu JJ (2010) A predictive ligand-based Bayesian model for human drug induced liver injury. Drug Metab Dispos 38:2302–2308
12. Zientek M, Stoner C, Ayscue R, Klug-McLeod J, Jiang Y, West M, Collins C, Ekins S (2010) Integrated in silico-in vitro strategy for addressing cytochrome P450 3A4 time-dependent inhibition. Chem Res Toxicol 23:664–676
13. Langdon SR, Mulgrew J, Paolini GV, van Hoorn WP (2010) Predicting cytotoxicity from heterogeneous data sources with Bayesian learning. J Cheminform 2:11

14. Clark RD, Wolohan PR, Hodgkin EE, Kelly JH, Sussman NL (2004) Modelling *in vitro* hepatotoxicity using molecular interaction fields and SIMCA. J Mol Graph Model 22:487–497
15. Cheng A, Dixon SL (2003) *In silico* models for the prediction of dose-dependent human hepatotoxicity. J Comput Aided Mol Des 17:811–823
16. Ung CY, Li H, Yap CW, Chen YZ (2007) *In silico* prediction of pregnane X receptor activators by machine learning approaches. Mol Pharmacol 71:158–168
17. Marechal JD, Yu J, Brown S, Kapelioukh I, Rankin EM, Wolf CR, Roberts GC, Paine MJ, Sutcliffe MJ (2006) *In silico* and *in vitro* screening for inhibition of cytochrome P450 CYP3A4 by co-medications commonly used by patients with cancer. Drug Metab Dispos 34:534–538
18. Ekins S, Waller CL, Swaan PW, Cruciani G, Wrighton SA, Wikel JH (2000) Progress in predicting human ADME parameters *in silico*. J Pharmacol Toxicol Methods 44:251–272
19. Boelsterli UA, Ho HK, Zhou S, Leow KY (2006) Bioactivation and hepatotoxicity of nitroaromatic drugs. Curr Drug Metab 7:715–727
20. Kassahun K, Pearson PG, Tang W, McIntosh I, Leung K, Elmore C, Dean D, Wang R, Doss G, Baillie TA (2001) Studies on the metabolism of troglitazone to reactive intermediates *in vitro* and *in vivo*. Evidence for novel biotransformation pathways involving quinone methide formation and thiazolidinedione ring scission. Chem Res Toxicol 14:62–70
21. Walgren JL, Mitchell MD, Thompson DC (2005) Role of metabolism in drug-induced idiosyncratic hepatotoxicity. Crit Rev Toxicol 35:325–361
22. Park BK, Kitteringham NR, Maggs JL, Pirmohamed M, Williams DP (2005) The role of metabolic activation in drug-induced hepatotoxicity. Annu Rev Pharmacol Toxicol 45:177–202
23. Schuster D, Laggner C, Langer T (2005) Why drugs fail—a study on side effects in new chemical entities. Curr Pharm Des 11:3545–3559
24. Xu JJ, Henstock PV, Dunn MC, Smith AR, Chabot JR, de Graaf D (2008) Cellular imaging predictions of clinical drug-induced liver injury. Toxicol Sci 105:97–105
25. Xia XY, Maliski EG, Gallant P, Rogers D (2004) Classification of kinase inhibitors using a Bayesian model. J Med Chem 47:4463–4470
26. Bender A (2005) Studies on molecular similarity. Ph.D. Thesis, University of Cambridge, Cambridge
27. Williams AJ, Ekins S (2012) A quality alert for chemistry databases. Towards a gold standard: regarding quality in public domain chemistry databases and approaches to improving the situation, Drug Discovery Today, Volume 17, Issues 13–14, Pages 685–701. Submitted for publication
28. Judson R (2010) Public databases supporting computational toxicology. J Toxicol Environ Health 13:218–231
29. Williams AJ, Tkachenko V, Lipinski C, Tropsha A, Ekins S (2009) Free online resources enabling crowd-sourced drug discovery. Drug Discov World 10(Winter):33–38
30. Richard AM, Williams CR (2002) Distributed structure-searchable toxicity (DSSTox) public database network: a proposal. Mutat Res 499:27–52
31. Judson R, Richard A, Dix D, Houck K, Elloumi F, Martin M, Cathey T, Transue TR, Spencer R, Wolf M (2008) ACToR—aggregated computational toxicology resource. Toxicol Appl Pharmacol 233:7–13
32. Overington J (2009) ChEMBL An interview with John Overington, team leader, chemogenomics at the European Bioinformatics Institute Outstation of the European Molecular Biology Laboratory (EMBL-EBI). Interview by Wendy A. Warr. J Comput Aided Mol Des 23:195–198
33. Richard AM (2006) DSSTox web site launch: Improving public access to databases for building structure-toxicity prediction models. Preclinica 2:103–108
34. Kortagere S, Krasowski MD, Reschly EJ, Venkatesh M, Mani S, Ekins S (2010) Evaluation of computational docking to identify pregnane × receptor agonists in the ToxCast™ database. Environ Health Perspect 118:1412–1417
35. Sanderson K (2011) It's not easy being green. Nature 469:18–20
36. Carroll JJ, Klyne G (2004) Resource description framework (RDF): concepts and abstract syntax. Tech rep, W3C
37. Prud'hommeaux E, Seaborne A (2008) SPARQL query language for RDF, W3C recommendation
38. Willighagen EL, Alvarsson J, Andersson A, Eklund M, Lampa S, Lapins M, Spjuth O,

Wikberg J (2011) Linking the resource description framework to cheminformatics and proteochemometrics. J Biomedical Semantics 2(Suppl 1):S1–S6

39. Chen B, Dong X, Jiao D, Wang H, Zhu Q, Ding Y, Wild DJ (2010) Chem2Bio2RDF: a semantic framework for linking and data mining chemogenomic and systems chemical biology data. BMC Bioinformatics 11:255
40. Ansell P (2011) Model and prototype for querying multiple linked scientific datasets. Future Generat Comput Syst 27:329–333
41. Belleau F, Nolin MA, Tourigny N, Rigault P, Morissette J (2008) Bio2RDF: towards a mashup to build bioinformatics knowledge systems. J Biomed Inform 41:706–716
42. Prud'hommeaux E (2007) Case study: FeDeRate for drug research. Tech Rep: 4–7
43. Wang Y, Xiao J, Suzek TO, Zhang J, Wang J, Bryant SH (2009) PubChem: a public information system for analyzing bioactivities of small molecules. Nucleic Acids Res 37: W623–W633
44. Crumb WJ Jr, Ekins S, Sarazan D, Wikel JH, Wrighton SA, Carlson C, Beasley CM (2006) Effects of antipsychotic drugs on I_{to}, I_{Na}, I_{sus}, I_{K1}, and hERG: QT prolongation, structure activity relationship, and network analysis. Pharm Res 23:1133–1143
45. Su BH, Shen MY, Esposito EX, Hopfinger AJ, Tseng YJ (2010) *In silico* binary classification QSAR models based on 4D-fingerprints and MOE descriptors for prediction of hERG blockage. J Chem Inf Model 50:1304–1318
46. Li Q, Jorgensen FS, Oprea T, Brunak S, Taboureau O (2008) hERG classification model based on a combination of support vector machine method and GRIND descriptors. Mol Pharm 5:117–127
47. Thai KM, Ecker GF (2009) Similarity-based SIBAR descriptors for classification of chemically diverse hERG blockers. Mol Divers 13:321–336
48. Ekins S, Williams AJ, Krasowski MD, Freundlich JS (2011) *In silico* repositioning of approved drugs for rare and neglected diseases. Drug Discov Today 16(7–8):298–310
49. Strachan RT, Ferrara G, Roth BL (2006) Screening the receptorome: an efficient approach for drug discovery and target validation. Drug Discov Today 11:708–716
50. O'Connor KA, Roth BL (2005) Finding new tricks for old drugs: an efficient route for public-sector drug discovery. Nat Rev Drug Discov 4:1005–1014
51. Roth BL, Lopez E, Beischel S, Westkaemper RB, Evans JM (2004) Screening the receptorome to discover the molecular targets for plant-derived psychoactive compounds: a novel approach for CNS drug discovery. Pharmacol Ther 102:99–110
52. Keiser MJ, Setola V, Irwin JJ, Laggner C, Abbas AI, Hufeisen SJ, Jensen NH, Kuijer MB, Matos RC, Tran TB, Whaley R, Glennon RA, Hert J, Thomas KL, Edwards DD, Shoichet BK, Roth BL (2009) Predicting new molecular targets for known drugs. Nature 462:175–181
53. Setola V, Dukat M, Glennon RA, Roth BL (2005) Molecular determinants for the interaction of the valvulopathic anorexigen norfenfluramine with the 5-HT2B receptor. Mol Pharmacol 68:20–33
54. Rothman RB, Baumann MH, Savage JE, Rauser L, McBride A, Hufeisen SJ, Roth BL (2000) Evidence for possible involvement of 5-HT(2B) receptors in the cardiac valvulopathy associated with fenfluramine and other serotonergic medications. Circulation 102:2836–2841
55. Chekmarev DS, Kholodovych V, Balakin KV, Ivanenkov Y, Ekins S, Welsh WJ (2008) Shape signatures: new descriptors for predicting cardiotoxicity *in silico*. Chem Res Toxicol 21:1304–1314
56. Zhu H, Tropsha A, Fourches D, Varnek A, Papa E, Gramatica P, Oberg T, Dao P, Cherkasov A, Tetko IV (2008) Combinatorial QSAR modeling of chemical toxicants tested against *Tetrahymena pyriformis*. J Chem Inf Model 48:766–784
57. Ekins S, Williams AJ (2010) Precompetitive preclinical ADME/Tox data: set It free on the web to facilitate computational model building to assist drug development. Lab Chip 10:13–22
58. Wishart DS, Knox C, Guo AC, Cheng D, Shrivastava S, Tzur D, Gautam B, Hassanali M (2008) DrugBank: a knowledgebase for drugs, drug actions and drug targets. Nucleic Acids Res 36:D901–D906
59. Hardy B, Douglas N, Helma C, Rautenberg M, Jeliazkova N, Jeliazkov V, Nikolova I, Benigni R, Tcheremenskaia O, Kramer S, Girschick T, Buchwald F, Wicker J, Karwath A, Gutlein M, Maunz A, Sarimveis H, Melagraki G, Afantitis A, Sopasakis P, Gallagher D, Poroikov V, Filimonov D, Zakharov A, Lagunin A, Gloriozova T, Novikov S, Skvortsova N, Druzhilovsky D, Chawla S, Ghosh I, Ray S, Patel H, Escher S (2010) Collaborative development of predictive toxicology applications. J Cheminform 2:7
60. Spjuth O, Alvarsson J, Berg A, Eklund M, Kuhn S, Masak C, Torrance G, Wagener J, Willighagen EL, Steinbeck C, Wikberg JE

(2009) Bioclipse 2: a scriptable integration platform for the life sciences. BMC Bioinformatics 10:397

61. Afzelius L, Arnby CH, Broo A, Carlsson L, Isaksson C, Jurva U, Kjellander B, Kolmodin K, Nilsson K, Raubacher F, Weidolf L (2007) State-of-the-art tools for computational site of metabolism predictions: comparative analysis, mechanistic insights, and future applications. Drug Metab Rev 39:61–86
62. Jolivette LJ, Ekins S (2007) Methods for predicting human drug metabolism. Adv Clin Chem 43:131–176
63. Crivori P, Poggesi I (2006) Computational approaches for predicting CYP-related metabolism properties in the screening of new drugs. Eur J Med Chem 41:795–808
64. Stjernschantz E, Vermeulen NP, Oostenbrink C (2008) Computational prediction of drug binding and rationalisation of selectivity towards cytochromes P450. Expert Opin Drug Metab Toxicol 4:513–527
65. Boyer S, Arnby CH, Carlsson L, Smith J, Stein V, Glen RC (2007) Reaction site mapping of xenobiotic biotransformations. J Chem Inf Model 47:583–590
66. Carlsson L, Spjuth O, Adams S, Glen RC, Boyer S (2010) Use of historic metabolic biotransformation data as a means of anticipating metabolic sites using MetaPrint2D and Bioclipse. BMC Bioinformatics 11:362
67. Rydberg P, Gloriam DE, Olsen L (2010) The SMARTCyp cytochrome P450 metabolism prediction server. Bioinformatics 26:2988–2989
68. Cruciani G, Carosati E, De Boeck B, Ethirajulu K, Mackie C, Howe T, Vianello R (2005) MetaSite: understanding metabolism in human cytochromes from the perspective of the chemist. J Med Chem 48:6970–6979
69. Spjuth O, Willighagen EL, Guha R, Eklund M, Wikberg JE (2010) Towards interoperable and reproducible QSAR analyses: exchange of datasets. J Cheminform 2:5
70. Floris F, Willighagen EL, Guha R, Rojas M, Hoppe C (2010) The blue obelisk descriptor ontology. Technical report
71. Wagener J, Spjuth O, Willighagen EL, Wikberg JE (2009) XMPP for cloud computing in bioinformatics supporting discovery and invocation of asynchronous web services. BMC Bioinformatics 10:279
72. Gupta RR, Gifford EM, Liston T, Waller CL, Bunin B, Ekins S (2010) Using open source computational tools for predicting human metabolic stability and additional ADME/TOX properties. Drug Metab Dispos 38:2083–2090
73. Steinbeck C, Hoppe C, Kuhn S, Floris M, Guha R, Willighagen EL (2006) Recent developments of the chemistry development kit (CDK)—an open-source java library for chemo- and bioinformatics. Curr Pharm Des 12:2111–2120
74. Brazma A (2001) On the importance of standardisation in life sciences. Bioinformatics 17:113–114
75. Brazma A, Hingamp P, Quackenbush J, Sherlock G, Spellman P, Stoeckert C, Aach J, Ansorge W, Ball CA, Causton HC, Gaasterland T, Glenisson P, Holstege FC, Kim IF, Markowitz V, Matese JC, Parkinson H, Robinson A, Sarkans U, Schulze-Kremer S, Stewart J, Taylor R, Vilo J, Vingron M (2001) Minimum information about a microarray experiment (MIAME)-toward standards for microarray data. Nat Genet 29:365–371

Chapter 11

Molecular Dynamics

Xiaolin Cheng and Ivaylo Ivanov

Abstract

Molecular dynamics (MD) simulation holds the promise of revealing the mechanisms of biological processes in their ultimate detail. It is carried out by computing the interaction forces acting on each atom and then propagating the velocities and positions of the atoms by numerical integration of Newton's equations of motion. In this review, we present an overview of how the MD simulation can be conducted to address computational toxicity problems. The study cases will cover a standard MD simulation performed to investigate the overall flexibility of a cytochrome P450 (CYP) enzyme and a set of more advanced MD simulations to examine the barrier to ion conduction in a human α7 nicotinic acetylcholine receptor (nAChR).

Key words: Molecular dynamics, Force field, Toxicity, Free energy, Enhanced sampling

1. Introduction

1.1. Overview of the Topic-How the Topic Fits into the Wider Scope of Computational Toxicology

Drug toxicity, an exaggerated pharmacological response, is one of the main reasons for high failure rates in drug discovery and development (1, 2). Major drug toxicity mechanisms can be divided into four general groups: *on-target* that is the result of a drug binding to the intended receptor, but at an inappropriate concentration, with suboptimal kinetics, or in unintended tissues; *off-target* that is caused by a drug binding to an unintended receptor; harmful immunological reactions; and idiosyncratic toxicity. The identification of drug toxicity is a complex task, and toxic effects are often not revealed until a compound has been selected for development, or has entered the clinic. However, with the structures of enzymes and receptors associated with chemical toxicity determined or characterized, structure-based studies will bring us closer to understanding the molecular mechanisms of drug toxicity and even predicting drug toxicity (1, 3). Drug metabolism through specialized enzymes, e.g., cytochrome P450, is a major consideration for drug clearance

Brad Reisfeld and Arthur N. Mayeno (eds.), *Computational Toxicology: Volume I*, Methods in Molecular Biology, vol. 929,
DOI 10.1007/978-1-62703-050-2_11,

and metabolite-induced toxicity (4, 5). Virtually all drug molecules are metabolized in the human body, which modifies the pharmacological properties of these drugs. Additionally, a number of hormone-mediated receptors are known to be related to toxicity, e.g., the epidermal growth factor receptor and the androgen receptor, towards which a variety of compounds show agonistic or antagonistic activities (6). Finally, the hERG potassium channels have become an important target for toxicity screening due to their possible involvement in the life-threatening drug-induced long QT syndrome (7, 13) in recent years.

Common approaches in computational toxicity involve the use of cheminformatics to determine if a new compound is toxic (8–11). Quantitative structure–activity relationship (QSAR) techniques have been widely used to correlate toxicity with a variety of physiochemical and structural properties for congeneric series of compounds. As toxicity data accumulates and expands in dimensions, data mining and machine learning techniques from computer science and statistics have also become popular in chemical toxicity predictions. In recent years, high-throughput screening (HTS) and "omic" technologies have made important strides in advancing our understanding of drug toxicity (12–14). In particular, systematic approaches to studying signaling networks and pathways have provided tremendous insights into the toxicity mechanisms at the molecular and cellular levels (1). HTS and "omic" studies produce enormous amount of data, which in turn requires computational approaches to analyze, interpret, and build more predictive models. Cheminformatic-based methods, however, do not provide insights into the molecular mechanisms of drug toxicity, and are nevertheless to predict which compound is likely to be toxic, and to estimate the relative margin of safety for a group of compounds. In addition to these more phenomenological approaches, molecular dynamics (MD) simulation is finding increasing applications in computational toxicity research, especially when the dynamic nature of enzymes or receptors involved in drug toxicity cannot be ignored. With ongoing developments of structural and systems biology and computational techniques, MD simulation has also become an essential component in the emerging multi-scale toxicity modeling approaches that incorporate structural and functional information at multiple scales (15, 16).

MD simulation based on the numerical integration of Newton's equation of motion can reveal the mechanisms of biological processes in their ultimate detail, and in general has been used in two different ways. The first is to use the simulation to probe the actual dynamics of the system. Thus MD simulation opens the possibility for direct observation of the motion of biomolecules at the atomic scale. For example, folding/unfolding of peptides or small proteins(17, 18), water molecules permeation through biological channels (19, 20) have been extensively studied with

MD, which shed light into the underlying mechanisms that would otherwise be difficult to obtain from experiments. The second way is to use simulation simply as a means of sampling. Then, in a statistical mechanics framework, a variety of equilibrium and kinetic properties of the systems can be derived and compared with experiments. Therefore, MD simulation not only allows direct visualization of the dynamics of biomolecules but also helps elucidate the underlying molecular mechanisms (driving forces) of the observed behavior. Since the 1970s, the method of MD simulation has gained popularity in biochemistry and biophysics (21–24). As the simulation capability increases in complexity and scale, MD has widely served as a computational microscope to investigate the molecular details of many complex biological processes, with application to protein folding (25), enzymatic catalysis (26, 27), molecular machines (28), etc. More recently, MD simulation has also been used to address drug toxicity problems in which the dynamic nature of proteins is essential.

1.2. Application Areas for the Techniques

In the context of computational toxicity, MD simulations have been used to help understand the molecular mechanisms of drug toxicity especially for the cases where the target structures are known, and to a less extent to predict drug toxicity. Biological systems are dynamic in nature; characterization of their dynamics at the atomistic level is therefore essential to understanding many biological phenomena including drug toxicity. MD simulation is an excellent computational tool for capturing dynamics of biological systems. When coupled with other computational tools, it opens the opportunity to address many fundamental mechanistic problems in drug toxicity. Current application areas for MD include the following:

First, when combined with the theory of thermodynamics and statistical mechanics, MD simulation can be used to compute many physicochemical properties of drug-like molecules, such as octanol/water partition coefficients, water solubility, solvent accessible area, and Henry's constant. The p*K*a values, chemical reactivity (such as redox potentials) and hydrolysis rate constants are also computable with a combined quantum mechanical/molecular mechanical (QM/MM) treatment (29, 30). Then, using QSAR-like or data modeling approaches, the computed physicochemical properties can be correlated with many toxicity-related properties, e.g., the distribution of a drug in blood and tissue or the kinetics of its metabolic activity.

Second, a few classes of enzymes or receptors are known to be involved in chemical toxicity for various reasons. MD simulations have been extensively applied to investigate the molecular mechanisms of these enzymes or receptors at an atomic level.

The cytochrome P450 superfamily (CYP) is a large and diverse group of enzymes, accounting for ~75% of the total metabolism

(31–33). They play a primary role in reducing drug toxicity through metabolic oxidation that then leads to the clearance of a drug. The CYP P450 enzymes are capable of recognizing a wide variety of chemically diverse substrates, the molecular details of this promiscuity, however, have remained elusive. Therefore, understanding the mechanism and specificity of substrate binding in the CYP enzymes is an important step toward explaining their key role in drug metabolism, toxicity, and xenobiotic degradation. A variety of experimental approaches have been employed to probe the dynamic nature of the enzymes, to study the substrate interaction with the active sites, and to identify potential residues involved in the catalytic mechanisms of the substrates. The availability of the structures of CYP P450 enzymes from X-ray crystallography allows MD simulations to be used to explore various aspects of the protein dynamics in substrate recognition and catalysis, complementing experimental findings (34).

Nuclear receptors comprise a family of ligand-mediated transcription factors within cells that are responsible for sensing hormones and other molecules (35–37). The binding of hormonally active compounds or endocrine disruptors to these receptors can elicit a variety of adverse effects in humans including promotion of hormone-dependent cancers and reproductive disorder. A number of receptors involved in drug or environmental toxicity are known, including thyroid hormone receptor (38), epidermal growth factor receptor (39), aryl hydrocarbon receptor (40, 41), androgen receptor, (42) and estrogen receptor (43). Ligands that bind to and activate nuclear receptors are typically lipophilic, such as endogenous hormones and xenobiotic endocrine disruptors. When the structures of the targets become available, usually from X-ray or by homology modeling, the most common computational approach for understanding the ligand–receptor interactions is molecular docking. However, MD simulations can help overcome many issues in molecular docking simulation, including optimization of the complex structures, accommodating flexibility of the receptors and improvement of scoring functions (44). MD simulations can also be used to provide an explanation for dynamic regulation and molecular basis of agonicity and antagonicity in nuclear receptors (45).

The hERG potassium channel is responsible for the electrical activity of the heart that coordinates the heart beating. hERG blockage has been linked to life-threatening arrhythmias and thus represents a major safety concern in drug development (7). Extensive experimental studies have greatly advanced our knowledge on the molecular basis of hERG-mediated arrhythmias. Although the crystal structure of hERG has not been determined, a few homology models have been developed, enabling MD simulations to be used to study these channels (46, 47). A variety of physiological processes are amenable to MD studies, e.g., how drug molecules are able to bind to hERG and then block the ion flow through the channel. MD simulations can be particularly valuable for membrane-bound hERG potassium channels as experimental characterization of their structural dynamics is very challenging.

1.3. How, When, and by Whom These Techniques or Tools Are Used in Practice

1.3.1. Cytochrome P450 Enzymes

MD simulations have been extensively employed to investigate the conformational dynamics of cytochrome P450 enzymes that is thought to play an important role in ligand binding and catalysis. Meharenna et al. have performed high-temperature MD simulations to probe the structural basis for enhanced stability in thermal stable cytochrome P450. The comparison of the MD trajectories at 500K suggests that the tight nonpolar interactions involving Tyr26 and Leu308 in the Cys ligand loop are responsible for the enhanced stability in CYP119, the most thermal stable P450 known (48). Using MD simulations at normal and high temperatures, Skopalík et al. have studied the flexibility and malleability of three microsomal cytochromes: CYP3A4, CYP2C9, and CYP2A6. MD simulations reveal flexibility differences between these three cytochromes, which appear to correlate with their substrate preferences (49). Hendrychová et al. have employed MD simulations and spectroscopy experiments to probe the flexibility and malleability of five forms of human liver CYP enzymes, and have demonstrated consistently from different techniques that CYP2A6 and CYP1A2 have the least malleable active sites while CYP2D6, CYP2C9, and CYP3A4 exhibit considerably greater degrees of flexibility (50). Lampe et al. have utilized MD simulations in conjunction with two-dimensional heteronuclear single quantum coherence NMR spectroscopy to examine substrate and inhibitor binding to CYP119, a P450 from *Sulfolobus acidocaldarius*. Their results suggest that tightly binding hydrophobic ligands tend to lock the enzyme into a single conformational substate, whereas weakly binding low-affinity ligands bind loosely in the active site, resulting in a distribution of localized conformers. Their MD simulation results further show that the ligand-free enzyme samples ligand-bound conformations of the enzyme, thus suggesting that ligand binding proceeds through conformational selection rather than induced fit (51). By means of MD simulations, Park et al. have examined the differences in structural and dynamic properties between CYP3A4 in the resting form and its complexes with the substrate progesterone and the inhibitor metyrapone (52).

The dynamics of the substrate binding site has also been a matter of extensive MD studies because of its crucial role in understanding the specificity and selectivity of the enzyme towards different substrates. Diazepam is metabolized by CYP3A4 with sigmoidal dependence kinetics, which has been speculated to be caused by the cooperative binding of two substrates in the active site. Fishelovitch et al. have performed MD simulations of the substrate-free CYP3A4 and the enzymes with one and two diazepam molecules bounded to understand the factors governing the cooperative binding (53). Seifert et al. performed extensive MD simulations to investigate the molecular basis of activity and regioselectivity of the human microsomal cytochrome CYP2C9 toward its substrate warfarin (54). The simulations suggest that the crystal structure of CYP2C9 with warfarin was captured in a nonproductive state, whereas in the productive states both 7- and 6-positions of warfarin (with position 7 markedly favored over position 6) are in contact with the heme, consistent with experimentally determined regioselectivity.

The crystal structure of CYP3A4 revealed that a small active site is buried inside the protein, extending to the protein surface through a narrow channel. Conventional MD simulations and MD simulations with enhanced sampling methods have been used to explore the dynamical entrance and exit of substrates/products into the active site through the access channel. Wade et al. have extensively investigated the ligand exit pathways and mechanisms of P450cam (CYP101), P450BM-3 (CYP102), and P450eryF (CYP107A1) by using random expulsion MD and conventional MD simulations (55), suggesting that the channel opening mechanisms are adjusted to the physicochemical properties of the substrates and can kinetically modulate the protein–substrate specificity (56). Steered molecular dynamics (SMD) simulations have been used by Li et al. and Fishelovitch et al. to pull metyrapone (57), temazepam, testosterone-6βOH (58) out of CYP3A4 respectively, in order to identify the preferred substrate/product pathways and their gating mechanism. Based on the simulation results, they concluded that product exit preferences in CYP3A4 are regulated by protein–substrate specificity. A quantitative assessment of the channel accessibility for various substrates would require the calculation of the potential of mean force (PMF) along the plausible ligand access pathways, but research along this line has not been reported to date due to the lack of efficient MD sampling techniques.

1.3.2. hERG Potassium Channel

No experimentally determined 3D structure of the hERG potassium channels is available, whereas the structures of several homologous voltage-gated potassium channels have been determined by X-ray crystallography. MD simulations have been previously used to refine, evaluate homology models and to acquire an atomic-level description of the channel pore and possible binding of channel blockers. Stary et al. have used a combination of geometry/packing/normality validation methods as well as MD simulations to obtain a consensus model of the hERG potassium channel (47). Subbotina et al. have used MD simulations to optimize a full model of the hERG channel including all transmembrane segments developed by using a template-driven de-novo design with ROSETTA-membrane modeling, leading to a structural model that is consistent with the reported structural elements inferred from mutagenesis and electrophysiology experiments (59). Masetti et al. have run MD simulations of the homology models of the hERG channel in both open and closed states, and showed that the combined use of MD and docking is suitable to identify possible binding modes of several drugs, reaching a fairly good agreement with experiments (46).

During the past decade, many aspects of channel functions (such as gating, ion permeation, and voltage sensing) of voltage-gated potassium channels have been extensively studied by means of MD simulations. In comparison, MD studies of the gating mechanisms and functions of the hERG channels have been rare. However, as our knowledge about the structural features of this important protein continues to advance, MD simulations are expected to be increasingly used in the study of hERG. Stansfeld et al. have studied the inactivation mechanisms in the hERG potassium channels, and have revealed that the carbonyl of Phe627, forming the S0 K^+ binding site, swiftly rotates away from the conduction axis in the wild-type channel while occurs less frequently in the non-inactivating mutant channels (60). MD simulations of a hERG model carried out by Kutteh et al. suggest that the fast inactivation might be caused by an unusually long S5-P linker in the outer mouth of hERG that moves closer to the channel axis, possibly causing a steric hindrance to permeating K^+ ions (61). Osterberg et al. have used MD simulations combined with docking and the linear interaction energy method to evaluate the binding affinities of a series of sertindole analogues binding to the human hERG potassium channel. The calculations reproduce the relative binding affinities of these compounds very well and indicate that both polar interactions near the intracellular opening of the selectivity filter as well as hydrophobic complementarity in the region around F656 are important for blocker binding (62).

2. Materials

2.1. Common Software and Methods Used in the Field

Many MD packages have been developed over the years, including CHARMM (63), AMBER (64), GROMOS (65), GROMACS (66), TINKER (67), MOLDY (68), DL_POLY (69), NAMD (70), LAMMPS (71), and Desmond (72). Some of them have their associated force fields, while others only provide an MD engine and require compatible force fields for running a simulation. For biomolecular simulation, CHARMM, AMBER, GROMOS, NAMD, and GROMACS are most widely used. CHARMM and AMBER have enjoyed the longest history of continuous development and offer a wide range of functionalities for advanced sampling as well as pre- and post-simulation analysis. They also carry their own force fields (73, 74) developed for proteins, nucleic acids, lipids, and carbohydrates. The development of NAMD and GOMACS has been focused on the performance. Recent versions of NAMD and GROMACS have shown remarkable parallel efficiency on both high-end parallel platforms and commodity clusters, thus being very attractive for large-scale MD simulations of biomolecular systems. GROMACS supports a variety of force fields, including all-atom GROMOS (75), AMBER (74), and CHARMM (73) as well as united atom and coarse-grained force fields. NAMD runs with standard AMBER, CHARMM, and GROMACS style ASCII topology and coordinates without the need of any format conversion. Desmond is a relatively new software package developed at D.E. Shaw Research to perform high-performance MD simulations of biological systems. Desmond supports several variants of the AMBER, CHARMM, and OPLS-AA force fields (76).

2.2. Special Requirements for the Software and Methods (e.g., Hardware, Computing Platform, and Operating System)

Most MD programs run under the Unix/Linux like operating systems. As MD simulations of biomolecules are computationally intensive, production runs are often performed on high-end parallel platforms or commodity clusters of many processors. Recently, with the emergence of special purpose processors such as field-programmable gate array (FPGA) and graphic processing unit (GPU) designed to speed up computing-intensive portions of applications, several MD codes have been adapted and ported to run on these platforms as well (77–79). Fortran or C/C++ compilers are required since most MD codes are written in Fortran or C/C++. Special parallel programming libraries are also required to run MD in parallel on multiple CPU cores or multiple nodes on a network, e.g., the MPI library for distributed memory systems, and POSIX threads and OpenMP for shared memory systems. Most MD codes use MPI, a message-passing application programmer interface, while NAMD uses Charm++ parallel objects for good performance on a wide variety of underlying hardware platforms. FFTW, a C subroutine library for computing the discrete Fourier transform, is extensively employed by state-of-the-art MD codes for treating long-ranged electrostatic interactions with the particle mesh Ewald (PME) (80) or particle–particle particle–mesh (P3M) (81) methods. Pre-compiled libraries, e.g., special math library, or script language library may also be required by some MD programs, such as the TCL library used in both NAMD (70) and VMD (82).

2.3. Preferred Software and Why

Several MD codes are currently used by the biomolecular simulation community, and each of them has its strength and weakness. Choosing which MD software for your simulation will depend on many factors, such as the force field compatibility, the speed/scalability, the support for simulation setup and post-simulation analysis, and the availability of special simulation techniques. It also depends on the problem/system under investigation and the computer resources at your disposal. NAMD (70) has been the software of preference for large-scale, conventional explicit solvent simulations for the following reasons: (1) superior scalability; (2) compatibility with AMBER and CHARMM force fields; (3) it uses the popular molecular graphics program VMD for simulation setup and trajectory analysis; (4) offers flexibility by providing a TCL scripting language interface so that the user can customize his/her own code for any special purpose without the need to modify the source code and recompile. Last but not least, NAMD provides excellent user support including online documentations, tutorials, and frequent workshops.

3. Methods

3.1. Molecule Building

3.1.1. Initial Coordinates

MD simulation starts with a 3D structure as the initial configuration of the system. This structure can be an NMR or X-ray structure obtained from the Brookhaven Protein Databank (http://www.rcsb.org/pdb/). If no experimentally determined structure is available, an atomic-resolution model of the "target" protein can be constructed from its amino acid sequence by homology modeling. Homology modeling can produce high-quality structural models when the target and the template, a homologous protein with an experimentally determined structure, are closely related. The choice of an initial configuration must be done carefully as this can influence the results of the simulation. When multiple PDB entries are available, which structure to choose usually depends on the quality of the structure, the state in which the structure was captured and the experimental condition under which the structure was determined. It is important to choose a configuration in a state best representing what one wishes to simulate.

3.1.2. Prune Protein Structure

With a 3D structure in hand, still a few things need to be sorted out before a simulation can get started. (1) *Removing redundant atoms*: X-ray structures may be captured in a multimer state; NMR may yield an ensemble of conformations; multiple conformations may exist for some flexible side chains; extra chemical agents may be added to facilitate the structure determination. All these redundant atoms should be removed prior to further structure processing. (2) *Add missing atoms*: depending on the quality of a PDB structure, some coordinates may be missing. It is important to check whether the missing coordinates are relevant to the question to be addressed. First, for those proteins which active form are multimeric, it is necessary to construct the multimer structures from the deposited monomer coordinates using associated symmetry and/or experimental constraints. Second, if only a few residues or atoms are missing, most MD building programs can reliably rebuild these missing parts. However, if the missing gap is large, e.g., a flexible loop that is often found missing or partially missing due to its dynamic nature, most MD building programs will fail in rebuilding a reliable structure. Ab initio loop modeling or homology modeling tools should be used instead for these cases (83). Third, since X-ray structures usually do not give the hydrogen positions, another necessary step is to add hydrogen atoms and assign appropriate protonation state to ionizable residues. The ionization states of residues such as glutamate, aspartate, histidine, lysine, and arginine can be extremely relevant for the function of a protein. So it is advisable to use p*Ka* calculations to aid in the assignment of protonation states. Several software packages and Web servers are available for this purpose, such as H++ (http://biophysics.cs.vt.edu/H++/), Karlsberg+ (http://agknapp.chemie.fu-berlin.de/karlsberg/), PROPKA (http://propka.ki.ku.dk/), and MCCE (http://134.74.90.158/). (3) *Replace mutated atoms*: For those proteins that do not form stable structures or only form transient (weak) structure complexes with their ligands, some of the residues might have been modified in order to obtain stable crystal structures. In these cases, the modified residues or atoms should be replaced by the native ones. (4) *Build the structure from multiple components*: when the complete structure of a protein–ligand complex or a multi-domain protein complex is not available, the molecular docking or protein–protein docking tools can be used to build an initial complex structure for simulation and refinement, from the available structures of its components.

3.1.3. Molecular Structure and/or Topology File

Given a refined PDB structure of the protein or protein complex, the next step is to generate the topology or parameter files for the system. The topology file contains the geometrical information of the system, e.g., bonds, angles, dihedral angles, and interaction list. Sometimes, topology files are combined with parameter files, thus may also contain the force field parameters, i.e., the functional forms and parameter sets used to describe the potential energy of the system. Various force fields have been developed for different types of biomacromolecules, including proteins, nucleic acids, lipids, and carbohydrates. The choice of an appropriate force field is of substantial importance, and will depend on the nature of the system (problem) of interest. In general, the chosen force field should be compatible with the MD engine, and the force fields for different components of the system should be consistent with each other. Most MD programs provide auxiliary utility programs for generating topology and parameter files from PDB files. The procedure is straightforward except for a few potentially confusing items, for instance, some special treatments (or patch) may be required if there exists a disulfide bond between a pair of cysteine residues, if the residue/atom name conventions in PDB and the topology generating programs are different, or if a nonstandard protonation state is assigned to a residue. If the system contains a nonstandard residue or novel ligand molecule, then its force field must be generated first. Also, the new force field must be compatible with that used for the rest of the system. Generalized force fields for drug-like molecules compatible with the AMBER (84) and CHARMM (85) all-atom force fields have become available.

3.1.4. Solvate the System

To simulate a biological system in an aqueous solution, a choice should be made between explicit and implicit solvent models. However, only explicit solvent models (such as the TIP3P, SPC/E water models) will be discussed in this review. For crystal waters, it is advisable to keep them, especially for those located in the active site or the interior of the protein that often play a structural or functional role. When necessary, additional water molecules can be placed inside or around the protein using programs such as DOWSER (http://hekto.med.unc.edu:8080/HERMANS/software/DOWSER/). The system is then solvated with a pre-equilibrated water box. Ions (usually Na^+, K^+, Cl^-) are added to neutralize the system, and to reach a desired ionic concentration. For membrane-associated systems, proteins need to be inserted to a pre-equilibrated lipid bilayer. The orientation and position of proteins in membrane can be determined by an online server OPM (http://opm.phar.umich.edu/), together with experiments and the modeler's intuition. The lipid composition is another issue worthy of consideration as accumulating evidence has shown it can have a significant and differentiate impact on the function of membrane-bound proteins. The CHARMM force field supports six types of lipids 1,2-dipalmitoyl-sn-phosphatidylcholine (DPPC), 1,2-dimyristoyl-sn-phosphatidylcholine (DMPC), 1,2-dilauroyl-sn-phosphatidylcholine (DLPC), 1-palmitoyl-2-oleoyl-sn-phosphatidylcholine (POPC), 1,2-dioleoyl-sn-phosphatidylcholine (DOPC), and 1-palmitoyl-2-oleoyl-sn-phosphatidylethanolamine (POPE), while VMD provides two types of pre-equilibrated POPC and POPE membrane patches. After everything is assembled together, the new structure/topology files can be built, followed by several rounds of energy minimization to remove bad van der Waals contacts.

3.2. Conducting Simulations

3.2.1. MD Theory and Algorithm

By integrating Newton's equations of motion, molecular dynamics allows the time evolution of a system of particles to be followed in terms of a dynamical trajectory (a record of all particle positions and momenta at discrete points in time over a time span T). For a system of N particles of masses m_i and positions $\{r_i\}$, the equations of motion are:

$$F_i = -\nabla_{\mathbf{r}_i} U = m_i \ddot{\mathbf{r}}_i, \quad i = 1, \ldots, N, \tag{1}$$

where the potential energy U acting on the atomic nuclei may be described, neglecting the electronic interactions, by simple functions of the ionic positions $U = U(\{\mathbf{r}_i\})$. The total potential energy function can be expressed as a sum of potentials derived from simple physical forces: van der Waals, electrostatic, mechanical strains arising from ideal bond length and angle deviations, and internal torsion flexibility. The forces can be separated into bonded and nonbonded terms.

$$\begin{aligned} U(\{\mathbf{r}_i\}) = & \sum_b K_b(l - l_0)^2 + \sum_a K_a(\theta - \theta_0)^2 + \sum_{\text{imp.}} K_\omega(\omega - \omega_0)^2 \\ & + \sum_{\text{dihed.}} K_\psi[(1 + \cos(n\psi - \delta)] + \sum_{i,j} \frac{q_i q_j}{4\pi\varepsilon_0\varepsilon_m r_{ij}} \\ & + \sum_{i,j} \left(\frac{A_{ij}}{r_{ij}^{12}} - \frac{B_{ij}}{r_{ij}^6} \right). \end{aligned} \tag{2}$$

In classical mechanics, the trajectory of the system is determined by its initial conditions, namely the initial atomic positions and velocities. The integration algorithm can be derived from the Taylor series expansion of the atomic positions with respect to the simulation time t:

$$r(t + \delta t) = r(t) + \dot{r}(t)\delta t + \frac{1}{2}\ddot{r}(t)\delta t^2 + \frac{1}{3}\dddot{r}\delta t^3 + \mathrm{O}(\delta t^4). \tag{3}$$

By summing the expansions for $r(t + \delta t)$and $r(t - \delta t)$, we obtain the Verlet algorithm:

$$r(t + \delta t) = 2r(t) - r(t - \delta t) + \frac{1}{2}a(t)\delta t^2 + \mathrm{O}(\delta t^4), \tag{4}$$

where $r(t)$ and a are the position vector and the acceleration vector, respectively; δt is the time step. The time step δt has to be chosen sufficiently small in order to ensure that continuity in the forces acting on the atoms and overall energy conservation.

The ensemble average of an observable A in a system characterized by the Hamiltonian H, in the classical limit of statistical mechanics ($\hbar \to 0$), in most of the cases can be considered equivalent to its time average:

$$\begin{aligned} <A> = & \frac{\int \mathrm{d}p^N \mathrm{d}r^N A(p^N, r^N) \mathrm{e}^{-\beta H(p^N, r^N)}}{\int \mathrm{d}p^N \mathrm{d}r^N \, \mathrm{e}^{-\beta H(p^N, r^N)}} \\ \cong & \lim_{t \to \infty} \frac{1}{t} \int \mathrm{d}t' A[p^N(t'), r^N(t')]. \end{aligned} \tag{5}$$

Here, $\beta = 1/k_\mathrm{B}T$, k_B is the Boltzmann constant and N is the number of degrees of freedom for the system under consideration. The above equivalence relation is known as the ergodic hypothesis, and for many systems similar to those contained in this thesis, the validity of the hypothesis has been confirmed. The ergodic hypothesis provides the rationale for the molecular dynamics method and a practical recipe that allows ensemble averages to be determined from time averaging over dynamical trajectories.

3.2.2. Interaction Treatment and Integration Method

How an MD simulation will be run is controlled by a set of input parameters contained in a configuration file, such as the number of steps and the temperature. The main options and values can be divided into three categories: (1) interaction energy treatment; (2) integration method; (3) ensemble specification. Additional sets of parameters may be used by advanced simulation techniques, e.g., enhanced sampling and free energy simulations. Most explicit solvent MD simulations employ periodic boundary conditions to avoid the boundary artifact. The most time consuming part of a simulation is the calculation of nonbonded terms in potential energy functions, e.g., the electrostatic and van der Waals forces. In principle, the nonbonded energy terms between every pair of atoms should be evaluated; in this case, the number of operations increases as the square of the number of atoms for a pair wise model (N^2). To speed up the computation, the nonbonded interactions, e.g., the electrostatic and van der Waals forces, are truncated if two atoms are separated greater than a predefined cutoff distance. The long-ranged electrostatic interactions typically use FFT-based PME (80) or particle–particle particle–mesh (P3M) (81) methods that reduce the computational complexity from N^2 to $N \log N$. MD simulation involves the numerical integration of Newton's equations of motion in finite time steps that must be small enough to avoid discretization errors. Typical time steps used in MD are in the order of 1 fs (i.e., smaller than the fastest vibrational frequency in biomolecular systems). This value may be increased by using constraint algorithms such as SHAKE (86), which fix the fastest vibrations of the atoms (e.g., hydrogens). Multiple-time-step methods are also available, which allow for extended times between updates of slowly varying long-range forces (87). The total simulation duration should be chosen to be long enough to reach biologically relevant time scales or allow sufficient sampling (barrier crossing), and should also account for the available computational resources so that the calculation can finish within a reasonable wall-clock time.

3.2.3. Temperature and Pressure Control

MD simulation is often performed on the following three thermodynamic ensembles: microcanonical (NVE), canonical (NVT), and isothermal–isobaric (NPT). In the NVE ensemble, the number of particles (N), the volume (V), and the total energy (E) of the system are held constant. In the canonical ensemble, N, V, and the temperature (T) are constant, where the temperature is maintained through a thermostat. In the NPT ensemble that corresponds most closely to laboratory conditions, N, T, and the pressure (P) are held constant, where a barostat is needed in addition to a thermostat. In the simulation of biological membranes, anisotropic pressure control is more appropriate, e.g., constant membrane area (P_A) or constant surface tension (P_γ). A variety of thermostat methods are available to control temperature, which include velocity rescaling (88), the Nosé-Hoover thermostat (89, 90), Nosé-Hoover chains (91), the Berendsen thermostat (92), and Langevin dynamics. Note that the Berendsen thermostat might cause unphysical translations and rotations of the simulated system. As with temperature control, different ways of pressure control are available for MD simulation, including the length-scaling technique of Berendsen (92) and the extended Nosé-Hoover (extended Lagrangian) formalism of Martyna et al. (93).

3.2.4. Equilibration

Before a production run, a multistage equilibration simulation is often necessary especially for a heterogeneous system composed of multiple components or phases. For a typical system of a protein embedded in an explicit solvent box, the equilibration often starts with fixing the protein and letting the waters move to adjust to the presence of the protein. After the waters are equilibrated, the constraints on the protein can be removed and let the whole system (protein + water) evolve with time. During the heating phase, initial velocities corresponding to a low temperature are assigned, and then the temperature is gradually brought up until the target temperature is reached. As the simulation continues, several properties of the system are routinely monitored, including the temperature, the pressure, the energies, and the structure. The production simulation can be started only until these properties become stable with respect to time. The purpose of the equilibration phase is to minimize nonequilibrium effects and avoid unphysical local structural distortions, thus leading to a more meaningful simulation.

3.3. Postprocessing and Analyzing the Results

During an MD simulation, the coordinates and velocities of every atom in the system can be saved at a prespecified frequency for later analysis. From the saved data, in principle, all the structural, thermodynamic (i.e., energy, temperature, pressure, velocity distributions) and dynamic (diffusion, time correlation functions) properties of the system can be computed. A variety of post-processing utility programs can be found in popular MD software packages. As MD simulations are often used to help visualize and understand conformational dynamics at an atomic level, after the simulation the MD trajectory is typically first loaded to molecular graphics programs to display possible structural changes of interest in a time-dependent way. Also, post-processing consists of more quantitative and detailed structural analysis, such as distance, angle, dihedral angle, contact, hydrogen bond, radius of gyration, radial distribution functions, protein secondary structure, sugar puckering, and DNA local curvature. Additional geometrical quantities that are routinely calculated from an MD simulation trajectory include the following: root mean square difference (RMSD) between two structures and RMS fluctuations (RMSF). The time trajectory of RMSD shows how a protein structure deviates from a reference structure as a function of time, while the time-averaged RMSF indicates the flexibility of different regions of a protein, which is related to the crystallographic *B*-factors.

Nowadays, large-scale MD simulations produce an immense quantity of data. Clustering and correlation analyses are standard mathematical tools that are used to group similar structures or detect correlations in large data sets. Moreover, the covariance matrix of atomic displacements can be diagonalized to obtain large-scale collective motions under a quasi-harmonic approximation, and to compute configurational entropies via various approximations(94, 95). Another valuable post-processing procedure for MD simulations is the (free) energy decomposition by the MM-PB (GB)SA method (96), or by the integration of individual force components. Time correlation functions can also be easily computed from an MD trajectory, which, in turn, can be used to relate the dynamics of atoms and electrons to various molecular spectroscopy data using the theory of nonequilibrium statistical mechanics. For nonequilibrium simulations or simulations using special sampling techniques, e.g., umbrella sampling (97), generalized ensemble algorithm (98), and Wang-Landau method (99). for achieving importance sampling, special post-processing, such as weighted histogram (100) or maximum-likelihood method, is required to recover equilibrium thermodynamic quantities from the biased simulations.

3.4. Validating the Results

The simulation results should be validated at two levels: first, to assess whether the simulation is conducted properly; second, to assess whether the model underlying the simulation sufficiently describes the problem to be probed. A variety of MD outputs can provide hints about whether the simulation is conducted properly, including the time-dependent thermodynamic quantities (i.e., temperature, pressure, and volume), their fluctuations, and the distribution of velocities in the system. For example, one would expect the conservation of the total energy in an NVE ensemble simulation, while any significant energy drift indicates possible problem of either the integration algorithm or the interaction force evaluation. Structural features of the system can be validated by visualization to rule out any unphysical (inappropriate) changes, contacts, or assembly; computer programs, e. g., Verify3D (http://nihserver.mbi.ucla.edu/Verify_3D/), Procheck phi/psi angle check (http://www.ebi.ac.uk/thornton-srv/software/PROCHECK/), WHAT_CHECK Packing 2 (http://swift.cmbi.ru.nl/gv/whatcheck/), Prosa2003 (http://www.came.sbg.ac.at/prosa_details.php), ModFOLD (http://www.reading.ac.uk/

bioinf/ModFOLD), and a local quality assessment method (101) are also available for assessing the overall quality of the structures sampled in simulation. The second level of validation is usually through comparing simulation results to experiments and/or results of other methods, which will be further discussed below. When no experimental result is available for comparison, one should run benchmark/test simulations first to validate on the quantities that have already been experimentally determined. Only upon this validation, can more trustworthy simulations be performed for other unknown but related properties. Finally, it is always advisable to run multiple or control simulations, since the behavior of the control simulations or the difference between the production and control simulations usually provides valuable insights into the reliability of the underlying simulations.

3.5. Comparing Simulation Results to Experiments and/or Results of Other Methods

MD simulation provides a window to investigate the ultimate details of biological processes. But it also suffers from a number of limitations as any theoretical models do, especially when applied to complex biological systems or processes. So whenever possible MD results should be compared to experiments. From an MD simulation, structural quantities can be easily calculated and most of them are directly comparable to experimental measurements. The average (or most populated) structure can be compared to X-ray crystallographic or nuclear magnetic resonance (NMR) structure. Positions of hydrogen atoms in the context of hydrogen bonding can be compared to neutron diffraction data. Distances in solution can be compared to the NMR nuclear Overhauser (NOE) experiments. Protein secondary structure features can be compared to circular dichroism, infrared, or Raman spectroscopic experiments. Orientation of molecular fragments can be compared to NMR order parameters.

It remains challenging for experimental techniques to probe the conformational dynamics of biomolecules at the atomic level. Most measurements are only able to capture one aspect of the dynamical changes, such as distance or overall shape. In this respect, MD simulations ideally complement the experiments. Time evolution of distance changes derived from simulation can be compared with fluorescence resonance energy transfer (FRET) experiment; slow conformational dynamics ($>10^{-9}$ s) captured in simulation can be compared to NMR residual dipolar coupling (RDC) experiment; correlated motions derived from simulation can be compared to quasi-elastic and inelastic neutron scattering, diffuse X-ray scattering, inelastic Mossbauer scattering, and dielectric spectroscopy.

Many thermodynamic (e.g., the free energy changes associated with solvation, ligand binding or conformational shift) or kinetic (e.g., the rate of an enzyme-catalyzed reaction or the single-channel conductance of a biological channel) quantities can also be calculated from an MD simulation and compared to experimental measurements. Even though a quantitative agreement between simulations and experiments is still a difficult task for many complex biological problems, an increasing number of successful examples are appearing in the literature. In practical applications, indirect comparison of the simulation results can also be made with many experiments that correlate the structural data with the functional measurements, such as mutagenesis and labeling experiments.

3.6. Interpreting the Results

The interpretation of MD simulation results are often straightforward since the simulation provides all necessary information in detail, i.e., the coordinates and velocities for every atom in the system for all the dynamic steps. However, experiments are often conducted for complex biological systems under complex conditions, so when interpreting the simulation results, one should always consider whether the simulation system and/or conditions adequately reflect or correspond to the experimental settings; otherwise the interpretation of the results could be irrelevant or even invalid. Furthermore, the interpretation of the simulation results can become complicated in the following two situations.

First, MD simulations are often limited in their abilities to investigate "long"-time scale motions. So in practical applications, it often involves the use of thermodynamic data to avoid the direct simulation of slow dynamics, such as the use of PMF computed with enhanced sampling techniques to help understand the ion conductance in biological channels; or the use of multiple short trajectories to approximate the long time dynamics; or the extrapolation of the limited data to project possible long time dynamics. The convergence of the thermodynamic quantities is the central question when interpreting large-scale biomolecular simulations (102). Unfortunately, assessing the errors introduced by these methods is very difficult to do and often is necessary to be validated by experiments as discussed above.

Second, when the experimental data is not directly comparable, some kind of assumption (or model) must be invoked to correlate the simulation data with the measurements, such as crystallographic *B*-factors, NMR *S* order parameters, and mutagenesis data. Techniques like quasi-elastic and inelastic neutron scattering can in principle provide information about correlated motions. However, the interpretation of neutron scattering at molecular level often requires a theoretical model. For these cases, one should bear in mind the underlying assumption and/or shortcomings of the model when interpreting the simulation results or comparing them to experiments. For example, thermal fluctuations are often related to crystallographic *B*-factors via the relationship $B = 8\pi^2 u^2$, where u is the time-averaged atomic fluctuation. However, this relationship assumes that the *B*-factors only approximate thermal motion within a well-ordered structure as a harmonic oscillator (i.e., isotropic vibration), while discounting the disorder of the protein.

3.7. Improving the Model

MD simulations suffer from several drawbacks, which are due to the empirical force fields, the limited simulation length and size, and the way the simulation models are built. The classical mechanical force field functions and parameters are derived from both experimental work and high-level quantum mechanical calculations based on numerous approximations (103, 104). Limitations in current force fields, such as inaccurate conformational preferences for small proteins and peptides in aqueous solutions have been known for years, which have led to a number of attempts to improve these parameters. Recent re-parameterization of the dihedral terms in AMBER (105) and the so-called CMAP correction in CHARMM (106) have significantly improved the accuracy of empirical force fields for protein secondary structure predictions. Moreover, many existing force fields based on fixed charge models that do not account for electronic polarization of the environment, although more sophisticated and expensive polarizable models have been shown to be necessary for accurate description of some molecular properties. A few polarizable force field models have been developed over the past several years (107–110). Recent systematic validations of AMOEBA force field have shown significant improvements over fixed charge models for a variety of structural and thermodynamic and dynamic properties (108). An increasing use of this next-generation of force field models in biomolecular simulations is anticipated within next few years. Furthermore, classical force fields are based on the assumption that quantum effects play a minor role or can be separated from classical Newtonian dynamics. A proper description of charge/electron transfer process or chemical bond breaking/forming requires quantum treatment that can be incorporated into simulations in different ways.

Another way of improving the model is to extend the time scales spanned by MD simulations. Currently attainable time scales are still about 3–4 orders of magnitude shorter than most biologically relevant ones. Methodologically, two general ways exist to extend the time scale of an MD simulation: to make each integration step faster (mainly) through parallelization, or to improve the exploration of phase space via enhanced sampling techniques. During recent years, tremendous efforts have been focused on improving parallel efficiency of the MD codes so that more CPU processors can be used, which has enabled many μs and even ms MD simulations of biological systems. One example is the use of a special hardware computer Anton to reach ms simulations of an Abl kinase (111), an NhaA antiporter (112), and a potassium channel (113). As the computing power continues to increase, we expect next generation of computer systems to significantly expand our capability to simulate more complex and realistic systems for longer times. However, the increase in computing power alone will not be sufficient, and the development of more efficient and robust enhanced sampling techniques will also be required to address many challenging thermodynamic and kinetic problems in biology.

A variety of enhanced sampling techniques have been developed, to obtain better converged equilibrium thermodynamic quantities, such as generalized ensemble methods (98, 114), Wang-Landau algorithm (99), meta-dynamics (115), and accelerated MD (116), or to obtain reaction pathways and reaction rates, such as transition path sampling (117) or Markov state models (118).

4. Examples

4.1. Cytochrome P450: A Simple Problem

We will first show how to carry out a standard MD simulation of the cytochrome P450 to investigate the overall flexibility of the protein. Three crystal structures of CYP3A4, unliganded, bound to the inhibitor metyrapone, and bound to the substrate progesterone, have been determined (119). The comparison of the three structures revealed little conformational change associated with the binding of ligand. So it will be interesting to investigate if any dynamical differences exist among the three structures. Here, we demonstrate how the protein dynamics can be probed by MD simulations using the NAMD package (70). The input and configuration files will be prepared with VMD (82) and TCL scripts.

4.1.1. Building a Structural Model of CYP3A4

Our simulation will start from an X-ray crystal structure of CYP3A4, which is an unliganded CYP3A4 soluble domain captured at a 2.80 Å resolution (119). The PDB file of CYP3A4 can be downloaded from the PDB database (PDB entries 1w0e). Given a PDB structure, the next step is to generate the PSF and PDB files using VMD and *psfgen* plugin. The script *protein.tcl* contains the detailed steps for the process, which can be executed by typing in a Linux terminal,

vmd -dispdev text –e protein.tcl > protein.log

4.1.2. Solvating and Ionizing the System

Now we will use VMD's *Solvate* plugin to solvate the protein. *Solvate* places the solute in a box of pre-equilibrated waters of a specified size, and then removes waters that are within a certain cutoff distance from the solute. After solvation, we will use VMD's *Autoionize* plugin to add ions to neutralize the system, which is important for Ewald-based long range electrostatic method such as particle Ewald mesh (PME) to work properly. It can also create a desired ionic concentration. What *Autoionize* does is to randomly replace water molecules with ions.

vmd -dispdev text –e solvate.tcl > solvate.log
vmd -dispdev text –e ionize.tcl > ionize.log

4.1.3. Running a Simulation of CYP3A4

The solvated system comprises about 40,000 atoms. We will run the simulations in a local Linux cluster computer using 32 cores (four nodes and each node with eight cores). We will first minimize

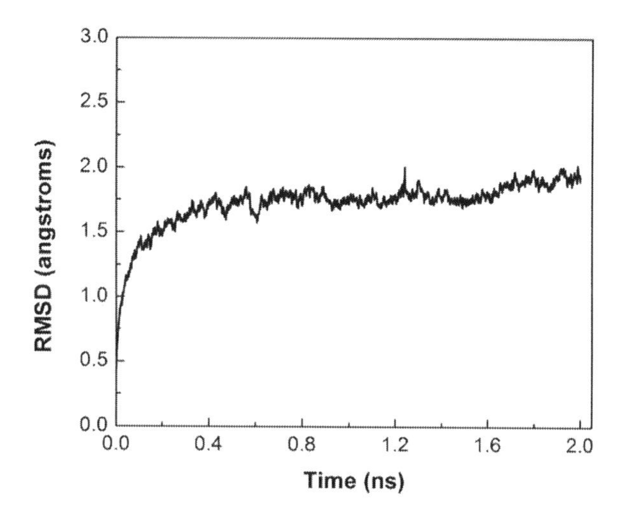

Fig. 1. Root-mean-square deviation (RMSD) as a function of time during the MD simulations of CYP3A4.

the system for 2,000 steps to remove bad contacts. The minimization run can be executed by submitting to the PBS batch scheduler with the command,

qsub runmin.pbs

The configuration file for minimization is *min.namd*, also shown below.

After minimization, an MD simulation starting from the minimized structure will be run by submitting to the PBS batch scheduler with the command,

qsub runmd.pbs

The PBS run script *runmd.pbs* is similar to *runmin.pbs*. The configuration file for the MD simulation *md.namd* is shown below.

4.1.4. Analyze the Results

The root mean square deviation (RMSD) is a frequently used measure of the differences between the structures sampled during the simulation and the reference structure. Using simulation trajectory as an input, RMSD can be calculated with VMD > Extensions > Analysis > RMSD Trajectory Tool for any selected atoms. Structure alignment will usually be performed to remove the translational and rotational movements. Backbone RMSD as a function of time for CYP3A4 is shown in Fig. 1. Overall, the CYP3A4 structure appears quite stable, with RMSD quickly reaching a plateau of 1.8 Å after about 0.4 ns of simulation.

The root mean square fluctuation (RMSF) measures the movement of a subset of atoms with respect to the average structure over the entire simulation. RMSF indicates the flexibility of different regions of a protein, which can be related to crystallographic *B* factors. Figure 2a illustrates the RMSFs of the Cα atoms from the simulation (red line) in comparison to those (black line)

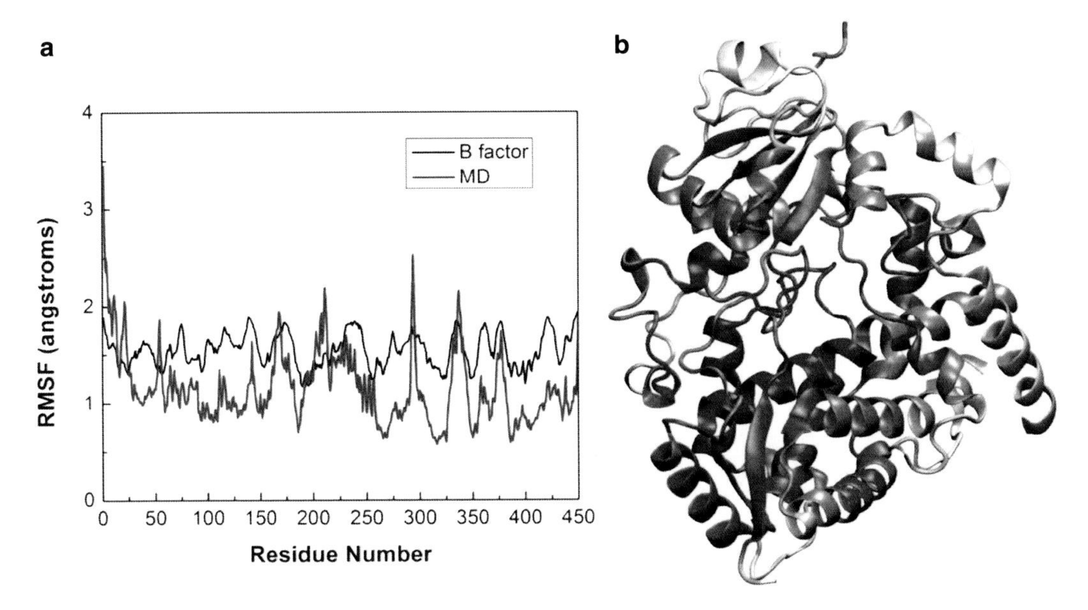

Fig. 2. RMSFs of the Cα atoms from the MD simulation as compared to the experimental data (*black line*), which were calculated from the *B*-factors of CYP3A4 (PDB code: 1w0e) using $\mathrm{RMSF} = \sqrt{(3/8)B_{\mathrm{factor}}}/\pi$. The computed RMSF values are color-coded onto a cartoon representation of the protein structure, with red corresponding to the most mobile region and blue corresponding to the most stable region.

obtained from crystallographic *B*-factors. Overall, the pattern of the computed RMSFs is moderately consistent with that obtained from crystallographic *B*-factors. The RMSF values can be filled in the Beta or Occupancy field of a PDB file so that the flexibility of different regions of a protein can be displayed in different colors with VMD. Figure 2b illustrates a cartoon representation of the CYP3A4 structure color-coded in the RMSF values. Clearly, the α helices and β-strands exhibit low flexibility, while regions that fluctuate most significantly are the loops connecting helices and strands.

4.2. Nicotinic Acetylcholine Receptor (nAChR): A More Complex Problem

For a more complex problem, we will use MD simulations to examine the barrier to ion conduction in a model of the cationic human α7 nicotinic acetylcholine receptor (nAChR)(Fig. 3). nAChR concentrates at synapses, where it responds to nerve-released acetylcholine to mediate excitatory transmission throughout the central and peripheral nervous systems (120). The binding of neurotoxins from snake venoms, such as bungarotoxin, to nAChR strongly blocks the channel conductance, thus leading to neuronal toxicity or even sudden death. The simulation strategies used for nAChR, a membrane integral ion channel, should be conveniently applicable to the study of hERG potassium channels.

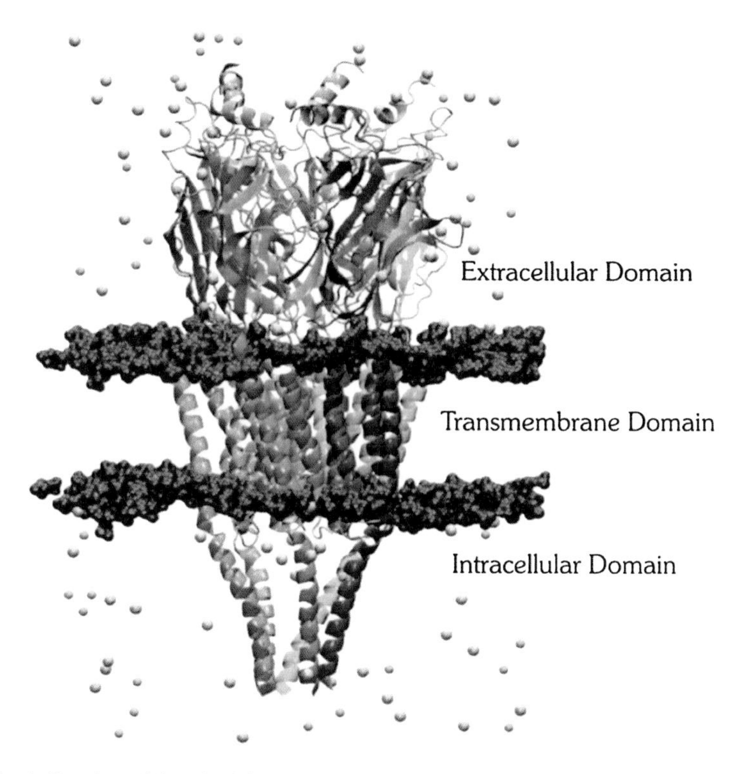

Fig. 3. Structure of the nicotinic acetylcholine receptor with the five subunits highlighted in different colors; chloride ions are shown in green, sodium ions in *yellow*, the head group region of the lipid bilayer in *dark blue*.

4.2.1. Building a Structural Model of nAChR

As no experimental 3D structure of human α7 nAChR is available, the first step is to construct a homology model based on the 4.0 Å resolution cryo-EM structure of nAChR from *T. marmorata* (PDB accession code: 2BG9) (121). The homology modeling will be conducted using the program Modeller 8 (83). The model used in the following simulation will be the lowest score models generated by Modeller and further evaluated with PROCHECK (122) and Prosa 2003 (123). Please refer to the reference (124) for more details of the homology modeling.

Given a PDB structure of nAChR, the script *protein.tcl* will be used to generate the PSF and PDB files with VMD and *psfgen*. The PDB structure by homology modeling is first divided into ten protein segments with each corresponding to a continuous chain and one segment for calcium ions. Five calcium ions in the β8–β9 loop regions are added during the homology modeling process, since it is known from experiments that the binding of these calcium ions help stabilize the otherwise flexible loop structures. The default protonation states are used for all the ionizable residues. Ten disulfide bond patches are applied.

vmd -dispdev text –e protein.tcl > protein.log

To simulate nAChR in a native-like environment, the next step is to place the protein in a fully hydrated membrane. We will first prepare a 120 Å × 120 Å palmitoyl-2-oleoyl-sn-glycerol-phosphatidylcholine (POPC) bilayer using the VMD *Membrane Builder* plugin.

vmd -dispdev text –e membrane.tcl > membrane.log

Then the structures of protein and membrane are aligned so that the channel axis overlays the membrane normal direction. After this is done, the protein can be placed into the membrane by running the *combine.tcl* script. What this script does is to combine the two PDB files, remove all the membrane atoms within 0.8 Å of the protein and write out a new set of PSF and PDB files for the combined protein/membrane system.

vmd -dispdev text –e combine.tcl > combine.log

At this point, we will use *solvate.tcl* to solvate, and *ionize.tcl* to ionize the combined protein/membrane system. Note the 10 Å padding in solvation is only applied for the membrane normal direction. Sometimes, the solvate procedure may put water molecules inside the membrane, which is undesired. So we will use the *delwat.tcl* script to remove those water molecules located within the membrane but not inside the channel pore.

vmd -dispdev text –e solvate.tcl > solvate.log
vmd -dispdev text –e ionize.tcl > ionize.log
vmd -dispdev text –e delwat.tcl > delwat.log

4.2.2. Minimization and Equilibration

Now that we have prepared all of our input files, we can start to run an MD simulation of the system. However, as the protein structure is built from homology modeling, and the final system consists of multiple components: protein, membrane, and water, we will use more sophisticated equilibration procedures to relax the system. The entire equilibration protocol consists of the following stages: minimization with fixed backbone atoms for 2,000 steps; minimization with restrained Cα atoms for 2,000 steps; Langevin dynamics with harmonic positional restraints on the Cα atoms for 100,000 steps (to heat the system to the target temperature); constant pressure dynamics (the ratio of the unit cell in the *x*–*y* plane is fixed) with decreasing positional restraints on the Cα atoms in five steps. The equilibration job will be submitted using

qsub runeq.pbs

4.2.3. Run a Free Energy Simulation Using the Adaptive Biasing Force (ABF) Method

Brute-force simulation of ion translocation through the nAChR channel is still a daunting task, often requiring specialized computer hardware. So here we will focus on understanding the intrinsic properties of the nAChR channel pore by computing the systematic forces, also known as the PMF, experienced by a sodium ion inside the channel. The adaptive biasing force (ABF) method as implemented in the NAMD package by Chipot and Henin will be used to

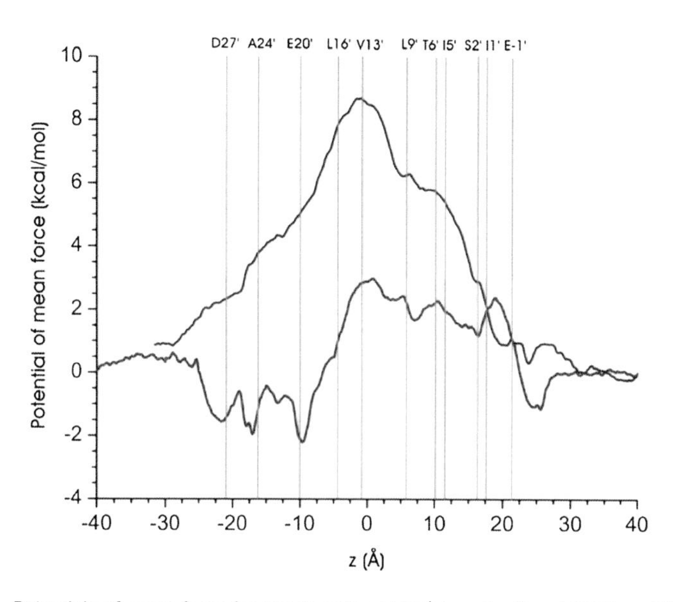

Fig. 4. Potentials of mean force for translocation of Na^+ ions (*red*) and Cl^- ions (*blue*) in nAChR. Positions of M2 pore-lining residues are shown with *gray lines* and labeled at the top of the graphs.

construct the PMF (125). The detail about this method has been given elsewhere. Briefly, a reaction coordinate ξ has to be selected. The average forces acting along ξ are accumulated in bins, providing an estimate of free energy derivative as the simulation progresses. Then the application of the biasing forces (the negative of the average forces) will allow the system to realize a free self-diffusion along ξ. In the following simulation, the reaction coordinate will be chosen as the normal to the bilayer surface (z). The simulations will carried out in ten windows of length 5 Å along this direction, which should be sufficient to cover the entire length of the transmembrane domain region of nAChR. The PBS submission script and the corresponding ABF simulation configuration file for one representative window are given below.

qsub runabf1.pbs

4.2.4. Analyze the Results

Combing the PMFs from all the windows by optimally matching the overlapping regions of two adjacent windows will produce the final PMF as displayed in Fig. 4 (red line). We only briefly summarize below the main finding of the PMF, and refer interested readers to the reference (126) for the detailed analysis of the PMF along with other calculations. The PMF for sodium inside the nAChR pore features two distinct areas of ion stabilization toward the extracellular end, corresponding to two distinct sets of negatively charged residues, D27′ and E20′. In both positions a sodium ion is stabilized by ~2 kcal/mol. Multiple ions can be accommodated at position D27′ due to the large pore radius at that position of the lumen. The PMF reaches an overall maximum at z ~ 0 Å. In this region the M2 helices expose primarily hydrophobic residues toward the interior of the receptor (Leu9′, Val13′, Phe14′, and Leu16′). This result implicates a hydrophobic nature of the gate. The effective free energy of sodium in the entire intracellular region of the pore (z between 0 and 20 Å) remains largely unfavorable compared to bulk solvent and goes through several minor peaks and troughs. Overall, the computed PMF provides a detailed thermodynamic description of a sodium ion inside the nAChR channel, such as equilibrium sodium distribution, location of ion binding site, or barrier. Moreover, when combined with a macroscopic or semi-microscopic diffusional theory, the PMF can be used to calculate ionic current, thus directly comparable to single-channel conductance measurements.

5. Notes

- Guidelines and best practices
- Recommendations and caveats about the tools and methods
- Common pitfalls and how these are mitigated

One of the significant challenges faced with MD simulation is the limitation on the time scales that we can simulate. So before starting an MD simulation, one should consider what timescale is expected for the biological process under investigation, and what timescale is affordable by the simulation. Timescale mismatches or unconverged simulations often lead to invalid conclusions or erroneous interpretation of the simulation results. If the biological timescale is indeed out of the reach for MD simulation with the available resources, one can consider using some advanced or alternative simulation techniques that enhance the sampling of phase space. Among many available enhanced sampling algorithms, it is advisable to choose one that maintains the rigorousness of thermodynamics and statistical mechanics, introduces minimum external perturbation to the system, and explores most efficiently the phase space.

The other major limitation of classical MD simulations is the underlying force field models. A molecular mechanical force field requires first the definition of a potential function form and then the fitting of a set of parameters to describe the interactions between atoms. During the development of a force field, various levels of approximations have to be introduced. So when choosing a force field and/or interpreting simulation results, one should be aware of how a particular force field is parameterized, what are potential limitations, and in what circumstance, the force field is applicable. For example, if the force field has been parameterized against thermodynamic properties or equilibrium structural data, then one would expect the simulation with this force field to be less accurate in reproducing the kinetics. In applications where electrostatic polarization or charge transfer is important, classical fixed charge force field models will not work very well.

In general, good practices of MD simulations involve the following: (1) a well-defined biological problem that can be addressed by MD simulations; (2) a high-quality starting structure that most closely represents the biological condition; (3) a good force field for the system and/or problem under investigation; (4) carefully monitored minimization and equilibration, especially for multicomponent systems or models generated by homology modeling; (4) always run multiple (control) simulations for self-consistency check. Finally, drug toxicity is a complex biological process: it is unlikely MD simulation alone will be able to identify the toxicity mechanism or predict the toxicity of a new compound; but when tightly coupled with other computational tools as well as a variety of experimental techniques, MD simulation is becoming a necessary tool for computational toxicity research. We anticipate more and more drug toxicity queries to be addressed by MD simulations in the near future.

6. Sample Input Files

Box 1
TCL script for building the protein structure

```
protein.tcl
################################
# Script to build the protein structure of 1WOE
# STEP 1: Build Protein

package require psfgen

# Use the specified CHARMM27 topology file.
topology /home/xc3/toppar/top_all27_prot_lipid.rtf

alias residue HIS HSD
alias atom ILE CD1 CD

# Build one segment
segment PROT {
 first ACE
 last CT3
 pdb 1w0e-prot.pdb
}

# Load the coordinates for each segment.
coordpdb 1w0e-prot.pdb PROT

# Guess the positions of missing atoms.
guesscoord

# Write out coor and psf file
writepdb protein.pdb
writepsf protein.psf

mol load psf protein.psf pdb protein.pdb
quit
```

Box 2
TCL scripts for solvating the protein structure

```
solvate.tcl
################################
# STEP 2: Solvate Protein

package require solvate
solvate protein.psf protein.pdb -t 10 -o solvated
quit

ionize.tcl
################################
# STEP 3: Ionize Protein

package require autoionize
autoionize -psf solvated.psf -pdb solvated.pdb -is 0.1
quit
```

Box 3
PBS script to run an energy minimization

```
runmin.pbs
################################
## Example PBS script to run a minimization on the linux cluster
#PBS -S /bin/bash
#PBS -j oe
#PBS -m ae
#PBS -N test
#PBS -l nodes=4:ppn=8,walltime=24:00:00
#PBS -V
#PBS -q md

source /share/apps/mpi/gcc/openmpi-1.2.8/bin/mpivars.sh
cd /home/xc3/data7/CYT450
/share/apps/mpi/gcc/openmpi-1.2.8/bin/mpiexec -np 32 /share/apps/
namd/NAMD_2.7b1_Source/Linux-amd64-MPI.arch/namd2 min.namd
> min.log
```

Box 4
NAMD configuration file for running an energy minimization

```
min.namd
################################
# minimization for 2000 steps

# molecular system
coordinates      ionized.pdb
structure        ionized.psf
firsttimestep    0
temperature      0
minimization     on
```

(continued)

Box 4
(continued)

```
numsteps          2000
# force field
paratypecharmm on
parameters        par_all27_prot.prm
exclude           scaled1-4
1-4scaling        1.0
switching         on
switchdist        8.5
cutoff            10
pairlistdist      12
#PME stuff
cellOrigin          57.56 77.37 10.48
cellBasisVector1    64.70 00.00 00.00
cellBasisVector2    00.00 93.59 00.00
cellBasisVector3    00.00 00.00 82.47
PME                 on
PmeGridsizeX        64
PmeGridsizeY        96
PmeGridsizeZ        81
margin 5
# output
outputname          min
outputenergies      1000
outputtiming        1000
restartname         min_restart
restartfreq         1000
restartsave         no
```

Box 5
PBS script to run an MD simulation

```
runmd.pbs
################################
## Example PBS script to run a minimization on the linux cluster
#PBS -S /bin/bash
#PBS -j oe
#PBS -m ae
#PBS -N test
#PBS -l nodes=4:ppn=8,walltime=24:00:00
#PBS -V
#PBS -q md
source /share/apps/mpi/gcc/openmpi-1.2.8/bin/mpivars.sh
cd /home/xc3/data7/CYT450
/share/apps/mpi/gcc/openmpi-1.2.8/bin/mpiexec -np 32 /share/apps/
namd/NAMD_2.7b1_Source/Linux-amd64-MPI.arch/namd2 md.namd
> md.log
```

Box 6
NAMD configuration file for running an MD simulation

```
md.namd
################################
# run md for 2000000 steps

# molecular system
coordinates     ionized.pdb
structure       ionized.psf
bincoordinates  min_restart.coor
binvelocities   min_restart.vel
extendedSystem  min_restart.xsc
firsttimestep   0
numsteps        2000000

# force field
paratypecharmm on
parameters      par_all27_prot_lipid.prm
exclude         scaled1-4
1-4scaling      1.0
switching       on
switchdist      8.5
cutoff          10
pairlistdist    12

# integrator
timestep        1.0
stepspercycle   20
nonbondedFreq   1

#PME stuff
cellOrigin          57.56 77.37 10.48
cellBasisVector1    64.70 00.00 00.00
cellBasisVector2    00.00 93.59 00.00
cellBasisVector3    00.00 00.00 82.47
PME                 on
PmeGridsizeX        64
PmeGridsizeY        96
PmeGridsizeZ        81
margin 5

# output
outputname      md
outputenergies  1000
outputtiming    1000
dcdfreq         1000
wrapAll         on
wrapNearest     on
restartname     md_restart
restartfreq     1000
restartsave     no
```

(continued)

Box 6
(continued)

```
# temperature & pressure
langevin                on
langevinDamping         10
langevinTemp            310
langevinHydrogen        on
langevinPiston          on
langevinPistonTarget    1.01325
langevinPistonPeriod    200
langevinPistonDecay     100
langevinPistonTemp      310
useGroupPressure        yes
useFlexibleCell         no
```

Box 7
TCL script files for building a protein structure embedded in membrane

```
protein.tcl
##############################
# STEP 1: Build Protein
# Script to build the protein structure of GA

package require psfgen

# Use the specified CHARMM27 topology file.
topology top_all27_prot_lipid.inp

alias atom ILE CD1 CD
alias residue HIS HSD

# Build two segments, one for each Chain.
segment GA1 {
 first ACE
 last CT3
 pdb Chain1.pdb
}
segment GA2 {
 first ACE
 last CT3
 pdb Chain2.pdb
}
segment GA3 {
 first ACE
 last CT3
 pdb Chain3.pdb
}
segment GA4 {
 first ACE
 last CT3
```

(continued)

Box 7 (continued)

```
 pdb Chain4.pdb
}
segment GA5 {
 first ACE
 last CT3
 pdb Chain5.pdb
}
segment GA6 {
 first ACE
 last CT3
 pdb Chain6.pdb
}
segment GA7 {
 first ACE
 last CT3
 pdb Chain7.pdb
}
segment GA8 {
 first ACE
 last CT3
 pdb Chain8.pdb
}
segment GA9 {
 first ACE
 last CT3
 pdb Chain9.pdb
}
segment GA10 {
 first ACE
 last CT3
 pdb Chain10.pdb
}
 segment GA11 {
 auto none
 pdb Chain11.pdb
}
# Add patches, for example disulphide bridges.
patch DISU GA1:128 GA1:142
patch DISU GA1:190 GA1:191
patch DISU GA3:495 GA3:509
patch DISU GA3:557 GA3:558
patch DISU GA5:862 GA5:876
patch DISU GA5:924 GA5:925
patch DISU GA7:1229 GA7:1243
patch DISU GA7:1291 GA7:1292
patch DISU GA9:1596 GA9:1610
patch DISU GA9:1658 GA9:1659

# Load the coordinates for each segment.
coordpdb Chain1.pdb GA1
```

(continued)

Box 7 (continued)

```
coordpdb Chain2.pdb GA2
coordpdb Chain3.pdb GA3
coordpdb Chain4.pdb GA4
coordpdb Chain5.pdb GA5
coordpdb Chain6.pdb GA6
coordpdb Chain7.pdb GA7
coordpdb Chain8.pdb GA8
coordpdb Chain9.pdb GA9
coordpdb Chain10.pdb GA10
coordpdb Chain11.pdb GA11
# Guess the positions of missing atoms.
guesscoord

# Write out coor and psf file
writepdb protein.pdb
writepsf protein.psf

mol load psf protein.psf pdb protein.pdb
quit
```

```
################################
# STEP 2: Building a Membrane Patch

package require membrane
membrane -l popc -x 120 -y 120

combine.tcl
################################
# STEP 3: Combine Protein and Membrane
##!/usr/local/bin/vmd

# need psfgen module and topology
package require psfgen
topology top_all27_prot_lipid.inp

# load structures
resetpsf
readpsf membrane.psf
coordpdb membrane.pdb
readpsf protein.psf
coordpdb protein.pdb

# write temporary structure
set temp "temp"
writepsf $temp.psf
writepdb $temp.pdb

# reload full structure (do NOT resetpsf!)
mol load psf $temp.psf pdb $temp.pdb

# select and delete lipids that overlap protein:any atom to any atom distance
under 0.8A
set sellip [atomselect top "resname POPC"]
```

(continued)

Box 7
(continued)

```
set lseglist [lsort -unique [$sellip get segid]]
foreach lseg $lseglist {
  # find lipid backbone atoms
  set selover [atomselect top "segid $lseg and within 0.8 of protein"]
  # delete these residues
  set resover [lsort -unique [$selover get resid]]
  foreach res $resover {
    delatom $lseg $res
  }
}
foreach res { } {delatom $LIP1 $res}
foreach res { } {delatom $LIP2 $res}
# select and delete waters that overlap protein:
set selwat [atomselect top "resname TIP3"]
set lseglist [lsort -unique [$selwat get segid]]
foreach lseg $lseglist {
  set selover [atomselect top "segid $lseg and within 3.8 of protein"]
  set resover [lsort -unique [$selover get resid]]
  foreach res $resover {
    delatom $lseg $res
  }
}
foreach res { } {delatom $WAT1 $res}
foreach res { } {delatom $WAT2 $res}
# write full structure
writepsf protein_and_membrane.psf
writepdb protein_and_membrane.pdb
file delete $temp.psf
file delete $temp.pdb
quit
```

Box 8
TCL script files for solvating the membrane protein structure

```
solvate.tcl
#################################
# STEP 4: Solvate Protein
package require solvate
solvate protein.psf protein.pdb -z 10 -o solvated
quit
ionize.tcl
#################################
# STEP 5: Ionize Protein
package require autoionize
```

(continued)

Box 8
(continued)

```
autoionize -psf solvated.psf -pdb solvated.pdb -is 0.1
quit

delwat.tcl
################################
# STEP 6: Delete water in the membrane

package require psfgen

# Load a pdb and psf file into both psfgen and VMD.
resetpsf
readpsf ionized.psf
coordpdb ionized.pdb
mol load psf ionized.psf pdb ionized.pdb

# Select waters located within the membrane but not inside the channel pore
set badwater [atomselect top "water and (z<14 and z>-14 and (((x>-50
and x<5) or (x>17 and x<75)) or ((y>-50 and y<5) or (y>17 and
y<75))))"]
# Delete the residues corresponding to the atoms we selected.
foreach segid [$badwater get segid] resid [$badwater get resid] {
delatom $segid $resid
}

# write out the new psf and pdb file
writepsf ionized_porewat.psf
writepdb ionized_porewat.pdb

quit
```

Box 9
NAMD configuration file for running an MD equilibration

```
equil.namd
################################
# STEP 7: Minimization and Equilibration

# molecular system
coordinates    ionized_porewat.pdb
structure      ionized_porewat.psf
temperature  0

# force field
paratypecharmm on
parameters       par_all27_prot_lipid.prm
exclude          scaled1-4
1-4scaling       1.0
switching        on
switchdist       8.5
cutoff           10
pairlistdist     12

# integrator
```

(continued)

Box 9
(continued)

```
timestep            1.0
stepspercycle       20
nonbondedFreq 1

#PME stuff
cellOrigin              11.53 12.51 -30.64
cellBasisVector1       123.90 000.00 000.00
cellBasisVector2       000.00 123.25 000.00
cellBasisVector3       000.00 000.00 136.78
PME                     on
PmeGridsizeX            128
PmeGridsizeY            128
PmeGridsizeZ            144
margin 5

# output
outputname          eq
outputenergies      1000
outputtiming        1000
dcdfreq             1000
dcdfile             eq.dcd
wrapAll             on
wrapNearest         on
fixedAtoms              on
fixedAtomsForces        on
fixedAtomsFile          fix_backbone.pdb
fixedAtomsCol           B

constraints   on
consRef       restrain_ca.pdb
consKFile     restrain_ca.pdb
consKCol      B

langevin                on
langevinDamping         10
langevinTemp            310
langevinHydrogen        no
langevinPiston          on
langevinPistonTarget 1.01325
langevinPistonPeriod 200
langevinPistonDecay  100
langevinPistonTemp   310
useGroupPressure        yes  # smaller fluctuations
useFlexibleCell                yes  # allow dimensions to fluctuate independently
useConstantRatio        yes  # fix shape in x-y plane

# run one step to get into scripting mode
minimize 0

# turn off until later
langevinPiston          off

# minimize nonbackbone atoms
minimize 2000
```

(continued)

Box 9
(continued)

```
output     min_fix

# min all atoms
fixedAtoms off
minimize            2000
output              min_all

# heat with CAs restrained
run        100000
output     heat

# equilibrate volume with CAs restrained
langevinPiston          on
constraintScaling       3.0
output                  equil_ca1
run                     200000
constraintScaling       1.0
output                  equil_ca2
run                     200000
constraintScaling       0.5
output                  equil_ca3
run                     200000
constraintScaling       0.25
output                  equil_ca4
run                     200000
constraintScaling       0
output                  equil_ca5
run                     1000000
```

Box 10
PBS script for running an MD equilibration

```
runeq.pbs
################################
## Example PBS script to run a minimization on the linux cluster
#PBS -S /bin/bash
#PBS -j oe
#PBS -m ae
#PBS -N test
#PBS -l nodes=12:ppn=8,walltime=24:00:00
#PBS -V
#PBS -q md

source /share/apps/mpi/gcc/openmpi-1.2.8/bin/mpivars.sh
cd /home/xc3/data7/nAChR
/share/apps/mpi/gcc/openmpi-1.2.8/bin/mpiexec -np 96 /share/apps/
namd/NAMD_2.7b1_Source/Linux-amd64-MPI.arch/namd2 equil.namd
> equil.log
```

Box 11
PBS script for running an MD production

```
runabf1.pbs
################################
## Example PBS script to run a minimization on the linux cluster
#PBS -S /bin/bash
#PBS -j oe
#PBS -m ae
#PBS -N abf1
#PBS -l nodes=12:ppn=8,walltime=24:00:00
#PBS -V
#PBS -q md
source /share/apps/mpi/gcc/openmpi-1.2.8/bin/mpivars.sh
cd /home/xc3/data7/nAChR
/share/apps/mpi/gcc/openmpi-1.2.8/bin/mpiexec -np 96 /share/apps/namd/
NAMD_2.7b1_Source/Linux-amd64-MPI.arch/namd2 abf1.namd > abf1.
log
```

Box 12
NAMD configuration file for running an ABF MD simulation

```
abf1.namd
################################
# STEP 7: ABF Simulation – window 1
# molecular system
# start from slightly modified equilibrated configuration
# the position of the permeating sodium ion is modified to be located within the
biasing window
coordinates     ionized_porewat-abf1.pdb
structure       ionized_porewat.psf
bincoordinates  eq_restart.coor
binvelocities   eq_restart.vel
extendedSystem  eq_restart.xsc
firsttimestep   0
temperature     310
numsteps        2000000
# force field
paratypecharmm on
parameters      par_all27_prot_lipid.prm
exclude         scaled1-4
1-4scaling      1.0
switching       on
switchdist      8.5
cutoff          10
pairlistdist    12
# integrator
timestep        1.0
stepspercycle   20
```

(continued)

Box 12 (continued)

```
nonbondedFreq 1

#PME stuff
cellOrigin           11.53 12.51 -30.64
cellBasisVector1     123.90 000.00 000.00
cellBasisVector2     000.00 123.25 000.00
cellBasisVector3     000.00 000.00 136.78
PME                  on
PmeGridsizeX         128
PmeGridsizeY         128
PmeGridsizeZ         144

# output
outputname           abf1
outputenergies       1000
outputtiming         1000
dcdfreq              1000
dcdfile              abf1.dcd
wrapAll              on
wrapNearest          on
restartname          abf1_restart
restartfreq          1000
restartsave          no

# restraints are applied to six Ca atoms on each subunit
# (three at the extracellular end and three at the intracellular end of the M2
helices)
constraints          on
consRef              restrain_ref.pdb
consKFile            restrain_ref.pdb
consKCol             B

langevin on
langevinDamping 5
langevinTemp 310
langevinHydrogen on
langevinPiston on
langevinPistonTarget 1.01325
langevinPistonPeriod 200
langevinPistonDecay 500
langevinPistonTemp 310
useGroupPressure yes
useFlexibleCell yes
useConstantArea yes

# ABF SECTION
colvars on
colvarsConfig Distance.in

Distance.in
################################
Colvarstrajfrequency 2000
Colvarsrestartfrequency 20000
```

(continued)

Box 12 (continued)

```
colvar {

  name COMDistance
  width 0.1
  lowerboundary -25.0
  upperboundary 20.0
  lowerwallconstant 10.0
  upperwallconstant 10.0

# distance along z axis between the ion and the 30 reference atoms
  distanceZ {
   group1 {
    atomnumbers { 245701 }
   }
   group2 {
    atomnumbers { 58205 58216 58223 58687 58705 58717
                  64078 64089 64096 64560 64577 64590
                  69951 69962 69969 70433 70450 70463
                  75824 75835 75842 76306 76323 76336
                  81697 81708 81715 82179 82196 82209 }
   }
  }
}
abf {
  colvars COMDistance
  fullSamples 800
  hideJacobian
}
```

References

1. Liebler DC, Guengerich FP (2005) Elucidating mechanisms of drug-induced toxicity. Nat Rev Drug Discov 4(5):410–420
2. Houck KA, Kavlock RJ (2008) Understanding mechanisms of toxicity: insights from drug discovery research. Toxicol Appl Pharmacol 227(2):163–178
3. Gillette JR, Mitchell JR, Brodie BB (1974) Biochemical mechanisms of drug toxicity. Annu Rev Pharmacol 14:271–288
4. Baillie TA (2008) Metabolism and toxicity of drugs. Two decades of progress in industrial drug metabolism. Chem Res Toxicol 21 (1):129–137
5. Guengerich FP (1999) Cytochrome P-450 3A4: regulation and role in drug metabolism. Annu Rev Pharmacol Toxicol 39:1–17
6. Gronemeyer H, Gustafsson J, Laudet V (2004) Principles for modulation of the nuclear receptor superfamily. Nat Rev Drug Discov 3:950–964
7. Sanguinetti MC, Tristani-Firouzi M (2006) hERG potassium channels and cardiac arrhythmia. Nature 440(7083):463–469
8. Cronin MT (2000) Computational methods for the prediction of drug toxicity. Curr Opin Drug Discov Dev 3(3):292–297
9. Dearden JC (2003) In silico prediction of drug toxicity. J Comput Aided Mol Des 17 (2–4):119–127
10. Valerio LG (2009) In silico toxicology for the pharmaceutical sciences. Toxicol Appl Pharmacol 241(3):356–370
11. Kavlock RJ et al (2008) Computational toxicology—a state of the science mini review. Toxicol Sci 103(1):14–27
12. Nicholson JD, Wilson ID (2003) Understanding 'Global' systems biology: metabonomics

and the continuum of metabolism. Nat Rev Drug Discov 2:668–676

13. Bugrim A, Nikolskaya T, Yuri Nikolsky Y (2004) Early prediction of drug metabolism and toxicity: systems biology approach and modeling. Drug Discov Today 9(3):127–135
14. Nicholson JK et al (2002) Metabonomics: a platform for studying drug toxicity and gene function. Nat Rev Drug Discov 1:153–161
15. Hunter PJ, Borg TK (2003) Integration from proteins to organs: the Physiome Project. Nat Rev Mol Cell Biol 4(3):237–243
16. Silva JR et al (2009) A multiscale model linking ion-channel molecular dynamics and electrostatics to the cardiac action potential. Proc Natl Acad Sci U S A 106(27):11102–11106
17. Dill KA et al (2008) The protein folding problem. Annu Rev Biophys 37:289–316
18. Zimmermann O, Hansmann UH (2008) Understanding protein folding: small proteins in silico. Biochim Biophys Acta 1784 (1):252–258
19. de Groot BL, Grubmüller H (2005) The dynamics and energetics of water permeation and proton exclusion in aquaporins. Curr Opin Struct Biol 15(2):176–183
20. Roux B, Schulten K (2004) Computational studies of membrane channels. Structure 12 (8):1343–1351
21. McCammon JA, Gelin BR, Karplus M (1977) Dynamics of folded proteins. Nature 267 (5612):585–590
22. Karplus M, McCammon JA (2002) Molecular dynamics simulations of biomolecules. Nat Struct Biol 9(9):646–652
23. van Gunsteren WF et al (2006) Biomolecular modeling: goals, problems, perspectives. Angew Chem Int Ed Engl 45 (25):4064–4092
24. Adcock SA, McCammon JA (2006) Molecular dynamics: survey of methods for simulating the activity of proteins. Chem Rev 106 (5):1589–1615
25. Scheraga HA, Khalili M, Liwo A (2007) Protein-folding dynamics: overview of molecular simulation techniques. Annu Rev Phys Chem 58:57–83
26. Warshel A (2002) Molecular dynamics simulations of biological reactions. Acc Chem Res 35(6):385–395
27. Garcia-Viloca M et al (2004) How enzymes work: analysis by modern rate theory and computer simulations. Science 303 (5655):186–195
28. Karplus M et al (2005) Protein structural transitions and their functional role. Philos Trans A Math Phys Eng Sci 363 (1827):331–355
29. Senn HM, Thiel W (2009) QM/MM methods for biomolecular systems. Angew Chem Int Ed Engl 48(7):1198–1229
30. Ridder L, Mulholland AJ (2003) Modeling biotransformation reactions by combined quantum mechanical/molecular mechanical approaches: from structure to activity. Curr Top Med Chem 3(11):1241–1256
31. Lewis DFV (2001) Guide to cytochromes P450: structure and function, 2nd edn. Informa Healthcare, London
32. Denisov IG et al (2005) Structure and chemistry of cytochrome P450. Chem Rev 105 (6):2253–2277
33. Wang JF, Chou KC (2010) Molecular modeling of cytochrome P450 and drug metabolism. Curr Drug Metab 11(4):342–346
34. Otyepka M et al (2007) What common structural features and variations of mammalian P450s are known to date? Biochim Biophys Acta 1770(3):376–389
35. Henley DV, Korach KS (2006) Endocrine-disrupting chemicals use distinct mechanisms of action to modulate endocrine system function. Endocrinology 147(6):S25–S32
36. Ankley GT et al (2010) Adverse outcome pathways: a conceptual framework to support ecotoxicology research and risk assessment. Environ Toxicol Chem 29(3):730–741
37. Jugan ML, Levi Y, Blondeau JP (2010) Endocrine disruptors and thyroid hormone physiology. Biochem Pharmacol 79(7):939–947
38. Pearce EN, Braverman LE (2009) Environmental pollutants and the thyroid. Best Pract Res Clin Endocrinol Metab 23(6):801–813
39. Prenzel N et al (2001) The epidermal growth factor receptor family as a central element for cellular signal transduction and diversification. Endocr Relat Cancer 8(1):11–31
40. Bock KW (1994) Aryl hydrocarbon or dioxin receptor: biologic and toxic responses. Rev Physiol Biochem Pharmacol 125:1–42
41. Bradshaw TD, Bell DR (2009) Relevance of the aryl hydrocarbon receptor (AhR) for clinical toxicology. Clin Toxicol (Phila) 47 (7):632–642
42. Gray LE Jr et al (2006) Adverse effects of environmental antiandrogens and androgens on reproductive development in mammals. Int J Androl 29(1):96–104
43. Roncaglioni A, Benfenati E (2008) In silico-aided prediction of biological properties of chemicals: oestrogen receptor-mediated effects. Chem Soc Rev 37(3):441–450

44. Lin JH et al (2002) Computational drug design accommodating receptor flexibility: the relaxed complex scheme. J Am Chem Soc 124(20):5632–5633
45. Cornell W, Nam K (2009) Steroid hormone binding receptors: application of homology modeling, induced fit docking, and molecular dynamics to study structure–function relationships. Curr Top Med Chem 9 (9):844–853
46. Recanatini M, Cavalli A, Masetti M (2008) Modeling the hERG potassium channel in a phospholipid bilayer: molecular dynamics and drug docking studies. J Comput Chem 29 (5):795–808
47. Stary A et al (2010) Toward a consensus model of the HERG potassium channel. ChemMedChem 5(3):455–467
48. Meharenna YT, Poulos TL (2010) Using molecular dynamics to probe the structural basis for enhanced stability in thermal stable cytochromes P450. Biochemistry 49 (31):6680–6686
49. Skopalík J, Anzenbacher P, Otyepka M (2008) Flexibility of human cytochromes P450: molecular dynamics reveals differences between CYPs 3A4, 2 C9, and 2A6, which correlate with their substrate preferences. J Phys Chem B 112(27):8165–8173
50. Hendrychováa T et al (2010) Flexibility of human cytochrome p450 enzymes: molecular dynamics and spectroscopy reveal important function-related variations. Biochim Biophys Acta 1814:58–68
51. Lampe JN et al (2010) Two-dimensional NMR and all-atom molecular dynamics of cytochrome P450 CYP119 reveal hidden conformational substates. J Biol Chem 285 (13):9594–9603
52. Park H, Lee S, Suh J (2005) Structural and dynamical basis of broad substrate specificity, catalytic mechanism, and inhibition of cytochrome P450 3A4. J Am Chem Soc 127 (39):13634–13642
53. Fishelovitch D et al (2007) Structural dynamics of the cooperative binding of organic molecules in the human cytochrome P450 3A4. J Am Chem Soc 129(6):1602–1611
54. Seifert A et al (2006) Multiple molecular dynamics simulations of human p450 monooxygenase CYP2C9: the molecular basis of substrate binding and regioselectivity toward warfarin. Proteins 64(1):147–155
55. Winn PJ et al (2002) Comparison of the dynamics of substrate access channels in three cytochrome P450s reveals different opening mechanisms and a novel functional role for a buried arginine. Proc Natl Acad Sci U S A 99(8):5361–5366
56. Lüdemann SK, Lounnas V, Wade RC (2000) How do substrates enter and products exit the buried active site of cytochrome P450cam? 1. Random expulsion molecular dynamics investigation of ligand access channels and mechanisms. J Mol Biol 303(5):797–811
57. Li W et al (2007) Possible pathway(s) of metyrapone egress from the active site of cytochrome P450 3A4: a molecular dynamics simulation. Drug Metab Dispos 35(4):689–696
58. Fishelovitch D et al (2009) Theoretical characterization of substrate access/exit channels in the human cytochrome P450 3A4 enzyme: involvement of phenylalanine residues in the gating mechanism. J Phys Chem B 113 (39):13018–13025
59. Subbotina J et al (2010) Structural refinement of the hERG1 pore and voltage-sensing domains with ROSETTA-membrane and molecular dynamics simulations. Proteins 78 (14):2922–2934
60. Stansfeld PJ et al (2008) Insight into the mechanism of inactivation and pH sensitivity in potassium channels from molecular dynamics simulations. Biochemistry 47 (28):7414–7422
61. Kutteh R, Vandenberg JI, Kuyucak S (2007) Molecular dynamics and continuum electrostatics studies of inactivation in the HERG potassium channel. J Phys Chem B 111:1090–1098
62. Osterberg F, Aqvist J (2005) Exploring blocker binding to a homology model of the open hERG K+ channel using docking and molecular dynamics methods. FEBS Lett 579:2939–2944
63. Brooks BR et al (2009) CHARMM: the biomolecular simulation program. J Comput Chem 30(10):1545–1614
64. Case DA et al (2005) The Amber biomolecular simulation programs. J Comput Chem 26 (16):1668–1688
65. Christen M et al (2005) The GROMOS software for biomolecular simulation: GROMOS05. J Comput Chem 26 (16):1719–1751
66. Van Der Spoel D et al (2005) GROMACS: fast, flexible, and free. J Comput Chem 26 (16):1701–1718
67. Ponder JW, Richards FM (1987) An efficient Newton-like method for molecular mechanics energy minimization of large molecules. J Comput Chem 8(7):1016–1024
68. Refson K (2000) Moldy: a portable molecular dynamics simulation program for serial and

parallel computers. Comput Phys Commun 126(3):310–329
69. Smith W, Forester TR (1996) DL_POLY_2.0: a general-purpose parallel molecular dynamics simulation package. J Mol Graph 14(3):136–141
70. Phillips JC et al (2005) Scalable molecular dynamics with NAMD. J Comput Chem 26:1781–1802
71. Plimpton SJ (1995) Fast parallel algorithms for short-range molecular dynamics. J Comp Phys 117:1–19
72. Bowers KJ et al (2006) Scalable algorithms for molecular dynamics simulations on commodity clusters. In: Proceedings of the ACM/IEEE conference on supercomputing (SC06). Tampa, FL.
73. MacKerell AD et al (1998) CHARMM: the energy function and its parameterization with an overview of the program. In: Schleyer PVR (ed) The encyclopedia of computational chemistry. Wiley, Chichester, pp 271–277
74. Cornell WD et al (1995) A second generation force field for the simulation of proteins, nucleic acids, and organic molecules. J Am Chem Soc 117(19):5179–5197
75. Oostenbrink C et al (2004) A biomolecular force field based on the free enthalpy of hydration and solvation: the GROMOS force-field parameter sets 53A5 and 53A6. J Comput Chem 25(13):1656–1676
76. Jorgensen WL, Maxwell DS, Tirado-Rives J (1996) Development and testing of the OPLS all-atom force field on conformational energetics and properties of organic liquids. J Am Chem Soc 118:11225–11236
77. Stone JE et al (2007) Accelerating molecular modeling applications with graphics processors. J Comput Chem 28(16):2618–2640
78. Friedrichs MS et al (2009) Accelerating molecular dynamic simulation on graphics processing units. J Comput Chem 30 (6):864–872
79. Davis JE et al (2009) Towards large-scale molecular dynamics simulations on graphics processors. Lecture Notes Comput Sci 5462:176–186
80. Darden T, York D, Pedersen L (1993) Particle mesh Ewald: an N log (N) method for Ewald sums in large systems. J Chem Phys 98:10089–10092
81. Hockney RW, Eastwood JW (1988) Computer simulation using particles. Taylor & Francis Croup, New York
82. Humphrey W, Dalke A, Schulten K (1996) VMD: visual molecular dynamics. J Mol Graph Model 14(1):33–38
83. Sali A et al (1995) Evaluation of comparative protein modeling by MODELLER. Proteins 23(3):318–326
84. Wang J et al (2004) Development and testing of a general amber force field. J Comput Chem 25(9):1157–1174
85. Vanommeslaeghe K et al (2010) CHARMM general force field: a force field for drug-like molecules compatible with the CHARMM all-atom additive biological force fields. J Comput Chem 31(4):671–690
86. Ryckaert JP, Ciccotti G, Berendsen HJC (1977) Numerical integration of the Cartesian equations of motion of a system with constraints: molecular dynamics of n-alkanes. J Comp Phys 23:327–341
87. Tuckerman M, Berne BJ, Martyna GJ (1992) Reversible multiple time scale molecular dynamics. J Chem Phys 97:1990–2001
88. Andersen HC (1980) Molecular dynamics at constant pressure and/or temperature. J Chem Phys 72:2384–2393
89. Nose S (1984) A unified formulation of the constant temperature molecular-dynamics methods. J Chem Phys 81(1):511–519
90. Hoover WG (1985) Canonical dynamics: equilibrium phase-space distributions. Phys Rev A 31(3):1695–1697
91. Martyna GL, Klein ML, Tuckerman M (1992) Nose-Hoover chains: the canonical ensemble via continuous dynamics. J Chem Phys 97(4):2635–2643
92. Berendsen HJC et al (1984) Molecular-dynamics with coupling to an external bath. J Chem Phys 81(8):3684–3690
93. Martyna GL, Tobias DJ, Klein ML (1994) Constant pressure molecular dynamics algorithms. J Chem Phys 101(5):4177–4189
94. Andricioaei I, Karplus M (2001) On the calculation of entropy from covariance matrices of the atomic fluctuations. J Chem Phys 115:6289–6292
95. Baron R, Hünenberger PH, McCammon JA (2009) Absolute single-molecule entropies from quasi-harmonic analysis of microsecond molecular dynamics: correction terms and convergence properties. J Chem Theory Comput 5(12):3150–3160
96. Kollman PA et al (2000) Calculating structures and free energies of complex molecules: combining molecular mechanics and continuum models. Acc Chem Res 33(12):889–897
97. Torrie GM, Valleau JP (1977) Nonphysical sampling distributions in Monte Carlo free-energy estimation—umbrella sampling. J Comput Phys 23:187–199

98. Mitsutake A, Sugita Y, Okamoto Y (2001) Generalized-ensemble algorithms for molecular simulations of biopolymers. Biopolymers 60(2):96–123
99. Wang F, Landau DP (2001) Efficient multiple range random walk algorithm to calculate density of states. Phys Rev Lett 86:2050
100. Kumar S et al (1993) The weighted histogram analysis method (WHAM) for free energy calculations on biomolecules: 1. The method. J Comput Chem 13:1011–1021
101. Fasnacht M, Zhu J, Honig B (2007) Local quality assessment in homology models using statistical potentials and support vector machines. Protein Sci 16(8):1557–1568
102. Grossfield A, Feller SE, Pitman MC (2007) Convergence of molecular dynamics simulations of membrane proteins. Proteins 67 (1):31–40
103. Guvench O, MacKerell AD Jr (2008) Comparison of protein force fields for molecular dynamics simulations. Methods Mol Biol 443:63–88
104. Ponder JW, Case DA (2003) Force fields for protein simulations. Adv Protein Chem 66:27–85
105. Hornak V et al (2006) Comparison of multiple Amber force fields and development of improved protein backbone parameters. Proteins 65(3):712–725
106. MacKerell AD Jr, Feig M, Brooks CL III (2004) Improved treatment of the protein backbone in empirical force fields. J Am Chem Soc 126(3):698–699
107. Patel S, Mackerell AD Jr, Brooks CL III (2004) CHARMM fluctuating charge force field for proteins: II protein/solvent properties from molecular dynamics simulations using a nonadditive electrostatic model. J Comput Chem 25(12):1504–1514
108. Ponder JW et al (2010) Current status of the AMOEBA polarizable force field. J Phys Chem B 114(8):2549–2564
109. Kaminski GA, Friesner RA, Zhou R (2003) A computationally inexpensive modification of the point dipole electrostatic polarization model for molecular simulations. J Comput Chem 24(3):267–276
110. Lopes PE, Roux B, Mackerell AD (2009) Molecular modeling and dynamics studies with explicit inclusion of electronic polarizability: theory and applications. Theor Chem Acc 124(1–2):11–28
111. Shan Y et al (2009) A conserved protonation-dependent switch controls drug binding in the Abl kinase. Proc Natl Acad Sci U S A 106(1):139–144
112. Arkin IT et al (2007) Mechanism of Na+/H+ antiporting. Science 317(5839):799–803
113. Jensen MØ et al (2010) Principles of conduction and hydrophobic gating in K+ channels. Proc Natl Acad Sci U S A 107 (13):5833–5838
114. Okamoto Y (2004) Generalized-ensemble algorithms: enhanced sampling techniques for Monte Carlo and molecular dynamics simulations. J Mol Graph Model 22(5):425–439
115. Laio A, Parrinello M (2002) Escaping free-energy minima. Proc Natl Acad Sci U S A 99 (20):12562–12566
116. Hamelberg D, Mongan J, McCammon JA (2004) Accelerated molecular dynamics: a promising and efficient simulation method for biomolecules. J Chem Phys 120:11919–11929
117. Bolhuis PG et al (2002) Transition path sampling: throwing ropes over rough mountain passes, in the dark. Annu Rev Phys Chem 53:291–318
118. Noé F et al (2007) Hierarchical analysis of conformational dynamics in biomolecules: transition networks of metastable states. J Chem Phys 126(15):155102
119. Williams PA et al (2004) Crystal structures of human cytochrome P450 3A4 bound to metyrapone and progesterone. Science 305 (5684):683–686
120. Sine SM, Engel AG (2006) Recent advances in Cys-loop receptor structure and function. Nature 440(7083):448–455
121. Unwin N (2005) Refined structure of the nicotinic acetylcholine receptor at 4A resolution. J Mol Biol 346(4):967–989
122. Laskowski RA et al (1993) PROCHECK: a program to check the stereochemical quality of protein structures. J Appl Cryst 26:283–291
123. Sippl MJ (1993) Recognition of errors in three-dimensional structures of proteins. Protein 17:355–362
124. Cheng X et al (2006) Channel opening motion of alpha7 nicotinic acetylcholine receptor as suggested by normal mode analysis. J Mol Biol 355(2):310–324
125. Chipot C, Henin J (2005) Exploring the free-energy landscape of a short peptide using an average force. J Chem Phys 123(24):244906
126. Ivanov IN et al (2007) Barriers to ion translocation in cationic and anionic receptors from the cys-loop family. J Am Chem Soc 129 (26):8217–24

Part IV

Pharmacokinetic and Pharmacodynamic Modeling

Chapter 12

Introduction to Pharmacokinetics in Clinical Toxicology

Pavan Vajjah, Geoffrey K. Isbister, and Stephen B. Duffull

Abstract

In clinical toxicology, a better understanding of the pharmacokinetics of the drugs may be useful in both risk assessment and formulating treatment guidelines for patients. Pharmacokinetics describes the time course of drug concentrations and is a driver for the time course of drug effects. In this chapter pharmacokinetics is described from a mathematical modeling perspective as applied to clinical toxicology. The pharmacokinetics of drugs are described using a combination of input and disposition (distribution and elimination) phases. A description of the time course of the input and disposition of drugs in overdose provides a basis for understanding the time course of effects of drugs in overdose. Relevant clinical toxicology examples are provided to explain various pharmacokinetic principles. Throughout this chapter we have taken a pragmatic approach to understanding and interpreting the time course of drug effects.

Key words: Pharmacokinetics, Clinical toxicology, Input, Disposition, Clearance, Volume of distribution, Compartmental models

1. Introduction

The primary tenet of pharmacology and toxicology alike is that the effects caused by a drug are related to the concentration of the drug, whether the effect is beneficial or toxic. Dose–effect relationships have been studied for decades and in these relationships dose is assumed to be an input or driving force for the effect. While such studies provide useful information about the size of the effect they lack information about the change of effect over time. Pharmacokinetics (PK) describes the time course of drug concentrations in the body and when used as the driver for the effect, it naturally adds time into the dose–effect relationship. Since pharmacodynamics (PD) is the relationship between concentration and effect then a primary purpose of studying PK is to incorporate the influence of time on effect. Figure 1 describes the purpose of pharmacokinetics in pharmacology and toxicology.

Brad Reisfeld and Arthur N. Mayeno (eds.), *Computational Toxicology: Volume I*, Methods in Molecular Biology, vol. 929,
DOI 10.1007/978-1-62703-050-2_12,

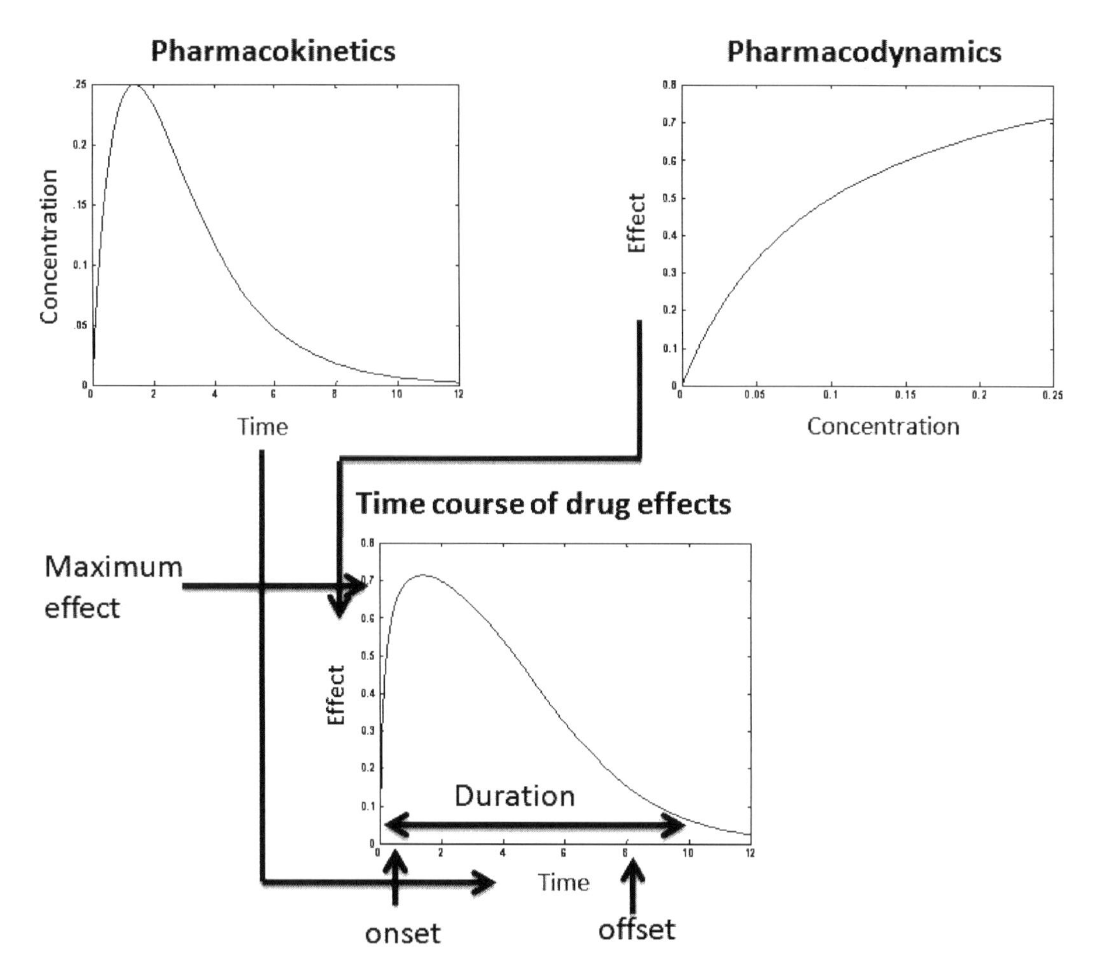

Fig. 1. A primary purpose of pharmacokinetics is to add time into the dose/concentration–effect relationship.

The combination of PK and PD therefore provides an understanding of the time course of drug effects, which means the time of onset of effects, the time for the effect to wear off, and the extent of the effect (as shown in Fig. 1).

In clinical toxicology, measurement of plasma concentrations of chemicals may be useful in making treatment decisions for chronic exposure to drugs/chemicals where there is reasonable access to assays in a clinically useful period of time. Examples include chronic lead and chronic pesticide exposure. Measurement of plasma concentrations of drugs in acute overdose or poisoning is rarely useful to guide treatment decisions because such assays are not routinely and rapidly available for most drugs. There are some exceptions for common or highly toxic drugs including acetaminophen, theophylline, iron, anticonvulsants, digoxin, lithium, and salicylates (1). It is therefore difficult to use drug concentration or PK to guide treatment decisions in clinical toxicology practice.

In contrast, studying the PK of drugs in overdose can be used to inform us about the treatment of overdose patients. Such research allows us to understand the dose–concentration–effect relationships in patients including the effect of various interventions such as decontamination (2). PKPD models can then be developed to inform guidelines regarding risk assessment and treatment with the aim of finding simple clinical determinants of outcomes such as dose or early clinical effects. Such guidelines have been developed for citalopram (3) from a PKPD model of citalopram in overdose (4).

In this chapter we will discuss the basic concepts of PK and use these to understand the time course and severity of drugs in overdose.

1.1. Toxic Dose

In this chapter we have taken a broad perspective of the word drug, where a drug is defined as any exogenously administered chemical that elicits an effect on the body. Hence from this perspective a drug could have therapeutic or toxic actions depending on the concentration in the target tissues, which is a function of dose and time. Based on this definition, the term toxicokinetics, which is used often used in the discipline of toxicology, is simply pharmacokinetics. We therefore use the term PK to refer to the concentration--time course of the concentration of any drug in the body whether it is used therapeutically or following inadvertent or deliberate self-poisoning.

2. Materials

2.1. Pharmacokinetics

Traditionally the PK of drugs has been described using ADME (Absorption, Distribution, Metabolism, and Excretion) principles. The acronym ADME suggests a serial approach to PK such that absorption occurs first followed by distribution then metabolism and then excretion. However these processes (ADME) are not temporally discrete and occur simultaneously even though at any point of time one of the processes may be predominant. For example the distribution, metabolism, and excretion of a drug usually continue to occur during the so-called absorption phase. Another way to consider PK processes is to categorize them into two components;

1. *Input* which describes the time course of drug movement from the site of administration to the site of measurement.
2. *Disposition* which describes the time course of drug distribution and elimination from the site of measurement.

Figures 2 and 3 show the input and disposition phases of the drug that is administered via an extravascular route. Here we

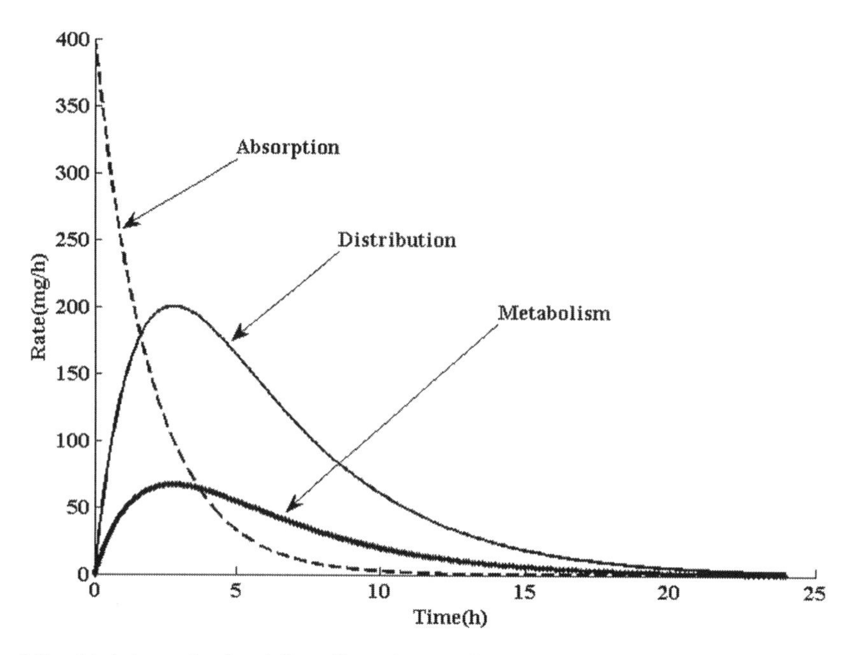

Fig. 2. The relationship between input and disposition rates over time.

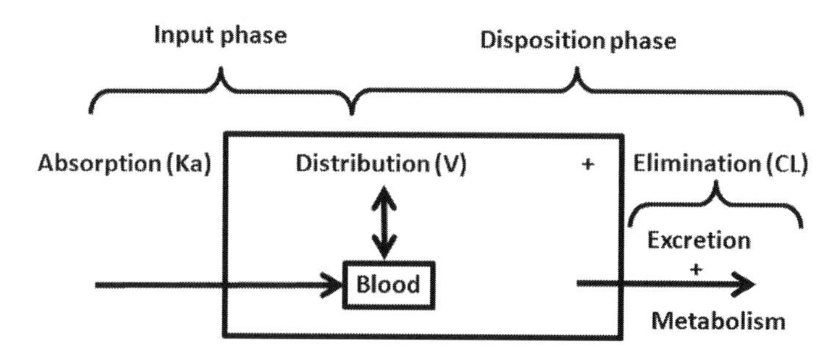

Fig. 3. The input and disposition phases of the PK process.

assume that the input and disposition rates are concentration dependent and hence described predominantly by a first-order process. A comparison of PK in terms of input and disposition versus ADME is discussed below.

PK when studied as ADME is useful in understanding the processes governing the fate of drug once it enters the body, i.e., we may know if the drug is absorbed through active/passive absorption or if the drug follows phase I or phase II metabolism. However, absence of time in the ADME concepts makes it a difficult paradigm for understanding the change of concentration over time and hence a driver for the time course of drug effects. In contrast, PK can be naturally divided into input and disposition since these processes are independent, i.e., the manner in which a drug is administered (e.g., percutaneous absorption, oral, intravenous) does not affect its disposition. Hence the overall time course of concentration in the body is the combination of input and disposition. In this sense ADME can be considered as a method to understand the mechanisms of drug movement and input–disposition as a method to understand the time course of drug movement.

2.2. Input

Input can be defined as the process by which unchanged drug travels from the site of administration to the site of measurement within the body. In the case of an intravenous bolus dose the entire amount of unchanged drug is available instantaneously in the body. In the case of the extravascular route of administration the input process usually involves more than one mechanism and there may be several possible sites where the drug may be irreversibly eliminated during the input process, hence absorption may be incomplete and variable in rate.

When the drugs are dosed orally loss may occur due to biological reasons and/or poor physicochemical properties of the drugs. Biological reasons include degradation of the drug in the gastrointestinal tract (e.g., insulin), pre-systemic metabolism by enzymes present in the gastrointestinal tract wall (e.g., cyclosporin) and transporters in the gastrointestinal tract (e.g., vincristine) wall providing a counter flux back into the gastrointestinal tract (5).

Most of the drugs that are substrates of the enzyme CYP3A4 (a class of cytochrome P 450 enzyme) may undergo pre-systemic metabolism. This is due to the presence of CYP3A4 in gut wall (6). A key example of a drug that undergoes pre-systemic metabolism through CYP3A4 is cyclosporine resulting in often highly erratic PK profiles (7). Terbutaline (8) also undergo pre-systemic sulphation. Drugs that are substrates of P-Glycoprotein (P-gp) also have poor bioavailability. P-gp is a glycoprotein present in the gut wall. The major function of P-gp in the gut wall is to transport the drug back into the gut. Examples of P-gp substrates include digoxin (9) and fexofenadine (10).

Physicochemical properties of the drug like poor solubility and/or permeability of the drug may also lead to loss of the drug. Some drugs like glibenclamide have low solubility in the gastrointestinal fluids and hence low bioavailability (11). Others like cimetidine have high solubility but low permeability (11). Once the drug gains entry passed the gastrointestinal tract and enters the portal vein, it may undergo first-pass metabolism in the liver. Figure 4 represents the absorption of the drug via the oral route.

2.2.1. Input and Drug Overdose

In most studies of overdose the input phase may be difficult to characterize. This is mainly because the patient does not take the overdose in the hospital and less information is available about the dose and the time at which the patient ingested the overdose (12). In addition, there is a lag between the time of ingestion of the dose and time at which the first plasma sample can be collected and the input phase of the drug may be partially or fully completed by this time.

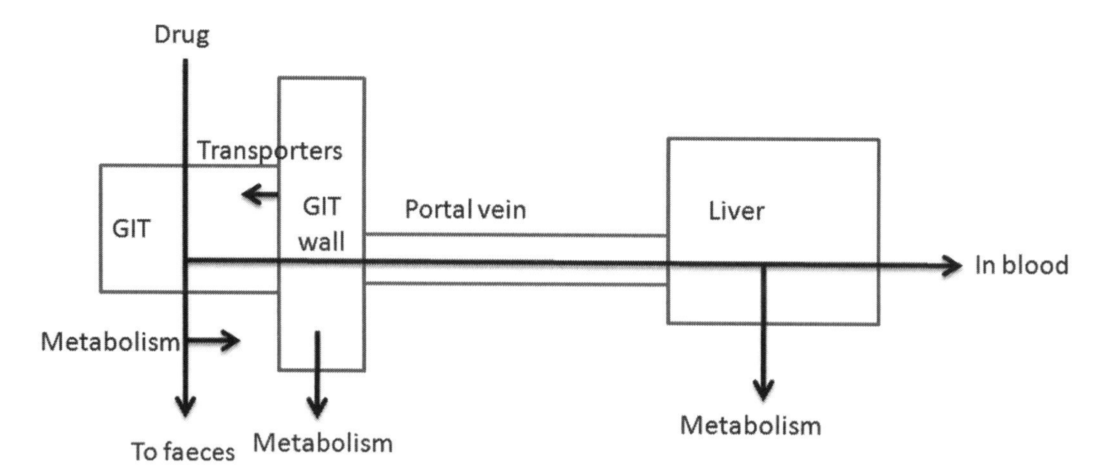

Fig. 4. Barriers that a drug encounters when given via the oral route. The drug can be degraded or poorly absorbed and eliminated through feces, metabolized by intestinal microsomes, affected by transporters in intestinal wall that may inhibit their passage through the GIT wall, and first pass metabolism in the liver.

Hence, the data collected may not provide much information about the input phase of the drug. Methods have been developed in order to account for this missing information in the input phase and account for some of the uncertainty which is particularly important for estimating the disposition parameters (e.g., clearance) (2, 12, 13).

It is often assumed that drug absorption takes longer when the drug is taken in overdose. However, in a number of pharmacokinetic studies of drugs in overdose the absorption appears to be rapid and complete, similar to the pharmacokinetics in therapeutic doses of the same drugs (2, 12–14). The input phase for drugs in overdose may therefore be similar to the drug in therapeutic doses in many cases. In some cases there is clearly prolonged absorption and this is assumed to be due to the tablets aggregating to form what are known as pharmacobezoars (15). Carbamazepine is an example of a drug in overdose that has a prolonged input phase (16), and this can be seen in Fig. 5 that suggests there is ongoing absorption for up to 24 h with increasing concentrations or a plateauing of drug concentrations. This may be due to the formation of pharmacobezoars or that carbamazepine affects its own absorption by perhaps reducing gastrointestinal motility.

2.3. Disposition

Disposition can be defined as the process by which the drug moves to and from the site of measurement. Once absorbed the drug molecules are distributed to the tissues of the body including to organs of elimination, leading to a decrease in concentration of the drug at the site of measurement. The decrease in the blood concentration could be due to reversible loss of drug from the blood to the tissues, defined as distribution, or the irreversible loss of drug

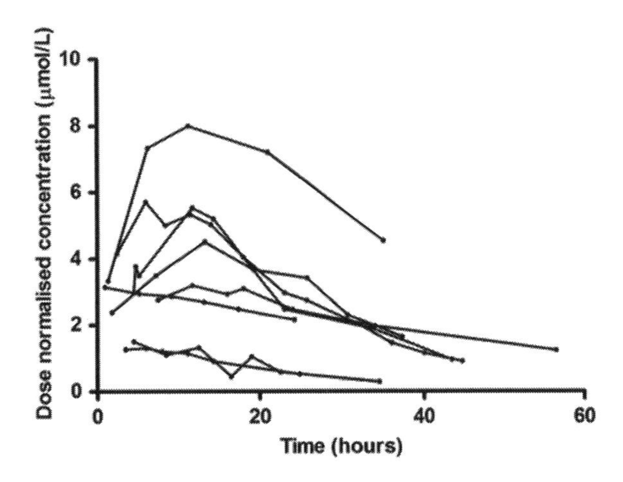

Fig. 5. Observed dose-normalized concentrations (to the therapeutic dose of 200 mg) versus reported time of administration curves for eight patients taking normal release carbamazepine overdoses with doses ranging from 3 to 36 g. Samples from the same dose event are connected.

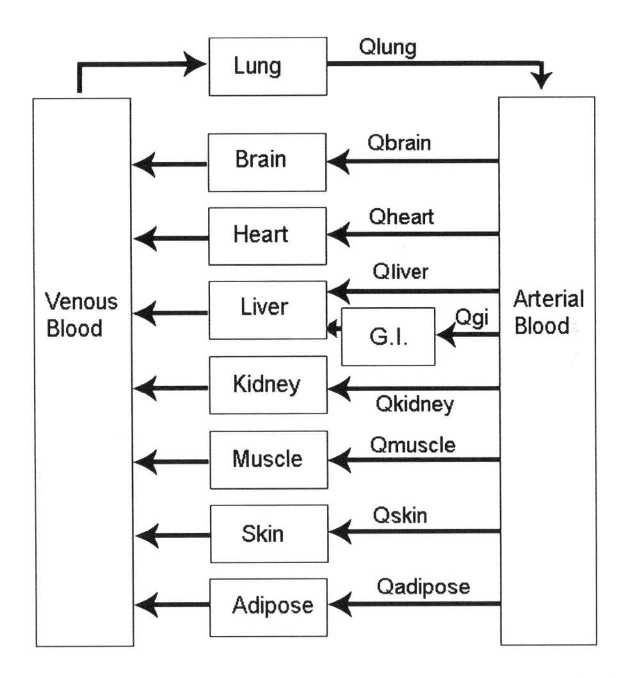

Fig. 6. A physiologically based pharmacokinetic model, describing the distribution of drug to various tissues.

from blood, defined as elimination. Disposition is therefore the combination distribution and elimination.

2.4. Distribution

Distribution can be defined as the reversible transfer of drug from the central compartment (blood) to the different tissues and is schematically represented in Fig. 6. The rate and extent of distribution of a drug to the different tissues is determined by the blood flow to the tissue, the ability of the drug to cross the tissue membrane, the solubility and the binding of the in the tissue (Fig. 6).

The clearance (CL) and volume of distribution (V) are the two most important disposition PK parameters. A compartment is a space in which the drug is assumed to distribute evenly and instantaneously. The *apparent* volume of a compartment may be much bigger than the physiological space since it is given by the product of the tissue weight and the partition coefficient (k_p) of the drug for that tissue.

In the model depicted in Fig. 6 the sum of the volumes of distribution of the different tissues is equivalent to the volume of distribution of the drug at steady state of the drug.

$$\begin{aligned} V_{ss} = {} & K_{\mathrm{pLung}} \times \mathrm{WT}_{\mathrm{Lung}} \times (1 - E_{\mathrm{Lung}}) \\ & + K_{\mathrm{pBrain}} \times \mathrm{WT}_{\mathrm{Brain}} \times (1 - E_{\mathrm{Brain}}) \\ & + K_{\mathrm{pHeart}} \times \mathrm{WT}_{\mathrm{Heart}} \times (1 - E_{\mathrm{Heart}}) + \cdots \\ & + K_{\mathrm{pAdipose}} \times \mathrm{WT}_{\mathrm{Adipose}} \times (1 - E_{\mathrm{Adipose}}), \end{aligned} \tag{1}$$

where K_p is the partition coefficient of the drug into the tissue which is a composite term that accounts for partitioning of drug and WT is the weight of the tissue and E is the extraction ratio of the tissue. If a tissue does not eliminate the drug then the extraction ratio (E) for that tissue will be equal to zero.

2.4.1. Volume of Distribution

The volume of distribution (V) of a drug is an important pharmacokinetic parameter and is defined as the ratio of the amount of drug in the body to the concentration in the compartment of interest (e.g., central compartment or plasma). For a one-compartment model, it is the ratio of the amount of drug in the body to the plasma concentration. It has units of volume.

$$V(\text{volume}) = \frac{A(\text{mass})}{C\left(\frac{\text{mass}}{\text{volume}}\right)} \tag{2}$$

The relationship between the first-order rate constant (k), clearance (CL) and apparent volume (V) of distribution is given by $k = \mathrm{CL}/V$. The elimination half-life, which is the length of time for the amount in the body to reduce by half can then be determined as $\ln 2/k$ in this mode (*see* Subheading 2.5.5 for a description of CL).

2.4.2. Distribution and Drug Overdose

Distribution is an important consideration in understanding drug toxicity. For some drugs the time course of distribution rather than the elimination appears to correlate well with the time course of toxicity. This is the case for tricyclic antidepressant overdose. Tricyclic antidepressants have long elimination half-lives (typically in the range of 8–80 h) that are not consistent with the time course of

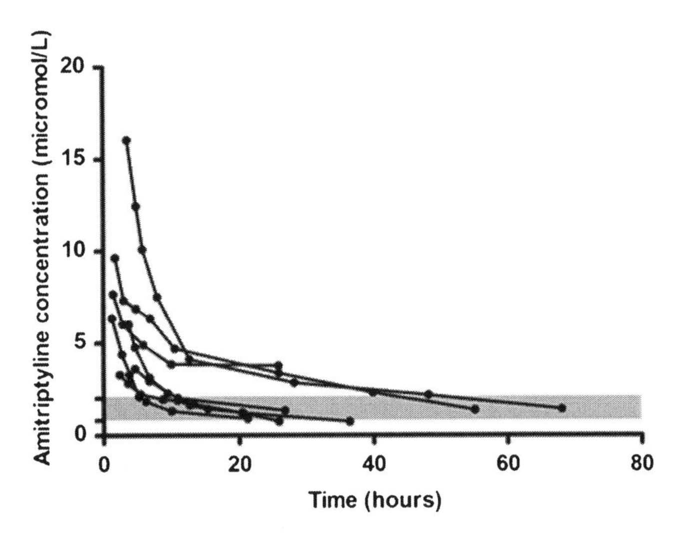

Fig. 7. Drug concentration time curves for seven patients with amitryptyline overdoses redrawn from Figure 3 in Hulten et al. (18) *Shaded area* is the therapeutic range of 0.8–2 μM.

toxicity in acute overdose. Tricyclic antidepressant toxicity develops rapidly and then resolves over a period of 6–24 h consistent with redistribution of drug to muscle and fat from the central nervous system and heart (17–19). Meineke et al. described the pharmacokinetics of imipramine in sheep given an intravenous overdose with a three-compartment model showing that there is a rapid decrease in plasma concentrations over the first few hours and hepatic metabolism is not an important factor in the initial rapid decrease in plasma drug concentrations. Studies of amitriptyline overdose in human patients demonstrate a similar rapid decrease in drug concentration in the first 6–24 h from toxic concentrations of 1,000–5,000 ng/mL back into the therapeutic range of 80–200 ng/mL (Fig. 7) (18). Tricyclic antidepressant overdose is characterized by central nervous system depression and coma, seizures, QRS widening, and cardiac arrhythmias. The clinical and electrocardiogram abnormalities (QRS widening) develop rapidly and resolve over 6–24 h consistent with the rapid changes in the plasma concentrations (17).

Lithium toxicity is another example of a drug where understanding its distribution is key to explaining acute versus chronic toxicity. Lithium is a small molecule that is distributed widely in extracellular fluid. The major site of toxicity is the central nervous system where it can cause cerebellar toxicity, confusion, coma, and death. However, this occurs only with chronic toxicity which develops over days to weeks (20). Figure 8 shows the time course of lithium concentrations in a patient with chronic toxicity who presented with QT prolongation, confusion, and cerebellar signs. In this case, the decline in lithium concentration was consistent with

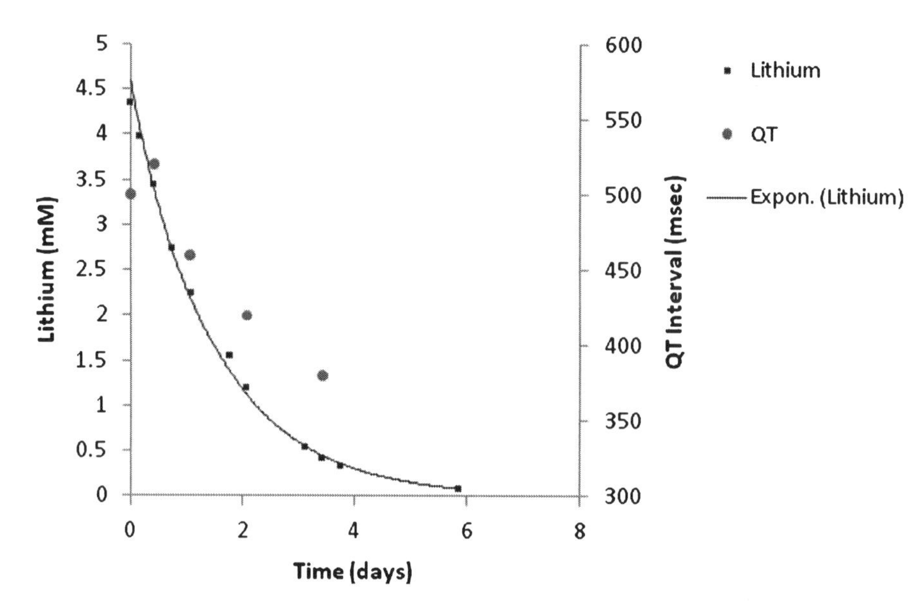

Fig. 8. Plasma lithium concentration (*black squares*) and QT interval (*gray circles*) versus time for a female patient with chronic lithium toxicity. The apparent half-life of lithium is 24.8 h assuming a one-compartmental model.

the clinical effects and this is shown with a similar decrease of the QT interval with the decline in lithium concentration (Fig. 8). In contrast, in acute lithium overdose toxicity occurs rarely because there is only transient exposure to high plasma lithium concentrations before it is distributed throughout the body. Uptake into the central nervous system is slow and for toxicity to occur there must be persistently high concentrations (above the upper level of the therapeutic range) for central nervous system toxicity to occur (20). In this setting, the distribution of lithium into the brain takes longer than its elimination, which is opposite to tricyclic antidepressants where the distribution to the brain is rapid and elimination slow.

The relationship between the first-order rate constant, clearance, and apparent volume of distribution is shown in Fig. 8 where there is mono-exponential decay of lithium with a first-order elimination constant of 0.028/h and an elimination half-life of 24.8 h. This apparent half-life of elimination is similar to that determined in pharmacokinetic studies of lithium therapeutically (21).

Methotrexate is another example of a drug which is similar to lithium in that its cellular uptake is slower than its renal elimination. Acute overdose of methotrexate orally has never been reported to cause severe toxicity compared to taking a therapeutic dose of methotrexate daily instead of weekly which can be life-threatening (22).

2.5. Elimination

Elimination is the irreversible loss of the drug from the site of measurement which occurs by excretion of unchanged drug, mainly through kidneys, or conversion of drug into a metabolite

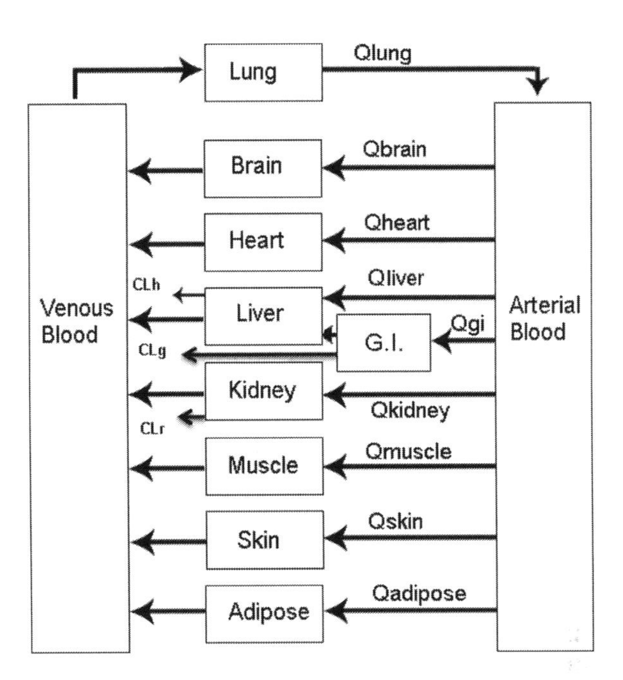

Fig. 9. A physiologically based pharmacokinetic model, describing the elimination of drug from various tissues.

via various metabolic pathways, mainly in the liver. Some drugs are excreted through bile after being metabolized in the liver, usually after phase II conjugation reactions. Uncommonly some drugs, mostly volatiles, may also be excreted through the lungs. We limit our discussion here to the two most common methods of drug elimination, namely the renal and the hepatic routes. Figure 9 replicates the distribution pathways in Fig. 6 and adds the major elimination pathways.

2.5.1. Hepatic Elimination

Metabolism is the predominant mechanism by which about 75% of drugs are eliminated from the body. The majority of drug metabolism occurs in the liver but can occur at other sites, including the gastrointestinal tract, the kidneys, and the lungs. Drug metabolites are not always inactive but are generally less active. There are special circumstances in which the metabolite is active while the parent is inactive. It is argued that codeine is a special case of this where codeine itself is thought to be inactive, and morphine, a metabolite, is active. In some cases drugs are designed specifically for the parent to be inactive and the metabolite active. These are termed prodrugs in which the chemical structure is modified so that absorption occurs more readily, e.g., dabigatran is given as the inactive dabigatran etexilate (23), and mycophenolate is given as the inactive mycophenolate mofetil (24). There are also less common circumstances in which the parent and metabolite are active and the metabolite itself is sometimes given as the active agent, e.g.,

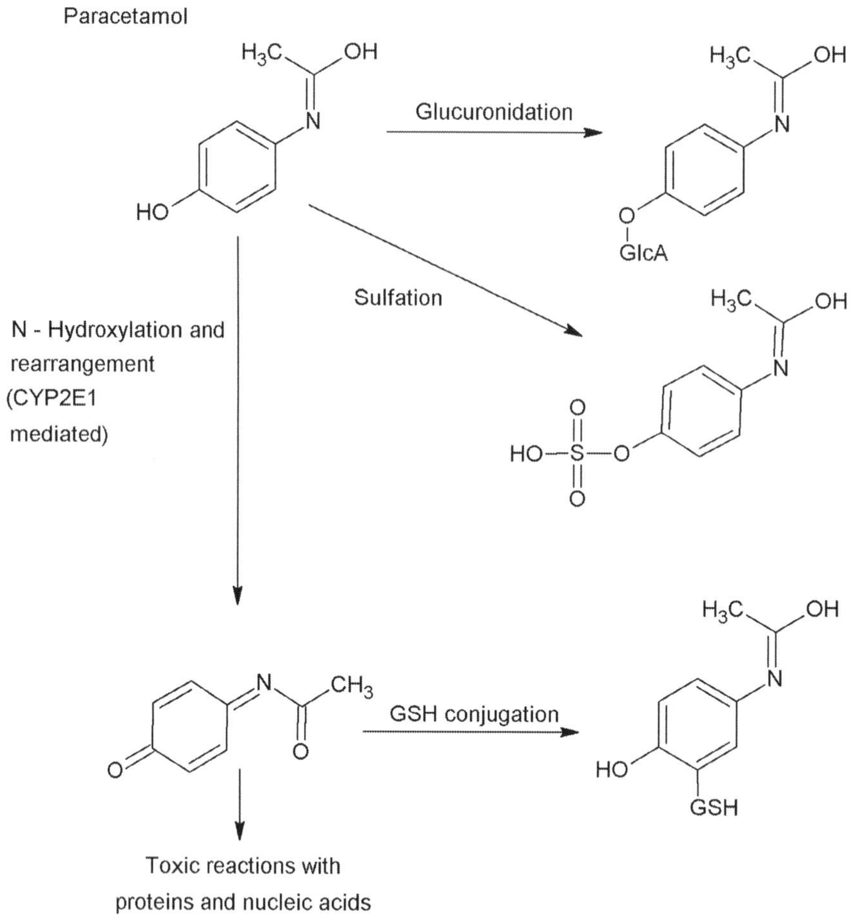

Fig. 10. Metabolic pathways for acetaminophen excluding the small amount excreted unchanged by the kidneys.

amitriptyline is metabolized to nortriptyline (25), hydroxyzine is metabolized to cetirizine (26).

There are two major pathways of hepatic metabolism referred to as phase I and phase II metabolic pathways. Phase I reactions generally convert the drug into a more polar species, which are then more easily eliminated by the kidneys, by either introducing a functional group or unmasking a functional group (e.g., –OH, $-NH_2$, –SH) and involve the oxidation, reduction or hydrolysis of the drug. The conversion of acetaminophen to *N*-acetyl-*p*-benzoquinone imine (NAPQI) by CYP2E1 (27) shown in Fig. 10 is an example of a phase I reaction. Phase II reactions are characterized by conjugation pathways and involve combining an endogenous substrate (e.g., glucuronide, sulfate) to a functional group on the drug. Figure 10 shows that the majority of acetaminophen is metabolized via conjugation pathways glucuronidation (40–80%) and sulfation (10–30%). Some drug fates include both phase I and II reactions. This occurs for acetaminophen where the toxic

Table 1
Examples of in vivo substrates, inhibitors, and inducers of various CYP isozymes relevant to clinical toxicology. This is not an exhaustive list

CYP	Substrate	Inhibitor	Inducer
1A2	Olanzapine, caffeine, amitriptyline	Fluvoxamine, fluoroquinolones	Smoking
2C9	Warfarin, phenytoin, ibuprofen, sulfonylureas	Fluconazole, valproate	Rifampicin
2C19	Omeprazole, citalopram, diazepam, imipramine	Omeprazole, fluvoxamine, moclobemide	Rifampicin
2D6	Imipramine, amitriptyline, fluoxetine, fluvoxamine, paroxetine, venlafaxine, oxycodone, tramadol, codeine, risperidone, metoprolol	Paroxetine, fluoxetine, bupropion	
2E1	Halothane and related anesthetics, acetaminophen, theophylline	Disulfiram	Ethanol
3A4/3A5	Midazolam, alprazoloam quetiapine, venlafaxine, methadone, mirtazapine, reboxetine, sertraline, diltiazem, many immunosuppressants, and chemotherapeutic agents	Clarithromycin, indinavir, ketoconazole	Rifampicin, carbamazepine, phenytoin, phenobarbitone

metabolite NAPQI is detoxified by conjugation to glutathione and then eliminated (Fig. 10).

The enzymes responsible for phase I metabolism belong to the family of cytochrome P 450 (CYP450). The CYP isoenzymes can be classified into various classes such as CYP2E1, CYP2C19, CYP2C9, CYP2D6, and CYP3A4/3A5 which are common and important ones for the metabolism of drugs taken in overdose. Generally all CYP enzymes are present in all metabolism tissues (e.g., gut, liver). CYP3A4 is the most abundant enzyme and CYP2D6 although less abundant is responsible for metabolism the largest fraction of drugs that undergo phase I metabolism. Drugs that undergo metabolism through a specific enzyme are called substrates for the enzyme. Some drugs may induce or inhibit the activity of the enzymes and this is important for both therapeutic use of drugs and in some cases the clearance of drugs in overdose (13). Table 1 provides examples of common drugs that are substrates, inducers and inhibitors of CYP isozymes relevant to clinical toxicology.

Phase II metabolic reactions are usually detoxification reactions. In these reactions the drug molecule is conjugated with a cofactor. Examples of cofactors include UDP-glucuronic acid and glutathione. The conjugation takes place in the presence of enzymes such as the UDP-glucuronosyltransferases, sulfotransferases, N-acetyltransferases, and glutathione S-transferases glucuronidases.

2.5.2. Metabolism and Drug Overdose

It is generally thought that in most overdoses these metabolic pathways are saturated. However, there is little evidence to support saturation in overdose and recent pharmacokinetics studies of citalopram, quetiapine, and venlafaxine in overdose suggest that metabolism is not saturated for these drugs despite drug concentrations 10- to 100-fold those seen with therapeutic use (12–14). Common examples of drugs where saturation occurs include ethanol, theophylline, and phenytoin; however saturation in these cases occurs with therapeutic doses, albeit for theophylline this is not noticeable clinically.

In the case of acetaminophen overdose, toxicity is due to the formation of the toxic metabolite NAPQI and there is some evidence that inhibition of CYP2E1 by ethanol decreases the formation of this metabolite and therefore toxicity, and conversely chronic alcohol use induces metabolism (28, 29). Although there is saturation of the sulfation pathway in acetaminophen overdose due to depletion of sulfate (30, 31), it is unclear if this changes the overall pharmacokinetics in overdose.

2.5.3. Renal Elimination

The renal elimination of the drugs is usually referred to as drug excretion and may be elimination of the parent drug or the drug metabolites. Excretion of drug and the metabolites into the urine involves three main mechanisms, glomerular filtration, active tubular secretion, and tubular reabsorption. About 25% of all the drugs undergo renal elimination as unchanged drug.

Renal elimination depends on various factors including the lipophilicity of the drug, plasma protein binding and the plasma drug concentration. The nephron is the basic anatomical unit of the kidney that is responsible for renal elimination of drugs. Like any substance eliminated by the kidneys, including endogenous substances, drugs can be eliminated by one of four major processes which occur at different sites along the nephron: glomerular filtration, active secretion, passive diffusion, and active reabsorption.

About 25% of cardiac output goes to the kidney and about 10% of it is filtered through the kidney. The glomerular filtration of a drug will depend on the glomerular blood flow and the concentration of unbound drug. Drugs that are bound to plasma proteins are not filtered. The glomerular filtration rate (GFR) is usually approximated by the clearance of creatinine, which is a catabolic product of amino acid metabolism in the muscle. Changes in the plasma protein binding of drugs will affect filtration, such as saturation of protein binding or changes in pH.

If the renal clearance of a drug is greater than GFR, then active secretion pathways are likely to be playing a role. It should be noted that only the net renal clearance can be determined and the individual influence of each process is generally not identifiable in clinical practice. Active secretion occurs via a carrier mechanism and tends to be sufficiently powerful to remove drug from plasma protein binding sites and hence is relatively unaffected by binding to plasma. Benzyl penicillin is 80% protein bound and almost completely removed from the blood by secretion into the proximal tubule. However, these efflux transporters can saturate at high concentrations of drugs. The P-glycoprotein and the organic anion transporting polypeptides are specific transport proteins for excretion of drugs. These are saturable and can lead to nonlinear pharmacokinetics (32).

Water is reabsorbed as the filtrate passes down the nephron so that only 1% of the original filtrate emerges as urine. Lipophilic drugs are more likely to permeate through the membrane and be reabsorbed. This is in contrast to polar drugs that have minimal reabsorption and their renal clearance will be similar to the GFR, e. g., gentamicin.

Drugs that resemble essential amino acids (levodopa, α-methyldopa, and thyroxine) are actively reabsorbed. Uric acid is also actively reabsorbed which is inhibited by probenecid, which competes with uric acid for the active transport system. This is the basis of the use of probenecid for the treatment of gout.

Renal clearance of a drug can be measured with timed collection of urine and analysis of the drug concentration in the urine using the following equation:

$$\mathrm{CL_r} = \frac{C \times Q}{C_{\mathrm{urine}}}, \tag{3}$$

where $\mathrm{CL_r}$ is renal clearance, C is the concentration of the drug in plasma, Q is the urine flow rate, and C_{urine} is the concentration of the drug in urine.

2.5.4. Renal Elimination and Drug Overdose

Understanding renal elimination is important for both acute overdose and poisoning with drugs that are mainly or completely eliminated by the kidneys. Chronic poisoning by drugs that are renally eliminated will occur in patients with abnormal renal function or acute renal failure, including digoxin toxicity, lithium toxicity, and metformin poisoning.

There are a number of drugs where renal elimination becomes important in overdose compared to therapeutic doses. An example of this is salicylate poisoning where the metabolic pathways in the liver are saturated in overdose leaving the major elimination pathway as renal clearance of salicylic acid (33). This explains why the apparent half-life of elimination of salicylate increases from 2 to 4 h in therapeutic doses (due to hepatic clearance) to approximately 20 h in overdose which is mainly due to renal elimination (34).

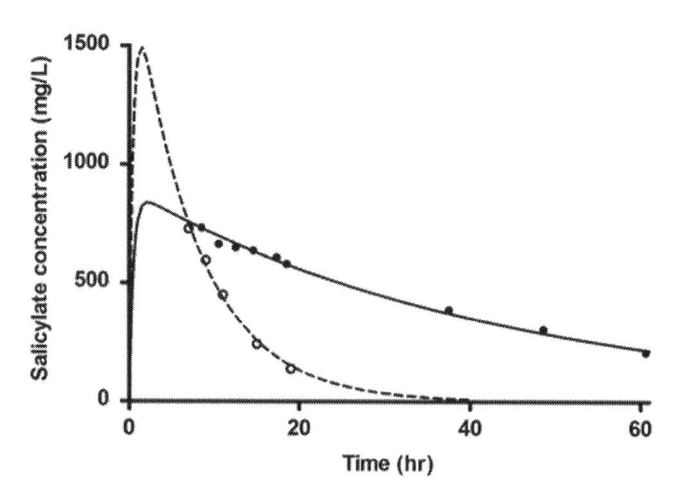

Fig. 11. Two patients with acute aspirin (acetylsalicylic acid) overdoses. The first patient (*filled circles*; *thick line*) ingested 36 g, was not treated with sodium bicarbonate and had a half-life of elimination of 29.8 h. The second patient (*open circles*; *dashed line*) ingested 15 g, was treated with a loading dose and infusion of bicarbonate and had a half life of 5.2 h. The observed concentrations (*filled* and *open circles*) have been fitted to a one-compartment model with first-order input.

Increasing the renal excretion of salicylic acidic is therefore important in the treatment of salicylate poisoning. Alkalinization of the urine, usually by administering a loading dose and infusion of bicarbonate, increases the dissociated anionic form of salicylic acid (i.e., salicylate anion) in the urine (35). Reabsorption is therefore reduced since the molecule is charged ("ion trapping") which increases the amount of salicylate excreted in the urine. The difference in rate of elimination is shown in Fig. 11 for two patients with aspirin overdose and only one treated with alkaline diuresis.

Toxic alcohols also have changed elimination in overdose. The metabolites of methanol and ethylene glycol (formic acid and oxalic acid) are the toxic species in toxic alcohol poisoning. Antidotal therapy aims to inhibit the enzyme alcohol dehydrogenase that metabolizes the alcohols either with ethanol or fomepizole. This then increases the half-life of elimination to about 50 h (methanol) or 17 h (ethylene glycol) since only the renal pathway remains. In the case of methanol this often necessitates the use of hemodialysis to speed the removal of methanol.

2.5.5. Clearance

The clearance of a drug from various tissues occurs in parallel as shown in Fig. 9 so the total body clearance of the drug (CL) is equal to the sum of the clearances of the individual tissues.

$$\mathrm{CL} = \mathrm{CL_h} + \mathrm{CL_g} + \mathrm{CL_r}. \tag{4}$$

Clearance is a constant that describes the relationship between drug concentration (C) in the body and the rate of elimination of the drug from the body and has units of volume per time. It is, however, perhaps most easily understood based on Fig. 9 as the product of the perfusion of the eliminating organ (Q) and the intrinsic ability of the organ to eliminate the drug termed extraction (E).

$$\mathrm{CL} = Q \times E. \tag{5}$$

Since E is unitless, then CL has the same units as perfusion (volume per time). The CL of the drug is always constant and is considered a primary parameter.

This is a physiologically appealing definition of CL in the sense that alterations in perfusion and extraction can be shown to change CL in a predictable manner.

3. Methods

3.1. Basic Model for PK

Given the many processes involved in the input and disposition of drugs, the mathematical models used to describe the PK of drugs can be complicated. While the processes can be, sometimes, exceedingly complex as depicted in Fig. 9, it is fortunate that a simple one-compartment model with first-order input provides a reasonable description of the time course for many drugs both therapeutically and in overdose (Fig. 12). Of course more complex PK models may be necessary.

In Fig. 12 we see that all the tissues of the body are lumped together as a single homogeneous compartment, the "body" (discussed in the next section). The arrows indicate the movement of the drug. The figure shows that part of the drug may be excreted

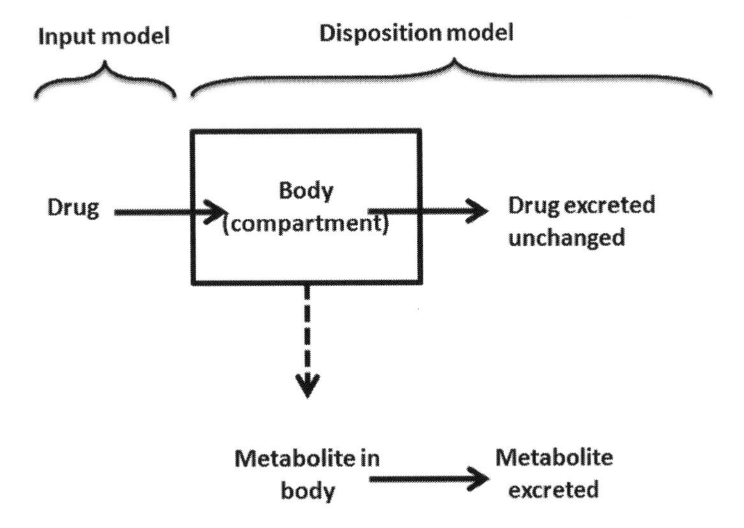

Fig. 12. A one compartment PK model with first-order input that can be used to describe the time course of drug concentration. Although two routes of elimination are illustrated the mathematical model describes total clearance.

unchanged and some part may be metabolized by enzymes in the liver. In the figure the time course of dose–concentration relationship may be described by combining input and disposition models. Conceptually this is provided by:

$$\text{Drug concentration}(t) = f[\text{input model } (t), \text{ disposition model}(t)]. \tag{6}$$

Here we just use f[input model (t), disposition model(t)] to signify that drug concentration is a function of input and disposition and these two models are essentially independent of one another. From a mass balance perspective the total amount of drug that enters the body is equal to the total amount of drug that is eliminated from the body. Hence

$$\begin{aligned} \text{Dose} = {} & \text{Amount of drug at absorption site} \\ & + \text{amount of drug in the body} \\ & + \text{amount of unchanged drug excreted} \\ & + \text{amount of metabolite in the body} \\ & + \text{amount of metabolite eliminated} \cdots \end{aligned} \tag{7}$$

We note here that this is a simplification of true mass balance and does not account for the change of mass due to metabolic process, but in a simple sense if mass did not change during these processes then this concept holds.

3.2. Rate of Movement of Drug Around the Body

Most drug movement processes are due to passive diffusion, therefore the driving factor is concentration. This means that the higher the concentration the greater the rate of the drug that diffuses across a membrane. This concentration-proportional rate is termed first order which holds for most drugs. If the rate of movement is constant and independent of concentration, then it is said to be zero order. If the rate of movement of drug is saturable (i.e., it changes from apparent first order to apparent zero order as concentration increases) then it is termed mixed order and most commonly described by a Michaelis–Menten process.

Note all the equations shown below are in terms of amount (A) but can be written in terms of concentrations (C).

Zero-order process:

$$\frac{dA}{dt} = -k_0 \tag{8}$$

A zero-order process is described by a constant rate (mass per time) alone and is independent of the amount of drug to be transferred.

First-order process:

$$\frac{dA}{dt} = -k \times A \tag{9}$$

A first-order process is described by rate constant (k) and the amount of drug to be transferred. If we describe the same equations as concentration instead of amount, the rate constant will be replaced by clearance (CL)

Michaelis–Menten (MM) process or mixed order process:

$$\frac{dA}{dt} = -\frac{V_{max}}{k_m + A/V} \times A/V \quad (10)$$

A Michaelis–Menten process is described by the Michaelis constant (k_m) and V_{max} the maximum rate at which the drug can be eliminated. k_m is defined as the amount for half the maximum rate or $V_{max}/2$ and has same units as A. Here V is the volume of distribution. When $A >> k_m$, k_m is negligible and the equation simplifies to,

$$\frac{dA}{dt} = -V_{max} \quad (11)$$

which is equivalent to a zero-order process (Eq. 8).

When $k_m >> A$, A is negligible and the equation becomes:

$$\frac{dA}{dt} = -\frac{V_{max}}{k_m} \times A, \quad (12)$$

where the ratio of V_{max} (e.g., mg/h) to k_m (e.g., mg) essentially provides k (1/h) and thus equivalent to a first-order process (Eq. 9).

3.3. Describing the Pharmacokinetics of Drugs

Three different approaches can be used to describe the PK of drugs:

1. Compartmental pharmacokinetic models
2. Non-compartmental pharmacokinetic models
3. Physiologically based pharmacokinetic (PBPK) models

In the compartmental approach the body is divided into one or more compartments the number of which is dictated by the data. PBPK involves the investigator defining a set of compartments based on physiology. The similarity of the behavior of the drug in various compartments is assessed based on data which is sampled from each of the compartments (Shown in Figs. 6 and 9).

The non-compartmental approach (NCA) does not require the assumption of any compartments for the purpose of analysis and includes the summary variables: peak or maximum drug concentration (C_{max}), time to peak drug concentration (T_{max}), area under the curve (AUC). However, they do require weak assumptions if the non-compartmental summary variables are converted into parameter values such as CL and V.

3.3.1. Compartmental Models

A commonly used approach to characterize the PK of a drug in the body is to represent the body as a series of compartments.

Assumptions are required when models are used to describe data. The compartment model assumes that each of the compartments is a kinetically homogenous unit. The model also assumes that the drug is instantaneously and evenly distributed throughout the compartment. Despite these being apparently strong assumptions it appears that most models perform well suggesting that these requirements are generally too subtle for clinically meaningful issues with the model.

Though not a requirement, it is often assumed when constructing a model that the rate of elimination of the drug from the compartment and the transfer of the drug between the compartments follows first order (linear kinetics). When naming a compartment model the number of compartments refers to the total number of disposition compartments. For example a one-compartment extravascular absorption model has two compartments, a gastrointestinal compartment and a compartment which incorporates all other tissues in the body. However, by convention this would be termed a one-compartment model based on the number of disposition compartments.

3.3.2. The One-Compartment PK Model

The one compartment model is the simplest of the models used to describe the PK of drugs. This model assumes that all the tissues in the body are lumped together into a single kinetically homogenous unit. In the below pharmacokinetic model, the whole body, except the gut, is assumed to be a single compartment. A schematic of a one-compartment PK model is given in Fig. 13.

When a dose D is administered extravascularly, it has to be absorbed through a biological barrier to enter the central compartment (blood) where it becomes systemically available. The process itself is complex and determined by factors such as, the route of administration, the amount administered, the formulation, and the physicochemical properties of the drug. This complex input process is often assumed to be a first-order input process governed by a single parameter k_a, the rate constant of absorption. This assumption of a first-order input process is not a requirement and the absorption rate may be described by any number of processes, including a zero-order input process (which would describe an intravenous infusion) or a mixture of these processes.

For this first-order input one-compartment model the rate of change of amount of drug over time (t) in both the absorption site and the body can be represented as a set of ordinary differential equations given by:

$$\frac{\mathrm{d}A(1)}{\mathrm{d}t} = -k_\mathrm{a} \times A(1) \tag{13}$$

$$\frac{\mathrm{d}A(2)}{\mathrm{d}t} = (k_\mathrm{a} \times A(1)) - (\mathrm{CL}/V \times A(2)), \tag{14}$$

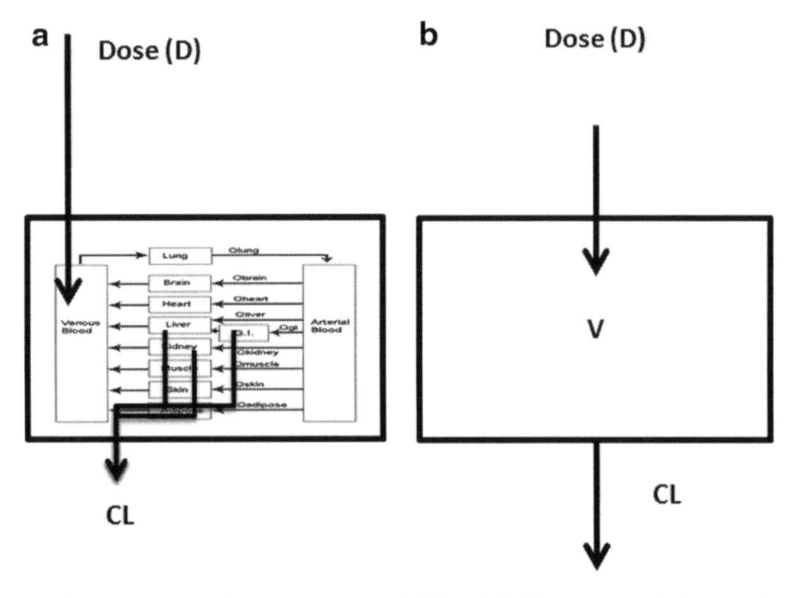

Fig. 13. (**a**) Schematic of a one-compartment PK model with intravenous bolus administration, showing all the tissues in the body. CL is the clearance and *V* is the volume of distribution of the drug. (**b**) Schematic of a one-compartment PK model with intravenous bolus administration. All the tissues are lumped together into a single homogeneous compartment. Both are the same; however, (**a**) shows the lumping assumptions.

with the initial conditions set to $A(1) = D$ and $A(2) = 0$ at $t = 0$.

It is important to understand that the one-compartment model does not imply that the concentration in the compartment (plasma or blood, the tissue most commonly used for measuring drug concentrations), is equal to the concentration in other tissues in the body. Rather the rate of change of concentration in the plasma or blood is identical to the rate of change of concentration in the tissues.

By simultaneously solving the above differential equations the amount of drug in plasma at any instance of time t is given by:

$$A(t) = D\frac{k_a}{k_a - \frac{CL}{V}}\left[\exp^{-\frac{CL}{V}\times t} - \exp^{-k_a\times t}\right]. \qquad (15)$$

The parameters in Eq. 11 are the absorption rate constant (k_a), clearance (CL) and volume of distribution (V). By dividing Eq. 15 by the volume of distribution (V), the concentration at any time can be obtained as shown in Eq. 16.

$$C(t) = D\frac{k_a}{V(k_a - \frac{CL}{V})}\left[\exp^{-\frac{CL}{V}\times t} - \exp^{-k_a\times t}\right]. \qquad (16)$$

Figure 14 gives an example of a concentration time data from a patient who ingested 70 g of amisulpride. The parameters k_a, V, and CL have been estimated using a first-order input one-compartment model and the half-life has been derived from the disposition

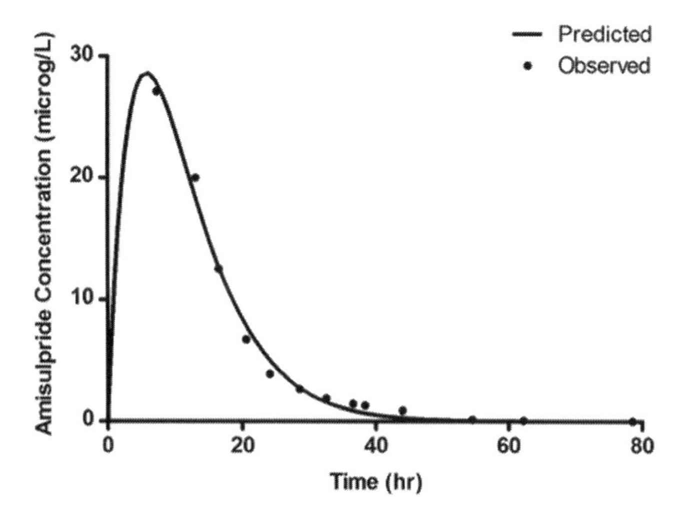

Fig. 14. Observed concentration time data from a female patient ingesting 14 g of amisulpride fitted with a first-order one-compartment model with parameter estimates of k_a, 0.173/h, V, 179 L and CL, 31.3 L/h and a half-life of 4 h.

parameters V and CL. The CL of amisulpride in healthy volunteer study at therapeutic doses was reported to be 31.2–41.6 L/h (36).

4. Notes

- Pharmacokinetics describes the time course of drug concentration in the body.
- Pharmacokinetic models are usually broken down into a model for drug input and a model for drug disposition.
- The primary purpose of pharmacokinetics is to add time into the concentration–effect relationship and thus allow us to understand the time course of drug effects.
- In clinical toxicology, a pharmacokinetic study can form the basis from which the pharmacokinetics after overdose can be compared to therapeutic doses to assess for any unexpected influence of very high doses. These studies also form the basis of determining the likely effectiveness of decontamination processes.

References

1. Dawson AH, Whyte IM (2001) Therapeutic drug monitoring in drug overdose. Br J Clin Pharmacol 52(Suppl 1):97S–102S
2. Isbister GK (2010) How do we use drug concentration data to improve the treatment of overdose patients? Ther Drug Monit 32:300–304
3. Isbister GK, Friberg LE, Duffull SB (2006) Application of pharmacokinetic-pharmacodynamic modelling in management

of QT abnormalities after citalopram overdose. Intensive Care Med 32:1060–1065

4. Friberg LE, Isbister GK, Duffull SB (2006) Pharmacokinetic-pharmacodynamic modelling of QT interval prolongation following citalopram overdoses. Br J Clin Pharmacol 61:177–190
5. Martinez MN, Amidon GL (2002) A mechanistic approach to understanding the factors affecting drug absorption: a review of fundamentals. J Clin Pharmacol 42:620–643
6. Lin JH, Chiba M, Baillie TA (1999) Is the role of the small intestine in first-pass metabolism overemphasized? Pharmacol Rev 51:135–158
7. Hoppu K, Koskimies O, Holmberg C, Hirvisalo EL (1991) Evidence for pre-hepatic metabolism of oral cyclosporine in children. Br J Clin Pharmacol 32:477–481
8. Pacifici GM, Eligi M, Giuliani L (1993) (+) and (−) terbutaline are sulphated at a higher rate in human intestine than in liver. Eur J Clin Pharmacol 45:483–487
9. Tanigawara Y, Okamura N, Hirai M, Yasuhara M, Ueda K, Kioka N, Komano T, Hori R (1992) Transport of digoxin by human P-glycoprotein expressed in a porcine kidney epithelial cell line (LLC-PK1). J Pharmacol Exp Ther 263:840–845
10. Cvetkovic M, Leake B, Fromm MF, Wilkinson GR, Kim RB (1999) OATP and P-glycoprotein transporters mediate the cellular uptake and excretion of fexofenadine. Drug Metab Dispos 27:866–871
11. Lindenberg M, Kopp S, Dressman JB (2004) Classification of orally administered drugs on the World Health Organization Model list of Essential Medicines according to the biopharmaceutics classification system. Eur J Pharm Biopharm 58:265–278
12. Friberg LE, Isbister GK, Hackett LP, Duffull SB (2005) The population pharmacokinetics of citalopram after deliberate self-poisoning: a Bayesian approach. J Pharmacokinet Pharmacodyn 32:571–605
13. Isbister GK, Friberg LE, Hackett LP, Duffull SB (2007) Pharmacokinetics of quetiapine in overdose and the effect of activated charcoal. Clin Pharmacol Ther 81:821–827
14. Kumar VV, Oscarsson S, Friberg LE, Isbister GK, Hackett LP, Duffull SB (2009) The effect of decontamination procedures on the pharmacokinetics of venlafaxine in overdose. Clin Pharmacol Ther 86:403–410
15. Buckley NA, Dawson AH, Reith DA (1995) Controlled release drugs in overdose: clinical considerations. Drug Saf 12:73–84
16. Brahmi N, Kouraichi N, Thabet H, Amamou M (2006) Influence of activated charcoal on the pharmacokinetics and the clinical features of carbamazepine poisoning. Am J Emerg Med 24:440–443
17. Hulten BA, Heath A, Knudsen K, Nyberg G, Starmark JE, Martensson E (1992) Severe amitriptyline overdose: relationship between toxicokinetics and toxicodynamics. J Toxicol Clin Toxicol 30:171–179
18. Hulten BA, Heath A, Knudsen K, Nyberg G, Svensson C, Martensson E (1992) Amitriptyline and amitriptyline metabolites in blood and cerebrospinal fluid following human overdose. J Toxicol Clin Toxicol 30:181–201
19. Meineke I, Schmidt W, Nottrott M, Schroder T, Hellige G, Gundert-Remy U (1997) Modelling of non-linear pharmacokinetics in sheep after short-term infusion of cardiotoxic doses of imipramine. Pharmacol Toxicol 80:266–271
20. Waring WS (2006) Management of lithium toxicity. Toxicol Rev 25:221–230
21. Sproule BA, Hardy BG, Shulman KI (2000) Differential pharmacokinetics of lithium in elderly patients. Drugs Aging 16:165–177
22. Balit CR, Daly FFS, Little M, Murray L (2006) Oral methotrexate overdose. Clin Toxicol 44:1
23. Sanford M, Plosker GL (2008) Dabigatran etexilate. Drugs 68:1699–1709
24. Goldblum R (1993) Therapy of rheumatoid arthritis with mycophenolate mofetil. Clin Exp Rheumatol 11(Suppl 8):S117–S119
25. Watson CP, Vernich L, Chipman M, Reed K (1998) Nortriptyline versus amitriptyline in postherpetic neuralgia: a randomized trial. Neurology 51:1166–1171
26. Tashkin DP, Brik A, Gong H Jr (1987) Cetirizine inhibition of histamine-induced bronchospasm. Ann Allergy 59:49–52
27. Manyike PT, Kharasch ED, Kalhorn TF, Slattery JT (2000) Contribution of CYP2E1 and CYP3A to acetaminophen reactive metabolite formation. Clin Pharmacol Ther 67:275–282
28. Schmidt LE, Dalhoff K, Poulsen HE (2002) Acute versus chronic alcohol consumption in acetaminophen-induced hepatotoxicity. Hepatology 35:876–882
29. Thummel KE, Slattery JT, Ro H, Chien JY, Nelson SD, Lown KE, Watkins PB (2000) Ethanol and production of the hepatotoxic metabolite of acetaminophen in healthy adults. Clin Pharmacol Ther 67:591–599
30. Levy G, Galinsky RE, Lin JH (1982) Pharmacokinetic consequences and toxicologic implications of endogenous cosubstrate depletion. Drug Metab Rev 13:1009–1020
31. Gelotte CK, Auiler JF, Lynch JM, Temple AR, Slattery JT (2007) Disposition of

acetaminophen at 4, 6, and 8 g/day for 3 days in healthy young adults. Clin Pharmacol Ther 81:840–848

32. Tirona RG, Kim RB (2002) Pharmacogenomics of organic anion-transporting polypeptides (OATP). Adv Drug Deliv Rev 54:1343–1352
33. Levy G, Tsuchiya T (1972) Salicylate accumulation kinetics in man. N Engl J Med 287:430–432
34. Done AK (1960) Salicylate intoxication. Significance of measurements of salicylate in blood in cases of acute ingestion. Pediatrics 26:800–807
35. Cumming G, Dukes DC, Widdowson G (1964) Alkaline diuresis in treatment of aspirin poisoning. Br Med J 2:1033–1036
36. Rosenzweig P, Canal M, Patat A, Bergougnan L, Zieleniuk I, Bianchetti G (2002) A review of the pharmacokinetics, tolerability and pharmacodynamics of amisulpride in healthy volunteers. Hum Psychopharmacol 17:1–13

Chapter 13

Modeling of Absorption

Walter S. Woltosz, Michael B. Bolger, and Viera Lukacova

Abstract

Absorption takes place when a compound enters an organism, which occurs as soon as the molecules enter the first cellular bilayer(s) in the tissue(s) to which is it exposed. At that point, the compound is no longer part of the environment (which includes the alimentary canal for oral exposure), but has become part of the organism. If absorption is prevented or limited, then toxicological effects are also prevented or limited. Thus, modeling absorption is the first step in simulating/predicting potential toxicological effects. Simulation software used to model absorption of compounds of various types has advanced considerably over the past 15 years. There can be strong interactions between absorption and pharmacokinetics (PK), requiring state-of-the-art simulation computer programs that combine absorption with either compartmental pharmacokinetics (PK) or physiologically based pharmacokinetics (PBPK). Pharmacodynamic (PD) models for therapeutic and adverse effects are also often linked to the absorption and PK simulations, providing PK/PD or PBPK/PD capabilities in a single package. These programs simulate the interactions among a variety of factors including the physicochemical properties of the molecule of interest, the physiologies of the organisms, and in some cases, environmental factors, to produce estimates of the time course of absorption and disposition of both toxic and nontoxic substances, as well as their pharmacodynamic effects.

Key words: Absorption, Permeability, Gastrointestinal, Dermal, Nasal/pulmonary, Ocular, Pharmacokinetic, Pharmacodynamic, Toxicology, Solubility, Dissolution, Precipitation, Supersaturation, Formulation, Transporters

1. Introduction

Because absorption is required before any compound can exert a toxicological influence, the field of computational toxicology requires not only the ability to predict whether a molecular structure is likely to be toxic to an organism, and in what ways, but also whether it can be expected to be absorbed at rates and in quantities sufficient to overcome the organism's natural defense mechanisms. If the molecule itself is nontoxic, but one or more of its metabolites are, then the absorption of the parent molecule will govern its availability at the sites of metabolism, but the rates and amounts

Brad Reisfeld and Arthur N. Mayeno (eds.), *Computational Toxicology: Volume I*, Methods in Molecular Biology, vol. 929,
DOI 10.1007/978-1-62703-050-2_13, © Springer Science+Business Media, LLC 2012

of metabolite formation will govern the toxicological effects, as seen for acetaminophen (1–3).

Absorption occurs when a compound enters an organism, which occurs as soon as the compound enters the first cellular bilayer(s) in the tissue(s) to which is it exposed. For oral absorption in the gastrointestinal tract, fraction absorbed (F_a) is defined as the fraction of the administered dose that gets absorbed into the apical membrane of the enterocytes (4). At that point, the compound is no longer part of the environment, but has become part of the organism. From there, the fate of the compound, and thereby the potential fate of the organism as a result of the exposure, will be determined by the pharmacokinetics and pharmacodynamics of the compound within the organism. Some compounds will be effluxed back out of absorbing cells by transporter proteins—one of nature's defense mechanisms (5). Some may be assisted into the cells by influx transporters, particularly if such transporters perceive them as nutrients (6). Some will be metabolized either in the absorbing cells or in other tissues within the organism. The resulting metabolite(s) may be therapeutic, toxic, or benign. These metabolites may be easily cleared out of the organism, or they may become bound to various tissues and clear very slowly. Toxicological effects can be from one or more metabolites in addition to, or in the absence of, toxicological effects of the original molecule that was absorbed.

Organisms absorb exogenous molecules through a variety of pathways. Some are ingested with food or liquids and are exposed to gastrointestinal tissues. Some are inhaled and are exposed to the tissues in the respiratory system. Some come in contact with the external surfaces of the organism, typically skin, eyes, and hair.

The rate and extent of absorption of molecules that come into contact with an organism depends on a number of factors that affect absorption through complex interactions. Typically, molecules must be in solution to be absorbed; however, endocytosis can result in very small solid particles being taken into some cells (7). When absorption requires molecules in solution, the solubility of the substance in the medium outside the membrane bilayer (intestinal fluids, mucus layer in the respiratory tract, tear layer in the eyes, etc.) will determine the concentration gradient across that first exposed bilayer, and hence the rate of absorption for passive diffusion.

Ionization of the invading molecules (when they are ionizable) can play an important role in affecting rate and extent of absorption into various tissues. The degree of ionization affects solubility for ionizable compounds, with solubility increasing with percent ionized. Transcellular permeability generally decreases with ionization (8), while observed paracellular permeability of protonated molecules was higher than permeability of their neutral forms (9). This results in a complex interplay between solubility and permeability that requires mechanistic simulations to solve. This interplay is

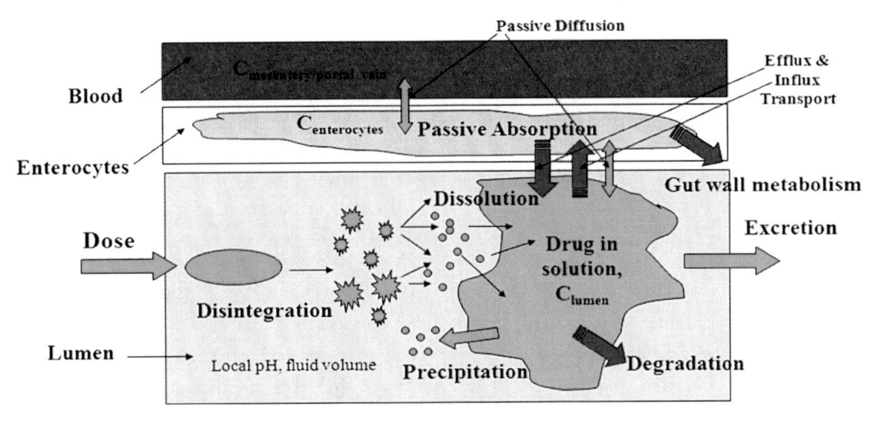

Fig. 1. Interacting processes involved in gastrointestinal absorption.

especially complex for absorption in the gastrointestinal tract, where pH is constantly changing, along with the fluid volume (10), surface area of the enterocytes (11), and the width of the tight junction gaps between enterocytes (12, 13). Food effects and biliary secretion at mealtime can result in changes to solubility and are often the cause for different absorption rates of the same dosage form of drugs between fasted and fed state (14). Many toxic substances would also be expected to show different absorption rates under fasted and fed conditions.

Thus, modeling of absorption of compounds via the various pathways requires knowledge of the physicochemical properties of the molecules, the physiology of the organism, and the biology of the tissues involved. The physical nature of the exposed compound (e.g., solution, powder, suspension) as well as the specific tissues exposed (e.g., skin, nasal/pulmonary passages, gastrointestinal tract, eyes) and duration of exposure (affected by transit time for some tissues) will determine the rate, and eventually the extent, of absorption into the organism. For some substances, degradation of the invading molecules by chemical or metabolic processes can reduce exposure, so that additional equations are required to calculate the degradation rate for such substances.

Thus, dissolution rate, solubility, permeability, transit time, degradation rate, and length of exposure must all be accounted for in an absorption model, and the interactions among these phenomena can be complex (15, 16). For example, a compound that is absorbed quickly produces a sink effect in the donor fluid, reducing the concentration, and allowing faster dissolution of any remaining solid particles, and perhaps also reducing degradation rate. Figure 1 shows the complex interactions involved in the gastrointestinal absorption of an oral dose of a pharmacological ingredient. These processes are essentially the same for ingested toxicants with the exception that a dose is not administered as a particular dosage form like a tablet or capsule, and so it may be introduced into the organism at a steady rate (e.g., inhalation of toxic fumes) or at multiple times with fixed or random spacing (e.g., via drinking water).

The simplest absorption model takes the form of (1)

$$dM_{abs}/dt = k_a \times M_{soln} \quad (1)$$

where M_{abs} is the amount absorbed, k_a is called an absorption rate constant (but is almost always not constant, but a time-varying coefficient), and M_{soln} is the mass of compound in solution at any point in time. This function is an oversimplification, and a more correct expression for movement of molecules across a membrane takes the form of (2), which is based on Fick's First Law for passive diffusion across a membrane (17).

$$dM_{abs}/dt = k_a \times V_{donor} \times (C_{donor} - C_{acceptor}) \quad (2)$$

where V_{donor} is the volume of fluid on the donor side of the membrane, C_{donor} is the concentration on the donor side, and $C_{acceptor}$ is the concentration on the acceptor side of the membrane. Notice that the product $V_{donor} \times C_{donor}$ is equal to M_{soln} in (1). Notice also that if the two concentration terms become equal, absorption stops. In fact, if $C_{acceptor}$ becomes greater than C_{donor}, the absorption rate would be negative and molecules would move back into the donor fluid. The presence of transporter proteins can override the concentration gradient and result in molecules moving from low concentration to high concentration against the gradient. Modern simulation software includes the ability to model the effects of transporters that both help (influx) and hinder (efflux) absorption of molecules.

The absorption rate coefficient, k_a, in (1) and (2) is influenced by different factors depending on the tissue into which absorption is taking place. The value of k_a to input into a simulation can be obtained from *in vitro* cell culture or artificial membrane experiments, from a variety of animal *in situ* experiments, or from an *in silico* prediction based on the molecular structure (18).

The most widely used absorption model for intestinal absorption in the pharmaceutical industry is the one incorporated into GastroPlus and known as the advanced compartmental absorption and transit (ACAT™) model (15). This model is based on the original compartmental absorption and transit (CAT) model developed by Yu (18). Figure 2 shows a diagram of the ACAT model in GastroPlus. A total of nine compartments are used to represent the gastrointestinal tract for humans and animals: stomach, six small intestine compartments, caecum, and colon. Each of these compartments will have a mean transit time, pH, permeability,

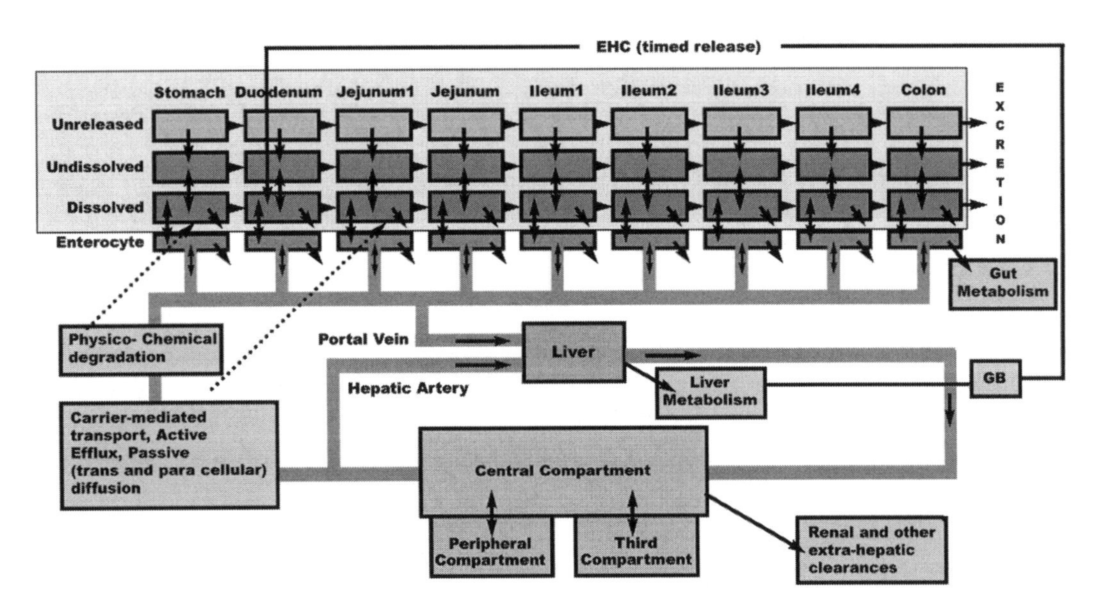

Fig. 2. Advanced compartmental absorption and transit (ACAT) model in GastroPlus.

concentration of bile salts, and fluid volume based on the physiology of the species being modeled. So each compartment will have its own absorption equation, as well as equations for dissolution of solid particles (which can have various sizes), metabolism in the absorbing cells (enterocytes), degradation of the substance in the intestinal lumen, and transit of absorbed compound through the enterocytes to the portal vein leading to the liver. Each compartment simulates all of the processes illustrated earlier in Fig. 1.

Simulating absorption with this program requires the following input parameters for the invading molecule as a minimum.

Molecular weight	pK_a (s)
Log P or log D at a specified pH	Ratio of ionized to un-ionized solubility
Solubility at a specified pH	Permeability
Dose amount	Dosage form (solution, solid, suspension)
Particle size(s) if solid particles exist	

When the species to be simulated is selected, built-in values for the numerous physiological and biological properties are automatically invoked to represent fasted or fed state. To obtain complete plasma concentration-time (Cp–time) predictions, PK parameters are also required inputs. These can be fitted against observed Cp–time data when such data are available.

Absorption modeling is intimately coupled with Pharmaco Kinetic/Toxico Kinetic (PK/TK), PBPK, and PD models in today's state-of-the-art software programs. By combining experimental data with *in silico* predictions for those parameters for which experimental data are not available, simulation programs like GastroPlus can be used not only to predict the absorption of the original (parent) compound but also to predict the rates of formation of metabolites (with known or predicted V_{max} and K_m), the distribution of parent and metabolites into various tissues (when metabolite structures are known or when metabolite concentration–time data are available), the clearance of parent and metabolites from the organism, and the pharmacodynamic (including toxicological) effects of parent and metabolites. The ability to predict specific metabolite PK and PD requires knowing the molecular structures of the metabolites as well as knowing their specific metabolizing enzymes.

Application areas for the techniques

Mechanistic simulation is the only way to quantitatively assess the complex interactions among the various mechanisms involved in absorption and eventual toxicity. A well-constructed simulation allows us to learn things we don't know from things that we do know. Simulation also affords the opportunity to test the sensitivities of the factors that affect absorption and toxicity, including variations in exposure (dose), dosage form, physiological factors, environment (e.g., wind velocity for inhaled compounds), and biology (e.g., changes in expression levels for transporters and enzymes in various tissues, which can change with age, gender, ethnicity, and disease state).

Applications for absorption modeling and simulation in computational toxicology can include the study of all forms of toxic exposure, including, but not limited to natural causes (solar exposure, water and air pollution as a result of natural events) (19), industrial chemicals (waste products, process materials, lubricants, high pressure fluids, coatings, etc.) (20, 21), vehicle exhaust products (aircraft, boats and ships, cars, and trucks), home products (cleaning supplies, automotive supplies, lawn and plant supplies, pet supplies, cosmetics) (22), agricultural chemicals (fertilizers, pesticides) (23), and pharmaceutical products (prescription and over-the-counter medications) (24).

How, when, and by whom these techniques or tools are used in practice

Absorption simulation and modeling tools for computational toxicology are used by environmental toxicologists, academic researchers, chemical industry scientists, clinical pharmacologists, and others to:

- Estimate pharmacokinetic effects (maximum concentration in plasma and various tissues) from potential exposure to toxic materials prior to actual exposure in individuals and populations (25, 26).

- Estimate safe levels of exposure to toxic substances (27, 28).
- Fit models that explain observed effects in subjects actually exposed to toxic materials (29, 30).
- From those models to anticipate actions needed to protect against further damaging exposure.

2. Materials

Simulation computer programs have been developed that employ mechanistic models that account for the interactions identified above through a series of differential equations that are integrated forward in time to provide complete absorption-versus-time profiles. These programs are commercially available from several sources, with the dominant ones used in the pharmaceutical industry listed below.

- GastroPlus™ (http://www.simulations-plus.com).
- PK-Sim™ (http://www.systems-biology.com).
- SimCYP™ (http://www.simcyp.com).

All of these programs are available on Microsoft Windows® platforms and should be able to run on any current version of the Windows operating system for standalone or server-based installations. GastroPlus, PK-Sim, and SimCYP are broad-based programs that offer built-in capabilities that include physiologically based pharmacokinetics (PBPK), drug–drug interactions (which can also be used for toxicant interactions with other molecules, including both drugs and other toxicants), stochastic simulations of variabilities expected in target populations, and model-fitting capabilities. The solution of as many as hundreds of differential equations over a typical exposure period is accomplished within seconds on modern personal computers, allowing scientists to quickly explore various "what if" scenarios as they attempt to explain observed behaviors and to predict outcomes for new conditions. To illustrate the importance of computer speed, the ratio of computer speed for a typical laptop computer in the year 2010 is roughly 15,000 times faster than the original IBM PC in 1981. Thus, a simulation that takes 10 s on the modern laptop computer would take on the order of 150,000 s, or approximately 42 h, on the original IBM PC. Fitting model parameters can require hundreds of such simulations, perhaps requiring an hour on the modern laptop (1.7 years on the 1981 computer). The tremendous improvement in speed and memory in today's inexpensive computers has made possible the sophisticated absorption/PK/PD software that is in common use today. One can only speculate how this will change in yet another 30 years.

For the greatest utility, absorption simulation software should be available for the most popular operating systems, especially Windows. It should provide accurate results using an appropriate level of detail in mechanistic models within the limitations of the state-of-the-art for such models and the required input parameters, i.e., limitations of *in vitro* and *in vivo* data available from which simulation parameters can be estimated. Values for critical model parameters should be built-in where possible. Software should run with reasonable speed and should incorporate built-in error checking that traps likely user mistakes and provides guidance for correcting them. Tutorials should be comprehensive and should use real-world data for examples. The software vendor should provide strong technical support with reasonable direct access to the actual software development science/engineering team.

3. Methods

During program development, complex simulation programs are tested repeatedly by running a variety of examples with known outcomes to ensure that the simulations duplicate observed results. Developers learn from outliers—situations that are not well-predicted at a certain stage of development and require more sophisticated mechanistic models and/or additional parameterization for proper simulation of those situations. With the continuous expansion of knowledge about physiology and biology, as well as the processes relevant in absorption and distribution of xenobiotics in the body, the models (especially the physiologically based ones) are improving with each newer version of these programs. The developers as well as end users need to check the reproducibility of the previous simulation results under the new assumptions. Developers ensure the repeatability of the simulation results from the previous versions of the programs. In cases where the repeatability of simulation results from previous versions is not possible, developers should provide clear explanations for the changes and what differences should the end user expect. In any case, developers should provide guidance on what changes should the end user make to their previous models to account for the latest body of knowledge.

Supporting data used during development and validation come from the scientific literature (including scientific posters and presentations at related meetings), the developer's own experimental data, and from *in silico* predictions when no experimental data are available. In addition to being one of the most important parts of developing new modeling capabilities, it is often the most tedious, requiring gathering, reading through, and filtering of hundreds of sometimes conflicting reports. Typical concerns with respect to literature data include, for example, consistency of the values reported by different labs, relevance of the *in vitro* experiment design to the *in vivo* situation (e.g., media selection in solubility or dissolution rate measurements), whether observations have been corrected for unbound fraction in cases where significant binding to different components in the *in vitro* assay may affect the results (e.g., *in vitro* measurement of clearance), whether separate data for different enantiomers are available. However, the gathering of supporting data is not limited to simply obtaining the values for obvious processes that would be affecting the absorption of the compound (e.g., solubility or permeability), but also gathering information about additional mechanisms that might play a role in a compound's absorption (e.g., designing an *in vitro* experiment to evaluate whether the compound might be a substrate for transporters).

3.1. Fitting Model Parameters

Regardless of the quantity and quality of data available during software development, fitting of one or more model parameters will be required in some, if not most, instances. Absorption simulation software should provide for robust numerical optimization of such parameters from as many types of experimental data as possible. The user should be provided a choice of optimization algorithms, objective functions, and weighting functions and should be able to set constraints on both fitted (optimized) parameters and on various results of the absorption simulation, such as maximum concentration in plasma (C_{max}) or other tissues, time of maximum concentration (T_{max}), area under the plasma concentration–time curve (AUC), fraction absorbed, and fraction bioavailable.

Scaling data from *in vitro* experiments or *in vivo* measurements in different species to human is a frequent challenge. For example, permeability in the gastrointestinal tract of rat or dog can be significantly different than that in human, even for compounds that obey simple passive diffusion (i.e., no transporters involved). Rat permeabilities tend to be around 3–4 times lower than human but are very well correlated after being scaled, while dog permeabilities tend to be around four times higher, but these are gross approximations and the actual reported ranges are much wider, with rat permeabilities as much as 15 times lower than human (31). It is also important to keep in mind the relevance of *in vitro* values to the *in vivo* situation. For example, aqueous solubility will generally not describe adequately dissolution of a compound in the gastrointestinal tract (14). The concentration of bile salts *in vivo* needs to be accounted for (32), together with the fact that it changes in different regions of the gastrointestinal tract as well as in relation to food intake (33). Of course, the effect of bile salts on solubility will not be relevant for ocular or pulmonary exposure where different types and amounts of surfactant may aid dissolution (34, 35).

3.2. Conducting Simulations and Validating Results

Conducting absorption simulations involves a series of steps:

- Gathering the data to be used as inputs and observations, and, if necessary, converting the values to units needed by the simulation software. Some programs provide built-in conversion tools that will accept data in a variety of commonly used units and convert them to units needed by the software.
- Examining the available data, both input parameters and expected outputs (observations) for likely challenges in building predictive models.
  - Incomplete data—missing values, aggregate rather than individual subject data.
  - Anomalies in the data that appear to be inconsistent.
  - Unknowns within the data (e.g., were *in vitro* results corrected for unbound fraction?).
  - Recognizable behaviors in observations that provide clues for using various model options (e.g., double peaks in plasma concentration–time data that might indicate mealtimes or enterohepatic circulation, or significant delay after dose/exposure before noticeable concentration of the molecule of interest is detected might indicate delayed gastric emptying when the exposure is by the oral route).
- If multiple data sets are available, dividing the available experimental data into two groups: that which will be used as to fit relevant model parameters for the simulations and that which will be used to test those models for generality.
- Running simulations with default parameter values to see if the simulated results are reasonably close to observations.
- Fitting model parameters as required to improve the ability of the absorption and pharmacokinetic models to predict observed concentration–time data. A very important part of this step is the selection of parameters that should be fitted. It is easy to match the observed concentration–time data by fitting many parameters, but one needs to keep in mind the final interpretability of the model. Only the parameters that are relevant to the mechanism of absorption for each particular compound should be fitted. One should also keep in mind the presence of measurement errors with each set of experimental data. These include errors in sampling as well as analytical measurement. Models can account for variability among subjects, but it is unreasonable to expect a model which was developed to describe the compound's absorption to account

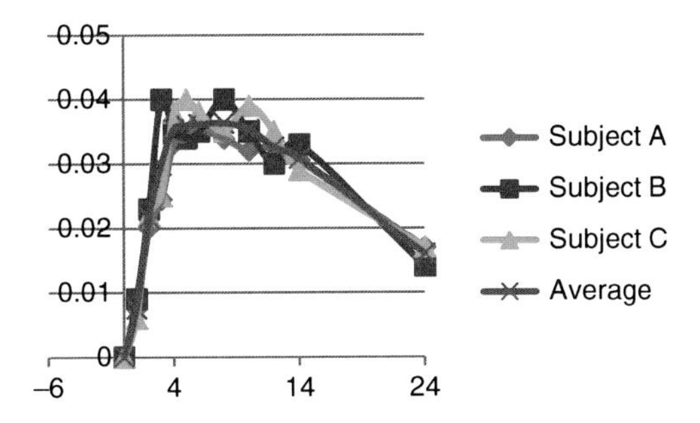

Fig. 3. Example of individual double peaks masked in mean data.

also for experimental errors. For example, the times at which samples are taken are typically reported in round numbers of minutes or hours, yet it is highly unlikely that multiple subjects were all sampled at exactly 5, 10, 15, and 30 min. The software has no way to correct these errors, and so the inputs are accepted as correct. Concern over the accuracies and variances of measured concentrations is typical, but concern over variances in sampling times at which the samples were taken is unusual, even though such variances add to the total variances observed for data among multiple subjects. Note, however, that simulation software can estimate variabilities in a population that are caused by known differences in physiology among subjects, known variations in the dosage form, and even estimated variabilities from measurement errors if the user can provide an estimate of those errors.

- Testing fitted models with additional data not used in the fitting process to assess whether the model will generalize to other conditions.
- Using validated models to predict outcomes from new dose/exposure conditions.

We advocate obtaining individual subject data rather than only mean or median data. Aggregated data can both hide certain behaviors such as double peaks in plasma concentration–time observations and can improperly indicate certain behaviors such as double peaks in plasma concentration–time observations. Figure 3 shows an example of double peaks that exist in individual subjects but that would be hidden if mean data were used.

- Interpreting the Results

Often, absorption models will provide results for which the interpretation is unambiguous. Occasionally, however, results will require interpretation. Such anomalous behaviors can be caused by environmental factors, disease states, different results in very young or very old subjects, interactions between substances, sleep/wake cycles, and mealtimes in general as well as specific food effects

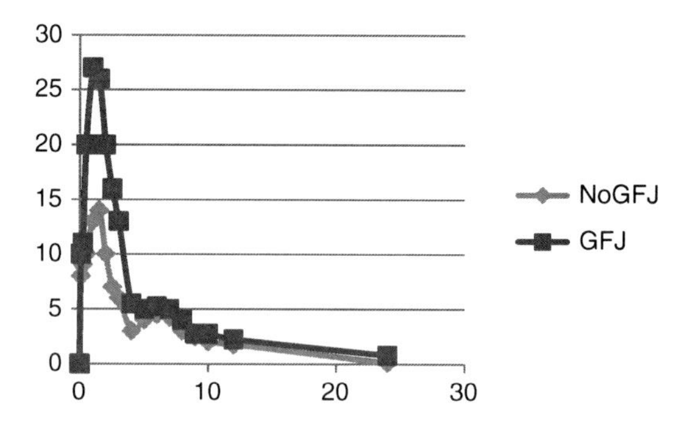

Fig. 4. Saquinavir plasma concentration–time without and after grapefruit juice.

(e.g., grapefruit juice effect on the metabolism of molecules metabolized by the cytochrome P450 3A4 enzyme), as shown in Fig. 4 for the difference in plasma concentration–time for the same 600 mg dose of saquinavir taken without and after grapefruit juice (36).

- Improving the Model

  Model quality is directly related to the quality and quantity of experimental data available upon which to build equations, and relationships that mechanistically describe how a system behaves. Models will typically improve as new data become available. In cases where the initial model is not able to describe the compound's behavior correctly, the model should be used to explore what additional mechanisms are missing. These may include involvement of transporters in absorption and their saturation, or possible precipitation of the compound in the gastrointestinal tract for oral exposure. Absorption of compound after inhalation exposure may not be limited to absorption from the respiratory system. The process of mucociliary clearance followed by swallowing may result in a significant portion of compound being absorbed from the gastrointestinal tract (37).

4. Examples

Consider the following relatively simple example from the world of pharmacology. A common drug used to treat hypertension, propranolol (Inderal), has both high permeability into intestinal cells and high solubility. An oral dose dissolves quickly and is absorbed quickly into the proximal small intestine. Using the GastroPlus software, the user need only enter a value for the dose, permeability, solubility, logP, and molecular weight, and with the default human fasted physiology will correctly predict rapid absorption of 100 % of the dose. This is all done without any calibration of the model.

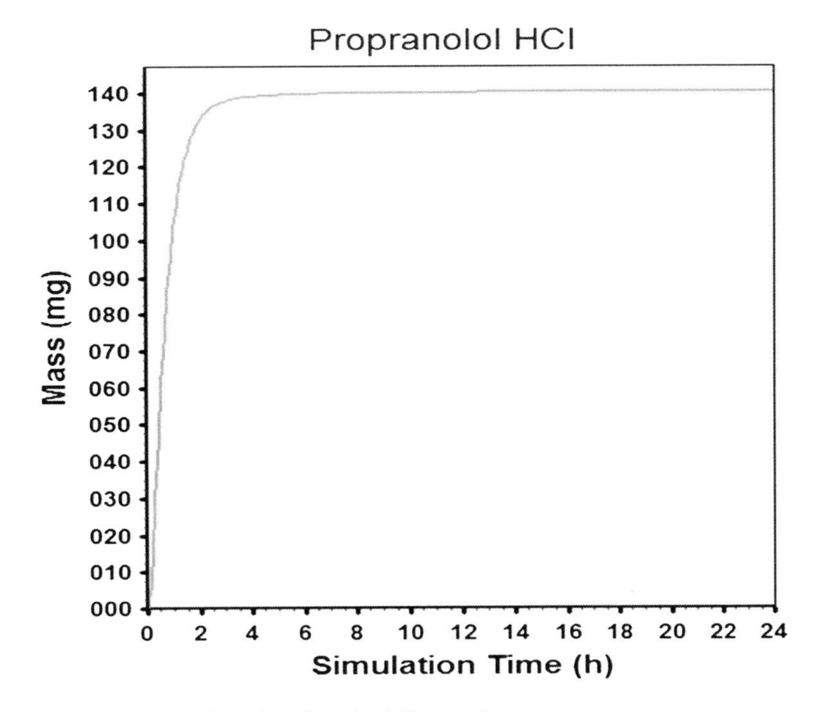

Fig. 5. Propranolol absorption–time simulation results.

Figure 5 shows the absorption–time plot predicted for propranolol by GastroPlus.

With the availability of intravenous (iv) data, pharmacokinetic parameters for a two-compartment PK model can be calculated. Figure 6 shows the plasma concentration–time plot produced using the same absorption model from above along with pharmacokinetic parameters fitted from intravenous data, and first pass extraction calculated from the difference in AUC/dose between intravenous and oral administration. Again, no calibration of the model was required to achieve a nearly perfect match to the observed plasma concentration–time data.

In a recent complex study that remains confidential at the time of this writing, a large amount of data were available for individual subjects, and the data showed that gastric emptying times had unusually high variance across subjects.

In Fig. 7, the unusual high variance in gastric emptying is evident within the same subject on different days. During a GastroPlus simulation study for this data, it was discovered that changing only gastric emptying time, while fixing the values of all other model parameters, provided an excellent fit to observed plasma concentration–time data for numerous subjects. This was an important discovery, as it would have been easy to adjust other model parameters (absorption rates and PK parameters) for different subjects, but doing so would have required different pharmacokinetic

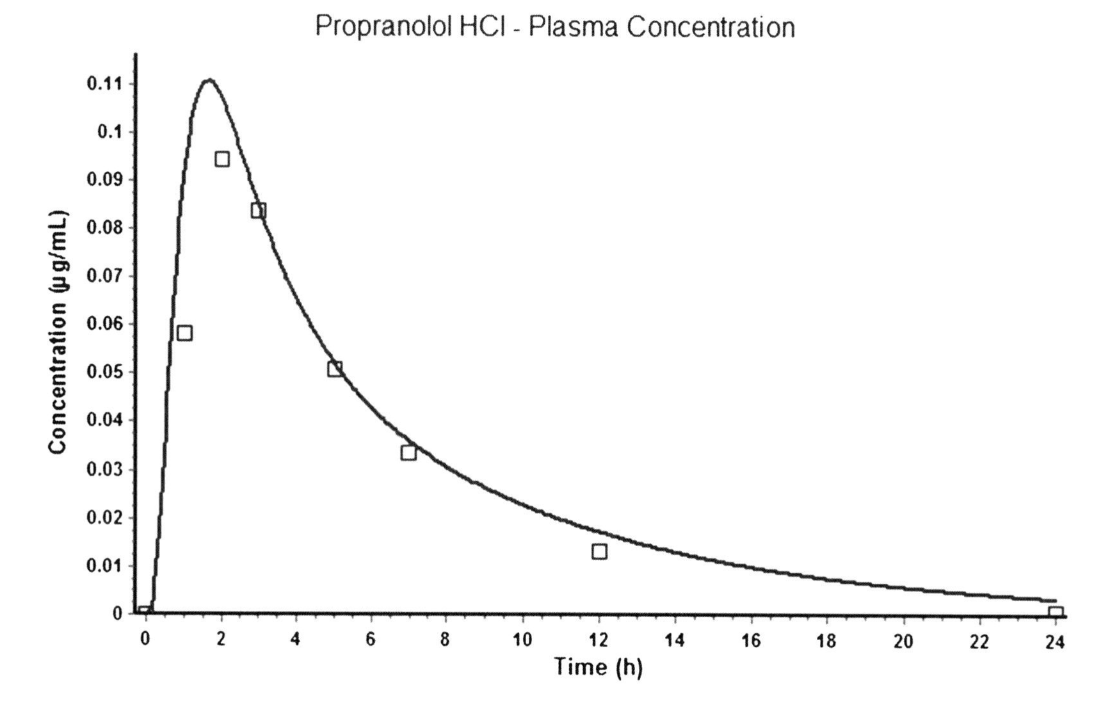

Fig. 6. Propranolol plasma concentration–time simulation results (*line* = simulation, *squares* = observations).

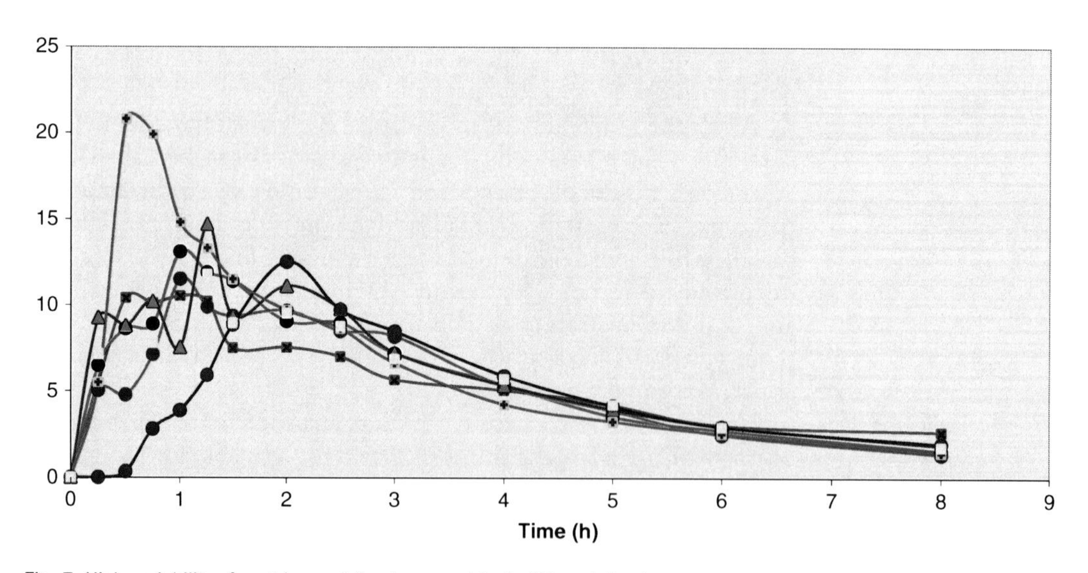

Fig. 7. High variability of gastric emptying (same subject, different days).

parameters for each subject—not a desirable way to fit data unless there is strong justification for doing so.

Another example of complex absorption behavior comes from a study of induced variability of gastric emptying in rats by alprazolam (38). In this study, theophylline and alprazolam were

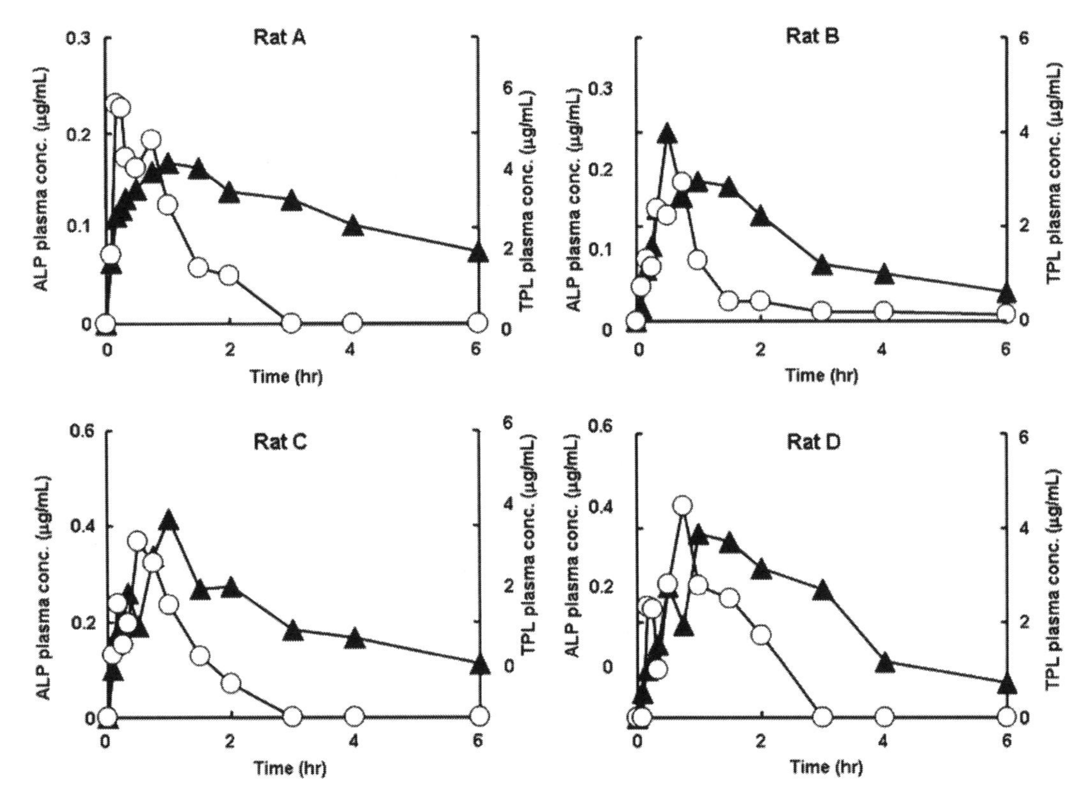

Fig. 8. Plasma concentration–time profiles for alprazolam (ALP) and theophylline (TPL) in four rats from Metsugi. Reprinted from (38) with permission from Springer.

coadministered. Plasma concentration–time data were obtained for independent intravenous doses and for coadministered oral solution doses. Figure 8 shows the plasma concentration–time profiles for four different rats following oral administration of alprazolam (12.5 mg/kg for rats A and B and 25 mg/kg for rats C and D) and theophylline (5 mg/kg administered as aminophylline for all rats) from the study. The double peaks in the oral data are evident for both alprazolam and theophylline in rats A, C, and D, while only alprazolam was observed to have a double peak in rat B. In order to match the data, the simulation must encompass the complete process of absorption through pharmacokinetics.

5. Notes

Using mechanistic simulation software requires an understanding of the various phenomena that interact and how they're being simulated. The expert user of such software must be a good generalist who understands each of the contributing areas of science well enough to communicate with the experts in each area to obtain the inputs needed for the simulations, and later to communicate the results back to them. There are many guidelines, best practices, and caveats, not all of which can be covered here. Advanced hands-on training workshops are typically several days to a full week long. Here we offer some suggestions that should help new users to avoid common pitfalls.

Some of the most common mistakes made with complex simulation software are

Failing to step back and look at the big picture before starting to run simulations.

Failing to have a simulation plan.

Providing incorrect or incomplete inputs.

Fitting the wrong model parameters, or too many of them.

Using oversimplified models that do not capture the actual governing mechanisms.

Having the wrong person run the simulation studies.

Failing to step back and look at the big picture before starting to run simulations. Before the simulation program is started up, it pays to review the data to be analyzed and the goals of the simulation study. For example, if the data consist of plasma concentration–time (Cp–time) data for various subjects, various dose levels, and various formulations, look over the data to see if it appears to be consistent. Do you have the same information for all individuals (species, age, body weight, gender, etc.). We find that quickly plotting the data in a spreadsheet and looking for trends, nonlinearities, or anomalies across different data sets often reveals clues as to the underlying key mechanisms affecting absorption and pharmacokinetics. For example, if double peaks are observed in mean Cp–time data, are double peaks seen in the individual data for all or most subjects, or are they simply due to different T_{max} values for single peaks in various subjects that only give the appearance of double peaks when averaged? Does AUC appear to be dose-proportional across different doses? If AUC is not dose-proportional, is it increasing or decreasing as the dose is increased? If it's decreasing, saturation of an absorption process should be investigated—saturation of solubility and saturation of an influx transporter system are often observed in such cases. If AUC/Dose increasing with dose amount, saturation of an efflux transporter, and/or saturation of a first pass extraction mechanism may be involved. The interaction of absorption and pharmacokinetics cannot be ignored—both must be simulated together. Knowing the most likely mechanisms provides guidance for the types of models to be investigated. Failure to employ a model with saturable transport or metabolism when they are involved can lead to much lost time and wasted simulations with a simpler model that simply cannot explain the data across all doses.

Our rule is if you have to change the model parameters for different doses, you don't have the right model–you need one where the dose amount is automatically accounted for so that the same model can be used for all doses.

Failing to have a simulation plan. What are the goals of the absorption/PK simulation study? What are the next decisions that need to be made to take the project forward? How are absorption/PK simulation results expected to affect those decisions?

Examples of the purposes for running absorption/PK simulations include

Analyzing animal data to assess a drug's behavior.

Estimating first dose in human.

Fitting models to try to understand unusual observations.

Testing theories to decide what steps to take next in animal or human studies.

Performing *in vitro–in vivo* correlations.

Each of these can involve different approaches to the simulation study. For example, if the goal is to develop an *in vitro–in vivo* correlation for a controlled release formulation, the first step should be to develop the pharmacokinetic and absorption models from data for iv and immediate release oral doses. Once those model parameters are set, then the only remaining factor for the controlled release formulation should be how it releases *in vivo*, which might be quite different than the dissolution–time data from an *in vitro* experiment. With modern IVIVC methods, the *in vivo* release can be fitted ("deconvoluted") to best match the simulated Cp–time curve to the observed Cp–time data. This allows direct comparison of the *in vivo* and *in vitro* release/dissolution–time profiles. A study of this type would have a different simulation plan than one designed to estimate first dose in human.

Always develop a plan for what the simulation study is intended to accomplish, and always organize and examine the data prior to running simulations. It doesn't take long, and you will save time and frustration later.

Incorrect or incomplete inputs. Simulation software is not intelligent—it does what you tell it to do. If you input water solubility at 25°C, then you are simulating a gastrointestinal tract filled with 25°C water, so don't blame the software if the solubility is too low and absorption is less than observed. If you don't provide a complete picture of ionization (all pK_as) then solubility-versus-pH and logD-versus-pH will be wrong, and your results might differ dramatically from that they should be. Figure 9 shows the difference in predicted absorption for a low-solubility monoprotic acid using pK_a values of 4, 4.5, and 5—at this pH, even small changes are critical. If you input Caco-2 permeability as human intestinal

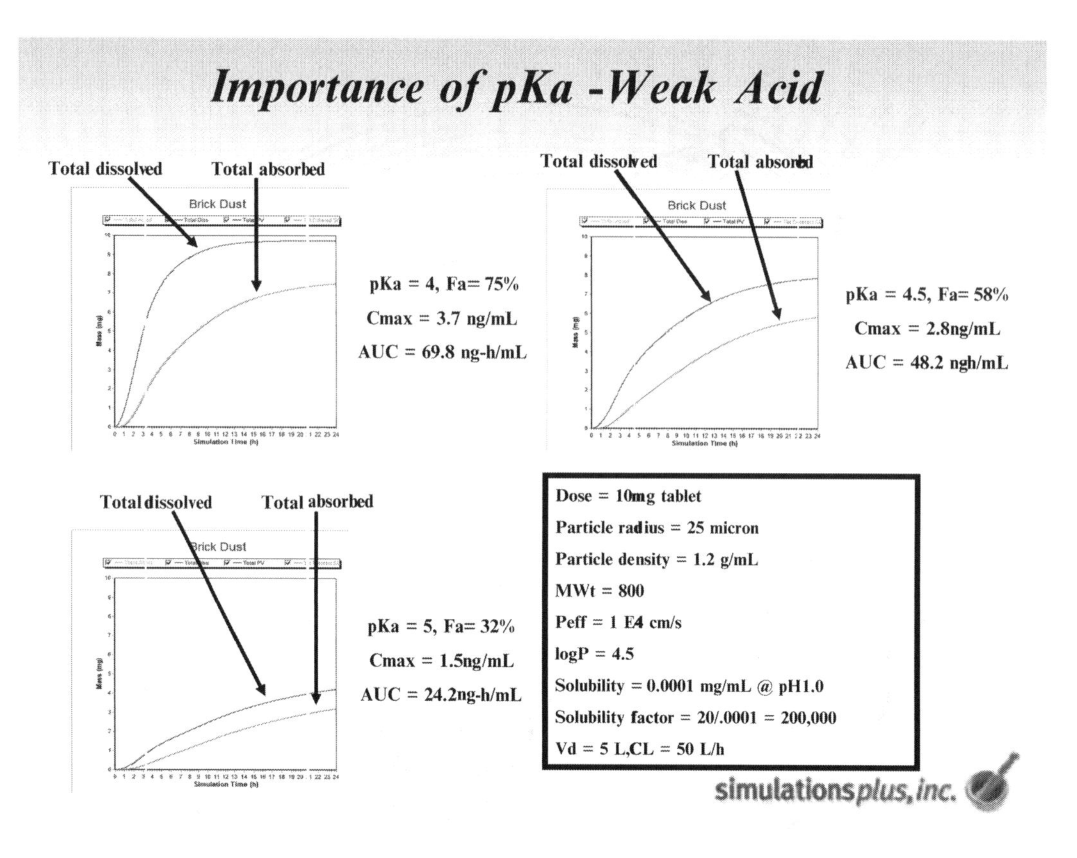

Fig. 9. Effects of small changes in pK_a on dissolution and absorption.

permeability without correcting it, you will get very little absorption, because Caco-2 permeabilities are typically 2–3 orders of magnitude lower than *in vivo* human permeability. If *in vitro* metabolism measurements are not properly scaled to the simulated species, body size, and enzyme expression levels, then the software will simulate something different. The old computer adage "garbage in—garbage out" applies!

Inputs to absorption/PK simulation software should represent *in vivo* conditions to the maximum possible extent. This will include, among others, ionization constants (pK_as—ALL of them for multiprotic compounds!), solubility versus pH in media that best represent *in vivo* conditions, log P (or log D at a specified pH—be sure you understand the difference!), permeability versus region in the intestinal tract, plasma protein binding, blood–plasma concentration ratio, enzyme and transporter expression levels in various tissues, and V_{max} and K_m values for metabolism and transport. Generally, V_{max} values for transporters will not be available and will need to be fitted; however, K_m values can often be gleaned from *in vitro* data. For oral doses, in addition to physiological and physicochemical inputs, the simulation program will also need an accurate description of the formulation(s) to be simulated—particle size distributions, shape factors for active pharmaceutical ingredient (API) particles, and any excipients that may affect solubility and dissolution.

Fitting the wrong model parameters, or too many of them. Modern complex mechanistic absorption/PK simulations allow fitting virtually any combination of model parameters to observed data. Often, many different combinations of fitted parameters can result in statistically similarly good fits to the data. The modeler should resist the temptation to force the simulation line through every data point—it's not about that! Statistical software may provide empirical functions that make pretty pictures, but with complex mechanistic simulations, the goal should be to understand the mechanisms that govern the behavior of the compound, not to create perfect plots. The fewest number of parameters should be used when fitting data to achieve an adequate (not perfect) fit.

Using oversimplified models that do not capture the actual governing mechanisms. It has been our experience that some scientists use very simple models for all of their work. Simple models can be adequate for some data sets, but not for all. Using a one-compartment PK model for a drug that actually undergoes two- or three-compartment (or more) distribution will lead to false conclusions and may lead to the wrong project decisions, wasting time, and money. Ignoring nonlinearities in absorption and/or PK can result in estimating incorrect doses for the next trial. Toxicity studies involve large doses for the animals involved—simulation can provide an idea of what dose levels achieve maximum exposure (so that animals are not wasted on higher doses that achieve no more exposure than lower ones), but only if the model used accounts for the true exposure–dose relationship.

Consider the behavior of the drug valacyclovir shown in Fig. 10 (39). The Cp–time data in the graph are for doses ranging from 100 to 1,000 mg. It is clear that the 1,000 mg dose does not result in ten times the C_{max} and AUC of the 100 mg dose, so right away we know that something is getting saturated. Could it be solubility? It is not likely for these dose levels as the solubility has been measured at 7 mg/mL at pH 8 and 39 mg/mL at pH 6 (personal communication, Richard Lloyd, GSK), so in 250 mL of fluid (the usual amount assumed in small intestine simulations), saturation at pH 6 would require 1,750 mg. *In vitro* data indicate that valacyclovir undergoes chemical degradation at $pH > 5$, with the rate of degradation increasing with pH, as shown in Fig. 11 (40). *In vitro* data also indicate that the molecule is a substrate for the PepT1 oligopeptide influx transporter (41) (in fact, valacyclovir was designed as a prodrug for acyclovir to take advantage of the increased permeability using the transporter). Thus, when concentrations are high enough to saturate the transporter, gut permeability will be a function of concentration, as shown in Fig. 12. Even at lower concentrations, permeability will be dependent on the

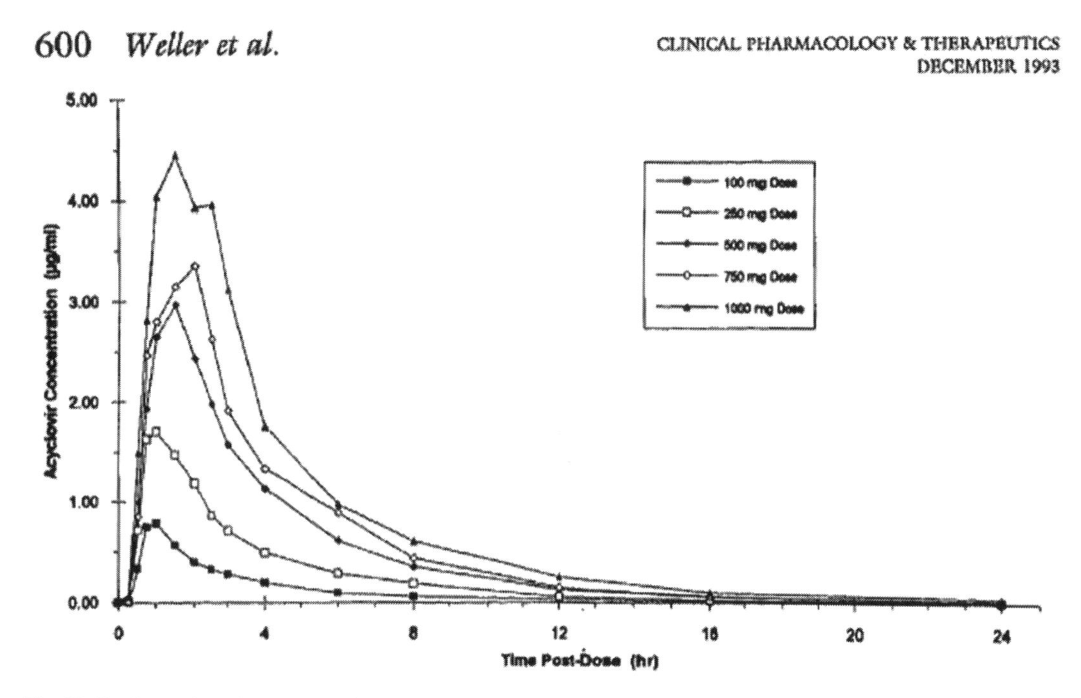

Fig. 10. Nonlinear dose-dependence of valacyclovir. Reprinted from (39) with permission from Macmillan Publishers Ltd.

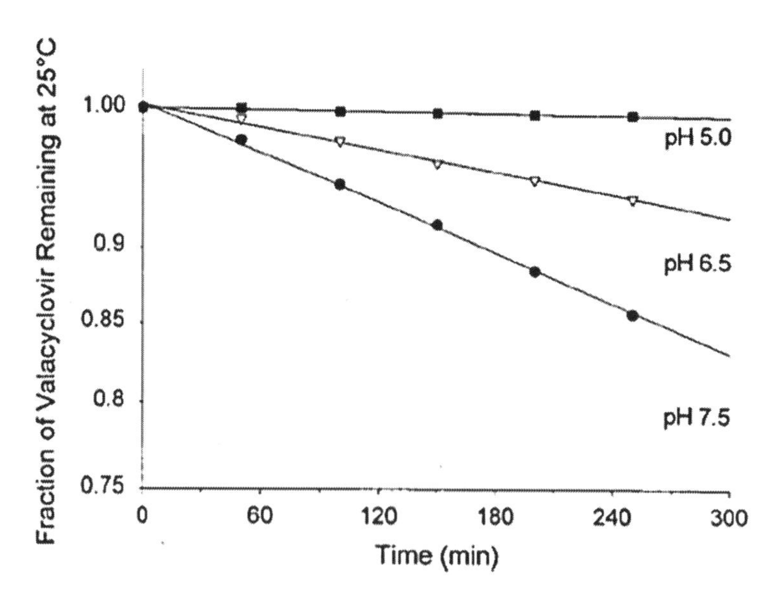

Fig. 11. Degradation of valacyclovir as a function of pH. Reprinted from (40) with permission from John Wiley & Sons, Inc.

amount (expression level) of influx transporter(s) in the enterocytes where the drug is in solution at the intestinal wall. *In vivo* data are available for both 5 and 10 mg iv doses, and fitting across these two doses simultaneously using the PKPlus™ module within GastroPlus shows the best fit (based on the lowest Akaike Information

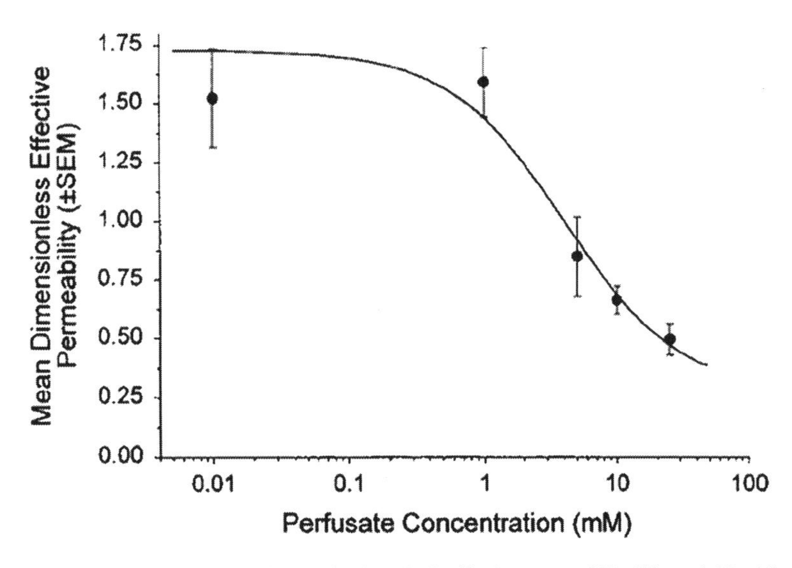

Fig. 12. Concentration dependence of valacyclovir effective permeability ($K_m = 1.22$ mM). Reprinted from (40) with permission from John Wiley & Sons, Inc.

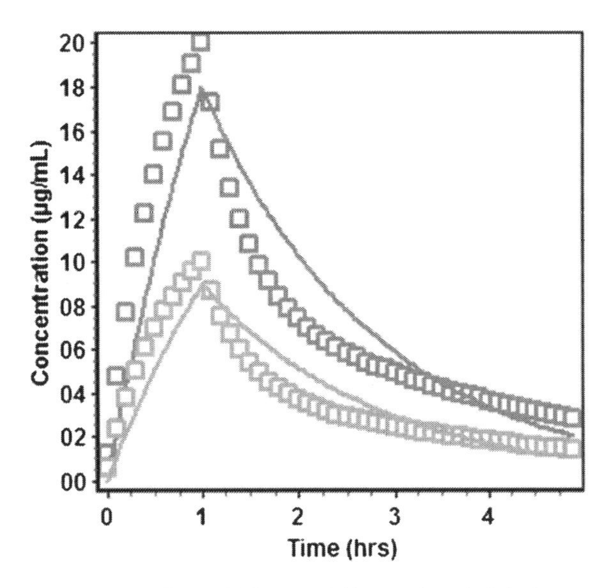

Fig. 13. One-compartment PK model for 5 and 10 mg valacyclovir iv doses.

Criterion) is a two-compartment PK model, as shown in Figs. 13 and 14. The effects of the aforementioned mechanisms of degradation and influx transport, as well as the multicompartment model, could be ignored with a simple linear one-compartment model. But the model parameters would be adjusted to cover up the inadequacies in the model, and it would only apply to one dose level at a time.

Having the wrong person run the simulation studies. In addition to how simulation software should be run, there is the question of

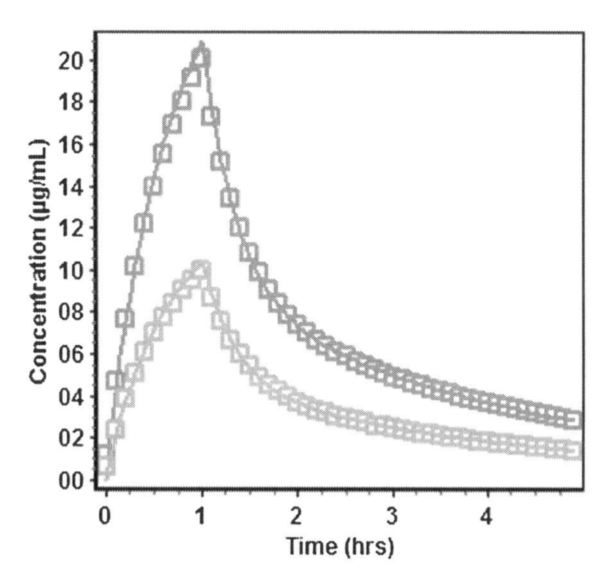

Fig. 14. Two-compartment PK model for 5 and 10 mg valacyclovir iv doses.

who should run it. The nature of simulation software is that it is highly multidisciplinary. Thus, there is a need for generalists who have an aptitude and a desire to learn a variety of areas of science, to communicate with the specialists in those areas, and to understand how to integrate their data into a cohesive model and interpret the results. Academia in the pharmaceutical sciences has not traditionally trained generalists. Instead, students of the pharmaceutical sciences tend to specialize—focusing narrowly on a particular area to become expert at it. The generalist is one who can go deep enough into any of the specialties to know what questions to ask, how to interpret and apply the data, and to have the insight and critical thinking skills when things are not going as expected to know where to look for resolution. Data from *in vivo* trials often exhibit behaviors that are not attributable to obvious explanations. Complex interplay among competing phenomena requires the generalist to experiment with the absorption/PK/PD simulations to determine which of multiple possible explanations best explain the data. Often, the best of a series of competing models cannot be identified, and the generalist should be able to guide experimentalists with respect to what experiments are needed to rule out incorrect models and home in on the one that truly represents what is happening *in vivo*.

To become such a generalist requires training, time, and practice, and the user's organization must provide the time and resources for these to take place. One cannot expect to be expert at solving complex problems using absorption/PK/PD software if it is only a small side activity. As with almost any tool, only through continued use can skills be kept at a high level. Incorrect use of a tool can cause more damage than good, so if the user is not provided appropriate time and resources to be good at what he/she is expected to do, then the organization should consider outsourcing for the required expertise. We have seen all too often that inexperienced users blame the software when things are going awry, only to discover that inputs did not represent reality, or the approach used modeling methods that were too simplified for the problem at hand.

References

1. Swaan PW, Marks GJ, Ryan FM et al (1994) Determination of transport rates for arginine and acetaminophen in rabbit intestinal tissues *in vitro*. Pharm Res 11(2):283–287
2. Slattery JT, Levy G (1979) Acetaminophen kinetics in acutely poisoned patients. Clin Pharmacol Ther 25(2):184–195
3. Clements JA, Heading RC, Nimmo WS et al (1978) Kinetics of acetaminophen absorption and gastric emptying in man. Clin Pharmacol Ther 24(4):420–431
4. Hogben CAM, Tocco DJ, Brodie BB et al (1959) On the mechanism of intestinal absorption of drugs. J Pharmacol Exp Ther 125:275–282
5. Tubic M, Wagner D, Spahn-Langguth H et al (2006) *In silico* modeling of non-linear drug absorption for the P-gp substrate talinolol and of consequences for the resulting pharmacodynamic effect. Pharm Res 23(8):1712–1720
6. Bolger MB, Lukacova V, Woltosz WS (2009) Simulations of the nonlinear dose dependence for substrates of influx and efflux transporters in the human intestine. AAPS J 11(2):353–363
7. Swaan PW (1998) Recent advances in intestinal macromolecular drug delivery via receptor-mediated transport pathways. Pharm Res 15 (6):826–834
8. Palm K, Luthman K, Ros J et al (1999) Effect of molecular charge on intestinal epithelial drug transport: pH- dependent transport of cationic drugs. J Pharmacol Exp Ther 291 (2):435–443
9. Adson A, Raub TJ, Burton PS et al (1994) Quantitative approaches to delineate paracellular diffusion in cultured epithelial cell monolayers. J Pharm Sci 83(11):1529–1536
10. Schiller C, Frohlich CP, Giessmann T et al (2005) Intestinal fluid volumes and transit of dosage forms as assessed by magnetic resonance imaging. Aliment Pharmacol Ther 22:971–979
11. Wilson JP (1967) Surface area of the small intestine in man. Gut 8:618–621
12. Fordtran JS, Rector FC Jr, Ewton MF et al (1965) Permeability characteristics of the human small intestine. J Clin Invest 44(12):1935–1944
13. Billich CO, Levitan R (1969) Effects of sodium concentration and osmolality on water and electrolyte absorption form the intact human colon. J Clin Invest 48(7):1336–1347
14. Parrott N, Lukacova V, Fraczkiewicz G et al (2009) Predicting pharmacokinetics of drugs using physiologically based modeling-application to food effects. AAPS J 11(1):45–53
15. Agoram B, Woltosz WS, Bolger MB (2001) Predicting the impact of physiological and biochemical processes on oral drug bioavailability. Adv Drug Deliv Rev 50(Suppl 1):S41–S67
16. Bolger MB, Agoram B, Fraczkiewicz R et al (2003) Simulation of absorption, metabolism, and bioavailability. In: Waterbeemd HVD, Lennernäs H, Artursson P (eds) Drug bioavailability. Estimation of solubility, permeability and bioavailability. Wiley, New York
17. Avdeef A (2001) Physicochemical profiling (solubility, permeability and charge state). Curr Top Med Chem 1(4):277–351
18. Yu LX, Amidon GL (1999) A compartmental absorption and transit model for estimating oral drug absorption. Int J Pharm 186 (2):119–125
19. Qiu Y, Kuo CH, Zappi ME (2001) Performance and simulation of ozone absorption and reactions in a stirred-tank reactor. Environ Sci Technol 35(1):209–215
20. Bogdanffy MS, Mathison BH, Kuykendall JR et al (1997) Critical factors in assessing risk from exposure to nasal carcinogens. Mutat Res 380(1–2):125–141
21. Fasano WJ, McDougal JN (2008) *In vitro* dermal absorption rate testing of certain chemicals of interest to the Occupational Safety and Health Administration: summary and evaluation of USEPA's mandated testing. Regul Toxicol Pharmacol 51(2):181–194
22. Nohynek GJ, Dufour EK, Roberts MS (2008) Nanotechnology, cosmetics and the skin: is

there a health risk? Skin Pharmacol Physiol 21 (3):136–149

23. Mansour SA, Gad MF (2010) Risk assessment of pesticides and heavy metals contaminants in vegetables: a novel bioassay method using Daphnia magna Straus. Food Chem Toxicol 48(1):377–389
24. Bohus E, Coen M, Keun HC et al (2008) Temporal metabonomic modeling of l-arginine-induced exocrine pancreatitis. J Proteome Res 7(10):4435–4445
25. Andersen ME, Krishnan K (1994) Physiologically based pharmacokinetics and cancer risk assessment. Environ Health Perspect 102 (Suppl 1):103–108
26. Dobrev ID, Andersen ME, Yang RSH (2002) *In silico* toxicology: simulating interaction thresholds for human exposure to mixtures of trichloroethylene, tetrachloroethylene, and 1,1,1-trichloroethane. Environ Health Perspect 110:1031–1039
27. Vinegar A, Jepson GW, Cisneros M et al (2000) Setting safe acute exposure limits for halon replacement chemicals using physiologically based pharmacokinetic modeling. Inhal Toxicol 12(8):751–763
28. Rao HV, Ginsberg GL (1997) A physiologically-based pharmacokinetic model assessment of methyl t-butyl ether in groundwater for bathing and showering determination. Risk Anal 17(5):583–598
29. Yang Y, Xu X, Georgopoulos P (2010) A Bayesian population PBPK model for multiroute chloroform exposure. J Expo Sci Environ Epidemiol 20(4):326–341
30. Campbell A (2009) Development of PBPK model of molinate and molinate sulfoxide in rats and humans. Regul Toxicol Pharmacol 53 (3):195–204
31. Fagerholm U, Johansson M, Lennernas H (1996) Comparison between permeability coefficients in rat and human jejunum. Pharm Res 13(9):1336–1342
32. Sugano K (2009) Computational oral absorption simulation for Low-solubility compounds. Chem Biodivers 6:2014–2029
33. Porter CJH, Trevaskis NL, Charman WN (2007) Lipids and lipid-based formulations: optimizing the oral delivery of lipophilic drugs. Nat Rev Drug Discov 6:231–248
34. Davies NM, Feddah MR (2003) A novel method for assessing dissolution of aerosol inhaler products. Int J Pharm 255 (1–2):175–187
35. Son YJ, McConville JT (2009) Development of a standardized dissolution test method for inhaled pharmaceutical formulations. Int J Pharm 2009(382):1–2
36. Kupferschmidt HH, Fattinger KE, Ha HR et al (1998) Grapefruit juice enhances the bioavailability of the HIV protease inhibitor saquinavir in man. Br J Clin Pharmacol 45 (4):355–359
37. Chilvers MA, O'Callaghan C (2000) Local mucociliary defence mechanisms. Paediatr Respir Rev 1(1):27–34
38. Metsugi Y, Miyaji Y, Ogawara K et al (2008) Appearance of double peaks in plasma concentration–time profile after oral administration depends on gastric emptying profile and weight function. Pharm Res 25(4):886–895
39. Weller S, Blum MR, Doucette M et al (1993) Pharmacokinetics of the acyclovir pro-drug valacyclovir after escalating single- and multiple-dose administration to normal volunteers. Clin Pharmacol Ther 54(6):595–605
40. Sinko PJ, Balimane PV (1998) Carrier-mediated intestinal absorption of valacyclovir, the L-valyl ester prodrug of acyclovir: 1. Interactions with peptides, organic anions and organic cations in rats. Biopharm Drug Dispos 19:209–217
41. Giacomini KM, Huang SM, Tweedie DJ et al (2010) Membrane transporters in drug development. Nat Rev Drug Discov 9(3):215–236

Chapter 14

Prediction of Pharmacokinetic Parameters

A.K. Madan and Harish Dureja

Abstract

In silico tools specifically developed for prediction of pharmacokinetic parameters are of particular interest to pharmaceutical industry because of the high potential of discarding inappropriate molecules during an early stage of drug development itself with consequent saving of vital resources and valuable time. The ultimate goal of the in silico models of absorption, distribution, metabolism, and excretion (ADME) properties is the accurate prediction of the in vivo pharmacokinetics of a potential drug molecule in man, whilst it exists only as a virtual structure. Various types of in silico models developed for successful prediction of the ADME parameters like oral absorption, bioavailability, plasma protein binding, tissue distribution, clearance, half-life, etc. have been briefly described in this chapter.

Key words: In silico models, Absorption, Distribution, Metabolism, Excretion, Bioavailability, Protein binding, Classification models, QSAR, Pharmacokinetics

1. Introduction

Drug discovery is a highly complex and expensive endeavor involving seven major steps: disease selection, target hypothesis, lead identification, lead optimization, preclinical trial, clinical trial, and pharmacogenomic optimization. Among the various techniques used to accelerate the drug discovery process, virtual (or in silico) ligand screening based upon the structure of known ligands or on the structure of the receptor is gradually emerging as a method of choice (1). The application of computational methodology during drug discovery process leads to significant reduction in number of experimental studies required for compound selection and development and for significantly improving the success rate. The in silico approaches are being widely used today to assess the absorption, distribution, metabolism, and excretion (ADME) properties of compounds at the early stages of drug discovery process (2). Study of ADME profiles is being widely used in drug discovery to understand

Brad Reisfeld and Arthur N. Mayeno (eds.), *Computational Toxicology: Volume I*, Methods in Molecular Biology, vol. 929, DOI 10.1007/978-1-62703-050-2_14, © Springer Science+Business Media, LLC 2012

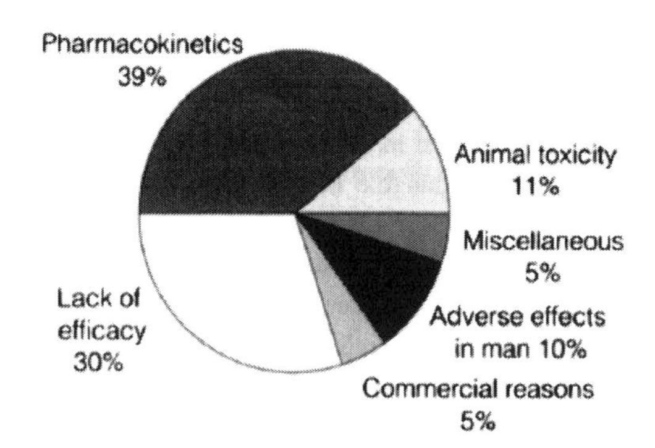

Fig. 1. Reasons for failure in drug development. Reproduced from ref. 6 with permission from Elsevier Limited, UK.

the properties necessary to convert lead structures into good medicines (3). Early consideration of ADME properties is also becoming increasingly important due to the implementation of combinatorial chemistry and high-throughput screening so as to generate vast number of potential lead compounds (4). The FDA has established a Simulation Working Group and information on current views can easily be found at the Center for Drug Development Science. In addition, other groups such as ECVAM (The European Centre for the Validation of Alternative Methods) have identified predictive pharmacokinetic modeling as a beneficial tool for reduction in the use of animals in drug discovery and development (5). As a result of studies in the late 1990s, indicating that the poor pharmacokinetics and toxicity were major causes of costly late-stage failures in drug development, there is increasing realization to consider these areas as early as possible in the drug discovery process (3).

Historically, inappropriate pharmacokinetic characteristics (Fig. 1) have been a major cause for the failure of compounds in the later stages of development (6). This was largely owing to an inability to rectify poor pharmacokinetic characteristics inherent in many lead series adopted for lead optimization (7). Poor pharmacokinetic properties are mainly responsible for terminating the development of drug candidates, with huge financial impact on the cost of R&D in the pharmaceutical industry (8). The failure rate due to pharmacokinetic problems may be greater than reported because poor pharmacokinetic properties such as lack of absorption, rapid metabolism or elimination, or unfavorable distribution may be clinically manifested as a lack of efficacy (5). There is an ever increasing need for good tools for predicting these properties to serve dual objectives—first, at the initial design stage of new compounds and compound libraries so as to minimize the risk of late-stage attrition; and second, to optimize the screening and testing by considering only the most promising compounds (9).

The significant failure/attrition rate of drug candidates in later developmental stages is the key driving force for the development of in vitro, in vivo, and in silico predictive tools that can eliminate inappropriate compounds before substantial time and money is invested in testing (10, 11). Firstly, a wide variety of in vitro assays have been automated through the use of robotics and miniaturization. Secondly, in silico models are being used to facilitate selection of appropriate assays, as well as selection of subsets of compounds to go through these screens. Thirdly, predictive models have been developed that might ultimately become sophisticated and reliable enough to replace in vitro assays and/or in vivo experiments (9).

Significant advances in automation technology and experimental ADME/Tox techniques, such as the Caco-2 permeability screening based on the 3-day Caco-2 culture system, the metabolic stability screening using microsomes or hepatocytes, and the P450 inhibition assay, have enabled the assaying of relatively much larger number of compounds when compared to traditional strategies (12). The approach of quickly predicting the ADME properties through computational means is of much greater importance because the experimental ADME testing is enormously expensive and arduous. Therefore, the use of computational models in the prediction of ADME parameters has been growing rapidly in the drug discovery process because of their immense benefits in throughput and early application in drug design (13). Compared to experimental approaches, these in silico methods have distinct advantage that they do not initially require the compounds to be synthesized and experimentally tested. Moreover, compound databases can be virtually screened rapidly in a high-throughput fashion if the calculations are computationally efficient. Until now, many computational methodologies have already been developed for the ADME/Tox properties which include aqueous solubility, bioavailability, intestinal absorption, blood–brain barrier (BBB) penetration, drug–drug interactions, transporter, plasma–protein binding, and toxicity (9, 11). The computational ADME has witnessed significant advances since early 1970s (14). In silico predictive methods are significantly influenced by the quality, quantity, sources, and generation of the measured data available for model development (15).

The early assessment of ADME characteristics will definitely help pharmaceutical scientists to select the best candidates for development as well as to discard those with low probability of success. The ultimate goal of the in silico models of ADME properties is the accurate prediction of the in vivo pharmacokinetic of a potential drug molecule in man, whilst it exists only as a virtual structure (2). The unexpectedly good prediction power of the simple computational models with high-throughput renders them vital tools in the early screening of drug candidates, whereas laborious cell culture models and animal studies can be beneficial in the later phases when comprehensive information about the transport mechanisms is needed (16). There is gradually emerging consensus that in silico predictions are no less predictive to what occurs in vivo

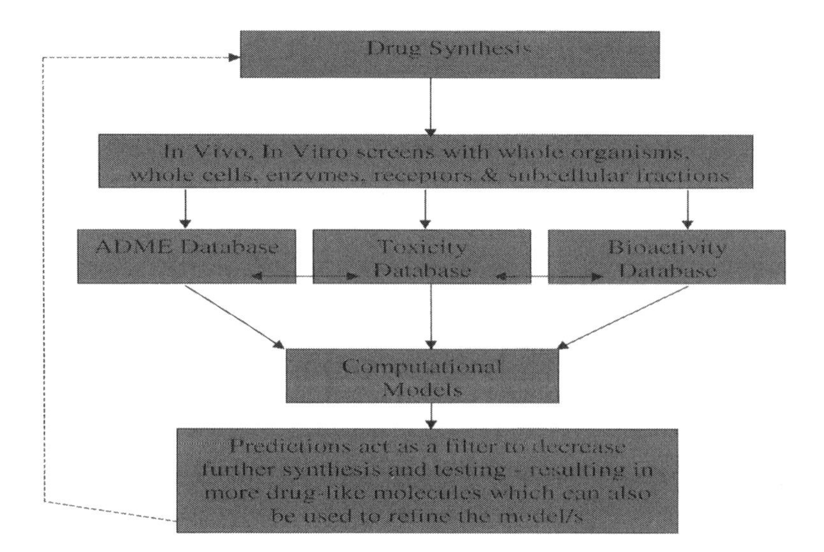

Fig. 2. The linkage between the data generation, databases, and model building. Reproduced from ref. 14 with permission from Elsevier Limited, UK.

than are in vitro tests, with the distinct advantage that far less investment in technology, resources, and time is needed (17). An amalgamation of in vitro experiments and in silico modeling will dramatically increase the insight and knowledge about the relevant physiological and pharmacological processes in drug discovery (3). The linkage between data generation and model building has been illustrated in Fig. 2 (14).

For data modeling, quantitative structure–property relationship (QSPR) approaches are generally applied. Based on appropriate descriptors, QSPR exploiting from simple linear regression to modern multivariate analysis techniques or machine-learning methods is now being extensively applied for the analysis of ADME data. Data modeling can be efficiently applied to large number of molecules, but require a significant quantity of high quality data to derive a relationship between the structures and the modeled property (11). Quantitative structure–activity relationship (QSAR) modeling is a well known and established discipline, where physicochemical and molecular descriptors are correlated with bioassay drug concentrations eliciting a standard pharmacological response such as EC_{50} or IC_{50}. The extension of the QSAR approach to pharmacokinetic parameters is similarly referred to as *quantitative structure–pharmacokinetic relationships* (QSPKR or QSPkR) modeling. A relatively less utilized technique, albeit just as empirical as the QSAR approach, is the direct mapping of molecular descriptors with the time course of plasma levels following administration of drug by various routes (18). Pharmacokinetics is the study of the time course of a drug within the body incorporating the processes of ADME. Pharmacokinetics can be defined as *the study of the time course of drug and metabolite levels in various fluids, tissues, and excreta of the body, and of the mathematical relationships required to describe them* (5). Pharmacokinetic data is highly beneficial in optimization of the dosage form design and establishing the dosage regimen (19).

Therefore, ADME parameters are of utmost importance and numerous computational approaches of diverse nature have been developed for these parameters, which includes bioavailability, human intestinal absorption, permeability, BBB penetration, half-life, volume of distribution, metabolism and clearance, etc. The present chapter deals with the in silico models for prediction of pharmacokinetic parameters.

2. Materials and Methods

The first phase of ADME computational models began in the 1960s with development of classical QSAR by Hansch. According to him quantitative relationships could be developed for the lipophilicity of the drugs as well as metabolic parameters such as microsomal hydroxylation, demethylation, CYP450–CYP420 conversion, and duration of drug action (20). The simplest ADME-concerned filter for short listing of potential drug molecule may be "rule of five" proposed by Lipinski et al. in 1997 (21). According to Lipinski, it is much easier to optimize pharmacokinetic properties in initial stage and to optimize the receptor binding affinity at a later stage of drug discovery process (22). Though the "rule of five" may be too simple approach but it has definitely generated/stimulated considerable interest in development of fast, generally applicable filters for ADME. This is now being suggested as a "*rule of thumb*" or guide rather than a definitive cutoff. However, "rule of five" and other general rules simply lay down minimum criterion of a molecule to be drug-like. The values of different desired characteristics proposed by various researchers in drug-like rules have been compiled in Table 1 (21, 23, 24). It is relatively easy for a molecule to fall within the "rule of five" but has no certainty to lead to a drug. As a matter of fact, 68.7% compounds of 2.4 million compounds in Available Chemical Directory (ACD) Screening Database and 55% compounds of 240,000 compounds in ACD have no violation of "rule of five" at all (11). *Therefore, much more stringent criteria need to be laid down so as to discriminate drug-like molecules from others.*

Various QSAR approaches ranging from simple linear regression to modern multivariate analysis techniques are now being applied to the analysis of ADME data. Data mining and machine-learning techniques which were originally developed and used in other fields are now being successfully used for this purpose. These techniques

Table 1
Drug-like rules proposed by various researchers

Research group	Data set employed for developing drug-like rules	Characteristics of drugs: Molecular weight	Octanol/water partition coefficient	Number of hydrogen-bond donors	Number of hydrogen-bond acceptors	Reference
Lipinski and coworkers	2,245 Drugs from World Drug Index	<500 Da	<5 (CLOGP) or 4.15 (MLOGP)	<5	<10	(21)
Ghose and coworkers	7,183 Compounds from Comprehensive Medicinal Chemistry database	160–480 (average value: 357)	–0.4 to 5.6 calculated log *P* (ALOGP) (average value: 2.52)	Molar refractivity: 40–130 (average value: 97)	Total number of atoms: 20–70 (average value: 48)	(22)
Wenlock and coworkers	594 Compounds from Physicians Desk Reference 1999	<473	<5 (calculated log *P*) or <4.3 (calculated log $D_{7.4}$)	<4	<7	(23)

include neural networks, classification and regression tree, self-organizing maps, support vector machine, and recursive partitioning (9). It is important to select an appropriate statistical and mathematical tool for the analysis of ADME data. Therefore, the data needs to be pre-analyzed so that linear versus nonlinear methods are correctly selected and utilized. Stringent model validation is an integral component of the successful development of any statistical/mathematical model. Without proper validation, the predictive ability of the model cannot be estimated (25).

Apart from the development of prediction of higher confidence, another major challenge is to develop an in silico ADME/Tox prediction software system and integrate the existing tools into a simplified single, consistent workflow environment. Number of companies active in the field of molecular modeling has now developed software (s) to assist in the estimation of ADME/Tox properties (11). These software/programs are basically computer simulation models developed and validated for prediction of ADME outcomes, such as rate and extent of absorption, using a limited number of in vitro data inputs. They are advanced compartmental absorption and transit models, in which physicochemical concepts, such as solubility and lipophilicity, are easily incorporated than physiological aspects involving transporters and metabolism (9). Commercially available software/programs used for prediction of ADME properties have been exemplified in Table 2.

2.1. Models for Prediction of ADME Parameters

Various in silico models have been developed for the prediction of pharmacokinetic parameters, which include oral absorption, bioavailability, plasma protein binding (PPB), tissue distribution, clearance, half-life, etc. Some of these predictive models have been exemplified in Table 3.

Absorption: The first step in the drug absorption process is the disintegration of the tablet or capsule and subsequent dissolution of the drug. Poor biopharmaceutical properties, i.e., poor aqueous solubility and slow dissolution rate can lead to poor and delayed oral absorption and hence low oral bioavailability. Consequently, low solubility is naturally detrimental to good and complete oral absorption, and so the early consideration of this property is of great significance in drug discovery process (9). Prediction of intestinal absorption is a major goal in the design, optimization, and selection of suitable candidates for development as oral drugs (28). Most of the computational approaches currently being used to predict absorption are based on the assumption that absorption is passive and can be predicted from various molecular descriptors of the compound. No account is taken care of active transport processes, including both uptake and efflux transporters. Presently, it is not known that how many compounds are actually actively transported in the gut (2). Drug absorption following oral administration is difficult to predict because of complex drug-specific parameters and physiological processes. These include drug release from the dosage form and dissolution, aqueous solubility, gastrointestinal (GI) motility and contents, pH, GI blood flow, active or passive transport systems, and pre-systemic and first-pass metabolism (18).

Permeability: Human intestinal permeability (important for the absorption of oral drugs) and BBB permeability (important for the distribution of CNS active agents and toxicity of non-CNS drugs) constitute important pharmacokinetic parameters (93). The hydrogen-bonding capacity of a drug solute is generally recognized as an important determinant of permeability. In order to penetrate through a membrane, a drug molecule needs to break/rupture hydrogen bonds with its aqueous environment. The more potential hydrogen bonds in a molecule will necessitate increased bond breaking costs. As a consequence high hydrogen-bonding potential is an unfavorable property that is often related to reduced permeability and absorption (9). Caco-2 cells permeability data still constitute a major target property for modeling, in spite of substantial inter- and even intra-laboratory variability in the data (94). The BBB represents a significant barrier towards entry of drugs into

Table 2
Some examples of commercially available software/program for prediction of ADME parameters

Name of the software/ program	Company	Website
ACD/ADME Suite	ACD Labs	www.acdlabs.com
C2.ADME	Accelrys	www.accelrys.com
CLOE PK	Cyprotex	www.cyprotex.com
Gastroplus	Simulations Plus	www.simlations-plus.com
KnowIAll ADME/Tox	Bio-Rad Laboratories	www3.bio-rad.com
META	Multicase	www.multicase.com
MetabolExpert	Compudrug	www.compudrug.com
QikProp	Schrodinger	www.schrodinger.com
Volsurf+	Molecular Discovery	www.moldiscovery.com
MetaCore	GeneGo	www.genego.com
MetaDrug	GeneGo	www.genego.com
Metasite	Molecular Discovery	www.moldiscovery.com
MEXAlert	Compudrug	www.compudrug.com
RetroMEX	Compudrug	www.compudrug.com
Moka	Molecular Discovery	www.moldiscovery.com
iDEA pkEXPRESS	Lion Biosciences; Biowisdom	www.lionbiosciences.com; www.biowisdom.com
MCASE/MC4PC	Multicase	www.multicase.com
METAPC	Multicase	www.multicase.com
ADMET Predictor	Simulations Plus	www.simlations-plus.com
METEOR	LHASA	www.lhasalimited.org
PK Solutions	Summit PK	www.summitpk.com
QMPRPlus	Simulations Plus	www.simlations-plus.com
PK-sim	Bayer Technology Services	www.bayertechnology.com

Table 3
Examples of predictive models for ADME

Pharmacokinetic parameter	Data set	Statistical/modeling technique	References
Oral bioavailability	188 Non-congeneric organic medicinals	Fuzzy adaptive least square	(26)
Intestinal drug absorption	6 β-Adrenoreceptor antagonists	Regression	(27)
Human intestinal absorption	67 Drugs and drug-like compounds (training set); 9 drugs (cross-validation set); and 10 drugs (external cross-validation set)	Artificial neural network	(28)
Intestinal drug absorption	6 β-Adrenoreceptor antagonists	Regression	(27)
Human intestinal absorption	67 Drugs and drug-like compounds (training set); 9 drugs (cross-validation set); and 10 drugs (external cross-validation set)	Artificial neural network	(28)
Tissue distribution	9 *n*-Alkyl-5-ethyl barbituric acids	Regression	(29)
Human jejunal permeability	22 Structurally diverse compounds (training set) and 34 compounds (external validation set)	Multivariate data analysis	(30)
Intestinal absorption	20 Drugs (training set); 74 drugs (prediction set)	Multilinear regression	(31)
Intestinal membrane permeability	10 Non-peptide endothelin receptor antagonists	Rule of five, molecular mechanics, and quantum mechanics	(32)
Intestinal absorption	20 Molecules	Artificial neural network based correlation (hashkey) model	(33)
Intestinal absorption	234 Compounds	Statistical pattern recognition model	(34)
Oral bioavailability	591 Structurally diverse compounds	Step-wise regression	(35)
Human intestinal absorption	38 Drugs (training set) and 131 drugs (prediction set)	Multilinear regression	(36)
Plasma protein binding	95 Diverse compounds	Genetic function approximation	(37)

(continued)

Table 3 (continued)

Pharmacokinetic parameter	Data set	Statistical/modeling technique	References
Oral bioavailability	21 Drugs and drug candidates	Graphical classification model	(38)
Oral bioavailability	1,100 Drug candidates	Regression	(39)
Clearance, volume of distribution, fractal clearance, and fractal volume	272 Structurally unrelated drugs	Principal component analysis and projection to latent structures	(40)
Human clearance	68 Drugs	Multiple linear regression, partial least squares method, and artificial neural network	(41)
Blood–brain barrier permeability	324 Drugs and drug-like molecules	Neural network and Support vector machine	(42)
Metabolic stability	631 Diverse chemicals (training set) and 107 chemicals (validation set)	k-Nearest neighbor	(43)
Human intestinal absorption	1,000 Drug-like compounds	Recursive partitioning analyses	(44)
Aqueous solubility, plasma protein binding, and human volume of distribution at steady state	202, 226, and 204 compounds (training set) and 442, 94, and 124 compounds (test set), respectively	Linear regression analysis	(45)
Clearance (oral)	87 Drugs	Multivariate regression analyses, multiple linear regression, and partial least squares analysis	(46)
Half-life, renal, and total body clearance, fraction excreted in urine, volume of distribution, and fraction bound to plasma proteins	20 Cephalosporins	Artificial neural network	(47)
Human intestinal absorption	82 Compounds (training set) and 127 drugs (prediction set) and 109 drugs (test set)	Topological substructural approach (TOPS-MODE)	(48)
Clearances, fraction bound to plasma proteins and volume of distribution	62 Structurally diverse compounds	Artificial neural network	(49)

(continued)

Table 3 (continued)

Pharmacokinetic parameter	Data set	Statistical/modeling technique	References
Blood–brain barrier penetration	150 Chemically diverse compounds	4D-molecular similarity and cluster analysis	(50)
Oral absorption	1,260 Compounds	Classification regression trees	(51)
Blood–brain barrier penetration	415 Drugs	Logistic regression, linear discriminant analysis, k-nearest neighbor, decision tree, probabilistic neural network, and support vector machine	(52)
Human serum albumin affinity	37 Structurally related interleukin-8 inhibitors	3D-QSAR	(53)
Metabolism	42 Derivatives	Comparative molecular field analysis	(54)
p-Glycoprotein inhibitors	Series of 1,4-dihydropyridines and pyridines	Forward inclusion coupled with multiple regression analysis and partial least square regression	(55)
Permeability	20 Drugs	Partial least square method	(56)
Steady-state volume of distribution	199 Compounds (human data) and 2,086 compounds (rat data)	Bayesian neural networks, classification and regression trees, and partial least squares	(57)
Oral drug absorption	22 Structurally diverse drugs (training set) and 169 drugs (prediction set)	Partial least square analysis	(58)
Blood–brain barrier permeability	191 Drugs (training set) and 50 drugs (test set)	Artificial neural network	(59)
Renal clearance	130 Diverse compounds (training set) and 20 compounds (test set)	Principal component analysis and partial least squares analysis	(60)
Drug clearance	398 Compounds (training set) and 105 compounds (validation set)	General regression neural network, support vector regression, and k-nearest neighbor	(61)
Blood–brain barrier permeability	28 Structurally diverse compounds (training set); 31 compounds (validation set) and 31 compounds (cross-validation set)	Moving average analysis based classification models	(62, 63)

(continued)

Table 3 (continued)

Pharmacokinetic parameter	Data set	Statistical/modeling technique	References
Human intestinal absorption	480 Structural diverse drug-like molecules (training set) and 98 molecules (test set)	Support vector machine based classification model	(64)
Human oral intestinal drug absorption	164 Compounds (training set) and 24 compounds (test set)	Membrane-interaction QSAR analysis	(65)
Blood–brain barrier permeability	136 Compounds	Approximate similarity matrices and partial least square	(66)
Plasma protein binding	686 Compounds	Partial least square regression	(67)
Oral bioavailability	30 Compounds	Regression	(68)
Oral bioavailability	768 Chemical compounds	Correlation	(69)
Oral bioavailability	250 Structurally diverse molecules (training set) and 52 molecules (test set)	Hologram-QSAR	(70)
Metabolism, tissue distribution, bioavailability	50 Structurally diverse compounds	Correlation	(71)
Human intestinal absorption	455 Compounds (training set) and 193 compounds (test set)	Genetic function approximation technique	(72)
Blood–brain barrier penetration	78 Compounds (training set) and 25 compounds (test set)	Multilinear regression	(73)
Blood–brain barrier permeability	159 Compounds (training set) and 99 drugs (external test set-1) and 267 organic compounds (external test set-2)	k-Nearest neighbors and support vector machine	(74)
Oral clearance	24 Compounds	Allometric approaches	(75)
Plasma protein binding and oral bioavailability	692 Compounds	Support vector machine combined with genetic algorithm	(76)
Half-life, renal, and total body clearance, fraction excreted in urine, volume of distribution, and fraction bound to plasma proteins	20 Cephalosporins	Random forest; decision tree; and moving average analysis	(77)

(continued)

Table 3 (continued)

Pharmacokinetic parameter	Data set	Statistical/modeling technique	References
T_{max}	28 Structurally diverse antihistamines	Decision tree and moving average analysis	(78)
Blood–brain barrier permeability	280 Compounds	Regression	(79)
Buccal permeability	15 Drugs (training set) and 13 compounds (test set)	Multiple linear regression and maximum likelihood estimations	(80)
Human volume of distribution at steady state	669 Drug compounds (training set) and 29 compounds (test set)	Linear and nonlinear statistical techniques, partial least squares, and random forest	(81)
Clearance	20,000 Unique compounds	Bayesian classification and extended connectivity fingerprints	(82)
Hepatic clearance	64 Drugs (training set) and 22 drugs(test set)	Multilinear regression analysis	(83)
Hepatic clearance	33 Drugs	Multilinear regression analysis	(84)
Drug distribution	93 Drugs	Partial least square and artificial neural network	(85)
Hepatic metabolic clearance	27,697 Compounds	Binary classification model	(86)
Hepatic clearance	50 Drugs	Multilinear regression analysis	(87)
Bioavailability	75 Compounds	Cluster analysis	(88)
Blood–brain barrier penetration	193 Compounds	Genetic approximation based regression model	(89)
Blood–brain barrier penetration and human intestinal absorption	1,093 Compounds (for BBB) and 480 compounds (for HIA)	Substructure pattern recognition and support vector machine	(90)
Clearance (total)	370 Compounds (training set) and 92 compounds (test set)	k-Nearest neighbors	(91)
Human hepatocyte intrinsic clearance	71 Drugs (training set); 18 drugs (test set-1); and 112 drugs (test set-2)	Artificial neural network	(92)

the brain and CNS (7). Brain penetration is of utmost importance for drug candidates where the site of action is within the CNS. However, BBB penetration needs to be minimized for drug molecules that target peripheral sites to reduce potential central nervous system-related side effects (95). Therefore, it is essential for the scientific community to develop models that can easily discriminate between molecules with high or low BBB permeability.

Bioavailability: Among the pharmacokinetic properties, a low and highly variable bioavailability is indeed the major cause for discontinuing further development of the drug. Oral bioavailability of a drug is dependent upon many factors, such as dissolution in the GI tract, intestinal membrane permeation, and intestinal/hepatic first-pass metabolism. The in silico prediction of oral bioavailability may be inspired by the Lipinski's "rule of five" (69). Membrane permeation is recognized as a vital requirement for oral bioavailability in the absence of active transport, and failure to achieve this will usually result in reduced oral bioavailability (38). Theoretical aspects on transporter proteins, especially P-glycoprotein (P-gp), have been vigorously studied during recent years. This is attributed to the reason that transport proteins are found in most organs involved in the uptake and elimination of endogenous compounds and xenobiotics, including drug molecules (10). In recent years, several prediction models for oral bioavailability based on QSPkR analysis have been reported.

Distribution: Tissue distribution is a vital determinant of the pharmacokinetic profile of a drug molecule. Currently, there are several techniques available for the prediction of tissue distribution. These techniques either predict tissue:plasma ratios or the volume of distribution at a steady state (2). The extent of distribution is defined by the volume of distribution, a proportionality factor relating the plasma level to the total amount of drug in the body. The volume of distribution, along with the clearance rate, determines the half-life of a drug and consequently its dosage regimen (18, 96). Hence, in drug development, the early prediction of both pharmacokinetic parameters would naturally be highly beneficial.

Plasma protein binding (PPB): Drugs frequently exhibit nonspecific binding in plasma and tissues to constituent proteins, such as albumin (primarily acidic drugs), α1-acid glycoprotein (basic drugs), and lipoproteins (neutral and basic drugs), as well as circulating cells including red blood cells and platelets (18). Drugs reveal equilibrium between their protein-bound and -free forms. Since only unbound free drug exhibits the intended therapeutic effect, therefore, the PPB affinity of drugs becomes a crucial property (93). For these reasons, development of the in silico models for the prediction of PPB is an active area of predictive ADME/Tox.

Metabolism: Drug metabolism and clearance also contribute towards the success of a drug, as these properties have a significant impact on bioavailability (oral or intravenous) (97). Chemical transformations of xenobiotics by liver (and other tissues and organs) are the key to the ADME profiling. It is extremely difficult to develop models for prediction of metabolic fate of drugs due to the complex nature of the biochemical processes involved. Of the two sets of metabolic transformations, oxidation(s) by CYP enzymes is/are critical (93). The rate and extent of metabolism influence clearance, whereas the involvement of particular enzyme(s) may lead to issues relating to the polymorphic nature of some of these enzymes as well as to drug–drug interactions (10). Metabolism, key factor influencing the excretion of drugs, has long been the target of computational models. Most efforts till date have mainly concentrated on modeling cytochrome P450 (CYP450) mediated metabolic stability and CYP450 inhibition. There are now examples of other drug metabolic end-points for which computational models have been developed include glucuronidation and enzymatic hydrolysis (94).

Excretion: Excretion (clearance) is the process through which the body eliminates the xenobiotics (i.e., "foreign" or "extraneous" compounds) and most of the biotransformations of these compounds take place in the liver in spite of the fact that this organ is not the exclusive site of metabolism. Another important route of excretion for a metabolite as well as the parent compound is renal clearance (94). There has been relatively little work conducted on the in silico modeling or prediction of excretion. Although most of the existing drugs are excreted to a variable extent as unchanged molecules via the kidneys or the bile, for only a few is urinary or biliary excretion a major route of elimination (2). Drug clearance is extremely difficult to correlate with physicochemical properties and molecular descriptors because of the complexity of the biological system, the influence of transporters, and the vast range of sites and mechanisms of biotransformation and elimination (18). The systemic clearance measures the efficiency for removal of a xenobiotic from the body. The hepatic clearance is the systemic clearance of those drugs which are exclusively metabolized by the liver. Apart from the evaluation of bioavailability, hepatic clearance also plays a vital role in the estimation and streamlining of early clinical trial doses/exposures in the clinic (93).

3. Example(s)

3.1. Prediction of Multiple Pharmacokinetic Parameters of Cephalosporins (77)

Multiple pharmacokinetic parameters of cephalosporins such as $t_{1/2}$, CL, CL_R, f_e, V, and f_b were predicted using random forest (RF), decision tree, and moving average analysis. The RFs were grown with the R program (version 2.1.0) using the RF library. R program along with the RPART library was used to grow decision tree.

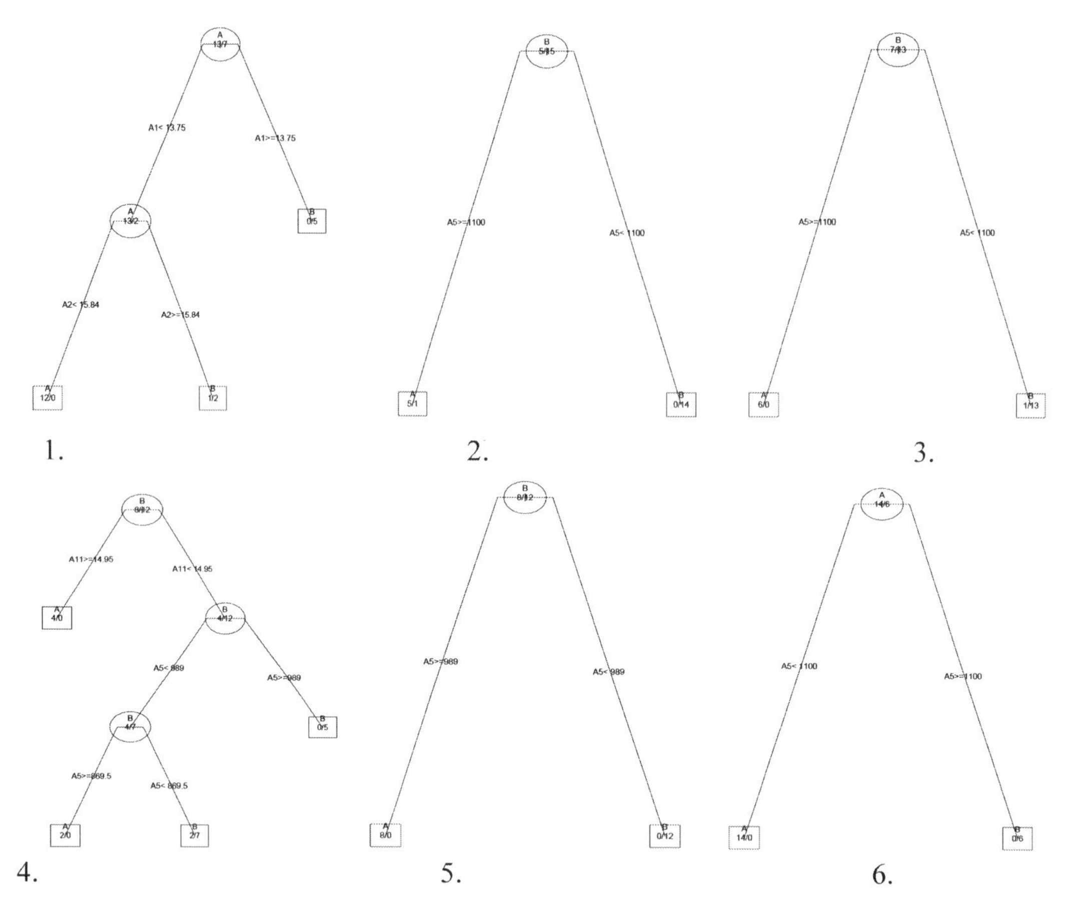

Fig. 3. The decision tree for distinguishing low value (A) from high value (B). (1) $t_{1/2}$; (2) CL; (3) CL_R; (4) f_e; (5) V; (6) f_b (A1, molecular connectivity topochemical index; A2, eccentric adjacency topochemical index; A5, eccentric connectivity topochemical index; A11, eccentric adjacency index). Reproduced from ref. 77 with permission from Austrian Pharmacists' Publishing House, Austria.

To construct a single topological index based model for predicting property/activity based ranges, moving average analysis of correctly predicted compounds was used. RF correctly classified the pharmacokinetic parameters into low and high ranges up to 95%. A decision tree was constructed for each pharmacokinetic parameter to determine the importance of topological indices. The decision tree learned the information from the input data with an accuracy of 95% and correctly predicted the cross-validated (tenfold) data with an accuracy of up to 90%. The classification of these pharmacokinetic parameters using decision tree is shown in Fig. 3. Three independent moving average based topological models were developed using a single range for simultaneous prediction of multiple pharmacokinetic parameters. The accuracy of classification of single index based models using moving average analysis varied from 65 to 100% (77).

3.2. Prediction of T_{max} of Antihistamines (78)

Models were developed for prediction of physicochemical property (octanol/water partition constant, log *P*), critical pharmacokinetic parameter (time to reach the maximum level of drug into the bloodstream, $T_{\max}$), and toxicological property (lethal dose, LD_{50}) of structurally diverse antihistaminic compounds using decision tree and moving average analysis. A decision tree was constructed for each property to determine the significance of topological descriptors. Single topological descriptor based models were developed using moving average analysis. The tree learned the information from the input data with an accuracy of >94% and subsequently predicted the cross-validated (tenfold) data with an accuracy of up to 71%. The classification of $T_{\max}$ using decision tree is shown in Fig. 4. The accuracy of prediction of single index based models using moving average analysis varied from ~63 to 80% (78).

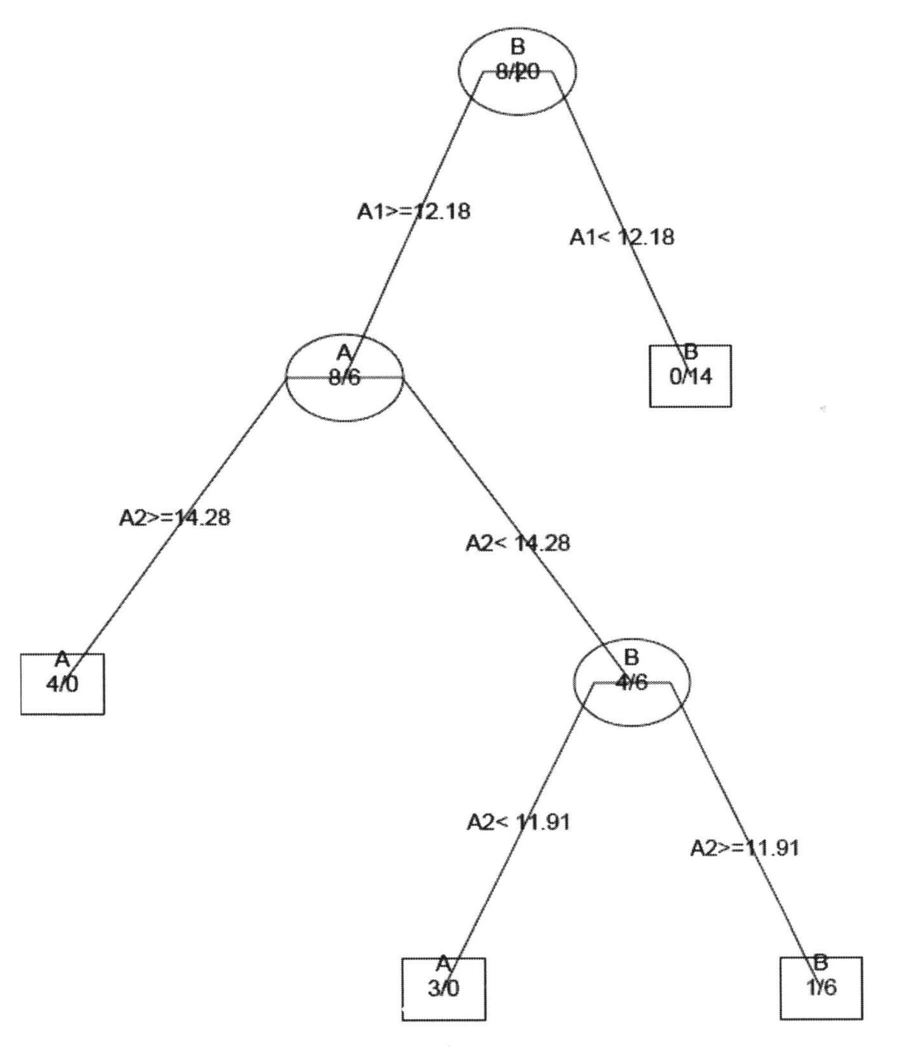

Fig. 4. A decision tree for distinguishing between low T_{max} and high T_{max}. A1, molecular connectivity topochemical index; A2, eccentric adjacency topochemical index; A, low T_{max}; B, high T_{max}. Reproduced from ref. 78 with permission from Inderscience Enterprises Limited, Switzerland.

4. Conclusion

In silico tools have not only accelerated drug discovery process but also have led to significant reduction in time, animal sacrifice, and expenditure. In silico tools specifically developed for prediction of ADME characteristics are of particular interest to pharmaceutical industry because of the high potential of discarding inappropriate molecules during an early stage of drug development with consequent saving of vital resources and valuable time. A well planned systematic integrated in silico, in vitro, and in vivo approach can discard inappropriate molecules at early stage and steeply accelerate drug discovery process at reduced cost.

References

1. Miteva MA, Violas S, Montes M et al (2006) FAF-drugs: free ADME/tox filtering of compound collections. Nucleic Acids Res 34: W738–W744
2. Boobis A, Gundert-Remy U, Kremers P et al (2002) In silico prediction of ADME and pharmacokinetics: report of an expert meeting organised by COST B15. Eur J Pharm Sci 17:183–193
3. Huisinga W, Telgmann R, Wulkow M (2006) The virtual lab approach to pharmacokinetics: design principles and concepts. Drug Discov Today 11:800–805
4. Hodgson J (2001) ADMET—turning chemicals into drugs. Nat Biotechnol 19:722–726
5. Grass GM, Sinko PJ (2002) Physiologically-based pharmacokinetic simulation modeling. Adv Drug Deliv Rev 54:433–451
6. Kennedy T (1997) Managing the drug discovery/development interface. Drug Disc Today 2:436–444
7. Spalding DJM, Harker AJ, Bayliss MK (2000) Combining high-throughput pharmacokinetic screens at the hits-to-leads stage of drug discovery. Drug Disc Today 5:S70–S76
8. Butina D, Segall MD, Frankcombe K (2002) Predicting ADME properties in silico: methods and models. Drug Disc Today 7:S83–S88
9. van Waterbeemd H, Gifford E (2003) ADMET in silico modelling: towards prediction paradise? Nat Rev Drug Disc 2:192–204
10. Hou T, Xu X (2004) Recent development and application of virtual screening in drug discovery: an overview. Curr Pharm Des 10:1011–1033
11. Hou T, Wang J, Zhang W et al (2006) Recent advances in computational prediction of drug absorption and permeability in drug discovery. Curr Med Chem 13:2653–2667
12. Li AP (2001) Screening for human ADME/Tox drug proteins in drug discovery. Drug Disc Today 6:357–366
13. Paul Y, Dhake AS, Singh B (2009) In silico quantitative structure pharmacokinetic relationship modeling of quinolones: apparent volume of distribution. Asian J Pharm 3:202–207
14. Ekins S, Waller CL, Swaan PW et al (2000) Progress in predicting human ADME parameters in silico. J Pharm Toxicol Methods 44:251–272
15. Goodwin JT, Clark DE (2005) In silico predictions of blood–brain barrier penetration: considerations to “keep in mind”. J Pharm Exp Ther 315:477–483
16. Linnankoski J, Ranta V-P, Yliperttula M et al (2008) Passive oral drug absorption can be predicted more reliably by experimental than computational models—fact or myth. Eur J Pharm Sci 34:129–139
17. Modi S (2004) Positioning ADMET in silico tools in drug discovery. Drug Disc Today 9:14–15
18. Mager DE (2006) Quantitative structure–pharmacokinetic/pharmacodynamic relationships. Adv Drug Del Rev 58:1326–1356
19. Gunaratna C (2001) Drug metabolism and pharmacokinetics in drug discovery: a primer for bioanalytical chemists, part II. Curr Sep 19:87–92
20. Hansch C (1972) Quantitative relationships between lipophilic character and drug metabolism. Drug Metab Rev 1:1–14
21. Lipinski CA, Lombardo F, Dominy BW et al (1997) Experimental and computational

approaches to estimate solubility and permeability in drug discovery and development settings. Adv Drug Deliv Rev 23:3–25

22. Lipinski CA (2000) Druglike properties and the causes of poor solubility and poor permeability. J Pharm Toxicol Methods 44:235–249
23. Ghose AK, Viswanadhan VN, Wendoloski JJ (1999) A knowledge-based approach in designing combinatorial or medicinal chemistry libraries for drug discovery. 1. A qualitative and quantitative characterization of known drug databases. J Comb Chem 1:55–68
24. Wenlock MC, Austin RP, Barton P, Davis AM, Leeson PD (2003) A comparison of physiochemical property profiles of development and marketed oral drugs. J Med Chem 46:1250–1256
25. Norinder U, Bergstrçm CAS (2006) Prediction of ADMET properties. ChemMedChem 1:920–937
26. Hirono S, Nakagome I, Hirano H et al (1994) Non-congeneric structure–pharmacokinetic property correlation studies using fuzzy adaptive least-squares: oral bioavailability. Biol Pharm Bull 17:306–309
27. Palm K, Luthman K, Ungel AL et al (1996) Correlation of drug absorption with molecular surface properties. J Pharm Sci 85:32–39
28. Wessel MD, Jurs PC, Tolan JW et al (1998) Prediction of human intestinal absorption of drug compounds from molecular structure. J Chem Inf Comput Sci 38:726–735
29. Nestorov I, Aarons L, Rowland M (1998) Quantitative structure-pharmacokinetics relationships II. Mechanistically based model for the relationship between the tissue distribution parameters and the lipophilicity of the compounds. J Pharmacokinet Biopharm 26:521–545
30. Winiwarter S, Bonham NM, Ax F et al (1998) Correlation of human jejunal permeability (in vivo) of drugs with experimentally and theoretically derived parameters. A multivariate data analysis approach. J Med Chem 41:4939–4949
31. Clark DE (1999) Rapid calculation of polar molecular surface area and its application to the prediction of transport phenomena. 1. Prediction of intestinal absorption. J Pharm Sci 88:807–814
32. Stenberg P, Luthman K, Ellens H et al (1999) Prediction of the intestinal absorption of endothelin receptor antagonists using three theoretical methods of increasing complexity. Pharm Res 16:1520–1526
33. Ghuloum AM, Sage CR, Jain AN (1999) Molecular hashkeys: a novel method for molecular characterization and its application for predicting important pharmaceutical properties of molecules. J Med Chem 42:1739–1748
34. Egan WJ, Merz KM, Baldwin JJ (2000) Prediction of drug absorption using multivariate statistics. J Med Chem 43:3867–3877
35. Andrews CW, Bennett L, Yu LX (2000) Predicting human oral bioavailability of a compound: development of a novel quantitative structure–bioavailability relationship. Pharm Res 17:639–644
36. Zhao YH, Le J, Abraham MH et al (2001) Evaluation of human intestinal absorption data and subsequent derivation of a quantitative structure–activity relationship (QSAR) with the Abraham descriptors. J Pharm Sci 90:749–784
37. Colmenarejo G, Alvarez-Pedraglio A, Lavandera JL (2001) Chemoinformatic models to predict binding affinities to human serum albumin. J Med Chem 44:4370–4378
38. Mandagere AK, Thompson TN, Hwang KK (2002) A graphical model for estimating oral bioavailability of drugs in humans and other species from their Caco-2 permeability and in vitro liver enzyme metabolic stability rates. J Med Chem 45:304–311
39. Veber DF, Johnson SR, Cheng H-Y et al (2002) Molecular properties that influence the oral bioavailability of drug candidates. J Med Chem 45:2615–2623
40. Karalis V, Tsantili-Kakoulidou A, Macheras P (2002) Multivariate statistics of disposition pharmacokinetic parameters for structurally unrelated drugs used in therapeutics. Pharm Res 19:1827–1834
41. Wajima T, Fukumura K, Yano Y et al (2002) Prediction of human clearance from animal data and molecular structural parameters using multivariate regression analysis. J Pharm Sci 91:2489–2499
42. Doniger S, Hofmann T, Yeh J (2002) Predicting CNS permeability of drug molecules: comparison of neural network and support vector machine algorithms. J Comput Biol 9:849–864
43. Shen M, Xiao YD, Golbraikh A et al (2003) Development and validation of k-nearest-neighbor QSPR models of metabolic stability of drug candidates. J Med Chem 46:3013–3020
44. Zmuidinavicius D, Didziapetris R, Japertas P et al (2003) Classification structure–activity relations (C-SAR) in prediction of human intestinal absorption. J Pharm Sci 92:621–633
45. Lobell M, Sivarajah V (2003) In silico prediction of aqueous solubility, human plasma

protein binding and volume of distribution of compounds from calculated pKa and AlogP98 values. Mol Divers 7:69–87
46. Wajima T, Fukumura K, Yano Y et al (2003) Prediction of human pharmacokinetics from animal data and molecular structural parameters using multivariate regression analysis: oral clearance. J Pharm Sci 92:2427–2440
47. Turner JV, Maddalena DJ, Cutler DJ et al (2003) Multiple pharmacokinetic parameter prediction for a series of cephalosporins. J Pharm Sci 92:552–559
48. Pérez MA, Sanz MB, Torres LR et al (2004) A topological sub-structural approach for predicting human intestinal absorption of drugs. Eur J Med Chem 39:905–916
49. Turner JV, Maddalena DJ, Cutler DJ (2004) Pharmacokinetic parameter prediction from drug structure using artificial neural networks. Int J Pharm 270:209–219
50. Pan D, Iyer M, Liu J et al (2004) Constructing optimum blood brain barrier QSAR models using a combination of 4D-molecular similarity measures and cluster analysis. J Chem Inf Comput Sci 44:2083–2098
51. Bai JP, Utis A, Crippen G et al (2004) Use of classification regression tree in predicting oral absorption in humans. J Chem Inf Comput Sci 44:2061–2069
52. Li H, Yap CW, Ung CY et al (2005) Effect of selection of molecular descriptors on the prediction of blood–brain barrier penetrating and nonpenetrating agents by statistical learning methods. J Chem Inf Model 45:1376–1384
53. Aureli L, Cruciani G, Cesta MC et al (2005) Predicting human serum albumin affinity of interleukin-8 (CXCL8) inhibitors by 3D-QSPR approach. J Med Chem 48:2469–2479
54. Rahnasto M, Raunio H, Poso A et al (2005) Quantitative structure–activity relationship analysis of inhibitors of the nicotine metabolizing CYP2A6 enzyme. J Med Chem 48:440–449
55. Zhou XF, Shao Q, Coburn RA et al (2005) Quantitative structure–activity relationship and quantitative structure–pharmacokinetics relationship of 1,4-dihydropyridines and pyridines as multidrug resistance modulators. Pharm Res 22:1989–1996
56. Jung SJ, Choi SO, Um SY et al (2006) Prediction of the permeability of drugs through study on quantitative structure–permeability relationship. J Pharm Biomed Anal 41:469–475
57. Gleeson MP, Waters NJ, Paine SW et al (2006) In silico human and rat Vss quantitative structure–activity relationship models. J Med Chem 49:1953–1963
58. Linnankoski J, Mäkelä JM, Ranta VP et al (2006) Computational prediction of oral drug absorption based on absorption rate constants in humans. J Med Chem 49:3674–3681
59. Garg P, Verma J (2006) In silico prediction of blood–brain barrier permeability: an artificial neural network model. J Chem Inf Model 46:289–297
60. Doddareddy MR, Cho YS, Koh HY et al (2006) In silico renal clearance model using classical Volsurf approach. J Chem Inf Model 46:1312–1320
61. Yap CW, Li ZR, Chen YZ (2006) Quantitative structure–pharmacokinetic relationships for drug clearance by using statistical learning methods. J Mol Graph Model 24:383–395
62. Dureja H, Madan AK (2006) Topochemical models for the prediction of permeability through blood brain barrier. Int J Pharm 323:27–33
63. Dureja H, Madan AK (2007) Validation of topochemical models for the prediction of permeability through blood brain barrier. Acta Pharm 57:451–467
64. Hou T, Wang J, Li Y (2007) ADME evaluation in drug discovery. 8. The prediction of human intestinal absorption by a support vector machine. J Chem Inf Model 47:2408–2415
65. Iyer M, Tseng YJ, Senese CL et al (2007) Prediction and mechanistic interpretation of human oral drug absorption using MI-QSAR analysis. Mol Pharm 4:218–231
66. Cuadrado MU, Ruiz IL, Gómez-Nieto MA (2007) QSAR models based on isomorphic and nonisomorphic data fusion for predicting the blood brain barrier permeability. J Comput Chem 28:1252–1260
67. Gleeson MP (2007) Plasma protein binding affinity and its relationship to molecular structure: an in-silico analysis. J Med Chem 50:101–112
68. Li C, Liu T, Cui X et al (2007) Development of in vitro pharmacokinetic screens using Caco-2, human hepatocyte, and Caco-2/human hepatocyte hybrid systems for the prediction of oral bioavailability in humans. J Biomol Screen 12:1084–1091
69. Hou T, Wang J, Zhang W et al (2007) ADME evaluation in drug discovery. 6. Can oral bioavailability in humans be effectively predicted by simple molecular property-based rules? J Chem Inf Model 47:460–463
70. Moda TL, Montanari CA, Andricopulo AD (2007) Hologram QSAR model for the prediction of human oral bioavailability. Bioorg Med Chem 15:7738–7745

71. De Buck SS, Sinha VK, Fenu LA et al (2007) The prediction of drug metabolism, tissue distribution, and bioavailability of 50 structurally diverse compounds in rat using mechanism-based absorption, distribution, and metabolism prediction tools. Drug Metab Dispos 35:649–659
72. Hou T, Wang J, Zhang W et al (2007) ADME evaluation in drug discovery. 7. Prediction of oral absorption by correlation and classification. J Chem Inf Model 47:208–218
73. Fu XC, Wang GP, Shan HL et al (2008) Predicting blood–brain barrier penetration from molecular weight and number of polar atoms. Eur J Pharm Biopharm 70:462–466
74. Zhang L, Zhu H, Oprea TI et al (2008) QSAR modeling of the blood–brain barrier permeability for diverse organic compounds. Pharm Res 25:1902–1914
75. Sinha VK, De Buck SS, Fenu LA et al (2008) Predicting oral clearance in humans: how close can we get with allometry? Clin Pharmacokinet 47:35–45
76. Ma CY, Yang SY, Zhang H et al (2008) Prediction models of human plasma protein binding rate and oral bioavailability derived by using GA-CG-SVM method. J Pharm Biomed Anal 47:677–682
77. Dureja H, Gupta S, Madan AK (2008) Topological models for prediction of pharmacokinetic parameters of cephalosporins using random forest, decision tree and moving average analysis. Sci Pharm 76:377–394
78. Dureja H, Gupta S, Madan AK (2009) Decision tree derived topological models for prediction of physico-chemical, pharmacokinetic and toxicological properties of antihistaminic drugs. Int J Comput Biol Drug Des 2:353–370
79. Lanevskij K, Japertas P, Didziapetris R et al (2009) Ionization-specific prediction of blood–brain permeability. J Pharm Sci 98:122–134
80. Kokate A, Li X, Williams PJ et al (2009) In silico prediction of drug permeability across buccal mucosa. Pharm Res 26:1130–1139
81. Berellini G, Springer C, Waters NJ et al (2009) In silico prediction of volume of distribution in human using linear and nonlinear models on a 669 compound data set. J Med Chem 52:4488–4495
82. McIntyre TA, Han C, Davis CB (2009) Prediction of animal clearance using naïve Bayesian classification and extended connectivity fingerprints. Xenobiotica 39:487–494
83. Li H, Sun J, Sui X, Yan Z et al (2009) Structure-based prediction of the nonspecific binding of drugs to hepatic microsomes. AAPS J 11:364–370
84. Emoto C, Murayama N, Rostami-Hodjegan A et al (2009) Utilization of estimated physicochemical properties as an integrated part of predicting hepatic clearance in the early drug-discovery stage: impact of plasma and microsomal binding. Xenobiotica 39:227–235
85. Paixão P, Gouveia LF, Morais JA (2009) Prediction of drug distribution within blood. Eur J Pharm Sci 36:544–554
86. Chang C, Duignan DB, Johnson KD (2009) The development and validation of a computational model to predict rat liver microsomal clearance. J Pharm Sci 98:2857–2867
87. Li H, Sun J, Sui X et al (2009) First-principle, structure-based prediction of hepatic metabolic clearance values in human. Eur J Med Chem 44:1600–1606
88. Grabowski T, Jaroszewski JJ (2009) Bioavailability of veterinary drugs in vivo and in silico. J Vet Pharmacol Ther 32:249–257
89. Fan Y, Unwalla R, Denny RA et al (2010) Insights for predicting blood–brain barrier penetration of CNS targeted molecules using QSPR approaches. J Chem Inf Model 50:1123–1133
90. Shen J, Cheng F, Xu Y et al (2010) Estimation of ADME properties with substructure pattern recognition. Chem Inf Model 50:1034–1041
91. Yu MJ (2010) Predicting total clearance in humans from chemical structure. J Chem Inf Model 50:1284–1295
92. Paixão P, Gouveia LF, Morais JA (2010) Prediction of the in vitro intrinsic clearance determined in suspensions of human hepatocytes by using artificial neural networks. Eur J Pharm Sci 39:310–321
93. Kharkar PS (2010) Two-dimensional (2D) in silico models for absorption, distribution, metabolism, excretion and toxicity (ADME/T) in drug discovery. Curr Top Med Chem 10:116–126
94. Lombardo F, Gifford E, Shalaeva MY (2003) In silico ADME prediction: data, models, facts and myths. Mini Rev Med Chem 3:861–875
95. Liu X, Tu M, Kelly RS et al (2004) Development of a computational approach to predict blood–brain barrier permeability. Drug Metab Dispos 32:132–139
96. Sui X, Sun J, Wu X et al (2008) Predicting the volume of distribution of drugs in humans. Curr Drug Metab 9:574–580
97. Ekins S, Boulanger B, Swaan PW et al (2002) Towards a new age of virtual ADME/TOX and multidimensional drug discovery. J Comput Aided Mol Des 16:381–401

Chapter 15

Ligand- and Structure-Based Pregnane X Receptor Models

Sandhya Kortagere, Matthew D. Krasowski, and Sean Ekins

Abstract

The human pregnane X receptor (PXR) is a ligand dependent transcription factor that can be activated by structurally diverse agonists including steroid hormones, bile acids, herbal drugs, and prescription medications. PXR regulates the transcription of several genes involved in xenobiotic detoxification and apoptosis. Activation of PXR has the potential to initiate adverse effects by altering drug pharmacokinetics or perturbing physiological processes. Hence, more reliable prediction of PXR activators would be valuable for pharmaceutical drug discovery to avoid potential toxic effects. Ligand- and protein structure-based computational models for PXR activation have been developed in several studies. There has been limited success with structure-based modeling approaches to predict human PXR activators, which can be attributed to the large and promiscuous site of this protein. Slightly better success has been achieved with ligand-based modeling methods including quantitative structure–activity relationship (QSAR) analysis, pharmacophore modeling and machine learning that use appropriate descriptors to account for the diversity of the ligand classes that bind to PXR. These combined computational approaches using molecular shape information may assist scientists to more confidently identify PXR activators. This chapter reviews the various ligand and structure based methods undertaken to date and their results.

Key words: Agonists, Alignment methods, Antagonists, Bayesian classification, Docking and Scoring, Pharmacophore, Pregnane Xenobiotic receptors, QSAR, Support vector machines

1. Introduction

Receptor mediated toxicity has received a lot of attention in the past decade with several studies commissioned to examine the deleterious effects of man-made chemicals (1–4). These chemicals include pesticides and other industrial chemicals that may elicit toxic endpoints by mimicking the action of endogenous hormones and neurotransmitters (3, 5). As a first line of investigation, the endocrine receptors have been implicated in receptor mediated toxicity although the role of other receptors and transporters has also been profiled (6, 7). One such class of endocrine receptors is the nuclear hormone receptors (NRs), which form the largest superfamily of ligand-dependent transcription factors. NRs are involved in a variety of functions including

Brad Reisfeld and Arthur N. Mayeno (eds.), *Computational Toxicology: Volume I*, Methods in Molecular Biology, vol. 929,
DOI 10.1007/978-1-62703-050-2_15, © Springer Science+Business Media, LLC 2012

Fig. 1. General architecture of nuclear hormone receptor family. DBD represents the DNA binding domain and LBD represents the ligand binding domain and AF2 represents the activation function region that binds to coactivator.

cell proliferation, differentiation, development, and metabolic homeostasis (8). Functionally, all NRs transcriptionally regulate the expression of target genes by the recruitment of coactivators and corepressors. The general architecture of all NRs consists of four distinct domains (Fig. 1): (a) an A/B domain present at the amino terminal region which contains the ligand-independent AF1 activation domain, (b) a DNA-binding domain (DBD) that contains two conserved C4-type zinc finger motifs, (c) a highly flexible hinge region ("D-domain"), and (d) the ligand binding domain (LBD, also known as the E domain) that connects to the second activation domain called the AF2 domain at the carboxy terminal of the protein. The DBD binds to target gene responsive elements that contain conserved hexameric sequences. The LBD among the NRs show much more structural diversity than the DBDs, which explains the diverse array of ligands that bind to NRs. Because of their diverse structural and functional properties and their ability to bind to a broad array of endogenous and exogenous molecules, NRs are involved in regulating a number of genes. Among the NR superfamily, the role of estrogen receptors (ERs), androgen receptor (AR), and peroxisome proliferator activated receptors (PPARs) have been well studied in terms of receptor mediated toxicity due to their direct role in mediating hormones (2, 9, 10). The federal agencies have brought in legislation to test for endocrine disrupting chemicals that interfere with these reproductive hormone regulators (11, 12).

The human PXR (hPXR) is a ligand dependent NR that plays a key role in regulating drug–drug interactions (13–15) by activating an essential class of cytochrome P450 enzymes that metabolize endobiotics and xenobiotics. PXR agonists include a wide range of structurally diverse endogenous and exogenous compounds such as bile acids, steroid hormones, dietary fat-soluble vitamins, prescription medications, and herbal drugs, as well as environmental chemicals such as pesticides, estrogens, and antiestrogens (16). PXR forms a heterotetramer with the retinoid X receptor (RXR) and a homodimer with itself. The homodimerization interface of PXR influences the coactivator binding site at AF2 by modulating long range motions of the LBD (17). These motions are responsible for the promiscuousness of the LBD in PXR that helps bind ligands of varying sizes and shapes. The role of PXR in various pathophysiological states indicates that PXR agonists could variably affect human and animal health. For example PXR agonists can impact cholesterol metabolism, cell cycle regulation, the endocrine system (18, 19) as well as potentiate the toxicity of other environmental contaminants as reviewed recently (20). Animal models may not reliably predict hPXR-related problems due to the diversity of PXRs across species (11, 14) resulting in differences in the ligand selectivity (21). The identification and characterization of hPXR agonists is thus important to human pharmacokinetics and toxicology of environmental chemicals. We (13–15, 22–28) and others (29–35) have tried to utilize a number of in silico methods to understand and characterize the agonist and antagonist binding modes of hPXR. In this chapter we summarize these *in silico* methods and the reader is urged to consult other references listed for a comprehensive overview of all the methods.

2. Results and Discussion

2.1. Ligand-Based Modeling of PXR

Human PXR agonist pharmacophore models have been shown to possess hydrophobic, hydrogen bond acceptor and hydrogen bond donor features (Table 1), consistent with the crystallographic structures of hPXR ligand–receptor complexes (17, 36–38). Several predictive computational models for hPXR have been developed to define key binding features of ligands (27, 28, 39). Most

Table 1
hPXR pharmacophore features and how to avoid being an agonist

	Pharmacophore features	How to avoid interaction with the protein
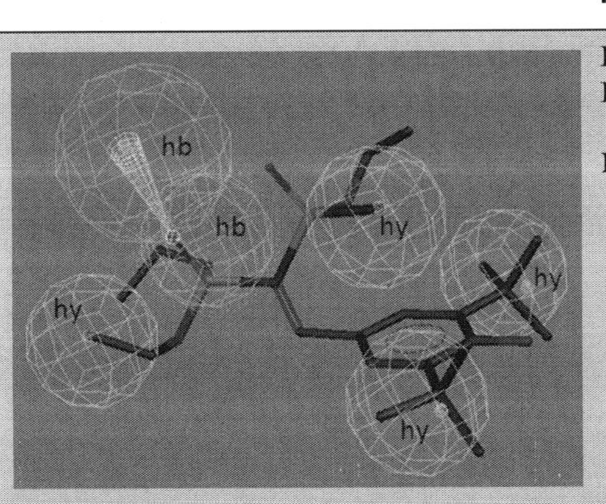	Hydrophobic (hy) Hydrogen bond acceptors (hb) Hydrogen bond donor (occasionally)	Attaching hydrogen bonding groups on one of the hydrophobic features, adding larger more rigid groups as well as removing central H-bond acceptors

pharmacophore models feature 4-5 hydrophobic features and at least 1–2 hydrogen bonding moieties.

One study has used hPXR activation data for 30 steroidal compounds (including 9 bile acids) to create a pharmacophore with four hydrophobic features and one hydrogen bond acceptor (13). This pharmacophore contained 5α-androstan-3β-ol (EC_{50} 0.8 μM) which contains one hydrogen bond acceptor, indicating that in contrast to the crystal structure of 17β-estradiol (published EC_{50} 20 μM) bound to hPXR with two hydrogen bonding interactions (37), hydrophobic interactions may therefore be more important for increased affinity. (13) This and other pharmacophores have been used to predict hPXR interactions for antibiotics (24) which were verified in vitro, suggesting one use for computational approaches in combination with experimental methods (24). The original hPXR pharmacophore (Table 1) consisted of 4 hydrophobes and a hydrogen bond acceptor feature and was found to map 5 of the antibiotics, one of which, RIF, was originally in the pharmacophore training set. Nafcillin, dicloxacillin, erythromycin, and troleandomycin mapped to the widely dispersed features well. The diverse ($n = 31$) pharmacophore consisted of 2 hydrophobes, a hydrogen bond acceptor and a hydrogen bond donor feature and mapped to 16 of the antibiotics, one of which, rifampin (RIF), was also in the pharmacophore training set. Tetracycline, sulfisoxazole, sulfmethazole, troleandomycin, and griseofulvin did not map to the features. Interestingly nafcillin fit well and dicloxacillin had the lowest fit value. The steroidal ($n = 30$) pharmacophore consisted of 4 hydrophobes and a hydrogen bond acceptor feature and surprisingly was found to map 4 of the antibiotics, dicloxacillin, troleandomycin, clindamycin, and griseofulvin. Twelve of the antibiotics were also docked (erythromycin failed to dock) into one of the hPXR crystal structures (Protein Data Bank (www.rcsb.org/pdb) accession # 1NRL), and were then scored using the docking program FlexX (40). The lower the FlexX score, the better the complementarity between ligand and receptor. All the penicillins and cephems docked and scored well apart from cefuroxime, tetracycline, doxycycline, and clindamycin which had poorer scores. Docking and scoring with FlexX scored the penicillins and cephems similarly and failed to dock the larger molecule erythromycin. We have also recently suggested that docking methods may need combination with quantitative structure–activity relationship (QSAR) or other computational methods, in order to improve predictions due to the flexibility of the protein and large binding site that could accommodate multiple pharmacophores (40).

The pharmacophore models have predominantly used structurally diverse ligands in the training set and have the limitation in most cases of compiling data from multiple laboratories using different experimental protocols, ultimately forcing the binary classification of ligands for the training sets (i.e., activating versus nonactivating). Others have described a statistical quantitative structure activity relationship (QSAR) model using VolSurf descriptors and partial least squares (PLS) for a training set of 33 hPXR ligands which identified hydrogen bond acceptor regions and amide responsive regions; however, no test set data were provided (41). A second statistical model using a recursive partitioning method has been used with 99 hPXR activators and nonactivators to predict the probability of aprepitant, L-742694, 4-hydroxytamoxifen and artemisinin binding to hPXR (26). Additional models based on machine learning methods using a set of hPXR activators and nonactivators (33) displayed overall prediction accuracies between 72 and ~80%, while an external test set of known activators had a similar level of accuracy.

To date there have been few attempts to build ligand-based models around a large structurally narrow set of hPXR activators. The absence of large sets of quantitative data for hPXR agonists has restricted QSAR models to a relatively small universe of molecules compared to the known drugs, drug-like molecules, endobiotics and xenobiotics in general (33). Various machine learning methods (e.g. support vector machines, recursive partitioning etc.) that can be used with binary biological data have been applied to hPXR. We have generated computational models for hPXR using recursive partitioning (RP), random forest (RF), and support vector machine (SVM) algorithms with VolSurf descriptors. Following 10-fold randomization the models correctly predicted 82.6–98.9% of activators and 62.0–88.6% of nonactivators. All models were validated with a test set ($N = 145$), and the prediction accuracy ranged from 63 to 67% overall (25). These test set molecules were found to cover the same area in a principal component analysis plot as the training set, suggesting that the predictions were within the applicability domain. A second study used the same training and test sets with SVM algorithms with molecular descriptors derived from two sources, Shape Signatures and the Molecular Operating Environment (MOE) application software. The overall test set prediction accuracy for hPXR activators with SVM was 72% to 81% (23).

A substantial amount of experimental hPXR data has recently been generated for classes of steroidal compounds, namely, androstanes, estratrienes, pregnanes, and bile salts (14) which was then used with an array of ligand-based computational methods including Bayesian modeling with 2D fingerprints methods. (15) All 115 compounds were used to generate a Bayesian classification model (42) using a definition of active as a compound having an EC_{50} for hPXR activation of less than 10 μM. Using FCFP_6 and 8 interpretable descriptors (AlogP, molecular weight, rotatable bonds, number of rings, number of aromatic rings, hydrogen bond acceptor, hydrogen bond donor, and polar surface area) a model was developed with a receiver operator characteristic for leave one out cross validation of 0.84. In addition to the leave one out cross validation, further validation methods were undertaken. After leaving 20% of the compounds out 100 times the ROC is 0.84, concordance 73.2%, specificity 69.1%, and sensitivity 84.1%. In comparison to molecular docking methods, ligand based models performed better in classifying the compounds. The Bayesian method appeared to have good model statistics for internal cross validation of steroids. We have additionally used this model to classify a previously used diverse molecule test set. The Bayesian hPXR model was used to rank 123 molecules (65 activators and 58 non activators). Out of the top 30 molecules scored and ranked with this model 20 (75%) were classified as activators (EC_{50} < 100 μM) (15). All hPXR positive contributing substructures were essentially hydrophobic, while hPXR negative contributing features possessed hydroxyl or other substitutions which are likely not optimally placed to facilitate interactions with hydrogen bonding features in the hPXR LBD. Therefore, possession of these hydrogen bond acceptor and donor features indicated in the steroidal substructures appears to be related to loss of hPXR activation. Among the alignment based methods, CoMFA and CoMSIA's performance in classification models was the least since they are based on rigid alignment of molecules, while hPXR binding site is very flexible. 4D and 5D QSAR methods are based on ensembles of ligand conformations and seem to perform better within a narrow chemical space (such as a sub-class of steroids), but the performance is limited when extended to a larger chemical class.

The Bayesian approach using fingerprints and 117 structural descriptors was also used recently with a large diverse training set comprising 177 compounds. The classifier was used to screen a subset of FDA approved drugs followed by testing of a few compounds (17 compounds from the top 25) with a cell-based luciferase reporter assay for evaluation of chemical-mediated hhPXR activation in HepG2 cells. The reporter assay confirmed nine drugs as novel hPXR activators: fluticasone, nimodipine, nisoldipine, beclomethasone, finasteride, flunisolide, megestrol, secobarbital, and aminoglutethimide (29). Such ligand-based Bayesian approaches with a diverse training set may be more useful than a narrow structural series of steroidal compounds that was previously used for database searching a set of pesticides and other industrial chemicals (22). These global models for hPXR could certainly be of value for selecting compounds for in vitro screening. We are not aware of an exhaustive screening of FDA approved drugs for potential hPXR agonists and then testing of compounds that score highly. Such an approach may be useful to understand hPXR mediated drug–drug interactions more comprehensively.

2.2. Structure-Based Modeling of hPXR

Currently at the time of writing there are five high-resolution crystal structures of the hPXR LBD in complex with a variety of ligands (17, 36–38) available in the PDB (and a sixth structure to be deposited (57)). The structures have provided atomic level details that have led to a greater understanding of the LBD and the structural features involved in ligand–receptor interactions (27, 28, 39). The cocrystallized ligands include the natural products hyperforin (active component of the herbal antidepressant St. John's wort) and colupulone (from hops), the steroid 17β-estradiol, and the synthetic compounds SR12813, T-0901317, and the antibiotic rifampicin (Fig. 2). These ligands span a range of molecular sizes (M.Wt range 272.38–713.81 Da, mean 487.58 ± 147.25 Da) and are predicted as generally hydrophobic (calculated ALogP (43) 3.54–10.11, mean 5.54 ± 2.41). The cavernous hPXR ligand binding pocket (LBP) with a volume >1,350 Å^3 accepts molecules of these widely varying dimensions and chemical properties, and is likely capable of binding small molecules in multiple orientations (44). This complicates overall prediction of whether a small molecule is likely to be classified as an hPXR agonist using traditional structure-based virtual screening methods like docking that treat the receptor as rigid for purposes of modeling (25, 45). With regard to this, we have previously shown that the widely used structure-based docking methods

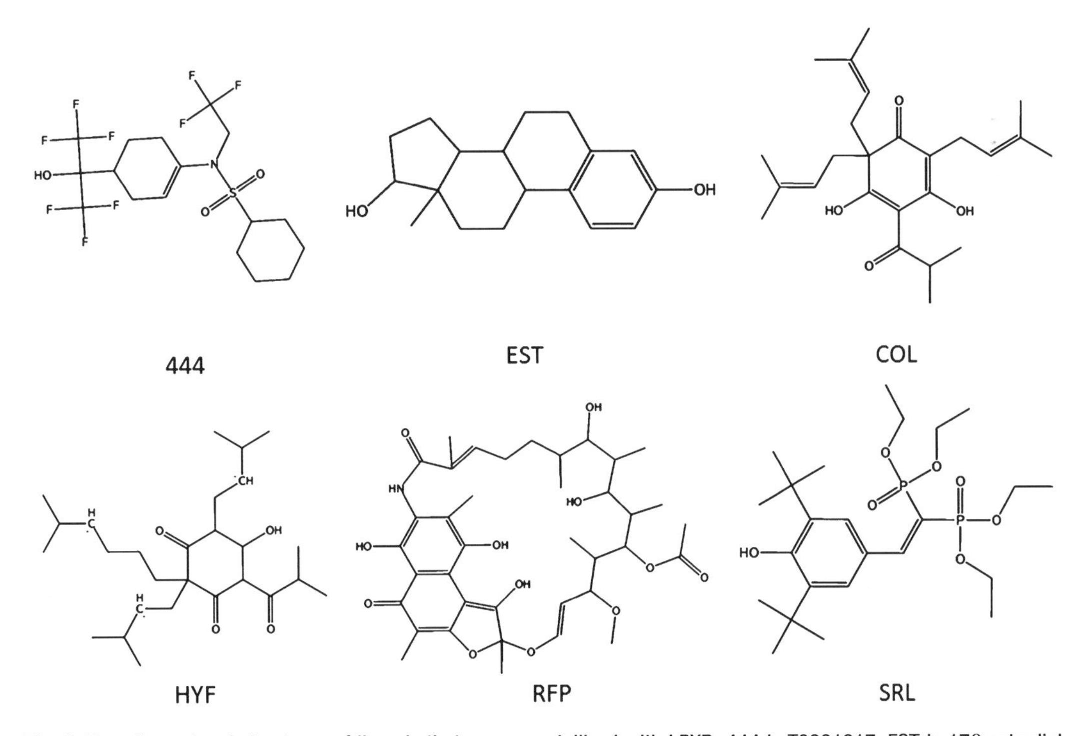

Fig. 2. Two dimensional structures of ligands that are cocrystallized with hPXR; 444 is T0901317, EST is 17β-estradiol, COL is colupulone, HYF is hyperforin, RFP is rifampicin, and SRL is SR12813.

FlexX and GOLD performed relatively poorly in predicting hPXR agonists and this is perhaps not surprising based on the observations described above.

In a recent paper we docked a series of 115 steroids to the six hPXR crystal structures using GOLD (15). To assess the diversity of the binding sites in the six published crystal structures, we superimposed the backbones. All six crystal structures superimposed with a backbone root mean squared deviation of 0.5 Å suggesting that they had very similar structures and their cocrystallized ligands bound to the same binding pocket. The docking scores for all 115 compounds were in the range of 36–77 (higher values are preferred) for all the crystal structures. In this study we designed a scoring scheme in which the docked scores were weighted based on their similarity to either the high affinity ligand (5α-androstan-3β-ol) or the cocrystal steroid ligand 17β-estradiol (see appendix for other scoring schemes). The corresponding Tanimoto similarity scores to 5α-androstan-3β-ol and the crystal ligand 17β-estradiol using MDL public keys were between 0.4 and 1 (with 1 being maximal similarity). We also chose an EC_{50} value of 10 μM as a cutoff to classify the compounds as hPXR activators and nonactivators. Based on these parameters the overall accuracy in classification using similarity weighted GoldScore ranged between 35% and 58%. Although the classification success rates were modest, these values were entirely dependent on the EC_{50} value chosen as cut off. The performance of docking based classification for individual classes of steroids was better than the overall performance. The classification of activators ranged from 66 to 100%. However, several of the nonactivators were classified as activators in the estratrienes and the reason being their high similarity scores towards the crystal structure ligand 17β-estradiol. Thus, similarity weighted scoring schemes have to be improved further to avoid this misclassification. Docking studies provided a number of insights into the mode of binding of steroids to hPXR. Analysis of the predicted binding mode of the various steroids showed that they all form hydrogen bonded interactions with His407 and also main chain atoms of Leu209 and Val211. The steroid core maintains hydrophobic interactions with Leu240, Leu239, Ile236, Trp299, Phe288, Leu411, and Met243 (Fig. 3). In a study by Gao et al. (30), molecular docking methods were used to derive the SAR of a known class of hPXR activators. It was found that the activation of hPXR could be attenuated by adding polar groups to portions of the ligand that would otherwise form favorable hydrophobic interactions with the protein. These results correlate well with our docking studies on other xenobiotics as well (45, 46).

In a recent study we used molecular docking and the similarity weighted GoldScore approach to classify activators of hPXR using the ToxCast™ database of pesticides and other industrial chemicals (22). Each ToxCast™ compound was docked into the 5 published

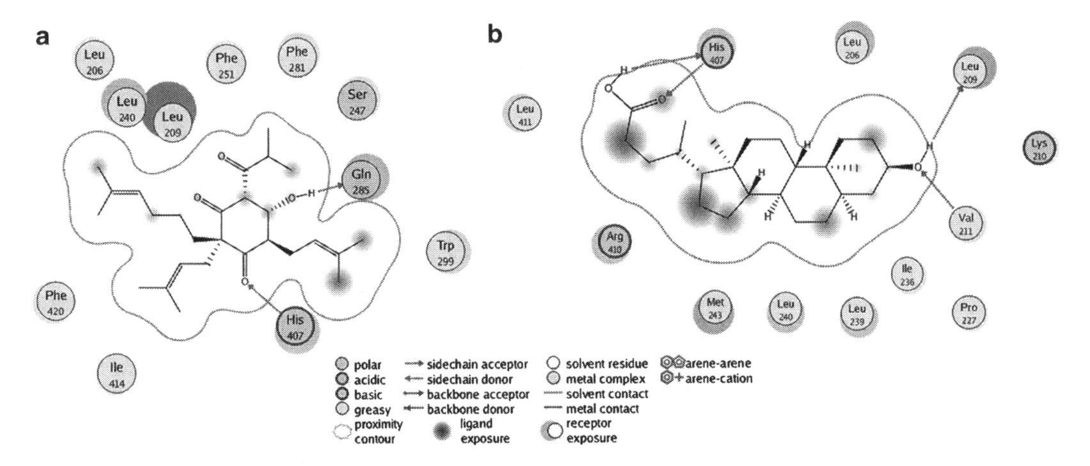

Fig. 3. Structural comparison of the binding modes of (**a**) hyperforin and (**b**) steroids compound in hPXR LBD is shown. The 2D interaction diagram was generated using LIGX module of program MOE.

crystallographic structures of hPXR and scored according to the hybrid scoring scheme and 11 compounds were selected based on their consensus docking scores for testing. The results from the docking studies were also correlated with a Bayesian model to classify the ToxCast^{TM} compounds into hPXR agonists and non-agonists. These chosen compounds were tested for hPXR activation using a luciferase-based reporter assays in the HepG2 and DPX-2 human liver cell lines. Among the eleven compounds tested, six were found to be strong agonists and two had weak agonist activity. Further, docking based classification of the entire Toxcast dataset consisting of 308 molecules was then compared to the experimental results. Docking based classification performed well with a prediction sensitivity of ~ 74% for the entire ToxCast^{TM} dataset published (47–49).

Another significant observation with reference to hPXR involves the presence of two splice variants of hPXR. A recent study has compared the splice variant hPXR.2 with hPXR.1, using homology based models. It was shown that hPXR.2 lacks the 37 amino acids that make up helix 2, and this might suggest why agonists do not activate this splice variant in recombinant expression systems (24).

2.3. hPXR Antagonists

There have been relatively few attempts to understand or develop in silico models of antagonism of hPXR (50). One computational approach focused on the LBD using the crystal structure of hPXR bound to T-0901317 (37), but this proved difficult (31). The list of hPXR antagonists is however steadily growing and even includes some compounds first characterized as weak hPXR agonists (15). For example, the antagonists ketoconazole (33), fluconazole, and enilconazole (51) have all been shown to inhibit the activation of hPXR in the presence of paclitaxel, while behaving as weak agonists on their own. Computational docking results revealed these hPXR antagonists partially occupied the same hydrophobic groove where the coactivator motif binds to receptor, antagonizing the essential protein–protein interaction ((13) and references therein). Ketoconazole was shown to inhibit the interaction of hPXR with the coactivator SRC-1 suggesting binding to the AF-2 site. This hypothesis was further confirmed with site-directed mutagenesis data (51), indicating ketoconazole behaved like the histidine residue of SRC-1 (51). Based on the three azole hPXR antagonists, a pharmacophore model was developed to elucidate the important structural features for binding. When this model was combined with docking studies and biological testing, it enabled discovery of several more potent nonazole hPXR antagonists which included commercially available synthetic compounds and the FDA approved prodrug leflunomide, confirmed experimentally in vitro (14). When ketoconazole was docked into the exterior site, the piperazine ring was predicted as solvent exposed. The pharmacophore model also indicated the minimum requirements of these azoles, suggesting the complementary nature of different computer-aided antagonist design methods (13). The new small molecule antagonists had good ligand efficiencies (as they were more potent on a per heavy atom basis) compared with ketoconazole when determined using the published approaches (52). This suggests that smaller molecules could also be effective antagonists and with optimal protein–ligand interactions. It is important to consider that antagonism of hPXR or other NRs could occur via interactions with other proteins that interact with hPXR or at other surface sites beyond those currently known. Novel hPXR antagonists provide a possible small molecule intervention to control drug metabolism and transport by reducing the activation of these genes during therapeutic treatment.

3. Conclusions

Long range motions of the hPXR LBD (17) and potential that the ligands may bind in multiple conformations within the LBD of hPXR may be responsible for the promiscuity of the hPXR LBD. Thus developing computational models to predict the binding mode and affinity of ligands for hPXR is challenging. In this study we have summarized the utility of ligand based and structure based models to predict agonists and antagonists of hPXR. Each of these methods has their inherent advantages and limitations specially when used on flexible proteins (22). Among the ligand based methods, some of them are based on ligand alignment while others are independent of alignments. The performance of alignment-dependent methods was limited in their ability to classify activators of hPXR. In the alignment independent category, the pharmacophore based models perform well within a given class of compounds by generating pharmacophoric features that are relevant to that class of molecules. However, their predictive potential decreases when applied to a diverse set of molecules. Machine learning methods such as Bayesian modeling and Support vector machines are also alignment independent methods that use 2D fingerprints and 2D or 3D molecular descriptors to build the models. In our studies we have found that they perform much better than the other in-silico methods; however their applicability domain is dependent on the training sets and consideration of molecular similar to molecules in the training set may be important. Some of the ligand based methods cannot deal with stereoisomers (unlike pharmacophore based methods) and need large training sets for improved performance. On the other hand structure based methods have their advantages of physically docking the molecules to the binding site and sampling both the ligand and protein conformational space, however they have their inherent limitations in scoring the best conformers and are computationally expensive for very large datasets. So the best option would be to utilize a combination of ligand and structure based methods that are complimentary in their approaches to better conquer flexible and promiscuous proteins such as hPXR. The methods described here are also generally applicable to other flexible proteins.

Acknowledgments

We would like to thank all our collaborators who have contributed to our studies on PXR. SK is supported by Scientist Development Grant awarded by American Heart Association.

Appendix

Structure-based Methods. Data collection: A comprehensive hPXR dataset consisting of a variety of chemical classes, namely, androstanes, pregnanes, estratrienes, bile acids, and regular xenobiotics is available from previously published studies. (53, 54). The Environmental Protection Agency has developed a dataset comprising of everyday chemicals, pesticides and other related chemicals aptly called Toxcast dataset (Toxcast^{TM}) which is available for use by researchers ((47, 49). For all these datasets, either the direct binding data to hPXR or relative fold induction changes have been published. However, the EC_{50} cutoff value used for classifying the molecules into hPXR activators and nonactivators is not always available or does not follow a strict pattern (25, 33), SMILES strings for compounds listed in datasets or 2D structures in mol or sdf formats can be obtained from the respective publications (25). Using these as input, single low energy three-dimensional structure of molecules can be generated using Molecular Operating Environment (MOE, Chemical Computing Group, Montreal, Canada) or CORINA (Molecular Networks GmbH, Nägelsbachstr. 25, 91052 Erlangen, Germany. http://www.mol-net.de) with partial charges assigned according to the Gasteiger-Marsili scheme (55).

Molecular Descriptors. hPXR binds a variety of ligands that differ in shape, size, and chemical composition. However, analysis of well known hPXR agonists show that the interactions are dominated by hydrophobic and hydrogen bonding features of the ligands. Hence to capture these properties, molecular descriptors that represent shape, size, flexibility and hydrogen bonding, and hydrophobic properties must be chosen. These include FCFP_6 fingerprints, volume, weight, KierA1-A3, Kier1-3, number of rotatable bonds, number of rings and KierFlex, electrostatic features like logP, TPSA, logs of lip_don, lip_acc, number of N atoms, and number of O atoms. The values for these specific molecular descriptors can be derived from MOE. In addition to analyze the specific role of shape and electrostatics, specific shape based descriptors such as shape signatures can be derived (56).

Molecular Docking. Five crystal structures of hPXR cocrystalized with a variety of ligands are available in the protein databank (PDB) under the codes 1 M13, resolution 2.00 Å (36), 1SKX, resolution 2.80 Å (38), 2O9I, resolution 2.80 Å (37), 1NRL, resolution 2.00 Å (36), and 2QNV, resolution 2.80 Å (37). In addition, another structure in complex with 17-β estradiol that is yet to be deposited in PDB was also used for docking studies (57). In all cases, the protein structure is first prepared by removing the crystal structure ligand and adding hydrogen atoms to the amino acids and the resulting structures are energy minimized to remove any steric contacts. All amino acids within 6 Å of the cocrystallized ligand are generally chosen as being part of the binding site. The docking program GOLD (ver 4) (58) or FlexX could be used for docking. GOLD uses a genetic algorithm to explore the various conformations of ligands and flexible receptor side chains in the binding pocket. In our studies, we have chosen to perform 20 independent docking runs for each ligand to sample the ligand and protein conformational space. The resulting docked complexes can be scored using GoldScore and ChemScore.

Scoring functions. One of the bottlenecks in docking studies is the scoring functions. Although most docking programs are capable of sampling the ligand in the binding pocket and generating solutions, the scoring functions are not sensitive enough to identify the truly best docked solutions. This is because, most scoring functions are empirical energy based schemes as shown below:

$$\Delta G_{\text{bind}} = \Delta G_{\text{solvent}} + \Delta G_{\text{conf}} + \Delta G_{\text{int}} + \Delta G_{\text{rot}} + \Delta G_{\text{t}}/r + \Delta G_{\text{vib}}$$

Where ΔG_{bind} is the binding free energy, $\Delta G_{\text{solvent}}$ is the penalty for desolvation, ΔG_{int} is the internal energy, ΔG_{rot} represents the energy contribution for bond rotation, and ΔG_{vib} represents the energy contribution for vibration component. However, the terms for solvation is approximated to a best fit model and the terms for entropy are most likely dropped from the energy equation due to the complexity involved in computing the entropy factors. Thus, in practice no single scoring function can work for every target and has to be customized to suit the needs of the target. In case of hPXR, given that the binding site is very promiscuous and binds a variety of ligands, developing a scoring scheme is challenging. We have used a number of methods to derive consensus scoring schemes including contact score, shape based scores, and molecular descriptor based schemes.

1. *Contact scoring scheme.* The docked receptor–ligand complexes were scored using a contact based scoring function. Accordingly, an in-house program was used to scan the docked complexes for contacts between the ligand and protein atoms (56). These contacts were scored based on a weighting scheme that was derived from the nature of interaction between the ligands cocrystallized with hhPXR. For example, hyperforin has hydrogen bond interactions with residues Gln285, His407, and Ser247 of the hhPXR protein in the crystal structure (PDB ID:1 M13) (Fig. 3). Thus the contact scoring function weighted all those docked protein–ligand complexes that featured the hydrogen bonding between the ligands and these three residues, higher than the rest of the interactions. Similarly, other nonbonded interactions were weighted based on the interactions of the ligands in the hPXR crystal structures. All interaction scores were then summed and normalized against all crystal structures. A consensus scoring scheme was developed for final classification based on the following rule: Only those compounds that had at least half the value of the highest GoldScore and a nonzero contact score were assigned as activators and the rest of the molecules were classified as nonactivators.
2. *Shape based scoring scheme.* In this scheme, the ligands were compared with the hPXR ligands from the five crystal structures for their shape based similarities using two different approaches. The first was based on the 2D similarity encoded

in MDL public fingerprint keys calculation using Discovery Studio 2.0 (Accelrys, San Diego, CA). The Tanimoto coefficient was used as the metric to compare the molecular fingerprints. The coefficients varied between 0 and 1, where 0 meant maximally dissimilar and 1 coded for maximally similar. The Tanimoto coefficient between fingerprints X and Y has been defined to be: [number of features in intersect (A,B)]/[number of features in union(A,B)], where A and B are two compounds.

In an approach, the 3D shapes of the molecules from the combined dataset were compared with the shapes of each of the four crystal structure ligands. This was achieved by comparing their corresponding 1D Shape Signatures and a dissimilarity score was computed for each ligand pair. The dissimilarity score was then converted to a similarity score, which was in turn used as weighting factor for the GoldScore. In all these scoring schemes the consensus score was calculated as shown below in Eq. (1).

3. *Molecular descriptor based scoring*. In this scheme, the molecular descriptors computed using MOE were used to calculate Euclidean distances from the crystal structure ligands. These Euclidean distances were used as weighting factors to GoldScore. Similarly, the values of the molecular descriptors were also used to calculate Tanimoto similarity indices (59) with reference to the cocrystal structure ligands. The values for the Tanimoto indices for each ligand in the combined set were calculated against each of the crystal structure ligands and then used as weighting factors to the GoldScores.

The weighted docking score of an active compound j with i conformations was described as

$$S_{i,j} = w_i s_{ij} \quad (1)$$

where s_{ij} was the original GoldScore for the compound i in its jth conformation and w_i is the weighting factor for compound i from either of the schemes described above.

References

1. Ruegg J, Penttinen-Damdimopoulou P, Makela S, Pongratz I, Gustafsson JA (2009) Receptors mediating toxicity and their involvement in endocrine disruption. EXS 99:289–323
2. Tabb MM, Blumberg B (2006) New modes of action for endocrine-disrupting chemicals. Mol Endocrinol 20:475–482
3. Safe S, Bandiera S, Sawyer T, Robertson L, Safe L, Parkinson A, Thomas PE, Ryan DE, Reik LM, Levin W et al (1985) PCBs: structure-function relationships and mechanism of action. Environ Health Perspect 60:47–56
4. Gustafsson JA (1995) Receptor-mediated toxicity. Toxicol Lett 82–83:465–470
5. Sewall CH, Lucier GW (1995) Receptor-mediated events and the evaluation of the Environmental Protection Agency (EPA) of dioxin risks. Mutat Res 333:111–122
6. Zollner G, Wagner M, Trauner M (2010) Nuclear receptors as drug targets in cholestasis and drug-induced hepatotoxicity. Pharmacol Ther 126:228–243

7. Lee EJ, Lean CB, Limenta LM (2009) Role of membrane transporters in the safety profile of drugs. Expert Opin Drug Metab Toxicol 5:1369–1383
8. Giguere V (1999) Orphan nuclear receptors: from gene to function. Endocr Rev 20:689–725
9. Ashby J, Houthoff E, Kennedy SJ, Stevens J, Bars R, Jekat FW, Campbell P, Van Miller J, Carpanini FM, Randall GL (1997) The challenge posed by endocrine-disrupting chemicals. Environ Health Perspect 105:164–169
10. Kelce WR, Gray LE, Wilson EM (1998) Antiandrogens as environmental endocrine disruptors. Reprod Fertil Dev 10:105–111
11. Melnick R, Lucier G, Wolfe M, Hall R, Stancel G, Prins G, Gallo M, Reuhl K, Ho SM, Brown T, Moore J, Leakey J, Haseman J, Kohn M (2002) Summary of the National Toxicology Program's report of the endocrine disruptors low-dose peer review. Environ Health Perspect 110:427–431
12. Goldman JM, Laws SC, Balchak SK, Cooper RL, Kavlock RJ (2000) Endocrine-disrupting chemicals: prepubertal exposures and effects on sexual maturation and thyroid activity in the female rat. A focus on the EDSTAC recommendations. Crit Rev Toxicol 30:135–196
13. Ekins S, Chang C, Mani S, Krasowski MD, Reschly EJ, Iyer M, Kholodovych V, Ai N, Welsh WJ, Sinz M, Swaan PW, Patel R, Bachmann K (2007) Human pregnane X receptor antagonists and agonists define molecular requirements for different binding sites. Mol Pharmacol 72:592–603
14. Ekins S, Kholodovych V, Ai N, Sinz M, Gal J, Gera L, Welsh WJ, Bachmann K, Mani S (2008) Computational discovery of novel low micromolar human pregnane X receptor antagonists. Mol Pharmacol 74:662–672
15. Biswas A, Mani S, Redinbo MR, Krasowski MD, Li H, Ekins S (2009) Elucidating the 'Jekyll and Hyde' nature of PXR: the case for discovering antagonists or allosteric antagonists. Pharm Res 26:1807–1815
16. Mnif W, Pascussi JM, Pillion A, Escande A, Bartegi A, Nicolas JC, Cavailles V, Duchesne MJ, Balaguer P (2007) Estrogens and antiestrogens activate PXR. Toxicol Lett 170:19–29
17. Teotico DG, Bischof JJ, Peng L, Kliewer SA, Redinbo MR (2008) Structural basis of human pregnane X receptor activation by the hops constituent colupulone. Mol Pharmacol 74:1512–1520
18. Wada T, Gao J, Xie W (2009) PXR and CAR in energy metabolism. Trends Endocrinol Metab 20:273–279
19. Gong H, Guo P, Zhai Y, Zhou J, Uppal H, Jarzynka MJ, Song WC, Cheng SY, Xie W (2007) Estrogen deprivation and inhibition of breast cancer growth in vivo through activation of the orphan nuclear receptor liver X receptor. Mol Endocrinol 21:1781–1790
20. Rotroff DM, Beam AL, Dix DJ, Farmer A, Freeman KM, Houck KA, Judson RS, LeCluyse EL, Martin MT, Reif DM, Ferguson SS (2010) Xenobiotic-metabolizing enzyme and transporter gene expression in primary cultures of human hepatocytes modulated by ToxCast chemicals. J Toxicol Environ Health B Crit Rev 13:329–346
21. Tirona RG, Leake BF, Podust LM, Kim RB (2004) Identification of amino acids in rat pregnane X receptor that determine species-specific activation. Mol Pharmacol 65:36–44
22. Kortagere S, Krasowski MD, Reschly EJ, Venkatesh M, Mani S, Ekins S (2010) Evaluation of computational docking to identify pregnane X receptor agonists in the ToxCast database. Environ Health Perspect 118:1412–1417
23. Ekins S, Kortagere S, Iyer M, Reschly EJ, Lill MA, Redinbo M, Krasowski MD (2009) Challenges Predicting Ligand-Receptor Interactions of Promiscuous Proteins: The Nuclear Receptor PXR. PLoS Comput Biol 5: e1000594
24. Yasuda K, Ranade A, Venkataramanan R, Strom S, Chupka J, Ekins S, Schuetz E, Bachmann K (2008) A comprehensive in vitro and in silico analysis of antibiotics that activate pregnane X receptor and induce CYP3A4 in liver and intestine. Drug Metab Dispos 36:1689–1697
25. Khandelwal A, Krasowski MD, Reschly EJ, Sinz MW, Swaan PW, Ekins S (2008) Machine learning methods and docking for predicting human pregnane X receptor activation. Chem Res Toxicol 21:1457–1467
26. Ekins S, Andreyev S, Ryabov A, Kirillov E, Rakhmatulin EA, Sorokina S, Bugrim A, Nikolskaya T (2006) A Combined approach to drug metabolism and toxicity assessment. Drug Metab Dispos 34:495–503
27. Bachmann K, Patel H, Batayneh Z, Slama J, White D, Posey J, Ekins S, Gold D, Sambucetti L (2004) PXR and the regulation of apoA1 and HDL-cholesterol in rodents. Pharmacol Res 50:237–246
28. Ekins S, Erickson JA (2002) A pharmacophore for human pregnane-X-receptor ligands. Drug Metab Dispos 30:96–99
29. Pan Y, Li L, Kim G, Ekins S, Wang H, Swaan PW (2011) Identification and validation of Novel hPXR activators amongst prescribed

drugs via ligand-based virtual screening. Drug Metab Dispos 39(2):337–44

30. Gao YD, Olson SH, Balkovec JM, Zhu Y, Royo I, Yabut J, Evers R, Tan EY, Tang W, Hartley DP, Mosley RT (2007) Attenuating pregnane X receptor (PXR) activation: a molecular modelling approach. Xenobiotica 37:124–138
31. Lemaire G, Benod C, Nahoum V, Pillon A, Boussioux AM, Guichou JF, Subra G, Pascussi JM, Bourguet W, Chavanieu A, Balaguer P (2007) Discovery of a highly active ligand of human pregnane x receptor: a case study from pharmacophore modeling and virtual screening to "in vivo" biological activity. Mol Pharmacol 72:572–581
32. Schuster D, Langer T (2005) The identification of ligand features essential for PXR activation by pharmacophore modeling. J Chem Inf Model 45:431–439
33. Huang H, Wang H, Sinz M, Zoeckler M, Staudinger J, Redinbo MR, Teotico DG, Locker J, Kalpana GV, Mani S (2007) Inhibition of drug metabolism by blocking the activation of nuclear receptors by ketoconazole. Oncogene 26:258–268
34. Lill MA, Dobler M, Vedani A (2005) In silico prediction of receptor-mediated environmental toxic phenomena-application to endocrine disruption. SAR QSAR Environ Res 16:149–169
35. Lin YS, Yasuda K, Assem M, Cline C, Barber J, Li CW, Kholodovych V, Ai N, Chen JD, Welsh WJ, Ekins S, Schuetz EG (2009) The major human pregnane X receptor (PXR) splice variant, PXR.2, exhibits significantly diminished ligand-activated transcriptional regulation. Drug Metab Dispos 37:1295–1304
36. Watkins RE, Davis-Searles PR, Lambert MH, Redinbo MR (2003) Coactivator binding promotes the specific interaction between ligand and the pregnane X receptor. J Mol Biol 331:815–828
37. Xue Y, Chao E, Zuercher WJ, Willson TM, Collins JL, Redinbo MR (2007) Crystal structure of the PXR-T1317 complex provides a scaffold to examine the potential for receptor antagonism. Bioorg Med Chem 15:2156–2166
38. Chrencik JE, Orans J, Moore LB, Xue Y, Peng L, Collins JL, Wisely GB, Lambert MH, Kliewer SA, Redinbo MR (2005) Structural disorder in the complex of human pregnane X receptor and the macrolide antibiotic rifampicin. Mol Endocrinol 19:1125–1134
39. Schuster D, Laggner C, Steindl TM, Palusczak A, Hartmann RW, Langer T (2006) Pharmacophore modeling and in silico screening for new P450 19 (aromatase) inhibitors. J Chem Inf Model 46:1301–1311
40. Khandelwal, A., Krasowski, M. D., Reschly, E. J., Sinz, M., Swaan, P. W., and Ekins, S. (2007) A Comparative Analysis of Quantitative Structure Activity Relationship Methods and Docking For Human Pregnane X Receptor Activation, *submitted.*
41. Jacobs MN (2004) In silico tools to aid risk assessment of endocrine disrupting chemicals. Toxicology 205:43–53
42. Hassan M, Brown RD, Varma-O'brien S, Rogers D (2006) Cheminformatics analysis and learning in a data pipelining environment. Mol Divers 10:283–299
43. Ghose AK, Viswanadhan VN, Wendoloski JJ (1998) Prediction of hydrophobic (lipophilic) properties of small organic molecules using fragmental methods: an analysis of ALOGP and CLOGP methods. J Phys Chem 102:3762–3772
44. Watkins RE, Wisely GB, Moore LB, Collins JL, Lambert MH, Williams SP, Willson TM, Kliewer SA, Redinbo MR (2001) The human nuclear xenobiotic receptor PXR: structural determinants of directed promiscuity. Science 292:2329–2333
45. Kortagere S, Chekmarev D, Welsh WJ, Ekins S (2009) Hybrid scoring and classification approaches to predict human pregnane X receptor activators. Pharm Res 26:1001–1011
46. Ekins S, Kortagere S, Iyer M, Reschly EJ, Lill MA, Redinbo MR, Krasowski MD (2009) Challenges predicting ligand-receptor interactions of promiscuous proteins: the nuclear receptor PXR. PLoS Comput Biol 5:e1000594
47. Judson RS, Houck KA, Kavlock RJ, Knudsen TB, Martin MT, Mortensen HM, Reif DM, Rotroff DM, Shah I, Richard AM, Dix DJ (2010) In vitro screening of environmental chemicals for targeted testing prioritization: the ToxCast project. Environ Health Perspect 118:485–492
48. Martin MT, Dix DJ, Judson RS, Kavlock RJ, Reif DM, Richard AM, Rotroff DM, Romanov S, Medvedev A, Poltoratskaya N, Gambarian M, Moeser M, Makarov SS, Houck KA (2010) Impact of environmental chemicals on key transcription regulators and correlation to toxicity end points within EPA's ToxCast program. Chem Res Toxicol 23:578–590
49. Knudsen TB, Martin MT, Kavlock RJ, Judson RS, Dix DJ, Singh AV (2009) Profiling the activity of environmental chemicals in prenatal developmental toxicity studies using the U.S. EPA's ToxRefDB. Reprod Toxicol 28:209–219
50. Mani S, Huang H, Sundarababu S, Liu W, Kalpana G, Smith AB, Horwitz SB (2005) Activation of the steroid and xenobiotic

receptor (human pregnane X receptor) by nontaxane microtubule-stabilizing agents. Clin Cancer Res 11:6359–6369

51. Chen Y, Tang Y, Wang MT, Zeng S, Nie D (2007) Human pregnane X receptor and resistance to chemotherapy in prostate cancer. Cancer Res 67:10361–10367
52. Reynolds CH, Tounge BA, Bembenek SD (2008) Ligand binding efficiency: trends, physical basis, and implications. J Med Chem 51:2432–2438
53. Ekins S, Reschly EJ, Hagey LR, Krasowski MD (2008) Evolution of pharmacologic specificity in the pregnane X receptor. BMC Evol Biol 8:103
54. Krasowski MD, Yasuda K, Hagey LR, Schuetz EG (2005) Evolution of the pregnane x receptor: adaptation to cross-species differences in biliary bile salts. Mol Endocrinol 19:1720–1739
55. Gasteiger JAMM (1980) Iterative partial equalization of orbital electronegativity—a rapid access to atomic charges. Tetrahedron 36:10
56. Kortagere S, Welsh WJ (2006) Development and application of hybrid structure based method for efficient screening of ligands binding to G-protein coupled receptors. J Comput Aided Mol Des 20:789–802
57. Xue Y, Moore LB, Orans J, Peng L, Bencharit S, Kliewer SA, Redinbo MR (2007) Crystal structure of the pregnane X receptor-estradiol complex provides insights into endobiotic recognition. Mol Endocrinol 21:1028–1038
58. Jones G, Willett P, Glen RC, Leach AR, Taylor R (1997) Development and validation of a genetic algorithm for flexible docking. J Mol Biol 267:727–748
59. Kogej T, Engkvist O, Blomberg N, Muresan S (2006) Multifingerprint based similarity searches for targeted class compound selection. J Chem Inf Model 46:1201–1213

Chapter 16

Non-compartmental Analysis

Johan Gabrielsson and Daniel Weiner

Abstract

When analyzing pharmacokinetic data, one generally employs either model fitting using nonlinear regression analysis or non-compartmental analysis techniques (NCA). The method one actually employs depends on what is required from the analysis. If the primary requirement is to determine the degree of exposure following administration of a drug (such as AUC), and perhaps the drug's associated pharmacokinetic parameters, such as clearance, elimination half-life, T_{max}, C_{max}, etc., then NCA is generally the preferred methodology to use in that it requires fewer assumptions than model-based approaches. In this chapter we cover NCA methodologies, which utilize application of the trapezoidal rule for measurements of the area under the plasma concentration–time curve. This method, which generally applies to first-order (linear) models (although it is often used to assess if a drug's pharmacokinetics are nonlinear when several dose levels are administered), has few underlying assumptions and can readily be automated.

In addition, because sparse data sampling methods are often utilized in toxicokinetic (TK) studies, NCA methodology appropriate for sparse data is also discussed.

Key words: Non-compartmental, NCA, AUC, Toxicokinetic, TK, λ_z

1. Non-compartmental Analysis

1.1. Non-compartmental Versus Regression Analysis

Most current approaches to characterize a drug's kinetics involve non-compartmental analysis, denoted *NCA*, and nonlinear regression analysis (1). The advantages of the regression analysis approach are the disadvantages of the non-compartmental approach and vice versa. *NCA* does not require the assumption of a specific compartmental model for either drug or metabolite. The method used involves application of the trapezoidal rule for measurements of the area under a plasma concentration–time curve. This method, which applies to first-order (linear) models, is rather assumption free and can readily be automated. Figure 1 gives a schematic picture of the *NCA* and nonlinear regression approaches. As can be seen, *NCA* deals with sums of areas whereas regression

Brad Reisfeld and Arthur N. Mayeno (eds.), *Computational Toxicology: Volume I*, Methods in Molecular Biology, vol. 929,
DOI 10.1007/978-1-62703-050-2_16, © Springer Science+Business Media, LLC 2012

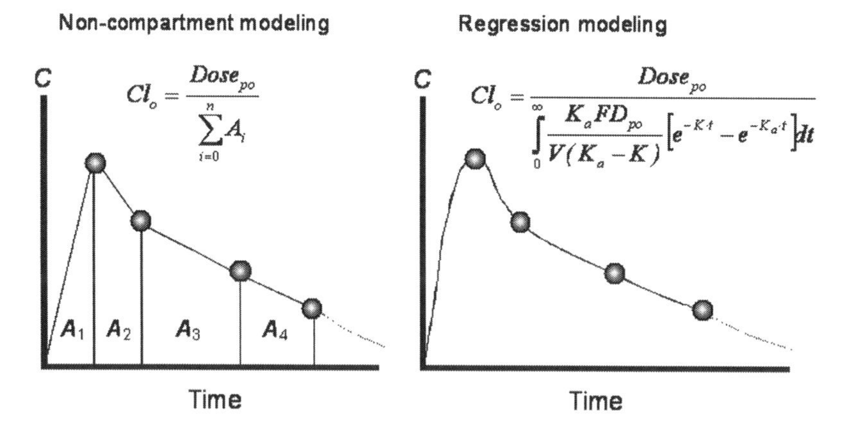

Fig. 1. Comparison of NCA (*left*) and nonlinear regression modeling (*right*). K_a, K, and V in the *right-hand panel* indicate the model parameters to be estimated by regressing the model to data.

modeling uses a function with regression parameters. Both methods are applied for the characterization of the kinetics of a compound.

The time course of drug concentration in plasma can usually be regarded as a statistical distribution curve. The area under a plot of the plasma concentration versus time curve is referred to as the area under the zero moment curve *AUC*, and the area under the product of the concentration and time versus time curve is then called the area under the first moment curve *AUMC*. Only the areas of the zero and first moments are generally used in pharmacokinetic analysis, because the higher moments are prone to an unacceptable level of computational error.

This section focuses on *NCA* with regard to computational methods, strategies for estimation of λ_z, pertinent pharmacokinetic estimates, issues related to steady state, and how to tackle situations where $t_{1/2}$ is much less than input time.

1.2. Computational Methods: Linear Trapezoidal Rule

The areas can either be calculated by means of the *linear trapezoidal rule* or by the *log-linear trapezoidal rule*. The total area is then measured by summing the incremental area of each trapezoid (Fig. 2).

The magnitude of the error associated with the estimated area depends on the width of the trapezoid and the curvature of the *true* profile. This is due to the fact that the linear trapezoidal rule overestimates the area during the descending phase assuming elimination is *first-order*, and underestimates the area during the ascending part of the curve (Fig. 3). This over/underestimation error will be more pronounced if the sampling interval Δt is large in relation to the half-life.

Using the linear trapezoidal method for calculation of the area under the zero moment curve *AUC* from 0 to time t_n, we have

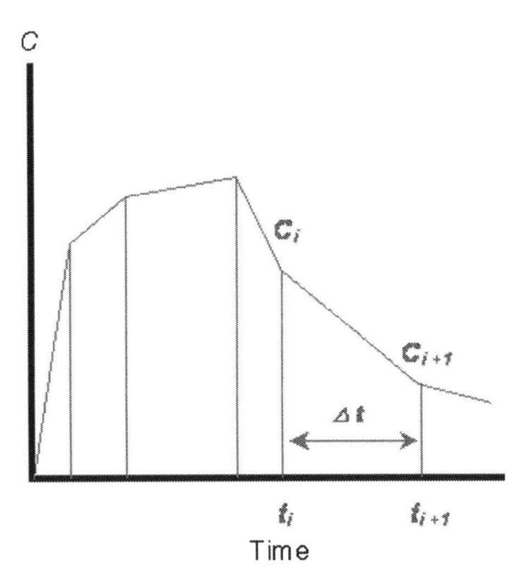

Fig. 2. Graphical presentation of the linear trapezoidal rule. $AUC_{t_i-t_{(i+1)}}$ is the area between t_i and t_{i+1}. C_i and C_{i+1} are the corresponding plasma concentrations, and Δt is the time interval. Note that Δt may differ for different trapeziums.

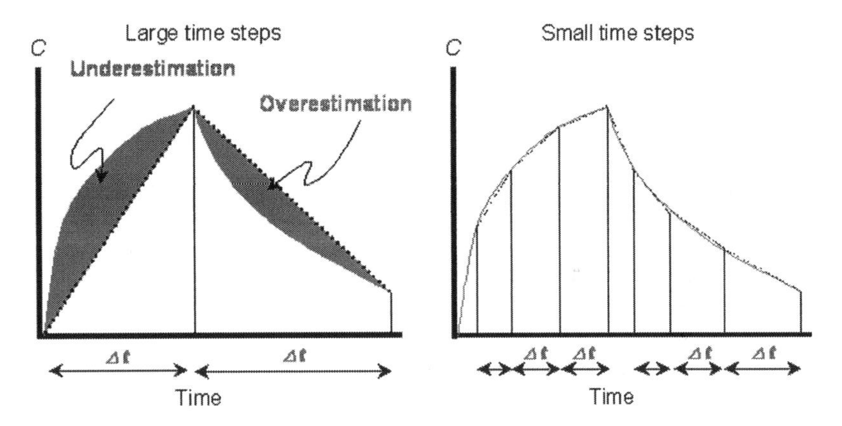

Fig. 3. Concentration versus time during and after a constant rate infusion. The *shaded area* represents underestimation of the area during ascending concentrations and overestimation of the area during descending concentrations. By decreasing the time step (Δt) between observations, this under- or overestimation of the area is minimized.

$$AUC_0^{t_{\text{last}}} = \sum_{i=1}^{n} \frac{C_i + C_{i+1}}{2} \cdot \Delta t, \tag{1}$$

where $\Delta t = t_{i+1} - t_i$ and t_{last} denotes the time of the last measurable concentration. Unless one has sampled long enough in time so that concentrations are negligible, the AUC as defined above will underestimate the *true* AUC. Therefore it may be necessary to extrapolate the curve out to t equal to infinity (∞). The extrapolated area under the zero moment curve from the last sampling time to infinity $AUC_{t_{\text{last}-\infty}}$ is calculated as

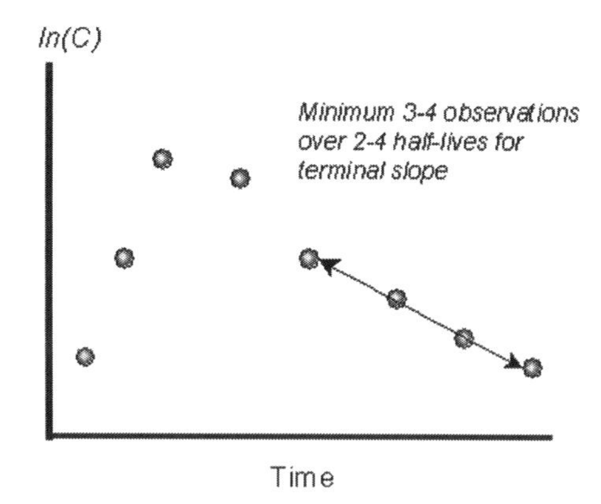

Fig. 4. Semilog plot demonstrating the estimation of λ_z. The terminal data points are fit by log-linear regression to estimate the slope.

$$AUC_{t_{\mathrm{last}}}^{\infty} = \int_{t_{\mathrm{last}}}^{\infty} C_{\mathrm{last}} \cdot \mathrm{e}^{-\lambda_z(t-t_{\mathrm{last}})} \mathrm{d}t = C_{\mathrm{last}} \left[\frac{\mathrm{e}^{-\lambda_z(t-t_{\mathrm{last}})}}{-\lambda_z} \right]_{t_{\mathrm{last}}}^{\infty}$$

$$= C_{\mathrm{last}} \left[0 - \frac{1}{-\lambda_z} \right] = \frac{C_{\mathrm{last}}}{\lambda_z}, \tag{2}$$

where C_{last} and λ_z are the last measurable nonzero plasma concentration and the terminal slope on a $\log_e$ scale, respectively. One may also use the predicted concentration at t_{last} if the observed concentration deviates from the terminal regression line (Fig. 4).

The λ_z parameter is graphically obtained from the terminal slope of the semilogarithmic concentration–time curve as shown in Fig. 4, with a minimum of 3–4 observations being required for accurate estimation. The Y axis $\ln(C)$ denotes the natural logarithm ($\log_e$) of the plasma concentration C.

The linear trapezoidal method for calculation of the area under the first moment curve $AUMC$ from 0 to time t_{last} is obtained from

$$AUMC_0^{t_{\mathrm{last}}} = \sum_{i=1}^{n} \frac{t_i \cdot C_i + t_{i+1} \cdot C_{i+1}}{2} \cdot \Delta t. \tag{3}$$

Remembering that $\int x \cdot \mathrm{e}^{-a \cdot x} \mathrm{d}x = -\frac{x \cdot \mathrm{e}^{-a \cdot x}}{a} - \frac{\mathrm{e}^{-a \cdot x}}{a^2}$, the corresponding area under the first moment curve from time t_{last} to infinity $AUMC_{t_{\mathrm{last}-\infty}}$ is computed as

$$AUMC_{t_{\mathrm{last}}}^{\infty} = \int_{t_{\mathrm{last}}}^{\infty} t \cdot C \mathrm{d}t = \int_{t_{\mathrm{last}}}^{\infty} t \cdot C_{\mathrm{last}} \mathrm{e}^{-\lambda_z(t-t_{\mathrm{last}})} \mathrm{d}t$$

$$= C_{\mathrm{last}} \cdot \mathrm{e}^{\lambda_z t_{\mathrm{last}}} \left[\frac{t \cdot \mathrm{e}^{-\lambda_z t_{\mathrm{last}}}}{-\lambda_z} + \frac{e^{-\lambda_z t_{\mathrm{last}}}}{-\lambda_z^2} \right]_{t_{\mathrm{last}}}^{\infty}$$

$$= \frac{C_{\mathrm{last}} \cdot t_{\mathrm{last}}}{\lambda_z} + \frac{C_{\mathrm{last}}}{\lambda_z^2}. \tag{4}$$

1.3. Computational Methods: Log-Linear Trapezoidal Rule

An alternative procedure that has been proposed is the log-linear trapezoidal rule. The underlying assumption is that the plasma concentrations decline mono-exponentially between two measured concentrations. However, this method applies only for descending data and fails when $C_i = 0$ or $C_{i+1} = C_i$. In these instances one would revert to the linear trapezoidal rule. The principal difference between the linear and the log-linear trapezoidal method is demonstrated in Fig. 5.

Remember that when the concentrations decline exponentially

$$C_{i+1} = C_i \cdot \mathrm{e}^{-K(t_{i+1}-t_i)} = C_i \cdot \mathrm{e}^{-K\Delta t}, \tag{5}$$

where $t_{i+1} - t_i$ is the time step Δt between two observations and K is the elimination rate constant for a one-compartment system. Otherwise, λ_z should be used as the slope. The above expression when rearranged gives the elimination rate constant K:

$$K = \frac{\ln(C_i/C_{i+1})}{\Delta t}. \tag{6}$$

The AUC within the time interval Δt is the difference between the concentrations divided by the slope K:

$$AUC_i^{i+1} = \frac{C_i - C_{i+1}}{K} = \frac{C_i - C_{i+1}}{\ln(C_i/C_{i+1})} \cdot \Delta t. \tag{7}$$

Using the log-linear trapezoidal method from time zero to t_n

$$AUC_0^{t_n} = \sum_{i=1}^{n} \frac{C_i - C_{i+1}}{\ln(C_i/C_{i+1})} \cdot \Delta t, \tag{8}$$

while the corresponding equation for $AUMC$ from time zero to t_n with this method yields

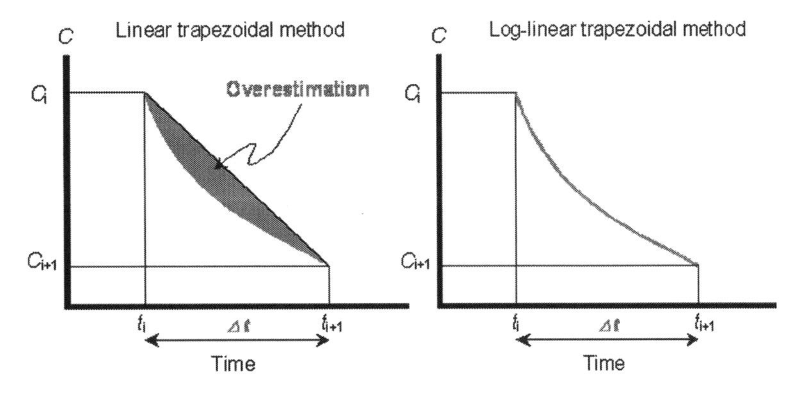

Fig. 5. The principal difference between the linear (*left*) and the log-linear (*right*) trapezoidal methods. The *shaded region* represents the over-predicted area with the linear trapezoidal rule. Note that the log-linear approximation is only true if the decay is truly mono-exponential between t_i and t_{i+1}.

$$AUMC_0^{t_n} = \sum_{i=1}^{n} \frac{t_i \cdot C_i - t_{i+1} \cdot C_{i+1}}{\ln(C_i/C_{i+1})} \cdot \Delta t - \frac{C_{i+1} - C_i}{[\ln(C_i/C_{i+1})]^2} \cdot \Delta t^2. \quad (9)$$

The extrapolated area under the zero moment curve from the last sampling time to infinity $AUC_{t_{\text{last-}\infty}}$ is calculated as

$$AUC_{t_{\text{last}}}^{\infty} = \frac{C_{\text{last}}}{\lambda_{\text{z}}}, \quad (10)$$

where C_{last} and λ_{z} are as defined earlier. The corresponding area under the first moment curve from time zero to infinity $AUMC_{t_{\text{last-}\infty}}$ is

$$AUMC_{t_{\text{last}}}^{\infty} = \frac{C_{\text{last}} \cdot t_{\text{last}}}{\lambda_{\text{z}}} + \frac{C_{\text{last}}}{\lambda_{\text{z}}^2}. \quad (11)$$

As previously pointed out, the linear trapezoidal method gives approximate estimates of *AUC* during both the ascending and descending parts of the concentration–time curve, although the bias is usually negligible for the upswing. The log-linear trapezoidal method may also give somewhat biased results, though to a lesser extent. Some people argue that the log-linear trapezoidal method may therefore be preferable for drugs with long half-lives relative to the sampling interval. From a practical point of view this still needs to be proven. However, our own experience is that the difference between the two methods is negligible as long as a reasonable sampling design has been used. We generally use a mixture of the two methods, which means that the linear trapezoidal method is applied for increasing and equal concentrations, e.g., at the peak or a plateau, and the log-linear trapezoidal method for decreasing concentrations. This is demonstrated in Fig. 6.

Note that *NCA* is often used in crossover studies comparing two formulations and 12–36 subjects. Thus, since the error associated

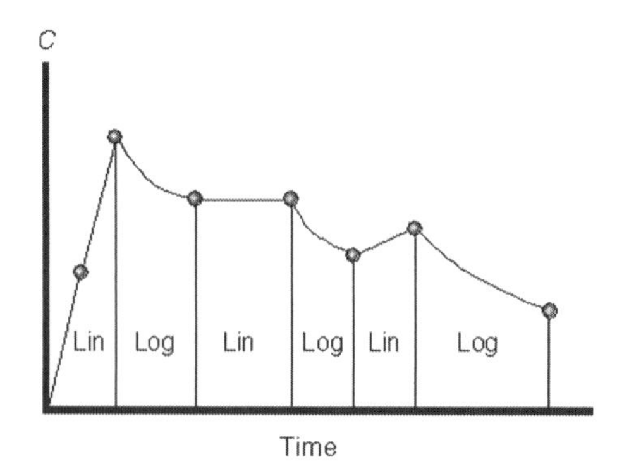

Fig. 6. NCA using a combination of the linear and log-linear trapezoidal methods. The linear method is used for consecutively increasing or consecutively equal concentrations. The log-linear method is used for decreasing concentrations.

with an individual patient's AUC is generally small, the (average) error associated with the average AUC for a formulation will generally be negligible, regardless of the method used. The choice of method is thus up to the discretion of the modeler, as long as one can explain why a particular method provides a more accurate estimate of AUC.

The linear trapezoidal method will work excellently in situations of zero-order kinetics since plasma concentrations decline linearly with time. Hence, even large sampling intervals will be acceptable. The log-linear trapezoidal rule may in some instances be more optimal within the first-order concentration range. The linear method will then overpredict the areas particularly when half-life is short relative to the sampling interval.

Direct integration of the function for the drug's kinetics in plasma is discussed under the introductory section on mono- and multi-exponential models and will therefore not be further elaborated here.

1.4. Strategies for Estimation of λ_z

When estimating λ_z, we recommend that data from each individual are first plotted in a semilog diagram. Ideally, to obtain a reliable estimate of the terminal slope, 3–4 half-lives would need to have elapsed. However, sometimes this is not possible. A minimum requirement is then to have 3–4 observations for the terminal slope (Fig. 7). By means of log-linear regression of those observations, the estimate of λ_z is obtained. This is then used for calculation of the extrapolated area as shown below:

$$AUC_{t_{\text{last}}}^{\infty}(\text{observed}) = \frac{C_{\text{last}}}{\lambda_z} \tag{12}$$

or

$$AUC_{t_{\text{last}}}^{\infty}(\text{predicted}) = \frac{\hat{C}_{\text{last}}}{\lambda_z}. \tag{13}$$

In Fig. 8 the last observed concentration C_{obs} deviates somewhat from the regression line. The extrapolated area, if based on

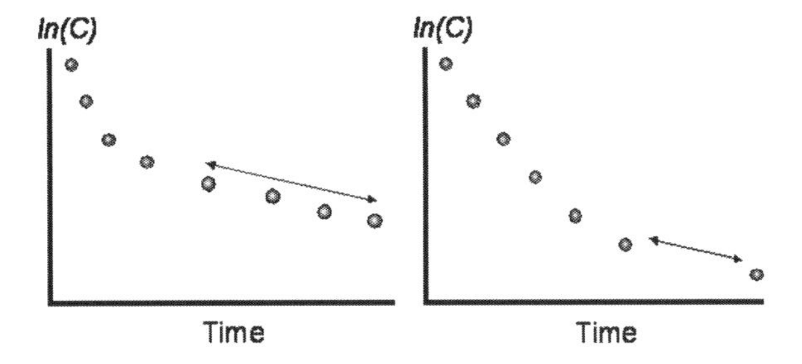

Fig. 7. The ideal situation (*left*) for estimation of the terminal slope λ_z. Another and perhaps more commonly encountered situation (*right*) is where one only has an indication of an additional slope.

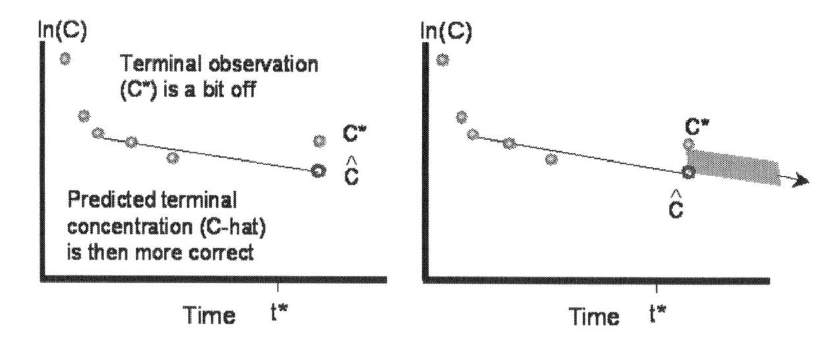

Fig. 8. Impact on the extrapolated area of using observed terminal concentration versus predicted concentration. The *shaded area* from t_{last} to infinity symbolizes the overestimation that would result. Note that if the observed terminal concentration lies below the predicted terminal concentration, then the extrapolated area would be underestimated. The *open circle* is the predicted concentration at t_{last}. The last observation is not included in the regression.

C_{obs}, would be disproportionately large as compared to the area based on the predicted concentration.

The total area is obtained by summing the individual areas obtained by means of the trapezoidal rule to the last time (t_{last}), and adding the extrapolated area according to

$$AUC_{total} = AUC_0^{\infty} = AUC_0^{t_{last}} + AUC_{t_{last}}^{\infty}. \tag{14}$$

The fraction of AUC_{extr} to $AUC_{0\text{-}\infty}$ is calculated as

$$\%\,\text{extrapolated area} = \frac{AUC_{t_{last}}^{\infty}}{AUC_0^{\infty}} \cdot 100. \tag{15}$$

The extrapolated area should ideally be as small as possible in comparison to the total area. We believe that $AUC_{t_{last\text{-}\infty}}$ should not exceed 20–25% of AUC_{total}, unless it is only used as a preliminary estimate for further study refinement.

1.5. Pertinent Pharmacokinetic Estimates

Moment analysis has been widely used in recent years as a noncompartmental approach to the estimation of clearance Cl, mean residence time MRT, steady-state volume of distribution V_{ss}, and volume of distribution during the terminal phase V_z (also called $V_{d\beta}$ for a bi-exponential system). A general treatment for the aforementioned parameters has been presented, which includes the possibility of input/exit from any compartment in a mammillary model (2, 3). This approach also defines *exit site-dependent* and *exit site-independent* parameters. We will, however, assume in the following examples that input/output occurs to the central compartment. Assuming a simple case with a one-compartment bolus system, the shape of the *concentration–time* and *t·concentration–time* profiles will take the form depicted in Fig. 9.

The extrapolated area from the last sample at t_{last} to infinity is in this case small. However, the corresponding area under the first

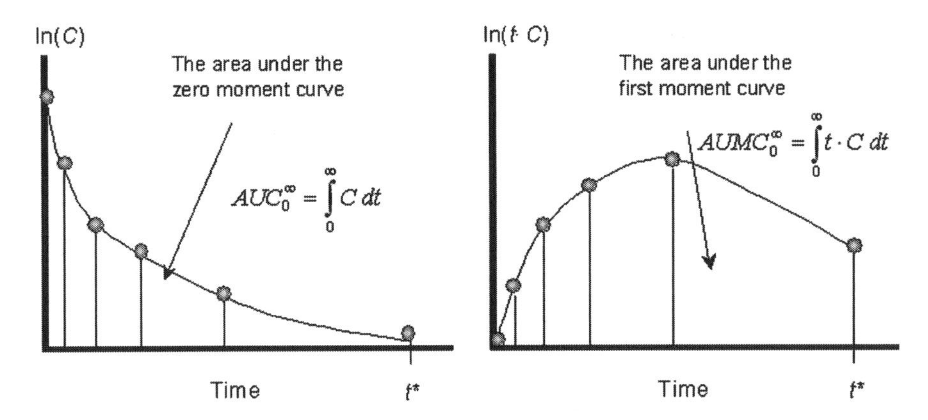

Fig. 9. Comparison of shape of area under the zero moment curve AUC and area under the first moment ($t{\cdot}C$) curve AUMC. The latter area contains usually an extensive extrapolated area as compared to AUC.

moment curve has an altogether different shape. Clearly, the extrapolated area from last sampling point to infinity will generally contribute to a much larger extent under the first moment curve as compared to the area under the zero moment curve.

Pharmacokinetics has moved almost completely from parameterizing elimination in terms of rate constants, with the more physiologically relevant use of clearance now being widely recognized. To put even more focus on clearance, Holford suggested that AUC no longer be used as a pharmacokinetic parameter. Clearance Cl or clearance over bioavailability Cl/F also denoted that Cl_0 is easily computed from AUC and dose, and Cl and CL/F can immediately be interpreted in a physiological context. On the other hand, AUC can be viewed as a parameter that confounds clearance and dose, and that has no intrinsic merit. While we agree with those ideas, AUC is still useful as a measure of exposure in toxicological studies and when dose is unknown.

Clearance is calculated from the dose and the area under the zero moment curve:

$$Cl = \frac{D_{\mathrm{iv}}}{AUC_0^\infty}. \tag{16}$$

Oral clearance Cl_0 or Cl/F is calculated from the oral dose and the area under the zero moment curve:

$$Cl_0 = \frac{Cl}{F} = \frac{D_{\mathrm{po}}}{AUC_0^\infty}. \tag{17}$$

Using the areas obtained from systemic, e.g., intravenous and extravascular, e.g., oral, dosing, the bioavailability F is calculated, after dose-normalization, according to

$$F = \frac{AUC_{\mathrm{ev}}}{AUC_{\mathrm{iv}}} \cdot \frac{D_{\mathrm{iv}}}{D_{\mathrm{ev}}}, \tag{18}$$

where AUC_{ev} and AUC_{iv} denote area under the extravascular and intravenous concentration–time profiles, respectively. D_{ev} and D_{ev} are the respective extravascular and intravenous doses.

If the drug is given at a constant rate over a period of T_{inf}, then one also needs to adjust *MRT* for the infusion time by means of subtracting $T_{inf}/2$ (infusion time/2) as follows:

$$MRT = \frac{AUMC_0^\infty}{AUC_0^\infty} - \frac{T_{inf}}{2}. \quad (19)$$

$T_{inf}/2$ originates from the average time a molecule stays in the infusion set (e.g., syringe, catheter, line). Half of the dose is infused when the piston has traveled half of the intended distance. $T_{inf}/2$ is the mean input time, *MIT*. Similarly for *first-order* input,

$$MRT = \frac{AUMC_0^\infty}{AUC_0^\infty} - \frac{1}{K_a}. \quad (20)$$

Remember that K_a is the apparent *first-order* absorption rate constant derived from plasma data. This parameter may also contain processes parallel to the true absorption step of drug in the gastrointestinal tract, e.g., chemical degradation (k_d). Consequently, the mean absorption time, *MAT*, is the sum of several processes including absorption and chemical degradation:

$$MAT = \frac{1}{K_{a(apparent)}} = \frac{1}{K_{a(true)} + k_d}. \quad (21)$$

The *MRT* of the central compartment *MRT*(1) is the sum of the inverse of the initial α and terminal β slopes corrected for the inverse of the sum of the exit rate constants from the peripheral compartment:

$$MRT_{iv}(1) = \frac{1}{\alpha} + \frac{1}{\beta} - \frac{1}{E_2}. \quad (22)$$

Assuming that there is only one exit rate constant from the peripheral compartment, which then is k_{21}, the MRT_{iv} is

$$MRT_{iv} = \frac{1}{\alpha} + \frac{1}{\beta} - \frac{1}{k_{21}}. \quad (23)$$

The observed *MRT* after extravascular dosing becomes

$$\frac{AUMC_{0\,measured}^\infty}{AUC_{0\,measured}^\infty} = MRT + MIT, \quad (24)$$

which is the sum of the true *MRT* and *MIT*. *MIT* can also be obtained from the input function according to Eq. 25 below:

$$MIT = \frac{\int_0^\infty \text{input function} \cdot t\,dt}{\int_0^\infty \text{input function}\,dt} = \frac{\int_0^\infty \text{input function} \cdot t\,dt}{F \cdot \text{Dose}}, \quad (25)$$

provided the input function is known,

$$A_{\text{gut}} = F \cdot D_{\text{po}} \cdot \mathrm{e}^{-K_a \cdot t}, \tag{26}$$

and MIT can be derived:

$$MIT = \frac{\int_0^{\infty} F \cdot D_{\text{po}} \cdot \mathrm{e}^{-K_a \cdot t} \cdot t \, \mathrm{d}t}{\int_0^{\infty} F \cdot D_{\text{po}} \cdot \mathrm{e}^{-K_a \cdot t} \, \mathrm{d}t} = \frac{F \cdot D_{\text{po}}/K_a^2}{F \cdot D_{\text{po}}/K_a} = \frac{1}{K_a}. \tag{27}$$

The volume of distribution at steady-state V_{ss} is computed as

$$V_{\text{ss}} = MRT \cdot Cl = \frac{AUMC_0^{\infty}}{AUC_0^{\infty}} \cdot \frac{D_{\text{iv}}}{AUC_0^{\infty}} = \frac{D_{\text{iv}} \cdot AUMC_0^{\infty}}{\left[AUC_0^{\infty}\right]^2}. \tag{28}$$

The volume of distribution during the terminal phase V_z is computed as

$$V_z = \frac{Cl}{\lambda_z} = \frac{D_{\text{iv}}}{AUC_0^{\infty}} \cdot \frac{1}{\lambda_z}. \tag{29}$$

The corresponding volume for a bi-exponential function is computed as

$$V_{d\beta} = \frac{Cl}{\beta} = \frac{D_{\text{iv}}}{AUC_0^{\infty}} \cdot \frac{1}{\beta}. \tag{30}$$

The terminal half-life $t_{1/2z}$ is readily estimated from the slope λ_z as

$$t_{1/2z} = \frac{\ln(2)}{\lambda_z}. \tag{31}$$

The half-life of the initial α-phase is

$$t_{1/2\alpha} = \frac{\ln(2)}{\alpha}. \tag{32}$$

The half-life of the terminal β-phase of a bi-exponential function is

$$t_{1/2\beta} = \frac{\ln(2)}{\beta}. \tag{33}$$

Note that the $t_{1/2z}$ parameter is referred to as $t_{1/2\beta}$ in a bi-exponential function and simply $t_{1/2}$ in a mono-exponential system.

1.6. NCA Approaches for Sparse Data

In some instances it may not be possible to obtain sufficient samples from each subject so as to completely characterize the plasma concentration–time curve. This may be due to the need to sacrifice the animal to obtain the samples, general concerns over blood loss (such as in human neonates or small rodents), or cost concerns. In these situations it is necessary to pool the data from multiple subjects to characterize the full time–plasma concentration curve. Generally these approaches are recommended only when the data are being collected from populations that do not exhibit extensive subject-to-subject variation, such as in highly inbred strains of animals.

One such approach is an extension of the NCA analysis for rich data described previously, and enables one to derive an estimated standard error (se) for AUC for sparse data (4–6). This procedure is implemented in Phoenix® WinNonlin®.

Another approach has also been proposed that involves nonlinear mixed effects modeling (also denoted as population modeling). In this instance a structural pk model is specified and fit to the data. This approach has the advantage of incorporating covariates (e.g., age, gender, body weight, etc.). That is, the ability (e.g.) to model changes in clearance as a function of age or body weight. It also has a limitation of possibly not being able to adequately identify the underlying structural model unless the sparse data can be pooled with rich data from some other cohort (7).

1.7. Suggested Reading

For further reading on basic pharmacokinetic principles, we refer the reader to Benet (8), Benet and Galeazzi (9), Gibaldi and Perrier (10), Nakashima and Benet (2, 3), Jusko (11), and Rowland and Tozer (12). Houston (13) and Pang (14) provide excellent texts on metabolite kinetics.

Benet and Galeazzi (9), Watari and Benet (15), and Nakashima and Benet (3) have elaborated on the theory of *NCA*, while Gillespie (16) discussed the pros and cons of *NCA* versus compartmental models.

References

1. Gabrielsson J, Weiner D (2006) Pharmacokinetic and pharmacodynamic data analysis: concepts and applications, 4th edn. Swedish Pharmaceutical Press, Stockholm
2. Nakashima E, Benet LZ (1988) General treatment of mean residence time, clearance and volume parameters in linear mamillary models with elimination from any compartment. J Pharmacokinet Biopharm 16:475
3. Nakashima E, Benet LZ (1989) An integrated approach to pharmacokinetic analysis for linear mammillary systems in which input and exit may occur in/from any compartment. J Pharmacokinet Biopharm 17:673
4. Bailer AJ (1988) Testing for the equality of area under the curve when using destructive measurement techniques. J Pharmacokinet Biopharm 16:303
5. Yeh C (1990) Estimation and significance tests of area under the curve derived from incomplete blood sampling. In: ASA Proceedings of the Biopharmaceutical Section, p 4
6. Nedelman JR, Jia X (1998) An extension of Satterthwaite's approximation applied to pharmacokinetics. J Biopharm Stat 8(2):317
7. Hing JP, Woolfrey SG, Wright PMC (2001) Analysis of toxicokinetic data using NONMEM: impact of quantification limit and replacement strategies for censored data. J Pharmacokinet Pharmacodyn 28(5):465
8. Benet LZ (1972) General treatment of linear mammillary models with elimination from any compartment as used in pharmacokinetics. J Pharm Sci 61:536
9. Benet LZ, Galeazzi RL (1979) Noncompartmental determination of the steady-state volume of distribution. J Pharm Sci 48:1071
10. Gibaldi M, Perrier D (1982) Pharmacokinetics. Revised and expanded, 2nd edn. Marcel Dekker Inc., New York, NY

11. Jusko WJ (1992) Guidelines for collection and analysis of pharmacokinetic data. In: Evans WE, Schentag JJ, Jusko WJ (eds) Applied pharmacokinetics: principles of therapeutic drug monitoring, 3rd edn. Applied Therapeutics, Spokane, WA
12. Rowland M, Tozer T (2010) Clinical pharmacokinetics and pharmacodynamics: concepts and applications, 4th edn. Lippincott Williams & Wilkins, Maryland
13. Houston JB (1994) Kinetics of disposition of xenobiotics and their metabolites. Drug Metab Drug Interact 6:47
14. Pang KS (1985) A review of metabolic kinetics. J Pharmacokinet Biopharm 13(6):633
15. Watari N, Benet LZ (1989) Determination of mean input time, mean residence time, and steady-state volume of distribution with multiple drug inputs. J Pharmacokinet Biopharm 17 (1):593
16. Gillespie WR (1991) Noncompartmental *versus* compartmental modeling in clinical pharmacokinetics. Clin Pharmacokinet 20:253

Chapter 17

Compartmental Modeling in the Analysis of Biological Systems

James B. Bassingthwaighte, Erik Butterworth, Bartholomew Jardine, and Gary M. Raymond

Abstract

Compartmental models are composed of sets of interconnected mixing chambers or stirred tanks. Each component of the system is considered to be homogeneous, instantly mixed, with uniform concentration. The state variables are concentrations or molar amounts of chemical species. Chemical reactions, transmembrane transport, and binding processes, determined in reality by electrochemical driving forces and constrained by thermodynamic laws, are generally treated using first-order rate equations. This fundamental simplicity makes them easy to compute since ordinary differential equations (ODEs) are readily solved numerically and often analytically. While compartmental systems have a reputation for being merely descriptive they can be developed to levels providing realistic mechanistic features through refining the kinetics. Generally, one is considering multi-compartmental systems for realistic modeling. Compartments can be used as "black" box operators without explicit internal structure, but in pharmacokinetics compartments are considered as homogeneous pools of particular solutes, with inputs and outputs defined as flows or solute fluxes, and transformations expressed as rate equations.

Descriptive models providing no explanation of mechanism are nevertheless useful in modeling of many systems. In pharmacokinetics (PK), compartmental models are in widespread use for describing the concentration–time curves of a drug concentration following administration. This gives a description of how long it remains available in the body, and is a guide to defining dosage regimens, method of delivery, and expectations for its effects. Pharmacodynamics (PD) requires more depth since it focuses on the physiological response to the drug or toxin, and therefore stimulates a demand to understand how the drug works on the biological system; having to understand drug response mechanisms then folds back on the delivery mechanism (the PK part) since PK and PD are going on simultaneously (PKPD).

Many systems have been developed over the years to aid in modeling PKPD systems. Almost all have solved only ODEs, while allowing considerable conceptual complexity in the descriptions of chemical transformations, methods of solving the equations, displaying results, and analyzing systems behavior. Systems for compartmental analysis include Simulation and Applied Mathematics, CoPasi (enzymatic reactions), Berkeley Madonna (physiological systems), XPPaut (dynamical system behavioral analysis), and a good many others. JSim, a system allowing the use of both ODEs and partial differential equations (that describe spatial distributions), is used here. It is an open source system, meaning that it is available for free and can be modified by users. It offers a set of features unique in breadth of capability that make model verification surer and easier, and produces models that can be shared on all standard computer platforms.

Brad Reisfeld and Arthur N. Mayeno (eds.), *Computational Toxicology: Volume I*, Methods in Molecular Biology, vol. 929, DOI 10.1007/978-1-62703-050-2_17, © Springer Science+Business Media, LLC 2012

Key words: Physiological and pharmacologic modeling, PKPD, Pharmacokinetics–pharmacodynamics, Compartmental systems, Systems biology, Physiome, JSim, CellML, SBML, Reproducible research, Unit checking, Verification, Validation, Ordinary and partial differential equations, Optimization, Confidence limits

1. Introduction

1.1. Overview of the Topic

Compartmental analysis implies the use of linear first-order differential operators as analogs for describing the kinetics of drug distribution and elimination from the body. Concentrations are measured in accessible fluids, usually the plasma, and the concentration–time curve is used to provide a measure of how long the drug concentration remains at a therapeutic level. This is the basis of *pharmacokinetics* (*PK*). The influences on efficacy and utility are considered by the term ADME, administration, distribution, metabolism, and elimination. Drugs are given in chemically significant amounts; they bind to enzymes, channels, receptors or transporters, changing reaction rates and fluxes in concentration-dependent fashion. The effects on the biological system is termed *pharmacodynamics* (*PD*). Precise mathematical statements about the kinetics and the body's responses comprise the combination *pharmacokinetics–pharmacodynamics* (*PKPD*).

Compartmental analysis had its historical start with the use of tracers. Tracer-labeled compounds were used in order to determine kinetics when the drug concentrations were too low to be measured chemically. Radioactive tracers were given in such low concentrations relative to those of native non-tracer mother substances that the kinetics were in fact linear. Consider a reaction rate, $k(C)$, that is dependent upon the concentration of the mother substance of concentration $C(t)$:

$$\text{Flux of mother substance} = k(C) \times C(t). \quad (1)$$

When tracer of concentration C' is added to the system, then

$$\text{Total flux, mother, and tracer} = k(C + C') \times [C(t) + C'(t)]. \quad (2)$$

When $C' << C$, then the rate constant is determined solely by C, as $k(C + C') \approx k(C)$, and the rate constant is independent of the tracer concentration:

$$\text{Tracer flux of } C'(t) = k(C) \times C'(t), \quad (3)$$

where the flux is first order in C' when the background non-tracer mother substance concentration is constant. When only the tracer is changing concentration, the $k(C)$ is constant and the system is first order and linear. In general then, one can look upon *tracers* in compartmental systems as being linear, first-order systems, though nowadays they can go far beyond that. The originators and later proponents of compartmental analysis (Berman (1); Jacquez (2, 3);

Cobelli et al. (4)) used this simplification, but were always aware of the greater possibilities of allowing nonlinear coefficients. Berman's classic 1963 article (1) provides much more than solutions to ordinary differential equations (ODEs) for he outlines an important philosophic approach to modeling in general. Jacquez' books and many articles, and the book by Cobelli et al. (4), give detailed mathematical approaches and explicit applications. The desire to use linear kinetics was not so much to avoid solving nonlinear equations as it was to use linear algebra to solve the differential equations. A system of linear differential equations can be solved by matrix inversion and can provide the much desired analytical solutions. As we shall see below, analytical solutions are still desired, for they serve as verification that the numerical solutions produced by modern simulation systems are correct in specific reduced cases, and thereby imply that the nonlinear system solutions in that neighborhood of parameter space are also correct. But, because most biological phenomena are nonlinear, such that the rate coefficients vary with the concentrations of one or usually more solutes, temperature, and pH, we have to acknowledge right at the start that using linear compartmental systems analysis is an approximation.

Compartmental analysis was mostly descriptive, not mechanistic. It was "Black Box," not attempting to define enzymatic reactions mechanistically, but to describe the time course, "White Box" modeling is where the innards of the operational analysis attempt to describe mechanism, not just the kinetics of a relationship. Nevertheless, the descriptive level was a success; in FDA reviews quantitative descriptions are valuable, for they distinguish groups of responses and allow categorization even when they cannot provide a physiological interpretation. The plasma concentration–time data are very useful in choosing methods of administration and in defining dosage regiments.

Modern molecular biology and emerging integrative multi-scale modeling analysis likewise are changing the game. Personalized medicine is pointedly mechanistic, with cell and molecular physiology dominating in the strategies of Administration, Distribution, Metabolism, and Elimination (ADME). Fortunately the huge increase in the rates of acquisition of data, causing a demand for detailed, informative but complex simulation analysis has been more than compensated for by the increases in computational speed and in improved software facilitating modeling analysis. Most importantly, software sharing is now relatively easy, and of much increased importance since comprehensive models may take years in development. Now, highly nonlinear complex systems, spatially distributed or lumped, are handled with faster computation, and can include kinetics and detailed physiological pharmacodynamics (the PD of PKPD), which is the systematic analysis of the body's responses to the drug or toxin. Given the relevance of physiological transport processes (diffusion, flow, transmembrane exchange, binding) in both administration and distribution, a relatively new term, PBPK, physiologically based pharmacokinetics, has arisen to recognize the importance of incorporating anatomy and physiology into PK.

Drugs are toxins. It is only a matter of dosage. The struggle to distinguish acceptable toxicity from unacceptable toxicity is the central conflict, not just for cancer therapy but also in defining drug usage in general. Aspirin, ibuprofen, oxycodone, sugar, and water, all create problems when in excess. Distinguishing "Therapeutic Dose" from "Toxic Dose" depends on the drug and on the particulars of the patient (size, age, body fat level, other drugs, physiological state and past history, genetic heritage). Many substances in our environment augment the difficulties, adding other sources of specific or general toxicity.

Drugs and toxins have many common features; for example the lipid solubility that allows easy permeation of cell membranes, so desirable in drugs, is the source of the problems with inhaled toxicants. While the body has evolved a rather general system for dealing with foreign toxicants, the P450 system in the liver that handles hundreds of different chemicals, it is also good at degrading and excreting drugs as well. As a corollary, hepatotoxicity can be a problem with drugs. Likewise, renal damage can be a risk from those drugs excreted in the urine, like ibuprofen.

Computer modeling includes all phases of ADME. The method of Administration by swallowing a pill is the commonest but is not as fast as intravenous (i.v.) injection. Intravenous is the method with the best defined administrative kinetics, followed by intramuscular (i.m.). There are a host of other local injection types, slow release subdermally, suppositories, inhalation, sublingual absorbance, etc., all of which have different rates of drug delivery into the circulation and to the target. Since the exposure of the target to the drug is most often measured in terms of the AUC (Area Under the Curve of the drug's plasma concentration versus time) the differences amongst methods of delivery are important. The AUC is influenced by dilution in the circulation (part of Distribution), degradation by hydrolysis or metabolism (the M of ADME), and elimination or excretion (the E of ADME), so a pharmacokinetic model must include all of these. See earlier chapters on Fundamentals of ADME and on Modeling of Absorption.

In the neighborhood of the target (enzyme, receptor, channel, transporter, or transcription regulator) much depends on physicochemical attributes of the drug or toxin. Does it bind to plasma proteins? What is the affinity of the target proteins compared to that of other competing proteins, or DNA or membrane-bound proteins? What are the on- and off-rates of any competing binding sites? What is the drug's tissue/blood partition coefficient or its solubility in body fat? One of the standard ways of getting clues on the fate of the drug is to do whole body distribution studies on rats to see where it is deposited at a succession of points in time. The distribution sites of positron-labeled drug may be shown with high resolution in reconstructed 3D imaging using MicroPET (Positron Emission Tomography) systems. Such observations give data on Distribution and Excretion and sometimes Metabolism. Since the retention times in specific tissue locations influence, and may dominate, the AUC, this information is crucial for optimizing effect on the target while minimizing toxic side effects. Modeling analysis then ideally provides specific information on all of these aspects of the kinetics, and further provides the information essential for a critical understanding of the pharmacodynamics. A by-product of a good understanding of the PK is that it allows one to consider using combined drug therapy wherein a second drug is used in advance to protect a sensitive binding site by binding to it harmlessly and preventing the toxicant drug from binding. Variants on this theme might be combining therapy with a drug preventing the activation of a receptor whose binding site accepts the drug of interest.

1.2. Model Types, Topologies, and Equation Types

1.2.1. In Terms of Input and Output Characteristics We May Classify Compartmental Systems as Closed or Open

1. *Closed system*: No sinks or sources, literally, all *fluxes* between inside and outside are zero, and *external driving forces* are all zero.
2. *Open system*: There are external sinks and/or sources for some of the constituents in some of the compartments or cells. (Sinks are defined as operations via which a substance vanishes from the system; sources are operations generating a substance.)

1.2.2. In Terms of Interconnections Amongst Compartments or Cells, the Topology of the Network Is Useful in Defining Mathematical or Analytical Approaches

1. *Catenary system*: Two or more compartments arranged in series.
2. *Cyclic system*: Three or more in series, with the last connected to the first allowing a circulating flux.
3. *Mammillary*: Two or more peripheral compartments connected to a central compartment and having no cyclic components, e.g., a blood compartment connected to each organ in the body.

Equating the circulatory system, for example, to a mammillary compartmental system raises questions. "How can this be a rational description?" Total circulatory mixing in humans requires many minutes. Is the rate of solute escape from blood so slow that mixing throughout the whole circulation is fast in comparison to the rate of exchange within the organs? Solute-binding to plasma proteins might so retard escape from the blood that the approximation is adequate. Alternatively, consider the mouse, where the circulatory mixing time is a few seconds, but permeabilities are similar to those in humans, so equilibration in tissues is slow compared to circulation times; here the idea that the system is mammillary is more reasonable. The basic compartmental premises, instantaneous mixing throughout the compartment, and therefore uniform internal compartmental concentrations are almost never truly valid; being alert to this implies the next step: evaluating the error due to failure to fulfill the requirements.

In terms of types of equations, models may be expressed in many forms, but physiological models set up for solving by numerical methods are mainly in the form of ODEs, Differential Algebraic Equations (DAEs), and partial differential equations (PDEs). First-order ODEs are central to compartmental modeling, and may be linear and nonlinear, for example a Michaelis–Menten equation. A set of first-order equations set up with a single variable to the left of each equal sign is said to be in state variable form. See the Background chapters on Linear Algebra and ODEs. ODEs and DAEs may be mixed together in a model, where the DAEs define variables used in the ODEs, either implicitly or explicitly. Seeking solutions sometimes requires prior analysis, but solvers like JSim and Matlab can handle many implicit forms. PDEs are becoming more common in PKPD problems now that computation is fast. Especially recently it is being recognized that there are usually concentration gradients along the lengths of capillaries for consumed substrates and for drugs during the uptake phase; the existence of gradients violates the compartmental assumption of uniform concentration (5, 6) and causes errors in the estimation of permeabilities and conductance parameters of all sorts. One-dimensional PDEs usually suffice for capillary–tissue exchange since in well-perfused organs the radial distances between blood and cells are so short that radial diffusional retardation is negligible compared to membrane permeation across endothelial cells or parenchymal cell membranes.

1.3. Distribution from the Site of Administration

Distribution, the D of ADME, is by convection, diffusion, and permeation, and must precede the drug's therapeutic and degradative reactions at target sites or metabolic sites. Distribution includes the reversible transfer of a drug between one compartment and another. Some factors affecting drug distribution include regional blood flow rates, volumes of interstitial and cellular spaces, molecular size, polarity, and binding to serum proteins or other nontarget sites, forming complexes. In using the AUC of plasma concentration versus time, one should keep in mind that plasma protein-bound drug is having no effect at the targeted site. Administration and Distribution are best treated together as it is the combination of processes that governs the effective concentration reaching the target.

Whatever the administration method (delivery on nanoparticles, laser-induced controlled release, ultrasound enhancement of intradermal diffusion, pill decomposition rates, location in gut, etc.), the next phases of Distribution are physiological processes that are fairly well understood (convection, permeation, diffusion, transport across membranes or intracellular microtubular transport), but not necessarily well characterized for the particular drug. Heterogeneities in regional flows, capillary densities, and tissue composition may have to be accounted for. These precede the reaction, binding, inhibition, or receptor-mediated responses that compose the desired pathophysiological responses. Metabolic reactions, sequestration, uptake, and excretion by epithelial cells (liver, kidney, saliva, skin, etc.), as by the liver's P450 system, or other reactions which inactivate (glucuronidation, glycosylation) are all a part of the ADME model and have to be considered in any PK modeling. For compounds which are degraded, the metabolic products have to be assessed for long-term effects. Compounds excreted by the liver or kidney become concentrated, even 100-fold, in the process of elimination in urine or bile, and if the drug is not conjugated or inactivated in some way there may be damage to the excretory organ, compromising the organ function and changing the pharmacokinetics after prolonged usage, and raising the AUC following each administration. Both renal glomerular filtration and tubular secretion and hepatic biliary excretion create high concentrations of the drug or metabolites.

At the level of the target, *Quantitative structure–activity relationship* (*QSAR*) (sometimes quantitative structure–property relationship: *QSPR*) comes into play. This is the process by which chemical structure is quantitatively correlated with biological or chemical reactivity. Biological activity can be expressed quantitatively as in the concentration of a substance required to give a certain biological response, but in the context of quantitative PK analysis, one would prefer that the PD (pharmacodynamics or biological response) also be expressed in mechanistic terms. Additionally, when the physicochemical properties are expressible in numbers, one can formulate the quantitative structure-activity relationships in the form of a mathematical model. The mathematical model may also predict the biological response to related chemical compounds. But it is the concentration of the bound agent complexed to the target site that determines the level and duration of the response.

There are a good many useful measures in considering the PK of ADME. These include, in addition to AUC (exposure), C_{max} (maximum concentration), T_{max} (Time to C_{max}), half-life, clearance route and rate, volume of distribution, and bioavailability in unbound form. In the steady state of long-term administration one assesses the plasma concentrations, total accumulation, linear or nonlinear PK, time-dependent changes in kinetics, and metabolites, their identity, and their PK. It is the combination of PK and PD that allows knowledgeable optimization of dosage regimens. A well-developed PKPD model will account for most of these considerations and thereby be predictive and advisory to the therapist. A clear description of the kinetics allows the planning of efficient dosage regimens.

2. Software for Systems Description, Simulation, and Data Analysis

2.1. Common Software and Methods Used in the Field

Pharmacokinetic models have been written in virtually every computer language that exists, and it is a field that has stimulated the development of a large set of relatively specialized simulation systems. A partial list of simulation software for compartmental analysis goes back half a century:

SAAM, Simulation, and Analysis Modeling, was the first, developed by Mones Berman at NIH for analyzing tracer kinetics (http://depts.washington.edu/saam2/). It exists still as SAAMII.

SIMCON, a general simulation control system (7), now evolved into JSim, was used to solve FORTRAN-based models of all sorts.

XPPAUT, from Bard Ermentrout (http://www.math.pitt.edu/~bard/xpp/xpp.html). XPPAUT is particularly good for bifurcation analysis of dynamical systems.

Gepasi, now CoPasi, from Pedro Mendez (http://www.softpedia.com/progDownload/Gepasi-Download-167140.html; http://www.copasi.org/tiki-view_articles.php). CoPasi is especially good for enzymatic reactions and biochemical systems, allowing a menu of choices for reaction types.

Modelica, http://www.openmodelica.org/, is excellent for linking operators and presenting the forms of model networks (http://www.ida.liu.se/labs/pelab/modelica/OpenSourceModelica-Consortium.html).

Jarnac is designed for symbolic or diagrammatic entry for biochemical and gene regulatory reactions (from Herbert Sauro (8, 9), http://sys-bio.org/jarnac/).

JSim: Developed from SIMCON and XSim (for X-windows linux systems, http://www.physiome.org/software/xsim/) into JSim (10) (http://www.physiome.org/jsim/). JSim was developed by Erik Butterworth (11); it provides automated unit balance checking and unit conversion, thus avoiding errors due to inconsistency in the units used in the code.

Non-MEM, nonlinear mixed effects modeling, is a commercial software package providing the capability to use a wide variety of pharmacokinetic models. It is particularly designed for the analysis of sparse data sets using combinations of single patient and population data, (http://www.iconplc.com/nonmem).

BioSPICE (derived from SPICE, for biology: http://sourceforge.net/projects/biospice) is designed for molecular biology. See also http://jigcell.cs.vt.edu/software.php for JigCell, based on BioSpice.

Cellular Open Resource (COR) is for a Windows environment: http://cor.physiol.ox.ac.uk/ for cellular level physiological systems. PCEnv for physiological systems is being developed from it.

Stella (http://www.iseesystems.com, a commercial system) for networks of operators such as in compartmental systems.

StochSim (http://www.ebi.ac.uk/~lenov/stochsim.html) is written explicitly for the treatment of molecular interaction when there are few molecules and interactions occur stochastically.

These simulation systems have one purpose in common: to make the programming of models simpler and to facilitate the analysis of experimental data in terms of the parameterized descriptions of the kinetics. They vary considerably with respect to their representation of physical/chemical mechanisms. Such modeling and analysis systems do not displace FORTRAN, C, C++, and Java as mainstream languages, but rather they replace the front-end entry to formulate the models and interface them to data sets. None of these have the general capabilities of Matlab or Mathematica, nor do they attempt algorithmic manipulation as in Maple, but are more directly tuned to the user's needs, as will be described.

2.2. A Preferred Simulation System, JSim

JSim is perhaps the most general of these simulation analysis systems, designed for the analysis of experimental data. It is built around a "project file, .proj," that may hold many data sets, several different models, and results of multiple analyses. JSim handles not only the ODEs around which traditional compartmental modeling is built, but also DAEs, implicit functions, PDEs, and stochastic equations. JSim, uniquely, and from its beginning in 1999, uses unit balance checking and automated unit conversion. (Unit balance checking assures that the units of the expressions on the left of the equal sign are the same as those on the right. Automated unit conversion means that when time is expressed in minutes a velocity expressed in cm/s will be converted to cm/min by multiplying cm/s by 60 s/min.) This pair of features is a great boon in programming since in the first phase of compilation it automates the first stage of *verification* of the model's mathematical implementation by making sure that every equation has unitary balance. The second phase of compilation parses the details of the equations, and sequences them for efficient computation. The run-time code is compiled into Java, which now runs almost as fast as FORTRAN and C. (On a cardiovascular–respiratory system model JSim ran exactly 300 times faster than a Matlab–Simulink version of the identical model.) JSim's advantages over the ODE-based systems listed above are the following:

1. Runs on Linux, Macintosh, and Windows.
2. Is free and downloadable from www.physiome.org. On the Macintosh it takes about 30 s to download and install, and another 10 s to bring up a model.

3. Is the only one that solves PDEs and offers an assortment of solvers for both PDEs (three available now) and ODEs (eight available).
4. Imports and exports both SBML and CellML archival forms.
5. Provides sensitivity analysis of two types, relative and absolute.
6. Graphical output is immediately available during the simulation run and setup in seconds.
7. Has seven built-in optimizers for excellent power in parameter adjustment to fit data.
8. Provides the covariance matrix giving the correlation among free parameters and estimates of parameter confidence limits.
9. Use project files that allow the analysis of many experimental data sets in one file.
10. Stores parameter sets so that individualized parameter sets for each data set can be stored.
11. Allows the use of several models within one project file so that competing hypotheses (models) can be compared and evaluated.
12. Is structured so that the front-end parameter control and graphical user interface (GUI) can be framed explicitly for any model.
13. Has linear and log line graphs, 2D contour plots representing 3D, and phase-plane plots.
14. Has "looping" capability, allowing discrete successive jumps of the values of one or two parameters at a time in order to explore system sensitivities visually and rapidly.
15. Uses a Mathematical Modeling Language, MML, in which one writes the equations directly, for simultaneous solution, and in which the order of the equations is not specified.

There are no special requirements for the JSim software or for its methods of use for model building and exploring or use in analysis with respect to hardware, computing platform, or operating system. It has important limitations, not being a procedural language but a declarative mathematical language. This means there is no equivalent of a FORTRAN DO-loop (or GO TOs or jumps). It cannot yet do matrix inversions (except through a special mechanism), and is in a continuing state of development. JSim 2.0, released in February 2011, is based on a new compiler providing many new features described at nsr.bioeng.washington.edu/JSim/.

The features listed above, and others not listed, have been implemented in JSim because the years of experience with a large variety of models, with teaching graduate and undergraduate classes, and postdoctoral and faculty workshops have led to a detailed understanding of how people use modeling in scientific research. Experiment design, hypothesis testing, and system parameterization are given priority in the conveniences provided.

Thus *JSim, since it is designed around the analysis of experimental data, is our preferred software* and will be used for the compartmental modeling shown next.

3. Compartmental Modeling

3.1. The Modeling Process

The overall process in the experiment/model hypothesis iteration loop of Platt (12) is as follows: (1) express the hypothesis in quantitative terms, as a mathematical model, with units on everything; (2) use the model to determine the best experiment that might contradict the predictions of the model, or, better yet, develop an alternative model that is seemingly as good but makes different predictions, and then design the experiment that clearly distinguishes between the models; and (3) do the experiment and analyze the data. One of the two competing models, maybe both, must be proven wrong, and so science is advanced.

The normal data analysis using models begins by putting the data to be analyzed in a "project file," modelname.proj, and displaying them on the JSim plot-pages. The second stage, coding and model verification in accord with standards (http://www.imagwiki.nibib.nih.gov/mediawiki/index.php?title=Working_Group_10), is building and testing the model, incorporating reference analytical solutions if appropriate to verify the solutions as being mathematically accurate, and representing the equations. The "project file" may contain two or more models so that alternative model forms can be compared directly by examining the solutions (changing parameters and rerunning, using "loops" to automatically change parameters, using behavioral analysis, plotting in various forms including phase-plane plots, contour plots). The verification stage is to show that the model solutions are computed correctly, done by testing different solvers, using different time-step sizes, and comparing with analytical solutions in special cases.

The validation stage is to test the fitting of the model solutions against the experimental data. The word "validation" is truly optimistic, because a good fit of the model solution to the data does not really validate the model, but merely fails to invalidate it. It is the failure of the model that leads to the scientific advances by forcing new ideas to be incorporated. Nevertheless, fitting of the model to the data provides characterization of the data, augments diagnostic acuity, assesses progress of disease or evidence of successful therapy, and is generally useful in reconciling the working hypothesis (model) with observations.

The final phase is preparing the model so that it can be reproduced by others. This is not only critical from a tutorial point of view but also in fact is a requirement for any scientific publication. Anything that cannot be reproduced is misleading and wastes time and money. Reproducible models can be tested by others or used as building blocks to advance the field when they pass muster.

3.2. A Simple Compartmental Model Implemented in JSim

The modeling code. For an introduction to JSim we use a two-compartment closed system with passive exchange between the compartments, and a conversion reaction of solute A to solute B in either or both compartments. This model has analytic solutions which could be used either to show the solutions or to provide verification of the accuracy of the numerical solution, but since these solutions run no faster than the numerical solutions, they will not be used here. Detailed instruction in JSim use is available at http://www.physiome.org/jsim/. This model is #246.

Many model programs are available at http://www.physiome.org/Models. One can search to find a model similar to what one might like to construct, e.g., from a tutorial list of compartmental models: http://www.physiome.org/jsim/models/webmodel/NSR/TUTORIAL/COMPARTMENTAL/index.html.

Open model #246, Comp2ExchReact, and you will be asked to allow the display on your computer, wait a moment for it to be compiled, and then click on "Source" at the bottom of the JSim page to show the source code, Table 1 for the model shown in Fig. 1. All the models on the Web site are archived to keep track of model changes (previous versions can be found under the "Model History" section on each model Web page). (Models edited over the Web cannot be saved on the Physiome Web server, so simply save what you want to your own directory. The JSim system can likewise be downloaded directly and the model worked on from your own computer.) The model for the code in Fig. 1 is diagrammed at the bottom of the figure; it is a two-compartment model for two substances, A and B. Both substances can passively move from one compartment to the other. A is irreversibly converted to B in either or both compartments. After the title a short description is provided. (Text enclosed by /* ...*/ is ignored by the compiler, as is comment text following // on any line.)

An important JSim feature is invoked next, the reading of a units file, nsrunit, which allows automated unit balance checking and automated unit conversion to common units during the compilation phase; MKS, CGS, and English units can all be used. (The unit conversion can be turned off, a feature useful

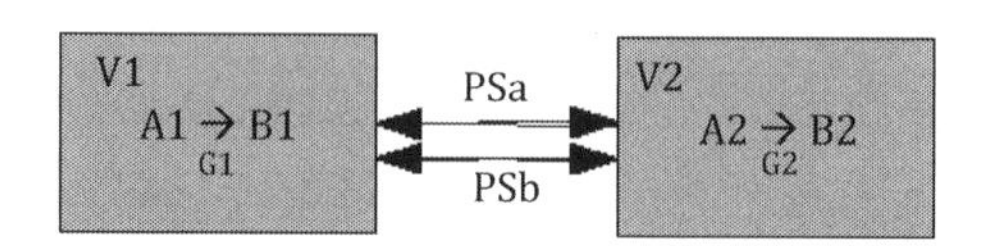

Fig. 1. JSim code for a two-compartment model in which a solute A can exchange across a membrane between volume V_1 and V_2 and can react irreversibly to form solute B in either compartment. This is available for download at www.physiome.org/jsim/models/webmodel/NSR/Comp2ExchangeReaction/index.html and is model #246.

Table 1
Code for two-compartmental model with passive bidirectional exchange

```
/*  MODEL NUMBER: 0246
    MODEL NAME: Comp2ExchangeReaction
    SHORT DESCRIPTION: Two comparment model with two substances,
    irreversibly converting A to B.
 */
 import  nsrunit; unit conversion on;

 math Comp2ExchangeReaction {

 // INDEPENDENT VARIABLE
 realDomain t sec; t.min=0.0; t.max=60; t.delta = 0.1;

 // PARAMETERS
 real V1 = 0.07 ml/g,                 // volume of compartment 1
      V2 = 0.15 ml/g,                 // volume of compartment 2
      PSa = 1 ml/(g*min),             // Permeability Surface area product
                                      //    for exchange of A between two compartments
      PSb = 1 ml/(g*min),             // Permeability Surface area product
                                      //    for exchange of A between two compartments
      G1   = 0.0 ml/(g*min),          // Conversion rate of A to B in compartment 1
      G2   = 1.0 ml/(g*min),          // Conversion rate of A to B in compartment 2
      A10 = 1 mM,                     // initial concentration of A in compartment 1
      A20 = 0 mM,                     // initial concentration of A in compartment 2
      B10 = 0 mM,                     // initial concentration of B in compartment 1
      B20 = 0 mM;                     // initial concentration of B in compartment 2

 // VARIABLES
 real A1(t) mM,                       // concentration of A in compartment 1
      A2(t) mM,                       // concentration of A in compartment 2
      B1(t) mM,                       // concentration of B in compartment 1
      B2(t) mM,                       // concentration of B in compartment 2
      Equilibrium mM;                 // Equilibrium concentration for B if G>0
                                      //    else for A if G=0
 // INITIAL CONDITIONS
 when (t=t.min) {A1 = A10; A2 = A20; B1 = B10; B2 = B20; }
 Equilibrium = (V1*(A10+B10)+V2*(A20+B20) )/(V1+V2);

 //ORDINARY DIFFERENTIAL EQUATIONS
 A1:t = (PSa/V1)*(A2-A1)-(G1/V1)*A1;
 B1:t = (PSb/V1)*(B2-B1)+(G1/V1)*A1;
 A2:t = (PSa/V2)*(A1-A2)-(G2/V2)*A2;
 B2:t = (PSb/V2)*(B1-B2)+(G2/V2)*A2;
 } // END OF MODEL
 /*
```

when importing models from CellML or SBML which sometimes have unit conversion factors hidden as dimensionless factors in their archived models, whereas they should be dimensioned, e.g., as 60 s/min.) The word "math" and the curly bracket designate the start of the model code, written in JSim's MML, which will be seen to be merely the equations for the model. The "realDomain" defines the *independent* variable as t, time in seconds, along with a starting and ending time and a time interval for graphing the solutions.

The parameters of the model are assigned units. We have put the units in physiological form in order to represent those for a perfused tissue, so much per gram of tissue. In general it is practical to avoid mistakes by using the same units as used for the experimental studies in which the data are acquired. The volumes of the two compartments are V_1 and V_2 ml/g. In this context, flow per gram of tissue (F), clearances (Cl), and permeability-surface area products (PS) all have the same units, ml/(g min). (The PS is defined as permeability, P cm/min, times membrane surface area per gram of tissue, S cm^2/g.) The reaction is described as a clearance, but here is given the symbol G, ml/(g min) for a gulosity or consumption. The solute A is converted from A to B. (The terminology used here is that used by the American Physiological Society (13), where an extensive set of terms for transport, flow, and electrophysiology are given.)

The model's variable are functions of the independent variable time, for example the concentration of A is defined as a real number and as a function of time by "real $A(t)$." In the equations $A(t)$ can be written without the "(t)," but in JSim's MML the use of the (t) initially is required to establish it as a function of time. Other languages such as Madonna and XPPaut do not demand this, but JSim users find that this reduces errors. The MML is fairly similar to that in those languages; the basic intention is just to write the equations directly.

The ODEs, used to describe compartmental models, need to be supplied with initial conditions, here provided as A10, A20, B10, and B20 where the first subscript digit refers the compartment and the second is "0" to refer to the initial time $t = t$ min, which can be negative or positive, but is usually $t = 0$.

In the ODEs, the derivative of $A(t)$, which you would expect to be $\mathrm{d}A/\mathrm{d}t$ is written A:t. The second derivative, $\mathrm{d}^2A/\mathrm{d}t^2$, would be A:t:t. The equations are written in state variable form in this model, for tidiness and simplicity, but this is not FORTRAN and one could equally well write the equation as "$V_1 \times A_1{:}t = A_2 \times \mathrm{PSa} - A_1 \times (\mathrm{PSa} + G_1)$;". In this latter form, the left-hand side, LHS, of the equation is the rate of change of mass of solute A, (mol/g)/min = mM (ml/(g min)) for A_1 in V_1.

3.2.1. Unit Balance Checking

The acute observer will have noticed that the independent variable, t, is in seconds: the differential equation therefore looks to have unbalanced units. It actually computes correctly since the phrase "unit conversion on" at the top of the program preceding the model code enlists the automated unit conversions so that in the compiled code the multiplier of the left side, "60 s/min," is inserted. Without unit conversion on, the best way to handle this is to put the independent variable t in minutes. In languages like Matlab there is no unit checking. In huge projects like the Mars Climate Orbiter mission (http://www.spaceref.com/news/viewpr.html?pid=2937), mixing units from the European and American programs led to the crash of the space vehicle and the termination of the billion dollar mission. There would have been no problem in JSim as long as the units were stated and unit conversion "on" (11). The *nsrunit* file may be viewed by clicking "Debug" (left, bottom) to drop down a menu including "View system units file." The four ODEs in Fig. 1 are a complete description of the system behavior after the initial moment. Following them is an algebraic equation for TotalC, also a function of time, to check the total mass at each point in time divided by the total volume and giving the average concentration. With the initial conditions and volumes given, the result is 0.33333333 mM for every time step. Run the model and then check the numbers for the plotted variables by clicking on "Text" instead of "Graph" at the bottom of the plot page, as in Fig. 2. The program end is marked by a right curly bracket, beyond which one can put notes, comments, key words, diagrams, references, etc.

3.2.2. Graphical User Interface

The JSim GUI for simulation control is shown in Fig. 2. The left panel has overlays for project contents: one or more models, data sets, parameters sets, setups for solvers for ODEs and PDEs, sensitivity analysis, optimization, and confidence limits. To start a simulation run one clicks "RUN" at the top of the run time window. On the right page one clicks on a "Message" panel for error messages, or the plot pages (1_Conc or 2_ReacSite, names chosen by the user), and at the page bottom choose "Graph" for displaying the output graphically or "Text" for seeing the numerical listings of the experimental data and the model solutions at each time point.

There is an extensive introduction to JSim at www.physiome.org/jsim giving precise detail to supplement this outline of usage. The JSim MML code is pretty easy for a beginner to use since it contains just the parameters with their units, the variables followed by (t) to indicate their dependence on the independent variable t for time, the initial conditions, and the equations for the model. The most common mistakes are to misapply the uses of commas and semicolons. Each of a sequence of events is usually comma separated, and a string of them is closed with a semicolon. The model code itself begins with the left curly bracket, "{" after the word "math" and ends with the right curly bracket,"}," after all of the equations have been written. Comments are preceded by a double slash, "//," or alternatively can be preceded by a "/*" followed by an "*/," without the quote signs. Equations end with a semicolon, including those in the initial condition statement.

3.2.3. Exploring Parameter Influences Using the "LOOP" Mode of Operation

In loop mode, the user can choose to enter a sequence of values (under Other Values) to explore model behavior widely, using comma separation, e.g., 2, 3, 5, 8, etc., and in the "auto" mode will do as many runs as there are values entered. One can also enter arithmetic changes such as @*2 or @ + 3 or more complicated expressions to indicate automatic changes in the starting value by

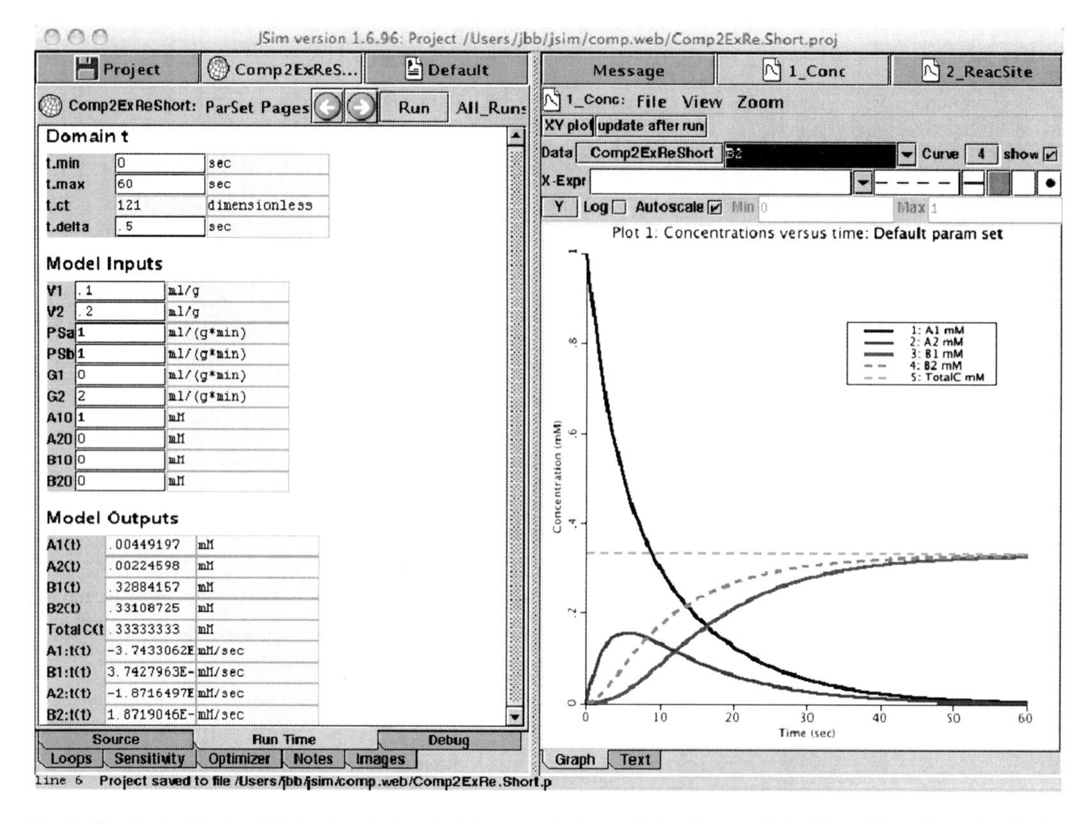

Fig. 2. Standard JSim Input/Output control and plot pages. *Left panel:* Runtime control: "Domain" is time, t, with $t_{\min}$ starting time, $t_{\max}$ ending, and t_{delta} the time step for plotting. "Model Inputs" gives parameter values and initial conditions. Model Outputs shows values of variables at the end of the run, at $t = 60$ s. The mass balance check, TotalC, is exact to eight decimals. *Right panel:* The time courses of concentrations and A and B in compartments 1 and 2 are shown for the parameters and initial conditions shown in the left panel and in the code in Fig. 1. The user chooses the variables to plot, and the colors and line type or point type. The title and labeling are user written and retained in the JSim project file. The legend for the graphics output is automated.

multiplication by 2 on each run, or the automatic addition of 3 on each run, for a chosen number of runs (Fig. 3).

3.2.4. Sensitivity Analysis

Sensitivity analysis is available to provide a quantitative measure of the effect of any chosen parameter on a particular variable. This extends the information gained by "looping." The sensitivity function $S(t)$ of a variable to a parameter is calculated by calculating the change in a variable such as $A(t)$ per fractional change in a parameter value, P. The linear sensitivity function is

$$S(t) = \partial A(t)/\partial P, \tag{4}$$

or alternatively, the log sensitivity function is

$$S_{\log}(t) = (\partial A(t)/A(t))/(\partial P/P) = \log(A(t))/\partial \log P. \tag{5}$$

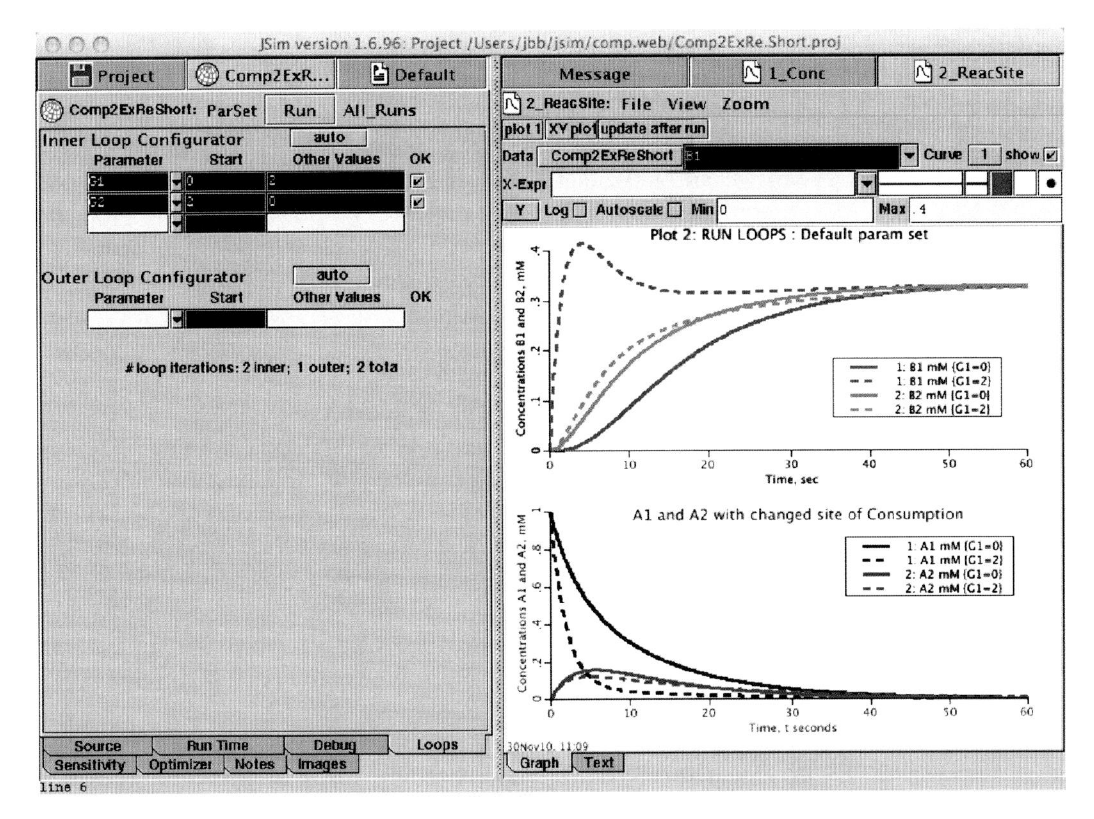

Fig. 3. Loop Mode Operation. The control panel for the looping operator is shown on the left. The starting values for G_1 and G_2 are those shown in the code (Fig. 1) and their solutions are given by the *solid lines*, and are the same as in Fig. 2, with conversion of A → B only in compartment 1. On the second run (*dashed lines* of same colors) the values entered by the user under Other Values (*left panel, top right*) are used, in this case setting $G_1 = 2$ and $G_2 = 0$, so that the conversion of A → B occurs now only in compartment 1 instead of in compartment 2.

Sensitivities are local linear approximations, and are calculated by computing two solutions, $A_1(t)$ and $A_2(t)$, the second having P changed by a small fraction, $0.001P$ or $0.01P$, so that

$$S(t) = (A_1(t) - A_2(t))/\Delta P, \quad (6)$$

and using the definition of the derivative as ΔP goes to zero gives the formal value.

3.2.5. Optimization in Data Fitting and Analysis

Optimization, either by manual parameter adjustment or by automated methods, is the procedure of adjusting parameters of the equations so that the solution provides a good fit to the data. The evaluation of the goodness of fit by minimizing the sum of squares of the differences between the model solution and the experimental data is based on an implicit assumption that the differences are Gaussian random. This is seldom correct, and it is important to appreciate that the choice of the distance function is personal, i.e., it is up to the investigator to characterize the noise in the data and to weight the influence of individual data points. See next section.

4. Applications

4.1. A Compartmental Approach to Aspirin Kinetics

Aspirin is a very old drug used to reduce fever and inflammation; only recently has its mechanism of action begun to be understood, the first being its action as a blocker of prostaglandin formation. Its kinetics have not been thoroughly worked out, so we present here our analysis of three sets of representative data from three different studies. Aspirin is acetylsalicylic acid, and its reactions are as follows:

$$\text{Acetylsalicylic acid} \rightarrow \text{salicylic acid} \rightarrow \text{salicyluric acid.} \quad (7)$$

Aspirin, acetylsalicylic acid, is hydrolyzed quickly via a plasma esterase to salicylate. Salicylate is pharmacologically active. Salicylate kinetics dominate the clearance. The modeling examines only salicylate's enzymatic conversion to product, where product is considered to be equivalent to excretion into the urine, a saturable process. This may produce salicyluric acid or a glucuronate. The model captures the kinetics of salicylate clearance over a 100-fold range of concentrations through consideration of one enzymatic reaction with parameters optimized to fit three very different data sets taken from the referenced papers.

The second reaction to form the excretable product is slow and is enzymatically facilitated. At high doses the enzyme becomes saturated, i.e., the reaction is limited by the fact that the enzyme is all in the form of the bound enzyme–substrate complex and raising the salicylate concentration does not accelerate the reaction. At low dosage the clearance is rapid; at medium or high therapeutic dosage the clearances are slower; at near-lethal toxic levels clearance is very slow and often requires treatment by infusion of alkaline salts and sometimes dialysis. For this example we have taken the data from three research studies. (Low dose data are from Fig. 1 Right of Benedek (14) Dose Period 1 (squares). Medium dose data are from Fig. 4 of Aarons (15) oral dosage, last nine points. High dose data are from Fig. 1 of Prescott (16) Control.) These particular data were chosen because the chemical methods and procedures appeared to be excellent, the data covered many hours, and, while we do not have the original data, conversion from the symbols in the figures to numerical representation was accomplished with good accuracy.

A reason for choosing aspirin as the subject for compartmental analysis is that the time course of the clearance, mainly by loss into the urine, is long compared to circulatory mixing times, so that the biochemical processes appeared to limit the clearance, and they could therefore be characterized. If the circulation and distribution times were long compared to the reaction processes, the latter would not be meaningfully determined from the observations. Another reason was that a comparison among the different data sets suggested that the clearance was enzymatically mediated: at high concentration the diminution was almost linear, at low concentrations it was almost exponential, and at intermediate concentrations the rate of clearance appeared to speed up with time. This fits the expectations for a *saturable enzymatic process*: at low concentrations well below the dissociation constant for the enzyme substrate complex almost all of the enzyme is free and available for the reaction and therefore the fractional transformation of substrate is at its highest. With all the enzyme free this is a first-order process, a single exponential. In contrast at high concentrations the enzyme is almost totally saturated and the conversion is at a maximum rate independent of the concentration; this is zero order kinetics, giving a linear diminution in concentration. Thus the shift from zero order kinetics to what occurs at intermediate concentration, a gradually increasing fractional rate of reaction, is what was suspected by looking at the middle-level concentrations.

Thus we hypothesize an enzyme conversion model for salicylic acid clearance, and then test the surmise by attempting to simultaneously fit the three independent data sets using one set of parameters. To do this in one program and to optimize the fit to the data for all three salicylate levels we coded three identical models in the one program (Table 2). These are computed simultaneously in order to fit the three independent data sets from the three research reports using *one common set of parameters*, and automated optimization was used to minimize the set of differences between data and model solutions.

The reactions, both the binding to the enzyme and the product formation, are reversible:

$$\mathrm{SA} + \mathrm{E} \underset{\leftarrow k_{\mathrm{off1}}}{\overset{k_{\mathrm{on1}}\rightarrow}{\longleftrightarrow}} \mathrm{SAE} \underset{\leftarrow k_{\mathrm{on2}}}{\overset{k_{\mathrm{off2}}\rightarrow}{\longleftrightarrow}} \mathrm{E} + \mathrm{P}. \tag{8}$$

The equations and parameters are identical for the three dosage levels. Here is the model code for the low dose, where the prefix L distinguishes this model equations from those for the medium dose, prefixed M and the high dose, prefixed H, neither of which are shown in Table 2.

In undertaking an analysis on a single enzymatic reaction we lack knowledge of the exact mechanism. In addition the compartmental approximation is certainly questionable for whole body studies. The hypothesis that a single enzymatic reaction dominates the clearance would be strengthened if the model provides good fits to the three data sets. There is no guidance from the literature on the dissociation constant, K_{D}, for our presumed enzymatic reaction, so that we are neither constrained nor aided.

We chose to use a simple enzymatic reaction, one that allowed characterizing the rates of binding and unbinding of substrate and enzyme, and a rate of the forward reaction to yield the product. We allow also a backward reaction, on the basis that *all reactions are*

Table 2
Model code for salicylate clearance (model downloadable from www.physiome.org/Models: search for model 280)

```
/* MODEL NUMBER 280
   MODEL NAME: Aspirin
   SHORT DESCRIPTION: Salicylic acid (SA) clearance for three different dose ranges is modeled as
   an enzyme reaction. This table is abbreviated by omitting the code for the Mid and High dose
   reactions.
*/

import nsrunit; unit conversion on;
math Aspirin {

// INDEPENDENT VARIABLE
realDomain t hour; t.min = 0; t.max = 16.0; t.delta = 0.05;

// PARAMETERS (SAME FOR ALL THREE MODELS)
real kon1 = 0.174 L*mg^(-1)/hour,   // On rate for SA + enzyme
   KD1 = 6.3 mg/L,                  // Dissociation constant for SA enzyme complex
   koff1 = KD1*kon1,                // Off rate for SA enzyme complex
   kon2 = 0.003 L/(mg*hour),        // On rate for Product + enzyme
   KD2 = 250 mg/L,                  // Dissociation constant for Product enzyme complex
   koff2 = KD2*kon2,                // Forward rate to form Product from complex
   Gp = 0.03 1/hour,                // Clearance rate from plasma
   Etot = 10 mg/L;                  // Total enzyme concn

// LOW DOSE MODEL: Data from Benedek (1995) Fig 1 Right Dose Period 1 (squares)
// LOW DOSE PARAMETER
real LSAtot = 8.2 mg/L,             // Total Low Dose concentration
// LOW DOSE MODEL VARIABLES
   LSA(t) mg/L,                     // Low dose SA
   LSAE(t) mg/L,                    // Low dose SA-enzyme complex
   LE(t) mg/L,                      // Low dose free enzyme
   LP(t) mg/L;                      // Low dose product
// LOW DOSE INITIAL CONDITIONS
when(t = t.min) {LSA = LSAtot; LSAE = 0; LP = 0;}

// LOW DOSE ORDINARY DIFFERENTIAL AND MASS BALANCE EQUATIONS
LSA:t = -kon1*LSA*LE + koff1*LSAE;
LSAE:t = kon1*LSA*LE-koff1*LSAE-koff2*LSAE + kon2*LE*LP;
LP:t = koff2*LSAE - kon2*LE*LP - Gp*LP;
LE = Etot - LSAE;                   // LSAtot = LSA + LSAE + LP; for an overall mass balance
  accounting
} //End of Low Dose Model. The omitted code for the medium and high dose models is identical
// but L is replaced with M or H. Copy and paste the Low Dose Model into the space preceding the
// right curly bracket, twice, and change the L to M for one copy, and L to H for the other; recompile.
```

thermodynamically reversible, at least in principle. This reaction setup allows reduction to the simpler, commonly used Michaelis--Menten reaction; this is accomplished by speeding up the binding and unbinding reactions of salicylate to enzyme and eliminating the

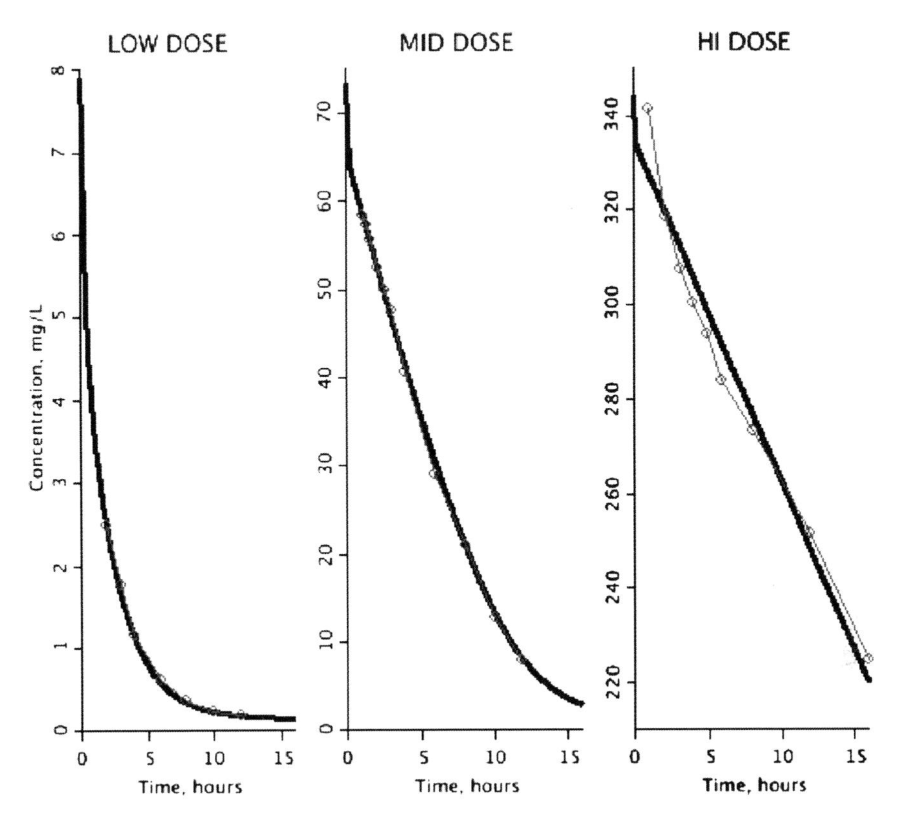

Fig. 4. Data on salicylate clearance from the three laboratories are fitted simultaneously with a single enzyme model to describe the clearance. Parameters are given in the model code in Table 2. Data are from Benedek (14) in the *left panel*, Aaron (15) in the *middle panel*, and Prescott (16) in the *right panel*. Note that the concentration ranges are markedly different. (http:/www.physiome.org/jsim/models/webmodel/NSR/Aspirin/ model 280).

reverse reaction from the product back to salicylate. So for the first level of testing all parameters are considered as open and adjustable, including the initial values of the concentrations in the system. In this analysis the system is considered to be a single well-stirred tank, as if the circulation were instantaneously mixed. This gross oversimplification also makes the assumption that the product either goes directly into the urine or to some other location in the body from which it does not return. Enzymatic conversion in the liver followed by excretion into the bile would be equivalent kinetically to conversion to a glucuronate followed by the circulation through the blood and clearance in the kidney. Ideally one would measure the urinary excretory rates simultaneously and model both the plasma clearance rates and the urinary clearance. This was not done here.

The results are shown in Fig. 4 where the curves at the low, middle, and high concentration ranges are shown to be fitted reasonably well by the model. The high concentration data (right panel) are fitted less well but do illustrate that the slope is

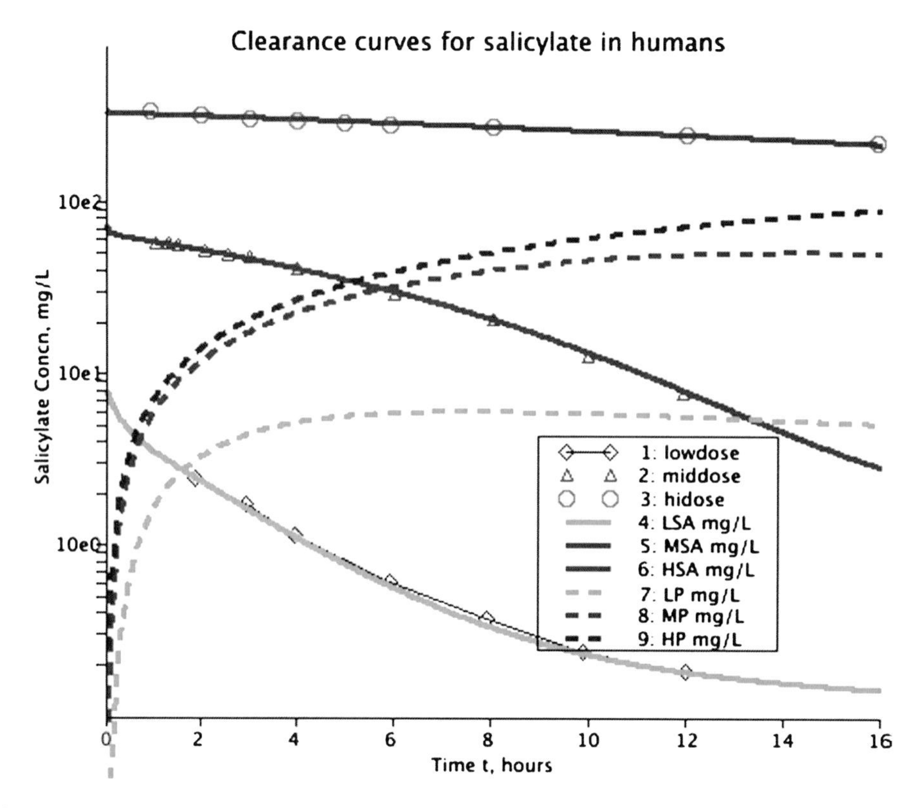

Fig. 5. Semilog plots of the data (symbols) fitted with the model solutions for the salicylic acid (LSA, MSA, and HSA, solids lines). All the data in both Figure 5 were fitted simultaneously with one parameter set for the enzyme as given in Table 2. The Product concentrations (dashed line for LP, MP, and HP) are merely predicted product concentrations. Since there are no data on product concentrations, the assumption that there is no degradation of Product must lead to some overestimation of its influence on the backward reaction.

approximately linear, as expected, whereas at the low concentrations (left panel) the curve is nearly exponential, as expected.

A different view is provided by Fig. 5, a semilog plot of the three sets of data each fitted by the model. The high concentration curve (open circles, purple line) appears to be almost linear on this plot, but the slope is shallow, and judgment based on Fig. 4 is better. The high dose concentrations are very much higher than the K_{D1} for substrate binding, estimated at 6.3 mg/l, so that there is no doubt that the enzyme was almost saturated. In fact, with the High Dose the concentrations are above apparent K_{D2} of 250 mg/l for product binding, so there is significant reversal of the reaction. At Mid Dose level (triangles, blue line) the slope diminishes as time progresses, that is to say the fractional clearance is increasing as the enzyme becomes less saturated. The Low Dose data (diamonds, green line) are fitted well and have considerable curvature on the semilog plot, the slope at late times being much diminished: this leads to the idea that there is some tendency for retention at the low concentrations, which could be either due to recirculation from other parts of the body where the concentrations were initially

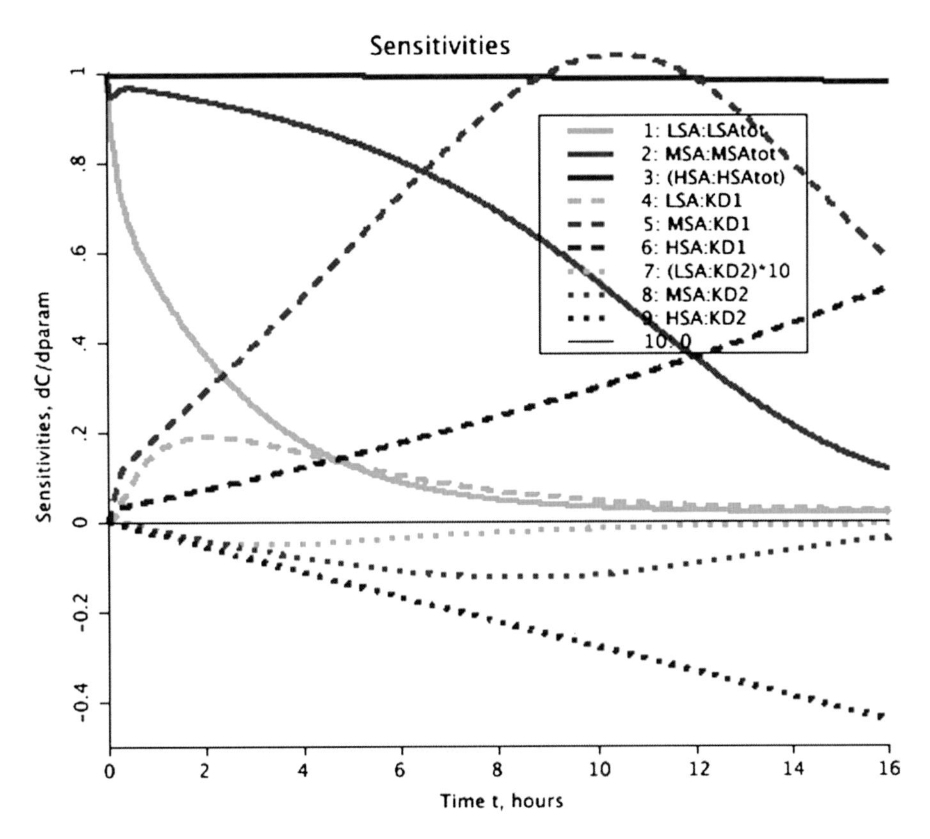

Fig. 6. Three sets of linear sensitivity functions versus time. *Solid lines* are sensitivities to the initial zero-time concentrations resulting from the doses. *Long dashed lines* are sensitivities to K_{D1}: the sensitivities are all positive. *Dotted lines* are sensitivities to K_{D2}, the dissociation constant for the reverse reaction: the sensitivities are all negative.

higher or due to "product pressure" to form salicylate by the reverse reaction. The reversibility is governed by the K_{D2} for the product, which here was about 250 mg/l. There must be even more reverse flux with the middle and higher level concentrations, HP and MP, a feature of product inhibition, and the estimate of this K_{D2} is determined from both of these, even though the effect is most evident from the curvature of the low salicylate dose, LSA.

4.1.1. Sensitivity Analysis

In Fig. 6 the linear or absolute sensitivity functions are shown for initial concentrations and for the two dissociation constants. The *solid lines* are sensitivities to the initial zero-time concentrations resulting from the doses; most of the sensitivity is at the earliest points. With the high dose the fractional clearance is so low that the high sensitivity extends throughout the 16 h of the study. For the middle dose, MSAtot, the sensitivity diminishes most steeply as a function of time at around 10 h when the concentration is close to K_{D1}, the dissociation constant for substrate binding. The *long dashed lines* are the sensitivities to K_{D1}; these are all positive, meaning that if K_{D1} were increased (decreasing the affinity of the enzyme for the substrate SA) the model solutions for all three doses would be at higher levels and the rate of disappearance would be diminished. Note that the time of peak sensitivity to K_{D1} is at early times for the low dose, at 10 h for the middle dose, and at late times for the high dose. The *dotted lines* are sensitivities to K_{D2}: the sensitivities are all negative, meaning that if K_{D2} were increased (decreasing the affinity of the enzyme for the product P) the model solutions would be at lower levels ***and the rate of disappearance would be increased*** because of reduced rates of reverse flux from product to salicylate.

Technically the sensitivity calculations are set up by a special mechanism: at the bottom of the left page is a button labeled "Sensitivity." Clicking on it takes one to the "Sensitivity Analysis Configurator." There in the leftmost column of the configurator table one types in, or chooses from the drag down menu, the parameter for which one wants to find the sensitivity. By clicking on the down arrowheads you bring up the choices. In the setup provided on the Web site at www.physiome.org, etc. the three starting values for the initial concentrations at $t = 0$ are listed: LSAtot, MSAtot, and HSAtot. Next on the list are the dissociation constants K_{D1} and K_{D2}.Their current values are automatically displayed under "value." The calculations of each $S(t)$ are made on the basis of the parameter change of 1% set under "delta" at 0.01. The tick marks in the OK column indicate that the calculation will be made as described earlier, namely, that the standard solution will be calculated and then another solution calculated for each of the five parameters listed, with this 1% change in parameter value. The $S(t)$ is the difference in the solution at each time point from the standard solution divided by the 1% change in the parameter value, Eq. 4.

4.1.2. Optimization

This is the process of fitting the model solutions as closely as possible to the data in order to guide one's thinking about and one's use of the model. When the fit is very close, then one has a descriptor of the fitted data sets, that is, the model and its parameter set provide a record of that description. Descriptions of many different studies, patients studies or experiments, allow comparisons and possible classification into categories having specific distinctions. Descriptive models are useful for diagnosis and possibly for prognosis or choosing modes of therapy. (If the model "explains" the data by defining the physical and chemical mechanisms, that is even better.)

When the fit is poor, then more exploration is needed. Was automated optimization used? A typical set of trajectories of parameter values during an optimization run using SENSOP is shown in Fig. 7. The values do not range widely; to assure one that they have not settled into local minima we also used other optimizers that search widely, e.g. simulated annealing. Try weighting the data differently: a simple sum of squares minimization may not be appropriate; it is almost never appropriate if the data range over one order of magnitude. For example when fitting a decay process that is exponential, one can use a reciprocal weighting so that points are weighted more evenly, or a weighting adjusted to the individual result such as "1/exp($-t$/(2 h))" for the low dose data set, and "1/exp($-t$/(8 h))" for the middle dose data. These choices of weighting are to be typed, without the quotes, into the appropriate line under Pwgt (point weighting) in the Data to Match Table on the Optimizer Configuration Page.

In the same table there is opportunity to even up the weightings for the three quite different curves by using Curve Weighting, Cwgt. In this case we weighted Low Dose data with 140 times the weight of the High Dose Data since the latter were about 140 times higher concentrations. This evens up the contributions to the sum of squares. Likewise the Middle dose data were weighted at 14, that is, about ten times higher than the High Dose data. This combination of Point weighting within each data curve and Curve weighting among the three data curves gives each point in the triple data set about the same weight. This makes the calculation of a root mean square error provided by the minimization of the sum of squares a reasonable strategy. This is probably close to what one would do if using "eyeball best fitting," examining the fitting to all the points. Both Figs. 1 and 2 plots are valuable in this regard since they weight the curves differently from the "eyeball" view. While arguing that the eyeball fitting is just as valid as that from automated optimization, the latter has the virtue that the weighting scheme is explicitly stated, and that the weighted sum of squares can be reproducibly reported when the weighting scheme is reported also.

There are situations where an optimizer fails to reach a good fit because the sum of squares has settled in a local minimum. Optimizers like Gridsearch and Simulated Annealing are designed to cover a wide range in state space so that even the corners get explored (Fig. 7). Others like SENSOP (17), GGOPT (18), and NL2SOL (19) make excursions, jumping out of the locale to test other regions for a better fit. To switch optimizers, simply pick another one from the drag down menu.

4.2. Multiple Sequential Dose Administration: Hepatic Function

Test of clinical functions evolves in a variety of ways, usually being designed long after the function has been clearly understood. Here is a counterexample, a case in which the idea of the clinical evaluation from the administration of a drug was evident right at the beginning. There was a coalescence of features that brought this about.

For the estimation of cardiac output using the indicator dilution technique Mayo Clinic's Earl Wood needed a dye that absorbed light at a wavelength of 800 nm, the isosbestic point at which oxyhemoglobin and reduced hemoglobin absorbed equally. This would allow optical detection and quantitation of the dye

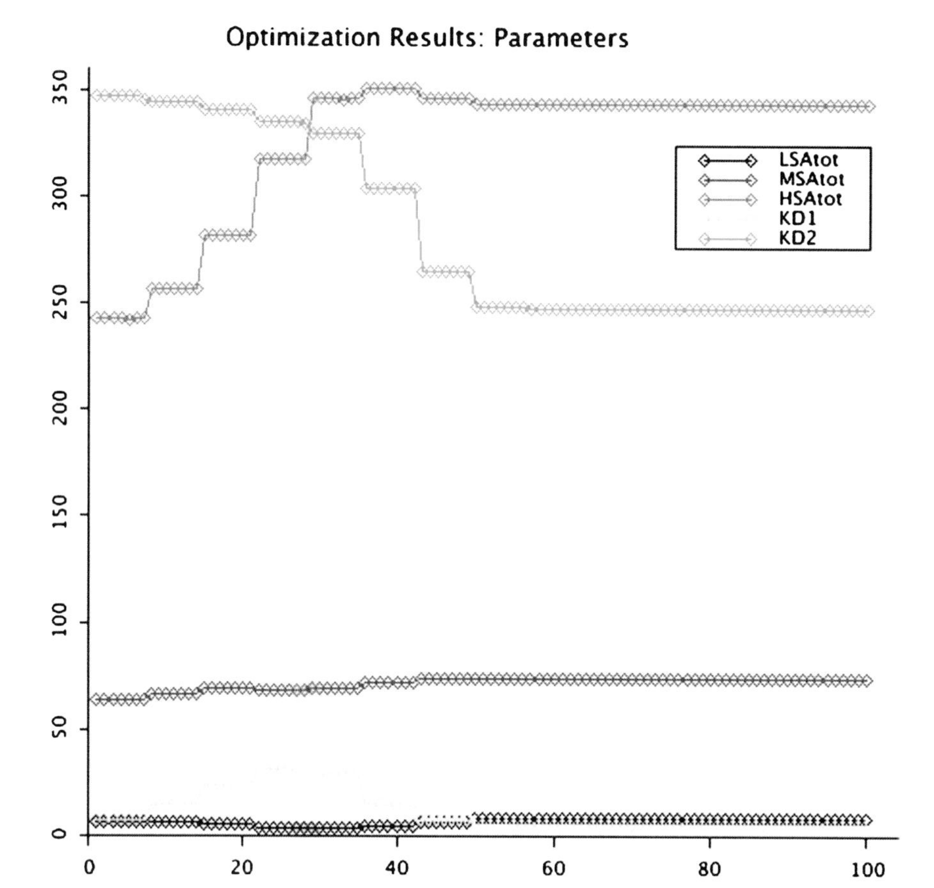

Fig. 7. Optimization: Trajectories of values for parameters being optimized, in this case the three initial concentrations and the two dissociation constants during 100 trials of fitting the model solutions to the salicylate data. The optimizer used was SENSOP (17); the staircase nature of the plotted values is that SENSOP reports the previous estimate of the parameter vector as it calculates each new sensitivity function, one for each parameter being optimized, and then reports and plots the new value of each parameter.

concentration independent of the blood oxygen level. The dye, indocyanine green, was found by Fox in Kodak's repository (20); it was not toxic; its absorbance peaked at 805 nm; it bound to albumin and so stayed in the circulation and did not color the body. The densitometer measuring the absorbance was developed (21), and a spin-off from it was an earpiece densitometer that could be used to detect blood concentration noninvasively. In early experiments to test the accuracy and reproducibility of the estimates of cardiac output, dye was injected repeatedly as a bolus into a vein and the arterial concentration–time curve recorded each time. The repeated injections led to a rise in the background arterial concentration, raising the question of whether the detector could be calibrated accurately over a large range of concentrations (22, 23). In the first studies, we found right away that the dye

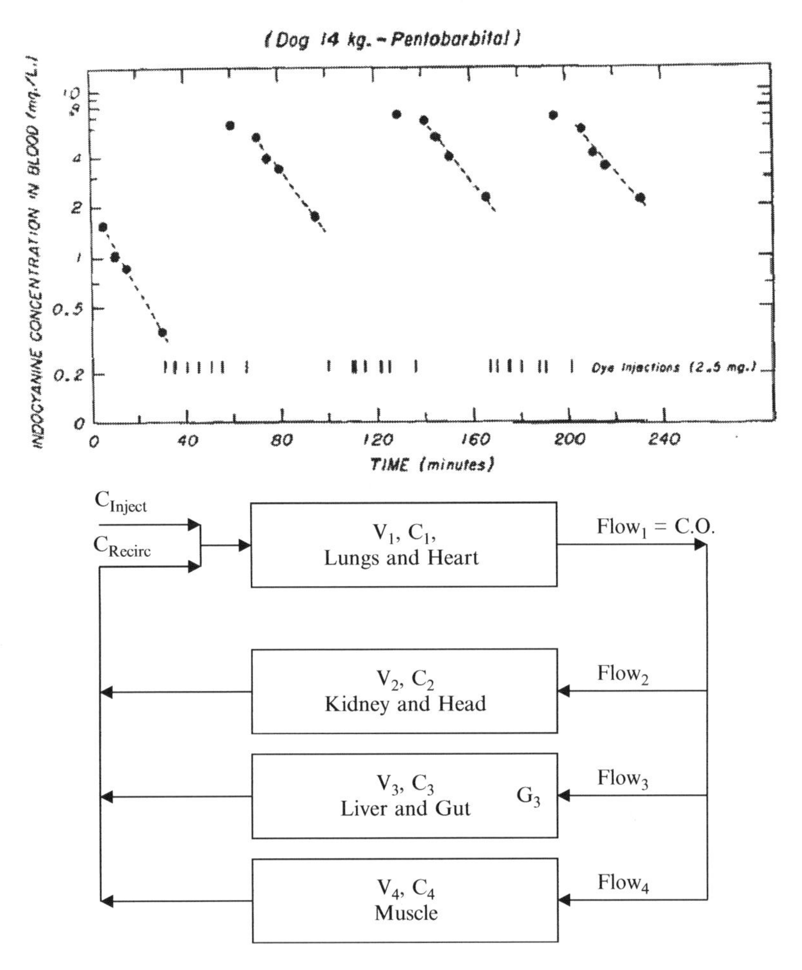

Fig. 8. Indocyanine Green Injection and Clearance. *Upper panel*: Blood concentrations in a 14 kg dog with 22 successive intravenous injections of 2.5 mg ICG (vertical pips). *Dashed lines* represent estimated single exponential decay after each series of injections. Data are from Edwards et al. (22). *Lower panel:* Circulatory model for hepatic clearance of ICG, a mammillary compartmental model. Model code, with all parameters and equations, is in Table 3. Modeling results are in Fig. 9.

was excreted via the bile: the feces turned green! Within a few years the Indocyanine Green clearance Test was used as a liver function test; it was a valued test even before the mechanisms of its hepatic excretion became known (24).

The data from one dog (Edwards, 1960), shown in Fig. 8 (top panel), invite analysis. After each series of injections a set of blood samples were obtained as the concentration diminished: the diminutions appeared as straight lines on the semilog plot shown in Fig. 8, top. This suggests a first-order clearance, i.e., a constant fraction of the dye was being removed per unit time. Now let us develop a simple model, and then test it against the data.

The anatomy and physiology provide the framework for the model. The dye distributes throughout the whole blood volume.

We hypothesize that it is removed at one point only, by the liver, and is excreted into the bile. We have no data on biliary concentration versus time, except that the dye ends up in the lower bowel within the duration of the experiment, 4 h. We model the quantitative data: the doses and their time of injection, the blood concentrations at the particular times, and use the dog's weight, 14 kg, as a constraint. The cardiac output is known from the areas of the dye–dilution curves: C.O. = Dose injected (mg)/Area under dye curve (min. mg/l). These averaged 1.5 l/min, with a standard deviation of about 10% (25).

A simple model of the whole body circulation is chosen as a compromise, three compartments to represent fast, moderate, and slow flows throughout the body and a single lumped compartment to represent blood in the heart and lungs, as in Fig. 8, lower panel. A mammillary model of this sort is far too crude to represent the indicator dilution curves used to measure the cardiac output; this requires higher temporal resolution (26, 27) and more precise models (28, 29). In this model there are up to three parameters for each of the four compartments, flow and volume, and in compartment 3, a consumption term G_3. The shape of the transport function is fixed by the assumption of a mixing chamber, so each has the impulse response $h(t) = (1/\tau)\exp(-t/\tau)$, where the time constant τ = volume/flow. The consumption influences the fraction leaving the compartment. If we were to try to use the observed concentrations in Fig. 8 upper as the input to the model we would see at once that we do not have enough data: the points sampled are too sparse.

But the doses and their times of injection were defined precisely, so we use the sequence of doses as the input. We represent the process of flow distribution and dilution of the injected ICG by the set of four first-order differential equations, stirred tanks, as in traditional compartmental analysis. Within each tank the concentration, C_{in}, is given by

$$dC_i/dt = (F_i \times (\mathrm{Cin_i} - C_i) - G_i \times C_i)/V_i, \tag{9}$$

where the index subscript i denotes the compartment, F is its flow, V is its volume, and G is the rate of consumption within the volume. $\mathrm{Cin_i}$ is the concentration in the blood entering the volume, and equals the concentration in the compartment just upstream. The concentration C_i is the same as the concentration flowing out because of the basic compartmental assumption that the tank is stirred instantaneously from entrance to exit. Via the test of fitting the model to the data we can assess whether or not this assumption is refuted by the data; if the data are fitted, then we do not prove that the assumption is correct, but only that it was not invalidated by the data. (This philosophical point underlies every assumption in formulating a model: a model can never be proven correct. It can only be demonstrated as adequate for the situation, and then can be used as a "working hypothesis," yet to be disproven. As T. H. Huxley put it: "The great tragedy of science: the slaying of a beautiful hypothesis by an ugly fact." The downfall leads to the advancement!)

The JSim code is given in Table 3. The equations for each compartment are closely similar. The recirculated dye and newly injected dye enter the heart and lung compartment with central blood volume V_1, and mix with the total cardiac output. The only clearance, G_3, is during passage through the liver where ICG is highly extracted and secreted into the bile; G_2 and G_4 are available for exploration, but are set to zero.

The modeling results are shown in Fig. 9. Each injection resulted in a sharp rise in concentration; when closely spaced the mean concentration rises in spite of the rapid clearance. In the absence of injections the concentration diminishes almost mono-exponentially as was suggested by the straight lines on the semilog plot in Fig. 8 (upper). If the system were a perfect single mixing chamber with first-order clearance the time constant for washout would be the volume divided by the clearance, V_{tot}/G_3, and in fact this is very close. The test is to plot the theoretical concentration diminution against the model and the data. The equation,

$$\text{Test} = 10 \times \exp(-(t - 58)/(V_{tot}/G_3)),$$

is the red dashed line in Fig. 9, and it does fit the phase where there are no injections after 64 min and before the next injection at 100 min. The multiplier, 10, and the delay, 58 min, in the equation simply position the curve. Test curves with longer delays (130 and 195 min for the purple and blue dashes) fit the data for the later phases without injections. The closeness of these fits suggests that approximating the whole blood volume as a single mixing chamber would give a fairly good fit too. But the exponential Test curves decay a little too rapidly compared to the model function and the data at the lowest concentrations. This systematically better fit of the model compared to the single exponential curve emphasizes that the circulation is not really instantaneously wholly mixed, and that a complete washout curve must be multi-exponential.

With respect to the physiological state of the animal, the fact that the same value for G_3 fits the data throughout the 4-h study says that the hepatic clearance was stable over this long period of anesthesia. We can also conclude that there is no evidence for the saturation of the clearance process over the range of ICG concentrations seen here, fairly high concentrations, nearing 10 mg/l. This implies that the processes for ICG clearance are not only very effective but must also have binding constants high compared to the concentrations found here. This dye soon found use as a clinical test of liver function (24), and is widely used clinically today (30–32). The hepatocyte's apical transporter has a very high capacity so that in the normal liver the transhepatic extraction is nearly 100% and the dye clearance can be used to estimate hepatic blood flow. The reason the extraction is so high is that there is an active ATP-supported extrusion of the dye from the hepatocyte into

Table 3
Compartmental model for hepatic ICG clearance

```
/* MODEL NUMBER: 0103
MODEL NAME: Comp4ICG
SHORT DESCRIPTION: Four-compartment whole body model with recirculation:
Repeated injections and first-order hepatic clearance of Indocyanine Green dye. */

import nsrunit; unit conversion on;
math Comp4ICG {

// INDEPENDENT VARIABLE
realDomain t min; t.min = 0; t.max = 30; t.delta = 0.1;

// PARAMETERS
real Flow1 = 1.5 L/min,          // Blood Flow through Heart/Lung = Cardiac Output
   Dose = 2.5 mg,                // Amt injected
   Vtot = 1. 256 L,              // Total Blood Volume
// First Compartmental unit (Heart/lung)
   Vfr1 = 0.25 dimensionless,    // fraction of Vtot in Comp 1
   V1 = Vfr1*Vtot,               // Volume of blood 1
   C10 = 0 mg/L,                 // Initial concentration
// Second Compartmental unit (Kidney and Head)
   Fr2 = 0.33,                   // Fraction of flow through comp 2
   Flow2 = Fr2*Flow1,            // Flow through Comp 2
   Vfr2 = 0.40,                  // fraction of Vtot in Comp 2
   V2 = Vfr2*Vtot,               // Volume of blood 2
   C20 = 0 mg/L,                 // Initial concentration
   G2 = 0 ml/min,                // consumption 2
// Third Compartmental unit (Liver and Gut)
   Fr3 = 0.25,                   // Fraction of Flow through Comp 3 (Liver)
   Flow3 = Fr3*Flow1,            // Flow through Comp 3
   Vfr3 = 0.25,                  // fraction of Vtot in Comp 3
   C30 = 0 mg/L,                 // Initial concentration
   G3 = 63.7 ml/min,             // consumption 3
// Fourth Compartmental unit (Muscle)
   Fr4 = 1 - Fr2 - Fr3,          // Fraction of Flow through Comp 4
   Flow4 = Fr4*Flow1,            // Flow through Comp 4
   Vfr4 = 1 -Vfr1 -Vfr2 -Vfr3,   // fraction of Vtot in Comp 4
   V4 = Vfr4*Vtot,               // Volume of blood 4
   C40 = 0 mg/L,                 // Initial concentration
   G4 = 0 ml/min;                // Consumption 4

// Total flow relationship: Flow4 = Flow1 - Flow2 - Flow3;

// EXTERNAL VARIABLE: The series of ICG injections, 2,5 mg each
extern real Cin1(t) 1/min;   // 60 second Pulse injection @ 0.1 min
extern real Cin2(t) 1/min;   // 60 second Pulse injection @ 30 min
extern real Cin3(t) 1/min;   // 60 second Pulse injection @ 34 min
extern real Cin4(t) 1/min;   // 60 second Pulse injection @ 39 min
extern real Cin5(t) 1/min;   // 60 second Pulse injection @ 44 min
extern real Cin6(t) 1/min;   // 60 second Pulse injection @ 49 min
extern real Cin7(t) 1/min;   // 60 second Pulse injection @ 54 min
extern real Cin8(t) 1/min;   // 60 second Pulse injection @ 64 min
```

(continued)

Table 3 (continued)

```
extern real Cin9(t) 1/min;   // 60 second Pulse injection @ 99 min
extern real Cin10(t) 1/min;   // 60 second Pulse injection @ 109 min
extern real Cin11(t) 1/min;   // 60 second Pulse injection @ 110 min
extern real Cin12(t) 1/min;   // 60 second Pulse injection @ 114 min
extern real Cin13(t) 1/min;   // 60 second Pulse injection @ 120 min
extern real Cin14(t) 1/min;   // 60 second Pulse injection @ 124 min
extern real Cin15(t) 1/min;   // 60 second Pulse injection @ 135 min
extern real Cin16(t) 1/min;   // 60 second Pulse injection @ 165.9 min
extern real Cin17(t) 1/min;   // 60 second Pulse injection @ 168.8 min
extern real Cin18(t) 1/min;   // 60 second Pulse injection @ 174.1 min
extern real Cin19(t) 1/min;   // 60 second Pulse injection @ 179.3 min
extern real Cin20(t) 1/min;   // 60 second Pulse injection @ 186.9 min
extern real Cin21(t) 1/min;   // 60 second Pulse injection @ 189.6 min
extern real Cin22(t) 1/min;   // 60 second Pulse injection @ 200.8 min

// DEPENDENT VARIABLES
real Crecirc(t) mg/L,                                  //Inflow + recirculation
   Qtot(t) mg,                                         // Total amt in 4 compartments
   Cinsum(t) 1/min,   // Sum of the string of pulse injections
// Comp1 Comp2 Comp3 Comp4
   C1(t) mg/L, C2(t) mg/L, C3(t) mg/L, C4(t) mg/L,    // concn in each region
   Q1(t) mg, Q2(t) mg, Q3(t) mg, Q4(t) mg;             // Quantity in each regions

// INITIAL CONDITIONS
when(t = t.min) {C1 = C10; C2 = C20; C3 = C30; C4 = C40;}

// INPUT CONCENTRATION FUNCTION
   real Qinjrate(t) mg/min;
   Cinsum = (Cin1+ Cin2 + Cin3 + Cin4 + Cin5 + Cin6 + Cin7 + Cin8+
   Cin9+ Cin10+ Cin11+ Cin12+ Cin13+ Cin14+ Cin15+
   Cin16+ Cin17+ Cin18+ Cin19+ Cin20+ Cin21+ Cin22);
   Qinjrate = Dose*Cinsum;
   Crecirc = (C2*Flow2+ C3*Flow3 + C4*Flow4)/Flow1;

// QUANTITY Retained at time t:
Qtot = V1*C1 + V2*C2 + V3*C3 + V4*C4                  // Qtot is the total amount of ICG in the
   body at time t.
real Area = Dose*(integral(t = t.min to t.max, Cinsum));   // check amt injected

// ORDINARY DIFFERENTIAL EQUATIONS
V1*C1:t = Flow1*(Crecirc-C1) + Cinject;               // Note that V1 can be left of = sign
C2:t = Flow2*(C1-C2)/V2 - G2*C2/V2;
C3:t = Flow3*(C1-C3)/V3 - G3*C3/V3;
C4:t = Flow4*(C1-C4)/V4 - G4*C4/V4;

}                                                      // END of program
/*

DETAILED DESCRIPTION:
This is a whole body model composed of a central blood volume from which flows the whole cardiac
   output. The C.O. is distributed into three organs labeled "kidney" for kidney and head, "liver" for
   liver and intestines, and "muscle" for the rest of the body.
```

(continued)

Table 3 (continued)

The four-compartment model has recirculation. The input function to compartment 1 (Heart and Lung) is the sum of the recirculated indicator plus the series of injection pulses into V1, Qinjrate, each injection at x mg/min for a short duration. Each injection, Cin#, is defined at run time using a separate function generator. Clearance of the injected dye, indocyanine green, is hepatic extraction from the blood via a saturable transporter on the hepatocyte sinusoidal membrane followed by ATP-dependent excretion across the hepatocyte apical membrane into the bile, but is represented here by a passive first-order loss, G3. This is adequate kinetically only at low concentrations of ICG, where the transporter is mainly uncomplexed. */

KEY WORDS: compartment, flow and exchange, mixing chamber, hepatic clearance. first-order consumption, washout, organ, multi-organ, recirculation

REFERENCES:

Edwards AWT, Bassingthwaighte JB, Sutterer WF, and Wood EH. Blood level of indocyanine green in the dog during multiple dye curves and its effect on instrumental calibration. Proc S M Mayo Clin 35: 747-751, 1960.

Edwards AWT, Isaacson J, Sutterer WF, Bassingthwaighte JB, and Wood EH. Indocyanine green densitometry in flowing blood compensated for background dye. J Appl Physiol 18: 1294-1304, 1963.

REVISION HISTORY:

Author: BEJ 06jan11

Revised by: JBB 09jan11 to combine the function generators, fgens, to speed computation

COPYRIGHT AND REQUEST FOR ACKNOWLEDGMENT OF USE:

Copyright (C) 1999-2011 University of Washington. From the National Simulation Resource, Director J. B. Bassingthwaighte, Department of Bioengineering, University of Washington, Seattle WA 98195-5061.

Academic use is unrestricted. Software may be copied as long as this copyright notice is included.

This software was developed with support from NIH grant HL073598.

Please cite this grant in any publication for which this software is used and send an e-mail with the citation and, if possible, a PDF file of the paper to staff@physiome.org.

*/ Table 3 is an example of a complete set of MML code in the format used in general for JSim project files, having the same sequence and format as those used in the repository of models at www.physiome.org/Models.

the bile, thus keeping the intra-hepatocyte concentration low. These features cannot be demonstrated at the low concentrations seen in Figs. 8 and 9, but one would expect that the model would have to be revised to include a saturable transporter if the concentrations were a lot higher.

5. Summary: The Processes Undertaken in Pharmacokinetics

In Subheading 4 we covered a standard approach to the steps in a modeling analysis of data. The order of the steps depends a little on the nature of the task. In the first model we performed no

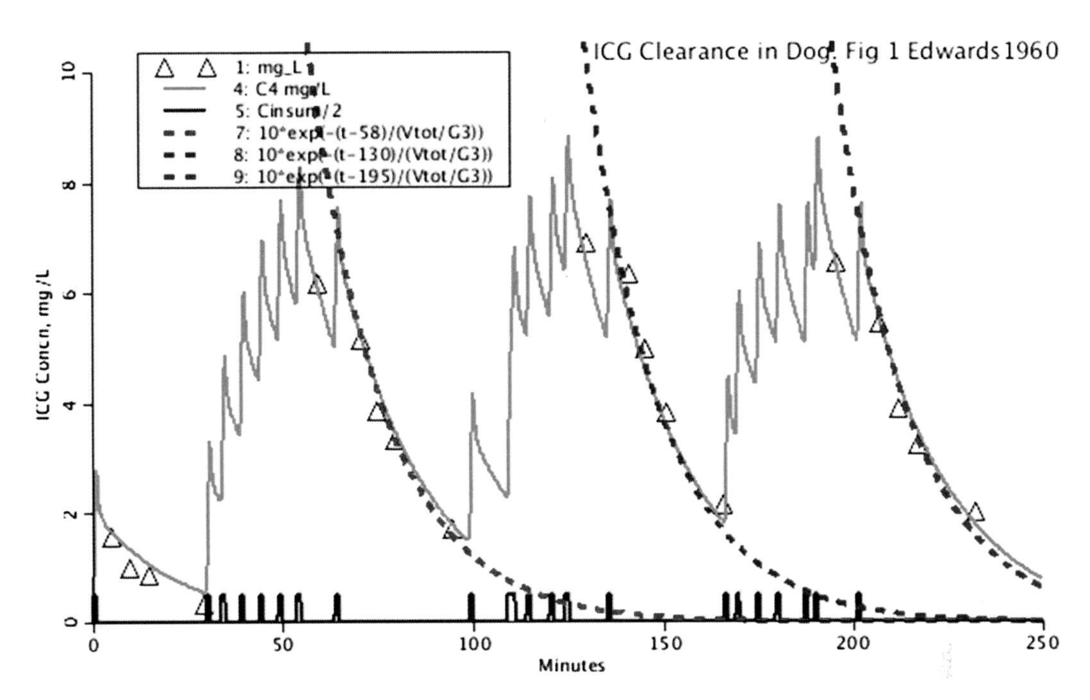

Fig. 9. Model solution to fit the ICG data in a 14 kg dog. The *triangles* are the data shown in Fig. 8 (*upper*). The parameters and initial conditions for the model are those given in the code in Table 3. The parameters are the result of optimization using NL2SOL from Dennis and Schnabel (19); the total blood volume, V_{tot}, and the hepatic clearance, G_3, were the only free parameters. The 2.5 mg injection pulses are shown along the abscissa. The dashed lines are mono-exponentials with time constant V_{tot}/G_3. The model is #103 at http:/www.physiome.org/jsim/models/webmodel/NSR/Comp4ICG/.

verification steps, and in the second the verification was done after the fitting of the model to the data. This is clearly in the wrong order: there is no point fitting a model to the data until it has been demonstrated to be computed correctly, so in the list that follows, the verification is done as soon as the draft model is constructed. One cannot argue that the verification is not needed until the model fits the data: a mathematically incorrect model might fit the data, and after all that work the effort would be shown to be a waste if the code had an error. Our failure to precede the data fitting by a formal verification in these two cases is based on prior observations that JSim's solutions for compartmental models provide four-digit accuracy compared to analytical solutions.

Taking a listing of the steps to a detailed level:

- Ideally, design the experimental protocol to be the best test of the model.
- Gathering supporting data, assess experimental accuracy.
- Obtain information on necessary parameters, a priori, and on possible constraints.
- Complete development of the model. List all the assumptions.

- Conduct simulations, compare the results with other methods of solution, perform verification tests, and find limiting cases for which there are analytical solutions.
- Determine how one should weight the data, for use in minimizing the sum of squares of differences between model solutions and data.
- Validate that the model is reasonable, that it provides good fits and that the, residual errors are not systematic or localized.
- Comparing simulation results to experiments and/or results of other methods.
- Post-processing: Analyzing the results for consistency with respect to physical and chemical and physiological expectations.
- Interpret the results scientifically. What does the model predict?
- Rethink the process. What are the weakest assumptions in the model? How might it be improved? The *model is always wrong*. Figure out new tests of it. Where might its predictions fail?

This overall process can be considered a success, if:

- Observations hitherto unexplained now fit a rational working hypothesis.
- The essence of the phenomenology has been captured. (A descriptive success only, perhaps.)
- Diagrams and schema of interactions aid understanding the model, the kinetic relationships, and the code.

6. Model Alternatives and Modifications: Interactive Hypothesis Revising

When the fit is not precise, outside of the limits of expectation relative to the noise in the data, despite all the attempts, then maybe the model is just wrong. Certainly it is not nicely descriptive, let alone explanatory! Given the philosophical premise that all models are wrong, in the sense of being incomplete, or incorrect mechanistically, every failure is a stimulus to find an alternative model. A most rewarding and successful strategy is that of Platt (12): he proposes that right from the outset one should have alternative hypotheses in mind, and that the experiment should be designed to distinguish between these hypotheses. "Strong Inference" is the title of his paper. We advocate that each hypothesis be expressed in terms of a computational model, since that means that it is described explicitly and is therefore testable. The strategy pays off because at least one of the hypotheses is proven wrong when the model fails to fit the data. Sometimes both are wrong! Regroup, rethink!

The least squares approach, minimization of the sums of squares of the differences between the data points and the model function, is a blunt tool, without specific information with respect to any misfitting. Displaying the residuals, plotting the point-by-point differences as a function of time (or in general, of the independent variable), is an excellent way to get insight into what to do next: a series of points above or below zero reveals a systematic misfitting. Comparing the time course of the deviations of the data from the model with the sensitivity functions of the various parameters could suggest that a particular parameter has not been optimized well, but this is most unlikely when the various options for optimization have been explored. It is more likely that the model lacks a feature that is needed. Back to the drawing board!

This is the usual iterative process: hypothesis in general terms, develop the model that represents a clear precise hypothesis, if possible design the model for an alternative hypothesis, design the distinguishing experiment while taking into account the accuracy of the data to be acquired, and disprove one or both hypotheses. The next step is to improve the model and repeat the series of steps until a satisfactory level of synchrony between model and experiment is achieved. This version of the model is usually then designated the working hypothesis. No working hypothesis is to be regarded as "the truth," though it is useful for practical purposes. And in fact it serves as the standing target to be disproved in order to advance the science.

7. What to Do When the Compartmental Representation Is not so Good?

What follows is a common example that applies in biophysics, physiology, and pharmacology. It is usual that drugs and substrates for metabolism and signaling molecules of molecular weight less than 1,000 Da are partially extracted during single passage through a capillary. Since capillaries are about a 1,000 μm long, but only 5 μm in diameter (as in Fig. 10), there is no possibility that they are instantaneously stirred tanks with uniform concentration from end to end: there must be gradients between capillary entrance and exit for any solute that is exchanging between blood and tissue. If the extraction is less than 5%, the gradient might be ignored, but for solutes of interest to us here the steady-state extractions are 30–90%, and so affect the estimates of the permeabilities and consumption rates. Consequently we now consider the computational differences between a stirred tank and an axially distributed capillary–tissues exchange unit.

Fig. 10. A venule and capillaries on the epicardium of a dog heart casted with microfil. Capillary diameters are 5.6 ± 1.3 μm, average intercapillary distances are 17–19 μm, and lengths are 800–1,000 μm. The distance between the long calibration lines is 100 μm. Modified from Bassingthwaighte et al. (33).

7.1. Capillary–Tissue Exchange: Convention, Permeation, Reaction, and Diffusion

A two-compartment system is here modified to incorporate flow, as shown in Fig. 11 (left panel), thus identifying V_1 as the vascular region, the membrane as the capillary barrier, and V_2 as the tissue. The system is considered as a homogeneously perfused organ with constant volumes and steady flow, F, in and out. Now, in order to put it into the context of substrate delivery and metabolism, we switch to standard physiological representation of the units, defining them per gram of organ mass. F, PS, and the consumption G have units ml/(g min), and the volumes have units ml/g. This notation normalizes flows, substrate use, etc. to be independent of organ mass. (The model is www.physiome.org/jsim/models/webmodel/NSR/Comp2FlowExchange/, #247.)

To keep the system simple so as to focus on the blood–tissue exchange, the intratissue consumption is considered to be a first-order process, as if the substrate concentration is far below the KD for any enzymatic reaction.

In the right panel of Fig. 11 is the equivalent axially distributed model that accounts for gradients in concentration along the length of the capillary. It is more general, but reduces to an exactly analogous model when the PS's are set to zero, so there is no entry into the cells. Other parameters for cellular permeation and reaction are as follows: for passage across endothelial cell luminal membrane (PS_{ecl}); endothelial cell abluminal membrane (PS_{eca}); and parenchymal cell membrane (PS_{pc}); G, intracellular consumption ml/(g min) or metabolism of solute by endothelial cells (G_{ec}) or by parenchymal cells (G_{pc}). The Vs, ml/g, are volumes of distribution in plasma (V_{pl}), endothelial cell (V_{ec}), interstitial (V_{isf}), and parenchymal cell (V_{pc}). With the cell permeabilities, PS_{ecl}, PS_{eca}, and

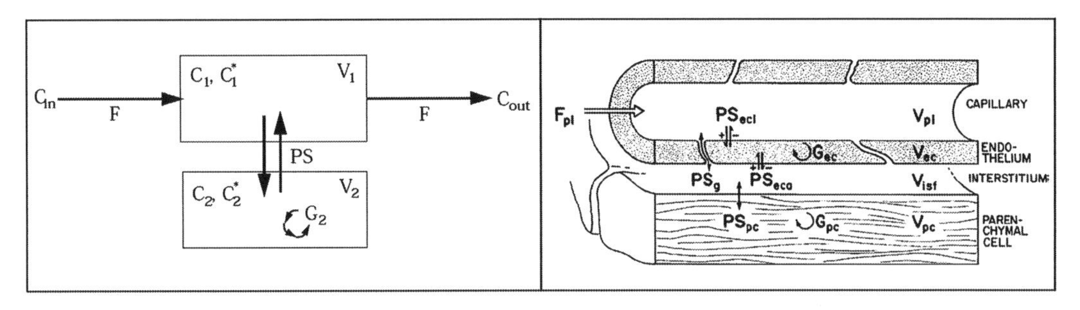

Fig. 11. Compartmental versus axially distributed models for capillary–tissue exchange. Exchange between capillary plasma and interstitial fluid regions can be regarded as providing two conceptually similar but mathematical distinguishable methods of representation. *Left panel:* Two compartment stirred tank for the exchange of solute C between flowing blood and surrounding stagnant tissue. Flow *F*, ml/(g min), carries in solute at concentration C_{in}, mM, and carries out a concentration $C_{out}t = C_p$, mM. The flux from capillary to ISF (compartment 2) is limited by the conductance PS ml/(g min), the permeability-surface area product of the membrane separating the two chambers, allowing bidirectional flux. G_2 ml/(g min) is a reaction rate for a transformation flux forming product at a rate $G2 \cdot C_2$ mmol/(g min). *Right panel*: Axially distributed model equivalent to the two-compartmental model when the solute does not enter the endothelial or parenchymal cells. PS_g, the conductance for permeation through the interendothelial clefts, is equivalent to the PS of the compartmental model in the *left panel*. *Right panel* from Gorman et al. (18) with permission from the American Physiological Society.

PS_{pc}, set to zero, PS_g is equivalent to the compartmental PS, and V_{pl} and V_{isf} are equivalent to V_1 and V_2. In each of the four regions there is an axial dispersion coefficient, equivalent to a diffusion coefficient, setting the rate of random spreading along the length. The full version of this model, GENTEX, a general multispecies model for blood-tissue exchange is available at: www.physiome.org/jsim/models/webmodel/NSR/gentex.proj.

Chinard (34) developed a technique to distinguish individual processes involved in blood–tissue exchange and reaction: the Multiple-Indicator Dilution (MID) technique (Fig. 12). He first used it for the purpose of estimating the volumes of distribution for sets of tracers of differing characteristics: the mean transit time volume, $V_{mtt} = F \times \bar{t}$ where $\bar{t}$ is the mean transit time through the system. He did not estimate permeabilities as his studies were on highly permeable solutes. Crone (35) analyzed the technique, showing how it could be used to estimate PS from the outflow curves for a simultaneous injection into the inflow of a solute and impermeable reference intravascular tracer as shown in Fig. 12. The figure diagrams an experimental setup for examining the uptake of D-glucose in an isolated perfused heart as by Kuikka et al. (36). L-Glucose, the stereoisomer, serves as an extracellular, non-metabolized reference. A more realistic diagram of a capillary–tissue exchange includes the endothelial cells and interstitial fluid (ISF), as shown in Fig. 11.

To determine capillary permeability, the relevant reference solute is one that does not escape from the capillary blood during single transcapillary passage; for example, albumin is the relevant reference solute to determine the capillary permeability to glucose. In this situation the albumin dispersion along the vascular space

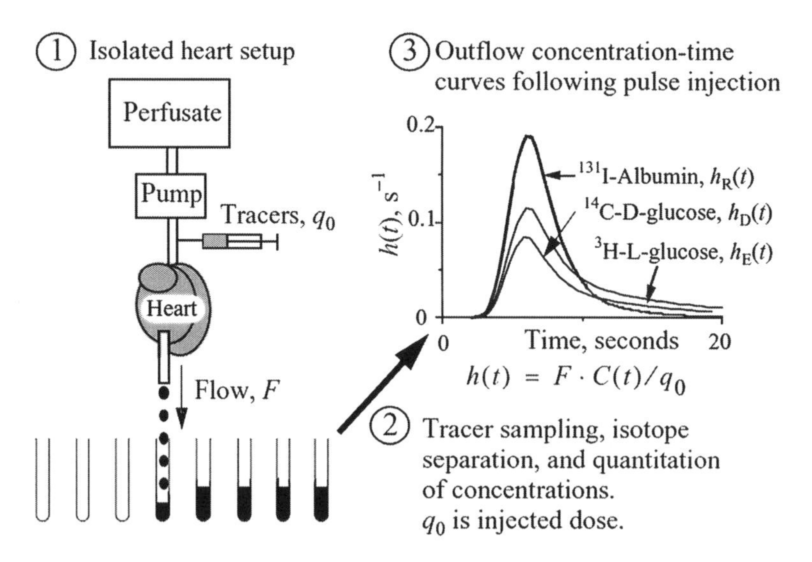

Fig. 12. Schematic overview of experimental procedures underlying the application of the multiple-indicator dilution technique to the investigation of multiple substrates passing through an isolated organ without recirculation of tracer. The approach naturally extends also to their metabolites.

may be assumed to be the same as that of the glucose; thus the shape of the albumin impulse response, $h_{\mathrm{alb}(t)}$, accounts for the intravascular transport of all the solutes. (L-Glucose, an extracellular reference tracer with the same molecular weight and diffusivity as D-glucose, is the extracellular reference for D-glucose, having the same capillary PS_g and the same interstitial volume of distribution, V_{isf}. Having simultaneous data on such reference tracers greatly reduces the degrees of freedom in estimating the parameters of interest for D-glucose.)

7.2. Model Equations for Tracer

The two diagrams in Fig. 11 look quite different, but the second can be reduced to the compartmental model, as we will show below. The essential difference is that the distributed model accounts for concentration gradients along the capillary length. Capillaries are about 1 mm long, and are 5 μm in diameter, an aspect ratio of 200. Diffusional relaxation times thus differ by a factor of 200 between radial and axial directions. Consequently, considering the capillary as a stirred tank is unreasonable.

The stirred tank expressions account for the flow through compartment 1, the permeation, and consumption terms G_2 ml/(g min) in the second compartment:

$$\frac{\mathrm{d}C_1}{\mathrm{d}t} = -\frac{\mathrm{PS}}{V_1}(C_1 - C_2) - \frac{F}{V_1}(\mathrm{Cin} - C_1),$$
$$\frac{\mathrm{d}C_2}{\mathrm{d}t} = +\frac{\mathrm{PS}}{V_2}(C_1 - C_2) - \frac{G_2}{V_2} \times C_2. \tag{10}$$

The use of these ODEs implies and builds into the calculations a discontinuity between the concentration of solute in the inflow and that in V_1. Because V_1 is assumed instantly mixed, there is no gradient along the capillary and the tracer entering the tank is immediately available to be washed out with the same probability as any molecule dwelling in there for a longer time.

Alternatively the capillary–tissue unit of Fig. 11 right, can be reduced to two regions represented by PDEs that allow a continuous gradient along the length of the capillary between entrance and exit, like those of Krogh and Erlang (37), Sangren and Sheppard (38), and Renkin (39). It improved upon these by incorporating axial dispersion in both the capillary and extravascular regions. Using the spatially distributed analogs for plasma, C_{pl}, or blood, and extravascular tissue, C_{isf}, instead of the lumped variables C_1 and C_2:

$$\frac{\partial C_{\mathrm{p}}(x,t)}{\partial t} = -\frac{F_{\mathrm{pl}}L}{V_{\mathrm{pl}}}\frac{\partial C_{\mathrm{pl}}}{\partial x} - \frac{\mathrm{PS_g}}{V_{\mathrm{pl}}}\left(C_{\mathrm{pl}} - C_{\mathrm{isf}}\right) + D_{x1}\frac{\partial^2 C_1}{\partial x^2},$$

$$\frac{\partial C_{\mathrm{isf}}(x,t)}{\partial t} = \frac{\mathrm{PS_g}}{V_{\mathrm{isf}}}\left(C_{\mathrm{pl}} - C_{\mathrm{isf}}\right) - \frac{G_{\mathrm{isf}}}{V_{\mathrm{isf}}} \times C_{\mathrm{isf}} + D_{x2}\frac{\partial^2 C_2}{\partial x^2}, \qquad (11)$$

where C_{pl} and C_{isf} are spatially distributed functions of both x and t, not just t. The axial position is denoted by x, where $0 < x < L$, the capillary length, cm. The analogy between this model, Eq. 11, and the compartmental version in Eq. 10 is $F_{\mathrm{p}} = F$, $V_{\mathrm{p}} = V_1$, $V_{\mathrm{isf}} = V_2$, and $\mathrm{PS_c} = \mathrm{PS}$, the permeability-surface area of the capillary wall, but we retain the two sets of names in order to allow comparisons between the estimated parameter values. The capillary length, L, is arbitrarily set to an average value such as 0.1 cm; in the computations what is being used is the dimensionless fractional length, x/L.

PDEs require boundary conditions. At the capillary entrance, in contrast to the compartmental model there is no discontinuity in the concentration profile, but there is a requirement for matching the diffusional and convective terms so that the influx is just the convected mass, $FC_{\mathrm{p}}(x = 0, t)$. The boundary conditions are written:

$$\text{when } (x = x_{\mathrm{min}})\{(-F_{\mathrm{p}} \times L/V_{\mathrm{p}}) \times (C_{\mathrm{p}} - C_{\mathrm{in}}) + D_{\mathrm{p}} \times \mathrm{d}C_{\mathrm{p}}/\mathrm{d}x = 0;\}$$
at inlet to capillary,
(12)

$$\text{when } (x = x_{\mathrm{max}})\{\mathrm{d}C_{\mathrm{p}}/\mathrm{d}x = 0;\ C_{\mathrm{out}} = C_{\mathrm{p}};\}$$
at exit from capillary. (13)

The form of the inlet condition is important when the diffusion is large, and we use it here for conceptual and practical accuracy. The outflow concentration C_{out} is set equal to the concentration just inside the exit, $C_{\mathrm{p}}(x = L, t)$, the same condition described by the ODEs.

The last term in each equation is the diffusion along the length of the capillary–tissue regions; the use of an anatomically correct length then makes using observed diffusion coefficients for D_{p} and D_{tiss}, cm^2/s, practical and meaningful. Gross exaggeration of the diffusion coefficients can be used in the equations to turn the distributed model into a de facto well-mixed, compartmental model. With both diffusion coefficients set to infinity this model behaves identically to the compartmental model of Eq. 10.

The negative sign on the flow term, the first term on the right of the equal sign in Eq. 11, merits further explanation. Consider the inflow to contain a bolus of solute: as it enters, the concentration at the capillary entrance rises. At this time, the slope of the curve of concentration versus position x, $\partial C/\partial x$, is negative as illustrated in the lower panel of Fig. 13 by the slope of $C(x, t)$ for the bolus shape at $t = 1.5$ s, at the capillary mid point, $x = 0.5$ L. The spatial slope has always the sign opposite to the temporal derivative $\partial C/\partial t$ at the same point, thus the negative sign on the term.

Functionally, therefore Eq. 11 is analogous to Eq. 10. But using the PDEs avoids the unrealistic discontinuity in the compartmental model at the entrance. Obnoxious, *unrealistic discontinuities at the entrances are the consequence of the instantaneous mixing within a compartment. Using the PDE allows continuity in concentrations* and concentration gradients along the capillary, and not only in concentration but also in the properties of the system such as axial gradients in transporter and enzyme densities that are evident in the liver sinusoid. For the following analysis, all parameters are assumed spatially uniform so as to minimize the difference from the compartmental models. There are many ways of representing axially distributed convecting systems, and two are shown in Fig. 13. One solves PDEs using a PDE solver (there are several choices within JSim), and here we used a Lagrangian method (40–42). A compartmental type of alternative is to approximate the capillary as a series of stirred tanks, each with the same volume and PS, as diagrammed in the upper part of the figure. With a large number of serial stirred tanks, the longitudinal concentration gradient is approximated increasingly accurately as the steps from one to the next are small. The intravascular transport process with serial stirred tanks is a Poisson process. In modeling, serial stirred tanks are convenient because the number of tanks, Ntanks, can be used as a free parameter. The relative dispersion RD over the length of the tube is determined by Ntanks such that the RD of the outflow curve (RD equals the coefficient of variation, the standard deviation divided by the mean transit time), induced during transit, is the reciprocal of the square root of Ntanks, so that with 100 tanks the RD is 10%.

Figure 13 (upper) shows that curves for the PDE solution and for the Poisson process are essentially similar, so that the dispersion coefficient D_p sufficed to create the same dispersion as occurred with the Poisson process using 109 tanks. The choice of 109 tanks is arbitrary, large that a plot of C(Ntanks) versus N would appear smooth. Because the capillary PS > 0, there is loss of solute as the

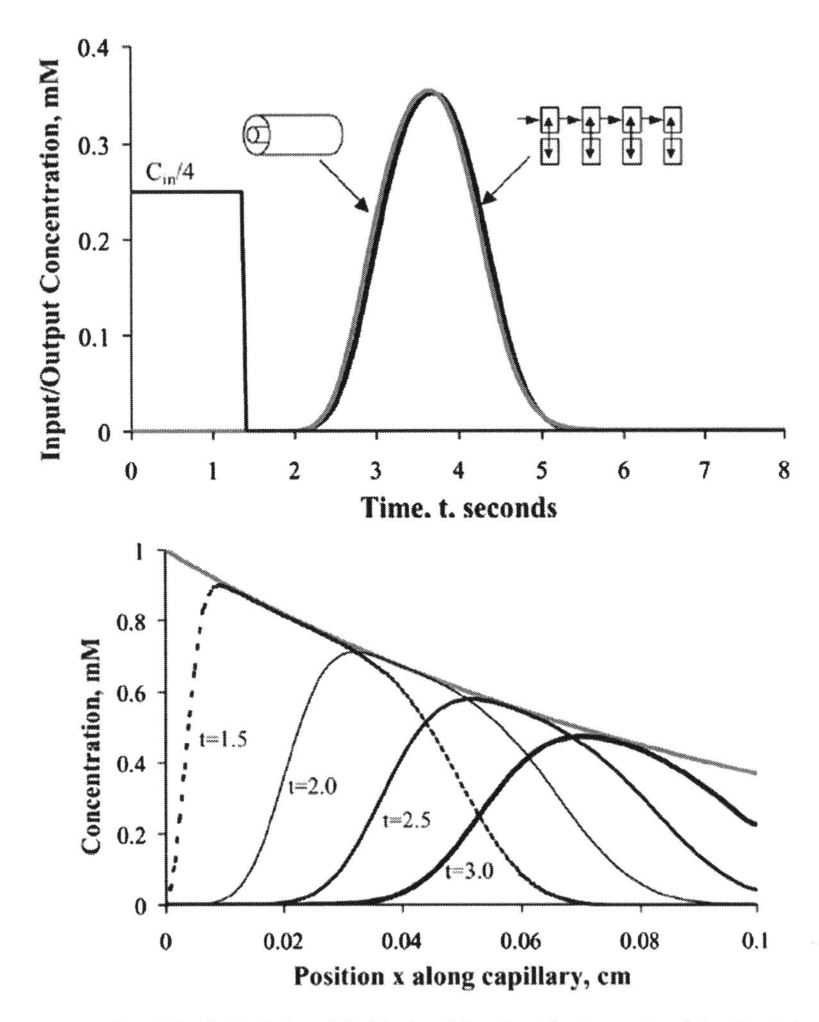

Fig. 13. Pulse responses in axially distributed models. The input function, C_{in}, is a pulse of duration 1.4 s. *Upper panel*: Outflow concentration–time curves for a partial differential equation solution using a Lagrangian sliding fluid element method and an intravascular dispersion coefficient, $D_p = 2.6 \times 10^{-5}$ cm^2/s (*gray curve*), and for a serial stirred tank algorithm representing a Poisson process with 109 stirred tanks (*black curve* almost superimposed on the *gray* one). *Lower panel*: Intracapillary spatial profiles in the distributed model (using the PDEs) at a succession of times, 1.5, 2.0, 2.5, and 3.0 s. The pulse slides and disperses due to the diffusion while some solute is lost from the vascular space by permeation of the capillary wall. Parameters were the same for the compartmental 109 tank Poisson model and the PDE: $F_p = 1$ ml/(g min), PSC = 2 ml/(g min), and tissue volume V_{tiss} was set to 10 ml/g so that there was negligible tracer flux from tissue back into the plasma space. (The model: http:/www.physiome.org/jsim/models/webmodel/NSR/Anderson_JC_2007/FIGURES/Anderson_JC_2007_fig11/index.html) Figure from Anderson (6).

bolus progresses along the capillary. The permeative loss is the same for both methods, with the result that the peak outflow concentrations are similar. Figure 13 (lower) shows the shape of the bolus as a function of position as it deforms continuously from its initial square pulse at the entrance to the capillary. The diminution in peak height is therefore due not only to the spreading but also to the loss. This loss is reflected of course in the reduction in the areas.

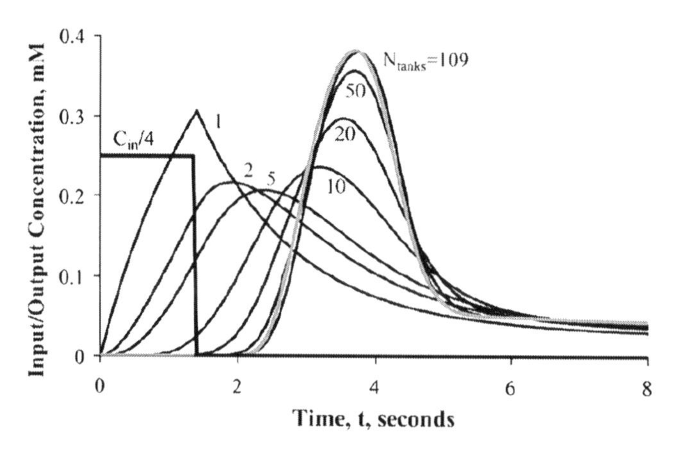

Fig. 14. Effect of reduction of Ntanks on the output $C(t)$. Responses of the *N*th order Poisson operator with Ntanks varied from 109 tanks in series down to 50, 20, 10, 5, 2, and finally to a single mixing chamber, Ntanks = 1. The *gray curve* is the Lagrangian solution to the PDEs as in Fig. 13. All of the Poisson operator outflow curves (*black*) have the same mean transit time, and the same parameters: $V_p = 0.05$ and $V_{tiss} = 0.15$ ml/g; $F_p = 1$ ml/(g min), PSg = 1 ml/(g min). Figure from Anderson (6). Model is #46, running loop mode to change Ntanks (http:/www.physiome.org/jsim/models/webmodel/NSR/Anderson_JC_2007/FIGURES/Anderson_JC_2007_fig12/index.html).

The grey curve touching to the top of each spatial profile is the theoretical curve from Crone (35) and Renkin (39), as in the model of Sangren and Sheppard (38):

$$C(x) = 1 - e^{-\frac{PS_g \bullet x}{F_p \bullet L}}, \tag{14}$$

where x/L is the fractional distance along the capillary, the abscissa in the lower panel.

Now that we know that the multi-compartmental serial tank representation can give results approximating the normal PDE representation, and that they differ basically only in the numerical method used, the question becomes: "Which of the methods produces the correct assessment of the parameters with the greatest efficiency?" The serial stirred tank model has the disadvantage that the waveforms are seriously distorted by reducing the number of stirred tanks, as is shown in Fig. 14.

While reducing the number of tanks in the stirred tank method has a dramatic effect on the shapes of the outflow curves, the problem is much less severe with the PDE representation, as shown in Fig. 15. Solutions are shown for Ngrid = 109, 51, 21, 11, and 7 for two methods of solving PDEs, one using a robust solver TOMS731 (43) and the other using a Lagrangian sliding fluid element algorithm (42).

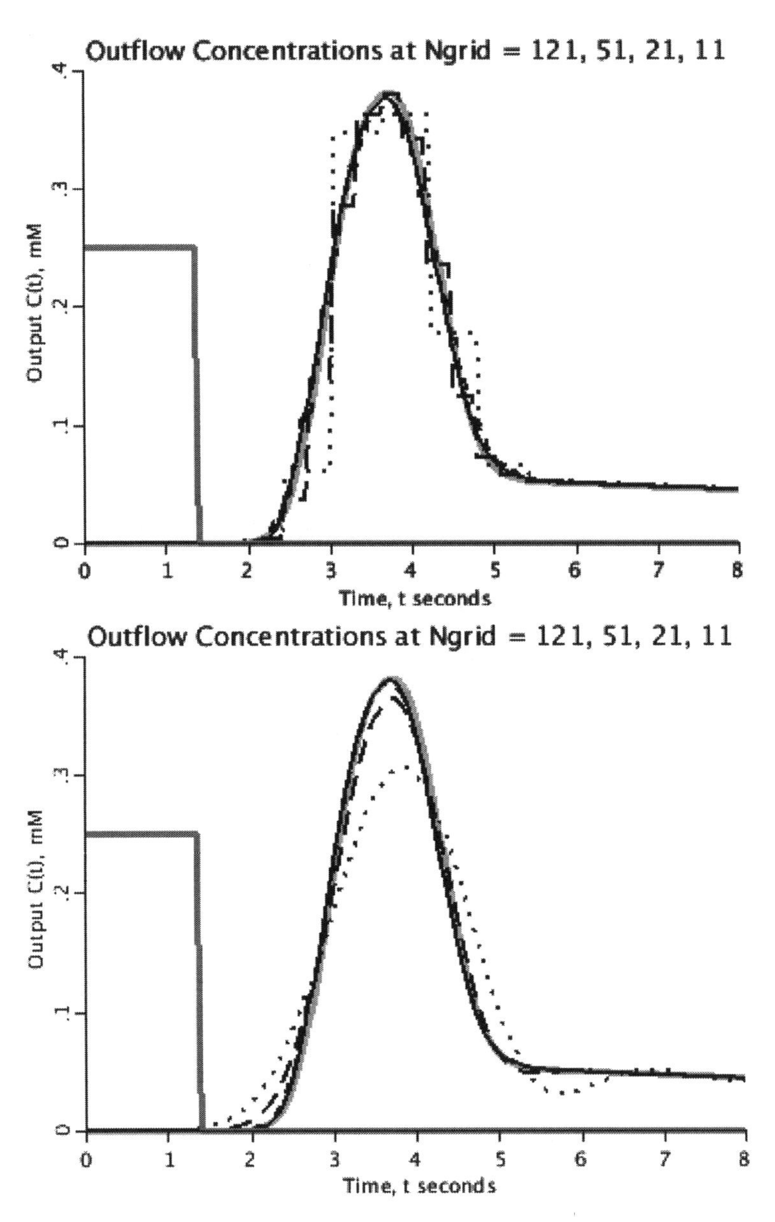

Fig. 15. Effect of reduction of Ngrid on the output $C(t)$ with two PDE solvers. The *green curve* is the serial compartmental model with 109 tanks. The *red rectangle* is the input function divided by 4. *Black curves* are the solutions to the PDE at varied resolutions. *Upper panel*: Lagrangian sliding fluid element algorithm, LSFEA, with Ngrid = 121 *black solid line*, 51 *short dashes* superimposed on the black solid line, 21 *long dashes*, and 11 *dotted*. This algorithm, while computationally fast, describes the outflow dilution curve as a series of square pulses; with a large number of segments the curves appear smooth. With fewer segments, e.g., Ngrid = 21 (*long dashes*) or 11 (*dotted line*), the steps are obvious but the approximation is reasonably good. *Lower panel*: TOMS731 solver. This slow, robust solver, like most PDE solvers, broadens the solution somewhat with reduced Ngrid but with Ngrid = 11 (*dotted line*) there is obvious spreading and oscillation in the solutions (www.physiome.org Model # 126).

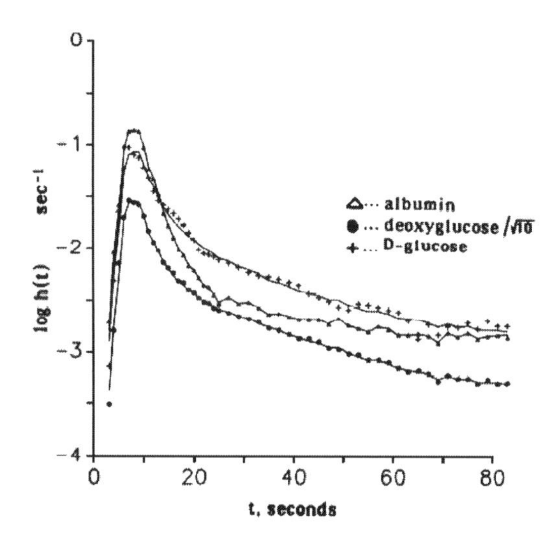

Fig. 16. Outflow dilution curves for D-glucose, 2-deoxy-D-glucose, and albumin (dog expt 4048-6) fitted with the model. The deoxyglucose curve has been shifted downward by half a logarithmic decade (ordinate values divided by 10 in.) to display it separately from D-glucose curve. Parameter estimates for D- and deoxyglucose were PS_c, 0.97 and 1.0; $PS_{pc} = 0.7$ and 0.5; $G_{pc} = 0.01$ and 0.05 ml/(g min); the volume ratio $V_{isf}/V_p = 6.5$ and $V_{pc}/V_p = 13.3$ ml/g, the same for both glucoses. Coefficients of variation were 0.19 and 0.09. From Kuikka (36) with permission from the American Physiological Society (Model is at www.physiome.org: Search on Kuikka, model 126).

Computation times for the PDE solvers are almost proportional to Ngrid, the number of spatial elements chosen for the computation. LSFEA is very much faster than TOMS731, and is also faster than the serial tanks model since fewer segments are needed. The key thing is that when the number of grid points is reduced the PDE algorithms do not change shape so much as the serial compartmental models. The result is that the PDEs are more efficient for fitting the data and provide a realistic solution.

Accounting correctly for smooth intracapillary gradients is important in the analysis of indicator dilution data for low-molecular-weight solutes using high-resolution techniques. Figure 16 is an example of modeling of the uptake of glucoses in the dog heart. The data from Kuikka (36) are outflow dilution curves from actively contracting blood-perfused dog hearts. Albumin is the intravascular reference data curve defining the dispersion and delay from the coronary inflow to the effluent veins for all of the tracers. The study was on D-glucose, a normal substrate, and 2-deoxy-D-glucose, an abnormal glucose that is transported into the cardiomyocyte, phosphorylated by hexokinase, and then cannot be further metabolized. The model is that illustrated in Fig. 11 (right panel) but without considering the negligible uptake by the endothelial cells. Parameters for the capillary wall permeation through the clefts and for the cellular uptake are given in the legend. The quality of the data is shown by the smoothness of the curves over time; the goodness of fit is a testimonial to the quality of the model defined through the PDEs.

8. Commentary

This presentation has emphasized issues important for new users of compartmental systems to contemplate. This reduced coverage neglects many important issues such as the identifiability of the models, distinguishing one option from another, the numbers and accuracy of the data points and their timing in obtaining parameter estimates, the consequences of model verification testing, and the basic principles of numerical methods and optimization techniques. These are all covered in the more general references listed in the Introduction.

We did however carefully express the models in physiological terms, trying to emphasize recognition of the nature of the processes of exchange and reaction. Thus Eq. 10 was deliberately not written in the common parlance used in compartmental modeling:

$$\frac{dC_1}{dt} = -k_1(C_1 - C_2) - k_2(\text{Cin} - C_1),$$
$$\frac{dC_2}{dt} = +k_3(C_1 - C_2) - k_4 C_2, \tag{15}$$

where $k_1 = \text{PS}/V_1$, $k_2 = F/V_1$, $k_3 = \text{PS}/V_2$, and $k_4 = G_2/V_2$. These equations look simpler, for it appears that there are fewer parameters, namely, four, while the original equations appear to have five: F, PS, V_1 and V_2, and G_2. But they are deceptively simple, quietly masking their true identities behind the fact that there are really only four independent parameters in the original Eq. 10. Putting $\tau = V_1/F$, $\gamma = V_2/V_1$, $\delta = G_2/F$, and $\varepsilon = \text{PS}/F$, we have $k_1 = \text{PS}/V_1 = \delta/\tau$; $k_2 = F/V_1 = 1/\tau$; $k_3 = \text{PS}/V_2 = \varepsilon/(\gamma\tau)$, and $k_4 = G_2/V_2 = \delta/(\gamma\tau)$; this illustrates that there were only four parameters in the original expressions, as expected, remembering that the exponential response of a single compartmental system has just one parameter $\tau = V_1/F$, its mean transit time. The point is that PS, G, F, and Vs have real anatomic and physiological meaning, and reminding us that when the flow is known, V_1 is the relevant unknown, and it is usually constrained by knowledge of the anatomy.

In general, "data" should include more than just a set of concentration–time curves, for one must recognize the underlying anatomy as data. Anatomic data constrain the estimates of water space or physical volumes as shown by Yipintsoi (44) and Vinnakota

(45). Further, large comprehensive models can be used to great advantage when built upon anatomic and physicochemical constraints combined into the physiological kinetics (46).

In this essay we have not emphasized the distinction between tracer and non-tracer kinetics except in the Introduction in Eqs. 1–3. We have however considered the nonlinearity of reactions in Application 4A, Aspirin Kinetics, where the dependence of the reaction flux on the concentration is evident. If the salicylate concentration were held constant at any particular value, then for a tracer, the tracer flux would be determined entirely by the concentration of the ambient non-tracer salicylate, and the system would be reduced to a linear one with constant coefficients (47). Likewise, when examining tracer fluxes in a situation where mother substance is varying, this should be taken into account by computing the dual model, for mother and tracer simultaneously, a consideration all too often forgotten in the common usage of compartmental analysis. The warning is: when using tracers, measure the background concentrations of mother substance often enough to assure its constancy.

The traditional compartmental analysis therefore has its greatest strength in situations where the overall system is in steady state for all substances related to the tracer substance, for then the power of linear systems analysis applies, convolution integration and stationarity apply, and matrix inversion can be used on sets of linear ODEs. Then all is well set up for linear systems analysis.

References

1. Berman M (1963) The formulation and testing of models. Ann N Y Acad Sci 108:182–194
2. Jacquez JA (1972) Compartmental analysis in biology and medicine. Kinetics of distribution of tracer-labeled materials. Elsevier Publishing Co, Amsterdam, 237 pp
3. Jacquez JA (1996) Compartmental analysis in biology and medicine, 3rd edn. BioMedware, Ann Arbor, MI, 514 pp
4. Cobelli C, Foster D, Toffolo G (2000) Tracer kinetics in biomedical research. From data to model. Kluwer Academic, New York
5. Zierler KL (1981) A critique of compartmental analysis. Annu Rev Biophys Bioeng 10:531–562
6. Anderson JC, Bassingthwaighte JB (2007) Tracers in physiological systems modeling. Chapter 8 Mathematical modeling in nutrition and agriculture. In: Mark D. Hanigan JN, Casey L Marsteller. Proceedings of the ninth international conference on mathematical modeling in nutrition, Roanoke, VA, 14–17 August 2006, Virginia Polytechnic Institute and State University Blacksburg, VA, pp 125–159
7. Knopp TJ, Anderson DU, Bassingthwaighte JB (1970) SIMCON–Simulation control to optimize man-machine interaction. Simulation 14:81–86
8. Sauro HM, Fell DA (1991) SCAMP: a metabolic simulator and control analysis program. Math Comput Model 15:15–28
9. Sauro HM, Hucka M, Finney A, Bolouri H (2001) The systems biology workbench concept demonstrator: design and implementation. Available via the World Wide Web at http://www.cds.caltech.edu/erato/sbw/docs/detailed-design/
10. Raymond GM, Butterworth E, Bassingthwaighte JB (2003) JSIM: free software package for teaching physiological modeling and research. Exp Biol 280.5:102. (www.physiome.org/jsim)
11. Chizeck HJ, Butterworth E, Bassingthwaighte JB (2009) Error detection and unit conversion. Automated unit balancing in modeling interface systems. IEEE Eng Med Biol 28(3):50–58
12. Platt JR (1964) Strong inference. Science 146:347–353

13. Bassingthwaighte JB, Chinard FP, Crone C, Goresky CA, Lassen NA, Reneman RS, Zierler KL (1986) Terminology for mass transport and exchange. Am J Physiol Heart Circ Physiol 250:H539–H545
14. Benedek IH, Joshi AS, Pieniazek JH, King S-YP, Kornhauser DM (1995) Variability in the pharmacokinetics and pharmacodynamics of low dose aspirin in healthy male volunteers. J Clin Pharmacol 35:1181–1186
15. Aarons L, Hopkins K, Rowland M, Brossel S, Thiercelin JF (1989) Route of administration and sex differences in the pharmacokinetics of aspirin, administered as its lysine salt. Pharm Res 6:660–666
16. Prescott LF, Balali-Mood M, Critchley JAJH, Johnstone AF, Proudfoot AT (1982) Diuresis or urinary alkalinisation for salicylate poisoning? Br Med J 285:1383–1386
17. Chan IS, Goldstein AA, Bassingthwaighte JB (1993) SENSOP: a derivative-free solver for non-linear least squares with sensitivity scaling. Ann Biomed Eng 21:621–631
18. Glad T, Goldstein A (1977) Optimization of functions whose values are subject to small errors. BIT 17:160–169
19. Dennis JE, Schnabel RB (1983) Numerical methods for unconstrained optimization and nonlinear equation. Prentice-Hall, New York
20. Fox IJ, Brooker LGS, Heseltine DW, Essex HE, Wood EH (1957) A tricarbocyanine dye for continuous recording of dilution curves in whole blood independent of variations in blood oxygen saturation. Proc Staff Meet Mayo Clin 32:478
21. Edwards AWT, Isaacson J, Sutterer WF, Bassingthwaighte JB, Wood EH (1963) Indocyanine green densitometry in flowing blood compensated for background dye. J Appl Physiol 18:1294–1304
22. Edwards AWT, Bassingthwaighte JB, Sutterer WF, Wood EH (1960) Blood level of indocyanine green in the dog during multiple dye curves and its effect on instrumental calibration. Proc Staff Meet Mayo Clin 35:747–751
23. Bassingthwaighte JB (1966) Plasma indicator dispersion in arteries of the human leg. Circ Res 19:332–346
24. Hunton DB, Bollman JL, Hoffman HN (1961) The plasma removal of indocyanine green and sulfobromophthalein: effect of dosage and blocking agents. J Clin Invest 30 (9):1648–1655 (PMCID PMC290858)
25. Bassingthwaighte JB, Edwards AWT, Wood EH (1962) Areas of dye-dilution curves sampled simultaneously from central and peripheral sites. J Appl Physiol 17:91–98
26. Stewart GN (1897) Researches on the circulation time and on the influences which affect it: IV. The output of the heart. J Physiol 22:159–183
27. Hamilton WF, Moore JW, Kinsman JM, Spurling RG (1932) Studies on the circulation. IV. Further analysis of the injection method, and of changes in hemodynamics under physiological and pathological conditions. Am J Physiol 99:534–551
28. Thompson HK, Starmer CF, Whalen RE, McIntosh HD (1964) Indicator transit time considered as a gamma variate. Circ Res 14:502–515
29. Bassingthwaighte JB, Ackerman FH, Wood EH (1966) Applications of the lagged normal density curve as a model for arterial dilution curves. Circ Res 18:398–415
30. Krenn CG, Krafft P, Schaefer B, Pokorny H, Schneider B, Pinsky MR, Steltzer H (2000) Effects of positive end-expiratory pressure on hemodynamics and indocyanine green kinetics in patients after orthotopic liver transplantation. Crit Care Med 28:1760–1765
31. Krenn CG, Pokorny H, Hoerauf K, Stark J, Roth E, Steltzer H, Druml W (2008) Non-isotopic tyrosine kinetics using an alanyl-tyrosine dipeptide to assess graft function in liver transplant recipients - a pilot study. Wien Klin Wochenschr 120(1–2):19–24
32. Kortgen A, Paxian M, Werth M, Recknagel P, Rauschfusz F, Lupp A, Krenn C, Muller D, Claus RA, Reinhart K, Settmacher U, Bauer M (2009) Prospective assessment of hepatic function and mechanisms of dysfunction in the critically ill. Shock 32(4):358–365
33. Bassingthwaighte JB, Yipintsoi T, Harvey RB (1974) Microvasculature of the dog left ventricular myocardium. Microvasc Res 7:229–249
34. Chinard FP, Vosburgh GJ, Enns T (1955) Transcapillary exchange of water and of other substances in certain organs of the dog. Am J Physiol 183:221–234
35. Crone C (1963) The permeability of capillaries in various organs as determined by the use of the indicator diffusion method. Acta Physiol Scand 58:292–305
36. Kuikka J, Levin M, Bassingthwaighte JB (1986) Multiple tracer dilution estimates of D- and 2-deoxy-D-glucose uptake by the heart. Am J Physiol Heart Circ Physiol 250: H29–H42
37. Krogh A (1919) The number and distribution of capillaries in muscles with calculations of the oxygen pressure head necessary for supplying the tissue. J Physiol (Lond) 52:409–415

38. Sangren WC, Sheppard CW (1953) A mathematical derivation of the exchange of a labeled substance between a liquid flowing in a vessel and an external compartment. Bull Math Biophys 15:387–394
39. Renkin EM (1959) Transport of potassium-42 from blood to tissue in isolated mammalian skeletal muscles. Am J Physiol 197:1205–1210
40. Bassingthwaighte JB (1974) A concurrent flow model for extraction during transcapillary passage. Circ Res 35:483–503
41. Bassingthwaighte JB, Wang CY, Chan IS (1989) Blood-tissue exchange via transport and transformation by endothelial cells. Circ Res 65:997–1020
42. Bassingthwaighte JB, Chan IS, Wang CY (1992) Computationally efficient algorithms for capillary convection-permeation-diffusion models for blood-tissue exchange. Ann Biomed Eng 20:687–725
43. TOMS./TOMS. Association of computing machinery: transactions on mathematical software. http://www.netlib.org/toms/index.html
44. Yipintsoi T, Scanlon PD, Bassingthwaighte JB (1972) Density and water content of dog ventricular myocardium. Proc Soc Exp Biol Med 141:1032–1035
45. Vinnakota K, Bassingthwaighte JB (2004) Myocardial density and composition: a basis for calculating intracellular metabolite concentrations. Am J Physiol Heart Circ Physiol 286: H1742–H1749
46. Bassingthwaighte JB, Raymond GR, Ploger JD, Schwartz LM, Bukowski TR (2006) GENTEX, a general multiscale model for *in vivo* tissue exchanges and intraorgan metabolism. Phil Trans Roy Soc A Math Phys Eng Sci 364(1843):1423–1442. doi: 10.1098/rsta.2006.1779
47. Bassingthwaighte JB, Goresky CA, Linehan JH (1998) Ch. 1 Modeling in the analysis of the processes of uptake and metabolism in the whole organ. In: Bassingthwaighte JB, Goresky CA, Linehan JH (eds) Whole organ approaches to cellular metabolism. Springer, New York, pp 3–27

Chapter 18

Physiologically Based Pharmacokinetic/Toxicokinetic Modeling

Jerry L. Campbell Jr., Rebecca A. Clewell, P. Robinan Gentry, Melvin E. Andersen, and Harvey J. Clewell III

Abstract

Physiologically based pharmacokinetic (PBPK) models differ from conventional compartmental pharmacokinetic models in that they are based to a large extent on the actual physiology of the organism. The application of pharmacokinetics to toxicology or risk assessment requires that the toxic effects in a particular tissue are related in some way to the concentration time course of an active form of the substance in that tissue. The motivation for applying pharmacokinetics is the expectation that the observed effects of a chemical will be more simply and directly related to a measure of target tissue exposure than to a measure of administered dose. The goal of this work is to provide the reader with an understanding of PBPK modeling and its utility as well as the procedures used in the development and implementation of a model to chemical safety assessment using the styrene PBPK model as an example.

Key words: PBPK, Styrene, Pharmacokinetics

1. Introduction

Pharmacokinetics is the study of the time course for the absorption, distribution, metabolism, and excretion of a chemical substance in a biological system. Implicit in any application of pharmacokinetics to toxicology or risk assessment is the assumption that the toxic effects in a particular tissue can be related in some way to the concentration time course of an active form of the substance in that tissue. Moreover, except for pharmacodynamic differences between animal species, it is expected that similar responses will be produced at equivalent tissue exposures regardless of animal species, exposure route, or experimental regimen (1–3). Of course the actual nature of the relationship between tissue exposure and response, particularly across species, may be quite complex, and exceptions to the rule of tissue dose equivalence are numerous.

Brad Reisfeld and Arthur N. Mayeno (eds.), *Computational Toxicology: Volume I*, Methods in Molecular Biology, vol. 929,
DOI 10.1007/978-1-62703-050-2_18, © Springer Science+Business Media, LLC 2012

Nevertheless, the motivation for applying pharmacokinetics is the expectation that the observed effects of a chemical will be more simply and directly related to a measure of target tissue exposure than to a measure of administered dose.

1.1. Compartmental Modeling

One of the first general descriptions of pharmacokinetic modeling was presented by Teorell (4, 5). The model consisted of a number of compartments representing specific tissues. The concentration of chemical in each compartment was described by a mass balance equation in which the rate of change of the amount of chemical in a compartment was determined from the rates at which the chemical entered and left the compartment in the blood as well as, when appropriate, the rate of clearance of the chemical in that compartment. Unfortunately, in order to obtain an analytical solution of the resulting system of differential equations, Teorell found it necessary to make a number of simplifying assumptions. These assumptions led to a solution in the form of a sum of exponential terms, and thus the "classical" compartmental modeling approach still used today was born.

Over the years, Teorell's association of the model compartments with specific tissues has to a large extent been lost, and compartmental modeling as currently practiced is largely an empirical exercise. In this empirical approach, data on the time course of the chemical of interest in blood (and perhaps other tissues, urine, etc.) are collected. Based on the behavior of the data, a mathematical model is selected which possesses a sufficient number of compartments (and therefore parameters) to describe the data. The compartments do not in general correspond to identifiable physiological entities but rather are described in abstract terms. An example of a simple two-compartment mathematical model of this type is shown in Fig. 1. This particular model consists of a "central" compartment, characterized by concentrations measured in the blood (but not considered to actually represent only the blood), and a "deep" compartment representing unspecified tissues communicating with the central compartment as described by kinetic parameters, k_{12} and k_{21}, which themselves have no obvious physiological or biochemical interpretation. Similarly, the volume of the central compartment and the uptake and clearance parameters (k_a and k_e) are empirically determined by the analysis or fitting of experimental data sets.

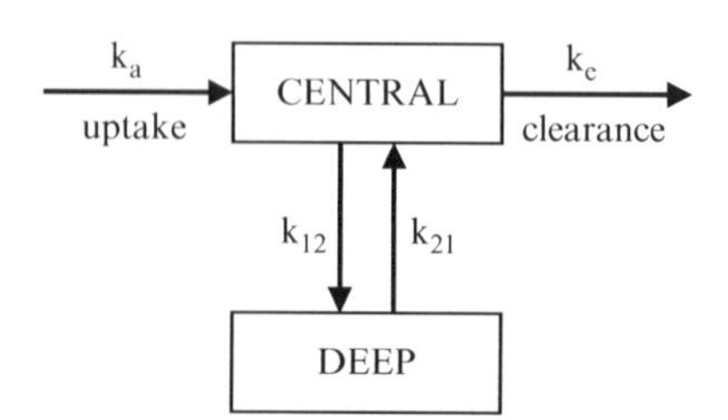

Fig. 1. Simple compartmental pharmacokinetic model.

The advantage of this modeling approach is that there is no limitation to fitting the model to the experimental data. If a particular model is unable to describe the behavior of a particular data set, additional compartments can be added until a successful fit is obtained. Since the model parameters do not possess any intrinsic meaning, they can be freely varied to obtain the best possible fit, and different parameter values can be used for each data set in a related series of experiments. Statistical tests can then be employed to compare the values of the parameters used, for example at two administered dose levels, in order to determine whether any apparent differences are statistically significant (6). Once developed, these models are useful for interpolation and limited extrapolation of the concentration profiles which can be expected as experimental conditions are varied. If the model parameters vary with dose, this information can provide evidence for the presence of nonlinearities in the animal system, such as saturable metabolism or binding. At this point, however, one of the serious disadvantages of the empirical approach becomes evident. Since the compartmental model does not possess a physiological structure, it is often not possible to incorporate a description of these nonlinear biochemical processes in a biologically appropriate context. For example, in the case of inhalation of chemicals subject to high-affinity, low-capacity metabolism in the liver, an important determinant of metabolic clearance at low inhaled concentrations is the fact that only the fraction of the chemical in the blood reaching the liver is available for metabolism (1). Without a physiological structure it is not possible to correctly describe the interaction between blood-transport of the chemical to the metabolizing organ and the intrinsic clearance of the chemical by the organ.

1.2. Physiologically Based Modeling

Physiologically based pharmacokinetic (PBPK) models differ from conventional compartmental pharmacokinetic models in that they are based to a large extent on the actual physiology of the organism. Instead of compartments defined solely by the experimental data, actual organ and tissue groups are described using weights and blood flows obtained from the literature. Moreover, instead of composite rate constants determined solely by fitting data, measured physical–chemical and biochemical constants of the compound can often be used. To the extent that the structure of the model reflects the important determinants of the kinetics of the chemical, PBPK models can predict the qualitative behavior of an experimental time course data set without having been based directly on it. Refinement of the model to incorporate additional insights gained from comparison with experimental data yields a model which can be used for quantitative extrapolation well beyond the range of experimental conditions on which it was based. In particular, a properly validated PBPK model can be used to perform

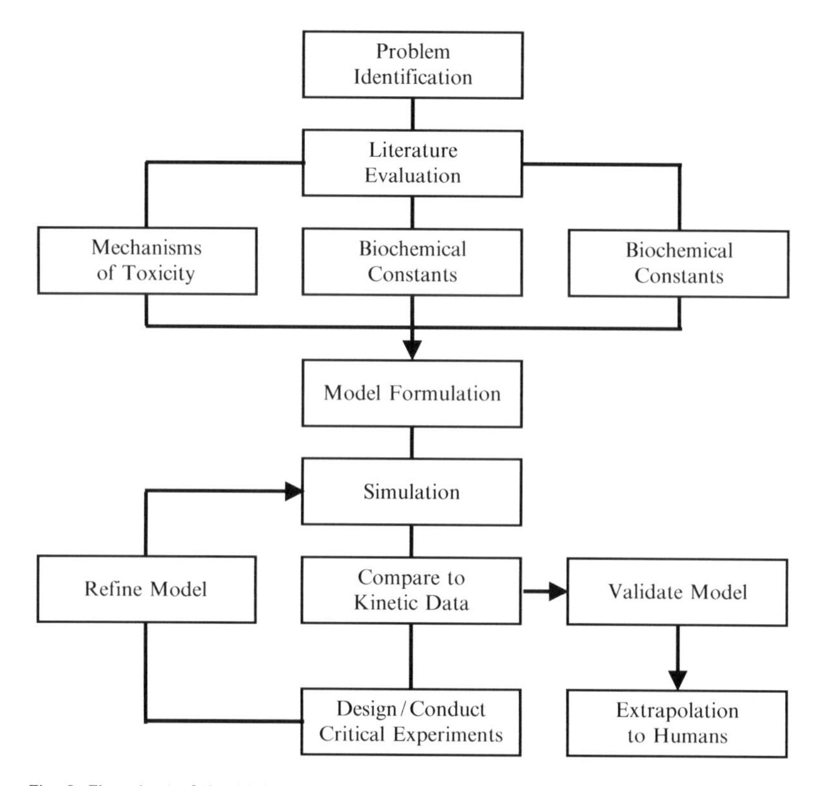

Fig. 2. Flowchart of the biologically motivated PBPK modeling approach to chemical risk assessment.

the high-to-low dose, dose-route, and interspecies extrapolations necessary for estimating human risk on the basis of animal toxicology studies (7–15).

A number of excellent reviews have been written on the subject of PBPK modeling (16–20). The basic approach is illustrated in Fig. 2. The process of model development begins with the definition of the chemical exposure and toxic effect of concern, as well as the species and target tissue in which it is observed. Literature evaluation involves the integration of available information about the mechanism of toxicity, the pathways of chemical metabolism, the nature of the toxic chemical species (i.e., whether the parent chemical, a stable metabolite, or a reactive intermediate produced during metabolism is responsible for the toxicity), the processes involved in absorption, transport and excretion, the tissue partitioning and binding characteristics of the chemical and its metabolites, and the physiological parameters (e.g., tissue weights and blood flow rates) for the species of concern (i.e., the experimental species and the human). Using this information, the investigator develops a PBPK model which expresses mathematically a conception of the animal–chemical system. In the model, the various time-dependent biological processes are described as a system of simultaneous differential equations. A mathematical model of this form can easily be written and exercised using commonly available computer software (21). The specific structure of the model is driven by the need to estimate the appropriate measure of tissue dose under the various exposure conditions of concern in both the experimental animal and the human. Before the model can be used in human risk assessment it has to be validated against kinetic, metabolic, and toxicity information and, in many cases, refined based on comparison with the experimental results. The model itself can frequently be used to help design critical experiments to collect data needed for its own validation.

The chief advantage of a PBPK model over an empirical compartmental description is its greater predictive power. Since fundamental biochemical processes are described, dose extrapolation over ranges where saturation of metabolism occurs is possible (22). Since known physiological parameters are used, a different species can be modeled by simply replacing the appropriate constants with those for the species of interest, or by allometric scaling (23–25). Similarly, the behavior for a different route of administration can be determined by adding equations which describe the nature of the new input function (21, 26). The extrapolation from one exposure scenario, say a single 6 h exposure, to another, e.g., a repetitive 6 h exposure, 5 days a week for the life of the animal, is relatively easy and only requires a little ingenuity in writing the equations for the dosing regimen in the kinetic model (27, 28).

Since measured physical–chemical and biochemical parameters are used, the behavior for a different chemical can quickly be estimated by determining the appropriate constants. An important result is the ability to reduce the need for extensive experiments with new chemicals (12). The process of selecting the most informative experimental data is also facilitated by the availability of a predictive pharmacokinetic model (29). Perhaps the most desirable feature of a physiologically based model is that it provides a conceptual framework for employing the scientific method in which hypotheses can be described in terms of biological processes, predictions can be made on the basis of the description, and the hypothesis can be revised on the basis of comparison with experimental data.

The trade-off against the greater predictive capability of physiologically based models is the requirement for an increased number of parameters and equations. However, values for many of the parameters, particularly the physiological ones, are already available in the literature (30–35), and in vitro techniques have been developed for rapidly determining the compound-specific parameters (36–39). An important advantage of PBPK models is that they provide a biologically meaningful quantitative framework within which in vitro data can be more effectively utilized (40). There is even a prospect that predictive PBPK models can someday be developed based almost entirely on data obtained from in vitro studies.

Some of the best examples of successful PBPK modeling efforts were performed to support the clinical use of chemotherapeutic drugs, e.g., methotrexate (41) and cisplatin (42) (see (43) for a review). There are also a large number of good examples of PBPK models which describe the kinetics of important environmental contaminants, including methylene chloride (8, 44, 45), trichloroethylene (46–48), chloroform (49, 50), 2-butoxyethanol (51), kepone (52), polybrominated biphenyls (53), polychlorinated biphenyls (54) and dibenzofurans (55), dioxins (56, 57), lead (58–62), arsenic (63, 64), methylmercury (65), atrazine (66, 67), acrylonitrile (68–70), perchlorate (71–76), and BTEX components (77–81). The U.S. EPA is currently compiling a compendium of PBPK models including source code. This should be available online within the next year.

2. Materials

Currently there exists a very diverse group of modeling software packages that vary in both complexity and range of application. Because of this diversity, there is a software package suitable for every level of user from the expert to the first-time modeler. However, not all modeling packages are created equal, and some of the more user-friendly software can lack the capabilities of the more complex programs. Consequently, no single software package available can meet all needs of all users, and the diversity and complexity of the programs can often make converting a model from one package to another rather difficult. Table 1 provides a list of some of the available software packages that may be useful for PBPK modeling (82).

An additional list of pharmacokinetic software is located at http://boomer.org/pkin/soft.html. However, not all of the software listed on this website is suitable for PBPK modeling.

The most commonly used software packages for PBPK modeling have included Advanced Continuous Modeling Language (ACSL) (now acslX), Berkeley Madonna, MATLAB, MATLAB/Simulink, ModelMaker, and SCoP. Table 2 provides a summary of the features of each of these followed by further information.

acslX is an updated version of the widely used ACSL software. It has graphical as well as text interface with automatic linkage to the integration algorithm. In particular, a Pharmacokinetic Toolkit in the graphic code interface makes it possible to build PBPK models by connecting predefined tissue code blocks. The software allows for the use of discrete blocks and script files and automatically sorts equations in the derivative block. The model may be compiled into either C/C++ or Fortran, although C++ is now the preferred compiler, and may be debugged interactively.

Table 1
Representative list of available software packages

Package	Source	Website
General-purpose high-level scientific computing software. These high-level programming language packages are very general modeling tools that are not specifically designed for PBPK modeling, but offer more complexity		
acslX	AEgis Technologies Group, Inc.	http://www.acslX.com
Berkeley Madonna	University of California at Berkeley	http://www.berkeleymadonna.com
GNU octave	University of Wisconsin	http://www.octave.org
MATLAB/Simulink	The MathWorks, Inc.	http://www.mathworks.com
MLAB	Civilized Software, Inc.	http://www.civilized.com
Biomathematical modeling software. Packages that were specifically designed for modeling biological systems and some are user-friendly. Their usefulness in PBPK modeling is determined by their graphical interfaces, computational speed, and language flexibility and may provide mixed-effects (population) capabilities allowing for the analysis of sparse data sets		
ADAPT II	Biomedical Simulations Resource, USC	http://bmsr.usc.edu
MCSim	INERIS	http://toxi.ineris.fr/activites/toxicologie_quantitative/mcsim/mcsim.php
ModelMaker	ModelKinetix	http://www.modelkinetix.com
NONMEM	University of California at San Francisco and Globomax Service Group	http://www.globomaxservice.com
SAAM II	SAAM Institute, Inc.	http://www.saam.com
SCoP	Simulation Resources, Inc.	http://www.simresinc.com
Stella	High Performance Systems, Inc.	http://www.hps-inc.com
WinNonlin	Pharsight Corp.	http://www.pharsight.com
WinNonMix	Pharsight Corp.	http://www.pharsight.com
Toxicokinetic software. These packages were designed specifically for PBPK and PBTK modeling and are extremely flexible. They are based on modeling languages developed in the aerospace industry for modeling complex systems		
SimuSolv	Dow chemical	Not maintained or subject to further development
Physiologically based custom-designed software. Custom-designed proprietary software programs specifically for biomedical systems or applications that provide a high level of biological detail but are not easily customized		
GastroPlus	Simulations Plus, Inc.	http://www.simulations-plus.com
Pathway prism	Physiome Sciences, Inc.	http://www.physiome.com
Physiolab	Entelos, Inc.	http://www.entelos.com
SimCYP	Simcyp, Ltd.	http://www.simcyp.com

Table 2
Comparison of modeling software features

Feature	acslX[c, d]	Berkeley Madonna[c, d]	MATLAB[c, d]	MATLAB/ Simulink[c, d]	Model maker[c, d]	SCoP[d]
Graphical interface	Y	Y	N	Y	Y	N
Text interface	Y	Y	Y	N	Y	Y
Automatic linkage to integration algorithm	Y	Y	N	Y	Y	Y
Discrete blocks	Y	N	N	Y	Y	N
Scripting	Y	N	Y	Y	N	N
Code sorting	Y	Y	N	Y	Y	N
Choice of target language	Y	N	N	N	N	N
Interactive model debugging	Y	Y	N	Y	N	N
Optimization	Y	Y	Y[a]	Y	Y	Y
Sensitivity analysis	Y[b]	Y	Y[b]	Y[b]	Y	Y
Monte Carlo	Y[b]	Y[b]	Y[b]	Y[b]	Y	N
Units checking	N	N	N	N	N	Y
Database of physiological values	N	N	N	N	N	N
Compiled (faster)	Y	Y	Y[a, e]	N	N	Y
Interpreted (more convenient)	N	N	Y	Y	Y	N

[a]Extra cost
[b]Can perform through the use of user-developed model code or script files
[c]Must contact vendor for price. Price may depend upon the type of license
[d]Student and/or academic licenses are available
[e]With separate compiler

An optimization program is also included. Sensitivity analysis and Monte Carlo analysis can currently be conducted with script files and there is some capability built into the package that allows these analyses to be conducted with the aid of a using a gui.

Berkeley Madonna has many of the same features as acslX; however, it does not allow for the use of discrete blocks or script files. It does currently have both an optimization and sensitivity analysis feature, but does not have a built-in Monte Carlo capability.

MATLAB has a text interface, but not a graphical interface. It does allow for the use of script files but not discrete blocks. It does not sort the code in the model so the user must be careful regarding the order of statements in the code. This statement order problem can be problematic in the case of PBPK models, which require simultaneous solution of multiple differential equations, and can complicate conversion from a software package that automatically sorts the model equations. An optimization package is available through an add-on toolbox, but sensitivity analysis and Monte Carlo analysis must be performed through the use of script files. A large variety of user-built model code blocks are available at the MATLAB web site. The model code may be compiled through the purchase and use of a MATLAB Compiler, but the user has no choice of the target language used.

Simulink is an add-on to MATLAB that offers a graphic interface but no text interface. The use of discrete blocks is also added with the use of Simulink. Since Simulink uses only a graphical interface, there is no code to be viewed. Simulink adds a graphical debugger to MATLAB and an optimizer.

ModelMaker has many of the same features as acslXtreme and Berkeley Madonna. It has both a graphical and a text interface with automatic linkage to the integration algorithm and allows for the use of discrete blocks but not the use of script files. ModelMaker also provides the capabilities for optimization, sensitivity analysis, and Monte Carlo analysis.

SCoP has only a text interface with automatic linkage to the integration algorithm. It does not allow for the use of discrete blocks or script files and it does not automatically sort the code. Optimization and sensitivity analysis capabilities are included, but not Monte Carlo analysis capabilities.

An additional modeling software package, not included in Table 2, is designed specifically to support Markov Chain Monte Carlo (MCMC) simulation. This program, MCSim, has been used to reestimate parameter distributions for PBPK models on the basis of the agreement between model predictions and measured data from kinetic studies. MCSim is available for free download from the website listed in Table 1. Its proper use requires expertise in programming and statistics.

Each of the software packages described above provides different features that would recommend them for a particular user. From the viewpoint of a risk assessor wanting to apply an existing PBPK model, as well as for a model developer seeking to have his model used in a risk assessment, the key requirement is the ability to readily evaluate (verify) the model, reproduce (validate) the capability of the model to simulate key data sets, and document its application for the risk assessment. The necessary documentation includes a model definition (preferably code-based) that can be reviewed to verify the mathematical correctness of the model, a description of the parameters that should be used to run the model for comparison with validation data, and a description of the parameters used to calculate the dose metrics for the risk assessment. The most important characteristics of a language for model evaluation are verifiable code, self-documentation, and ease of use. The feature that is most important with regard to self-documentation is scripting, which allow the model developer to create procedures consisting of sequences of commands that, for example, set model parameters, run the model, and plot the model predictions against the appropriate data set. Other features which contribute to ease of model evaluation include viewable model definition code, code sorting, and automatic linkage to integration algorithms.

The modeling language that has seen the most widespread use in PBPK modeling is the ACSL, which is currently implemented as acslX. The ACSL language has also served as the basis for a variety of older packages, including SimuSolv (from Dow Chemical, no longer supported), ACSL/Tox (from Pharsight, no longer supported), and ERDM (currently used by the USEPA). The automatic code sorting provided by ACSL allows code to be grouped functionally (liver, lung, fat, etc.) rather than in program order, greatly simplifying model development and improving readability of the code. The graphic code block capability in acslX is particularly attractive for PBPK modeling because it provides the ability to create a model by connecting functional units (e.g., tissues) in a graphic environment, while at the same time creating a model definition in ACSL code that can be reviewed for model verification. The scripting capability, which permits both MATLAB-like m-files, greatly expedites the comparison of the model with multiple data sets that is generally required for PBPK modeling of data from multiple species and routes of exposure. The scripting capability also makes it possible to document the use of a model in a risk assessment, since m-scripts or command file procedures can be written that set the model parameters and run the model for each of the dose metrics required for the risk assessment, as well as for each of the data sets used for model evaluation/validation.

Berkeley Madonna provides a particularly intuitive, flexible platform for model development, and has been very popular in academic settings. The software provides automatic code sorting, the ability to automatically convert between code and graphic model descriptions, and automatic compilation, greatly simplifying and expediting model development, debugging, and verification. Conversion of models between ACSL and Berkeley Madonna is relatively straightforward. However, the lack of a scripting capability makes comparison of the model with data fairly cumbersome, particularly in situations where a large number of data sets are being modeled. The lack of scripting also makes it more difficult to document the actual use of a model in a risk assessment and greatly complicates model evaluation.

MATLAB is a very powerful and flexible software package that is particularly attractive in research activities. Its main drawback is that its use requires significant expertise in programming. MATLAB/Simulink avoids this drawback by providing a graphical interface, and is very popular with engineers in the automotive and electronic industries. However, verification of a Simulink model by a nonengineer is hampered by the lack of any model definition code. That is, the model is specified only by a "wiring diagram" that shows the connections between blocks built up from basic mathematical functions (adders, multipliers, integrators, etc.). Conversion of a model between MATLAB and Simulink, or between one of these programs and another software package, can be very difficult.

ModelMaker, which is popular in Europe, provides surprisingly broad functionality at a relatively low price. Its only serious drawback is the lack of a scripting capability.

SCoP, which along with ACSL was one of the first languages to be used for PBPK modeling, continues to be used due to its familiarity and low cost. However, its lack of scripting or code sorting, and its DOS-based, menu-driven run-time interface can make model evaluation more difficult.

3. Methods

3.1. General Concepts

The methods will begin with a description of the seminal PBPK model published by Ramsey and Andersen in 1984 and lead into the elements necessary for successful model development, refinement and validation. The experience of Ramsey and Andersen serves as a useful example of the advantages of the PBPK modeling approach. In this case, blood and tissue time-course curves of styrene had been obtained for rats exposed to four different concentrations of 80, 200, 600, and 1,200 ppm (83). Data were obtained during a 6 h exposure period and for 18 h after cessation of the exposure. The initial analysis of these data had been based on a simple compartmental model, similar to the model shown in Fig. 1, which had a zero-order input related to the amount of styrene inhaled, a two-compartment description of the rat, and linear metabolism in the central compartment. The compartmental model was successful with lower concentrations but was unable to account for the more complex behavior at higher concentrations (note the different behavior of the data at the two concentrations shown in Fig. 3).

In an attempt to provide a more successful description, a PBPK model was developed with a realistic equilibration process for pulmonary uptake and Michaelis–Menten saturable metabolism in the

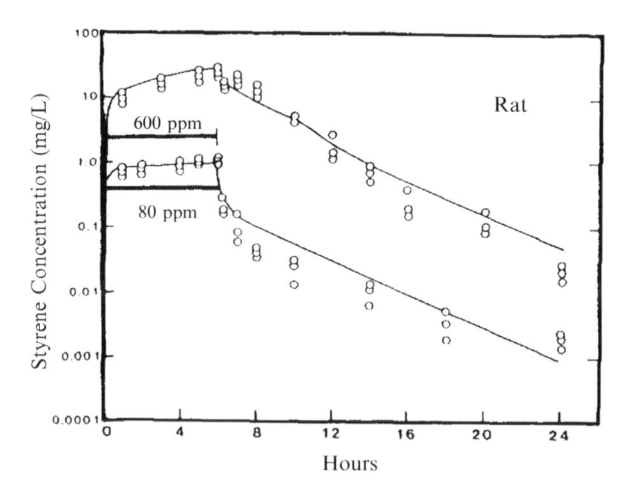

Fig. 3. Model predictions (*solid lines*) and experimental blood styrene concentrations in rats during and after 6 h exposures to 80 and 600 ppm styrene. The *thick bars* represent the chamber air concentrations of styrene and are shown to highlight the nonlinearity of the relationship between administered and internal concentrations. The model (Fig. 1.3) contains sufficient biological realism to predict the very different behaviors observed at the two concentrations.

liver. A diagram of the PBPK model that was used by Ramsey and Andersen (1984) (22) to describe styrene inhalation in both rats and humans is shown in Fig. 4. In this diagram, the boxes represent tissue compartments and the lines connecting them represent blood flows. The model contained several "lumped" tissue compartments: fat tissues, poorly perfused tissues (muscle, skin, etc.), richly perfused tissues (viscera), and metabolizing tissues (liver). The fat tissues were described separately from the other poorly perfused tissues due to their much higher partition coefficient for styrene, which leads to different kinetic properties, while the liver was described separately from the other richly perfused tissues due to its key role in the metabolism of styrene. Each of these tissue groups was defined with respect to their blood flow, tissue volume, and their ability to store (partition) the chemical of interest. Although the model diagram in Fig. 4 shows a lung compartment, a steady-state approximation for the equilibration of lung blood with alveolar air was used in the mathematical formulation of the model to eliminate the need for an actual lung tissue compartment. This simple model structure, with realistic constants for the physiological, partitioning, and metabolic parameters, very accurately predicted the behavior of styrene in both fat and blood of the rat at all concentrations. Fig. 3 compares the model-predicted time course in the blood with the experimental data for the highest and lowest exposure concentrations in the rat studies.

The structure of the PBPK model for styrene reflects the generic mammalian architecture. Organs are arranged in a parallel system of blood flows with total blood flow through the lungs.

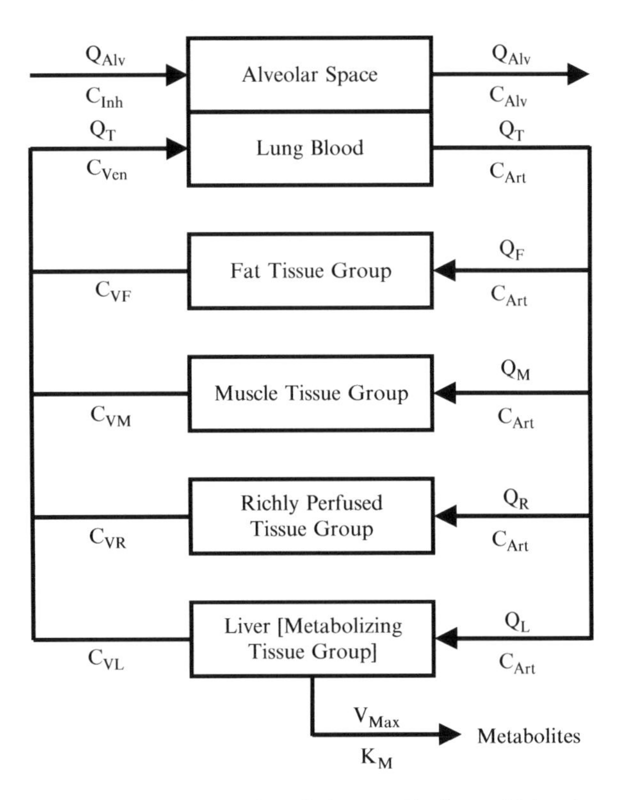

Fig. 4. Diagram of a physiologically based pharmacokinetic model for styrene. In this description, groups of tissues are defined with respect to their volumes, blood flows (Q), and partition coefficients for the chemical. The uptake of vapor is determined by the alveolar ventilation (Q_{ALV}), cardiac output (Q_T), blood–air partition coefficient, and the concentration gradient between arterial and venous pulmonary blood (C_{ART} and C_{VEN}). Metabolism is described in the liver with a saturable pathway defined by a maximum velocity (V_{max}) and affinity (K_m). The mathematical description assumes equilibration between arterial blood and alveolar air as well as between each of the tissues and the venous blood exiting from that tissue.

This model can easily be scaled-up to examine styrene kinetics for other mammalian species. In the case of styrene, exposure experiments had also been conducted with human volunteers (84). In order to model this data, the PBPK model parameters were changed to human physiological values, the human blood–air partitioning was determined from human blood samples, and the metabolism was scaled allometrically so that capacity (V_{max}) was related to basal metabolic rate (body weight raised to the 0.7 power) and affinity (K_m) was the same in the human as in the rat, 0.36 mg/L. Ramsey (84) measured both venous blood and exhaled air concentrations in these human volunteers. Although the rat PBPK model was developed for blood and fat, not for exhaled air, the physiologically based description automatically provides information on expected exhaled air concentrations. It was straightforward then to predict expected

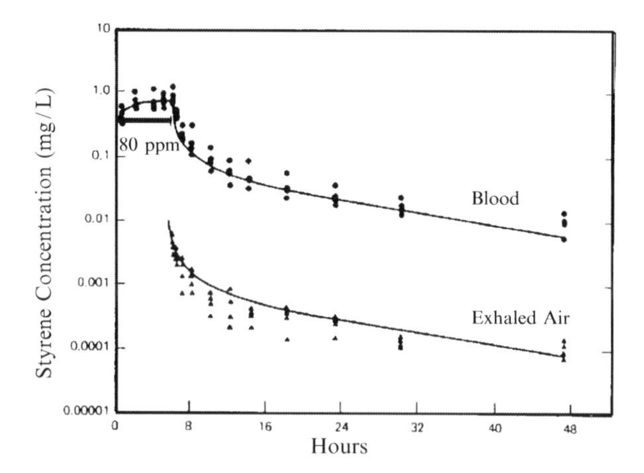

Fig. 5. Model predictions and experimental blood and exhaled air concentrations in human volunteers during and after 6 h exposures to 80 ppm styrene. The model is identical to that used for rats (Fig. 1.4). The model parameters have been changed to values appropriate for humans on the basis of physiological and biochemical information, and have not been adjusted to improve the fit to the experimental data.

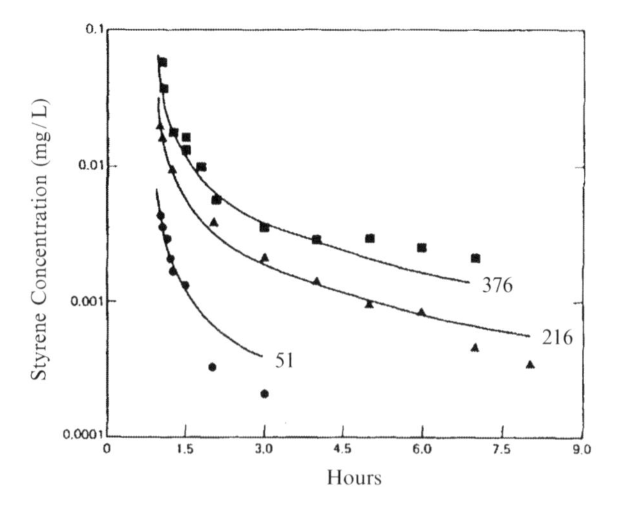

Fig. 6. Model predictions and experimental exhaled air concentrations in human volunteers following 1 h exposures to 51, 216, and 376 ppm styrene. The model is the same as Fig. 1.5.

exhaled air concentrations in humans and compare the predictions with the concentrations measured during the experiments (Fig. 5). A similar comparison of the model's predictions with another human data set from (85) also demonstrated the ability of the PBPK structure to support extrapolation of styrene kinetics from the rat to the human (Fig. 6).

3.2. Modeling Philosophy

This basic PBPK model for styrene has several tissue groups which were lumped according to their perfusion and partitioning characteristics. In the mathematical formulation, each of these several compartments is described by a single mass-balance differential equation. It would be possible to describe individual tissues in each of the lumped compartments, if necessary. This detail is usually unnecessary unless some particular tissue in a lumped compartment is the target tissue. One might, for example, want to separate brain from other richly perfused tissues if the model were for a chemical that had a toxic effect on the central nervous system (86–88). Other examples of additional compartments include the addition of placental and mammary compartments to model pregnancy and lactation (89–91). The interactions of chemical mixtures can even be described by including compartments for more than one chemical in the model (92–94). Increasing the number of compartments does increase the number of differential equations required to define the model. However, the number of equations does not pose any problem due to the power of modern desktop computers.

On the other hand, as the number of compartments in the PBPK model increases, the number of input parameters increases correspondingly. Each of these parameters must be estimated from experimental data of some kind. Fortunately, the values of many of these can be set within narrow limits from nonkinetic experiments. The PBPK model can also help to define those experiments which are needed to improve parameter estimates by identifying conditions where the sensitivity of the model to the parameter is the greatest (95). The demand that the PBPK fit a variety of data also restricts the parameter values that will give a satisfactory fit to experimental data. For example, the styrene model (described above) was required to reproduce both the high and low concentration behaviors, which appeared qualitatively different, using the same parameter values. If one were independently fitting single curves with a model, the different parameter values obtained under different conditions would be relatively uninformative for extrapolation.

As the renowned statistician George Box has said, "All models are wrong, and some are useful." Even a relatively complex description such as a PBPK model will sometimes fail to fit reliable experimental data. When this occurs, the investigator needs to think how the model might be changed, i.e., what extra biological aspects must be added to the physiological description to bring the predictions in line with experimental observation? In the case of the work with styrene cited above, continuous 24 h styrene exposures could not be modeled with a time-independent maximum rate of metabolism, and induction of enzyme activity had to be included to yield a satisfactory representation of the observed kinetic behavior (96).

When a PBPK model is unable to adequately describe kinetic data, the nature of the discrepancy can provide the investigator with additional insight into time dependencies in the system. This insight can then be utilized to reformulate the biological basis of the model and improve its fidelity to the data. The resulting model may be more complicated, but it will still be useful if the pertinent kinetic constants can be estimated for human tissues. Indeed, as long as the model maintains its biological basis the additional parameters can often be determined directly from separate experiment, rather than estimated by fitting the model to kinetic data. As the models become more complex, they necessarily contain larger numbers of physiological, biochemical, and biological constants. The crucial task during model development is to keep the description as simple as possible and to ensure the identifiability of new parameters that are added to the model; every attempt should be made to obtain or verify model parameters from experimental studies separate from the modeling exercises themselves (97).

The following section explores some of the key issues associated with the development of PBPK models. It is meant to provide a general understanding of the basic design concepts and mathematical forms underlying the PBPK modeling process, and is not meant to be a complete exposition of the PBPK modeling approach for all possible cases. It must be understood that the specifics of the approach can vary greatly for different types of chemicals, e.g., volatiles, nonvolatiles, and metals, and for different applications.

Model building is an art, and is best understood as an iterative process in the spirit of the scientific method (97). The literature articles cited in the introductory section include examples of successful PBPK models for a wide variety of chemicals and provide a wealth of insight into various aspects of the PBPK modeling process. They should be consulted for further detail on the approach for applying the PBPK methodology in specific cases.

3.3. Tissue Grouping

The first aspect of PBPK model development that will be discussed is determining the extent to which the various tissues in the body may be grouped together. Although tissue grouping is really just one aspect of model design, which is discussed in the next section, it provides a simple context for introducing the two alternative approaches to PBPK model development: "lumping" and "splitting" (Fig. 7). In the context of tissue grouping, the guiding philosophy in the lumping approach can be stated as follows: "Tissues which are pharmacokinetically and toxicologically indistinguishable may be grouped together." In this approach, model development begins with information at the greatest level of detail that is practical, and decisions are made to combine physiological elements (tissues and blood flows) to the extent justified by their similarity. The common grouping of tissues into richly (or rapidly) perfused and poorly (or slowly) perfused on the basis of their perfusion rate (ratio of blood flow to tissue volume) is an example of the lumping approach. The contrasting philosophy of splitting is

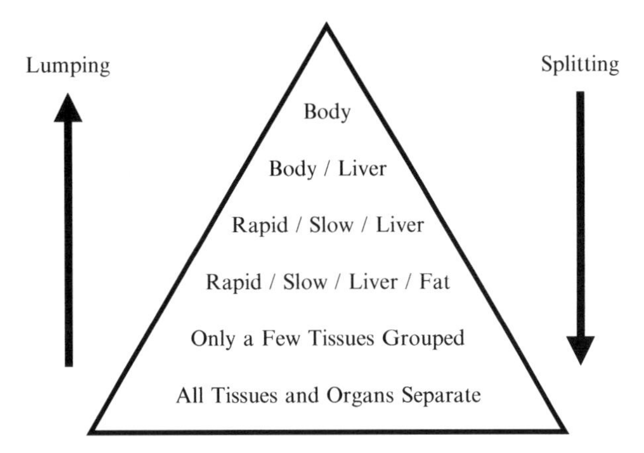

Fig. 7. The role of lumping and splitting processes in PBPK model development.

as follows: "Tissues which are pharmacokinetically or toxicologically distinct must be separated." This approach starts with the simplest reasonable model structure and increases the model's complexity only to the extent required to reproduce data on the chemical of concern for the application of interest. Splitting requires the greater initial investment in data collection and, if taken to the extreme, could paralyze model development. Lumping, on the other hand, is more efficient but runs a greater risk of overlooking chemical-specific determinants of chemical disposition.

The description of fat tissue in the PBPK model of styrene described in the previous section can be used to provide an example of the different approach associated with the two philosophies. In the splitting approach, which is the approach used by Ramsey and Andersen (22), a single fat compartment was used initially with volume, blood flow and partitioning parameters selected to represent all adipose tissues in the body. Clearly, there are actually a number of distinguishable adipose tissues, including inguinal, perirenal, and brown fat, among others, which may have different partitioning and kinetic characteristics for styrene. However, since this single-compartment treatment provided an adequate description of the available data on the kinetics of styrene in the fat and blood, no attempt was made to split the fat tissue group into multiple compartments. For a more lipophilic chemical, polychlorotrifluoroethylene oligomer, on the other hand, it was not possible to adequately reproduce fat and blood kinetic data using a single fat compartment (98); therefore, the fat compartment was split into two parts: "perirenal fat" and "other fat tissues," resulting in an acceptable simulation of the observed kinetic behavior.

The splitting process just described can be contrasted with a lumping approach, in which the PBPK model would initially be designed to include separate compartments for all physiologically distinguishable fat tissues. Partition coefficients for each of the fat tissues would be determined experimentally, and the volume and blood flow for each fat tissue would be estimated. If it were then determined that the kinetic characteristics of the various fat tissues were not sufficiently different to justify retaining separate compartments, they would be lumped together by appropriately combining the individual parameter values (adding the volumes and blood flows and averaging the partition coefficients).

3.3.1. Criteria for Grouping Tissues

There are two alternative approaches for determining whether tissues are kinetically distinct or should be lumped together. In the first approach, the tissue rate-constants are compared. The rate-constant (k_T) for a tissue is similar to the perfusion rate except that the partitioning characteristics of the tissue are also considered:

$$k_T = Q_T/(P_T \times V_T),$$

where Q_T = the blood flow to the tissue (L/h), P_T = the tissue–blood partition coefficient for the chemical, V_T = the volume of the tissue (L).

Thus the units of the tissue rate-constant are the same as for the perfusion rate, h^{-1}, but the rate-constant more accurately reflects the kinetic characteristics of a tissue for a particular chemical. It was the much smaller rate-constant for fat in the case of a lipophilic chemical such as styrene that required the separation of the fat compartment from the other poorly perfused tissues (muscle, skin, etc.) in the PBPK model for styrene (22).

The second, less rigorous, approach for determining whether tissues should be lumped together is simply to compare the performance of the model with the tissues combined and separated. This approach is essentially the reverse of the example given above for splitting of the fat compartment. The reliability of this approach depends on the availability of data under conditions where the tissues being evaluated would be expected to have an observable impact on the kinetics of the chemical. Sensitivity analysis can sometimes be used to determine the appropriate conditions for such a comparison (95).

3.4. Model Design Principles

There is no easy rule for determining the structure and level of complexity needed in a particular modeling application. The wide variability of PBPK model design for different chemicals can be seen by comparing the diagram of the PBPK model for methotrexate (41), shown in Fig. 8, with the diagram for the styrene PBPK model shown in Fig. 3. Model elements which are important for a volatile, lipophilic chemical such as styrene (lung, fat) do not need to be considered in the case of a nonvolatile, water soluble compound such as methotrexate. Similarly, while kidney excretion and enterohepatic recirculation are important determinants of the kinetics of methotrexate, only metabolism and exhalation are

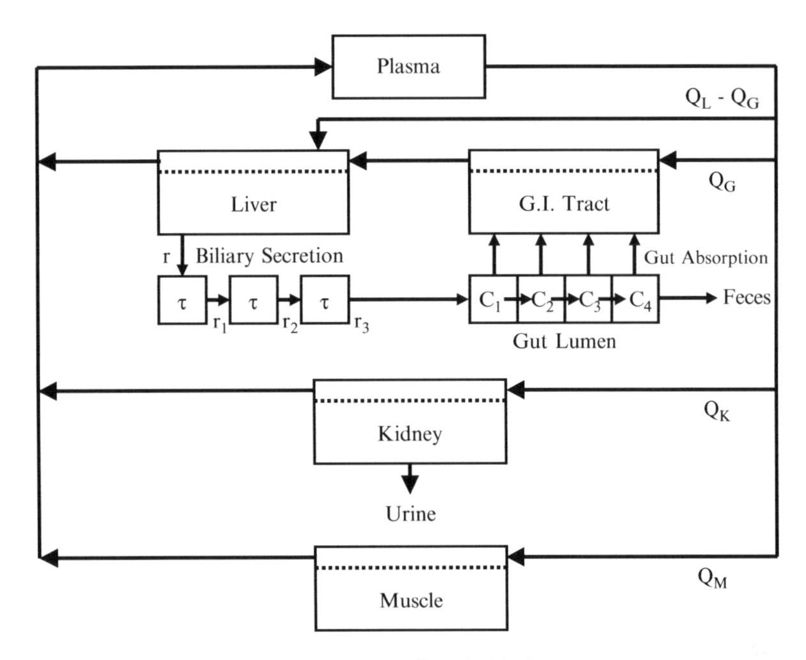

Fig. 8. PBPK model for methotrexate (Bischoff et al. 1971).

significant for styrene. The decision of which elements to include in the model structure for a specific chemical and application draws on all of the modeler's experience and knowledge of the animal–chemical system.

The alternative approaches to tissue grouping discussed above are actually just reflections of the two competing criteria which must be balanced during model design: parsimony and plausibility. The principle of parsimony simply states that a model should be as simple as possible for the intended application (but no simpler). This "splitting" philosophy is related to that of Occam's Razor: "Entities should not be multiplied unnecessarily." That is, structures and parameters should not be included in the model unless they are needed to support the application for which the model is being designed. For example, if a model is developed to describe inhalation exposure to a chemical over periods from hours to years, as in the case of the styrene model discussed earlier, it is not necessary to describe transient, breath-by-breath behavior of chemical uptake and exhalation in the lung. On the other hand, if the model is being developed to predict initial inhalation uptake of the chemical at times on the order of minutes, this level of detail clearly might be justified (98, 99).

The desire for parsimony in model development is driven not only by the desire to minimize the number of parameters whose values must be identified, but also by the recognition that as the number of parameters increases, the potential for unintended interactions between parameters increases disproportionately. A generally accepted rule of software engineering warns that it is relatively easy to design a computer program which is too complicated to be completely comprehended by the human mind. As a model becomes more complex, it becomes increasingly difficult to validate, even as the level of concern for the trustworthiness of the model should increase.

Countering the desire for model parsimony is the need for plausibility of the model structure. As discussed in the introduction, it is the physiological and biochemical realism of PBPK models that gives them an advantage for extrapolation. The credibility of a PBPK model's predictions of kinetic behavior under conditions different from those under which the model was validated rests to a large extent on the correspondence of the model design to known physiological and biochemical structures. In general, the ability of a model to adequately simulate the behavior of a physical system depends on the extent to which the model structure is homomorphic (having a one-to-one correspondence) with the essential features determining the behavior of that system. For example, if the model of styrene had not included a description of saturable metabolism, it would not have been able to adequately simulate the kinetics of styrene at both low and high doses using a single parameterization.

3.4.1. Model Identification

The process of model identification begins with the selection of those model elements which the modeler considers to be minimum essential determinants of the behavior of the particular animal–chemical system under study, from the viewpoint of the intended application of the model. Comparison with appropriate data, relevant to the intended purpose of the model, then can provide insights into defects in the model which must be corrected either by reparameterization or by changes to the model structure. Unfortunately, it is not always possible to separate these two elements. In models of biological systems, estimates of the values of model parameters will always be uncertain, due both to biological variation and experimental error. At the same time, the need for biological realism unavoidably results in models that are "overparameterized"; that is, they contain more parameters than can be identified from the kinetic data the model is used to describe.

As an example of the interaction between model structure and parameter identification, the two metabolic parameters, V_{max} and K_m, in the model for styrene discussed earlier could both be identified relatively unambiguously in the case of the rat. Indeed, as pointed out previously, the inclusion of capacity-limited metabolism in the model was necessary in order to reproduce the available data at both low and high exposure concentrations. In the case of the human, however, data was not available at sufficiently high concentrations to saturate metabolism. Therefore, only the ratio, V_{max}/K_m, would actually be identifiable. The use of the same model structure, including a two-parameter description of metabolism, in the human as in the rat was justified by the knowledge that similar enzymatic systems are responsible for the metabolism of chemicals such as styrene in both species. However, if the model were to be used to extrapolate to higher concentrations in the human, the potential impact of the uncertainty in the values of the individual metabolic parameters would have to be carefully considered.

Model identification is the selection of a specific model structure from several alternatives, based on conformity of the models' predictions to experimental observations. The practical reality of model identification in the case of biological systems is that regardless of the complexity of the model there will always be some level of "model error" (lack of homomorphism) which will result in systematic discrepancies between the model and experimental data. This model structural deficiency interacts with deficiencies in the identifiability of the model parameters, potentially leading to misidentification of the parameters or misspecification of structures. This most dangerous aspect of model identification is exacerbated by the fact that, in general, adding equations and parameters to a model increases the model's degrees of freedom, improving its ability to reproduce data, regardless of the validity of the underlying structure. Therefore, when a particular model structure improves the agreement of the model with kinetic data, it can only be said that the model structure is "consistent" with the kinetic data; it cannot be said that the model structure has been "proved" by its consistency with the data. In such circumstances, it is imperative that the physiological or biochemical hypothesis underlying the model structure is tested using nonkinetic data.

3.5. Elements of Model Structure

The process of selecting a model structure can be broken down into a number of elements associated with the different aspects of uptake, distribution, metabolism, and elimination. In addition, there are several general model structure issues that must be addressed, including mass balance and allometric scaling. The following section treats each of these elements in turn.

3.5.1. Storage Compartments

Naturally, any tissues which are expected to accumulate significant quantities of the chemical or its metabolites need to be included in the model structure. As discussed earlier, these storage tissues can be grouped together to the extent that they have similar time constants. Three storage compartments were included in the styrene model described above: fat tissues, richly perfused tissues, and poorly perfused tissues. The generic mass balance equation for storage compartments such as these is (Fig. 9):

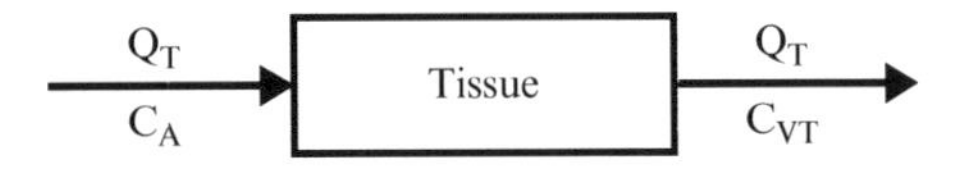

Fig. 9. Blood flow through a storage compartment.

$$dA_T/dt = Q_T \times C_A - Q_T \times C_{VT},$$

where A_T = the mass of chemical in the tissue (mg), Q_T = the blood flow to (and from) the tissue (L/h), C_A = the concentration of chemical in the arterial blood reaching the tissue (mg/L), C_{VT} = the concentration of the chemical in the venous blood leaving the tissue (mg/L).

Thus this mass balance equation simply states that the change in the amount of chemical in the tissue with respect to time (dA_T/dt) is equal to the difference between the amount of chemical entering the tissue and the amount leaving the tissue. We can then calculate the concentration of the chemical in the storage tissue (C_T) from the amount in the tissue and the tissue volume (V_T):

$$C_T = A_T/V_T.$$

In PBPK models, it is common to assume "venous equilibration"; that is, that in the time that it takes for the blood to perfuse the tissue, the chemical is able to achieve its equilibrium distribution between the tissue and blood. Therefore, the concentration of the chemical in the venous blood can be related to the concentration in the tissue by the equilibrium tissue–blood partition coefficient (P_T):

$$C_{VT} = C_T/P_T.$$

Therefore we obtain a differential equation in A_T:

$$dA_T/dt = Q_T \times C_A - Q_T \times A_T/(P_T \times V_T).$$

If desired, we can reformulate this mass balance equation in terms of concentration:

$$dA_T/dt = d(C_T \times V_T)/dt = C_T \times dV_T/dt + V_T \times dC_T/dt.$$

If (and only if) V_T is constant (i.e., the tissue does not grow during the simulation), $dV_T/dt = 0$, and:

$$dA_T/dt = V_T \times dC_T/dt,$$

so we have the alternative differential equation:

$$dC_T/dt = Q_T \times (C_A - C_T/P_T)/V_T.$$

This alternative mass balance formulation, in terms of concentration rather than amount, is popular in the pharmacokinetic literature. However, in the case of models with compartments that change volume over time it is preferable to use the formulation in terms of amounts in order to avoid the need for the additional term reflecting the change in volume ($C_T \times dV_T/dt$).

Depending on the chemical, many different tissues can potentially serve as important storage compartments. The use of a fat storage compartment in the styrene model is typical of a lipophilic chemical. The gut lumen can also serve as a storage site for chemicals subject to enterohepatic recirculation, as in the case of methotrexate. Important storage sites for metals, on the other hand, can include the kidney, red blood cells, intestinal epithelial cells, skin, bone, and hair. Transport to and from a storage compartment does not always occur via the blood, as described above; for example, in some cases the storage is an intermediate step in an excretion process (e.g., hair, intestinal epithelial cells). As with methotrexate, it may also be necessary to use multiple compartments in series, or other mathematical devices, to model plug flow (i.e., a time delay between entry and exit from storage).

3.5.2. Blood Compartment

The description of the blood compartment can vary considerably from one PBPK model to another depending on the role the blood plays in the kinetics of the chemical being modeled. In some cases the blood may be treated as a simple storage compartment, with a mass balance equation describing the summation (Σ) of the venous blood flows from the various tissues and the return of the total arterial blood flow (Q_C) to the tissues, as well as any urinary clearance (Fig. 10):

$$\mathrm{d}A_B/\mathrm{dt} = \sum (Q_T \times C_T/P_T) - Q_C \times C_B - K_U \times C_B,$$

where A_B = the amount of chemical in the blood (mg), Q_C = the total cardiac output (L/h), C_B = the concentration of chemical in the blood (mg/L), K_U = the urinary clearance (L/h).

For some chemicals, such as methotrexate, all of the chemical is present in the plasma rather than the red blood cells, so plasma flows and volumes are used instead of blood. For other chemicals it may be necessary to model the red blood cells as a storage compartment in communication with the plasma via diffusion-limited transport. Note that if the blood is an important storage compartment for a chemical, it may be necessary to carefully evaluate data on tissue concentrations, particularly the richly perfused tissues, to determine whether chemical in the blood perfusing the tissue could be contributing to the measured tissue concentration.

For still other chemicals, such as styrene, the amount of chemical actually in the blood may be relatively unimportant. In this case, instead of having a true blood compartment, a steady-state approximation can be used to estimate the concentration in the blood at any time. Assuming the blood is at steady-state with respect to the tissues:

$$\mathrm{d}A_B/\mathrm{dt} = 0.$$

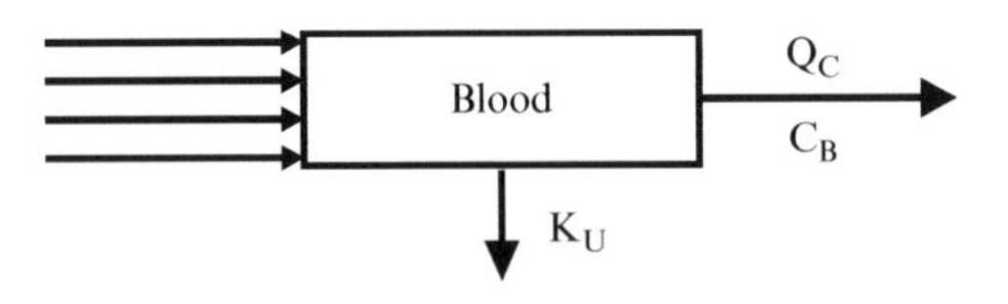

Fig. 10. Blood flow to tissue and urinary clearance.

Therefore, solving the blood equation for the concentration:

$$C_B = \sum \frac{(Q_T \times C_T/P_T)}{Q_C}.$$

3.5.3. Metabolism/ Elimination

The liver is frequently the primary site of metabolism for a chemical. The following equation is an example of the mass balance equation for the liver in the case of a chemical which is metabolized by two pathways (Fig. 11):

$$\mathrm{d}A_L/\mathrm{dt} = Q_L \times (C_A - C_L/P_L) - k_F \times C_L \times V_L/P_L - V_{max} \times C_L/P_L/(K_m + C_L/P_L).$$

In this case, the first term on the right-hand side of the equation represents the mass flux associated with transport in the blood and is identical to the case of the storage compartment described previously. The second term describes metabolism by a linear (first-order) pathway with rate constant k_F (h^{-1}) and the third term represents metabolism by a saturable (Michaelis–Menten) pathway with capacity V_{max} (mg/h) and affinity K_m (mg/L). If it were desired to model a water soluble metabolite produced by the saturable pathway, an equation for its formation and elimination could be added to the model (Fig. 12):

$$\mathrm{d}A_M/\mathrm{dt} = R_{stoch} \times V_{max} \times C_L/P_L/(K_m + C_L/P_L) - k_e \times A_M$$

$$C_M = A_M/V_D,$$

where A_M = the amount of metabolite in the body (mg), R_{stoch} = the stoichiometric yield of the metabolite times the ratio of its molecular, Weight to that of the parent chemical, k_e = the rate

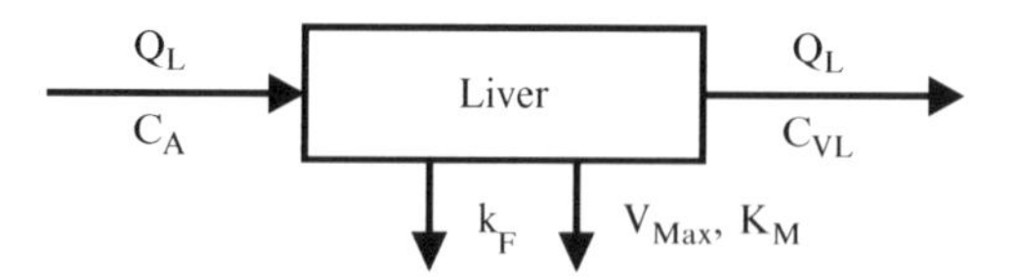

Fig. 11. Liver compartment metabolizing a chemical by two pathways.

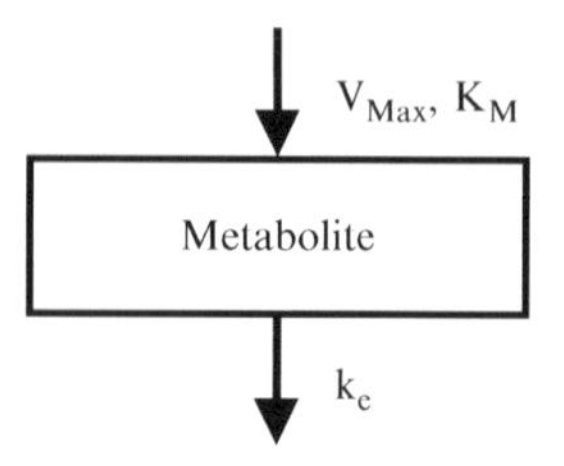

Fig. 12. Metabolite formation and elimination compartment.

constant for the clearance of the metabolite from the body (h^{-1}), C_M = the concentration of the metabolite in the plasma (mg/L), V_D = the apparent volume of distribution for the metabolite (L).

3.5.4. Metabolite Compartments

In principle, the same considerations which drive decisions regarding the level of complexity of the PBPK model for the parent chemical must also be applied for each of its metabolites, and their metabolites, and so on. As in the case of the parent chemical, the first and most important consideration is the purpose of the model. If the concern is direct parent chemical toxicity and the chemical is detoxified by metabolism, then there is no need for a description of metabolism beyond its role in the clearance of the parent chemical. The models for styrene and methotrexate discussed above are examples of parent chemical models. Similarly, if reactive intermediates produced during the metabolism of a chemical are responsible for its toxicity, as in the case of methylene chloride, a very simple description of the metabolic pathways might be adequate (8). The cancer risk assessment model for methylene chloride described the rate of metabolism for two pathways: the glutathione conjugation pathway, which was considered responsible for the carcinogenic effects, and the competing P450 oxidation pathway, which was considered protective.

On the other hand, if one or more of the metabolites are considered to be responsible for the toxicity of a chemical, it may be necessary to provide a more complete description of the kinetics of the metabolites themselves. For example, in the case of teratogenicity from all-*trans*-retinoic acid, both the parent chemical and several of its metabolites are considered to be toxicologically active; therefore, in developing the PBPK model for this chemical it was necessary to include a fairly complete description of the metabolic pathways (100). Fortunately, the metabolism of xenobiotic compounds often produces metabolites which are relatively water soluble, simplifying the description needed. In many cases, such as the production of trichloroacetic acid from trichloroethylene (46–48), a classical one-compartment description may be adequate for describing the metabolite kinetics. An example of such a description was provided earlier. In other cases, however, the description of the metabolite (or metabolites) may have to be as complex as that of the parent chemical. An example of such a case is the PBPK model for parathion (88), in which the model for the active metabolite, paraoxon, is actually more complex than that of the parent chemical.

3.5.5. Target Tissues

Typically, a PBPK model used in toxicology or risk assessment applications will include compartments for any target tissues for the toxic action of the chemical. The target tissue description may in some cases need to be fairly complicated, including such features as in situ metabolism, binding, and pharmacodynamic processes in order to provide a realistic measure of biologically effective tissue exposure (57). For example, whereas the lung compartment in the styrene model was represented only by a steady-state description of alveolar vapor exchange, the PBPK model for methylene chloride that was applied to perform a cancer risk assessment (8) included a two-part lung description in which alveolar vapor exchange was followed by a lung tissue compartment with in situ metabolism. This more complex lung compartment was required to describe the dose–response for methylene chloride induced lung cancer, which was assumed to result from the metabolism of methylene chloride in lung clara cells.

In other cases, describing a separate compartment for the target tissue may be unnecessary. For example, the styrene model described above could be used to relate acute exposures associated with neurological effects without the necessity of separating out a brain compartment. Instead, the concentration or AUC of styrene in the blood could be used as a metric, on the assumption that the relationship between brain concentration and blood concentration would be the same under all exposure conditions, routes, and species, namely, that the concentrations would be related by the brain–blood partition coefficient. In fact, this is probably a reasonable assumption across different exposure conditions in a given species. However, while tissue–air partition coefficients for volatile lipophilic chemicals appear to be similar in dog, monkey, and man (101), human blood–air partition coefficients appear to be roughly half of those in rodents (102). Therefore, the human brain–blood partition would probably be about twice that in the rodent. Nevertheless, if the model were to be used for extrapolation from rodents to humans, this difference could easily be factored into the analysis as an adjustment to the blood metric, without the need to actually add a brain compartment to the model.

A fundamental issue in determining the nature of the target tissue description required is the need to identify the toxicologically active form of the chemical. In some cases, a chemical may produce a toxic effect directly, either through its reaction with tissue constituents (e.g., ethylene oxide) or through its binding to cellular control elements (e.g., dioxin). Often, however, it is the metabolism of the chemical that leads to its toxicity. In this case, toxicity may result primarily from reactive intermediates produced during the process of metabolism (e.g., chlorovinyl epoxide produced from the metabolism of vinyl chloride) or from the toxic effects of stable metabolites (e.g., trichloroacetic acid produced from the metabolism of trichloroethylene).

The specific nature of the relationship between tissue exposure and response depends on the mechanism of toxicity, or mode of action, involved. Some toxic effects, such as acute irritation or acute neurological effects, may result primarily from the current concentration of the chemical in the tissue. Other toxic effects, such as tissue necrosis and cancer, may depend on both the concentration and duration of the exposure. For developmental effects, the chemical time course may also have to be convoluted with the window of susceptibility for a particular gestational event. The selection of the dose metric, that is, the active chemical form for which tissue exposure should be determined and the nature of the measure to be used—e.g., peak concentration (C_{max}) or area under the concentration (AUC)–time profile—is the most important step in a pharmacokinetic analysis and a principal determinant of the structure and level of detail that will be required in the PBPK model.

3.5.6. Uptake Routes

Each of the relevant uptake routes for the chemical must be described in the model. Often there are a number of possible ways to describe a particular uptake process, ranging from simple to complex. As with all other aspects of model design, the competing goals of parsimony and realism must be balanced in the selection of the level of complexity to be used. The following examples are meant to provide an idea of the variety of model code which can be required to describe the various possible uptake processes.

Intravenous Administration

$$A_{B0} = \text{Dose} \times \text{BW},$$

where A_{B0} = the amount of chemical in the blood at the beginning of the simulation ($t = 0$), Dose = administered dose (mg/kg), BW = animal body weight (kg).

or, in the case where a steady-state approximation has been used to eliminate the blood compartment:

$$C_B = (Q_L \times C_{VL} + \cdots + Q_F \times C_{VF} + k_{IV})/\text{QC},$$

where

$$\begin{aligned} k_{IV} &= \text{Dose} \times \text{BW}/t_{IV}, \quad (t < t_{IV}) \\ &= 0 \quad (t > t_{IV}) \end{aligned}$$

t_{IV} = the duration of time over which the injection takes place (h).

In the latter case, the model code must be written with a "switch" to change the value of k_{IV} to zero at $t = t_{IV}$.

Drinking Water (Fig. 13)

$$k_0 = \text{Dose} \times \text{BW}/24$$

$$dA_L/dt = Q_L \times (C_A - C_L/P_L) - k_F \times C_L \times V_L/P_L + k_0,$$

where Dose = the daily ingestion rate of chemical in drinking water (mg/kg/day) and the liver compartment as shown includes only first-order metabolism

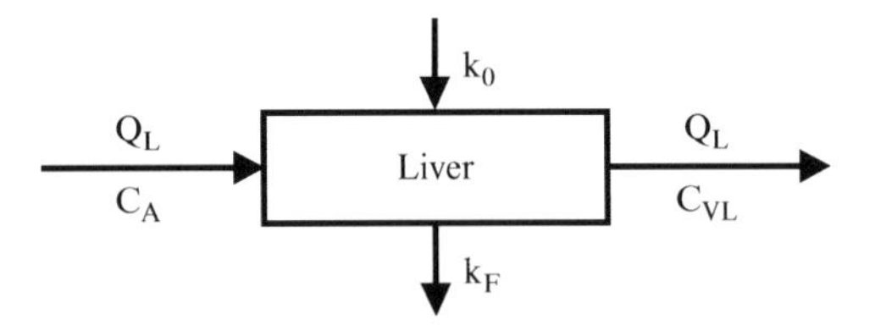

Fig. 13. Ingestion of chemical through drinking water.

Oral Gavage

For a chemical which is not excreted in the feces (Fig. 14):

$$A_{\mathrm{ST0}} = \mathrm{Dose} \times \mathrm{BW}$$

$$\mathrm{d}A_{\mathrm{ST}}/\mathrm{dt} = -k_{\mathrm{A}} \times A_{\mathrm{ST}}$$

$$\mathrm{d}A_{\mathrm{L}}/\mathrm{dt} = Q_{\mathrm{L}} \times (C_{\mathrm{A}} - C_{\mathrm{L}}/P_{\mathrm{L}}) - k_{\mathrm{F}} \times C_{\mathrm{L}} \times V_{\mathrm{L}}/P_{\mathrm{L}} + k_{\mathrm{A}} \times A_{\mathrm{ST}},$$

where Dose = the gavage dose (mg/kg), BW = the animal body weight (kg), A_{ST0} = the amount of chemical in the stomach at the beginning of the simulation, A_{ST} = the amount of chemical in the stomach at any given time, k_{A} = the oral absorption rate (h^{-1}).

For a chemical which is excreted in the feces (Fig. 15):

$$A_{\mathrm{ST0}} = \mathrm{Dose} \times \mathrm{BW}$$

$$\mathrm{d}A_{\mathrm{ST}}/\mathrm{dt} = -k_{\mathrm{A}} \times A_{\mathrm{ST}}$$

$$\mathrm{d}A_{\mathrm{I}}/\mathrm{dt} = k_{\mathrm{A}} \times A_{\mathrm{st}} - k_{\mathrm{I}} \times A_{\mathrm{I}} - K_{\mathrm{F}} \times A_{\mathrm{I}}/V_{\mathrm{I}}$$

$$\mathrm{d}A_{\mathrm{L}}/\mathrm{dt} = Q_{\mathrm{L}} \times (C_{\mathrm{A}} - C_{\mathrm{L}}/P_{\mathrm{L}}) - k_{\mathrm{F}} \times C_{\mathrm{L}} \times V_{\mathrm{L}}/P_{\mathrm{L}} + k_{\mathrm{I}} \times A_{\mathrm{I}},$$

where A_{I} = the amount of chemical in the intestinal lumen (mg), k_{I} = the rate constant for intestinal absorption (h^{-1}), K_{F} = the fecal clearance (L/h), V_{I} = the volume of the intestinal lumen (L).

The rate of fecal excretion of the chemical is then:

$$\mathrm{d}A_{\mathrm{F}}/\mathrm{dt} = K_{\mathrm{F}} \times A_{\mathrm{I}}/V_{\mathrm{I}}.$$

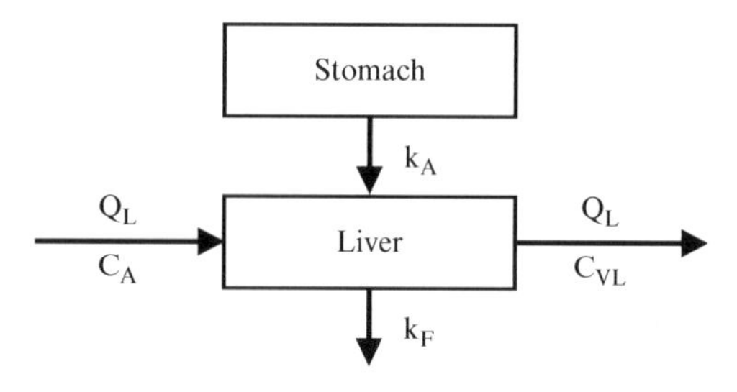

Fig. 14. Chemical ingested through oral gavage and not excreted in the feces.

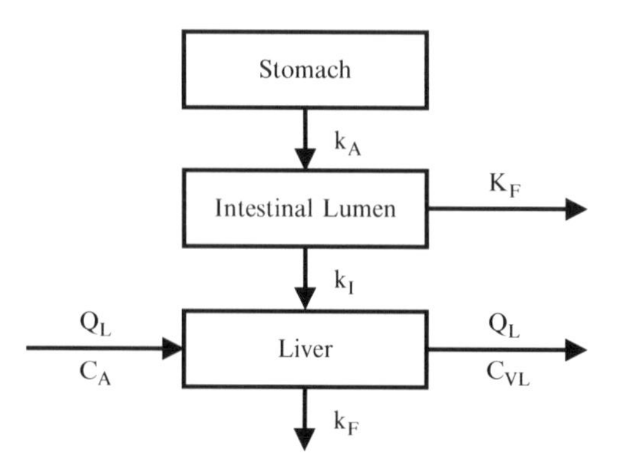

Fig. 15. Chemical being excreted through feces.

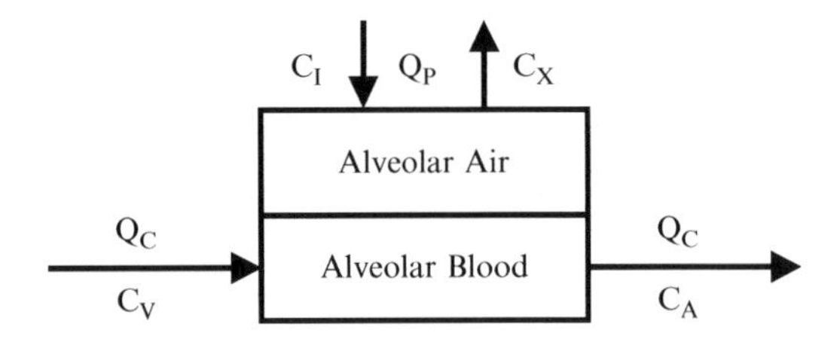

Fig. 16. Inhalation of chemical through the lung compartment.

Note, however, that this simple formulation does not consider the plug flow of the intestinal contents and will not reproduce the delay which actually occurs in the appearance of the chemical in the feces. Such a delay could be added using a delay function available in common simulation software, or multiple compartments could be used to simulate plug flow, as shown in the diagram of the methotrexate model (Fig. 8).

Inhalation (Fig. 16)

$$dA_{AB}/dt = Q_C \times (C_V - C_A) + Q_P \times (C_I - C_X),$$

where A_{AB} = the amount of chemical in the alveolar blood (mg), Q_C = the total arterial blood flow (L/h), C_V = the concentration of chemical in the pooled venous blood (mg/L), C_A = the concentration of chemical in the alveolar (arterial) blood (mg/L), Q_P = the alveolar (not total pulmonary) ventilation rate (L/h), C_I = the concentration of chemical in the inhaled air (mg/L), C_X = the concentration of chemical in the exhaled air (mg/L).

Assuming the alveolar blood is at steady-state with respect to the other compartments:

$$dA_{AB}/dt = 0.$$

Also, assuming lung equilibration (i.e., that the blood in the alveolar region has reached equilibrium with the alveolar air prior to exhalation):

$$C_X = C_A/P_B.$$

Substituting into the equation for the alveolar blood, and solving for C_A:

$$C_A = \frac{(Q_C \times C_V + Q_P \times C_I)}{(Q_C + Q_P/P_B)}.$$

This steady-state approximation is used in the styrene model described earlier. Note that the rate of elimination of the chemical by exhalation is just $Q_P \times C_X$. The alveolar ventilation rate, Q_P, does not include the "deadspace" volume (the portion of the inhaled air which does not reach the alveolar region), and is therefore roughly 70% of the total respiratory rate. The concentration C_X represents the "end-alveolar" air concentration; in order to

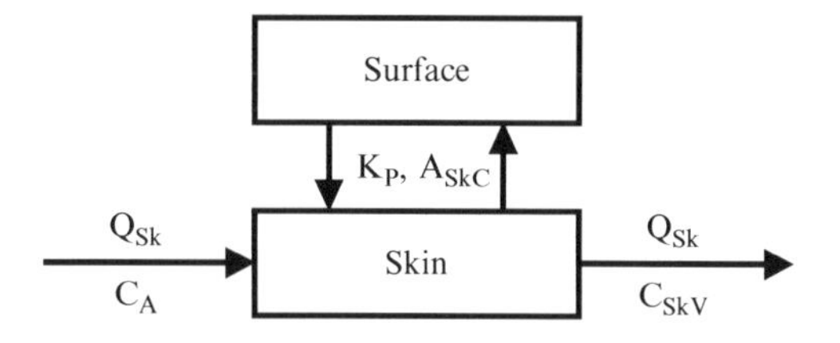

Fig. 17. Diffusion of chemical through the skin.

estimate the average exhaled concentration (C_{EX}), the dead-space contribution must be included:

$$C_{EX} = 0.3 \times C_I + 0.7 \times C_X.$$

Dermal (Fig. 17)

$$dA_{SK}/dt = K_P \times A_{SFC} \times (C_{SFC} - C_{SK}/P_{SKV})/\,1{,}000 + Q_{SK} \times (C_A - C_{SK}/P_{SKB}),$$

where A_{SK} = the amount of chemical in the skin (mg), K_P = the skin permeability coefficient (cm/h), A_{SFC} = the skin surface area (cm^2), C_{SFC} = the concentration of chemical on the surface of the skin (mg/L), C_{SK} = the concentration of the chemical in the skin (mg/L), P_{SK} = the skin–vehicle partition coefficient (i.e., the vehicle containing the chemical on the surface of the skin), Q_{SK} = the blood flow to the skin region (L/h), C_A = the arterial concentration of the chemical (mg/L), P_{SKV} = the skin–blood partition coefficient.

Note that due to adding this compartment, the equation for the blood in the model must also be modified to add a term for the venous blood returning from the skin ($+Q_{SK} \times C_{SK}/P_{SKB}$), the blood flow and volume parameters for the slowly perfused tissue compartment must be reduced by the amount of blood flow and volume for the skin.

3.5.7. Experimental Apparatus

In some cases, in addition to compartments describing the animal–chemical system, it may also be necessary to include model compartments that describe the experimental apparatus in which measurements were obtained. An example of such a case is modeling a closed-chamber gas uptake experiment. In a gas uptake experiment, several animals are maintained in a small, enclosed chamber while the air in the chamber is recirculated, with replenishment of oxygen and scrubbing of carbon dioxide. A small amount of a volatile chemical is then allowed to vaporize into the chamber, and the concentration of the chemical in the chamber air is monitored over time. In this design, any loss of the chemical from the chamber air reflects uptake into the animals (103). In order to simulate the change in the concentration in the chamber air as the chemical is taken up into the animals, an equation is required for the chamber itself (Fig. 18):

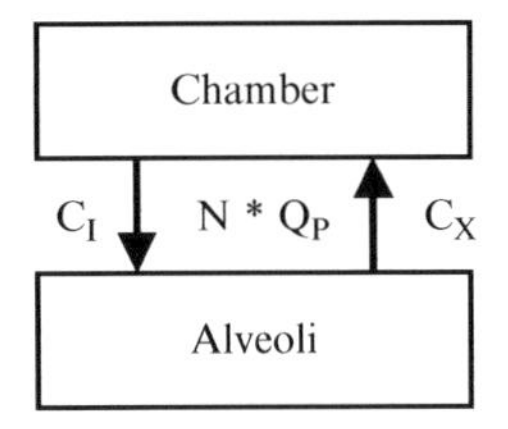

Fig. 18. The change in concentration in the chamber air as the chemical is absorbed.

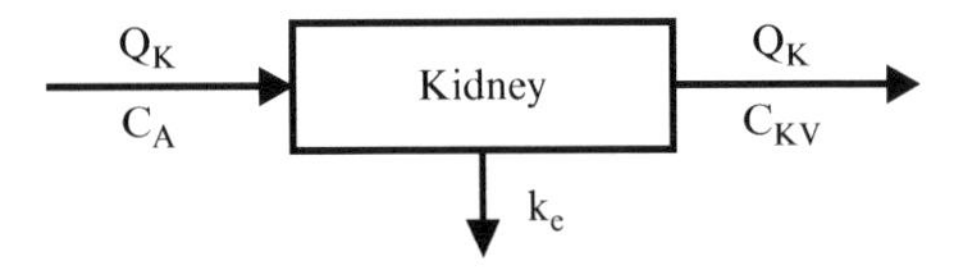

Fig. 19. Saturable binding in the kidney.

$$\mathrm{d}A_{\mathrm{CH}}/\mathrm{dt} = N \times Q_{\mathrm{P}} \times (C_{\mathrm{X}} - C_{\mathrm{I}})$$

$$C_{\mathrm{I}} = A_{\mathrm{CH}}/V_{\mathrm{CH}},$$

where A_{CH} = the amount of chemical in the chamber (mg), N = the number of animals in the chamber, C_X = the concentration of chemical in the air exhaled by the animals (mg/L), Q_P = the alveolar ventilation rate for a single animal (L/h), C_I = the chamber air concentration (mg/L), V_{CH} = the volume of air in the chamber (L).

3.5.8. Distribution/Transport

There are a number of issues associated with the description of the transport and distribution of the chemical that must be considered in the process of model design. Examples of a few of the more common ones are included here.

Binding

When there is evidence that saturable binding is an important determinant of the distribution of a chemical (such as an apparent dose-dependence of the tissue partitioning), a description of binding can be added to the model. For example, in the case of saturable binding in the kidney (Fig. 19):

$$\begin{aligned}\mathrm{d}A_{\mathrm{KT}}/\mathrm{dt} &= V_{\mathrm{K}}\mathrm{d}C_{\mathrm{KT}}/\mathrm{dt} \\ &= Q_{\mathrm{K}} \times (C_{\mathrm{A}} - C_{\mathrm{KF}}/P_{\mathrm{KF}}) - k_{\mathrm{e}} \times C_{\mathrm{KF}}/P_{\mathrm{KF}},\end{aligned}$$

where A_{KT} = the total amount of chemical in the kidney (mg), V_K = the volume of the kidney (L), C_{KT} = the total concentration of chemical in the kidney (mg/L), Q_K = the blood flow to the kidney, C_{KF} = the concentration of free (unbound) chemical in the kidney (mg/L), P_{KF} = the kidney–blood partition coefficient for free chemical, k_e = the urinary excretion rate constant (h^{-1}).

The apparent complication in adding this equation to the model is that the total movement of chemical is needed for the mass balance, but the determinants of the kinetics are in terms of the free concentration. To solve for free in terms of total, we note that:

$$C_{KT} = C_{KF} + C_{KB},$$

where C_{KB} = the concentration of bound chemical.

We can describe the saturable binding with an equation similar to that for saturable metabolism:

$$C_{KB} = B \times C_{KF}/(K_B + C_{KF}),$$

where B = the binding capacity (mg/L), K_B = the binding affinity (mg/L).

Substituting this equation in the previous one:

$$C_{KT} = \frac{C_{KF} + B \times C_{KF}}{(K_B + C_{KF})}.$$

Rewriting this equation to solve for the free concentration in terms of only the total concentration would result in a quadratic equation, the solution of which could be obtained with the quadratic formula. However, taking advantage of the iterative algorithm by which these PBPK models are exercised (as will be discussed later), it is not necessary to go to this effort. Instead, a much simpler implicit equation can be written for the free concentration (i.e., an equation in which the free concentration appears on both sides):

$$C_{KF} = C_{KT}/(1 + B/(K_B + C_{KF})).$$

In an iterative algorithm, this equation can be solved at each time step using the previous value of C_{KF} to obtain the new value! A new value of C_{KT} is then obtained from the mass balance equation for the kidney and the process is repeated.

Diffusion Limitation

Most of the PBPK models in the literature are flow-limited models; that is, they assume that the rate of tissue uptake of the chemical is limited only by the flow of the chemical to the tissue in the blood. While this assumption appears to be reasonable in general, for some chemicals and tissues uptake may instead be diffusion-limited. Examples of tissues for which diffusion-limited transport has often been described include the skin, placenta, brain, and fat. The model compartments described thus far have all assumed flow-limited transport. If there is evidence that the movement of a chemical between the blood and a tissue is limited by diffusion, a two-compartment description of the tissue can be used with a "shallow" exchange compartment in communication with the blood and a diffusion-limited "deep" compartment (Fig. 20):

$$dA_S/dt = Q_S \times (C_A - C_S) - K_{PA} \times (C_S - C_D/P_D)$$

$$dA_D/dt = K_{PA} \times (C_S - C_D/P_D),$$

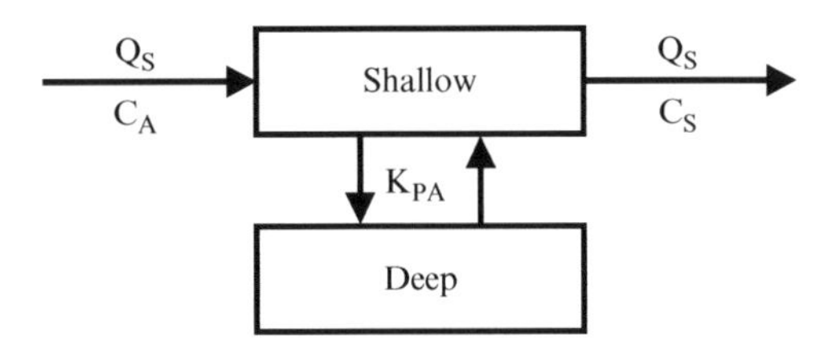

Fig. 20. A two-compartment model describing the movement of a chemical between the "shallow" and diffusion-limited "deep" compartment.

where A_S = the amount of chemical in the shallow compartment (mg), Q_S = the blood flow to the shallow compartment (L/h), C_S = the concentration of chemical in the shallow compartment (mg/L), K_{PA} = the permeability-area product for diffusion-limited transport (L/h), C_D = the concentration of chemical in the deep compartment (mg/L), P_D = the tissue–blood partition coefficient, A_D = the amount of chemical in the deep compartment (mg).

3.6. Model Parameterization

Once the model structure has been determined, it still remains to identify the values of the input parameters in the model.

3.6.1. Physiological Parameters

Estimates of the various physiological parameters needed in PBPK models are available from a number of sources in the literature, particularly for the human, monkey, dog, rat, and mouse (30–35). Estimates for the same parameter often vary widely, however, due both to experimental differences and to differences in the animals examined (age, strain, activity). Ventilation rates and blood flow rates are particularly sensitive to the level of activity (31, 33). Data on some important tissues is relatively poor, particularly in the case of fat tissue. Table 3 shows typical values of a number of physiological parameters in several species.

3.6.2. Biochemical Parameters

For volatile liquids, the type of chemicals which are common environmental contaminants, tissue partition coefficients can be determined by a simple in vitro technique called vial equilibration (36, 38) and tissue metabolic constants by a modification of the same technique (37) or other in vitro methods (105). Alternatively, rapid in vivo approaches for determining metabolic constants can be used either based on steady-state (96) or gas uptake experiments (102–104, 106, 107). Determination of the total amount of chemical metabolized in a particular exposure situation can also be used to estimate metabolic parameters (109). In addition, determination of stable end-product metabolites after exposure can be a particularly attractive technique in some cases (108, 110). Similar approaches can be used with nonvolatile chemicals (39, 111) and metals (112).

Table 3
(II-1) Typical physiological parameters for PBPK models

Species		Mouse	Rat	Monkey	Human
Ventilation					
Alveolar	(L/(h - 1 kg))	29.0	15.0	15.0	15.0
Blood flows					
Total	(L/(h - 1 kg))	16.5	15.0	15.0	15.0
Muscle	(Fraction)	0.18	0.18	0.18	0.18
Skin	(Fraction)	0.07	0.08	0.06	0.06
Fat	(Fraction)	0.03	0.06	0.05	0.05
Liver (arterial)	(Fraction)	0.035	0.03	0.065	0.07
Gut (portal)	(Fraction)	0.165	0.18	0.185	0.19
Other organs	(Fraction)	0.52	0.47	0.46	0.45
Tissue volumes					
Body weight	(kg)	0.02	0.3	4.0	80.0
Body water	(Fraction)	0.65	0.65	0.65	0.65
Plasma	(Fraction)	0.04	0.04	0.04	0.04
RBCs	(Fraction)	0.03	0.03	0.03	0.03
Muscle	(Fraction)	0.34	0.36	0.048	0.33
Skin	(Fraction)	0.17	0.195	0.11	0.11
Fat	(Fraction)	0.10	0.07	0.05	0.21
Liver	(Fraction)	0.046	0.037	0.027	0.023
Gut tissue	(Fraction)	0.031	0.033	0.045	0.045
Other organs	(Fraction)	0.049	0.031	0.039	0.039

3.6.3. Allometry

The different types of physiological and biochemical parameters in a PBPK model are known to vary with body weight in different ways (23). Typically, the parameterization of PBPK models is simplified by assuming standard allometric scaling (33, 113), as shown in Table 4, where the scaling factors, b, can be used in the following equation:

$$Y = aX^b,$$

where Y = the value of the parameter at a given body weight, X (kg), a = the scaled parameter value for a 1 kg animal.

While standard allometric scaling provides a useful starting point, or hypothesis, for cross-species scaling, it is not sufficiently accurate for some applications, such as risk assessment. In the case of the physiological parameters, the species-specific parameter values are generally available in the literature (30–35), and can be used directly in place of the allometric estimates. This is often not the case for the other model parameters, however. The use of the allometric scaling convention provides a useful way to estimate reasonable initial values for parameters across species. It also provides a reasonable method for intraspecies scaling.

Table 4
Standard allometric scaling for physiologically based pharmacokinetic model parameters

Parameter type (units)	Scaling (power of body weight)
Volumes	1.0
Flows (volume per time)	0.75
Ventilation (volume per time)	0.75
Clearances (volume per time)	0.75
Metabolic capacities (mass per time)	0.75
Metabolic affinities (mass per volume)	0
Partition coefficients (unit less)	0
First-order rate constants (inverse time)	−0.25

3.6.4. Parameter Optimization

In many cases, important parameters values needed for a PBPK model may not be available in the literature. In such cases it is necessary to measure them in new experiments, to estimate them by QSAR techniques, or to identify them by optimizing the fit of the model to an informative data set. Even in the case where an initial estimate of a particular parameter value can be obtained from other sources, it may be desirable to refine the estimate by optimization. For example, given the difficulty of obtaining accurate estimates of the fat volume in rodents, a more reliable estimate may be obtained by examining the impact of fat volume on the kinetic behavior of a lipophilic compound such as styrene. Of course, being able to uniquely identify a parameter from a kinetic data set rests on two key assumptions: (1) that the kinetic behavior of the compound under the conditions in which the data was collected is sensitive to the parameter being estimated, and (2) that other parameters in the model which could influence the observed kinetics have been determined by other means, and are held fixed during the estimation process.

The actual approach for conducting a parameter optimization can range from simple visual fitting, where the model is run with different values of the parameter until the best correspondence appears to be achieved, or by a quantitative mathematical algorithm. The most common algorithm used in optimization is the least-squares fit. To perform a least-squares optimization, the model is run to obtain a set of predictions at each of the times a data point was collected. The square of the difference between the model prediction and data point at each time is calculated and the results for all of the data points are summed. The parameter being estimated is then modified, and the sum of squares is recalculated. This process is repeated until the smallest possible sum of squares is obtained, representing the best possible fit of the model to the data.

In a variation on this approach, the square of the difference at each point is divided by the square of the prediction. This variation, known as relative least squares, is preferable in the case of data with an error structure which can be described by a constant coefficient of variation (that is, a constant ratio of the standard deviation to the mean). The former method, known as absolute least squares, is preferable in the case of data with a constant variance. From a practical viewpoint, the absolute least squares method tends to give greater weight to the data at higher concentrations and results in fits that look best when plotted on a linear scale, while the relative least squares method gives greater weight to the data at lower concentrations and results in fits that look best when plotted on a logarithmic scale.

A generalization of this weighting concept is provided by the extended least squares method, available in a number of optimization packages including ACSL/Opt (MGA Software, Concord, MA). In the extended least squares algorithm, the heteroscedasticity parameter can be varied from 0 (for absolute weighting) to 2 (for relative weighting), or can be estimated from the data. In general, setting the heteroscedasticity parameter from knowledge of the error structure of the data is preferable to estimating it from a data set.

A common example of identifying PBPK model parameters by fitting kinetic data is the estimation of tissue partition coefficients from experiments in which the concentration of chemical in the blood and tissues is reported at various time points. Using an optimization approach, the predictions of the model for the time course in the blood and tissues could be optimized with respect to the data by varying the model's partition coefficients. There is really little difference in the strength of the justification for estimating the partition coefficients in this way as opposed to estimating them directly from the data (by dividing the tissue concentrations by the simultaneous blood concentration). In fact, the direct estimates would probably be used as initial estimates in the model when the optimization was started.

A major difficulty in performing parameter optimization results from correlations between the parameters. When it is necessary to estimate parameters which are highly correlated, it is best to generate a contour plot of the objective function (sum of squares) or confidence region over a reasonable range of values of the two parameters. Generation of a contour plot with ACSL/Opt is relatively straightforward. An example of a contour plot for two of the metabolic parameters in the PBPK model for methylene chloride is

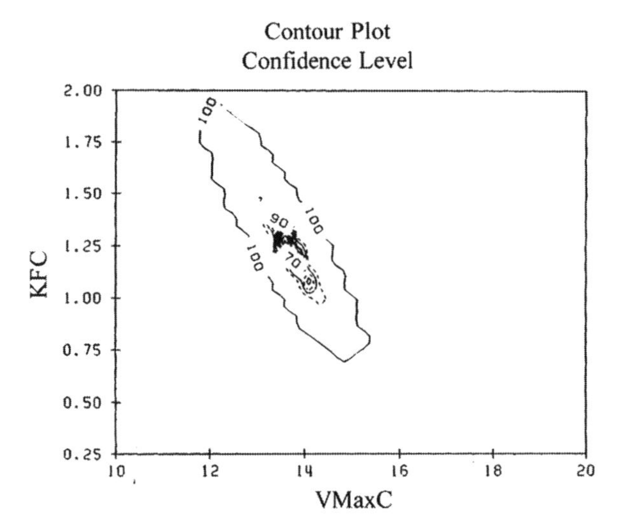

Fig. 21. Contour plot for correlated metabolic parameters in the PBPK model for methylene chloride.

shown in Fig. 21. The contours in the figure outline the joint confidence region for the values of the two parameters, and the fact that the confidence region is aligned diagonally reflects the correlation between the two parameters.

3.7. Mass Balance Requirements

One of the most important mathematical considerations during model design is the maintenance of mass balance. Simply put, the model should neither create nor destroy mass. This seemingly obvious principle is often violated unintentionally during the process of model development and parameterization. A common violation of mass balance, which typically leads to catastrophic results, involves failure to exactly match the arterial and venous blood flows in the model. As described above, the movement of chemical in the blood (in units of mass per time) is described as the product of the concentration of chemical in the blood (in units of mass per volume) times the flow rate of the blood (in units of volume per time). Therefore, to maintain mass balance, the sum of the blood flows leaving any particular tissue compartment must equal the sum of the blood flows entering the compartment. In particular, to maintain mass balance in the blood compartment (regardless of whether it is actually a compartment or just a steady-state equation), the sum of the venous flows from the individual tissue compartments must equal the total arterial blood flow leaving the heart:

$$\sum Q_{\mathrm{T}} = Q_{\mathrm{C}}.$$

Another obvious but occasionally overlooked aspect of maintaining mass balance during model development is that if a model is modified by splitting a tissue out of a lumped compartment, the blood flow to the separated tissue (and its volume) must be subtracted from that for the lumped compartment. Moreover, even though a model may initially be designed with parameters that meet the above requirements, mass balance may unintentionally be violated later if the parameters are altered during model execution. For example, if the parameter for the blood flow to one compartment is increased, the parameter for the overall blood flow must be increased accordingly or an equivalent reduction must be made in the parameter for the blood flow to another compartment. Particular care must be taken in this regard when the model is subjected to sensitivity or uncertainty analysis; inadvertent violation of mass balance during Monte Carlo sampling has lead in the past to the publication of erroneous sensitivity results (95).

A similar mass balance requirement must be met for transport other than blood flow. For example, if the chemical is cleared by biliary excretion, the elimination of chemical from the liver in the bile must exactly match the appearance of chemical in the gut lumen in the bile. Put mathematically, the same term for the transport must appear in the equations for the two compartments, but with opposite signs (positive vs. negative). For example, if the following equation were used to describe a liver compartment with first-order metabolism and biliary clearance.

$$\mathrm{d}A_{\mathrm{L}}/\mathrm{dt} = Q_{\mathrm{L}} \times (C_{\mathrm{A}} - C_{\mathrm{L}}/P_{\mathrm{L}}) - k_{\mathrm{F}} \times C_{\mathrm{L}} \times V_{\mathrm{L}}/P_{\mathrm{L}} - K_{\mathrm{B}} \times C_{\mathrm{L}},$$

where K_{B} = the biliary clearance rate (L/h).

The equation for the intestinal lumen would then need to include the term: $+K_{\mathrm{B}} \times C_{\mathrm{L}}$.

As a model grows in complexity, it becomes increasingly difficult to assure its mass balance by inspection. Therefore, it is a worthwhile practice to check for mass balance by including an equation in the model that adds up the total amount of chemical in each of the model compartments, including metabolized and excreted chemical, for comparison with the administered or inhaled dose.

3.8. Model Diagram

As described in the previous sections, the process of developing a PBPK model begins by determining the essential structure of the model based on the information available on the chemical's toxicity, mechanism of action, and pharmacokinetic properties. The results of this step can usually be summarized by an initial model diagram, such as those depicted in Figs. 3 and 8. In fact, in many cases a well-constructed model diagram, together with a table of the input parameter values and their definitions, is all that an accomplished modeler should need in order to create the mathematical equations defining a PBPK model. In general, there should be a one-to-one correspondence of the boxes in the diagram to the mass balance equations (or steady-state approximations) in the model. Similarly, the arrows in the diagram correspond to the transport or metabolism processes in the model. Each of the arrows connecting the boxes in the diagram should correspond to one of the terms in the mass balance equations for both of the compartments it connects, with the direction of the arrow pointing from the compartment in which the term is negative to the compartment in which it is positive. Arrows only connected to a single compartment, which represent uptake and excretion processes, are interpreted similarly. The model diagram should be labeled with the names of the key variables associated with the compartment or process represented by each box and arrow. Interpretation of the model diagram is also aided by the definition of the model input parameters in the corresponding table. The definition and units of the parameters can indicate the nature of the process being modeled (e.g., diffusion-limited vs. flow-limited transport, binding vs. partitioning, saturable vs. first-order metabolism, etc.).

3.9. Elements of Model Evaluation

One of the key advantages of PBPK models is their ability to perform extrapolations across species, routes of exposure, and exposure conditions. The reliability of a particular PBPK model for the purpose of extrapolation depends not only on the adequacy of its structure, but also on the correctness of its parameterization. The following section discusses some of the key issues associated with evaluating the adequacy of a model to predict chemical kinetics under conditions different from those for which experimental data are available.

3.9.1. Model Documentation

In cases where a model previously developed by one investigator is being evaluated for use in a different application by another investigator, adequate model documentation is critical for evaluation of the model. The documentation for a PBPK model should include sufficient information about the model so that an experienced modeler could accurately reproduce its structure and parameterization. Usually the suitable documentation of a model will require a combination of one or more "box and arrow" model diagrams together with any equations which cannot be unequivocally derived from the diagrams. Model diagrams should clearly differentiate blood flow from other transport (e.g., biliary excretion) or metabolism, and arrows should be used where the direction of transport could be ambiguous. All tissue compartments, metabolism pathways, routes of exposure, and routes of elimination should be clearly and accurately presented. All equations should be dimensionally consistent and in standard mathematical notation. Generic equations (e.g., for tissue "*i*") can help to keep the description brief but complete. The values used for all model parameters should be provided, with units. If any of the listed parameter values are based on allometric scaling, a footnote should provide the body weight used to obtain the allometric constant as well as the power of body weight used in the scaling.

3.9.2. Model Validation

Internal validation consists of the evaluation of the mathematical correctness of the model (114). It is best accomplished on the actual model code, but if necessary can be performed on appropriate documentation of the model structure and parameters, as described above (Assuming, of course, that the actual model code accurately reflects the model documentation). A more important issue regards the provision of evidence for external validation (sometimes referred to as verification). The level of detail incorporated into a model is necessarily a compromise between biological accuracy and parsimony. The process of evaluating the sufficiency of the model for its intended purpose, termed model verification, requires a demonstration of the ability of the model to predict the behavior of experimental data different from that on which it was based.

Whereas a simulation is intended simply to reproduce the behavior of a system, a model is intended to confirm a hypothesis concerning the nature of the system (115). Therefore, model validation should demonstrate the ability of the model to predict the behavior of the system under conditions which test the principal aspects of the underlying hypothetical structure. While quantitative tests of goodness of fit may often be a useful aspect of the verification process, the more important consideration may be the ability of the model to provide an accurate prediction of the general behavior of the data in the intended application.

Where only some aspects of the model can be verified, it is particularly important to assess the uncertainty associated with the aspects which are untested. For example, a model of a chemical and its metabolites which is intended for use in cross-species extrapolation to humans would preferably be verified using data in different species, including humans, for both the parent chemical and the metabolites. If only parent chemical data is available in the human, the correspondence of metabolite predictions with data in several animal species could be used as a surrogate, but this deficiency should be carefully considered when applying the model to predict human metabolism. One of the values of biologically based modeling is the identification of specific data which would improve the quantitative prediction of toxicity in humans from animal experiments.

In some cases it is necessary to use all of the available data to support model development and parameterization. Unfortunately, this type of modeling can easily become a form of self-fulfilling prophecy: models are logically strongest when they fail, but psychologically most appealing when they succeed (116). Under these conditions, model verification can particularly difficult, putting an additional burden on the investigators to substantiate the trustworthiness of the model for its intended purpose. Nevertheless, a combined model development and verification can often be successfully performed, particularly for models intended for interpolation, integration, and comparison of data rather than for true extrapolation.

3.9.3. Parameter Verification

In addition to verifying the performance of the model against experimental data, the model should be evaluated in terms of the plausibility of its parameters. This is particularly important in the case of PBPK models, where the parameters generally possess biological significance, and can therefore be evaluated for plausibility independent of the context of the model. The source of each model input parameter value should be identified, whether it was obtained from prior literature, determined directly by experiment, or estimated by fitting a model output to experimental data. Parameter estimates derived independently of tissue time course or dose-response data are preferred. To the extent feasible, the degree of uncertainty regarding the parameter values should also be evaluated. The empirically derived "Law of Reciprocal Certainty" states that the more important the model parameter, the less certain will be its value. In accordance with this principle, the most difficult, and typically most important, parameter determination for PBPK models is the characterization of the metabolism parameters.

When parameter estimation has been performed by optimizing model output to experimental data, the investigator must assure that the parameter is adequately identifiable from the data (114). Due to the confounding effects of model error, overparameterization, and parameter correlation, it is quite possible for an optimization algorithm to obtain a better fit to a particular data set by modifying a parameter which in fact should not be identified on the basis of that data set. Also, when an automatic optimization routine is employed it should be restarted with a variety of initial parameter values to assure that the routine has not stopped at a local optimum. These precautions are particularly important when more than one parameter is being estimated simultaneously, since the parameters in biologically based models are often highly correlated, making independent estimation difficult. Estimates of parameter variance obtained from automatic optimization routines should be viewed as lower bound estimates of true parameter uncertainty since only a local, linearized variance is typically calculated. In characterizing parameter uncertainty, it is probably more instructive to determine what ranges of parameter values are clearly inconsistent with the data than to accept a local, linearized variance estimate provided by the optimization algorithm.

It is usually necessary for the investigator to repeatedly vary the model parameters manually to obtain a sense of their identifiability and correlation under various experimental conditions, although some simulation languages include routines for calculating parameter sensitivity and covariance or for plotting confidence region contours. Sensitivity analysis and Monte Carlo uncertainty analysis techniques can serve as useful methods to estimate the impact of input parameter uncertainty on the uncertainty of model outputs (95, 117). However, care should be taken to avoid violation of mass balance when parameters are varied by sensitivity or Monte Carlo algorithms (95, 114), particularly where blood flows are affected.

3.9.4. Sensitivity Analysis

To the extent that a particular PBPK model correctly reflects the physiological and biochemical processes underlying the pharmacokinetics of a chemical, exercising the model can provide a means for identifying the most important physiological and biochemical parameters determining the pharmacokinetic behavior of the chemical under different conditions (95). The technique for obtaining this information is known as sensitivity analysis and can be performed by two different methods. Analytical sensitivity coefficients are defined as the ratio of the change in a model output to the change in a model parameter that produced it. To obtain a sensitivity coefficient by this method, the model is run for the exposure scenario of interest using the preferred values of the input parameters, and the resulting output (e.g., hair concentration) is recorded. The model is then run again with the value of one of the input parameters varied slightly. Typically, a 1% change is appropriate. The ratio of the resulting incremental change in the output to the change in the input represents the sensitivity coefficient. For example, if a 1% increase in an input parameter resulted in a 0.5% decrease in the output, the sensitivity coefficient would be −0.5. Sensitivity coefficients >1.0 in absolute value represent amplification of input error and would be a cause for concern. An alternative approach is to conduct a Monte Carlo analysis, as described below, and then to perform a simple correlation analysis of the model outputs and input parameters. Both methods have specific advantages. The analytical sensitivity coefficient most accurately represents the functional relationship of the output to the specific input under the conditions being modeled. The advantage of the correlation coefficients is that they also reflect the impact of interactions between the parameters during the Monte Carlo analysis.

3.9.5. Uncertainty Analysis

There are a number of examples in the literature of evaluations of the uncertainty associated with the predictions of a PBPK model using the Monte Carlo simulation approach (10, 117, 118). In a Monte Carlo simulation, a probability distribution for each of the PBPK model parameters is randomly sampled, and the model is run using the chosen set of parameter values. This process is repeated a large number of times until the probability distribution for the desired model output has been created. Generally speaking, 1,000 iterations or more may be required to ensure the reproducibility of the mean and standard deviation of the output distributions as well as the 1st through 99th percentiles. To the extent that the input parameter distributions adequately characterize the uncertainty in the inputs, and assuming that the parameters are reasonably independent, the resulting output distribution will provide a useful estimate of the uncertainty associated with the model outputs.

In performing a Monte Carlo analysis it is important to distinguish uncertainty from variability. As it relates to the impact of pharmacokinetics in risk assessment, uncertainty can be defined as the possible error in estimating the "true" value of a parameter for a representative ("average") person. Variability, on the other hand, should only be considered to represent true interindividual differences. Understood in these terms, uncertainty is a defect (lack of certainty) which can typically be reduced by experimentation, and variability is a fact of life which must be considered regardless of the risk assessment methodology used. An elegant approach for separately documenting the impact of uncertainty and variability is "two-dimensional" Monte Carlo, in which distributions for both uncertainty and variability are developed and multiple Monte Carlo runs are used to convolute the two aspects of overall uncertainty. Unfortunately, in practice it is often difficult to differentiate the contribution of variability and uncertainty to the observed variation in the reported measurements of a particular parameter (118).

Due to its physiological structure, many of the parameters in a PBPK model are interdependent. For example, the blood flows must add up to the total cardiac output and the tissue volumes (including those not included in the model) must add up to the body weight. Failure to account for the impact of Monte Carlo sampling on these mass balances can produce erroneous results (95, 117). In addition, some physiological parameters are naturally correlated, such as cardiac output and respiratory ventilation rate, and these correlations should be taken into account during the Monte Carlo analysis (118).

3.9.6. Collection of Critical Data

As with model development, the best approach to model evaluation is within the context of the scientific method. The most effective way to evaluate a PBPK model is to exercise the model to generate a quantitative hypothesis; that is, to predict the behavior of the system of interest under conditions "outside the envelope" of the data used to develop the model (at shorter/longer durations, higher/lower concentrations, different routes, different species, etc.). In particular, if there is an element of the model which remains in question, the model can be exercised to determine the experimental design under which the specific model element can best be tested. For example, if there is uncertainty regarding whether uptake into a particular tissue is flow or diffusion limited, alternative forms of the model can be used to compare predicted tissue concentration time courses under each of the limiting assumptions under various experimental conditions. The experimental design and sampling time which maximizes the difference between the predicted tissue concentrations under the two assumptions can then serve as the basis for the actual experimental data collection. Once the critical data has been collected, the same model can also be used to support a more quantitative experimental inference. In the case of the tissue uptake question just described, not only can the a priori model predictions be compared with the observed data to test the alternative hypotheses, but the model can also be used a posteriori to estimate the quantitative extent of any observed diffusion limitation (i.e., to estimate the relevant model parameter by fitting the data). If, on the other hand, the model is unable to reproduce the experimental data under either assumption, it may be necessary to reevaluate other aspects of the model structure. The key difference between research and analysis is the iterative nature of the former. It has wisely been said, "If we knew when we started what we had to do to finish, they'd call it search, not research."

4. Example

The previous sections have focused on the process of designing the PBPK model structure needed for a particular application. At this point the model consists of a number of mathematical equations: differential equations describing the mass balance for each of the compartments and algebraic equations describing other relationships between model variables. The next step in model development is the coding of the mathematical form of the model into a form which can be executed on a computer. The discussion in this section will be couched in terms of a particular software package, acslX.

4.1. Mathematical Formulation

Mathematically, a PBPK model is represented by a system of simultaneous linear differential equations. The model compartments are represented by the differential equations that describe the mass balance for each one of the "state variables" in the model. There may also be additional differential equations to calculate other necessary model outputs, such as the area under the AUC in a particular compartment, which is simply the integral of the concentration over time. The resulting system of equations is referred to as simultaneous because the time courses of the chemical in the various compartments are so interdependent that solving the equations for any one of the compartments requires information on the current status of all the other compartments; that is, the equations for all of the compartments must be solved at the same time. The equations are considered to be linear in the sense that they include only first-order derivatives, not due to any pharmacokinetic considerations. This kind of mathematical problem, in which a system is defined by the conditions at time zero together with differential equations describing how it evolves over time, is known as an initial value problem, and matrix decomposition methods are used to obtain the simultaneous solution.

A number of numerical algorithms are available for solving such problems. They all have in common that they are step-wise approximations; that is, they begin with the conditions at time zero and use the differential equations to predict how the system will change over a small time step, resulting in an estimate of the conditions at a slightly later time, which serves as the starting point for the next time step. This iterative process is repeated as long as necessary to simulate the experimental scenario.

The more sophisticated methods, such as the Gear algorithm (named after the mathematician, David Gear, who developed it) use a predictor–corrector approach, in which the corrector step essentially amounts to "predicting backwards" after each step forward, in order to check how closely the algorithm is able to reproduce the conditions at the previous time step. This allows the time step to be increased automatically when the algorithm is performing well, and to be shortened when it is having difficulty, such as when conditions are changing rapidly. However, due to the wide variation of the time constants (response times) for the various physiological compartments (e.g., fat vs. richly perfused), PBPK models often represent "stiff" systems. Stiff systems are characterized by state variables (compartments) with widely different time constants, which cause difficulty for predictor–corrector algorithms. The Gear algorithm was specifically designed to overcome this difficulty. It is therefore generally recommended that the Gear algorithm be used for executing PBPK models. An implementation of the Gear algorithm is available in most of the advanced software packages.

Regardless of the specific algorithm selected, the essential nature of the solution, as stated above, will be a step-wise approximation. However, all of the algorithms made available in computer software are convergent; that is, they can stay arbitrarily close to the true solution, given a small enough time step. On modern personal computers, even large PBPK models can be run to more than adequate accuracy in a reasonable timeframe.

4.2. Model Coding in ACSL

The following sections contain typical elements of the ACSL code for a PBPK model, interspersed with comments, which will be written in *italics* to differentiate them from the actual model code. The first section describes the model definition file, which by convention in ACSL is given a filename with the extension CSL.

The model used as an example in the following sections is a simple, multiroute model for volatile chemicals, similar to the styrene model discussed earlier, except that is also has the ability to simulate closed-chamber gas uptake experiments.

4.3. Typical Elements in a Model File

An acslX source file follows the structure defined in the Standard for Continuous Simulation Languages (just like there is a standard for C++). Thus, for example, there will generally be an INITIAL block defining the initial conditions followed by a DYNAMIC block which contains DISCRETE and/or DERIVATIVE sub-blocks that define the model. In addition, conventions which have been generally adopted by the PBPK modeling community (most of which started with John Ramsey at Dow Chemical during the development of the styrene model) help to improve the readability of the code. The following file shows typical elements of "Ramseyan code."

The first line in the code must start with the word PROGRAM (a remnant of ACSL's derivation from FORTRAN).

```
PROGRAM
```

Lines starting with an exclamation point (and portions of lines to the right of one) are ignored by the ACSL translator and can be used for comments:

```
! Developed for ACSL Level 10
! by Harvey Clewell (KS Crump Group, ICF Kaiser Int'l., Ruston, LA)
! and Mel Andersen (Health Effects Research Laboratory, USEPA,
RTP, NC)
```

The first section of an ACSL source file is the INITIAL block, which is used to define parameters and perform calculations that do not need to be repeated during the course of the simulation:

```
INITIAL ! Beginning of preexecution section
```

Only parameters defined in a CONSTANT statement can be changed during a session using the SET command:

```
LOGICAL CC ! Flag set to .TRUE. for closed chamber runs

! Physiological parameters (rat)
CONSTANT QPC = 14. ! Alveolar ventilation rate (L/hr)
CONSTANT QCC = 14. ! Cardiac output (L/hr)
CONSTANT QLC = 0.25 ! Fractional blood flow to liver
CONSTANT QFC = 0.09 ! Fractional blood flow to fat
CONSTANT BW = 0.22 ! Body weight (kg)
CONSTANT VLC = 0.04 ! Fraction liver tissue
CONSTANT VFC = 0.07 ! Fraction fat tissue
!——————Chemical specific parameters (styrene)
CONSTANT PL = 3.46 ! Liver/blood partition coefficient
CONSTANT PF = 86.5 ! Fat/blood partition coefficient
CONSTANT PS = 1.16 ! Slowly perfused tissue/blood partition
CONSTANT PR = 3.46 ! Richly perfused tissue/blood partition
CONSTANT PB = 40.2 ! Blood/air partition coefficient
CONSTANT MW = 104. ! Molecular weight (g/mol)
CONSTANT VMAXC = 8.4 ! Maximum velocity of metabolism
(mg/hr-1kg)
CONSTANT KM = 0.36 ! Michaelis–Menten constant (mg/L)
CONSTANT KFC = 0. ! First order metabolism (/hr-1kg)
CONSTANT KA = 0. ! Oral uptake rate (/hr)
```

```
!————————Experimental parameters
CONSTANT PDOSE = 0. ! Oral dose (mg/kg)
CONSTANT IVDOSE = 0. ! IV dose (mg/kg)
CONSTANT CONC = 1000. ! Inhaled concentration (ppm)
CONSTANT CC = .FALSE.! Default to open chamber
CONSTANT NRATS = 3. ! Number of rats (for closed chamber)
CONSTANT KLC = 0. ! First order loss from closed chamber (/hr)
CONSTANT VCHC = 9.1 ! Volume of closed chamber (L)
CONSTANT TINF = .01 ! Length of IV infusion (hr)
```

It is an understandable requirement in ACSL to define when to stop and how often to report. The parameter for the reporting frequency ("communication interval") is assumed by the ACSL translator to be called CINT unless you tell it otherwise using the CINTERVAL statement. The parameter for when to stop can be called anything you want, as long as you use the same name in the TERMT statement (see below), but the Ramseyan convention is TSTOP:

```
CONSTANT TSTOP = 24. ! Length of experiment (hr)
```

The following parameter name is generally used to define the length of inhalation exposures (the name LENGTH is also used by some):

```
CONSTANT TCHNG = 6. ! Length of inhalation exposure (hr)
```

The INITIAL block is a useful place to perform logical switching for different model applications, in this case between the simulation of closed-chamber gas uptake experiments and normal inhalation studies. It is also sometimes necessary to calculate initial conditions for one of the integrals ("state variables") in the model (the initial amount in the closed chamber in this case):

```
IF (CC) RATS = NRATS ! Closed chamber simulation
IF (CC) KL = KLC

IF (.NOT.CC) RATS = 0. ! Open chamber simulation
IF (.NOT.CC) KL = 0.
! (Turn off chamber losses so concentration in chamber remains constant)

IF (PDOSE.EQ.0.0) KA = 0. ! If not oral dosing, turn off oral uptake

    VCH = VCHC-RATS*BW ! Net chamber air volume (L)
    AI0 = CONC*VCH*MW/24450. ! Initial amount in chamber (mg)
```

After all the constants have been defined, calculations using them can be performed. In contrast to the DERIVATIVE block (of which more later), the calculations in the INITIAL block are performed in

the order written, just like in FORTRAN, so a variable must be defined before it can be used.

Note how allometric scaling is used for flows (QC, QP) and metabolism (VMAX, KFC). Also note how the mass balance for the blood flows and tissue volumes is maintained by the model code. Run-time changes in the parameters for fat and liver are automatically balanced by changes in the slowly and richly perfused compartments, respectively. The fractional blood flows add to one, but the fractional tissue volumes add up to only 0.91, allowing 9% of the body weight to reflect nonperfused tissues:

```
!————Scaled parameters
    QC = QCC*BW**0.74 ! Cardiac output
    QP = QPC*BW**0.74 ! Alveolar ventilation
    QL = QLC*QC ! Liver blood flow
    QF = QFC*QC ! Fat blood flow
    QS = 0.24*QC-QF ! Slowly perfused tissue blood flow
    QR = 0.76*QC-QL ! Richly perfused tissue blood flow
    VL = VLC*BW ! Liver volume
    VF = VFC*BW ! Fat tissue volume
    VS = 0.82*BW-VF ! Slowly perfused tissue volume
    VR = 0.09*BW-VL ! Richly perfused tissue volume
    VMAX = VMAXC*BW**0.7 ! Maximum rate of metabolism
    KF = KFC/BW**0.3 ! First-order metabolic rate constant
    DOSE = PDOSE*BW ! Oral dose
    IVR = IVDOSE*BW/TINF ! Intravenous infusion rate
```

An END statement is required to delineate the end of the initial block:

```
END ! End of initial se-ction
```

The next (and often last) section of an ACSL source file is the DYNAMIC block, which contains all of the code defining what is to happen during the course of the simulation:

```
DYNAMIC ! Beginning of execution section
```

ACSL possesses a number of different algorithms for performing the simulation, which mathematically speaking consists of solving an initial value problem for a system of simultaneous linear differential equations. (Although it is easier to just refer to it as integrating.) Available methods include the Euler, Runga-Kutta, and Adams-Moulton, but the tried and true choice of most PBPK modelers is the Gear predictor–corrector, variable step-size algorithm for stiff systems, which PBPK models often are. (Stiff, that is):

```
ALGORITHM IALG = 2        ! Use Gear integration algorithm
NSTEPS NSTP = 10          ! Number of integration steps in
                          communication interval
```

```
MAXTERVAL MAXT = 1.0e9    ! Maximum integration step size
MINTERVAL MINT = 1.0e-9   ! Minimum integration step size
CINTERVAL CINT = 0.01     ! Communication interval
```

One of the structures which can be used in the DYNAMIC block is called a DISCRETE block. The purpose of a DISCRETE block is to define an event which is desired to occur at a specific time or under specific conditions. The integration algorithm then keeps a lookout for the conditions and executes the code in the DISCRETE block at the proper moment during the execution of the model. An example of a pair of discrete blocks which are used to control repeated dosing in another PBPK model are shown here as an example:

```
DISCRETE DOSE1 ! Schedule events to turn exposure on and off daily
  INTERVAL DOSINT= 24. ! Dosing interval
    !(Set interval larger than TSTOP to prevent multiple exposure)
  IF (T.GT.TMAX) GOTO NODOSE
  IF (DAY.GT.DAYS) GOTO NODOSE
      CONC= CI ! Start inhalation exposure
      TOTAL= TOTAL + DOSE ! Administer oral dose
      TDOSE= T ! Record time of dosing
  SCHEDULE DOSE2 .AT. T + TCHNG ! Schedule end of exposure
  NODOSE..CONTINUE
  DAY= DAY + 1.
  IF (DAY.GT.7.) DAY= 0.5
END ! of DOSE1

DISCRETE DOSE2
    CONC= 0. ! End inhalation exposure
END ! of DOSE2
```

Within the DYNAMIC block, a group of statements defining a system of simultaneous differential equations is put in a DERIVATIVE block. If there is only one it does not have to be given a name:

```
DERIVATIVE ! Beginning of derivative definition block
```

The main function of the derivative block is to define the "state variables" which are to be integrated. They are identified by the INTEG function. For example, in the code below, AI is defined to be a state variable which is calculated by integrating the equation defining the variable RAI, using an initial value of AI0. For most of the compartments, the initial value is zero.

```
!——————————CI = Concentration in inhaled air (mg/L)
  RAI = RATS*QP*(CA/PB-CI) - (KL*AI) ! Rate equation
    AI = INTEG(RAI,AI0) ! Integral of RAI
    CI = AI/VCH*CIZONE ! Concentration in air
CIZONE = RSW((T.LT.TCHNG).OR.CC,1.,0.)
  CP = CI*24450./MW ! Chamber air concentration in ppm
```

Any experienced programmer would shudder at the code shown above, because several variables appear to be used before they have been calculated (for example, CIZONE is used to calculate CI and CI is used to calculate RAI. However, within the derivative block, writing code is almost too easy because the translator will automatically sort the statements into the proper order for execution. That is, there is no need to be sure that a variable is calculated before it is used.

The down side of the sorting is that you cannot be sure that two statements will be calculated in the order you want just because you place them one after the other. Also, because of the sorting (as well as the way the predictor–corrector integration algorithm hops forward and backward in time), IF statements will not work right. The RSW function above works like an IF statement, setting CIZONE to 1. whenever T (the default name for the time variable in ACSL) is Less Than TCHNG, and setting CIZONE to 0. (and thus turning off the exposure) whenever T is greater than or equal to TCHNG.

The following blocks of statements each define one of the compartments in the model. These statements can be compared with the mathematical equations described in the previous sections of the manual. One of the advantages of models written in ACSL following the Ramseyan convention is that they are easier to comprehend and reasonably self-documenting.

```
!——MR = Amount remaining in stomach (mg)
    RMR = -KA*MR
    MR = DOSE*EXP(-KA*T)
```

Note that the stomach could have been defined as one of the state variables:

MR = INTEG(RMR,DOSE)

But instead the exact solution for the simple integral has been used directly.

Similarly, instead of defining the blood as a state variable, the steady-state approximation is used:

```
!——CA = Concentration in arterial blood (mg/L)
    CA = (QC*CV + QP*CI)/(QC + (QP/PB))
    AUCB = INTEG(CA,0.)

!——AX = Amount exhaled (mg)
    CX = CA/PB ! End-alveolar air concentration (mg/L)
    CXPPM = (0.7*CX+0.3*CI)*24450./MW ! Average exhaled
    air concentration (ppm)
    RAX = QP*CX
    AX = INTEG(RAX,0.)

!——AS = Amount in slowly perfused tissues (mg)
    RAS = QS*(CA-CVS)
    AS = INTEG(RAS,0.)
    CVS = AS/(VS*PS)
```

```
    CS = AS/VS

!——AR = Amount in rapidly perfused tissues (mg)
    RAR = QR*(CA-CVR)
    AR = INTEG(RAR,0.)
    CVR = AR/(VR*PR)
    CR = AR/VR

!——AF = Amount in fat tissue (mg)
    RAF = QF*(CA-CVF)
    AF = INTEG(RAF,0.)
    CVF = AF/(VF*PF)
    CF = AF/VF

!——AL = Amount in liver tissue (mg)
    RAL = QL*(CA-CVL)-RAM + RAO
    AL = INTEG(RAL,0.)
    CVL = AL/(VL*PL)
    CL = AL/VL
    AUCL = INTEG(CL,0.)

!——AM = Amount metabolized (mg)
    RAM = (VMAX*CVL)/(KM + CVL) + KF*CVL*VL
    AM = INTEG(RAM,0.)

!——AO = Total mass input from stomach (mg)
    RAO = KA*MR
    AO = DOSE-MR

!——IV = Intravenous infusion rate (mg/h)
    IVZONE = RSW(T.GE.TINF,0.,1.)
    IV = IVR*IVZONE

!——CV = Mixed venous blood concentration (mg/L)
    CV = (QF*CVF + QL*CVL + QS*CVS + QR*CVR + IV)/
    QC

!——TMASS = mass balance (mg)
    TMASS = AF + AL + AS + AR + AM + AX + MR

!——DOSEX = Net amount absorbed (mg)
DOSEX = AI + AO + IVR*TINF-AX
```

Last, but definitely not least, you have to tell ACSL when to stop:

```
TERMT(T.GE.TSTOP) ! Condition for terminating simulation

END ! End of derivative block
END ! End of dynamic section
```

Another kind of code section, the TERMINAL block, can also be used here to execute statements that should only be calculated at the end of the run.

```
END ! End of program
```

4.4. Model Evaluation

The following section discusses various issues associated with the evaluation of a PBPK model. Once an initial model has been developed, it must be evaluated on the basis of its conformance with experimental data. In some cases, the model may be exercised to predict conditions under which experimental data should be collected in order to verify or improve model performance. Comparison of the resulting data with the model predictions may suggest that revision of the model will be required. Similarly, a PBPK model designed for one chemical or application may be adapted to another chemical or application, requiring modification of the model structure and parameters. It is imperative that revision or modification of a model is conducted with the same level of rigor applied during initial model development, and that structures are not added to the model with no other justification than that they improve the agreement of the model with a particular data set.

In addition to comparing model predictions to experimental data, model evaluation includes assessing the plausibility of the model input parameters, and the confidence which can be placed in extrapolations performed by the model. This aspect of model evaluation is particularly important in the case of applications in risk assessment, where it is necessary to assess the uncertainty associated with risk estimates calculated with the model.

4.5. Model Revision

An attempt to model the metabolism of allyl chloride (119) serves as an excellent example of the process of model refinement and validation. As mentioned earlier, in a gas uptake experiment several animals are maintained in a small, enclosed chamber while the air in the chamber is recirculated, with replenishment of oxygen and scrubbing of carbon dioxide. A small amount of a volatile chemical is then allowed to vaporize into the chamber, and the concentration of the chemical in the chamber air is monitored over time. In this design, any loss of the chemical from the chamber air reflects uptake into the animals. After a short period of time during which the chemical achieves equilibration with the animals' tissues, any further uptake represents the replacement of chemical removed from the animals by metabolism. Analysis of gas uptake data with a PBPK model has been used successfully to determine the metabolic parameters for a number of chemicals (103).

In an example of a successful gas uptake analysis, (108) described the closed chamber kinetics of methylene chloride using a PBPK model which included two metabolic pathways: one saturable, representing oxidation by Cytochrome P450 enzymes, and one linear, representing conjugation with glutathione (Fig. 22). As can be seen in this figure, there is a marked concentration dependence of the observed rate of loss of this chemical from the chamber. The initial decrease in chamber concentration in all of the experiments results from the uptake of chemical into the animal

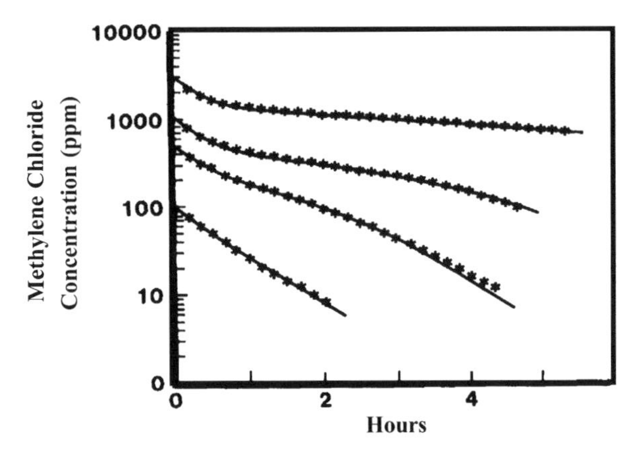

Fig. 22. Gas uptake experiment. Concentration (ppm) of methylene chloride in a closed, recirculated chamber containing three Fischer 344 rats. Initial chamber concentrations were (*top to bottom*) 3,000, 1,000, 500, and 100 ppm. *Solid lines* show the predictions of the model for a V_{max} of 4.0 mg/h/kg, a K_m of 0.3 mg/L, and a first-order rate constant of 2.0/h/kg, while symbols represent the measured chamber atmosphere concentrations.

tissues. Subsequent uptake is a function of the metabolic clearance in the animals, and the complex behavior reflects the transition from partially saturated metabolism at higher concentrations to linearity in the low concentration regime. The PBPK model is able to reproduce this complex behavior with a single set of parameters because the model structure appropriately captures the concentration dependence of the rate of metabolism.

A similar analysis of gas uptake experiments with allyl chloride using the same model structure was less successful. The smooth curves shown in Fig. 23 are the best fit that could be obtained to the observed allyl chloride chamber concentration data assuming a saturable pathway and a first-order pathway with parameters that were independent of concentration. Using this model structure there were large systematic errors associated with the predicted curves. The model predictions for the highest initial concentration were uniformly lower than the data, while the predictions for the intermediate initial concentrations were uniformly higher than the data. A much better fit could be obtained by setting the first-order rate constant to a lower value at the higher concentration; this approach would provide a better correspondence between the data and the model predictions, but would not provide a basis for extrapolating to different exposure conditions.

The nature of the discrepancy between the PBPK model and the data for allyl chloride suggested the presence of a dose-dependent limitation on metabolism not included in the model structure. This indication was consistent with other experimental evidence indicating that the conjugative metabolism of allyl chloride depletes glutathione, a necessary cofactor for the linear

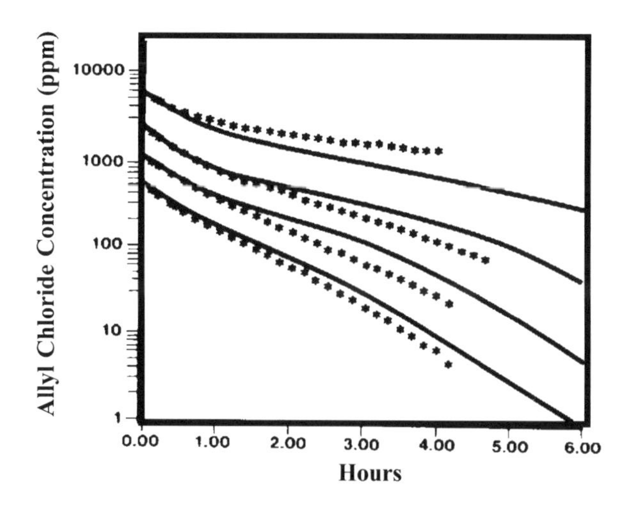

Fig. 23. Model failure. Concentration (ppm) of allyl chloride in a closed, recirculated chamber containing three Fischer 344 rats. Initial chamber concentrations were (*top to bottom*) 5,000, 2,000, 1,000, and 500 ppm. Symbols represent the measured chamber atmosphere concentrations. The curves represent the best result that could be obtained from an attempt to fit all of the data with a single set of metabolic constants using the same closed chamber model structure as in Fig. 4.1.

conjugation pathway. The conjugation pathway for reaction of methylene chloride and glutathione regenerates glutathione, but in the case of allyl chloride glutathione is consumed by the conjugation reaction. Therefore, to adequately reflect the biological basis of the kinetic behavior, it was necessary to model the time dependence of hepatic glutathione. To accomplish this, the mathematical model of the closed chamber experiment was expanded to include a more complete description of the glutathione-dependent pathway. The expanded model structure used for this description (120) included a zero-order production of glutathione and a first-order consumption rate that was increased by reaction of the glutathione with allyl chloride; glutathione resynthesis was inversely related to the instantaneous glutathione concentration. This description provided a much improved correspondence between the data and predicted behavior (Fig. 24). Of course, the improvement in fit was obtained at the expense of adding several new glutathione-related parameters to the model. To ensure that the improved fit is not just a consequence of the additional parameters providing more freedom to the model for fitting the uptake data, a separate test of the hypothesis underlying the added model structure (depletion of glutathione) was necessary. Therefore the expanded model was also used to predict both allyl chloride and hepatic glutathione concentrations following constant concentration inhalation exposures. Model predictions for end-exposure hepatic glutathione concentrations compared very favorably with actual data obtained in separate experiments (Table 5).

Table 5
Predicted glutathione depletion caused by inhalation exposure to allyl chloride

	Depletion (μM)	
Concentration (ppm)	Observed	Predicted
0	7,080 ± 120	7,088
10	7,290 ± 130	6,998[a]
0	7,230 ± 80	7,238[a]
100	5,660 ± 90	5,939
0	7,340 ± 180	7,341[a]
1,000	970 ± 10	839
0	6,890 ± 710	6,890[a]
2,000	464 ± 60	399

Note: Glutathione depletion data were graciously supplied by John Waechter, Dow Chemical Co., Midland, Michigan

[a]For the purpose of this comparison, the basal glutathione consumption rate in the model was adjusted to obtain rough agreement with the controls in each experiment. This basal consumption rate was then used to simulate the associated exposure

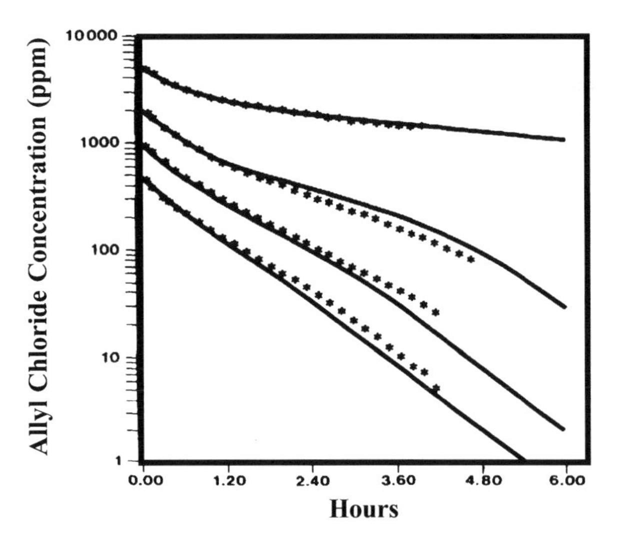

Fig. 24. Cofactor depletion. Symbols represent the same experimental data as in Fig. 23. The curves show the predictions of the expanded model, which not only included depletion of glutathione by reaction with allyl chloride, but also provided for regulation of glutathione biosynthesis on the basis of the instantaneous glutathione concentration, as described in the text.

To reiterate the key points of this example:

1. A PBPK model which had successfully described experimental results for a number of chemicals was unable to reproduce similar kinetic data on another chemical.
2. A hypothesis was developed that depletion of a necessary cofactor was affecting metabolism. This hypothesis was based on:
   (a) The nature of the discrepancy between the model predictions and the kinetic data.
   (b) Other available information about the nature of the chemical's biochemical interactions.
3. The code for the PBPK model was altered to include additional mass balance equations describing the depletion of this cofactor, and its resynthesis, as well as the resulting impact on metabolism.
4. The modification to the model was then tested in two ways:
   (a) By testing the ability of the new model structure to simulate the kinetic data that the original model was unable to reproduce.
   (b) By testing the underlying hypothesis regarding cofactor depletion against experimental data on glutathione depletion from a separate experiment.

Both elements of testing the model, kinetic validation and mechanistic validation, are necessary to provide confidence in the model. Unfortunately, there is a temptation to accept kinetic validation alone, particularly when data for mechanistic validation are unavailable. It should be remembered, however, that the simple act of adding equations and parameters to a model will, in itself, increase the flexibility of the model to fit data. Therefore, every attempt should be made to obtain additional experimental data to provide support for the mechanistic hypothesis underlying the model structure.

References

1. Andersen ME (1981) Saturable metabolism and its relation to toxicity. Crit Rev Toxicol 9:105–150
2. Monro A (1992) What is an appropriate measure of exposure when testing drugs for carcinogenicity in rodents? Toxicol Appl Pharmacol 112:171–181
3. Andersen ME, Clewell HJ, Krishnan K (1995) Tissue dosimetry, pharmacokinetic modeling, and interspecies scaling factors. Risk Anal 15:533–537
4. Teorell T (1937) Kinetics of distribution of substances administered to the body. I. The extravascular mode of administration. Arch Int Pharmacodyn 57:205–225
5. Teorell T (1937) Kinetics of distribution of substances administered to the body. I. The intravascular mode of administration. Arch Int Pharmacodyn 57:226–240
6. O'Flaherty EJ (1987) Modeling: an introduction. National Research Council. In: Pharmacokinetics in risk assessment. Drinking water and health, vol 8. National Academy Press, Washington DC, pp. 27–35
7. Clewell HJ, Andersen ME (1985) Risk assessment extrapolations and physiological modeling. Toxicol Ind Health 1(4):111–131

8. Andersen ME, Clewell HJ, Gargas ML, Smith FA, Reitz RH (1987) Physiologically based pharmacokinetics and the risk assessment for methylene chloride. Toxicol Appl Pharmacol 87:185–205
9. Gerrity TR, Henry CJ (1990) Principles of route-to-route extrapolation for risk assessment. Elsevier, New York
10. Clewell HJ, Jarnot BM (1994) Incorporation of pharmacokinetics in non-carcinogenic risk assessment: Example with chloropentafluorobenzene. Risk Anal 14:265–276
11. Clewell HJ (1995) Incorporating biological information in quantitative risk assessment: an example with methylene chloride. Toxicology 102:83–94
12. Clewell HJ (1995) The application of physiologically based pharmacokinetic modeling in human health risk assessment of hazardous substances. Toxicol Lett 79:207–217
13. Clewell HJ, Gentry PR, Gearhart JM, Allen BC, Andersen ME (1995) Considering pharmacokinetic and mechanistic information in cancer risk assessments for environmental contaminants: examples with vinyl chloride and trichloroethylene. Chemosphere 31:2561–2578
14. Clewell HJ, Andersen ME (1996) Use of physiologically-based pharmacokinetic modeling to investigate individual versus population risk. Toxicology 111:315–329
15. Clewell HJ III, Gentry PR, Gearhart JM (1997) Investigation of the potential impact of benchmark dose and pharmacokinetic modeling in noncancer risk assessment. J Toxicol Environ Health 52:475–515
16. Himmelstein KJ, Lutz RJ (1979) A review of the application of physiologically based pharmacokinetic modeling. J Pharmacokinet Biopharm 7:127–145
17. Gerlowski LE, Jain RK (1983) Physiologically based pharmacokinetic modeling: principles and applications. J Pharm Sci 72:1103–1126
18. Fiserova-Bergerova V (1983) Modeling of inhalation exposure to vapors: uptake distribution and elimination, vol 1 and 2. CRC, Boca Raton
19. Bischoff KB (1987) Physiologically based pharmacokinetic modeling. National Research Council. In: Pharmacokinetics in Risk Assessment. Drinking water and health, vol 8. National Academy Press, Washington, DC, pp. 36–61
20. Leung HW (1991) Development and utilization of physiologically based pharmacokinetic models for toxicological applications. J Toxicol Environ Health 32:247–267
21. Clewell HJ, Andersen ME (1986) A multiple dose-route physiological pharmacokinetic model for volatile chemicals using ACSL/PC. In: Cellier FD (ed) Languages for continuous system simulation. Society for Computer Simulation, San Diego, pp 95–101
22. Ramsey JC, Andersen ME (1984) A physiological model for the inhalation pharmacokinetics of inhaled styrene monomer in rats and humans. Toxicol Appl Pharmacol 73:159–175
23. Adolph EF (1949) Quantitative relations in the physiological constitutions of mammals. Science 109:579–585
24. Dedrick RL (1973) Animal scale-up. J Pharmacokinet Biopharm 1:435–461
25. Dedrick RL, Bischoff KB (1980) Species similarities in pharmacokinetics. Fed Proc 39:54–59
26. McDougal JN, Jepson GW, Clewell HJ, MacNaughton MG, Andersen ME (1986) A physiological pharmacokinetic model for dermal absorption of vapors in the rat. Toxicol Appl Pharmacol 85:286–294
27. Paustenbach DJ, Clewell HJ, Gargas ML, Andersen ME (1988) A physiologically based pharmacokinetic model for inhaled carbon tetrachloride. Toxicol Appl Pharmacol 96:191–211
28. Vinegar A, Seckel CS, Pollard DL, Kinkead ER, Conolly RB, Andersen ME (1992) Polychlorotrifluoroethylene (PCTFE) oligomer pharmacokinetics in Fischer 344 rats: development of a physiologically based model. Fundam Appl Toxicol 18:504–514
29. Clewell HJ, Andersen ME (1989) Improving toxicology testing protocols using computer simulations. Toxicol Lett 49:139–158
30. Bischoff KB, Brown RG (1966) Drug distribution in mammals. Chem Eng Prog Symp 62 (66):33–45
31. Astrand P, Rodahl K (1970) Textbook of work physiology. McGraw-Hill, New York
32. International Commission on Radiological Protection (ICRP) (1975) Report of the task group on reference man. ICRP Publication 23
33. Environmental Protection Agency (EPA) (1988) Reference physiological parameters in pharmacokinetic modeling. EPA/600/6-88/004. Office of Health and Environmental Assessment, Washington, DC
34. Davies B, Morris T (1993) Physiological parameters in laboratory animals and humans. Pharm Res 10:1093–1095
35. Brown RP, Delp MD, Lindstedt SL, Rhomberg LR, Beliles RP (1997) Physiological parameter values for physiologically based

pharmacokinetic models. Toxicol Ind Health 13(4):407–484
36. Sato A, Nakajima T (1979) Partition coefficients of some aromatic hydrocarbons and ketones in water, blood and oil. Br J Ind Med 36:231–234
37. Sato A, Nakajima T (1979) A vial equilibration method to evaluate the drug metabolizing enzyme activity for volatile hydrocarbons. Toxicol Appl Pharmacol 47:41–46
38. Gargas ML, Burgess RJ, Voisard DE, Cason GH, Andersen ME (1989) Partition coefficients of low-molecular-weight volatile chemicals in various liquids and tissues. Toxicol Appl Pharmacol 98:87–99
39. Jepson GW, Hoover DK, Black RK, McCafferty JD, Mahle DA, Gearhart JM (1994) A partition coefficient determination method for nonvolatile chemicals in biological tissues. Fundam Appl Toxicol 22:519–524
40. Clewell HJ (1993) Coupling of computer modeling with in vitro methodologies to reduce animal usage in toxicity testing. Toxicol Lett 68:101–117
41. Bischoff KB, Dedrick RL, Zaharko DS, Longstreth JA (1971) Methotrexate pharmacokinetics. J Pharm Sci 60:1128–1133
42. Farris FF, Dedrick RL, King FG (1988) Cisplatin pharmacokinetics: application of a physiological model. Toxicol Lett 43:117–137
43. Edginton AN, Theil FP, Schmitt W, Willmann S (2008) Whole body physiologically-based pharmacokinetic models: their use in clinical drug development. Expert Opin Drug Metab Toxicol 4:1143–1152
44. Andersen ME, Clewell HJ III, Gargas ML, MacNaughton MG, Reitz RH, Nolan R, McKenna M (1991) Physiologically based pharmacokinetic modeling with dichloromethane, its metabolite carbon monoxide, and blood carboxyhemoglobin in rats and humans. Toxicol Appl Pharmacol 108:14–27
45. Andersen ME, Clewell HJ, Mahle DA, Gearhart JM (1994) Gas uptake studies of deuterium isotope effects on dichloromethane metabolism in female B6C3F1 mice in vivo. Toxicol Appl Pharmacol 128:158–165
46. Fisher J, Gargas M, Allen B, Andersen M (1991) Physiologically based pharmacokinetic modeling with trichloroethylene and its metabolite, trichloroacetic acid, in the rat and mouse. Toxicol Appl Pharmacol 109:183–195
47. Fisher JW, Allen BC (1993) Evaluating the risk of liver cancer in humans exposed to trichloroethylene using physiological models. Risk Anal 13:87–95
48. Allen BC, Fisher J (1993) Pharmacokinetic modeling of trichloroethylene and trichloroacetic acid in humans. Risk Anal 13:71–86
49. Corley RA, Mendrala AL, Smith FA, Staats DA, Gargas ML, Conolly RB, Andersen ME, Reitz RH (1990) Development of a physiologically based pharmacokinetic model for chloroform. Toxicol Appl Pharmacol 103:512–527
50. Reitz RH, Mendrala AL, Corley RA, Quast JF, Gargas ML, Andersen ME, Staats DA, Conolly RB (1990) Estimating the risk of liver cancer associated with human exposures to chloroform using physiologically based pharmacokinetic modeling. Toxicol Appl Pharmacol 105:443–459
51. Johanson G (1986) Physiologically based pharmacokinetic modeling of inhaled 2-butoxyethanol in man. Toxicol Lett 34:23–31
52. Bungay PM, Dedrick RL, Matthews HB (1981) Enteric transport of chlordecone (Kepone) in the rat. J Pharmacokinet Biopharm 9:309–341
53. Tuey DB, Matthews HB (1980) Distribution and excretion of 2,2′,4,4′,5,5′-hexabromobiphenyl in rats and man: pharmacokinetic model predictions. Toxicol Appl Pharmacol 53:420–431
54. Lutz RJ, Dedrick RL, Tuey D, Sipes IG, Anderson MW, Matthews HB (1984) Comparison of the pharmacokinetics of several polychlorinated biphenyls in mouse, rat, dog, and monkey by means of a physiological pharmacokinetic model. Drug Metab Dispos 12(5):527–535
55. King FG, Dedrick RL, Collins JM, Matthews HB, Birnbaum LS (1983) Physiological model for the pharmacokinetics of 2,3,7,8-tetrachlorodibenzofuran in several species. Toxicol Appl Pharmacol 67:390–400
56. Leung HW, Ku RH, Paustenbach DJ, Andersen ME (1988) A physiologically based pharmacokinetic model for 2,3,7,8-tetrachlorodibenzo-p-dioxin in C57BL/6J and DBA/2J mice. Toxicol Lett 42:15–28
57. Andersen ME, Mills JJ, Gargas ML, Kedderis L, Birnbaum LS, Norbert D, Greenlee WF (1993) Modeling receptor-mediated processes with dioxin: implications for pharmacokinetics and risk assessment. Risk Anal 13 (1):25–36
58. O'Flaherty EJ (1991) Physiologically based models for bone seeking elements. I. Rat skeletal and bone growth. Toxicol Appl Pharmacol 111:299–312
59. O'Flaherty EJ (1991) Physiologically based models for bone seeking elements. II. Kinetics

of lead disposition in rats. Toxicol Appl Pharmacol 111:313–331

60. O'Flaherty EJ (1991) Physiologically based models for bone seeking elements. III. Human skeletal and bone growth. Toxicol Appl Pharmacol 111:332–341
61. O'Flaherty EJ (1993) Physiologically based models for bone seeking elements. IV. Kinetics of lead disposition in humans. Toxicol Appl Pharmacol 118:16–29
62. O'Flaherty EJ (1995) Physiologically based models for bone seeking elements. V. Lead absorption and disposition in childhood. Toxicol Appl Pharmacol 131:297–308
63. Mann S, Droz PO, Vahter M (1996) A physiologically based pharmacokinetic model for arsenic exposure. I. Development in hamsters and rabbits. Toxicol Appl Pharmacol 137:8–22
64. Mann S, Droz PO, Vahter M (1996) A physiologically based pharmacokinetic model for arsenic exposure. II. Validation and application in humans. Toxicol Appl Pharmacol 140:471–486
65. Farris FF, Dedrick RL, Allen PV, Smith JC (1993) Physiological model for the pharmacokinetics of methyl mercury in the growing rat. Toxicol Appl Pharmacol 119:74–90
66. McMullin TS, Hanneman WH, Cranmer BK, Tessari JD, Andersen ME (2007) Oral absorption and oxidative metabolism of atrazine in rats evaluated by physiological modeling approaches. Toxicology 240:1–14
67. Lin Z, Fisher JW, Ross MK, Filipov NM (2011) A physiologically based pharmacokinetic model for atrazine and its main metabolites in the adult male C57BL/6 mouse. Toxicol Appl Pharmacol 251:16–31
68. Kirman CR, Hays SM, Kedderis GL, Gargas ML, Strother DE (2000) Improving cancer dose-response characterization by using physiologically based pharmacokinetic modeling: an analysis of pooled data for acrylonitrile-induced brain tumors to assess cancer potency in the rat. Risk Anal 20:135–151
69. Sweeney LM, Gargas ML, Strother DE, Kedderis GL (2003) Physiologically based pharmacokinetic model parameter estimation and sensitivity and variability analyses for acrylonitrile disposition in humans. Toxicol Sci 71:27–40
70. Takano R, Murayama N, Horiuchi K, Kitajima M, Kumamoto M, Shono F, Yamazaki H (2010) Blood concentrations of acrylonitrile in humans after oral administration extrapolated from in vivo rat pharmacokinetics, in vitro human metabolism, and physiologically based pharmacokinetic modeling. Regul Toxicol Pharmacol 58:252–258
71. Clewell RA, Merrill EA, Robinson PJ (2001) The use of physiologically based models to integrate diverse data sets and reduce uncertainty in the prediction of perchlorate and iodide kinetics across life stages and species. Toxicol Ind Health 17:210–222
72. Clewell RA, Merrill EA, Yu KO, Mahle DA, Sterner TR, Mattie DR, Robinson PJ, Fisher JW, Gearhart JM (2003) Predicting fetal perchlorate dose and inhibition of iodide kinetics during gestation: a physiologically-based pharmacokinetic analysis of perchlorate and iodide kinetics in the rat. Toxicol Sci 73:235–255
73. Clewell RA, Merrill EA, Yu KO, Mahle DA, Sterner TR, Fisher JW, Gearhart JM (2003) Predicting neonatal perchlorate dose and inhibition of iodide uptake in the rat during lactation using physiologically-based pharmacokinetic modeling. Toxicol Sci 74:416–436
74. Merrill EA, Clewell RA, Gearhart JM, Robinson PJ, Sterner TR, Yu KO, Mattie DR, Fisher JW (2003) PBPK predictions of perchlorate distribution and its effect on thyroid uptake of radioiodide in the male rat. Toxicol Sci 73:256–269
75. Merrill EA, Clewell RA, Robinson PJ, Jarabek AM, Gearhart JM, Sterner TR, Fisher JW (2005) PBPK model for radioactive iodide and perchlorate kinetics and perchlorate-induced inhibition of iodide uptake in humans. Toxicol Sci 83:25–43
76. McLanahan ED, Andersen ME, Campbell JL, Fisher JW (2009) Competitive inhibition of thyroidal uptake of dietary iodide by perchlorate does not describe perturbations in rat serum total T4 and TSH. Environ Health Perspect 117:731–738
77. Haddad S, Charest-Tardif G, Tardif R, Krishnan K (2000) Validation of a physiological modeling framework for simulating the toxicokinetics of chemicals in mixtures. Toxicol Appl Pharmacol 167:199–209
78. Haddad S, Beliveau M, Tardif R, Krishnan K (2001) A PBPK modeling-based approach to account for interactions in the health risk assessment of chemical mixtures. Toxicol Sci 63:125–131
79. Dennison JE, Andersen ME, Yang RS (2003) Characterization of the pharmacokinetics of gasoline using PBPK modeling with a complex mixtures chemical lumping approach. Inhal Toxicol 15:961–986
80. Dennison JE, Andersen ME, Clewell HJ, Yang RSH (2004) Development of a

physiologically based pharmacokinetic model for volatile fractions of gasoline using chemical lumping analysis. Environ Sci Technol 38:5674–5681

81. Campbell JL, Fisher JW (2007) A PBPK modeling assessment of the competitive metabolic interactions of JP-8 vapor with two constituents, m-xylene and ethylbenzene. Inhal Toxicol 19:265–273
82. Rowland M, Balant L, Peck C (2004) Physiologically based pharmacokinetics in drug development and regulatory science: a workshop report (Georgetown University, Washington, DC, May 29-30, 2002). AAPS PharmSci 6:56–67
83. Ramsey JC, Young JD (1978) Pharmacokinetics of inhaled styrene in rats and humans. Scand J Work Environ Health 4:84–91
84. Ramsey JC, Young JD, Karbowski R, Chenoweth MB, Mc Carty LP, Braun WH (1980) Pharmacokinetics of inhaled styrene in human volunteers. Toxicol Appl Pharmacol 53:54–63
85. Stewart RD, Dodd HC, Baretta ED, Schaffer AW (1968) Human exposure to styrene vapors. Arch Environ Health 16:656–662
86. Gearhart JM, Clewell HJ, Crump KS, Shipp AM, Silvers A (1995) Pharmacokinetic dose estimates of mercury in children and dose-response curves of performance tests in a large epidemiological study. Water Air Soil Pollut 80:49–58
87. Gearhart JM, Jepson GW, Clewell HJ, Andersen ME, Conolly RB (1990) Physiologically based pharmacokinetic and pharmacodynamic model for the inhibition of acetylcholinesterase by diisopropylfluorophosphate. Toxicol Appl Pharmacol 106:295–310
88. Gearhart JM, Jepson GW, Clewell HJ, Andersen ME, Connoly RB (1995) A physiologically based pharmacokinetic model for the inhibition of acetylcholinesterase by organophosphate esters. Environ Health Perspect 102(11):51–60
89. Fisher JW, Whittaker TA, Taylor DH, Clewell HJ, Andersen ME (1989) Physiologically based pharmacokinetic modeling of the pregnant rat: a multiroute exposure model for trichlorethylene and its metabolite, trichloroacetic acid. Toxicol Appl Pharmacol 99:395–414
90. Fisher JW, Whittaker TA, Taylor DH, Clewell HJ, Andersen ME (1990) Physiologically based pharmacokinetic modeling of the lactating rat and nursing pup: a multiroute exposure model for trichlorethylene and its metabolite, trichloroacetic acid. Toxicol Appl Pharmacol 102:497–513
91. Luecke RH, Wosilait WD, Pearce BA, Young JF (1994) A physiologically based pharmacokinetic computer model for human pregnancy. Teratology 49:90–103
92. Andersen ME, Gargas ML, Clewell HJ, Severyn KM (1987) Quantitative evaluation of the metabolic interactions between trichloroethylene and 1,1-dichloroethylene by gas uptake methods. Toxicol Appl Pharmacol 89:149–157
93. Mumtaz MM, Sipes IG, Clewell HJ, Yang RSH (1993) Risk assessment of chemical mixtures: biological and toxicologic issues. Fundam Appl Toxicol 21:258–269
94. Barton HA, Creech JR, Godin CS, Randall GM, Seckel CS (1995) Chloroethylene mixtures: pharmacokinetic modeling and in vitro metabolism of vinyl chloride, trichloroethylene, and trans-1,2-dichloroethylene in rat. Toxicol Appl Pharmacol 130:237–247
95. Clewell HJ, Lee T, Carpenter RL (1994) Sensitivity of physiologically based pharmacokinetic models to variation in model parameters: methylene chloride. Risk Anal 14:521–531
96. Andersen ME, Gargas ML, Ramsey JC (1984) Inhalation pharmacokinetics: evaluating systemic extraction, total *in vivo* metabolism and the time course of enzyme induction for inhaled styrene in rats based on arterial blood: inhaled air concentration ratios. Toxicol Appl Pharmacol 73:176–187
97. Andersen ME, Clewell HJ, Frederick CB (1995) Applying simulation modeling to problems in toxicology and risk assessment—a short perspective. Toxicol Appl Pharmacol 133:181–187
98. Vinegar A, Winsett DW, Andersen ME, Conolly RB (1990) Use of a physiologically based pharmacokinetic model and computer simulation for retrospective assessment of exposure to volatile toxicants. Inhal Toxicol 2:119–128
99. Vinegar A, Jepson GW (1996) Cardiac sensitization thresholds of halon replacement chemicals predicted in humans by physiologically-based pharmacokinetic modeling. Risk Anal 16:571–579
100. Clewell HJ, Andersen ME, Wills RJ, Latriano L (1997) A physiologically based pharmacokinetic model for retinoic acid and its metabolites. J Am Acad Dermatol 36:S77–S85
101. Fiserova-Bergerova V (1975) Biological—mathematical modeling of chronic toxicity.

AMRL-TR-75-5. Aerospace Medical Research Laboratory, Wright-Patterson Air Force Base

102. Gargas ML, Andersen ME (1989) Determining kinetic constants of chlorinated ethane metabolism in the rat from rates of exhalation. Toxicol Appl Pharmacol 97:230–246
103. Gargas ML, Andersen ME, Clewell HJ (1986) A physiologically-based simulation approach for determining metabolic constants from gas uptake data. Toxicol Appl Pharmacol 86:341–352
104. Gargas ML, Clewell HJ, Andersen ME (1990) Gas uptake techniques and the rates of metabolism of chloromethanes, chloroethanes, and chloroethylenes in the rat. Inhal Toxicol 2:295–319
105. Reitz RH, Mendrala AL, Guengerich FP (1989) *In vitro* metabolism of methylene chloride in human and animal tissues: use in physiologically-based pharmacokinetic models. Toxicol Appl Pharmacol 97:230–246
106. Filser JG, Bolt HM (1979) Pharmacokinetics of halogenated ethylenes in rats. Arch Toxicol 42:123–136
107. Andersen ME, Gargas ML, Jones RA, Jenkins LH Jr (1980) Determination of the kinetic constants of metabolism of inhaled toxicant *in vivo* based on gas uptake measurements. Toxicol Appl Pharmacol 54:100–116
108. Gargas ML, Clewell HJ, Andersen ME (1986) Metabolism of inhaled dihalomethanes *in vivo*: differentiation of kinetic constants for two independent pathways. Toxicol Appl Pharmacol 82:211–223
109. Watanabe P, Mc Gown G, Gehring P (1976) Fate of [^{14}] vinyl chloride after single oral administration in rats. Toxicol Appl Pharmacol 36:339–352
110. Gargas ML, Andersen ME (1982) Metabolism of inhaled brominated hydrocarbons: validation of gas uptake results by determination of a stable metabolite. Toxicol Appl Pharmacol 66:55–68
111. Lam G, Chen M, Chiou WL (1981) Determination of tissue to blood partition coefficients in physiologically-based pharmacokinetic studies. J Pharm Sci 71(4):454–456
112. O'Flaherty EJ (1995) PBPK modeling for metals. Examples with lead, uranium, and chromium. Toxicol Lett 82–83:367–372
113. Environmental Protection Agency (EPA) (1992) EPA request for comments on draft report of cross-species scaling factor for cancer risk assessment. Fed Reg 57:24152
114. Carson ER, Cobelli C, Finkelstein L (1983) The mathematical modeling of metabolic and endocrine systems model formulation, identification, and validation. Wiley, New York
115. Rescigno A, Beck JS (1987) The use and abuse of models. J Pharmacokinet Biopharm 15:327–340
116. Yates FE (1978) Good manners in good modeling: mathematical models and computer simulations of physiological systems. Am J Physiol 234:R159–R160
117. Clewell HJ (1995) The use of physiologically based pharmacokinetic modeling in risk assessment: a case study with methylene chloride. In: Olin S, Farland W, Park C, Rhomberg L, Scheuplein R, Starr T, Wilson J (eds) Low-dose extrapolation of cancer risks: issues and perspectives. ILSI, Washington, DC
118. Allen BC, Covington TR, Clewell HJ (1996) Investigation of the impact of pharmacokinetic variability and uncertainty on risks predicted with a pharmacokinetic model for chloroform. Toxicology 111:289–303
119. Clewell HJ, Andersen ME (1994) Physiologically-based pharmacokinetic modeling and bioactivation of xenobiotics. Toxicol Ind Health 10:1–24
120. D'Souza RW, Francis WR, Andersen ME (1988) Physiological model for tissue glutathione depletion and increased resynthesis after ethylene dichloride exposure. J Pharmacol Exp Ther 245:563–568

Chapter 19

Interspecies Extrapolation

Elaina M. Kenyon

Abstract

Interspecies extrapolation encompasses two related but distinct topic areas that are germane to quantitative extrapolation and hence computational toxicology—dose scaling and parameter scaling. Dose scaling is the process of converting a dose determined in an experimental animal to a toxicologically equivalent dose in humans using simple allometric assumptions and equations. In a hierarchy of quantitative extrapolation approaches, this option is used when minimal information is available for a chemical of interest. Parameter scaling refers to cross-species extrapolation of specific biological processes describing rates associated with pharmacokinetic (PK) or pharmacodynamic (PD) events on the basis of allometric relationships. These parameters are used in biologically based models of various types that are designed for not only cross-species extrapolation but also for exposure route (e.g., inhalation to oral) and exposure scenario (duration) extrapolation. This area also encompasses in vivo scale-up of physiological rates determined in various experimental systems. Results from *in vitro* metabolism studies are generally most useful for interspecies extrapolation purposes when integrated into a physiologically based pharmacokinetic (PBPK) modeling framework. This is because PBPK models allow consideration and quantitative evaluation of other physiological factors, such as binding to plasma proteins and blood flow to the liver, which may be as or more influential than metabolism in determining relevant dose metrics for risk assessment.

Key words: Scaling, Extrapolation, *In vitro* scale-up, Allometry, Cross-species, *In vitro* to in vivo extrapolation (IVIVE)

1. Introduction

Interspecies extrapolation, also referred to as cross-species scaling, includes the vast topic of allometry. Allometry is the study of the usual variation in measurable characteristics of anatomy and physiology as a function of overall body size (1, 2). It has been an area of active research interest for over a century, originally inspired by a desire to explain the general observation that smaller mammals have higher rates of metabolism and shorter life spans compared to larger mammals (3–5). In more recent years, renewed interest in the study of allometry has been based on the need to determine

Brad Reisfeld and Arthur N. Mayeno (eds.), *Computational Toxicology: Volume I*, Methods in Molecular Biology, vol. 929, DOI 10.1007/978-1-62703-050-2_19, © Springer Science+Business Media, LLC 2012

toxicologically equivalent doses to extrapolate the results of toxicity studies in experimental animals to equivalent doses in humans (6–8). This process is pivotal to the development of health protective guidance values that are used to establish limits on pollutant levels in various environmental media (9, 10).

Physiologically based pharmacokinetic (PBPK) models are extensively used as a tool for multiple risk assessment applications and are the preferred method to address cross-species dose extrapolation. The reliability of PBPK models is directly related to the accuracy of the chemical-specific parameters used as model inputs. These chemical-specific parameters include metabolic rate constants, partition coefficients, diffusion constants, and parameters describing dermal or gastrointestinal absorption rates (11). Metabolic rate parameters are often estimated using data derived from *in vitro* experimental techniques and then scaled up for in vivo use in PBPK models. Estimation of metabolic parameters using *in vitro* data is increasingly necessary due to the number of chemicals for which data are needed, the trend towards minimizing laboratory animal use, and very limited opportunity to collect data in human subjects (12).

This chapter deals with two related but distinct topic areas that are germane to quantitative interspecies extrapolation and hence computational toxicology—dose scaling and parameter scaling. Dose scaling is the process of converting a dose determined in an experimental animal to a toxicologically equivalent dose in humans. Parameter scaling refers to cross-species extrapolation of specific biological processes describing rates associated with pharmacokinetic (PK) or pharmacodynamic (PD) events on the basis of allometric relationships. These parameters are used in biologically based models of various types that are designed for not only cross-species extrapolation but also for exposure route (e.g., inhalation to oral) and exposure scenario (duration) extrapolation. This area also encompasses in vivo scale-up of physiological rates determined using various *in vitro* experimental systems, i.e., *in vitro* to in vivo extrapolation (IVIVE).

2. Materials

There are a range of tools to assist in the analysis of the types of data to which allometric scaling and *in vitro* scale-up procedures may be applied. Because the calculations themselves tend to be numerically simple, many commercially available spreadsheet and graphical software (with curve fitting capabilities) packages are suitable for these types of analyses. These software packages are typically able to run on most desktop or laptop personal computers.

Experimentally determined enzyme kinetic data are often used to estimate metabolism parameters that are scaled from the *in vitro* to the *in vivo* situation for use in PBPK models. Graphical analyses are often used as a means to easily and quickly estimate rate parameters such as K_M, the Michaelis–Menten constant, and V_{max}, the maximum reaction velocity, for enzymes exhibiting saturable (Michaelis–Menten) kinetic behavior. Graphical analyses using the classical Lineweaver–Burk plot and more recently nonlinear regression analyses or alternative linear forms of the Michaelis–Menten equation (e.g., Eadie–Hofstee or Hanes–Woolf) plot can be accomplished with a variety of graphical packages specifically adapted for this purpose (e.g., SigmaPlot, SAAM II). Specific applications and limitations of various methods for analysis of kinetic data are covered in depth in a number of texts (e.g., (13–15)).

3. Methods

3.1. Dose Scaling

As used in this chapter, the term dose scaling is the process of directly converting a dose determined in an experimental animal model to a toxicologically equivalent dose in humans at a gross or default level. The scientific basis for this procedure is the generalized allometric equation

$$\Upsilon = a(\mathrm{BW})^b, \quad (1)$$

where Υ is the physiological variable of interest, BW is body weight, a is the y-intercept, and b is the slope of the line obtained from a plot of log Υ vs. log BW.

This relationship was originally studied in regard to energy utilization or basal metabolism by Kleiber whose analyses suggested that basal metabolism scales to the 3/4 (0.75) power of body weight across species (3–5). This was in contrast to the generally accepted "surface law" at that time (16), i.e., the concept that basal metabolism scales across species according to body surface area or the 2/3 power of body weight (3, 4, 16). A variety of theories have been advanced to explain this allometric relationship, including "elastic similarity" of skeletal and muscular structures (17), "the fractal nature of energy distributing vascular networks" (18, 19), and others (20, 21). It should be noted that agreement on the scaling exponent, i.e., 0.75, is not universal and some analyses have been published which suggest 2/3 is the more appropriate value (22). Dodds et al. (23) performed a reanalysis of data from the published literature and concluded that available information did not allow one to distinguish between an exponent of 2/3 vs. 3/4 as being more predictive.

In practice, $BW^{0.75}$ is more widely accepted as a "default" method for scaling oral doses across species for both cancer and noncancer health effects compared to other exponents (2, 9, 10). The meaning of default as used here applies to the situation where insufficient chemical-specific information is available to use a more data-informed approach (e.g., as outlined in ref. 24) or a PBPK model. In this spectrum of approaches, an appropriately documented and evaluated PBPK model would be considered the optimal choice for cross-species dose extrapolation (see next section for discussion of interspecies scaling for PBPK model parameters). It should be noted that for inhalation exposure, the US EPA has an established framework for cross-species dosimetric adjustment that does not depend on the availability of a PBPK model. This categorical methodology is based on physical state (gas, particulate) as well as reactivity and solubility for gases and regionally deposited dose for particulates (25) and is conceptually analogous to the default scaling of oral doses illustrated here (10).

For a default-type dosimetric adjustment, an oral dose (in mg/kg/day) for an experimental animal is converted to a "human equivalent dose" (HED) by multiplying the animal dose by a dosimetric adjustment factor (DAF):

$$DAF = \left(\frac{BW_a}{BW_h}\right)^{0.25}. \quad (2)$$

Or the mathematically equivalent alternative,

$$DAF = \left(\frac{BW_h}{BW_a}\right)^{-0.25}, \quad (3)$$

where the subscripts "a" and "h" denote animal and human, respectively, and the ¼ exponent results from the application of $BW^{0.75}$ scaling to exposure in units of mg/kg/day (rather than mg/day) such that

$$\frac{BW^{0.75}}{BW^{1/1}} = BW^{-0.25}. \quad (4)$$

Therefore the HED is calculated as

$$HED(mg/kg/day) = animal\ dose(mg/kg/day) \times DAF. \quad (5)$$

The empirical and theoretical basis for the development of a generalized cross-species scaling factor inclusive of pharmacokinetic processes has been detailed elsewhere (6, 7, 26), including limitations and caveats which are described in greater detail in Subheading Note 2 of this chapter.

3.2. Parameter Scaling

As used in this chapter, the term parameter scaling is the process of scaling a physiological variable from an experimental animal value to a human value for use in a PBPK model. Examples of parameters typically scaled in PBPK models include alveolar ventilation rate, cardiac output, and rates describing absorption and metabolism (27). This is a more specific or refined use (compared to dose scaling) of the allometric relationship denoted by equation (1) proposed by Kleiber (3, 4) to describe energy utilization.

Analyses have been conducted to derive equations to describe the relationship of body weight to a variety of physiological characteristics and functions (e.g., (18, 28)). Examples of the coefficients conforming to the equation form, $y = a(\mathrm{BW})^{0.75}$, are shown in Table 1. Examination of Table 1 reveals characteristics such as blood volumes and organ weight increase in roughly direct proportion to body weight. Rate processes, e.g., clearances and outputs, typically scale in proportion to approximately $(\mathrm{BW})^{0.75}$. When a physiological parameter that varies in proportion to $\mathrm{BW}^{0.75}$ is normalized against a characteristic that varies directly ($\mathrm{BW}^{1/1}$), scaling will approximate $\mathrm{BW}^{-0.25}$, i.e., $\mathrm{BW}^{0.75}/\mathrm{BW}^{1/1} = \mathrm{BW}^{-0.25}$ (10).

On the basis of allometry, cross-species scaling by $\mathrm{BW}^{0.75}$ power of body weight would be expected to yield reasonable initial estimates for physiological quantities measured in volume per unit time (e.g., flows, ventilation, clearances) or mass per unit time (e.g., metabolic capacity, i.e., $V_{\max}$). In the case of first-order rate processes (units of inverse time, e.g., h^{-1}), cross-species scaling on the basis of $\mathrm{BW}^{-0.25}$ would be used. In PBPK modeling, allometric scaling of parameter values is generally used as a first approximation when appropriate species-specific or chemical-specific parameter values are not available (11). Numerical and statistical methods for parameter estimation and optimization using pharmacokinetic data are discussed in greater detail in Part VI of Computational Toxicology, Volume II.

In practice, species-specific parameter values are generally available for physiological parameters such as blood flows and organ volumes which also usually scale directly on the basis of body weight ($\mathrm{BW}^{1/1}$). Chemical-specific parameters associated with metabolism (e.g., $V_{\max}$) or binding ($B_{\max}$) that follow Michaelis–Menten (saturable) kinetics are usually scaled by $\mathrm{BW}^{0.75}$ when extrapolating from experimental animals to humans using *in vivo* data (11).

Alternatively, metabolic rates may be determined for the chemical of interest *in vitro* and scaled to the *in vivo* case as described in Subheading 3.3 and illustrated in Subheading 4.3 of this chapter. Note that body weight scaling is not applied in the case of affinity constants (e.g., K_{M}, B_{M}) measured in units of mass per volume (e.g., mg/L, μM). Tissue partition coefficients (PC), a unitless, chemical-specific measure of tissue solubility, are typically normalized against the chemical-specific blood-to-air partition coefficient for humans as illustrated in Subheading 4.2. The underlying basis for this correction is that tissue lipid and water content are important determinants of tissue solubility and tissue composition is highly conserved across mammalian species (27).

Table 1
Selected physiological characteristics and their body weight (BW) scaling coefficients[a, b]

Physiological characteristic (*Y*)	Units of *Y*	*a* (*y*-intercept)	*b* (slope)
Basal O_2 consumption	mL STP/h	3.8	0.734
Water intake	mL/h	0.01	0.88
Urine output	mL/h	0.0064	0.82
Ventilation rate	mL/h	120	0.74
Tidal volume	mL	6.2×10^{-3}	1.01
Urea clearance	mL/h	1.59	0.72
Inulin clearance	mL/h	1.74	0.77
Creatinine clearance	mL/h	4.2	0.69
Hippurate clearance	mL/h	5.4	0.80
Heartbeat duration	h	1.19×10^{-5}	0.27
Breath duration	h	4.7×10^{-5}	0.28
Peristaltic (gut beat) duration	h	9.3×10^{-5}	0.31
Total nitrogen output	g/h	7.4×10^{-5}	0.735
Endogenous nitrogen output	g/h	4.2×10^{-5}	0.72
Creatinine nitrogen output	g/h	1.09×10^{-6}	0.9
Sulfur output	g/h	1.71×10^{-6}	0.74
Kidneys weight	g	0.0212	0.85
Brain weight	g	0.081	0.7
Heart weight	g	6.6×10^{-3}	0.98
Lungs weight	g	0.0124	0.99
Liver weight	g	0.082	0.87
Thyroids weight	g	2.2×10^{-4}	0.80
Adrenals weight	g	1.1×10^{-3}	0.92
Pituitary weight	g	1.3×10^{-4}	0.76
Stomach weight	g	0.112	0.94
Blood weight	g	0.055	0.99
Number of nephrons	None	2,600	0.62

[a] The values correspond to the equation $Y = a(\mathrm{BW})^b$, where Y is the physiological variable of interest, BW is body weight, a is the y-intercept, and b is the slope of the line obtained from a plot of log Y vs. log BW
[b] Adapted from (16, 26, 29)

3.3. In Vitro to In Vivo Scale-Up

Data from various *in vitro* systems are commonly used to estimate metabolic rate parameters that may then be scaled to the level of the whole tissue and organism. Systems used include precision cut organ slices, whole cell preparations (e.g., hepatocytes), subcellular tissue fractions (i.e., microsomes and cytosol), and recombinantly expressed enzymes. The performance, advantages, and limitations of these systems have been compared and reviewed in a number of publications (e.g., (12, 30–32)). Figure 1 provides a schematic illustration of the scaling procedures used with *in vitro* systems derived from liver tissue for hepatocytes and subcellular fractions (microsomes and

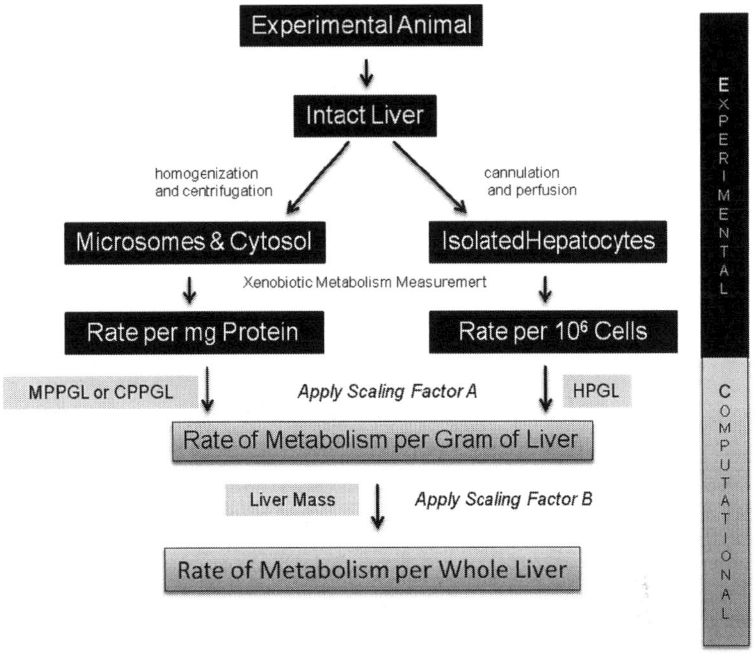

Fig. 1. Schematic representation of experimental and computational steps necessary for IVIVE based on the use of hepatic subcellular fractions (microsomes, cytosol) or hepatocytes. Microsomal protein per gram of liver (MPPGL), cytosolic protein per gram of liver (CPPGL), and hepatocytes per gram of liver (HPGL) are ideally determined in the specific experiment, but default values may also be obtained from the literature. While IVIVE for rate of metabolism is illustrated here, other parameters such as K_M and intrinsic clearance (V_{max}/K_M) are also estimated utilizing these experimental systems.

cytosol), the two most commonly used experimental systems. While the liver is generally the tissue of interest in metabolism studies, conceptually the same principles apply to other tissues (e.g., kidney, lung, gut) that can be metabolically active or toxicologically important due to toxicity being caused by metabolism in the target organ.

The normalizing basis or units for rates of metabolism (V_{max}) reported in the literature will vary depending upon the experimental system used. When using subcellular fractions, V_{max} is typically reported in units of mass per time per mg of microsomal or cytosolic protein (e.g., nmol/min/mg protein). Specifically, it is the rate of product formed (sometimes measured as disappearance of parent compound) per unit time normalized to mg of microsomal protein or cytosolic protein. To scale this rate to the whole liver, it is necessary to know the mg of microsomal protein per gram of liver (MMPGL) or mg of cytosolic protein per gram of liver (MCPGL) and the liver weight (LW) in grams. The overall rate of the whole liver (LR) is given as

$$\mathrm{LR} = V_{max}(\mathrm{mass/time/mg\,protein}) \times \mathrm{MMPGL} \times \mathrm{LW(g)}. \quad (6)$$

To convert this rate (in units of mass/time/liver) for use in a PBPK model, it is necessary to divide this figure by $\mathrm{BW}^{0.75}$ to yield units of mass/time/kg. This figure is often referred to as $V_{max}C$ in the biological modeling literature. It is optimal to use figures for MMPGL (or MCPGL) reported in the original source in which the metabolism rate was reported. In practice, these data are often not reported, thus necessitating the use of "default" or "typical" values for MMPGL. Also, whole organ weights are generally estimated on the basis of the average percentage of organ weight as a function of body weight (12). For example, in humans, the liver is assumed to be 2.6% of body weight. These figures vary depending upon the species and their physiological status (age, gender, strain). Compilations of physiological data, including organ weights as a percentage of body weight, are available (e.g., (33)).

When V_{max} is derived from a recombinantly expressed enzyme system (very often a cytochrome P450 isoenzyme or CYP, e.g., mass of product/time/mass CYP protein), it is necessary to know the mg of the particular enzyme (protein) per gram of liver (MEGL). In this case, the overall rate of metabolism in the whole liver (LR) is

$$\mathrm{LR} = V_{max}(\mathrm{mass/time/mg\,enzyme\,protein}) \times \mathrm{MEGL} \times \mathrm{LW(g)}. \tag{7}$$

The same scaling procedure referenced above (division by $BW^{0.75}$) is applied to derive a $V_{max}C$ for use in a PBPK model.

If V_{max} is determined using hepatocytes, data are typically expressed in units of product formed per unit time per million cells (e.g., nmol/min/10^6 cells). In this instance, the scaling factor needed to estimate the rate for the whole liver is hepatocellularity per gram of liver (HPGL). Thus, the overall rate of metabolism for the whole liver (LR) is

$$\mathrm{LR} = V_{max}(\mathrm{mass/time/10^6\,cells}) \times \mathrm{HPGL(cells/g\,liver)} \times \mathrm{LW(g)}. \tag{8}$$

This calculation yields a rate for the whole liver in units of mass/time which is converted to $V_{max}C$ for use in a PBPK model by division by $BW^{0.75}$ as described previously (12).

Although generally estimated from *in vitro* studies, the Michaelis–Menten constant, K_M, is often used directly in PBPK models after conversion to appropriate units as necessary. Usually studies in the published literature report K_M values that were determined in aqueous solution, but for use in a PBPK model it is more biologically relevant and accurate to express K_M in terms of the concentration in venous blood at equilibrium with liver (assuming liver is the tissue of interest). This is accomplished by dividing the K_M determined in aqueous suspension by the liver:blood partition coefficient for the chemical that is used in the PBPK model, i.e.,

$$\mathrm{PBPK}\ K_M = \frac{\mathrm{aqueous\ suspension}\ K_M}{\mathrm{liver:blood\ PC}}. \tag{9}$$

This procedure assumes that the concentration determined in the *in vitro* suspension adequately represents the concentration in liver that results in the half-maximal rate of metabolism (34).

In all cases where metabolism parameters are determined *in vitro* using tissues from experimental animals, these parameters are extrapolated from the experimental animals to humans using the procedures outlined in Subheading 3.2 and illustrated in Subheading 4.2 of this chapter. While the above discussion has focused on point estimates which may be thought of in the context of measures of central tendency, the impact of interindividual variability in metabolic rate parameters on PBPK model predictions used in risk assessment has been explored and illustrated in several publications (35–37).

4. Examples

4.1. Dose Scaling

Table 2 below illustrates the results of $BW^{0.75}$ scaling of a hypothetical dose of 10 mg/kg in each of the four species of experimental animal to a "toxicologically equivalent" dose in a 70 kg human using equations (2) or (3) and (5) from Subheading 3.1. The significant assumptions and limitations inherent in this procedure are discussed in Subheading 5.1.

4.2. Parameter Scaling

Example 1—Cross-Species Extrapolation of Partition Coefficients: Tissue-to-blood partition coefficients for humans are typically estimated by dividing the tissue-to-air partition coefficient obtained using rodent tissues by the blood-to-air partition coefficient using human blood which is readily obtainable compared to human tissue.

Table 2
Use of the Dosimetric Adjustment Factor (DAF) in derivation of a human equivalent dose (HED) from an oral animal exposure

Animal species	BW_a (kg)	BW_h (kg)	Animal dose (mg/kg/day)	DAF Eq. (2)	DAF Eq. (3)	HED (mg/kg-day) Eq. (5)
Mouse	0.025	70	10	0.137	0.137	1.37
Rat	0.25	70	10	0.244	0.244	2.44
Rabbit	3.5	70	10	0.473	0.473	4.73
Dog	12	70	10	0.643	0.643	6.43

Note that Eqs. (2) and (3), i.e., $DAF = (BW_a/BW_h)^{0.25}$ and $DAF = (BW_h/BW_a)^{-0.25}$, yield the same result

Table 3
Calculation of human tissue-to-blood partition coefficients (PC) for toluene[a]

Tissue	Rat experimental tissue:air PC	Calculate human tissue:blood PC
Liver[b]	42.3	42.3/13.9 = 3.04
Liver[c]	83.6	83.6/13.9 = 6.01
Muscle[b]	27.8	27.8/13.9 = 2.00
Muscle[c]	27.7	27.7/13.9 = 1.99
Brain[b]	36.1	36.1/13.9 = 2.60
Lung[b]	23.9	23.9/13.9 = 1.72
Stomach[b]	55.1	55.1/13.9 = 3.96
Skin[b]	22.9	22.9/13.9 = 1.65

[a]Partition coefficients were measured using the vial equilibration technique. Thrall et al. (38) reported a human blood:air PC of 13.9
[b]Reported in Thrall et al. (38)
[c]Reported in Gargas et al. (39)

$$\text{Human tissue : blood PC} = \frac{\text{animal tissue : air PC}}{\text{human blood : air PC}}. \quad (10)$$

This is illustrated for the volatile organic chemical toluene in Table 3 below. Assumptions and limitations of this approach are discussed in Subheading 5.2.

Example 2—Cross-Species Scaling of Michaelis–Menten V_{max} and First-Order Rate (k_a, k_f) Constants: Michaelis–Menten rate parameters can be estimated from in vivo experimental animal pharmacokinetic data using a PBPK model. In this case, $BW^{0.75}$ scaling is generally used, such that

$$\begin{aligned} &\text{Human } V_{\max}(V_{\max}\,\text{h in mass/time}) \\ &\quad = \text{animal } V_{\max}(\text{mass/time/kg BW}) \times (BW_h)^{0.75}. \quad (11) \end{aligned}$$

In the case of first-order processes for absorption (k_a) or metabolism (k_f) that have units of reciprocal time (e.g., h^{-1}), scaling is to the -0.25 power such that (Table 4)

$$\begin{aligned} &\text{Human } k_a \text{ or } k_f(k_f\,\text{h or } k_a\,\text{h in time}^{-1}) = \text{animal } k_a \text{ or } K_f \\ &\quad \times (BW_h)^{-025}/(BW_a)^{-0.25}. \quad (12) \end{aligned}$$

4.3. In Vitro to In Vivo Scale-Up

Example: Scaling V_{max} and K_M for Chloroform ($CHCL_3$) from In Vitro Data Obtained Using Human Microsomes (34):

$$\begin{aligned} LR = &\text{ in vitro } V_{\max} \times \text{MEGL} \times \text{LW} = 5.24\,\text{pmole CHCl}_3/\,\text{min}\,/ \\ &\text{pmol CYP 2E1} \times 2,562\,\text{pmole CYP2E1/g liver} \times 1,820\,\text{g liver} \\ &= 24,433,281.6\,\text{pmoles CHCl}_3/\,\text{min}\,/\text{whole liver}. \end{aligned}$$

Table 4
Examples of cross-species scaling of rate parameters from a variety of in vivo pharmacokinetic data

Chemical and reference	Experimental data and animal parameter	Scaled human parameter
BDCM (40)	$V_{max}a = 12.8$ mg/h/kg Data used for estimation was Br ion concentration in blood of rats at the end of 4-h continuous exposure to 50, 100, 150, 200, 400, 800, 1,200, 1,600, or 3,200 ppm bromodichloromethane (BDCM)	Assuming a 70 kg human, $V_{max}h = 12.8 \times (70)^{0.75}$ $= 310$ mg/h
Benzene (41)	$V_{max}a = 14.0$ μmol/h/kg Data used for estimation was a set of curves for disappearance of benzene from a closed vapor uptake chamber for male $B6C3F_1$ mice exposed to initial benzene concentrations of 440, 1,250, or 2,560 ppm	Assuming a 70 kg human, $V_{max}h = 14.0 \times (70)^{0.75}$ $= 339$ μmol/h
Methyl chloroform (42)	$k_f = 7.8$/h for a 0.225 kg rat Data used for estimation was a set of curves for disappearance of benzene from a closed vapor uptake chamber for male F344 rats exposed to initial concentrations of 0.2, 1.0, 10, or 210 ppm methyl chloroform	Assuming a 70 kg human, $k_f h = k_f a \times (BW_h)^{-0.25}/(BW_a)^{-0.25}$ $= 7.8 \times (70)^{-025}/(0.225)^{-025}$ $= 1.86$/h

Note: If V_{max} estimated from in vivo animal data is given in units of mass/time in a study, then the conversion is $V_{max}h = V_{max}a \times (BW_h)^{0.75}/(BW_a)^{0.75}$

For this example the units of time, mass of chloroform, organ or body mass, and volume used in the PBPK model were h, mg, kg, and L, respectively. Thus to convert the above figure to units appropriate for the model (Table 5),

$$\text{LR} = 24{,}433{,}281.6\,\text{pmoles}\,\text{CHCl}_3/\text{min} \times 60\ \text{min}/\text{h} \times 119\,\text{pg}\ \text{CHCl}_3/\text{pmole} \times 10^{-9}\,\text{mg/pg} = 175\,\text{mg/h/liver}.$$

Converting this LR to $V_{max}C$,

$$V_{max}C = 175\,\text{mg/h/liver}/(70\,\text{kg})^{0.75} = 8.9\,\text{mg/h/kg}.$$

Table 5
Data used for in vitro to in vivo extrapolation of V_{max} and K_M for $CHCl_3$

Item	Value and units	Source
In vitro V_{max}	5.24 pmole $CHCl_3$/min/pmol CYP 2E1	(43)
In vitro K_M	18.27 μg/L	(43)
CYP2E1 concentration	2,562 pmole CYP2E1/g liver	(35)
Body mass	70 kg	Assumed value
Liver % of BW	2.6%	(33)
Liver:blood PC	1.6	(34)
Time	60 min/h	
FW $CHCl_3$	119.4 pg $CHCl_3$/pmole	
Liver weight (LW)	70 kg × 0.026 = 1.82 kg (1,820 g)	Calculated value

To convert the *in vitro* K_M determined in aqueous solution to a value for use in PBPK model,

$$\begin{aligned} \text{PBPK } K_M &= \text{aqueous suspension } K_M/\text{liver:blood PC} \\ &= 18.27\ \mu g/L/1.6 \\ &= 11.4\ \mu g/L \times mg/10^3\ \mu g \\ &= 0.012\, mg/L. \end{aligned}$$

An important consideration for IVIVE of this type is how well they compare to estimates derived using other methods. For example, the $V_{max}C$ calculated above (8.9 mg/h/kg) compares well (within twofold) with other $V_{max}C$ values developed for chloroform of 7 mg h/kg (44) and 15.7 mg/h/kg (45).

5. Notes

1. A fundamental, but easy-to-overlook, principle if one is focused on numerical analysis is that the quality of any parameter estimate is only as reliable as the data on which it is based. For this reason, the underlying biochemical, physiological, or toxicological data used for parameter estimation or cross-species scaling should be evaluated carefully for appropriateness of

experimental design and validity of methodology. The latter includes but is not limited to *in vitro* and in vivo experimental techniques, analytical chemistry methodology, and statistical analysis.

2. Dose Scaling - Pharmacokinetic dose scaling based on $BW^{0.75}$ is a scientifically sound default option when it is necessary to perform cross-species scaling to estimate "toxicologically equivalent doses" in the absence of sufficient chemical-specific data. If the dose scaled is one that elicited a toxic response in the experimental animal, this may be a function of either or both pharmacokinetic and pharmacodynamic processes. Rhomberg and Lewandowski (7) have noted that allometric relationships for cross-species scaling of pharmacodynamic processes have received relatively little study and thus represent an important source of uncertainty that merits further study. Some analyses have demonstrated that there are certain conditions under which default allometrically based dose scaling may yield clearly erroneous estimates and other instances where significant uncertainties exist.
3. Dose Scaling - Overall, $BW^{0.75}$ scaling is most reliable for situations in which the chemical itself or its stable metabolite is the putative toxic agent and clearance follows first-order processes, i.e., when the appropriate dose metric is area under the curve (6, 46). Conversely, when the putative toxic agent is a highly reactive chemical or a metabolite that reacts with cellular components at the site of formation, $BW^{0.75}$ scaling is less reliable (8, 46, 47).
4. Dose Scaling - Scaling on the basis of delivered dose per unit surface area is preferable to $BW^{3/4}$ scaling in cases where toxic effects are mediated by direct action of a chemical or its metabolites on tissues at the site of first contact. These are often referred to as "portal-of-entry" effects. This could apply to the respiratory tract for inhalation exposures, skin in the case of dermal exposures, and epithelial tissues lining the gastrointestinal tract for oral exposures (10, 25).
5. Dose Scaling - Evidence also suggests that $BW^{0.75}$ scaling is not applicable in cases where toxicity is a consequence of an acute high-level exposure that suddenly overwhelms normal physiological processes without opportunity for repair or regeneration. Analysis by Rhomberg and Wolff (48) suggests that this is the case among small laboratory rodents for lethality as a consequence of acute oral exposure. Further analysis by Burzala-Kowalczyk and Jongbloed (49) extended this observation to larger species (e.g., monkeys, dogs, rabbits) and parenteral routes of exposure (e.g., intramuscular, intravenous, intraperitoneal). In both cases direct body weight scaling ($BW^{1/1}$) fits the data better than $BW^{0.75}$ scaling (48, 49).

6. Dose Scaling - BW scaling across life stages can engender considerable uncertainty. Within species, dose scaling for effects observed in mature organisms to immature organisms is most reliable when there is mechanistic data on the physiological processes affected with corresponding data on the maturity of these systems in the life stage of interest. Some recent analyses suggest that $BW^{0.75}$ scaling is descriptive of various toxicokinetic differences observed for pharmaceuticals across ages in humans down to about 6 months of age (50–52). Toxicodynamic differences across life stages are a largely unexplored research area (7). In addition, differences across species in patterns of development can have considerable impact on interspecies dose scaling from experimental animals to immature humans (53–55).
7. Parameter Scaling - The reliability of parameter scaling will be highly dependent on the reliability of the experimental technique used to obtain the data and its applicability to the chemical in question. For example, for volatile chemicals the *in vitro* vial equilibration technique is well established as an experimental method to obtain estimates of tissue partitioning in a variety of media (39). Analogous *in vitro* experimental techniques for estimating partitioning of nonvolatile chemicals (e.g., using equilibrium dialysis or ultrafiltration) are fewer and not as widely used (56, 57). It is also feasible to use in vivo pharmacokinetic data to estimate partition coefficients for nonvolatile chemicals, e.g., the area method of Gallo et al. (58), provided that measurements are obtained under steady-state conditions.
8. Parameter Scaling - An important caveat in the interpretation of all experimental data used to estimate tissue partitioning is that the estimates themselves may be compromised if other physiological effectors (e.g., metabolism, binding, presence of residual blood) are not accounted for in the experimental design or analysis of the data (59, 60). In the case of partition coefficients, there are also a number of computational algorithms that have been developed to estimate them on the basis of tissue composition and physicochemical characteristics in the absence of direct experimental data (e.g., (61, 62)). Application and limitations for some of these approaches are discussed in Chapter 6.
9. Parameter Scaling - In the case where rate constants are estimated from in vivo animal pharmacokinetic data, reliable parameter estimation is only possible when the parameter is sensitive to changes in the type of experimental endpoints being measured. Vapor uptake data, typically a series of curves over a range of initial starting concentrations measuring a decline in air concentration in a chamber due to metabolism by live rodents, are a good example of this. Vapor uptake data

are very sensitive to changes in V_{max}, but typically much less so to changes in K_M. Techniques, such as sensitivity analysis (see Chapter 17 and 18), are useful to determine whether a particular type of data (response) is sufficiently sensitive (under a given set of experimental conditions) to a change in the value of a parameter to be useful to estimate that parameter.

10. Parameter Scaling - Another important aspect of metabolic parameter estimation using in vivo data (that also applies to *in vitro* data) is matching what is measured experimentally to the scale or level of detail at which it is desired to describe metabolism in the PBPK model. For example, experimental techniques that measure the disappearance of a parent chemical can provide a good estimate of overall metabolic clearance. However, these same data would yield relatively less accurate rate estimates of individual downstream metabolic reactions.
11. IVIVE - Reliable *in vitro* to *in vivo* scale-up requires accurate scaling factors and thus a thorough understanding of the biological and experimental factors that can influence their estimation. This is particularly important if the parameter being estimated from *in vitro* data is subject to variation that may impact model predictions for pharmacokinetic responses (i.e., dose metrics such as AUC) to be used in quantitative risk assessment. The discussion here focuses on some more general factors that impact IVIVE as well as the scaling factors themselves—microsomal protein (MPPGL) and cytosolic protein (CPPGL) per gram of liver and HPGL.
12. IVIVE - It is a generally recognized principle that *in vitro* pharmacokinetic measurements intended for extrapolation to the level of the intact or whole organism (animal or human) should be done under experimental conditions that mimic the in vivo situation as closely as possible (35). Thus, it is critical that the concentrations used for studies are within the range of tissue concentrations that are either (1) observed in the intact organism or (2) predicted on the basis of a kinetic model under realistic exposure conditions. Another basic principle is that the experimental conditions under which enzyme activity was determined must be carefully evaluated in terms of biological reality and the impact this may have on numerical values reported. This will help ensure that methodology-specific limitations are appropriately identified to distinguish experimental artifacts from inherent biological characteristics. This is necessary because experimental conditions are usually designed to mitigate limiting factors, i.e., pH, buffering (ionic strength), temperature, oxygen tension, cofactors and substrate concentration, etc., are adjusted to optimize metabolic transformation (12).

13. IVIVE - Ideally, scaling factors (MPPGL, CPPGL, and HPGL) used should be obtained from the same source (publication) or matched to the same experimental model (e.g., a specific strain, age, gender of rat) from which the enzymatic data are obtained. In practice, such data are rarely provided in published manuscripts because either they were not directly relevant to the goals of the original experiment or were not feasible to determine in the *in vitro* system used (e.g., vendor-purchased hepatocytes). This necessitates the use of average or default scaling factors from other sources and requires an understanding of factors which may impact their value. In general, MPPGL seems to exhibit greater variability compared to HPGL. For example, Carlile et al. (31) reported that in rats treated with different enzyme inducers (e.g., phenobarbital, dexamethasone), hepatocellularity was essentially unchanged compared to untreated rats, whereas recovery of microsomal protein varied by twofold. It is also worthy of note that scaling to the level of the whole organism from other experimental systems, e.g., using data from recombinantly expressed enzyme systems, especially the cytochrome P450 enzymes, is complicated by differences in lipid matrix in this *in vitro* system compared to intact hepatocytes or microsomes. These issues are more fully discussed in ref. 12.
14. IVIVE - Barter et al. (63) reviewed the literature with the goal of establishing average consensus values for MPPGL and HPGL in adult human liver and estimates of variability as well as significant covariates. On the basis of a meta-analytic approach, they recommended weighted mean values of 32 mg/g liver (95% CI 29–34 mg/g) and 99×10^6 cells/g liver (95% CI 74–131 $\times 10^6$ cells/g liver) for MPPGL and HPGL, respectively. The authors reported a weak, but statistically significant, negative correlation between age and both MMPGL and HPGL, whereas no statistically significant relationship was reported for other covariates, such as gender, smoking status, or alcohol consumption.
15. IVIVE - Some studies have been conducted with the goal of evaluating the predictive value of different *in vitro* systems compared to in vivo data for metabolic clearance. These studies illustrate some of the issues which arise with various methodologies. For example, Houston(64) reported that for low-clearance drugs both hepatocyte and microsomal systems were reasonably predictive of in vivo metabolic clearance in rats. For high-clearance drugs, rat hepatocytes were also predictive of in vivo metabolic clearance whereas estimates based on rat microsomal data tended to underestimate in vivo clearance. Based on their review of the literature, these authors also reported values of MMPGL and HPGL of 45 mg/gram liver and 1.35×10^8 hepatocytes/gram liver for the rat.

16. IVIVE - Andersson et al. (30) compared the metabolic clearance of the antiarrhythmic drug, Almokalant, predicted from multiple *in vitro* systems (microsomes, tissue slices, hepatocytes, and cDNA-expressed P450 enzymes) with in vivo data from human subjects; they reported that all the *in vitro* systems evaluated under-predicted in vivo clearance. The authors suggested that for drugs that are extensively conjugated (like Almokalant which is predominantly glucuronidated) under-prediction could be accounted for by differences in access to UDP-glucuronic acid, latency of the enzyme in the endoplasmic reticulum, or hydrolysis by other enzymes in the incubation system. In studies with Diazepam (an antianxiety drug), Carlile et al. (31) presented evidence that the under-prediction of in vivo clearance using rat liver microsomes was attributable to product inhibition, i.e., accumulation of initially formed metabolites in the medium effectively inhibited further metabolism of the parent compound. Other investigators have demonstrated the importance of quantifying and accounting for factors in the experimental system such as binding to proteins or nonbiological components in the experimental system (e.g., (65–67)). Finally it should also be noted that results from *in vitro* metabolism studies are most useful for extrapolation purposes when integrated into a PBPK modeling framework. This is because PBPK models allow consideration and quantitative evaluation of other physiological factors such as binding to plasma proteins and consideration of blood flow to the liver which may be as or more influential than metabolism in determining relevant dose metrics for purposes of risk analysis (12, 67, 68).

Disclaimer

This manuscript has been reviewed in accordance with the policy of the National Health and Environmental Effects Research Laboratory, US Environmental Protection Agency, and approved for publication. Approval does not signify that the contents necessarily reflect the views and policies of the Agency, nor does mention of trade names or commercial products constitute endorsement or recommendation for use.

References

1. Dedrick RL (1973) Animal scale-up. J Pharmacokinet Biopharm 1:435–461
2. U.S. EPA (U.S. Environmental Protection Agency) (1992) Draft report: a cross-species scaling factor for carcinogen risk assessment based on equivalence of $mg/kg^{3/4}/day$. Notice Fed Reg 57:24152–24173
3. Kleiber M (1932) Body size and metabolism. Hilgardia 6:315–353

4. Kleiber M (1947) Body size and metabolic rate. Physiol Rev 27:511–541
5. Kleiber M (1961) The fire of life: an introduction to animal energetics. Wiley, New York, NY
6. O'Flaherty EJ (1989) Interspecies conversion of kinetically equivalent doses. Risk Anal 9:587–598
7. Rhomberg LR, Lewandowski TA (2006) Methods for identifying a default cross-species scaling factor. Hum Ecol Risk Assess 12:1094–1127
8. Travis CC, White RK (1988) Interspecies scaling of toxicity data. Risk Anal 8:119–125
9. U.S. EPA (2005) Guidelines for carcinogen risk assessment. EPA/630/P-03/001F Risk Assessment Forum, Washington, DC
10. U.S. EPA (2011) Harmonization in interspecies extrapolation: use of body weight$^{3/4}$ as the default method in derivation of the oral reference dose. EPA/100/R11/0001 Risk Assessment Forum, Washington, DC
11. Clewell HJ, Reddy MB, Lave T, Andersen ME (2008) Physiologically based pharmacokinetic modeling. In: Gad SC (ed) Preclinical development handbook: ADME and biopharmaceutical properties. Wiley, New York, NY, pp 1167–1227
12. Lipscomb JC, Poet TS (2008) *In vitro* measurements of metabolism for application in pharmacokinetic modeling. Pharmacol Ther 118:82–103
13. Matthews JC (1993) Fundamentals of receptor, enzyme and transport kinetics. CRC, Boca Raton, FL
14. Cornish-Bowden A (1995) Analysis of enzyme kinetic data. Oxford University Press, Oxford
15. Cornish-Bowden A (2004) Fundamentals of enzyme kinetics, 3rd edn. Portland Press, London
16. Rubner M (1883) Uber den einfluss der korpergrosse auf stoff- und kraftwechsel. Zeit Biol 19:536–562
17. McMahon TA (1975) Using body size to understand the structural design of animals: quadrupedal locomotion. J Appl Physiol 39:619–627
18. West GB, Brown JH, Endquist BJ (1997) A general model for the origin of allometric scaling laws in biology. Science 276:122–126
19. West GB, Woodruff WH, Brown JH (2002) Allometric scaling of metabolic rate from molecules and mitochondria to cells and mammals. Proc Natl Acad Sci U S A 99(Suppl 1):2473–2478
20. Banavar JR, Maritan A, Rinaldo A (1999) Size and form in efficient transportation networks. Nature 399:130–131
21. Bejan A (2000) Shape and structure, from engineering to nature. Cambridge University Press, Cambridge
22. White CR, Seymour RS (2003) Mammalian basal metabolic rate is proportional to body mass$^{2/3}$. Proc Natl Acad Sci U S A 100:4046–4049
23. Dodds PS, Rothman DH, Weitz JS (2001) Re-examination of the "3/4-law" of metabolism. J Theor Biol 209:9–27
24. IPCS (International Programme on Chemical Safety) (2005) Guidance document for the use of data in development of chemical-specific adjustment factors (CSAFs) for interspecies differences and human variability in dose/concentration-response assessment. World Health Organization, Geneva
25. U.S. EPA (1994) Methods for derivation of inhalation reference concentrations and application of inhalation dosimetry. EPA/600/8-90/066F. Environmental Criteria and Assessment Office, Washington, DC
26. Boxenbaum H (1982) Interspecies scaling, allometry, physiological time, and the ground plan of pharmacokinetics. J Pharmacokinet Biopharm 10:201–227
27. Reddy MB, Yang RSH, Clewell HJ, Andersen ME (2005) Physiologically based pharmacokinetic modeling—science and applications. Wiley Interscience, Hoboken, NJ
28. Adolph EF (1949) Quantitative relations in the physiological constitutions of mammals. Science 109:579–585
29. Mordenti J (1986) Man versus beast: pharmacokinetic scaling in mammals. J Pharm Sci 75:1028–1040
30. Andersson TB, Sjoberg H, Hoffman K-J, Boobis AR, Watts P, Edwards RJ, Lake BJ, Price RJ, Renwick AB, Gomez-Lechon MJ, Castell JV, Ingelman-Sundberg M, Hidestrand M, Goldfarb PS, Lewis DFV, Corcos L, Guillouzo A, Taavitsainen P, Pelkonen O (2001) An assessment of human liver-derived *in vitro* systems to predict the *in vivo* metabolism and clearance of almokalant. Drug Metab Dispos 29:712–720
31. Carlile DJ, Zomorodi K, Houston JB (1997) Scaling factors to relate drug metabolic clearance in hepatic microsomes, isolated hepatocytes and the intact liver—studies with induced livers involving diazepam. Drug Metab Dispos 25:903–911
32. Tang W, Wang RW, Lu AYH (2005) Utility of recombinant cytochrome P450 enzymes: a drug metabolism perspective. Curr Drug Metab 6:503–517
33. Brown RP, Delp MD, Lindstedt SL, Rhomberg LR, Beliles RP (1997) Physiological parameter values for physiologically based pharmacokinetic models. Toxicol Ind Health 13:407–484
34. U.S. EPA (Lipscomb JC, Kedderis GL) (2005) Use of physiologically based pharmacokinetic

models to quantify the impact of human age and interindividual differences in physiology and biochemistry pertinent to risk: final report for cooperative agreement ORD/NCEA Cincinnati, OH EPA/600/R-06-014A

35. Lipscomb JC, Teuschler LK, Swartout JC, Popken D, Cox T, Kedderis GL (2003) The impact of cytochrome P450 2E1-dependent metabolic variance on a risk relevant pharmacokinetic outcome in humans. Risk Anal 23:1221–1238
36. Lipscomb JC, Kedderis GL (2002) Incorporating human interindividual biotransformation variance in health risk assessment. Sci Total Environ 288:12–21
37. Lipscomb JC (2004) Evaluating the relationship between variance in enzyme expression and toxicant concentration in health risk assessment. Hum Ecol Risk Assess 10:39–55
38. Thrall KD, Gies RA, Muniz J, Woodstock AD, Higgins G (2002) Route-of-entry and brain tissue partition coefficients for common superfund contaminants. J Toxicol Environ Health Part A 65:2075–2086
39. Gargas ML, Burgess RJ, Voisard DE, Cason GH, Andersen ME (1989) Partition coefficients of low-molecular-weight volatile chemicals in various liquids and tissues. Toxicol Appl Pharmacol 98:87–99
40. Lilly PD, Andersen ME, Ross TM, Pegram RA (1997) Physiologically based estimation of *in vivo* rates of bromodichloromethane metabolism. Toxicology 124:141–152
41. Kenyon EM, Kraichely RE, Hudson KT, Medinsky MA (1996) Differences in rates of benzene metabolism correlate with observed genotoxicity. Toxicol Appl Pharmacol 136:649–656
42. Gargas ML, Andersen ME, Clewell HJ (1986) A physiologically based simulation approach for determining metabolic constants from gas uptake data. Toxicol Appl Pharmacol 86:341–352
43. Lipscomb JC, Barton H, Tornerol-Velez R (2004) The metabolic rate constants and specific activity of human and rat hepatic cytochrome P450 2E1 toward chloroform. J Toxicol Environ Health 67:537–553
44. Delic JI, Lilly PD, MacDonald AJ, Loizou GD (2000) The utility of PBPK in the safety assessment of chloroform and carbon tetrachloride. Reg Toxicol Pharmacol 32:144–155
45. Corley RA, Mendrala AL, Smith FA et al (1990) Development of a physiologically based pharmacokinetic model for chloroform. Toxicol Appl Pharmacol 103:512–527
46. Beck BD, Clewell HJ III (2001) Uncertainty/safety factors in health risk assessment: opportunities for improvement. Hum Ecol Risk Assess 7:203–207
47. Travis CC (1990) Tissue dosimetry for reactive metabolites. Risk Anal 10:317–321
48. Rhomberg LR, Wolff SK (1998) Empirical scaling of single oral lethal doses across mammalian species base on a large database. Risk Anal 18:741–753
49. Burzala-Kowalczyk L, Jongbloed G (2011) Allometric scaling: analysis of LD_{50} data. Risk Anal 31:523–532
50. Ginsberg G, Hattis D, Sonawane B, Russ A, Banati P, Kozlak M, Smolenski S, Goble R (2002) Evaluation of child/adult pharmacokinetic differences from a database derived from the therapeutic drug literature. Toxicol Sci 66:185–200
51. Ginsberg G, Hattis D, Miller R, Sonawane B (2004) Pediatric pharmacokinetic data: implications for environmental risk assessment for children. Pediatrics 113(Suppl):973–983
52. Hattis D (2004) Role of dosimetric scaling and species extrapolation in evaluating risks across life stages IV pharmacodynamic dosimetric considerations. Report to the U.S. Environmental Protection Agency under RFQ No DC-03-00009
53. Finlay BL, Darlington RB (1995) Linked regularities in the development and evolution of mammalian brains. Science 268:1578–1584
54. Renwick AG, Lazarus NR (1998) Human variability and noncancer risk assessment—an analysis of the default uncertainty factor. Regul Toxicol Pharmacol 27:3–20
55. Clancy B, Darlington RB, Finlay BL (2001) Translating developmental time across mammalian species. Neuroscience 105:7–17
56. Krishnam K, Andersen ME (1994) Physiologically based pharmacokinetic modeling in toxicology. In: Hayes AW (ed) Principles and methods of toxicology, 3rd edn. Raven Press, New York, NY, pp 149–188
57. Jepson GW, Hoover DK, Black RK, McCafferty JD, Mahle DA, Gearhart JM (1994) A partition coefficient determination method for nonvolatile chemicals in biological tissues. Fundam Appl Toxicol 22:519–524
58. Gallo JM, Lam FC, Perrier DG (1987) Area method for the estimation of partition coefficients for physiological pharmacokinetic models. J Pharmacokinet Biopharm 15:271–280
59. Teo SKO, Kedderis GL, Gargas ML (1994) Determination of tissue partition coefficients for volatile tissue-reactive chemicals: acrylonitrile and its metabolite 2-cyanoethylene oxide. Toxicol Appl Pharmacol 128:92–96
60. Khor SP, Mayersohn M (1991) Potential error in the measurement of tissue to blood distribution

coefficients in physiological pharmacokinetic modeling residual tissue blood I theoretical considerations. Drug Metab Dispos 19:478–485

61. Poulin P, Krishnan K (1995) An algorithm for predicting tissue:blood partition coefficients or organic chemicals from n-octanol:water partition coefficient data. J Toxicol Environ Health 46:117–129
62. Poulin P, Krishnan K (1996) A mechanistic algorithm for predicting blood:air partition coefficients of organic chemicals with the consideration of reversible binding in hemoglobin. Toxicol Appl Pharmacol 136:131–137
63. Barter ZE, Bayliss MK, Beaune PH, Bobbis AR, Carlile DJ, Edwards RJ, Houston JB, Lake BG, Lipscomb JC, Pelkonen OR, Tucker GT, Rostami-Hodjegan A (2007) Scaling factors for the extrapolation of *in vivo* metabolic drug clearance from *in vitro* data: reaching a consensus on values of human microsomal protein and hepatocellularity per gram of liver. Curr Drug Metab 8:33–45
64. Houston JB (1994) Utility of *in vitro* drug metabolism data in predicting *in vivo* metabolic clearance. Biochem Pharmacol 47:1469–1479
65. Blaauboer BJ (2010) Biokinetic modeling and *in vitro-in vivo* extrapolations. J Toxicol Environ Health Part B 13:242–252
66. Howgate EM, Yeo KR, Proctor NJ, Tucker GT, Rostami-Hodjegan A (2006) Prediction of *in vivo* drug clearance from *in vitro* data I impact of inter-individual variability. Xenobiotica 36:473–497
67. Obach RS (1999) Prediction of human clearance of twenty-nine drugs from hepatic microsomal intrinsic clearance data: an examination of *in vitro* half-life approach and nonspecific binding to microsomes. Drug Metab Dispos 27:1350–1359
68. Kedderis GL (1997) Extrapolation of *in vitro* enzyme induction data to human *in vivo*. Chem-Biol Interact 107:109–121

Chapter 20

Population Effects and Variability

Jean Lou Dorne, Billy Amzal, Frédéric Bois, Amélie Crépet, Jessica Tressou, and Philippe Verger

Abstract

Chemical risk assessment for human health requires a multidisciplinary approach through four steps: hazard identification and characterization, exposure assessment, and risk characterization. Hazard identification and characterization aim to identify the metabolism and elimination of the chemical (toxicokinetics) and the toxicological dose–response (toxicodynamics) and to derive a health-based guidance value for safe levels of exposure. Exposure assessment estimates human exposure as the product of the amount of the chemical in the matrix consumed and the consumption itself. Finally, risk characterization evaluates the risk of the exposure to human health by comparing the latter to with the health-based guidance value. Recently, many research efforts in computational toxicology have been put together to characterize population variability and uncertainty in each of the steps of risk assessment to move towards more quantitative and transparent risk assessment. This chapter focuses specifically on modeling population variability and effects for each step of risk assessment in order to provide an overview of the statistical and computational tools available to toxicologists and risk assessors. Three examples are given to illustrate the applicability of those tools: derivation of pathway-related uncertainty factors based on population variability, exposure to dioxins, dose–response modeling of cadmium.

Key words: Population variability, Risk assessment, Dose–reponse modeling, Benchmark dose, Toxicokinetics, Toxicodynamics, Physiologically based models, Cadmium, Dioxins, Pathway-related uncertainty factors, Meta-analysis, Systematic review

1. Introduction

Population variability is a critical quantitative matrix to all sciences that have been applied to many disciplines to model systems in biological, medical, environmental, and social sciences, economy, and help risk analysts and risk assessors in the decision making process. In the context of chemical risk assessment, a multidisciplinary approach at the cross road between medicine, biochemistry, toxicology, ecology, analytical chemistry, applied mathematics, and bioinformatics is applied to protect humans, animals, and the

Brad Reisfeld and Arthur N. Mayeno (eds.), *Computational Toxicology: Volume I*, Methods in Molecular Biology, vol. 929, DOI 10.1007/978-1-62703-050-2_20, © Springer Science+Business Media, LLC 2012

environment from chemical hazards. According to *Regulation (EC) No 178/2002 of the European Parliament and of the Council of 28 January 2002*, "Hazard" is defined as a biological, chemical, or physical agent in, or condition of, food and "Risk" is defined a function of the probability of an adverse health effect and the severity of that effect, consequential to a hazard (1). The WHO International Program on Chemical Safety (IPCS) within its program dealing with the harmonization of chemical risk assessment methodologies has defined hazard as "the inherent property of an agent or situation having the potential to cause adverse effects when an organism, system or (sub)population is exposed to that agent" and risk as "the probability of an adverse effect in an organism, system or (sub)population caused under specified circumstances by exposure to an agent" (2). To *qualify and quantify such hazard and risk*, the application of the four pillars of risk assessment, namely, hazard identification and hazard characterization, exposure assessment, risk characterization have enabled scientists and public health agencies to protect consumers from adverse health effects that may result from acute and chronic chemical exposure (3).

Hazard identification and hazard characterization can be underpinned as the toxicological dimension of chemical risk assessment: the toxicokinetics (TK) translates a chemical external dose to an internal dose leading to overall elimination from the body, i.e., absorption from the gastrointestinal tract, distribution in body fluids/tissues, metabolism, and ultimately excretion in the urine/feces. The toxicodynamics (TD) expresses the toxicity once the toxic species, either as the parent compound or an active/toxic metabolite reaches its target (receptor/cell/organ) to derive a health-based guidance value (4).

Exposure assessment is defined as the qualitative and/or quantitative evaluation of the likely intake of biological, chemical or physical agents via food as well as exposure from other sources if relevant (2). Because of the absence of adequate biomarkers, the external exposure is often the only information available to be calculated and compared with toxicological thresholds. There is an overall agreement that the exposure assessment should be a stepwise approach, each step being more conservative that the following one but also less costly in terms of time and resources. In the field of food safety the exposure from food is often assumed to be the unique source of exposure. In a vast majority of cases, its assessment consists in a deterministic approach combining the mean and high levels of occurrence for the chemical under consideration with the consumption of an average and a high consumer for the contaminated food. The deterministic approach is quite well described in various guidelines (2, 5) and is used routinely for its simplicity and its conservatism. However, it does not represent an accurate or even a realistic picture of the true exposure over the population. In particular it does not account either for the day to day variability of the food consumption on long term or for the distribution of hazard within foodstuffs.

Risk characterization consists in comparing the exposure with one or another threshold for safety concern. The exposure could first be compared with the Health based guidance values (e.g., provisional tolerable weekly intake) defined as: *The estimate of the amount of a chemical in food or drinking-water, expressed on a body weight basis, that can be ingested [weekly] over a lifetime without appreciable health risk to the consumer.* The derivation of health-based guidance value can use, as the point of departure, the Benchmark Dose Limit (BMDL) which is the lower boundary of the confidence interval on a dose of a substance associated with a specified low incidence of risk, of a health effect (2). The BMDL accounts for the uncertainty in the estimate of the dose–response that is due to characteristics of the experimental design, such as sample size. Such comparison allows setting up safe levels of exposure and consequently safe level of food consumption and/or hazard occurrence. When health effects in humans are well characterized, the exposure could also be compared with the level at which these effects are likely to occur or with the benchmark dose. Such type of comparison allows estimating a burden of disease attributable to the hazard under consideration. In a number of cases it is necessary to develop statistical methodologies to reconcile the dietary exposure with the results of PB-TK models describing the fate of toxic compounds within the body. Ultimately, consistent outputs from the two exposure models (external and internal) should allow a robust estimation of health effects.

Beyond classes of chemicals and the risk assessment pillars, regulators have relied upon two basic mechanistic differences to assess human health risks related to chemical exposure, i.e., whether the chemicals are genotoxic carcinogens or nongenotoxic carcinogens. Such classification constitutes the basis for the derivation of health-based guidance values in humans and has been reviewed in detail elsewhere (3, 4, 6). In essence, genotoxic carcinogens and their metabolites are assumed to act via a mode of action that involves a direct and potentially irreversible DNA-covalent binding with a linear dose–response relationship over a chronic to life time exposure with no threshold or dose without a potential effect. To deal with such genotoxic carcinogens, the Margin Of Exposure (MOE) approach has been recently used by the World Health Organization (WHO) and the European Food Safety Authority (EFSA). The MOE is determined as the ratio of a defined point on a dose–response curve for adverse effects obtained in animal experiments (in absence of human epidemiological data) and human intake data. Two reference points describing the dose–response relationship have been proposed, namely, the preferred Benchmark Dose (BMD) and BMDL or the T25. For the interpretation of MOEs, the Scientific Committee of EFSA considered that an

MOE of 10,000 or more, based on a BMDL10 derived from animal cancer bioassay data and taking into account the uncertainties in the interpretation, "*would be of low concern from a public health point of view and might reasonably be considered as a low priority for risk management actions*" (7). Recent examples of risk assessments performed by the Joint FAO/WHO Expert Committee on Food additives (JECFA) and EFSA using this approach have included ethyl carbamate, polycyclic aromatic hydrocarbons, and arsenic (8–11).

In contrast, nongenotoxic carcinogens and their metabolites are assumed to act via an epigenetic mode of action without covalent binding to DNA. It is assumed a threshold level of exposure below which no significant effects are induced implying that homeostatic mechanisms can counteract biological perturbations produced by low levels of intake, and that structural or functional changes leading to adverse effects, that may include cancer, would be observed only at higher intakes (4). For such thresholded toxicants, health-based guidance values are derived and have been defined as "without appreciable health risk" when consumed every day or weekly for a lifetime such as the "Acceptable and Tolerable Daily Intakes (ADI/TDI)" or provisional tolerable weekly intake (PTWI) in Europe and the "Reference dose" in the United States. For thresholded chemicals with acute toxic effects, the acute reference dose approach (ARfD) has been defined as by the Joint FAO/WHO Meeting on Pesticide Residues (JMPR) as "an estimate of the amount a substance in food and/or drinking water, normally expressed on a body weight basis, that can be ingested in a period of 24 h or less without appreciable health risk to the consumer on the basis of all known facts at the time of the evaluation" (12). Recent risk assessment for which ARfD have been derived for humans based on the consumption of shellfish contaminated with marine biotoxins or mushrooms contaminated with nicotine (13, 14). These health-based guidance values are derived using surrogates for the threshold such as the no-observed-adverse-effect-level (NOAEL) or the BMD or BMDL from laboratory animal species used in risk assessment (mice, rat, rabbit, dog) and an uncertainty factor of a 100-fold, to allow for interspecies differences (10) and human variability (10).

Over the years, the scientific basis for the uncertainty factors has been challenged and considerable research efforts have aimed to further split the products of such factors to include TK (4.0 for interspecies differences and 3.16 human variability) and TD aspects (2.5 for interspecies differences and 3.16 human variability) to ultimately replace them with chemical-specific adjustment factors (CSAF) when chemical-specific data are available and physiologically based models can be developed (Fig. 1) (15, 16). Application of such adjustment factors is a substantial refinement in quantitative risk assessment as it integrates information specific to the chemicals or the studied populations, even from limited data. This integration

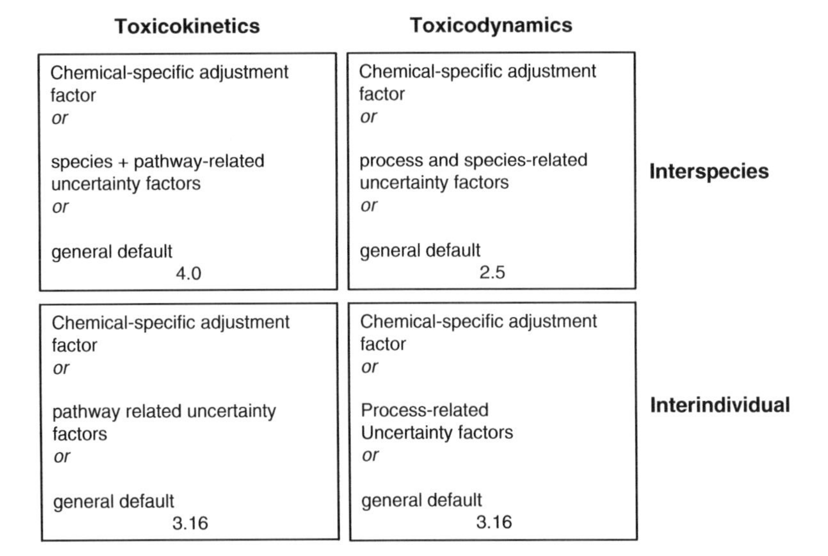

Fig. 1. Uncertainty factors, chemical-specific adjustment factors (CSAF), pathway-related uncertainty factors and the general default uncertainty factors. Based on ref. 15.

is made through statistical analysis of specific data with ad hoc modeling of the population variability. An example of adjustment factor derivation from physiological-based model for both TK and TD aspects is the recent risk assessment of cadmium performed by EFSA (17, 18). Intermediate options using simpler models have also been applied. For example, when the metabolic route is known in humans, the default uncertainty factor of 3.16 can be replaced by pathway-related uncertainty factors which have been developed based on meta-analyses of human variability in TK for phase I enzymes [cytochrome P-450 isoforms (CYP1A2, CYP2E1, CYP2C9, CYP2C19, CYP2D6, CYP3A4), hydrolysis...], phase II enzymes (glucuronidation, sulfation, *N*-acetylation), and renal excretion using the pharmaceutical database. These pathway-related uncertainty factors have been derived for subgroups of the population for whom data were available such as healthy adults from different ethnic backgrounds including genetic polymorphisms (CYP2D6, CYP2C9, CYP2C19, *N*-acetylation, infants, children and neonates, and the elderly) (19).

These examples which are detailed further in this chapter highlight that considerable efforts move have been made to develop population and hierarchical models using advanced statistical modeling to quantify population variability in the critical processes driving the setting of health standards, namely, exposure, TK, and TD or dose–response. This chapter focuses such population models in computational toxicology applied to TK, TD (dose–response), and exposure assessment. The first part focuses on the basic description of such models together with the computer software and packages available. Mathematical and practical aspects to develop the models are then given and three models are given as examples: the development of pathway-related uncertainty factors, dose–response modeling of cadmium in humans, and probabilistic exposure assessment of dioxins in humans. Finally, critical conclusions and limitations of this growing discipline conclude the chapter.

2. Materials: Main Approaches and Software Packages

2.1. Hazard Identification and Characterization

2.1.1. Toxicokinetic Modeling

After a chemical compound penetrates into a living mammalian organism (following intentional administration or unintentional exposure), it is distributed to various tissues and organs by blood flow (20). Following its distribution to tissues, the substance can bind to various proteins and receptors, undergo metabolism, or can be eliminated unchanged. The concentration versus time profiles of the xenobiotic in different tissues, or the amount of metabolites formed, are often used as surrogate markers of internal dose or biological activity (21). Such information can be predicted with the help of toxicokinetic (TK)/pharmacokinetic (PK) modeling for which two general approaches have been applied :

- Noncompartmental analyses, consisting in a statistical regression analysis of TK measurements versus time and a number of covariates.
- Compartmental analyses, consisting in the modeling of TK profiles accounting for the distribution and metabolism of the toxic compounds throughout the body.

Like for any modeling exercise, the choice of method depends on the objectives and scope of the analysis, the data available, and the resources and constrains attached to the analysis. Usually, noncompartmental approaches are used for exploratory analyses, the screening of potentially influential factors, and the determination of sample size for further TK studies. Conversely, compartmental analyses are more suitable for refined analyses and predictions. The methods and related tools are further detailed and compared below.

Compartmental TK Analysis

Compartmental analyses assume that the substance of interest distributes in the body as if it was made of homogeneous well-stirred compartments. Empirical compartmental models assign not particular meaning to the compartment identified (i.e., the substance behaves "as if" it was distributed in two or three compartments in the body, without looking for interpretations of what these compartments might be) (22). That type of analysis can only be performed usefully in data-rich situations and does not lend itself to necessary interpolations and extrapolations. It is less and less used, and survives only in heavily regulated and slowly changing contexts. On the

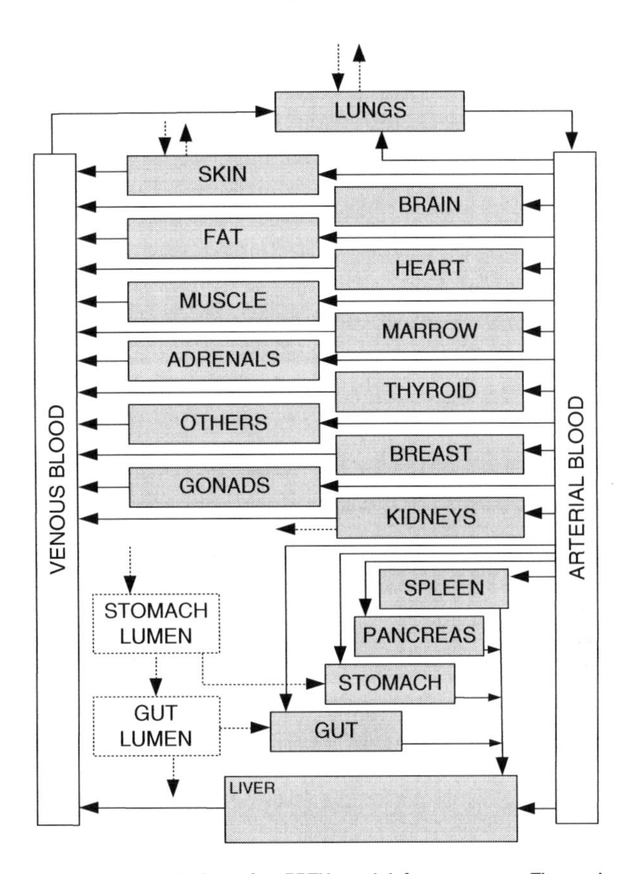

Fig. 2. Schematic representation of a PBTK model for a woman. The various organs or tissues are linked by blood flow. In this model, exposure can be through the skin, the lung or *per os*. Elimination occurs through the kidney, the GI tract, and the lung. The parameters involved are compartment volumes, blood flows, tissue affinity constants (or partition coefficients), and specific absorption, diffusion and excretion rate constants. The whole life of the person can be described, with time-varying parameters. The model structure is not specific of a particular chemical (see http:/www.gnu.org/software/mcsim/, also for a pregnant woman model).

contrary, physiological-based TK models (PBTK models) insist in assigning physiological meaning to the compartments they define. Their parameter values can be determined on the basis of in vitro data, in vivo data in humans or animals, quantitative structure–property relationship (QSPR) models, or the scientific literature (basically reporting and summarizing past experiments of the previous types) (23). PBTK models have evolved into sophisticated models which try to use realistic biological descriptions of the determinants of the disposition of the substance in the body (24). Those models describe the body as a set of compartments corresponding to specific organs or tissues (e.g., adipose, bone, brain, gut, heart, kidney, liver, lung, muscle, skin, and spleen, etc.) as illustrated by Fig. 2. Between compartments, the transport of substances is dictated by various physiological flows (blood, bile, pulmonary ventilation, etc.) or by diffusions (20). Perfusion-rate-limited kinetics applies when the tissue membrane presents no barrier to distribution. Generally, this condition is likely to be met by small lipophilic substances. In contrast, permeability-rate kinetics applies when the distribution of the substance to a tissue is limited by the permeability of a compound across the tissue membrane. That condition is more common with polar compounds and large molecular structures. PBTK models may therefore exhibit different degrees of complexity. In such a model, any of the tissues can be a site of metabolism or excretion, if that is biologically justified.

Building a PBTK model requires gathering a large amount of data which can be categorized in three groups:

- Model structure, which refers to the arrangement of tissues and organs included in the model.
- System data (physiological, anatomical, biochemical data).
- Compound-specific data.

Additional details on PBTK modeling and applications can be found in (20, 25–27) or in this book (Bayesian inference-Chapter 25). Indeed, such descriptions of the body are approximate but a balance has to be found between precision (which implies complexity) and simplicity (for ease of use). Yet, the generic structure of a PBTK model facilitates its application to any mammalian species as long as the corresponding system data are used. Therefore, the same structural model can be used for a human, a rat or a mouse, or for various individuals of the same species. That is why such models are widely used for interspecies extrapolations, and also for interindividual and intraindividual extrapolations.

Interindividual or intraindividual extrapolations refer to the fact that a given exposure may induce different effects in the individuals of a population, and that the same individual may respond differently to the same exposure at different times in his/her lifetime. These extrapolations are performed by setting parameter values to those of the subpopulation or individual of interest, and are mainly used to predict the differential effects of chemicals on sensitive populations such as children, pregnant women, the elderly, the obese, and the sick, taking into account genetic variation of key metabolic enzymes, etc. The toxicokinetic characteristics of a compound can also be studied under special conditions, such as physical activity. Understanding the extent and origins of interindividual differences can be done predictively, on the basis of Monte Carlo simulations for example. However, that requires a large amount of system information, and in particular statistical distributions for the systems parameters. An alternative is to estimate the variability in parameters from experimental or clinical data. This form of inference is well developed and known as population or multilevel pharmacokinetic analyses (21).

Besides such extrapolations, PBTK models are also used to perform interdose and interroute *extrapolations.* Interdose extrapolations are achieved by capturing both the linear and nonlinear components of the biological processes (e.g., transport and metabolism) governing the kinetics of the chemical of interest. Interroute variations can also been captured by PBTK models to be used to extrapolate or predict TK of a given compound from one route to another (e.g., intravenous infusion, oral uptake, dermal absorption, inhalation). The use of numerical integration also allows arbitrary forms of time-varying inputs (e.g., air concentration) to be used. Combinations of exposure routes can also be modeled.

Note that PBTK models have historically been used primarily to estimate internal exposures following some defined external exposure scenario. As with any model, PBTK models can be run backwards (numerically at least) to reconstruct exposures from the TK measurements. That application, even if obvious, is a bit more recent (28).

The level of the complexity structuring PBTK models (e.g., number of compartments or differential equations) needs to be commensurate with the available data and the primary objective of the analysis. In chemical risk assessment, the evaluation of interindividual variability may be of primary importance. In this case, compartmental models may be reduced to a simpler one-compartment model describing only the first-order magnitude of TK variations over time, but integrating a subject-specific random effect (typically lognormal) to describe the population variability. Generally speaking, multicompartment PBTK models describe much better the metabolic pathways as a whole and allow the calculation of chemicals' concentrations in the main organs in the body. On the other hand, the numerous parameters may require a substantial amount of information on parameters which makes more difficult any statistical evaluation. It also generally requires thorough sensitivity analysis and model validation. Alternatively, a one-compartment model only focuses on the overall elimination of the toxicant from the body, making rough and global assumptions on the involved pathways. In case of poor prior knowledge on these pathways, it allows a simplified and parsimonious description of toxicant elimination, hence easier statistical evaluations (such as the evaluation of population variability). However, in some cases where the simple TK modeling assumptions are not met (like zero-order absorption, dose linearity), such an approach could lead to poor fits and inflated residuals. Moreover, by definition, one-compartment models do not allow for the evaluation of toxicant concentrations in each organ where it distributes. Typically, risk assessors may face the choice between implementing an extended PBTK model without population variability versus fitting a one-compartment model with a subject-specific random effect to enable the evaluation of interindividual variability of the overall toxicant half-time. The choice will then be made depending on the available data and on the objective of the modeling analysis. A discussion on this choice exemplified by the Cadmium example is detailed in ref. 17.

The simple one-compartment, first order, TK models can be use to relate external exposure to internal exposure (29, 30). Since one-compartment TK models are widely used especially for chronic risk assessment, we detail hereafter their general form.

One-compartment TK models are represented by a differential equation describing the change of the chemical body concentration x over time:

$$\frac{dx}{dt}(t) = (I \times \text{ABS}) - (k \times x(t)), \quad k = \frac{\ln(2)}{\text{HL}} \tag{1}$$

where I is the daily intake, ABS the fraction of chemical absorbed after oral exposure (absorption fraction) and k the elimination rate described by the biological half-life HL.

After a certain period of time, the chemical concentration in body lipids settles to an equilibrium regime or steady-state. In this regime, a balance is obtained between ingested dose of chemical and its elimination. The differential equation 1 is therefore equals to 0 and the burden in body at steady state situation can be simply calculated as a linear relation:

$$x(t) = \frac{I \times \text{ABS}}{k}. \tag{2}$$

This equation at steady-state is often used to relate external and internal exposure (29, 30). Using the above equation requires that the daily intake and the absorption fraction are constant over the period of time needed to reach steady-state (31). No variability on the chemical daily intake can be considered. However, for chemicals present in food not daily consumed such as methylmercury in fishery products a high variability between individual intakes over lifetime can occur. Pinsky and Lorber (32) found that TK model integrating time-varying exposure profile produced more reliable predictions of body burden than compared with the equation at steady-state. Moreover, the Eq. 2 cannot be used to relate external exposure to measurements on population which has not reached the steady-state, which is the particular case of chemicals with long half-life.

For chemicals with long half-life the absorption time can be considered as insignificant compared to the elimination time. In such situation, there is no absorption process as in the case of an intravenous injection of the chemical. A simplified equation differential (33, 34) obtained by removing the first part related to the intake dose of the Eq. 1 is thus used to define the elimination process between two intakes time

$$\frac{dx}{dt}(t) = -k \times x(t). \tag{3}$$

Any software package able to solve systems of differential equations can be used to build a PBTK model and run simulations of it. They include GNU Octave (http://www.gnu.org/software/octave), Scilab (http://www.scilab.org/), GNU MCSim (free software), Mathematica (http://www.wolfram.com/mathematica), Matlab (http://www.mathworks.com/products/matlab), and acslX (http://www.acslx.com). A list of software packages used for PBTK modeling is detailed in Table 1. Most of packages specifically developed for compartmental analysis were tailored for PBPK and applications in clinical pharmacology, though they can be used equally in the context of toxicological assessments. Only a small number of them allow user-specific integration of random effects to evaluate population variability on PK parameters. As a matter of fact, such nonlinear mixed effect models require powerful algorithms for their statistical inference, such as simulation-based algorithms. This can be achieved using Bayesian inference and Monte Carlo Markov chains as implemented, e.g., in GNU MCSim (http://www.gnu.org/software/mcsim) or using the SAEM algorithm, a stochastic version of the EM algorithm (35) maximum likelihood estimation. The Monolix application developed as a free toolbox for Matlab is a powerful and flexible user-friendly platform to implement SAEM-based estimation of PBPK and PBTK models with population variability. It is increasingly used in particular in drug development and pharmacological research as it combines flexibility, statistical performance and reliability with minimal coding required from users. From this perspective, it shows substantial advantage compared to the NONMEM software, the gold standard of pharmacological modeling used in the pharmaceutical world, mostly based on gradient optimization and first-order approximation of the likelihood.

Even more specialized software using pathway-specific information is available and has devoted considerable effort to building databases of parameter values for various species and human populations: notable examples are Simcyp Simulator (http://www.simcyp.com), PK-Sim (http://www.pk-sim.com), GastroPlus (http://www.simulations-plus.com), and Cyprotex's CloePK (https://www.cloegateway.com).

Noncompartmental TK Analysis

Noncompartmental methods are in fact a misnomer, because they still consider that the body as one compartment. In fact they are simpler, often nonparametric compared with methods that aim at estimating general kinetic properties of a substance (for example, the substance biodisponibility, total clearance, etc.) (36). Noncompartmental models are convenient for exploratory analyses and easy to use; they produce estimates of common parameters of interest. They can be used when very little information is available on the

Table 1
List of the main software packages for compartmental PBPK/PBTK modeling, in alphabetic order, with the indication that they include built-in features for analyzing population variability or not

Software (Web site)	Description	Built-in population variability
asclX http://www.acslxtreme.com/	Built-in compartmental modeling including pharmacodynamics, toxicity studies, Monte Carlo simulations tools	No
Berkeley Madonna www.berkeleymadonna.com	Generic powerful ODE solver and Monte Carlo simulations	No
Boomer/Multi-Forte www.boomer.org	Estimate and simulate from compartmental models, using a range of possible algorithms including Bayesian inference	No
CloePK https://www.cloegateway.com	PBPK predictive tools for animals and humans including open-source database on metabolic pathways	No
ERDEM http://www.epa.gov/heasd/products/erdem/erdem.html	PBPK/PD solver and simulator for risk assessment used by EPA and exposure assessment using backwards simulations	No
GastroPlus™ http://www.simulations-plus.com	Popular PBPK solver and simulator, including trial simulations	No
GNU Octave http://www.gnu.org/software/octave	Generic ODE similar to Matlab	No
Matlab with MONOLIX http://www.monolix.org/	Powerful PBPK solver integrating population variability, with GUI interface for graphical analyses and predictions. It uses MCMC and SAEM algorithms	**Yes**
Mathematica with Biokmod http://web.usal.es/~guillermo/biokmod/mathjournal.pdf	ODE solver specific to compartmental biokinetic systems, includes design analysis	No
MCSim http://www.gnu.org/software/mcsim	Flexible PBPK solver with population variability, it gives examples of detailed generic PBPK models	**Yes**
NONMEM http://www.iconplc.com	Gold standard of the population PK compartmental analysis in the pharmaceutical industry	**Yes**
Phoenix® NLME™ (formerly WinNonMix) http://www.pharsight.com	Population PK modeling tool with similar algorithms as in Nonmem but with better graphical features	**Yes**
PKBugs www.mrc-bsu.cam.ac.uk/bugs/.../pkbugs.shtml	Efficient and user-friendly Bayesian software built in the WinBUGS, for analysis of population PK models	**Yes**

(continued)

Table 1 (continued)

Software (Web site)	Description	Built-in population variability
PK-Sim http://www.systems-biology.com/products/pk-sim.html	User-friendly tool for PBPK modeling and simulations in humans and animals, it includes a population database	Yes
PK solution http://www.summitpk.com	Easy-to-use Excel-based analysis tool for noncompartmental and simple compartmental models	No
R with PKfit http://cran.r-project.org/web/packages/PKfit/index.html	Free R routine for the analysis of compartmental models	No
SAAM II and Popkinetics http://depts.washington.edu/saam2	Compartmental modeling PK software with population analysis using Popkinetics	Yes
S-Adapt http://bmsr.usc.edu/Software/ADAPT/ADAPT.html	Compartmental PK/PD modeling platform including simple population models fitted with the EM algorithm	Yes
Simcyp http://www.simcyp.com/	Major and powerful PBPK software suite with built-in PK and genetic database for predictions in animals and humans	Yes

metabolic pathways and few data are to be analyzed. But they have assumptions, first and foremost linearity.

As the methods involved are mainly simple calculations or statistical regressions, most standard statistical software packages can be used. Population variability is then assessed using linear mixed-effect models built in most statistical packages. Nevertheless, more specific computer-based tools are available; the main ones are listed in Table 2.

2.1.2. Toxicodynamics and Dose–Response Modeling

Toxicodynamic (TD) modeling quantifies the plausible causal link between chemical exposure and health or environmental consequences. TD assessments are typically based on the development of dose–response models or dose–effect relationships. Like for TK models, a proper quantification of the interindividual variability of dose–response is essential for quantitative risk assessment, and more specifically for the derivation of health-based guidance values.

TK/TD models provide a number of advantages over more descriptive methods to analyze toxicity/safety data. In particular, these models make better use of the available data as all effects observations over time are used to estimate model parameters.

Table 2
List of the main software packages for noncompartmental TK modeling, in alphabetic order, with the indication that they include built-in features for analyzing population variability or not

Software (Web site)	Main features	Population variability
ADAPT II http://bmsr.usc.edu/Software/ADAPT/ADAPT.html	PKPD software with Monte Carlo simulations capacities	No
BE/BA for R http://pkpd.kmu.edu.tw/bear/	Freeware using linear and ANOVA models, include bioequivalence analysis and study design analysis	No
PK solution http://www.summitpk.com	Easy-to-use Excel-based analysis tool for noncompartmental and simple compartmental models	No
WinNonLin http://www.pharsight.com	Industry standard for noncompartmental analyses. Includes design analysis	**Yes**

As a consequence, TK/TD models can also successfully be used to extrapolate and compare toxic effects between different exposure scenarios. Finally, TK/TD models facilitate a mechanism-based comparison or extrapolation of effects from different substances, species, doses, and life stages.

Similarly to PBPK in TK assessments, biologically based dose–response (BBDR) models have been developed TD analyses in order to incorporate information on biological processes at the cellular and molecular level to link external exposure to an adverse effect. Although BBDR models provide a framework for testing mechanistic hypotheses (37), their role in risk assessment, e.g., for low-dose extrapolation appears to be limited (38) as the population variability is then only expressed at cellular level rather than at subject levels without much improvements on toxicity predictivity.

In contrast with mechanism-based BBDR models, empirical modeling can be used to describe the patterns relating adverse effects to increasing exposure to a contaminant. Empirical dose–response modeling hence attempts to find a simple mathematical model that adequately describes this pattern. In general, empirical models have minimal or even no direct linkage to the underlying mechanisms driving such adverse effect. Instead, they focus on flexible and often simpler mathematical forms that can fit a broad spectrum of dose–response data. This mathematical form is then by nature strongly dependent on the dose range observed so that extrapolation, e.g., to low dose should be carefully approached in light of information available on the biology underlying mechanisms of action. Furthermore, empirical models can generally not be used to extrapolate hazard characterization of mixtures from dose–response of individual compounds without a strong assumption on how the individual dose–response curves combine into the overall dose–response for the mixture (e.g., assuming effect additivity). Typical empirical models include linear, log-linear, Poisson, and Hill models. Empirical models can then be fitted using any standard statistical software and then used to estimate a point of departure.

In toxicological assessment where population variability needs to be accounted for, the integration of TK and TD models into a consistent mathematical description is essential. Indeed, misspecification of population variability in one of the TK and TD models would then make the overall assessment inaccurate. Moreover, such integrated approaches allow for correlating subject-specific TK and TD parameters, hence avoiding over-estimation of population variability in risk assessments.

There are very few dedicated software packages specific to population TD assessments. In general, such package are integrated either within a PK software or associated to a benchmark dose evaluation tool (see below). This is the case, e.g., for commercial packages like Toxtools which allow for dose–response model evaluations among a set of other features.

Reference Dose Evaluation: NOAEL, LOAEL and BMD

Risk assessment of thresholded toxicants (typically noncarcinogenic contaminants) requires the evaluation of a health-based exposure limit values or Reference Dose (RfD) which are used for regulatory purposes. Exposure to chemicals below such RfD will then generally be determined to be safe and exposures above the RfD will generally be determined to be unsafe. Many variations are observed between risk assessment bodies, but a general approach to the evaluation of RfD is:

- To evaluate the dose–effect or dose–response relationship.
- To calculate a point of departure from this TD characterization.
- To apply an uncertainty factor to the point of departure, in order to account for possible interspecies and/or interindividual variability.

Points of departure are typically one of these:

- No-observed-adverse-effect-level (NOAEL);
- Lowest-observed-adverse-effect-level (LOAEL); or
- Benchmark dose (BMD).

To determine these points of departure for a given compound, specific prospective toxicological (laboratory) studies can be conducted, generally on animals due to obvious ethical, cost and regulatory hurdles. Alternatively, retrospective analyses of human data can also be performed, with the evaluation of interindividual variability as far as possible. This variability estimate can support the derivation of evidence-based uncertainty factors, as illustrated further in the next sections of this chapter.

NOAEL is defined by the US EPA as the level of exposure at which there is no statistically or biologically significant increase in the risk or severity of any adverse affect in the exposed population as compared to a control or nonexposed population. It corresponds to the highest level of exposure without adverse effects. Closely related is the LOAEL, defined as the lowest level of exposure at which adverse effect or early precursor of adverse effect is observed. An obvious advantage of using NOAEL/LOAEL is that they are simple to evaluate based on published data. Even toxicological studies typically involving ten identical rats by dose groups are rather straightforward to design and to analyze, essentially because interindividual variability is by essence eliminated by the animal study population and therefore ignored in the NOAEL/LOAEL assessment. Using NOAEL/LOAEL as a basis for the evaluation of RfD implies therefore that evidence-based uncertainty factors can be derived to account for such population variability. In addition, a typical concern of NOAEL/LOAEL is the fact that the uncertainty attached to their evaluation is usually underestimated and their values can be highly dependent on the study designs and sample size. NOAEL/LOAEL generally larger with smaller sample sizes which makes their use nonconservative in case of sparse data.

As an alternative the BMD approach have developed and increasingly used to better account for population variability and properly assess estimation uncertainty using a statistical assessment of the dose–response model (see e.g., refs. 39–41 for a recent overview of the state-of-the-art). The BMD is defined as the dose needed to achieve an excess risk of a given adverse effect compared to the background exposure, usually defined as level of a biomarker, above a given threshold. If P(d) denotes the probability for an individual to reach the defined threshold for adverse effect, when exposed to dose d, then the BMD can be defined in two different ways:

- As the dose leading to "Additional risk" of $X\%$ (often called benchmark response (BMR)):

  $P(\mathrm{BMD}) = P(\mathrm{Background}) + X\%$

  where $P(\mathrm{Background})$ stands for the probability of adverse effect at background exposure, and $X\%$ (BMR) for the additional prevalence.

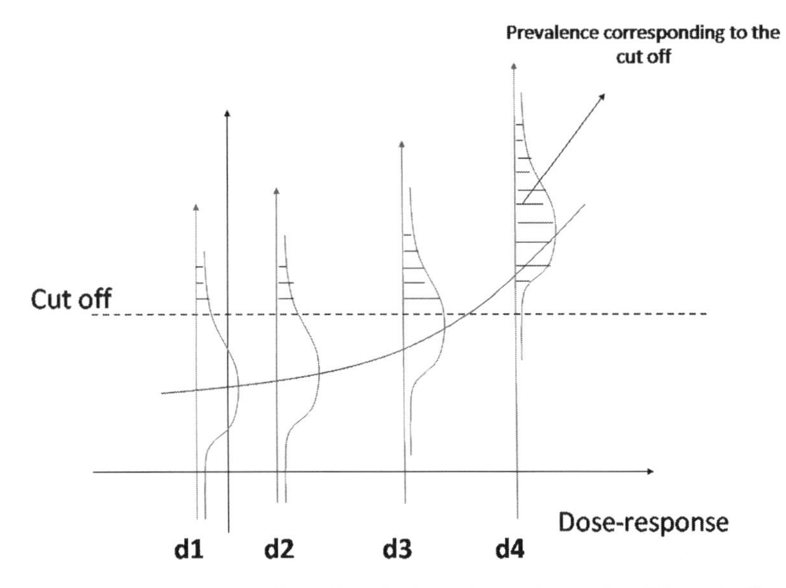

Fig. 3. Hybrid approach used for BMD evaluation, using continuous data dichotomized into quanta data with a predefined cut-off.

- As the dose leading to "Extra risk" of X%:
  $P(\text{BMD}) = P(\text{Background}) + X\%(1 - P(\text{Background}))$
  where P(Background) stands for the probability of adverse effect at background exposure, and X% for the probability to observe adverse effect at BMD given it was not observed at background exposure.

BMDs can be derived using dose–response model evaluated on quantal data, corresponding to the proportion of population reaching a given threshold effect at a given dose. This standard approach is the one proposed in the available software such as the EPA BMD software and the PROAST package (www.epa.gov/ncea/bmds and www.rivm.nl/proast). However, the so-called hybrid approach (41–43) uses a dose–effect model describing the continuous relationship between dose and effect level, hence allowing for risk calculation without dichotomizing the outcome, using all information available from continuous data. Risks or prevalence can then be derived with respect to any given biologically relevant threshold. This approach is valid under the assumption that effect levels are log-normally distributed over the population at a given dose. The main idea of this hybrid approach is to model the population variability around the mean dose–effect curve using a statistical (log-normal) distribution at each given dose, as illustrated by Fig. 3. Then, for any cut-off, one could derive the corresponding prevalence and dose–response curve.

Once a dose–response model fitted to appropriate populated data, BMD can be derived according to:

$$\text{BMD} = \exp\left(\mu^{-1}\left(\log(\text{cutoff}) - \sigma^{-1}(p)\right)\right),$$

with:

$$p = (1 - \mathrm{BMR})\left(\frac{\log(\mathrm{cutoff}) - \mathrm{background}}{\sigma}\right),$$

for extra risk

$$p = \left(\frac{\log(\mathrm{cutoff}) - \mathrm{background}}{\sigma}\right) - \mathrm{BMR},$$

for additional risk

where σ stands for the population of effect, BMR for the benchmark response, p for the cumulative distribution function of the standardized normal distribution, μ for the dose–effect function.

By construction, a BMD includes the population variability which implies that no uncertainty factors will be subsequently required account for it. Furthermore, since BMDs are the results of a parametric statistical estimation procedure, a confidence interval can be derived. The lower bound of such interval (BMDL) can be chosen as a more conservative point of departure, hence accounting for estimation uncertainty.

In general, BMD analysis requires individual data are available for a number of dose groups, which may sometimes be hurdle to the implementation of BMD approach. However, approaching dose–effect curve estimation in the context of meta-analysis where only aggregated data from the literature are available can be made under additional assumptions (e.g., log-normality of population distributions). This should also be accounted for in the statistical modeling and in the use of adjustment factors for any BMD evaluation. Examples and discussion of analysis of aggregated epidemiological data for dose–response analysis can be found in (44). There is generally not a unique dose–response model that can be chosen to meet the purpose. Therefore, it is necessary to assess the sensitivity of results with respect to the modeling assumption. This can be achieved, e.g., by comparing results from different possible models or using model averaging techniques (45).

Although the BMD approach provides a powerful framework to derive health-based guidance values for both carcinogenic and noncarcinogenic compounds, it is far from being systematically implemented by regulatory bodies. One of the main hurdles is the substantial need for statistical modeling expertise even if specific software packages have been developed. Implementation and interpretation of outputs still require an expert modeler to be involved in the assessment. In a way this just acknowledges that chemical risk assessments remains by nature a multidisciplinary task.

As usual, the choice between NOAEL/LOAEL and BMD approaches should be made in relation with the data available, the resources available and the objectives to be met. Table 3 summarizes the main differences between NOAEL and BMD-based assessments, while Table 4 lists the pros and cons for the two approaches.

Table 3
Summary table comparing requirements and characteristics of NOAEL versus BMD evaluations

Comparative table	NOAEL	BMD
Data/design requirements	NOAEL must be one of the doses used in the studies Small sample size reduces statistical power	Precision on doses should be good (if not, it should be know and accounted for)
Effect of Sample size per dose	Smaller sample size leads to larger NOAEL	Smaller sample size leads to smaller BMD
Effect of number of dose groups (or number of publication)	Fewer doses leads to larger NOAEL	Fewer doses leads to smaller BMD
Additional assumptions	Need for an uncertainty factor in most cases	Dose–response model needed (sensitivity analysis often required) Need for an uncertainty factor in some cases

Table 4
Summary table comparing pros and cons in the application of NOAEL versus BMD in toxicological risk assessment

Summary table	NOAEL	BMD
PROS	Easy/faster to understand and implement	BMD calculated for any risk levels decided upon
	Intuitive appeal	Uncertainty accounted for in a robust manner
	Do not require strong assumptions on dose–response models	The whole dose–response curve accounted for
	More robust when rare effects are observed	Allows better adjustment for covariates/clusters
CONS	NOAEL must be of the doses tested	Data requirements not always achieved to allow modeling (especially with human epidemiological data)
	Fewer subjects/doses give higher NOAEL (rewards poor designs)	Few software packages available
	Does not provide a measure of risk	Except for ideal cases, requires statistical/modeling expertise
	Uncertainty around NOAEL is dealt with uncertainty factors, often arbitrary or poorly robust	Less robust when doses are widely spread
		Less robust with complex or shallow dose–response

The evaluation of BMD can be implemented by direct coding of dose–response models in generic modeling platform such as R, Matlab or S-Plus. Since the BMD methodology has been increasingly used by risk assessors worldwide, national agencies have developed their own software. It is in particular the case of the BMDS software developed by the US EPA and the PROAST software developed by the Dutch agency RIVM. Version 2.1.2 of the U.S. EPA software BMDS can be downloaded free of charge at: http://www.epa.gov/NCEA/bmds/ while version 26.0 of the RIVM software PROAST can be obtained free of charge at: http://www.rivm.nl/en/foodnutritionandwater/foodsafety/proast.jsp. PROAST is a subroutine of the R package hence requires preliminary knowledge of R commands.

The same exponential family of models can be fitted in BMDS and in PROAST. However some important features differentiate these two software packages:

- BMDS uses a window driven environment and is therefore more user-friendly than PROAST which requires basic knowledge of the R-software environment.
- PROAST uses the lognormal distribution as the default while BMDS has the normal distribution as the default setting.
- BMDS does not at the moment allow for covariates to be included in the analysis while PROAST does.
- BMDS is not suitable for studies with a large number of individual data points as there is a limit in the number of rows in the data file; the software is therefore of limited use for human studies.
- The BMDS software gives the outcomes for each model of the exponential family, leaving the user to select the "best" model, while PROAST selects it automatically.

Both packages are still regularly being improved and these features may substantially change with the upcoming versions.

Other routines are available for more advanced or more specific analyses. Considering the relatively strong sensitivity of BMD estimate with respect to the modeling assumptions, it is worth mentioning routines available for BMD evaluation with model averaging, as averaging over a list of potential models can generally increase the robustness of results (45). The Model Averaging for Dichotomous Response Benchmark Dose (MADr-BMD) has been developed for this purpose by the US NIOSH. MADr-BMD fits the same quantal response models used in the BMDS software with the same estimation algorithms, but generates BMDs based on model-averaged dose–response. It is available free of charge as open-source and standalone package on http://www.jstatsoft.org/v26/i05. Note that Bayesian model averaging approach to BMD can also advantageously be applied to a wider range of models, possibly including covariates (46). However no specific user-friendly built-in routine have been developed to our knowledge, so that ad hoc Bayesian modeling would need to be implemented, e.g., in WinBUGS or similar generic Bayesian package.

Among commercial alternatives, none can really make significant differences from the freeware packages already named. Note that Toxtools is similar to the BMDS software with more practical and user-friendly outputs, and offering GEE models for repeated measurements to account, e.g., for intraindividual variability.

2.2. Exposure Assessment

The exposure to chemicals may occur by consumption of food and drinking water, inhalation of air and ingestion of soil so in theory exposure modeling should cover all these routes. As an example, in the European Union System for the Evaluation of Substances (47), the overall human exposure from all sources could be estimated by EUSES. In practice, such an exposure modeling is strongly dependent of the quality and accuracy of data available and default assumptions. For example, for the populations leaving close to a source of pollution, the exposure assessment is based on typical worst case scenario since all food items, water and air are derived from the vicinity of a point source. When dietary exposure assessment is performed in isolation, quantitative data (i.e., food consumption and chemical occurrence in food) are generally available allowing more accurate estimates than those based on scenarios. Unfortunately similar data on the distribution of chemical occurrence in air are rarely available and therefore dietary exposure assessment cannot be compared with exposure scenarios to avoid an underestimation of the contribution of exposure from food relative to other media.

As mentioned in the introduction the deterministic dietary exposure assessment represents the most common approach currently used in the area of chemical risk from food. The practices are detailed in the recent guidelines of the World Health Organization (2) and are not in the scope of this paragraph. In a number of cases the deterministic assessment fails to conclude on an exposure below the threshold of safety concern. It is therefore necessary to develop statistical methodologies to refine the exposure assessment and to bridge the gaps between the external exposure and the potential health effects.

A further refinement of the deterministic approach is therefore to use basic Monte-Carlo simulations. In such a stochastic model the amounts of food consumed, the concentrations of the hazard in food and the individual body weights of exposed populations are assumed to arise from probability distributions. For this kind of assessments, many software are commercially available, e.g.:

- @ risk software, Copyright ©2011 Palisade Corporation

  http://www.palisade.com/risk/
- crystalball

  http://www.oracle.com/us/products/applications/crystalball/crystalball-066563.html

- Creme Software Ltd. Dublin

  http://www.biotechnologyireland.com/pooled/profiles/BF_COMP/view.asp?Q=BF_COMP_45078

More interestingly, for chemicals with long half-life, the burden accumulated in the body can be completely different from the dose exposure estimated from dietary intake at a fixed time or considering a mean intake over lifetime. In that case, it is necessary to integrate in the dietary exposure assessment the TK of the chemical. For this last model described below, no software or routines currently exists.

2.2.1. Dynamic Modeling of Dietary Exposure and Risk

The dynamic exposure process, mathematically described in ref. 48, is determined by the accumulation phenomenon due to successive dietary intakes and by the elimination process in between intakes. Verger et al. (49) propose a Kinetic Dietary Exposure Model (KDEM) which describes the dynamic evolution of the dietary exposure over time. It is assumed that at each eating occasion T_n, $n \in N$ the dynamic exposure process jumps with size equal to the chemical intake U_n related to this eating occasion. Between intakes the exposure process decreases exponentially according the Eq. 3. The value of the total body burden X_{n+1} at intake time T_{n+1} is thus defined as:

$$X_{n+1} = X_n e^{-k\Delta T_{n+1}} + U_{n+1}, \quad n \in \mathrm{N} X_0 = x_0 \tag{4}$$

with $\Delta T_{n+1} = T_{n+1} - T_n, \quad n > 1$, is the time between two intakes (interintake time), U_n the intake of at time T_n, $\ln(2)/k$ is the half-life of the compound, and x_0 the initial body burden.

The exposure process X of a single individual over lifetime is not available. Indeed, chemical occurrences in food ingested by an individual and his consumption behavior over a long period of time have never been surveyed. Thus, computations of the exposure process are conducted from chemical concentration data and dietary data reported over a short period of time (described in Subheading 20.3). Starting from a population-based model, exposure process simulations over lifetime can be refined to approach an individual-based model. To perform simulations, values for model input variables are needed. The level of information that could be included in the model depends on the available data (see section3.1 gathering adequate supporting data for risk assessment). For some variables, the information is available only at population level and for others at individual level. For the latter, inter- and intraindividual variability can be included in the model. Each of these input variables is further characterized and discussed hereafter.

2.2.2. Intake Time

Depending on the available information from consumption survey and on the consumption frequency of the contaminated products, intake times can be the eating occasion, the day, the week, etc. Another way is to consider as in ref. (49) that the consumption occasions occur randomly so that interintake times are independently drawn from the same probability distribution such as a right censored exponential distribution (fitted on the observed interintake times).

2.2.3. Chemical Intake

Chemical dietary intakes U are estimated in combining data on consumed food quantities to data on chemical occurrence in food. The intake U_n of an individual at time T_n is therefore the sum of each multiplication of the chemical concentration by the consumed quantity of product p:

$$U_n = \frac{\sum_{p=1}^{P} Q_p C_{n,p}}{W_n},$$

where Q_p is the concentration in total chemical of the consumed product p at time T_n, $C_{n,p}$ is the consumed quantity of the product p at time T_n, P is the total number of products containing the chemical consumed at time T_n, and W_n is the body weight of the individual at time T_n.

Probabilistic methods to compute dietary intakes are described and compared with deterministic approaches in ref. 50. The use of the different methods depends on the nature of available data. The increased number of data permits to better account for the variability in the intake assessment. Indeed, when empirical or parametric distributions are available, single or double Monte Carlo simulations can be performed to combine chemical concentration and food consumption.

2.2.4. Half-Life

The half-life is the time required for the total body burden of a chemical to decrease by half in absence of further intake. Many studies on rats and humans have been conducted to determine half-lives of various chemicals (33, 51, 52). The half-life value depends on the toxicological properties of the chemical but also on the individual's personal characteristics. Variability between individual half-lives can be integrated in the modeling using the estimate of particular half-life for the different population groups (children, women, men). Some authors have characterized a linear relation between half-life and personal characteristics, allowing generating individual half-life. Moreover, changing individual's personal characteristics with time, intraindividual variability of half-life over the lifetime can be integrated in the model. However, level of half-life variability which can be included in dynamic exposure modeling depends on the available data. When it is possible, the impact on model output of using a population half-life or an individual half-life varying with time has to be tested and discussed.

2.2.5. Initial Body Burden

Under mathematical properties of KDEM, at steady-state situation, the dynamic of the exposure process does not depend on its initial value X_0 anymore. Considering that the aim of the exposure modeling is to estimate quantities at steady-state situation, there is therefore no need to accurately define the initial body burden. Thus, the convergence of the dynamic process to the steady-state has to be checked before evaluating steady-state quantities (cf. Subheading "Validation of Exposure Models").

When the steady state is not reached within a reasonable horizon of time compared to human life, the initial body burden is required. It corresponds to the concentration in chemical present in the body at the starting time of the exposure process. Often, simulations are done over life-time and the initial body burden required is the one of newborn children aged of few months. Chemical body burden of newborn children comes from their mother during pregnancy and breast-feeding. Given one compartmental kinetic, a half-life in the order of several years and a constant exposure during breast-feeding, one may expect approximate linear accumulation kinetics of chemicals in children body. Therefore, body burden of newborn children can be easily estimated combining data on daily breast milk consumption and data on burden of breast milk. In both cases, the impact of X_0 on the estimates of quantities of interest has to be tested based on simulations starting with different initial values.

2.2.6. Model Computation

At steady-state situation, the exposure process X_t is described by a governing steady-state distribution $\mu(\mathrm{d}x) = f(x)\mathrm{d}x$ (48). This stationary distribution is crucial to determine steady-state or long-term quantities such as the steady state mean exposure,

$$E_\mu[X] = \int x\mu(x)\mathrm{d}x = \lim_{T\to\infty} \frac{1}{T}\int X_t \mathrm{d}t \tag{5}$$

In some rare circumstances, the steady state distribution can be analytically determined. For example, with exponential distributions for both intakes and interintake times, the resulting steady state distribution is a Gamma distribution (49). In most cases, quantities of interest and time to reach steady state are rather determined through computer simulations. Computer simulations are also useful to estimate quantities at time before steady-state.

2.2.7. Simulation of Individual Exposure from Population Exposure Models

Verger et al. (49) describe a population-based approach of the KDEM applied to the occurrence of methylmercury in fish and seafood consumed by women aged over 15 years. The intakes and interintake times are considered independent and drawn from two exponential distributions fitted from the French national consumption survey INCA (53). A fixed half-life is used to define the elimination process between intakes over lifetime. A trajectory of the exposure process is computed in randomly selecting an intake and an interintake time in their respective distribution. The corresponding body burden is calculated from Eq. 4 using the previous selected values and the fixed value of the half-life. Each trajectory is therefore computed from intakes of the whole women population of INCA and represents the dynamic of the exposure process of this population. In such way, estimated quantities are the ones of a "mean individual" computed with information from the whole population.

Based on this population-based model, individual intakes can be randomly selected from the population distribution without accounting for the change of intake levels with the increasing age. However, since food consumption behaviors, personal characteristic, and chemical occurrence vary with age, simulations over lifetime must integrate the intake and half-life variability with time. Simulating individual trajectories over time faces a lack of individual data. One solution can be to compute simulations combining children and adults intakes, according to the following procedure. For each intake time T_n over lifetime, an intake value can be randomly selected from the intake distribution given the age of the simulated individual at T_n. The time window can be defined considering a sufficient number of observed intakes to compute an exposure distribution. For example, if the time window is of 3 years, the exposure distribution changes every 3 years with the increasing intake time T_n. Intake time can be fixed to a day or a week, or else randomly selected in the interintake distribution of the corresponding time window.

2.3. Risk Characterization: The Case of Dynamic Modeling of Exposure from Food

Risk characterization is usually the step of risk assessment which combines outputs from risk characterization and exposure into one final assessment.

In the case of the dynamic exposure model described above, some more specific risk characterization can be undertaken. As a matter of fact, an interesting risk measure is the probability of the exposure process exceeds a threshold d_{ref} given by:

$$P_\mu\{X > d_{\mathrm{ref}}\} = \mu([d_{\mathrm{ref}}, \infty]) = \lim_{T\to\infty} \frac{1}{T} \int \mathrm{I}_{\{X(t) \geq d_{\mathrm{ref}}\}} \mathrm{d}t.$$

Another interesting quantity is the amount of chemical over the d_{ref}, calculated as:

$$E_\mu[X - d_{\mathrm{ref}} | X > d_{\mathrm{ref}}] = \mu([d_{\mathrm{ref}}, \infty])^{-1} \int (x - d_{\mathrm{ref}})\mu(\mathrm{d}x).$$

The threshold d_{ref}, can be the extension in a dynamic context of the tolerable daily intake (TI) of a chemical to which consumers could be exposed all along their life without health risk. The tolerable intake can be the health based guidance values defined by international instances (JECFA, EFSA) such as acceptable daily intake (ADI), the provisional tolerable weekly intake (PTWI) of the chemical of interest. According to the definition of the tolerable intake, a Kinetic Tolerable Intake (KTI) can be built (49). The KTI is constructed considering that the tolerable intake is the dose of chemical ingested per time and eliminated between two intake times with the kinetic model (3). Above that condition, the dynamic evolution of the process is

$$x_{n+1} = x_n e^{-k\Delta T_{n+1}} + \mathrm{TI}.$$

At the steady-state, the reference process stabilizes at a safe level, obtained from the TI and the half-life:

$$\lim_{n\to\infty} x_n = \frac{\mathrm{TI}}{1 - e^{-k}}.$$

Note that when k is close to 0, the function $1 - e^{-k}$ can be approximated to k with the Taylor series and thus the previous equation equals to Eq. (2).

Another value to define threshold d_{ref} can be the body burden of chemical (endpoints) corresponding to possible health adverse effects. According to that definition the probability of the exposure process to be above d_{ref} corresponds to an incidence of disease. Endpoints are derived from dose–response relationship provided by laboratory animal experiments. They can be the lowest chemical dose which can induce an observed effect of disease (LOAEL) or the dose associated with a level of this effect (benchmark dose BMD). For example, a BMD related to effects on reproductive toxicity can be the percentage of decrease in sperm production. To extrapolate human dose–response relationship from animal dose–response, interspecies (variability between man and rat) and intraspecies (variability between humans) factors have to be used. Each factor can be defined by a single value or by a probabilistic distribution.

3. Methods: Practical Implementation and Best Practices for Quantitative Modeling of Population Variability

3.1. Gathering Adequate Supporting Data for Risk Assessment

Various methodologies have been developed to gather available data for the quantification of population variability with respect to hazard identification and characterization [toxicokinetics, toxicodynamics (dose–response)], occurrence and consumption (exposure assessment) for a particular chemical.

3.1.1. Gathering Evidence for Chemical Hazard Characterization in Humans

Prior to any hazard characterization (dose-dependent TK or dose--response modeling), the collection of relevant information is obviously a key step which essentially drives the approach to undertake for the data modeling. In particular, it will condition the way population variability will or can be accounted for. Data collection, either prospectively through ad hoc studies or retrospectively based on literature reviews or SR (54), is described hereafter, with a focus on how such population variability can be best synthesized.

For hazards with long term effects, the exposure should be compared to the health based guidance values (e.g., provisional tolerable weekly intake, see above) or with the benchmark dose level at which these effects are likely to occur. In practice the comparisons are commonly done without considering the kinetic of the substance and assuming its complete elimination within the period of reference of the health based guidance value (e.g., 1 week for a tolerable weekly intake). For chemicals with a long half-life this assumption is not valid and the temporal evolution of the [weekly] exposure should be compared dynamically with the [weekly] consumption of the tolerable amount of the substance which can then be called the kinetic tolerable intake or KTI (49). In that case the model for dietary exposure should combine the distributions of consumption and occurrence with the distribution of time intervals between exposure occasions so that interindividual variability in half-lives can be taken into account and allow to move from a population-based to an individual-based model.

Design and Analysis of Randomized Studies

Most of human and animal randomized studies can be used for toxicological analyses. By nature, randomized studies are designed so that the analysis of primary outcomes is simply restricted to group comparisons. Such randomized studies are typically designed to assess a given effect in relation with chemical exposure. In the context of risk assessment, these designs are therefore more suitable for hazard identification rather than hazard characterization. The simplest design would then be a two-arm study comparing exposed rats versus nonexposed rats. This assumes that the exposure level is predefined. Responses to exposure are then compared between the two groups using typical statistical testing such as Fischer tests for quantal responses. In cases where a range of doses needs to be tested, multiple group comparisons can be analyzed with ANOVA-type of analyses.

By definition, the randomization structure is deemed to balance all factors that could affect primary outcomes, so that no adjustments for covariates are needed in the final group comparison analysis, which can be performed by any standard statistical package. As a consequence, the main statistical challenge lies in the randomization structure which should account for the variability structure and clustering of the data to be collected.

Design and Analysis of Observational Studies

The variability structure of data is more critical in the analysis of observational human data, as influential factors have generally not been balanced by the study design. The list of potential covariates affecting human data can be vast (e.g., age, gender, body weight, etc.). Confounding factors often interfere with the outcome variable of interest. For example, in the case of dose–response assessment, socioeconomic factors may confound with exposure when high exposure is correlated with poorer living conditions, which may also increase the risk for the ill health. This difficulty is commonly (partly) addressed by including the confounding factors as covariates in a multiple regression model.

An extreme particular case of design unbalance is the absence of comparators, which includes the case of absence of control. For example, in the case of exposure assessment, it can be that zero exposure data are not available. Therefore the response at zero exposure is, in fact, estimated by low-dose extrapolation based on a dose–response model, resulting in uncertain (model-dependent) estimates of the response in an unexposed population, and hence the outcome of the analysis is likely to be strongly model-dependent. Such weaknesses of the analysis need to be acknowledged and possibly quantified in the discussion of the results (e.g., using sensitivity analyses).

Although covariates may explain a part of interindividual variations, the variability structure present in human or environmental data is often more complex at the large scale on which public health or risk assessment questions are raised (including, e.g., variations between population ethnic subgroups or temporal and spatial variations). Moreover, some clusters often underlie such data, such as the country, the slaughterhouse, or the field from where those data come. Some of this complexity may or may not be captured and handled by appropriate study design. Where the data are available, hierarchical models can often help to account for such complexity which is an important aspect to be evaluated in risk assessment.

A common particular case of variability pattern is the time and space variations of the data collected which often requires a particular care in food and feed safety questions. Cross-sectional studies can be problematic regarding time variations such as seasonal or periodic effects or time trends. For example, in the case of hazard characterization with long-term effects, exposure at the time of the study may not reflect the long-term exposure. Similarly, cohort studies may not render the spatial variation. As a result, a careful data selection is necessary prior to any analysis. If data are not expected to give unbiased picture, conservative choices should be made (e.g., focusing on vulnerable subgroups). Epidemiological studies are often conducted on a larger scale than experimental studies, hence with much larger sample sizes allowing for data exclusion when necessary to improve relevance and quality of data. The level of precision and accuracy expected from the toxicological assessment should be commensurate to the size of the time and space scale of relevance.

Missing or censored data are classical statistical issues. They can be of large importance in meta-analysis especially for nonrandomized studies and for national surveys where the proportion of missing or censored data can be above 50% (e.g., up to more than 80% in chemical exposure studies) and where the missingness is usually far from being at random. Statistical approaches have been developed to handle such missing data (e.g., multiple imputation or mixed effect models) and censored data (e.g., using adequate maximum likelihood approaches, or Kaplan Meier estimators). However, it is often useful for risk managers to compare results using more naïve imputations based on worst and best cases scenarios.

Gathering Evidence from Multiple Published Studies

The systematic review (SR) methodology combined with a meta-analysis can be applied in the context of chemical risk assessment, when a number of studies for that particular chemical are available. A thorough account of the SR methodology is beyond the aim of this chapter and its potential applicability in risk assessment applied to food and feed safety has been explored by EFSA in a guidance document. However, a few basic concepts are worth mentioning. SR has been developed by the cochrane library collaboration and applied in human medicine, epidemiology and ecology for a number of evidence-based syntheses (cost-effectiveness of treatments, reporting of side effects, relative risk of disease, meta-analysis of biodiversity or abundance data in ecology). SR has been defined as "an overview of existing evidence pertinent to a clearly formulated question, which uses prespecified and standardized methods to identify and critically appraise relevant research, and to collect, report, and analyze data from the studies that are included in the review" (54).

When applied to risk assessment, the starting point of an SR is to identify the question type and how to frame the question, using the Cochrane collaboration methodology, four key elements frame the question, namely, population, exposure, comparator, and outcome. Such a question can be close-framed or open-framed. Typically, SR can be appropriate for a close-framed question since primary research study design can be envisaged to answer the question.

Dose-dependent toxicokinetic (i.e., half-life, clearance) or dose–response in humans or in a test species (rat, mouse, dog, rabbit..) can be taken as a generic example, the population can be the human population or in the absence of relevant human data, the population of interest would become the test species (rat, mouse...), the exposure is the chemical, the comparator would correspond to different dose levels and the outcome can be either a toxicokinetic parameter (half-life, clearance, C_{max}...) or toxicity in a specific target organ (liver, lung, kidney, heart, bladder...). Ideally, once the primary studies in humans or test species referring to the dose-dependent toxicokinetics or dose–response for a specific chemical have been collected, a meta-analysis can be performed so that modeling of the population variability is possible (54).

Whether a question is suitable for systematic review or not does not necessarily mean that a systematic review would be worthwhile or practically feasible. Considerations include the following: prioritization of risk assessment model parameters for which refinement of the parameter estimates is considered most critical; the quantity and quality of available evidence; the source and potential confidentiality of the evidence; the need for transparency and/or for integrating conflicting results (54). General aspects of the methodology of meta-analysis are given below and specific examples with regards to analysis of (1) toxicokinetic variability for a number of metabolic routes in humans and (2) human toxicodynamics of cadmium are described in Chapter 4.

Meta-analysis is a statistical procedure to review and summarize data from several past studies. By combining information from relevant studies, meta-analyses may provide more precise estimates than those derived from the individual studies included within a review. Typically, such compilation of literature data is used in risk assessment for hazard characterization in most cases, but also sometimes to identify unknown or hard-to-study hazards such as for Genetically Modified Organisms (55). Meta-analyses also facilitate investigations of the consistency of evidence across studies and the exploration of differences across studies.

Generally speaking, a proper meta-analysis should allow to:

- Adjust for all identified sources of bias, such as selection bias or design bias.
- Account for all significant and measurable sources of variability.
- Weight each study according to the evidence it provides: weighting of evidence is a classical issue of meta-analysis, and one of its primary objective. It includes weighting between study designs and a weighting between studies of the same type, e.g., based on their sample size, e.g., according to their sample size or to the precision of estimates each of them provide.

More specifically, a number of issues are worth being listed when carrying out such analyses. Each of them may require a dedicated statistical method to be handled in the analysis.

- The data collected often result from aggregation or averaging of different individual values (e.g., regional averages or age groups averages). Depending on the level of heterogeneity and differences within the aggregated data, this aggregation can translate into a simple precision problem or into a severe fallacy issue in the results interpretation (e.g., for ecological designs) if not accounted for by the statistical models (56).
- Heterogeneity between studies is also a typical issue to be addressed in meta-analysis. Study heterogeneity can arise from

many sources such as the use of different study population, different environmental conditions or different analytical techniques employed in different studies. In such cases specific statistical tools are needed to address the related specific issues, such as the variation across studies or the publication bias (57, 58). These tools include, e.g., the use of random effect models to account for interstudy variability. Statistical tests for heterogeneity have commonly been used especially in the case of clinical trials (59), but they should be utilized and interpreted very cautiously in most other types of study as they are often too weak to detect heterogeneity in large scale food safety problems where variability sources are often large, numerous and complex and in which data quality is often poor.

Although the chi-square test for homogeneity is often used in meta-analyses (60), its power is known to be very low (61) especially with low number of studies. It is therefore a good statistical practice to investigate the post-hoc power of such a test to detect a difference of the size of the claimed effect size. Moreover, the Higgins test usually complements the assessment of homogeneity, testing for inconsistency between studies (59). In general, the Higgins test is therefore preferable as it is less sensitive to the number of studies included.

3.1.2. Gathering Evidence for Exposure Assessment

To perform the probabilistic assessment of human dietary exposure, the basic available data are the distribution of food consumed and the distribution of hazard occurrence in food. Most of the data available are collected at national level and are assumed to be representative for the country. For an international perspective the various data sets should be combined together. This chapter emphasizes on this step and aim to describe the data preparation.

Food Consumption Data

Food consumption data are collected at national level based on various survey methodologies like food record or recall (62). Moreover, the year of conduct, the population groups and the age categories differ greatly between countries, therefore, it is not possible to use directly these data for an international probabilistic assessment (63). The approach used currently consists in comparing national distributions and using the worst case, i.e., the highest consumption for consumers only observed in one country, for the comparison with health based guidance value. This means in practice that the percentage of risk or the percentage of individuals exposed above the health based guidance values is extrapolated from the national to the international population level and is likely to be considerably higher than in the reality depending of the ratio between the considered populations. For example: if the risk is estimated for the 95th percentile of exposure in Netherlands (16.6 million inhabitants), it represents about 828,000 people.

When extrapolated to the European Union (501 million inhabitants) it represents 25 million people. Assuming that the population at risk within the European Union is only 5% of the Dutch population would correspond to a risk assessment at about the 99.8th percentile. This represents a large source of uncertainty in the context of establishing an appropriate level of protection.

Possible improvements could be envisaged in the absence of harmonized data collection. At first national food consumption data based on individuals can be combined together using a common food classification. A single distribution for food consumption of each food category could be simulated based on national data weighted as a function of the country population. In that case for each individual, a single day should be used to increase the comparability between survey results (62). However, a number of assumptions should be made:

- All surveys were performed at the same time.
- Recall and record provide similar results.
- The individuals in the survey fully represent the whole population of the country.
- The consumption of each food category is assumed to be independent of the ones of other food groups.

Another approach should consist in identifying similar dietary patterns across countries or regions. This would have the advantage to allow estimating the exposure based on various consumer behaviors rather than on the citizenship to one or another country. Moreover this approach would account for the dependency between various food groups.

The clustering techniques are widely applied to reach this goal and to identify the similarities and differences in dietary patterns between countries and regions. As an example the World Health Organization (WHO) developed this approach to describe the various diets around the world and resulted in 13 so-called cluster diets (64). In that case the clustering was based on the economical data collected by the Food and Agricultural Organization and known as the FAO Food Balance Sheets. A more recent study (65) is applying nonnegative matrix factorization (NMF) techniques to food consumption data in order to understand these combinations. This paper proposes a different modeling of population consumption that can be directly applied to the consumed food quantities. Even though a very large number of different foods are involved in individual consumption patterns, all possible food combinations are not observed in practice. Certain foods are preferentially combined or substituted as a function of hedonic choices and/or sociocultural habits. One may then realistically expect that the vast majority of consumption data can be described by a few patterns, which are linear combinations of consumption vectors of specific foods. These underlying factors can be interpreted as latent variables. Therefore, according to this modeling, an individual diet must be seen as a linear superposition of several consumption systems. When identified, the consumption systems could be used for risk assessment and their optimal combination should be identified for risk/benefit analysis.

Occurrence Data

The main difference between occurrence and consumption data is that food market is generally assumed to be global and therefore, the country in which food is sampled is not considered to be the only or even the main source of variability.

Because of the inherent variability between samples regarding the occurrence of hazards in food, the uncertainty in the results is likely to increase when the number of sample decreases. Therefore in such cases, exposure assessment should be questioned and when necessary a deterministic estimate of the range of exposure should be performed (66). The current chapter is assuming a dataset with a sufficient number of samples (e.g., > 50 samples) for each food category to be considered, allowing a probabilistic approach with a reasonable level confidence.

After combining all analytical results together using a common food classification, the whole dataset should be described (e.g., with histograms/density plot/cumulative distributions) and analyzed with a particular focus on its potential sources of heterogeneity (Country of origin, year of sampling, laboratory characteristics, analytical techniques, etc.).

One of the main issues is therefore very often to handle concentration data described as being below the limit of reporting (analytical limit of detection or limit of quantification). These data are often known as nondetects, and the resulting occurrence distribution is left-censored. As a first step the impact of left censored data could be addressed imputing nondetects with values equal to zero and to the limit of reporting according respectively to a lower- and upper-bound scenario. The effects of the substitution method could be evaluated on the mean, and/or the high percentiles. If the effect is negligible then exposure assessment should be based on substitution of nondetects by the limit of reporting (upper bound approach). This approach has the disadvantage of hiring the variability between samples and overestimating the exposure but can be used in that particular case because of its low impact on the overall result.

In other cases, depending on the percentage of censored data, parametric or nonparametric modeling could be used. When the percentage of censoring is high (e.g., >50%), it has been observed in the literature (66, 67) than a parametric approach is performing better than the nonparametric ones. A set of candidate parametric models should be defined, by inspecting the density plots. In recent guidelines, EFSA (66) proposes to select the best parametric model on the basis of a goodness of fit statistics (AIC or BIC) and to implement the lack-of-fit test (Hollander–Proschan test) of the selected model. When the percentage censoring is lower and when datasets are very heterogeneous with multiple limits of reporting, the (nonparametric) Kaplan–Meier method can also be used (50). This method spreads the weight of censored data, those below limits of reporting, over all lower uncensored data and zero.

The number of chemical measurements in a specific product is generally too small to estimate high percentiles of the occurrence distribution. To improve estimation of such percentiles, the upper tails of the chemical occurrence distribution can be modeled using extreme value theory (67, 68). This method could also be directly applied on the exposure distribution (69).

3.2. Specific Good Modeling Practices

3.2.1. Population Model Building

The purpose of this section is not to detail general "best modeling practices" that would apply to any statistical modeling. Instead, we propose to emphasize the good practices that are specific to population models applied to chemical risk assessment. Note that a thorough dose–response model-building guidance document prepared by the WHO International Program for Chemical at Safety is available in ref. 70 for the purpose of risk assessment.

Generally speaking the key element in structuring a population model is the definition of its stochastic component and how to balance it with the deterministic one. This balance necessarily depends on the objective of the modeling exercise. In case individual predictions are of primary interest (e.g., when using the model to define a maximum tolerable dose of a compound for a specific individual), the deterministic structure should be developed to its largest extent allowed by the data available. Indeed, such deterministic model could then better account for individual specificities and covariates. Conversely, when the population predictions are of primary interest (e.g., when using the model to assess the response of 95th population percentile), the variability component becomes essential and the stochastic structure should be developed to best capture such population variability.

This choice between population one-compartment model (with population variability of TK parameters) versus larger PBTK models (without population variability) with more compartments to refine the description of metabolism is a typical illustration of such balance between deterministic and stochastic structures as exemplified in the cadmium example below.

The way to construct the population or hierarchical structure of a stochastic model relies on the identification of the main patterns found in the toxicological data to be used and of importance in the risk assessment to be made. Those patterns can be captured or not depending on the design of the data collection. A common way to grasp and describe this variability structure is made by considering the hierarchy of the different levels or scales at which data are

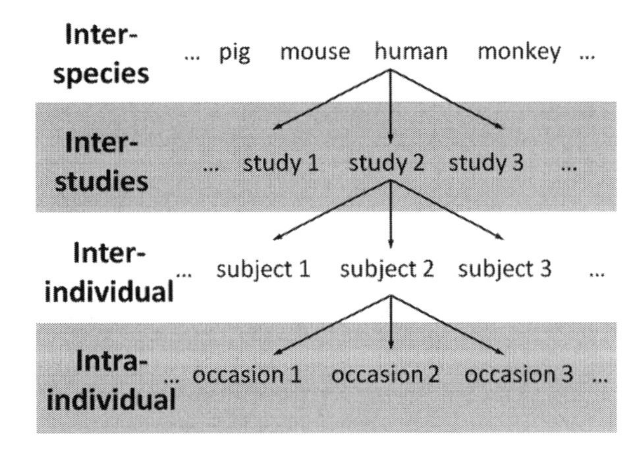

Fig. 4. Hierarchical levels commonly found in toxicological data, from interspecies variability down to intraindividual variations. Such patterns should drive the variability compound of statistical models used for the data analysis.

collected, starting from the wider scale (ecological or animal data) down to the individual scale (data from one individual) as exemplified by Fig. 4. Each level would then be described by a specific random effect in a regression model. This description is often meaningful from both statistical and biological standpoints. Accounting for interspecies variability applies for large scale assessment (e.g., ecological evaluation) involving a large number of species. Interstudy variability is accounted for in a context of meta-analysis as it is often the case in risk assessment (see Subheading "Gathering Evidence from Multiple Published Studies"). Interindividual variability is the most common source of variability captured by toxicological data and of primary importance. It can sometimes be reduced or "explained" by covariates inclusion. Finally, intraindividual variations such as interoccasion variability or seasonal effects can be assessed in case of longitudinal or repeated measurements over time.

The critical modeling assumption to be set at each hierarchical level is obviously the population distribution chosen for each random effect. Biological assessments have shown to be sensitive with respect to this choice and could therefore be a weak point in a quantitative toxicological or risk assessment. The standard choice of lognormal distribution is usually applied when the main outcome modeled is a chemical concentration. It is a good practice to plot the empirical distributions of such outcomes to visualize and support the modeling assumptions. Furthermore, it is essential to check that residual variances are not correlated with the random effects. Otherwise, it is likely that the distribution chosen is not adequate.

A way to include more determinism in a model is to refine the underlying structural model by describing in more details the biological mechanisms involved. In PBPK models, this translates in adding more compartments or in sophisticating the differential equations describing the metabolism (e.g., by adding saturation effects or elimination lag times).

Another way to add more determinism in a population model is to include more predictors or covariates. These can explain the population variability, and therefore reduce the residual variability component of the model. Covariate selection is therefore a critical step in population modeling. When using frequentist inference and the standard paradigm of statistical testing, there are two common and widely used approaches to select covariates on statistical grounds:

1. For nested models, which is typically the case for nonlinear mixed effects regression models used in TK or TD assessments, likelihood ratio tests are particularly suited to compare models with versus without a set of covariates. Their comparisons is then based on the difference of their respective objective function ($-2 \times$ Log-Likelihood) which asymptotically follows a chi-square distribution with degree of freedom equals to the difference in number of model parameters. NONMEM software reports as a default output such objective functions for each fitted model to allow the users for model comparisons.
2. For nonnested models, information criteria such as AIC or BIC can be used and seen as an extension of log-likelihood ratio tests. However, it should be noticed that degrees of freedom of fixed effects in the context of mixed-effect models are often debatable (71). Deviance Information Criteria (DIC) used in a Bayesian framework are therefore often more robust for hierarchical models to be compared (72).

In the area of population PK/TK, the forward/backward selection is often used when the number of potential covariates is so large that all possible pair-wise model comparisons are not feasible (e.g., when the number of potential covariates is above 5). This empirical and pragmatic way of selecting a set of covariates does not ensure theoretically full optimality of the final model but it generally performs well and provides close-to-optimal solutions. In this context, prior to any investigation, a base model (with no covariate) is first fitted to the data. Subsequently, covariates are screened using a two-step forward–backward approach in which covariates arte added into or removed from the base model based on likelihood ratio tests with a predetermined significance level. More specifically:

- The forward (selection) step consists in testing, for each single factor independently, the effect on the main outcome. At the end of this one-by-one selection process, a full model can be built, including all selected factors.

- The backward (elimination) step consists in testing one-by-one whether each selected factor could be removed or not from the full model. At the end of this elimination process, a final model can be built, which integrated all remaining significant factors.

One obvious theoretical flaw in this approach is the multiple testing aspect: the final model building is based on the outcome of a series of tests at a given significance level. For this reason, it usually advised to either adjust for multiplicity (which can be tricky), or to set a low level of significance for each test e.g., 1% instead of the too conventional 5%.

Aside all statistical considerations, model structure can always be motivated and justified by biological rationales based e.g., on mechanism plausibility or on the relevance of the predictions derived from the model in the dose range of interest.

3.2.2. Model Validation

Validation of Population TK/TD Models

Population models are typically evaluated with respect to how the model simulations can reflect past and/or future observations, or more specifically a function of observations of highest interest for the assessment such as the 95th population percentile of a concentration.

The prediction error is often evaluated on the observations compared either to individual or to population predictions. Population predictions are obtained after setting all random effects to zero, i.e., the first-order approximation PRED in NONMEM. Individual predictions are obtained after setting random effects to their subject-specific values as estimated by the model. These are typically evaluated using post hoc Bayesian estimates as with the IPRED command of NONMEM. Bayesian individual predictions are straightforward outputs from any Bayesian software like WinBUGS or MCSim.

Another measure of the predictive performance can be derived from evaluating the likelihood of the data given the model estimated with fixed population parameters.

Model validation can be done using an external dataset or alternatively using a cross-validation approach. In the latter case, the dataset is split into two subsets: one for the fitting (the larger) and one for the validation (the smaller). The split should be done randomly, and could be repeated.

Critical validation steps of population models include the graphical checks of residual errors variances and of random effect distributions. Visualize checks can be done using boxplots and should be centralized around zero without systematic bias or patterns.

In the context of chemical risk assessment, sensitive assumptions in population modeling are typically:

- The choice of random effects' distributions
- The choice of dose–response models

Models can be compared using Akaike or Bayes information criteria (AIC, BIC). The deviance information criterion (DIC) can be considered as a generalization of AIC for hierarchical models. The advantage of DIC over other criteria in the case of Bayesian model selection is that the DIC is easily calculated from the samples generated by a Markov chain Monte Carlo simulation. AIC and BIC require calculating the maximum likelihood, which is not readily available from the MCMC simulations. WinBUGS provides DIC calculations in its standard tool menu.

Validation of Exposure Models

The KDEM model predicts for each simulated trajectory the mean body burden of chemical at steady-state. The predictions from dietary exposure can be compared with internal exposure (i.e., biomarkers) to validate the model. Measurements on internal exposures to chemical are usually sampled in urine, hair, blood, or breast milk. The chemical body burden predicted by KDEM has to be converted in same unit of the measurements with conversion factors. For example, predicted body burden have to be converted in concentration in body fat to compare with measurements in breast milk. The conversion factors such as percentage of fat depend of personal characteristics of the individuals and its variability could also be included.

Often, internal measurements are not available for the population for which the dynamic modeling of exposure process has been computed, especially when the population of interest is the whole population of a country. In that case, validation of the exposure process is done at population level in comparing both the distributions of predicted body burden and internal measurements. When studies coupling biological measurements with frequency consumption questionnaire have been carried out, the interest is to link the body burden predicted from the questionnaire with the associated internal measurement for each individual. Due to high uncertainty on past contamination and consumption, for an individual the point estimate with one trajectory can be far from its internal measurement value. In such a case, for an individual, several trajectories can be simulated under different scenarios regarding contamination and consumption using probabilistic distributions. A confidence interval for the predicted exposure can then be constructed and compared with the internal exposure value. Sometimes, measurements have been performed on a specific population that has certainly not reached the steady-state situation, an example is pregnant women. Using a well-defined initial body burden X_0 and computing a sensitivity analysis to this value, estimation of the body burden can however be conducted based on the external exposure and compared with internal measurements.

3.2.3. Numerical Considerations

Convergence of Fitting Algorithms of Population Models

Aside specific model validation steps, the fitting of nonlinear mixed-effect models involves implementation computational methods such as stochastic or iterative algorithms. As a consequence, outputs from parameter estimation rely on valid convergence of such algorithms. In maximum likelihood estimation, fitting algorithms implemented in e.g., NONMEM are based on first order approximations of likelihood. Those approximations may not be valid in case of highly skewed likelihoods and nonnormal random effects, or they may be highly sensitive to initial values. As a consequence, a good practice is to test a range of different approximation methods and initial values. To gain in flexibility and reliability, stochastic approximation of EM algorithms (SAEM) have been developed and used in PBTK modeling with the MONOLIX software. Although SAEM is more powerful gradient algorithms for nonlinear population models, only visual convergence check is available in MONOLIX. Finally, for Bayesian inference, convergence of MCMC needs also to be checked using as for any Bayesian modeling exercise. WinBUGS provides ready-to-use tools for convergence check such as the Gelman-Rubin statistics.

Convergence of KDEM Simulations

Time to reach steady-state situation can be determined through simulations. For example, a set of 1,000 trajectories is computed for several initial values X_0. For each trajectory of the different set the mean exposure over increasing periods of time $[0,T]$ is calculated using the eq. 5. For each set, the mean of the 1,000 means is calculated and plotted (49). After a period of time which corresponds to the time to reach the steady-state situation, the trajectories of the means converge to the steady-state mean exposure value. This value corresponds to the mean body burden of the population at steady-state situation. In the KDEM application to occurrence of methylmercury (49), the steady-state distribution is simply a Gamma distribution with parameters depending of the two exponential distributions used to model intakes and intertime intakes. Therefore the steady-state mean exposure value is the one of the Gamma distribution.

4. Examples

Three examples are described below to explore population modeling in computational toxicology: (1) Systematic review and meta-analysis of human variability data in toxicokinetics for the metabolic routes and derivation of pathway-related uncertainty factors using the pharmaceutical database, (2) population modeling of toxicokinetics and toxicodynamics of cadmium in humans using Bayesian inference for aggregated data, and (3) exposure assessment in human populations to dioxins using probabilistic estimates of exposure and toxicokinetic parameters.

4.1. Example 1: Systematic Review and Meta-analysis of Human Variability Data in Toxicokinetics for Major Metabolic Routes and Derivation of Pathway-Related Uncertainty Factors

Systematic review and meta-analysis of TK variability for the major human metabolic routes, (1) phase I metabolism (CYP1A2, CYP2A6, CYP2C9, CYP2C19, CYP2D6, CYP2E1, CYP3A4, hydrolysis, alcohol dehydrogenase), (2) phase II metabolism (*N*-acetyltransferases, glucuronidation, glycine conjugation, sulfation) and (3) renal excretion; *have been performed* following a number of methodological steps using the pharmaceutical database. The purpose was to derive *pathway-related uncertainty factors using human variability data in TK as an intermediate between default uncertainty factors and chemical-specific adjustement factors* (i.e., *based on PB-TK models*) *for chemical risk assessment* (for more details on uncertainty factors see Subheading 20.1) (4, 6, 19, 73–75).

– Selection of probe substrates and systematic review of the TK literature:
  Probe substrates were identified by searching the literature (MEDLINE, PUBMED and TOXLINE depending on the pathway). The selection of probe substrates followed a number of specific criteria: (1) oral absorption complete, (2) a single pathway (phase I, phase II metabolism and renal excretion) responsible for the elimination of the compound (60–100% of the dose), (3) intravenous data used for compounds for which absorption was variable. The specific metabolic pathway was identified using quantitative biotransformation data from in vitro (microsomes, cell lines and primary cell cultures) and in vivo (urinary and fecal excretion) studies.
– Systematic review and meta-analysis of TK data.

Selection of TK studies and data ranking for population variability analysis

For each metabolic route, a systematic review for TK studies in humans was performed for each probe substrate selected for each pathway for each subgroup of the population: general healthy adults (16-70 years from different ethnic backgrounds (Caucasian, Asian, African) and genetic polymorphisms (CYP2C9, CYP2D6, CYP2C19, NAT-2) and other subgroups of the population [elderly: healthy adults older than 70 years, children (>1 year to <16 years), infants (>1 month to <1 year) and neonates (<1 month)] African, effects of genetic polymorphisms, effects of age (elderly, children, infants, neonates), effects of ethnicity (Caucasian, Asian, African) using two different types of TK parameters: (1) TK parameters reflecting chronic exposure [metabolic and total clearances, area under the plasma concentration–time curve (AUC)]. (2) TK parameters reflecting acute exposure (C_{max}) markers of chronic (clearances and AUCs) and acute (C_{max}) exposure.

A number of meta-analysis were then performed in order to quantify human variability in TK for each compound, marker (acute, chronic), subgroup of the population and metabolic route in humans including phase I metabolism (CYP1A2, CYP2A6, CYP2C9, CYP2C19, CYP2D6, CYP2E1, CYP3A4, hydrolysis, alcohol dehydrogenase), phase II metabolism (*N*-acetyltransferases, glucuronidation, glycine conjugation, sulfation), and renal excretion.

Variability in TK for each probe substrate/subgroup/parameter was estimated using ranking of the parameters, as described previously in ref. 19:

1. Data were selected from peer-reviewed studies using specific and sensitive analytical methods (i.e., HPLC).
2. Data for the oral route were preferred to intravenous data for the purpose of application in chemical risk assessment since humans are exposed to environmental contaminants and nutrients mainly in food and drinking water.
3. TK parameters were abstracted preferentially as metabolic or total plasma clearance (CL) adjusted to body weight (ml/min/kg), then unadjusted metabolic or total plasma clearance (ml/min), the area under the plasma concentration–time curve (AUC).
4. AUC and C_{max} values were corrected for body weight (mg/kg) using the published mean adult body weight, or 70 kg (males), 60 kg (females), 65 kg (mixed males and females) for adults when the weights were not reported.
5. Data were analyzed such that no individual would contribute more than once to each analysis.
6. TK data were assumed to be linear at dose studied and to follow a log-normal distribution. However, data from individual kinetic studies were usually reported as arithmetic means (X) and standard deviations (SD) or coefficient of variation assuming a normal distribution (CV_N), and were transformed into geometric mean (GM), geometric standard deviation (GSD) and the corresponding coefficient of variation (CV_{LN}) (73, 74) using:

$$\mathrm{GM} = \frac{X}{\sqrt{(1 + \mathrm{CV}_N^2)}} \quad (6)$$

$$\mathrm{GSD} = \exp\left\{\sqrt{\ln(1 + \mathrm{CV}_N^2)}\right\}. \quad (7)$$

The coefficient of variation for the normal distribution (CV_N) is given by

$$\mathrm{CV}_N = \frac{\mathrm{SD}}{X}. \quad (8)$$

The coefficient of variation for the log-normal distribution (CV_{LN}) is given its relationship with the appropriate measure of variation (geometric standard deviation (GSD))

$$CV_{LN} = \sqrt{\exp\{[\ln(GSD)]^2\} - 1}. \quad (9)$$

7. CV are assumed to represent interindividual differences in the TK of compound and its metabolic pathway (full oral absorption and >60% metabolized by this route) either after chronic (clearance, AUC) or acute exposure (C_{max}) rather than measurement errors or random analytical errors since TK parameters are derived from multiple measurements.

Meta-analyses of TK studies
As described above, meta-analyses of the TK data were performed first o estimate population variability at the level of the compound for each subgroup of the population, route of exposure (oral, intravenous) and TK parameter (AUCs/clearances or Cmax) using a weighted mean approach (73). Each individual TK study reporting data for the same kinetic parameter, compound and subgroup of the population were then combined using the weighted mean method described previously using the number of subjects in each study as the weight (73, 74, 76, 77). The overall coefficients of variation (CV_N and CV_{LN}) for each parameter/subgroup of the population were then combined for all probe substrates for a particular pathway as an average on the log-scale to derive the pathway-related variability for each metabolic pathway.

Published data related to subgroups of the population were also analyzed to quantify differences in internal dose (for both means and variability) compared to healthy adults:

The difference in mean level of internal dose (based on clearance, AUC or C_{max}) was expressed as the ratio of the GM for the healthy to the GM for the subgroup. The GM ratio was expressed as the magnitude of any increase in the internal dose in the subpopulation compared to healthy adults (i.e., ratio of 2 would arise from a twofold lower clearance or twofold higher AUC or C_{max} in the subgroup).

The difference in variability was expressed as the ratio CV_{LN} for healthy adults to CV_{LN} for subgroup and expressed as the magnitude of any increase in the variability of the subgroup compared to healthy adults (a ratio of 2 would indicate a twofold greater variability in the subgroup).

These differences in internal dose and variability for subgroups of the population were also averaged on the log-scale to define pathway-related differences in internal dose and variability.

As an example of subgroup analysis, for polymorphic metabolic routes such as CYP2D6, CYP2C19 and *N*-acetyltransferases 2 (NAT-2), the interphenotypic differences in internal dose were determined in the different subgroups of the population under two different scenarios depending on data availability (74):

1. TK data available only in nonphenotyped subgroups of the population: TK data for different probe compounds of CYP2D6, CYP2C19, and NAT-2 in nonphenotyped were compared with the corresponding data for nonphenotyped healthy adults.
2. TK data available in phenotyped subgroups of the population [extensive metbolizers (EM), SEM (slow extensive metabolizers) and poor metabolizers (PM) (CYP2D6, CYP2C19), fast acetylators (FA) and slow acetylators(Nas) (NAT-2)] were compared with the corresponding data for EM (CYP2D6, CYP2C19) or FA healthy adults(NAT-2) respectively in each subgroup as an average on the log-scale between the compounds.

4.1.1. Derivation of Pathway-Related Uncertainty Factors

Pathway-related uncertainty factors for markers of chronic and acute exposure were derived as the *Z* scores of the pathway-related variability to cover the 95th, 97.5th and 99th centiles of the healthy adult population and subgroups of the population (including pathway-related ratio of internal dose and pathway-related variability in the specific subgroup). The pathway-related uncertainty factors were calculated as the difference between each percentile (95th, 97.5th and 99th centiles) of the subgroup compared with healthy adults, without taking into account the incidence of the subgroup in the overall population. These values assume that higher circulating concentrations of the parent compound would result in increased risk, i.e., that the parent compound is the toxic chemical species (6, 19, 73–77).

4.2. Example 2: Population TK/TD of Cadmium Using Aggregated Data

This example is extracted from a recent human risk assessment of Cadmium in food by the European Food Safety Authority (18).

Cadmium is a widespread environmental pollutant that has been shown to exert toxic effects on kidney and bones in humans after long-term exposure. The recent health risk assessment of Cadmium in food performed by the European Food Safety Authority (EFSA), illustrates the whole process of TK/TD model-based risk assessment. A population TK and dose-effect models (TD) were developed and linked together in order to evaluate a “safe dose” (“health-based guidance value”) for the Caucasian population.

4.2.1. Toxicokinetics

Urinary cadmium (U-Cd) concentration is considered a good biomarker of accumulated cadmium in kidney, and diet is the main source of cadmium among nonsmokers.

The TK assessment involved the comparison of an eight-compartment toxicokinetic model and a one-compartment population TK model, based on a cohort study of 680 Swedish women, over a 20-year-long period.

Second, an alternative one-compartment model (78) was considered. Such a simpler model focuses on the kidney accumulation and urinary excretion, making rough and global assumptions on the other pathways. In case of poor prior knowledge on the physiological parameters involved in cadmium kinetics, it allows a simplified and parsimonious description of cadmium excretion, hence facilitating further statistical evaluations such as for example the evaluation of population variability or integration of intraindividual variability.

The one-compartment model considered here is a standard first-order elimination model with bolus administration; see ref. 22. It can be described as follows: For a given intake of cadmium (d_0) at time 0, the accumulated amount of cadmium in the kidney at time t is calculated as:

$$\mathrm{Cd}_{\mathrm{kidney}}(t, d_0) = f_{\mathrm{k}} \frac{d_0 t_{1/2}}{\log(2)} \exp\left(-\frac{\log(2)t}{t_{1/2}}\right),$$

where $t_{1/2}$ is the cadmium half-time and f_{k} is a factor aggregating several physiological and cadmium-related constants:

$$f_{\mathrm{k}} = \frac{\mathrm{Abs} \times \mathrm{frac}_{\mathrm{kidney}} \times \mathrm{coef}_{\mathrm{cortex}}}{\mathrm{Weight}_{\mathrm{kidney}}},$$

with Abs = gastrointestinal absorption coefficient (in %); $\mathrm{frac}_{\mathrm{kidney}}$ = fraction of cadmium transported to kidney (in %); $\mathrm{coef}_{\mathrm{cortex}}$ = coefficient translating cadmium in the whole kidney into cadmium in the kidney cortex; $\mathrm{Weight}_{\mathrm{kidney}}$ = kidney weight (in g) assumed to be proportional to body weight.

Repeated exposure to cadmium is considered as daily bolus doses, thus:

$$\mathrm{Cd}_{\mathrm{kidney}}(t) = f_{\mathrm{k}} \sum_i \frac{d_i t_{1/2}}{\log(2)} \exp\left(-\frac{\log(2)(t-i)}{t_{1/2}}\right).$$

The urinary cadmium concentration is assumed to be proportional to the cadmium concentration in the kidney cortex, thus the urinary cadmium concentration (in μg/g creatinine) at day t is obtained by:

$$\mathrm{Cd}_{\mathrm{urine}}(t) = f_{\mathrm{u}} \times \mathrm{Cd}_{\mathrm{kidney}}$$

where f_{u} is the ratio between cadmium in urine (μg/g creatinine) and in kidney cortex (μg/kg kidney cortex). Note that, in the case where dose per kg body weight is assumed to be constant over the lifetime, the equation model can be simplified into the more classical one where urinary cadmium concentration is just proportional to $1 - \exp[(t/t_{1/2})\log(2)]$.

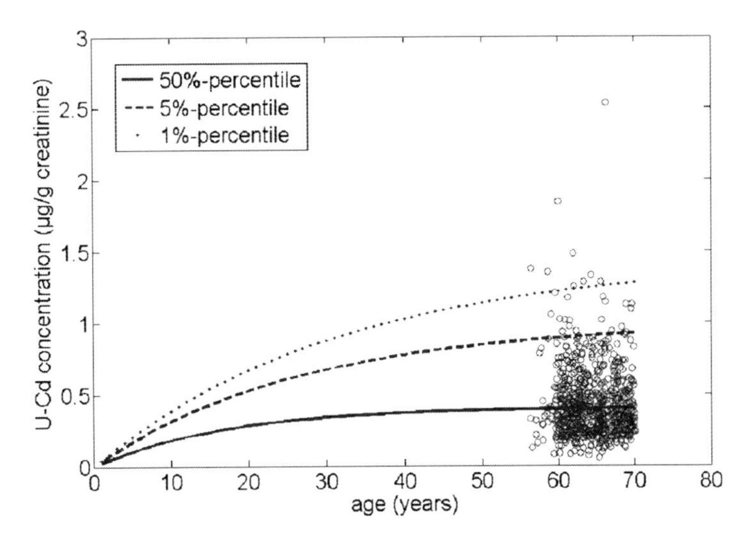

Fig. 5. Measured individual urinary cadmium concentrations (U-Cd) versus redicted individual concentrations using the population one-compartment TK model.

A Bayesian approach was used to perform the statistical inference. A uniform prior between 5 and 35 years was set for individual $t_{1/2}$ in order to constrain the estimation only within biologically plausible values. An informative prior was set over ($f_k \times f_u$) in order to integrate the prior knowledge on the cadmium-related and physiological parameters described previously in ref. 17. This prior was set to a normal distribution (truncated at zero) centered on the central value derived from the literature, and with CV = 30% allowing to cover the range of possible values for ($f_k \times f_u$).

Based on the estimated parameters, Monte Carlo simulations were run in order to predict the population variation in urinary cadmium as a function of lifetime exposure, for a given daily intake. The predicted urinary cadmium concentrations corresponding to a daily cadmium intake of 0.3 µg/kg body weight over 70 years in the 50th, 95th, and 99th percentiles of the population are shown in Fig. 5. The upper percentiles represent the individuals at most risk for high urinary cadmium concentration (mainly), because of long retention time in the body.

Based on the model, the population distribution in the daily cadmium intake corresponding to a given level of urinary cadmium can also be obtained. Thus, we calculated the population variation in dietary cadmium intake corresponding to urinary cadmium concentrations of 0.5, 1, 2 and 3 µg/g creatinine, in a 50-year-old individual with 70 kg body weight (Fig. 6). The functions show for each population percentile, the maximum dietary cadmium intakes allowed in order not to exceed the predefined urinary cadmium concentrations. Thus, in order to remain below e.g., 1 µg cadmium/g creatinine in urine in 50% of the population by the age of 50 (average urinary cadmium is 1 µg/g), the average daily

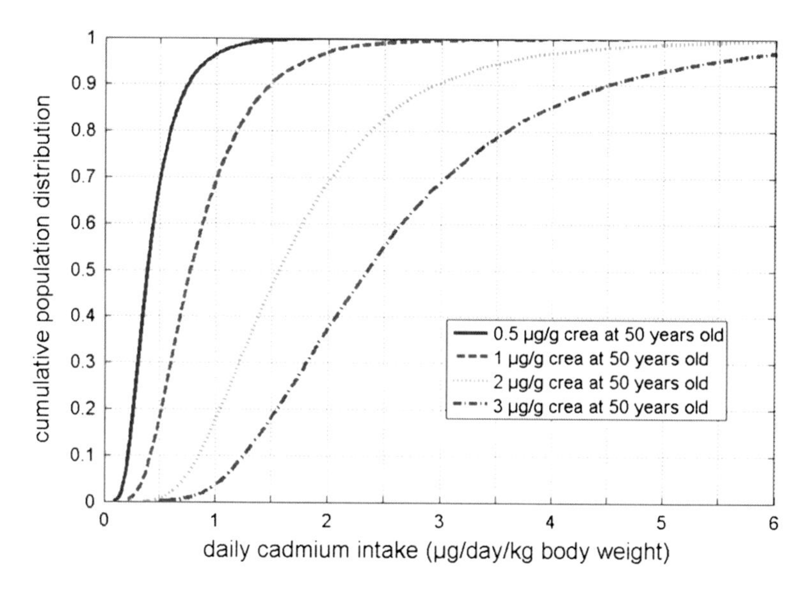

Fig. 6. Cumulative population distribution of daily Cd intake needed to achieve 0.5, 1, 2, and 3 μg/g creatinine of U-Cd concentration.

dietary cadmium intake should not exceed 0.8 μg Cd/kg body weight. The corresponding intake for the 95th percentile of the population remaining below 1 μg/g is 0.4 μg Cd/kg body weight and day.

4.2.2. Modeling Population Variability in the Dose–Response: Toxicodynamics and BMD Assessment

For cadmium renal effects, β2-microglobulinuria (β_2-MG) was the most commonly reported biomarker and was therefore chosen as biomarker of effect. On the basis of a systematic review of the literature, 35 epidemiological studies that measured both biomarkers of exposure and effect in urine were compiled into an aggregated dataset made up of 165 groups of matching urinary cadmium levels and β2-microglobulinuria.

A meta-analysis was then performed to determine the overall relationship between urinary cadmium and β2-microglobulin for subjects over 50 years of age (with purely occupational studies excluded) and for the whole population with all studies.

The consolidated dataset of U-Cd dose groups versus β2-MG effect groups is displayed on Fig. 7, illustrating dose and effect population variability, as well as the interstudy variations. It exhibits an S-shaped dose–effect relationship in the log-scale of both U-Cd and β2-MG, which can be described by a Hill model, with equation:

$$\text{Effect}\,(d) = \text{background} + \text{amplitude} \times [d^{\eta}/(d^{\eta} + \text{ed}_{50}{}^{\eta})]$$

Where: d stands for the dose, i.e., urinary cadmium (in log scale), "amplitude" corresponds to the difference between the two plateaus of the S-shape, ed_{50} corresponds to the dose where 50% of the maximal effect is achieved, and η corresponds to the shape parameter defining the steepness of the S curve.

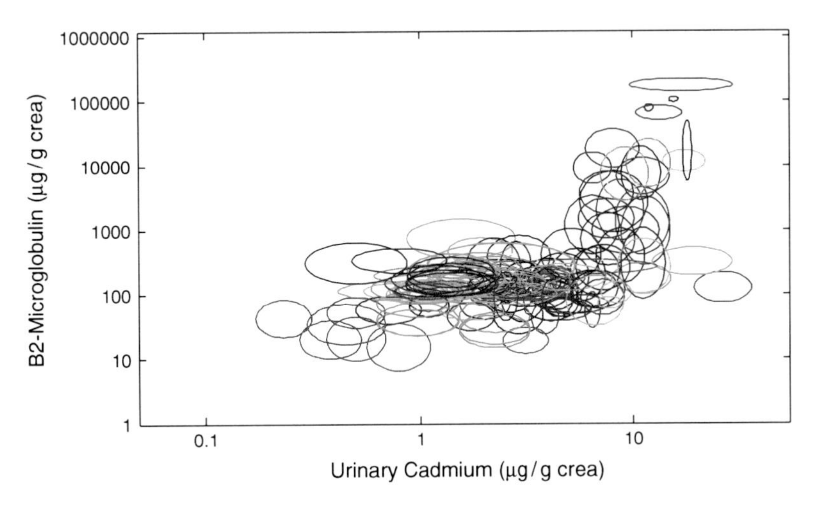

Fig. 7. Scatter plot of data from all studies linking urinary cadmium to β_2-microglobulinuria, using a different color for each study and illustrating within group variability. Each dose group is represented by an ellipse on the log scale, with log(GSD) as radium.

Especially in such cases of large interstudy variability due to many uncontrolled factors, data homogeneity cannot be assumed and variation between studies should be accounted for (79). This is usually made using a random study effect in the statistical model (80, 81). Therefore, the group geometric means and variances were meta-analyzed using a mixed-effect Hill dose-effect model, hence accounting for study heterogeneity and the group sample sizes.

If $\Upsilon_i^{(k)}$ stands for a measurement of e.g., β_2-MG for individual (i) in the study (k), then empirical means and variances ($S^{(k)2}$) of log ($\Upsilon_i^{(k)}$) (as derived from the recorded geometric means and standard deviations in the data) are assumed to follow the following statistical distributions:

$$\sum_i \frac{\log(\Upsilon_i^{(k)})}{n^{(k)}} \sim N\left(\mu^{(k)}, \frac{\sigma^2}{n^{(k)}}\right) S^{(k)2} \sim \frac{\sigma^2}{n^{(k)}} \chi^2(n^{(k)} - 1)$$

where $n^{(k)}$ is the sample size of study (k), $\mu^{(k)}$ is the population subgroup mean effect (in log scale) and σ^2 is the interindividual variance of the effect at a given dose. This statistical model naturally accounts for interstudy and interindividual variability and weights studies according to their sample sizes. It is valid under the assumption that individual doses and effect levels are log-normally distributed within each group, which were generally what was reported in the original publications from the grouped data were collected. The population mean $\mu^{(k)}$ was adjusted for ethnicity to differentiate Caucasian from Asian data. The resulting model fit to the data is reported by Fig. 8, showing that Asian subjects were estimated to have more than twofold higher β_2-microglobulinuria than Caucasians with the same exposure ($p < 0.01$). Further

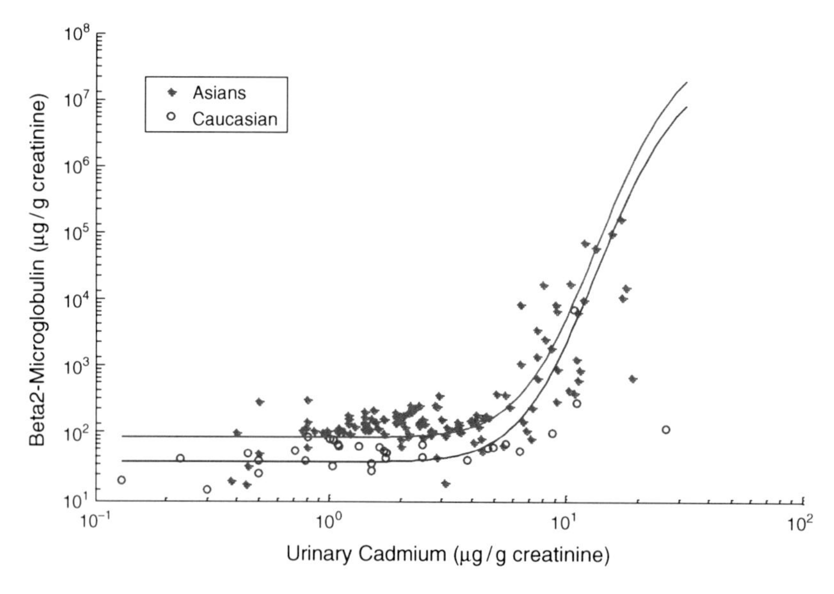

Fig. 8. Hill model fitted to the Caucasian data (open circles) versus the Asian data , using the complete dataset.

investigations are however required to evaluate possible sources of bias and confounding issues which may temper this finding.

Based on the interindividual variability of effect (σ^2) estimated from this analysis, a benchmark dose was derived first assuming no variability of dose within the groups. This benchmark dose was defined as the urinary cadmium level at which an extra 5% of the Caucasian population at 50 years of age is expected to show a biomarker response (e.g., β2-microglobulinuria greater than 300 μg/g creatinine in the case of the EFSA assessment) as compared to the background exposure. To account for estimation uncertainty, the lower bound of the one-sided 95%-confidence interval was used instead of a central estimate, hence defining a BMDL rather than a BMD.

The statistical model was fitted via Bayesian inference as it particularly suited to hierarchical models fitting. Bayesian setup also offers an integrated and robust framework to derive BMDs and BMDLs by Monte Carlo simulations using posterior samples. The Bayesian evaluation was made using WinBUGS (version 1.4, (82)). For each fitted model, three Monte Carlo Markov chains were simultaneously run until convergence. Convergence was assessed using the Gelman–Rubin test available in the WinBUGS software, and by visual inspection of the chains. Posterior means were reported as statistical estimates of model parameters. Prior distributions were chosen as "noninformative," i.e., flat normal distributions for mean parameters and flat gamma distributions for precision parameters.

4.2.3. Final Risk Assessment and TDI Determination

The overall approach taken by EFSA for this assessment is summarized by Fig. 9 showing how the various data, analyses and outputs of analyses were sequenced and combined together

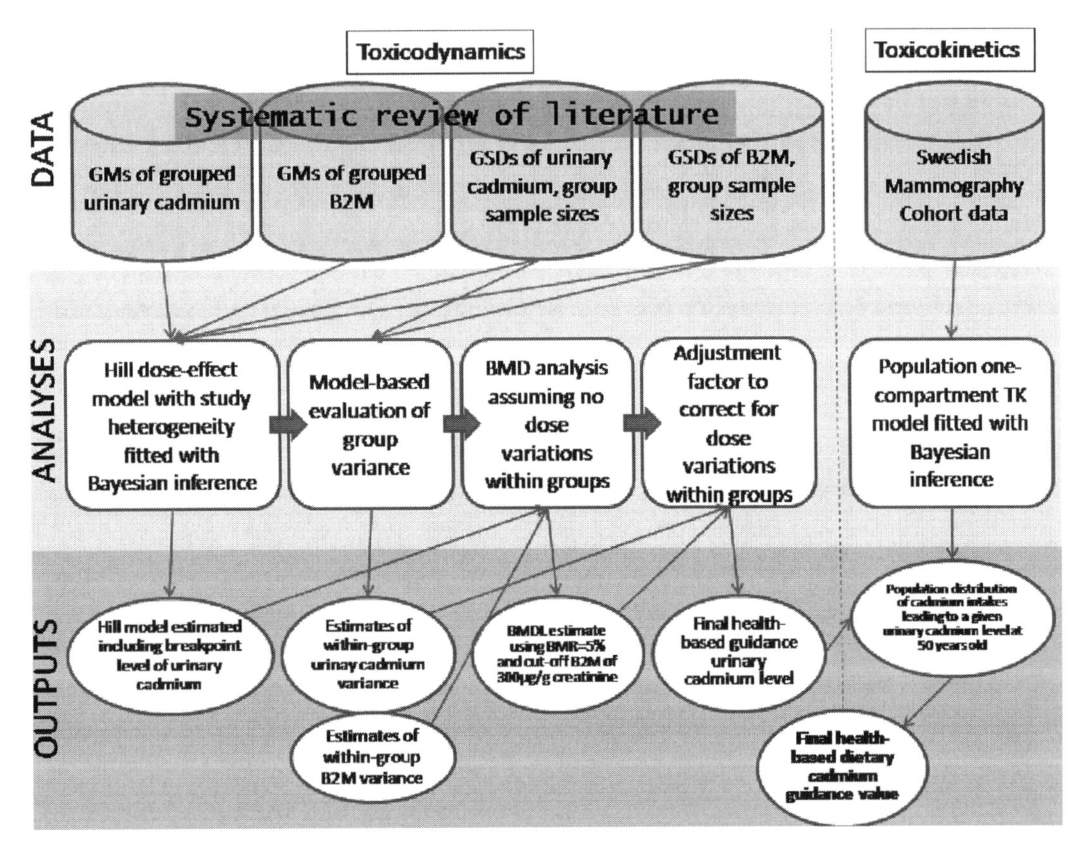

Fig. 9. Graphic representation of step-wise TK/TD assessment performed by EFSA to derive the final HBGV on dietary cadmium.

to derive the final health based guidance value (HBGV) on dietary cadmium. To account for dose variations within groups, the CONTAM Panel then corrected this benchmark dose by an adjustment factor based on estimated variance of exposure within groups. After adjustment, the CONTAM Panel estimated that 1 μg/g creatinine of urinary cadmium was the critical exposure level on which health-based reference of cadmium dietary intake should be based. Subsequently, the TK model described above showed that a food intake of about 2.5 μg/kg body weight (b.w.) per week would prevent 95% of the Caucasian population from being above the threshold of 1 μg/g creatinine of urinary cadmium at 50 years of age (18).

The dose–effect model showed that a urinary cadmium level above 1 μg/g creatinine leads to an excess risk of 5% for the Caucasian population of being with β2-microglobulinuria above the critical cut-off of 300 μg/g creatinine. The TK model showed that a food intake of about 2.5 μg/kg body weight and per week would prevent 95% of the Caucasian population from being above the threshold of 1 μg/g creatinine of urinary cadmium.

A thorough discussion of this EFSA approach, including a critical review of the model assumptions and robustness can be found in ref. 83. Furthermore, recent BMD analyses using the same biomarkers of dose and effect but with individual data confirmed the findings of this TK/TD evaluations (84) on the Japanese population.

4.3. Example 3: Population Exposure to Dioxins over Time Using Individual Data

In this section, KDEM is applied to the French population exposure to dioxin and dioxins like compounds (DLCs). Several scenarios for the model input variables have been performed to test their impact on model output.

In the case of dioxin and related compounds, the total dietary intake estimation should be performed for a group of congeners with similar toxicological properties. The TEF approach, (51) are based on an additive model and are used to convert concentrations of each congener relatively to the compound defined as a reference. The converted concentrations are then summed to obtain concentration in total chemical expressed in toxic equivalents (TEQs).

Dietary exposure data are the individual exposures estimated by (85) in combining concentration data from French monitoring programs (2001–2004) with individual consumption data provided by the first French Individual Consumption Survey (53). The daily exposure to total dioxins expressed in toxic equivalents (TEQs) was estimated to be on average 1.8 and 2.8 pg TEQ_{98}/kg b.w./day for respectively the adult and the children population. The 95th percentile of exposure is 4.5 and 6.6 pg TEQ_{98}/kg b.w/day for respectively the adult and the children population. The empirical exposure distributions of adults and children are used as input values in the dynamic exposure process.

Both the population-based model and the individual-based model presented in the Subheading 3.2 have been computed using different values of initial body burden X_0. Variability between congeners' half-lives is considered to be included in the TEFs and the half-life value of the TCDD which is the reference compound is used (51). Two scenarios considering fixed or individual half-lives were tested for the population-based model. With the first scenario, the half-life of the adults' population was fixed to 7.1 years and the one of the children population to 2.8 years (31). Some authors (86) have observed for dioxins and dioxins like compounds a linear relationship between half-lives and personal characteristics such as age, body fat, smoking, and breast-feeding. With the second scenario, half-lives for each individual were estimated according to the equation of (86). Trajectories were simulated over 90 years. The interintake times for both model was set to a week and the time window of the individual-based model was set to 3 years.

Figure 10 shows the individual variability of the dynamic process in plotting 30 exposure trajectories computed with the individual-based model. The jumps of the dynamic exposure

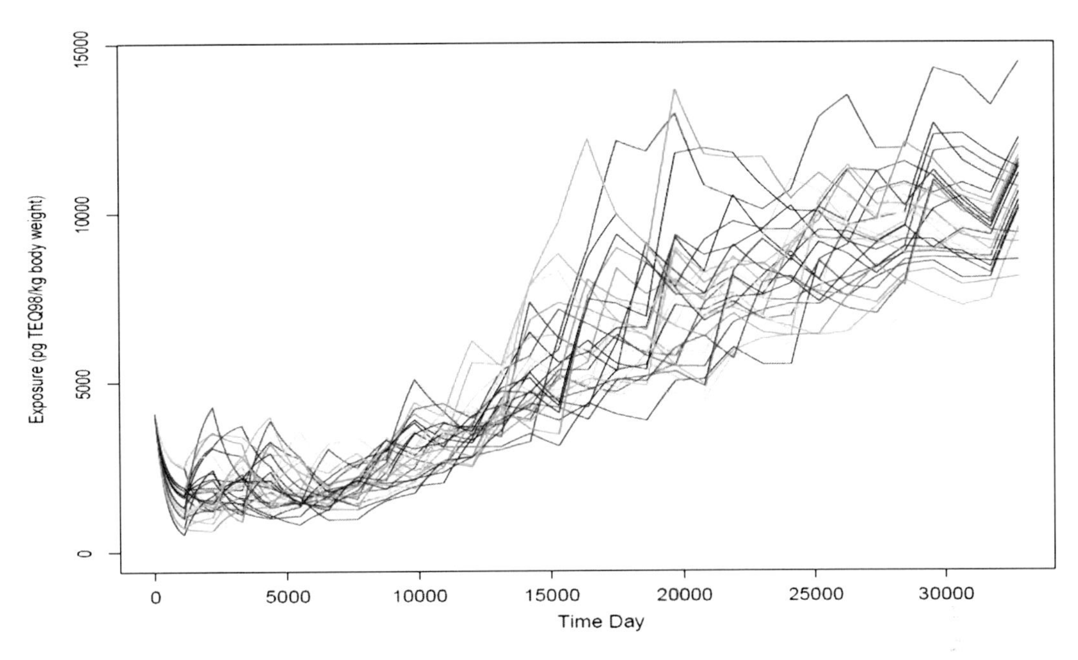

Fig. 10. An example of 30 trajectories of the exposure process over 90 years computed with the individual-based model.

process at each intake-time and its decreases between two intakes can be observed.

The convergence to the steady-state of the dynamic exposure process performed with the population-based model using a fixed half-life can be visually checked in plotting the mean exposure trajectories for different initial body burdens (Fig. 11, see section 3.2.3 on numercail considerations- convergence of fitting algorithms of population model). Simulations were conducted over 90 years for both the children and adult population separately. With increasing time, the trajectories converge to similar values of body burden (between 3.8 and 4.2 ng TEQ/kg b.w. for the children population).

The mean trajectories of the exposure process over different scenarios are compared in Fig. 12. To simulate mean trajectories over lifetime with the population-based model, the trajectories of the children and the adults were combined creating a break in the trajectories at 15 years of age. Using individual half-lives leads with lower predicted body burden. Indeed, because half-lives calculated with the equation of (86) are lower (values range between 0.6 and 2.4 years) than the fixed one for children aged over 15 years, the exposure process decreases rapidly during the first years of life. The individual-based model leads with lower mean exposures than that computed with the population-based model during 64 years of life. This phenomenon is due to the fact that the calculated half-lives for individuals over 65 years are longer than the fixed half-life. The trajectories of the mean exposure performed with the individual-

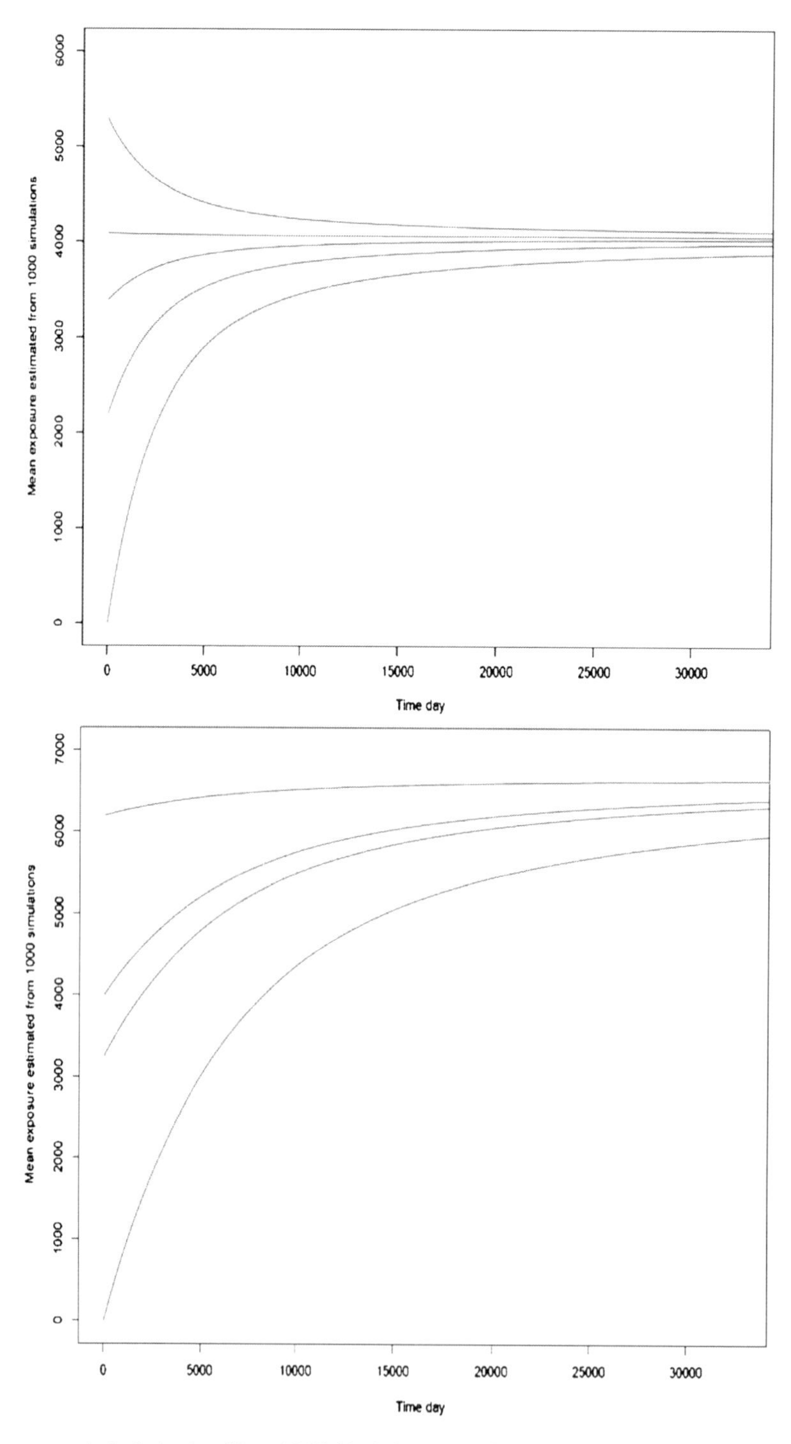

Fig. 11. Mean exposure trajectories for different initial body burden performed with the population-based model using a fixed half-life. A mean exposure trajectory is represented by the mean of the computed mean exposures over increasing intervals $[0, T]$ for a set of 1,000 trajectories.

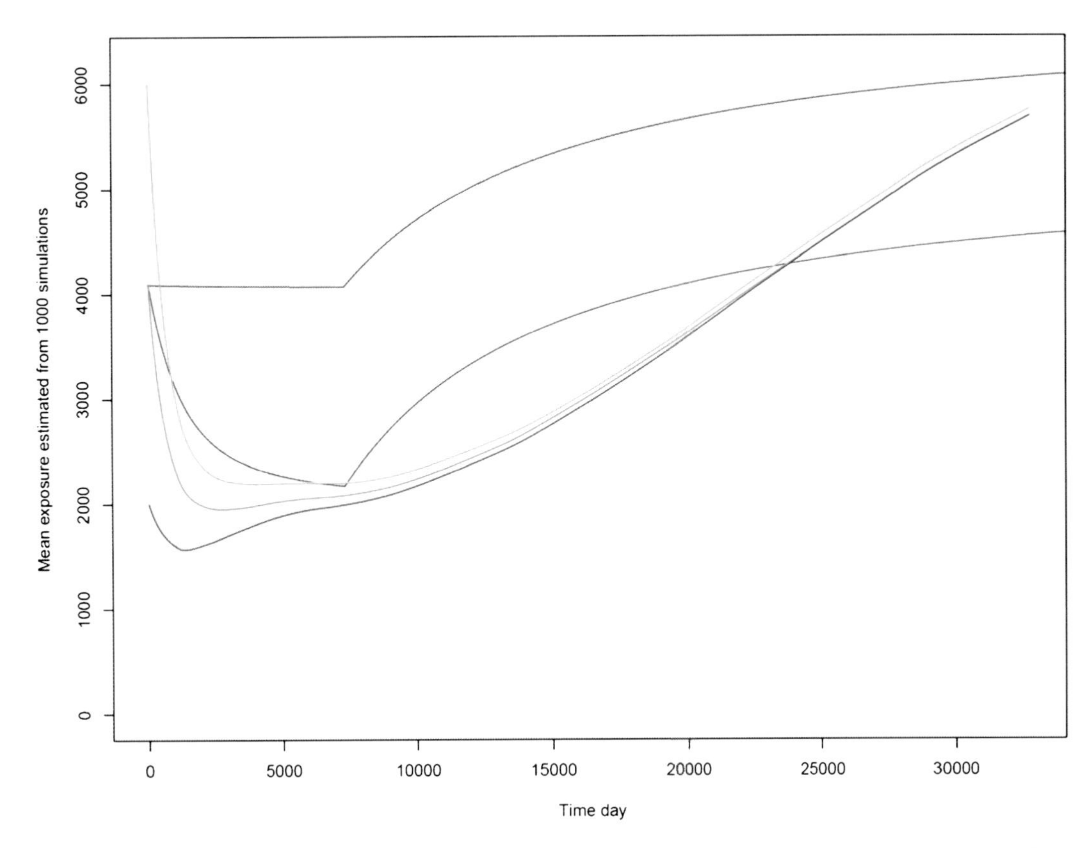

Fig. 12. Mean exposure trajectories for different initial body burden performed with the population-based model using a fixed half-life. A mean exposure trajectory is represented by the mean of the computed mean exposures over increasing intervals $[0, T]$ for a set of 1,000 trajectories.

based model for three different initial body burdens are close after a period of around 30 years.

The test of different scenarios results in that the dynamic modeling of the exposure process is sensitive to the half-life values and to the choice between the population-based model and individual one. The individual-based model seems to be less sensitive to the starting point X_0. This phenomenon can be explained by the fact that the exposure variability is lower within an age group than among the whole population. Therefore, exposures performed with two set of 1,000 trajectories using the individual-based model will be closer than the ones performed with the population-based model. Note that for both models, the steady-state situation is reached after a very long time of exposure, more than a life with the individual-based model.

To give example of risk measures, human BMD was extrapolated from the rat to human combining animal BMD determined by (87) and intraspecies factor modeled by a lognormal distribution with a geometric mean of 1 and a geometric standard-deviation of

2 (88). The BMD is related to 5% of decrease in sperm production. The probabilities that the mean exposure body burden, calculated over 40 and 70 years of life, exceeds the extrapolated BMD were computed with the individual-based model. Over 40 years, the estimate of the probability that the mean exposure exceeds the BMD is of 0.5%. The 95% confidence interval related to the variability of the BMD is of [0.1;1]%. As body burden increases with age, over 70 years, the estimate of the probability that the mean exposure exceeds the BMD is of 3.2% with a 95% confidence interval of [2.2%;4.3%].

Another measure of risk is the percentage of time (as the probability) that the exposure process exceeds the BMD. This probability is estimated for each individual trajectory and therefore a distribution of that probability is computed. The median estimated over 40 years and 70 years are of 0%. This means that for the 50% of the population the probability to exceed the BMD is null. The 97.5th of the distribution is respectively of 7% and 49% for 40 years and 70 years of exposure. This latter result means that 2.5% of the population is exposed over the lower BMDL during around 50% of their life time over 70 years and that the overshoot of exposure is more important between 40 and 70 years old.

5. Notes/ Conclusions and Future Perspectives

Generally speaking, there are limitations for the various in vitro toxicokinetic assays which have an impact on the predictive accuracy of PB-TK models. Difficulties in predicting metabolism, dermal absorption, renal excretion and active transport are foremost in that respect, and improvements will proceed at the pace adopted to solve these problems. For instance, checking the validity of PB-TK models is much easier when such models have a stable and well-documented physiological structure. In this context a number of generic PB-TK models have been developed, i.e., Simcyp, Bayer Technology Services, Cyprotex (https://www.cloegateway.com), or Simulation Plus. However, modelers are still facing the need to validate QSAR submodels or in vitro assays used to assign a PB-TK model's parameter values. Obviously, the quality of those inputs conditions the validity of the PB-TK model using them. The validation of those submodels and in vitro assays requires particular attention and follow the relevant procedures combining sensitivity and uncertainty analyses to understand and quantify which model parameters or assumptions are the most critical to develop relevant, stable and quantitatively informative toxicokinetic/toxicodynamic models for risk assessment (89).

In any case, the major challenge will probably be the coupling of PB-TK models to predictive toxicity models, at the cellular and at the organ level. For example, liver models are being developed (90, 91), but their predictive power is far from established for chronic repeated dose toxicity.

Additionnally, many improvements can still be implemented within exposure models; such as the KDEM; to integrate the variability of input variables in the modeling. As an example, in the dynamic exposure modeling developed by (49), the absorption fraction (ABS) has been considered to be equal to 1. However, experimental studies have shown that after oral exposure, the fraction of the chemical absorbed can be variable and less than full absorption (91, 92). This parameter could be included in the modeling through the introduction of a probabilistic distribution reflecting population variability and move away from point estimates. A potential challenge is the fact that modeling of exposure over time faces a lack of data on past contamination and consumption. Indeed, for a number of chemicals related to environmental and food contamination, exposure has decreased over the past two/three decades in some countries. In contrast, in countries for which environmental and food contamination is not controlled, such exposure can potentially increase and become a public health concern. In terms of food consumption, behaviors have changed with the evolution of taste and food cultural practices and past and future histories of exposures to chemicals can be reconstructed by combining data from studies past and current studies. For example, some countries have surveyed food consumption or purchases over several years and such data can be combined to simulate individual consumption over years. TK models used in KDEM can be extended to PB-TK models with additional compartments representing the different tissues or physiological regions of the body (93–95). The distribution of the chemical between the different tissues or regions is represented by differential equations. The large number of parameters required in PB-TK models limits their use in statistical modeling to predict internal exposure from external exposure.

5.1. Overall Perspective for Chemical Risk Assessment

Over the last decade, risk assessment methodologies have evolved considerably and such evolution has been stimulated by the scientific community as a whole, specialists in the areas of toxicology, pharmacokinetic modeling, applied statistics, systems biology to cite but a few, as well as international public health agencies moving towards "evidence-based" quantitative approaches. Historically, an important historical factor of this trend has been the rise of the bioinformatics in the postgenomic era serving as an interface between the biomedical sciences (molecular biology, pharmacology, toxicology, and epidemiology), mathematics/statistics, and computer sciences. Additionally, high-performance computational platforms as tools have favored such interdisciplinary approaches to develop, such as the integration of population variability data from the molecular, cellular, clinical, and epidemiological scales thus providing powerful tools to quantify interindividual and interspecies variations in physiological processes and toxicological responses (6).

Currently, another trend is the tremendous growth of the application of systems biology using OMICs methodologies (e.g., genomics, proteomics, metabolomics, metabonomics) both from an experimental and modeling aspects to provide overall views on gene, protein or metabolic profile changes in humans, animals, and plants. These global views combined with appropriate population modeling can provide not only an understanding of toxic mode of action of a specific chemical in a specific species but opportunities to develop specific new biomarkers with the associated population variability. Together with systems biology, quantitative structure activity relationship (QSAR) and molecular modeling, biologically based or physiologically based models to quantify variability and uncertainty have been central to these developments to progressively more towards data-based risk assessment instead of default assumptions, potentially used only in the case of total absence of data. Indeed, a number of conceptual, experimental, and modeling tools are still under development and will allow in the future to integrate in a full quantitative manner physiological and toxicological knowledge together with population variability to move towards fully integrated risk assessment. Again, combining knowledge on mode of action (MOA) at the organ and cellular level together with genome-wide and functional measurements (transcriptomics, proteomics, metabolomics...) can provide an global of the biological basis of MOA. In practice for contaminants, the identification of epidemiological biomarkers in humans, when available, reflecting target organ toxicity expressed in molecular terms, can provide a reliable way to determine quantitative levels of exposure with a high level of protection for human subpopulations (health-based guidance value) and at the same time considerably reduce in vivo animal toxicity testing (71). In addition, intelligent or integrated hierarchical testing strategies are being developed in Europe for the implementation of the legislation for Registration, Evaluation, and Authorization of Chemical Substances (REACH). These strategies include the use of human cells/tissues to potentially eliminate interspecies extrapolation, increase efficiencies in testing and reduce the use of animals. Importantly, the validation of in vitro testing methods can specifically address particular modes of action for specific toxicological endpoints.

Finally, large chemical databases dealing specifically with physicochemical properties, toxicokinetics and toxicity in humans, mammals and species used for ecological risk assessments (daphnia, fish, bacteria, soil invertebrates) are being developed around the world including the echem-portal of the OECD and the European Chemical Agency under REACH, the chemical hazards database of EFSA and the TOX 21 program of the National toxicology program to cite but a few. In the near future, these public, free-access database will constitute a considerable step forward for researchers and risk assessors to exchange data, reduce uncertainty in the risk assessment process through a deeper mechanistic understanding and the integration of weight of evidence in toxicokinetics and toxicodynamics, and integrate them at the individual and population variability level using the appropriate statistical methodologies, e.g., Bayesian methods efficient to integrate complex data and quantify uncertainties in the data.

Another Challenge is the integration of new methods and tools to provide quantitative descriptors regarding variability and uncertainty for the risk assessment of chemical mixtures (toxicokinetics, toxicodynamics) in humans and the environment. A number of frameworks are currently available to deal with the human risk assessment of chemical mixtures. These methods will prove useful to risk assessors in ecological and human risk assessment from the public or private sector so that the relevant tools and modeling techniques can provide science-based risk assessment that are more transparent.

Acknowledgments

The views reflected in this review are the authors' only and do not reflect the views of the European Food Safety Authority, the Technological university of Compiegne, the French Agency for Food, Environment, and Occupational Health Safety, the French National Institute of Agronomical Research (INRA), or the World Health organization.

References

1. European Commission (EC) (2002) Regulation (EC) No 178/2002 of the european parliament and of the council laying down the general principles and requirements of food law, establishing the European Food Safety Authority and laying down procedures in matters of food safety. http://eur-lex.europa.eu/LexUriServ/LexUriServ.do?uri=OJ:L:2002:031:0001:0024:EN:PDF
2. WHO (2009) Principles and methods for the risk assessment of chemicals in food, Environmental health criteria 240. http://www.who.int/foodsafety/chem/principles/en/index1.html
3. Svendsen C, Ragas AM, Dorne JLCM (2008) Contaminants in organic and conventional food: the missing link between contaminant levels and health effects. Book comparing organic vs non-organic food at the nutritional, microbiological and toxicological level. In: Givens DI et al (eds) Health benefits of organic

foods: effects of the environment. Chapter 6, vol 119. CABI, Wallingford

4. Dorne JLCM, Bordajandi LR, Amzal B, Ferrari P, Verger P (2009) Combining analytical techniques, exspoure assessment and biological effects for risk assessment of chemicals in food. Trends Anal Chem 2009(28):695
5. Kroes R, Müller D, Lambe J, Löwik MR, van Klaveren J, Kleiner J, Massey R, Mayer S, Urieta I, Verger P, Visconti A (2002) Assessment of intake from the diet. Food Chem Toxicol 40(2–3):327–385
6. Dorne JCM (2010) Metabolism, variability and risk assessment. Toxicology 268 (3):156–164
7. EFSA (European Food Safety Authority) (2005) Opinion of the Scientific Committee on a request from EFSA related to a harmonised approach for risk assessment of substances which are both genotoxic and carcinogenic. EFSA J 282:1–31. http://www.efsa.europa.eu/EFSA/Scientific_Opinion/sc_op_ej282_gentox_en3.pdf
8. FAO/WHO (Food and Agriculture Organisation of the United Nations/World Health Organization) (2006) Safety evaluation of certain contaminants in food. Prepared by the Sixty-fourth meeting of the Joint FAO/WHO Expert Committee on Food Additives (JECFA). FAO Food Nutr Pap 82:1–778
9. EFSA (European Food Safety Authority) (2007) Opinion of the Scientific Panel on Contaminants in the Food chain on a request from the European Commission on ethyl carbamate and hydrocyanic acid in food and beverages. EFSA J 551:1–44. :http://www.efsa.europa.eu/cs/BlobServer/Scientific_Opinion/Contam_ej551_ethyl_carbamate_en_rev.1.pdf?ssbinary=true
10. EFSA (European Food Safety Authority) (2008) Scientific Opinion of the Panel on Contaminants in the Food Chain on a request from the European Commission on Polycyclic Aromatic Hydrocarbons in Food. EFSA J 724:1–114. http://www.efsa.europa.eu/cs/BlobServer/Scientific_Opinion/contam_ej_724_PAHs_en,1.pdf?ssbinary=true
11. EFSA (European Food Safety Authority) (2009) Scientific opinion on arsenic in food. EFSA J 7(10):1051. http://www.efsa.europa.eu/en/efsajournal/doc/1351.pdf
12. JMPR (Joint FAO/WHO Meetings on Pesticide Residues) (2002) Report of the JMPR, FAO Plant Production and Protection Paper, 172, 4. FAO, Rome
13. EFSA (2009) Scientific opinion on marine biotoxins in shellfish—Palytoxin group. EFSA J 7 (12):1293. http://www.efsa.europa.eu/en/efsajournal/doc/1393.pdf
14. EFSA (2009) Potential risks for public health due to the presence of nicotine in wild mushrooms. EFSA J RN-286:2–47. http://www.efsa.europa.eu/en/efsajournal/doc/286r.pdf
15. Renwick AG, Lazarus NR (1998) Human variability and noncancer risk assessment—an analysis of the default uncertainty factor. Regul Toxicol Pharmacol 27:3–20
16. WHO (2005) International Programme on Chemical Safety: chemical-specific adjustment. Factors for interspecies differences and human variability: guidance document for use of data in dose/concentration response assessment. World Health Organization, Geneva. http://www.who.int/ipcs/methods/harmonization/areas/uncertainty/en/index.html
17. Amzal B, Julin B, Vahter M, Johanson G, Wolk A, Åkesson A (2009) Population toxicokinetic modeling of cadmium for health risk assessment. Environ Health Perspect 117 (8):1293–1301
18. EFSA (European Food Safety Authority) (2009) Scientific Opinion of the Panel on Contaminants in the Food Chain on a request from the European Commission on cadmium in food. EFSA J 980:1–139. http://www.efsa.europa.eu/cs/BlobServer/Scientific_Opinion/contam_op_ej980_cadmium_en_rev.1.pdf?ssbinary=true
19. Dorne JLCM, Walton K, Renwick AG (2005) Human variability in xenobiotic metabolism and pathway-related uncertainty factors for chemical risk assessment: a review. Food Chem Toxicol 43:203–216
20. Gerlowski LE, Jain RK (1983) Physiologically based pharmacokinetic modeling: principles and applications. J Pharm Sci 72:1103–1127
21. Bois F, Jamei M, Clewell HJ (2010) PBPK modelling of inter-individual variability in the pharmacokinetics of environmental chemicals. Toxicology 278:256–267
22. Gibaldi M, Perrier D (1982) Pharmacokinetics, 2nd edn, revised and expanded ed. Marcel Dekker, New York
23. Jamei M, Marciniak S, Feng KR, Barnett A, Tucker G, Rostami-Hodjegan A (2009) The Simcyp population-based ADME simulator. Expert Opin Drug Metab Toxicol 5:211–223
24. Bouvier d'Yvoire M, Prieto P, Blaauboer BJ, Bois FY, Boobis A, Brochot C, Coecke S, Freidig A, Gundert-Remy U, Hartung T, Jacobs MN, Lavé T, Leahy DE, Lennernäs H, Loizou GD, Meek B, Pease C, Rowland M, Spendiff M, Yang J, Zeilmaker M (2007) Physiologically-based kinetic modelling (PBK

modelling): meeting the 3Rs agenda—the report and recommendations of ECVAM Workshop 63a. Altern Lab Anim 35:661–671

25. Edginton AN, Schmitt W, Willmann S (2006) Development and evaluation of a generic physiologically based pharmacokinetic model for children. Clin Pharmacokinet 45:1013–1034
26. Luecke RH, Pearce BA, Wosilait WD, Slikker W, Young JF (2007) Postnatal growth considerations for PBPK modeling. J Toxicol Environ Health A 70:1027–1037
27. Jones HM, Gardner IB, Watson KJ (2009) Modelling and PBPK simulation in drug discovery. AAPS J 11:155–166
28. Allen BC, Hack CE, Clewell HJ (2007) Use of Markov chain Monte Carlo analysis with a physiologically-based pharmacokinetic model of methylmercury to estimate exposures in US women of childbearing age. Risk Anal 27:947–959
29. Lorber M (2008) Exposure of Americans to polybrominated diphenyl ethers. J Expo Sci Environ Epidemiol 18(1):2–19
30. Fromme H, Korner W et al (2009) Human exposure to polybrominated diphenyl ethers (PBDE), as evidenced by data from a duplicate diet study, indoor air, house dust, and biomonitoring in Germany. Environ Int 35 (8):1125–1135
31. US-EPA (2003) Exposure and human health reassessment of 2,3,7,8-tetrachlorodibenzo-*p*-dioxin (TCDD) and related compounds National Academy Sciences (NAS) review draft. Part III. EPA, Washington, DC
32. Pinsky PF, Lorber MN (1998) A model to evaluate past exposure to 2,3,7,8-TCDD. J Expo Anal Environ Epidemiol 8(2):187–206
33. Smith JC, Farris FF (1996) Methyl mercury pharmacokinetics in man: a reevaluation. Toxicol Appl Pharmacol 137(2):245–252
34. Albert I, Villeret G et al (2010) Integrating variability in half-lives and dietary intakes to predict mercury concentration in hair. Regul Toxicol Pharmacol 58(3):482–489
35. Delyon B, Lavielle M, Moulines E (1999) Convergence of a stochastic approximation version of the EM algorithm. Ann Stat 27(1):94–128
36. Rowland M, Benet LZ, Graham GG (1973) Clearance concepts in pharmacokinetics. J Pharmacokinet Biopharm 1:123–136
37. Shuey DL, Lau C, Logsdon TR, Zucker RM, Elstein KH, Narotsky MG, Setzer RW, Kavlock RJ, Rogers JM (1994) Biologically based dose-response modeling in developmental toxicology: biochemical and cellular sequelae of 5-fluorouracil exposure in the developing rat. Toxicol Appl Pharmacol 126(1):129–144
38. Crump KS, Chen C, Chiu WA, Louis TA, Portier CJ et al (2010) What role for biologically based dose-response models in estimating low-dose risk? Environ Health Perspect 118(5)
39. Crump KS (1984) A new method for determining allowable daily intakes. Fundam Appl Toxicol 4:854–871
40. Budtz-Jørgensen E, Keiding N, Grandjean P (2001) Benchmark dose calculation from epidemiological data. Biometrics 57:698–706
41. Sand S et al (2008) The current state of knowledge on the use of the benchmark dose concept in risk assessment. J Appl Toxicol 28 (4):405–421
42. Crump KS (2002) Critical issues in benchmark calculations from continuous data. Crit Rev Toxicol 32:133–153
43. Suwazono Y et al (2006) Benchmark dose for cadmium-induced renal effects in humans. Environ Health Perspect 114(7):1072–1076
44. Ryan L (2008) Combining data from multiple sources, with applications to environmental risk assessment. Stat Med 27:698–710
45. Wheeler MW, Bailer AJ (2007) Properties of model-averaged BMDLs: a study of model averaging in dichotomous response risk estimation. Risk Anal 27:659–670
46. Morales KH, Ibrahim JG, Chen CJ, Ryan LM (2006) Bayesian model averaging with applications to benchmark dose estimation for arsenic in drinking water. J Am Stat Assoc 101 (473):9–17
47. EC (2004) European Union System for the Evaluation of Substances 2.0 (EUSES 2.0). Prepared for the European Chemicals Bureau by the National Institute of Public Health and the Environment (RIVM), Bilthoven, The Netherlands (RIVM Report no. 601900005). Available via the European Chemicals Bureau. http://ecb.jrc.it
48. Bertail P, Clémençon S et al (2010) Statistical analysis of a dynamic model for dietary contaminant exposure. J Biol Dyn 4(2):212–234
49. Verger P, Tressou J, Clémençon S (2007) Integration of time as a description parameter in risk characterisation: application to methyl mercury. Regul Toxicol Pharmacol 49 (1):25–30
50. Tressou J, Leblanc JCh, Feinberg M, Bertail P (2004) Statistical methodology to evaluate food exposure to a contaminant and influence of sanitary limits: application to ochratoxin A. Regul Toxicol Pharmacol 40(3):252–263
51. Van den Berg M, Birnbaum L et al (1998) Toxic equivalency factors (TEFs) for PCBs, PCDDs, PCDFs for humans and wildlife. Environ Health Perspect 106:775–792

52. Thuresson, Höglund et al (2000) In: Medicine and health policy. New York: Marcel Dekker
53. AFSSA (2009) Etude individuelle Nationale des consommations Alimentaires 2 (INCA 2) (2006-2007), Rapport AFSSA, 228p, http://www.anses.fr/Documents/PASER-Ra-INCA2.pdf
54. EFSA (2010) Application of systematic review methodology to food and feed safety assessments to support decision making. EFSA J 8 (6):1637. http://www.efsa.europa.eu/en/efsajournal/doc/1637.pdf
55. Marvier M, McCreedy C, Regetz J, Kareiva P (2007) A meta-analysis of effects of Bt cotton and maize on nontarget invertebrates. Science 316(5830):1475–1477
56. Greenland S, Robins J (1994) Invited commentary: ecologic studies—biases, misconceptions, and counterexamples. Am J Epidemiol 139:747–60
57. Terrin N, Schmidt CH, Lau J, Olkin I (2003) Adjusting for publication bias in the presence of heterogeneity. Stat Med 22:2113–2212
58. Stangl D, Berry DA (eds) Meta-analysis
59. Higgins JP, Thompson SG, Deeks JJ, Altman DG (2003) Measuring inconsistency in meta-analyses. BMJ 327:557–560
60. Egger M et al (2001) Systematic reviews in health care. BMJ books, London
61. Egger M, Smith GD, Schneider M, Minder C (1997) Bias in meta-analysis detected by a simple, graphical test. BMJ 315:629–634
62. Biro G, Hulshof K, Ovesen L, Amorim Cruz JA (2002) Selection of methodology to assess food intake. Eur J Clin Nutr 56(Suppl 2): S25–S32. doi:10.1038/sj/ejcn/1601426
63. Verger P, Ireland J, Møller A, Abravicius JA, De Henauw S, Naska A (2002) Improvement of comparability of dietary intake assessment using currently available individual food consumption surveys. Eur J Clin Nutr 56(Suppl 2):S1–S7. doi:10.1038/sj/ejcn/1601425
64. Wirfält E, Hedblad B, Gullberg B, Mattisson I, AndrénC RU, Janzon L, Berglund G (2001) Food patterns and components of the metabolic syndrome in men and women: A cross-sectional study within the Malmö diet and cancer cohort. Am J Epidemiol 154(12):1150–1159
65. Zetlaoui M, Feinberg M, Verger P, Clémencon S (2011) Extraction of food consumption systems by non-negative matrix factorization (NMF) for the assessment of food choices. Biometrics (in press). http://hal.archives-ouvertes.fr/docs/00/48/47/94/PDF/NMF_food.pdf
66. EFSA (2010) European Food Safety Authority; management of left-censored data in dietary exposure assessment of chemical substances. EFSA J 8(3):1557. http://www.efsa.europa.eu/en/efsajournal/doc/1557.pdf
67. Helsel DR (2005) Nondetects and data analysis. Wiley, New York
68. Kennedy MC, Roelofs VJ et al (2011) A hierarchical Bayesian model for extreme pesticide residues. Food Chem Toxicol 49(1):222–232
69. Tressou J, Bertail P et al (2003) 709 Evaluation of food risk exposure using extreme value theory-application to heavy metals for sea products consumers. Toxicol Lett 144(Supplement 1):s190
70. WHO (2009) Principles for modelling dose-response for the risk assessment of chemicals. Environmental Health Criteria. http://www.who.int/tipcs/methods/harmonization/dose_response/en/
71. Spilke J, Piepho HP, Hu X (2005) A simulation study on tests of hypotheses and confidence intervals for fixed effects in mixed models for blocked experiments with missing data. J Agric Biol Environ Stat 10:374–389
72. Spiegelhalter DJ, Best NG et al (2002) Bayesian measures of model complexity and fit. J R Stat Soc Series B Stat Methodol 64:583–640
73. Dorne JLCM, Walton K, Renwick AG (2001) Uncertainty factors for chemical risk assessment: human variability in the pharmacokinetics of CYP1A2 probe substrates. Food Chem Toxicol 39:681–696
74. Dorne JLCM, Walton K, Slob W, Renwick AG (2002) Human variability in polymorphic CYP2D6 metabolism: is the kinetic default uncertainty factor adequate? Food Chem Toxicol 40:1633–1656
75. Dorne JLCM, Renwick AG (2005) The refinement of uncertainty/safety factors in risk assessment by the incorporation of data on toxicokinetic variability in humans. Toxicol Sci 86:20–26
76. Dorne JLCM, Walton K, Renwick AG (2003) Human variability in CYP3A4 metabolism and CYP3A4-related uncertainty factors for risk assessment. Food Chem Toxicol 41:201–224
77. Dorne JLCM, Walton K, Renwick AG (2003) Polymorphic CYP2C19 and N-acetylation: human variability in kinetics and pathway-related uncertainty factors. Food Chem Toxicol 41:225–245
78. Kjellström T (1971) A mathematical model for the accumulation of cadmium in human kidney cortex. Nord Hyg Tidskr 52:111–119
79. Sutton AJ, Higgins JPT (2008) Recent developments in meta-analysis. Stat Med 27:625–650

80. Berry D, Strangl DK (eds) (2001) Meta-analysis in medicine and health policy. Biostatistics, New York
81. Morales KH, Ryan LM (2005) Benchmark dose estimation based on epidemiologic cohort data. Environmetrics 16:435–447
82. Lunn DJ, Thomas A, Best N, Spiegelhalter D (2000) WinBUGS—a Bayesian modelling framework: concepts, structure and extensibility. Stat Comput 10:325–337
83. EFSA (2011) Comparison of the approaches taken by EFSA and JECFA to establish a HBGV for cadmium. http://www.efsa.europa.eu/en/efsajournal/doc/2006.pdf.
84. Suwazono Y, Nogawa K, Uetani M et al (2011) Application of hybrid approach for estimating the benchmark dose of urinary cadmium for adverse renal effects in the general population of Japan. J Appl Toxicol 31(1):89–93
85. Tard A, Gallotti S, Leblanc JC, Volatier JL (2007) Dioxins, furans and dioxin-like PCBs: occurrence in food and dietary intake in France. Food Addit Contam 24(9):1007–1017
86. Milbrath MO, Wenger Y, Chang CW, Emond C, Garabrant D, Gillespie BW, Jolliet O (2009) Apparent half-lives of dioxins, furans, and polychlorinated biphenyls as a function of age, body fat, smoking status, and breast-feeding. Environ Health Perspect 117(3):417–425
87. Gray LE, Ostby JS et al (1997) A dose-response analysis of the reproductive effects of a single gestational dose of 2,3,7,8-tetrachlorodibenzo-p-dioxin in male Long Evans Hooded rat offspring. Toxicol Appl Pharmacol 146(11–20)
88. Bokkers, B. G. H., M. J. Zeilmaker, et al. (2009). RIVM report on framework and integration methods. The application of animal toxicity data in risk-benefit analysis: 2,3,7,8-TCDD as an example.
89. Bernillon P, Bois FY (2000) Statistical issues in toxicokinetic modeling: a Bayesian perspective. Environ Health Perspect 108(Suppl 5):883–893
90. Yan L, Sheihk-Bahaei S, Park S, Ropella GE, Hunt CA (2008) Predictions of hepatic disposition properties using a mechanistically realistic, physiologically based model. Drug Metab Dispos 36(4):759–768
91. Lerapetritou MG, Georgopoulos PG, Roth CM, Androulakis LP (2009) Tissue-level modeling of xenobiotic metabolism in liver: an emerging tool for enabling clinical translational research. Clin Transl Sci 2(3):228–237
92. McDonald TA (2005) Polybrominated diphenylether levels among United States residents: daily intake and risk of harm to the developing brain and reproductive organs. Integr Environ Assess Manag 1(4):343–354
93. Van der Molen GW, Kooijman SALM et al (1996) A generic toxicokinetic model for persistent lipophilic compounds in humans: an application to TCDD. Fundam Appl Toxicol 31(1):83–94
94. Verner MA, Ayotte P et al (2009) A physiologically based pharmacokinetic model for the assessment of infant exposure to persistent organic pollutants in epidemiologic studies. Environ Health Perspect 117(3):481–487
95. Lu C, Holbrook CM et al (2010) The implications of using a physiologically based pharmacokinetic (PBPK) model for pesticide risk assessment. Environ Health Perspect 118 (1):125–130

Chapter 21

Mechanism-Based Pharmacodynamic Modeling

Melanie A. Felmlee, Marilyn E. Morris, and Donald E. Mager

Abstract

Pharmacodynamic modeling is based on a quantitative integration of pharmacokinetics, pharmacological systems, and (patho-) physiological processes for understanding the intensity and time-course of drug effects on the body. Application of such models to the analysis of meaningful experimental data allows for the quantification and prediction of drug–system interactions for both therapeutic and adverse drug responses. In this chapter, commonly used mechanistic pharmacodynamic models are presented with respect to their important features, operable equations, and signature profiles. In addition, literature examples showcasing the utility of these models to adverse drug events are highlighted. Common model types that are covered include simple direct effects, biophase distribution, indirect effects, signal transduction, and irreversible effects.

Key words: Adverse drug effects, Exposure–response relationships, Mathematical modeling, Pharmacodynamics, Pharmacokinetics

1. Introduction

Pharmacodynamics represents a broad discipline that seeks to identify drug- and system-specific properties that regulate acute and long-term biological responses to drugs. The term is typically used in the context of therapeutic effects, whereas toxicology or toxicodynamics relates to adverse drug reactions. In contrast to classical conceptualizations whereby beneficial and adverse responses occur via distinct mechanisms, it is increasingly clear that diseases and both types of drug responses may emerge from perturbations of singular complex interconnected networks (1). Thus, mechanism-based pharmacodynamic models, by definition, should be multipurpose and readily adapted to understand the extent and time-course of adverse drug effects.

In the mid-1960s, Gerhard Levy was the first to mathematically demonstrate a link between pharmacokinetics (factors controlling drug exposure) and the rate of decline of in vivo pharmacological

Brad Reisfeld and Arthur N. Mayeno (eds.), *Computational Toxicology: Volume I*, Methods in Molecular Biology, vol. 929,
DOI 10.1007/978-1-62703-050-2_21, © Springer Science+Business Media, LLC 2012

responses (2, 3). Since that landmark discovery, pharmacodynamic modeling has evolved into a quantitative field that aims to mathematically characterize the temporal aspects of drug effects via emulating mechanisms of action (4). The application of mathematical models to describe drug–system interactions allows for the quantification and prediction of subsequent interactions within the system. The major goals of pharmacodynamic modeling are to integrate known system components, functions, and constraints, generate and test competing hypotheses of drug mechanisms and system responses under new conditions, and estimate system-specific parameters that may be inaccessible (5). These models are applicable to a wide range of disciplines within the biological sciences including pharmacology and toxicology, wherein there is a critical need to understand and predict desired and adverse responses to xenobiotic exposure, which together define the clinical utility or therapeutic index.

The main objectives of this chapter are to illustrate commonly used mechanistic pharmacodynamic models, providing important model features, operable equations, and signature profiles, as well as examples of the application of these models to the analysis of drug-induced adverse reactions.

2. Modeling Requirements

Useful pharmacodynamic models are based on plausible mathematical and pharmacological exposure–response relationships. Basic model components encompassing a range of pharmacodynamic systems are illustrated in Fig. 1. For most drug effects, both pharmacological mechanisms, often characterized by sensitivity-grounded capacity-limited effector units, and physiological turnover processes need to be integrated with drug disposition when constructing a PK/PD model.

The construction and evaluation of relevant PK/PD models require suitable pharmacokinetic data, an appreciation for molecular and cellular mechanisms of pharmacological/toxicological responses, and a range of quantitative experimental measurements of meaningful biomarkers within the causal pathway between drug–target interactions and clinical effects. Good experimental designs are essential to ensure that sensitive and reproducible data are collected. These data should cover a reasonably wide dose/concentration range and appropriate study duration to ascertain net drug exposure and the ultimate fate of the biomarkers or outcomes under investigation. A wide range of systemic drug concentrations is also typically required for the accurate and precise estimation of pharmacodynamic parameters. Typically studies should involve a minimum of two to three doses to adequately estimate the

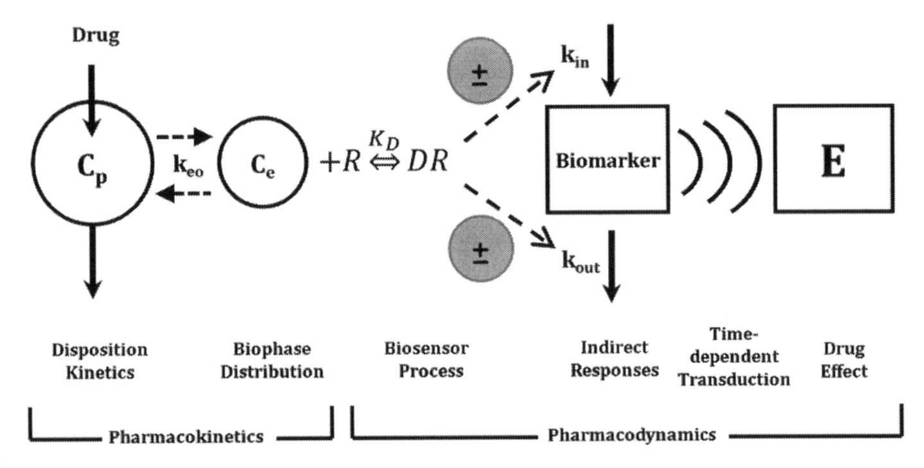

Fig. 1. Basic components of pharmacodynamic models. The time-course of drug concentrations in a relevant biological fluid (e.g., plasma, C_p) or the biophase (C_e) is characterized by a mathematical function that serves to drive PD models. The biosensor process involves the interaction between the drug and the pharmacologic target (R), and may be described using various receptor-occupancy models, may require equations that consider the kinetics of the drug–receptor complex formation and dissociation, or may encompass irreversible drug–target interactions. Many drugs act via indirect mechanisms and the biosensor process may serve to stimulate or inhibit the production (k_{in}) or loss (k_{out}) of endogenous mediators. These altered mediators may not represent the final observed drug effect (E) and further time-dependent transduction processes may occur, thus requiring additional modeling components. System complexities such as drug interactions, functional adaptation, changes with pathophysiology, and other factors may play a role in regulating drug effects after acute and long-term drug exposure (adapted from ref. 33).

nonlinear parameters of most pharmacodynamic models with simultaneous collection of concentration and response data. For more complex systems (and therefore models), more extensive datasets are required as these models typically incorporate multiple nonlinear processes and pharmacodynamic endpoints. Models are typically defined using ordinary differential equations and include both drug- and system-specific parameters. This separation of terms provides a platform for translational research, whereby relationships with in vitro bioassays and preclinical experiments can be identified.

Once a structural model has been selected, unknown parameter values can be estimated using nonlinear regression techniques. It is beyond the scope of this chapter to review the vast array of software programs and algorithms available, and the best tool and approach will often be defined by the characteristics of the experimental data, the familiarity of the end user with specific programs, and the goals and objectives of the analysis. The type of model (e.g., data-driven versus systems models), the nature of the biomarker (e.g., continuous versus categorical), the degree of inter-subject variability, and complexities within a dataset (e.g., missing variables, data above or below a limit of quantification, and availability of covariates) are just a few considerations when selecting an approach to develop and qualify PK/PD models.

3. Practical Modeling Approaches

The first steps in any modeling endeavor are to define the objectives of the analysis and to perform a careful graphical analysis of raw data. Both efforts should facilitate selection of appropriate techniques and conditions for model construction and evaluation. A good graphical analysis (along with a priori knowledge of drug mechanisms) may be used to narrow down the number of structural models being considered as a base model and also help in calculating initial parameter estimates. Despite progress in computational algorithms, good initial parameter estimates can reduce the likelihood of falling into local minima and can also be used as a reality check when compared to final parameter estimates or literature reported values. Next, an appropriate drug/toxin pharmacokinetic/toxicokinetic function is derived from fitting a model to concentration–time profiles in relevant biological fluids. Depending on the complexity of the pharmacodynamic model/system, the pharmacokinetic model and associated parameters are often fixed to serve as a driving function for the pharmacodynamic model relating drug exposure to pharmacological/toxicological effects. Although simultaneous PK/PD modeling is desirable, this can still be a formidable challenge for complex models. Objective model-fitting criteria (e.g., diagnostic and goodness-of-fit plots) are frequently compared to select a final model, and a variety of techniques are available to verify or qualify models, which can range in complexity depending on the modeling approach (e.g., population versus pooled data). Ideally, an external dataset, not used in the construction of the model, could be used to determine whether the model is generalizable; however, internal validation steps are far more common as most model-builders will attempt to incorporate all available experimental data. In any event, final models should reasonably recapitulate the data used to derive the model, generate new insights and testable hypotheses of factors controlling drug responses, and provide guidance for subsequent decisions in drug discovery, development, and pharmacotherapy. Subsequent sections will highlight commonly used pharmacodynamic models with increasing degrees of complexity, as well as provide literature examples on the application of such models to the analysis of drug-induced adverse events.

3.1. Simple Direct Effect Models

The Hill equation assumes that drugs effects (E) are directly proportional to receptor occupancy (i.e., linear transduction), assumes that plasma drug concentrations are in rapid equilibrium with the effect site, and represents a fundamental pharmacodynamic relationship (6):

$$E = E_0 \pm \frac{E_{max} \times C_{\mathrm{p}}}{\mathrm{EC}_{50} + C_{\mathrm{p}}}. \quad (1)$$

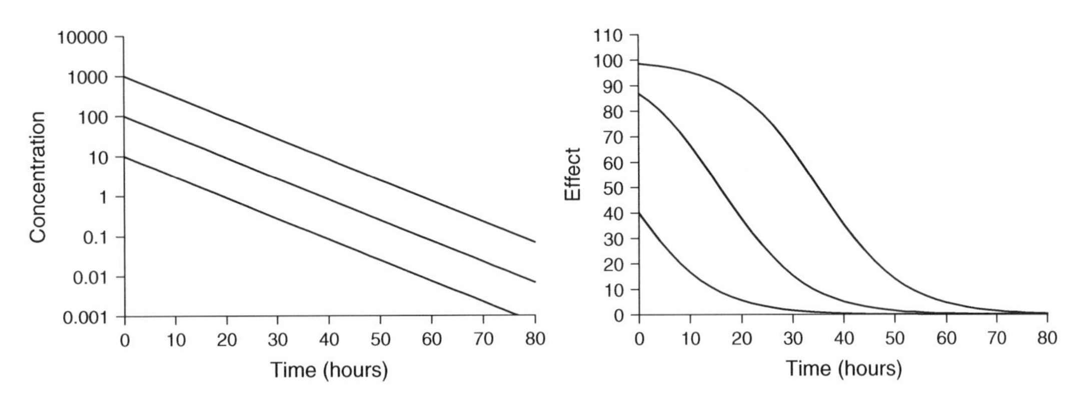

Fig. 2. Simulated drug concentrations (*left*) and response curves (*right*) using a simple E_{max} model (Eq. 1). Drug concentrations follow monoexponential disposition: $C_p = C^o e^{(-kt)}$. C^o was set to 10, 100, or 1,000 units to achieve increasing dose levels. Parameter values were $k = 0.12$/h, $E_{max} = 100$ units, and $EC_{50} = 15$ units.

This equation, also known as the E_{max} model, describes the concentration–effect relationship in terms of a baseline effect or E_0 (if applicable), the maximum possible effect (E_{max}), and the drug concentration producing half maximal effect (EC_{50}). These parameters can be visualized easily from a plot of effect versus log-concentration where E_{max} is the plateau at relatively high concentrations and EC_{50} is the drug concentration associated with $E = 0.5 \times E_{max}$. Signature temporal profiles for simple direct effects for a compound with monoexponential disposition are shown in Fig. 2. The effect versus time curves appear saturated at high dose levels, decline linearly and in parallel over a range of doses, and the peak response time corresponds with the time of peak drug concentrations.

If a sufficient range of concentrations is not achieved, or cannot be obtained for safety reasons, the Hill equation can be reduced to simpler functions. For concentrations significantly less than the EC_{50}, C_p in the denominator of Eq. 1 is negligible, and drug effect is directly proportional to plasma drug concentrations:

$$E = E_o \pm S \times C_p, \tag{2}$$

with S as the slope of the relationship. When the effect is between 20 and 80% maximal, according to Eq. 1, the effect is directly proportional to the log of drug concentrations:

$$E = E_0 \pm m \log C_p, \tag{3}$$

with m as the slope of the relationship. These reduced functions are only valid within certain ranges of drug concentrations relative to drug potency, and hence cannot be extrapolated to identify the maximal pharmacodynamic effect of a compound.

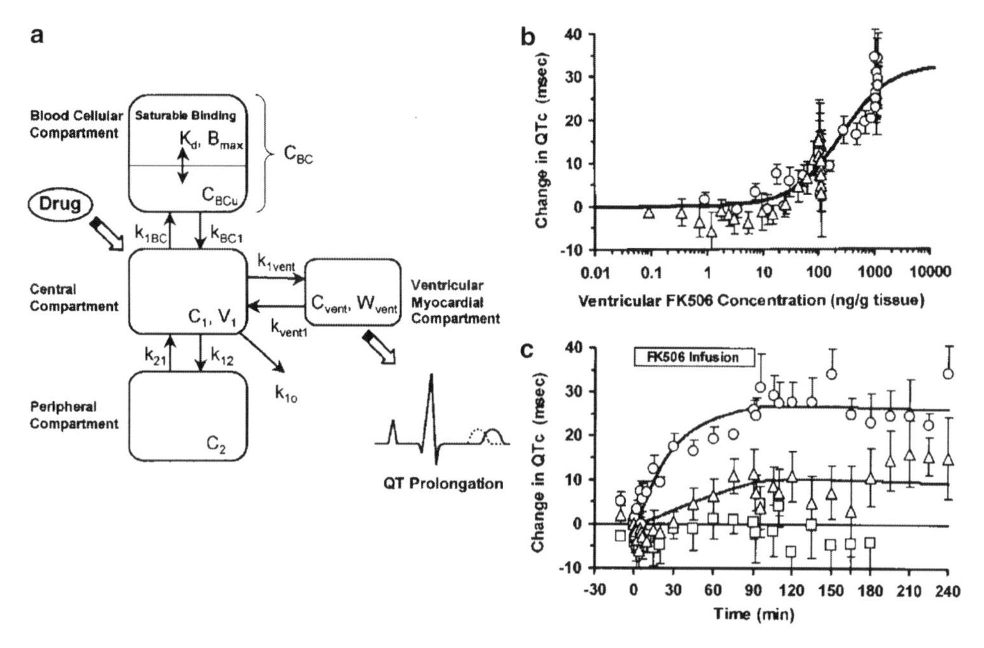

Fig. 3. Direct effect model of tacrolimus-induced changes of QTc intervals in guinea pigs. The pharmacokinetic model includes both plasma and ventricular myocardial drug concentrations (**a**), and the latter are associated with changes in QTc according to Eq. 4 (**b**). The PK/PD relationship results in the time-course of changes in QTc (**c**). Reprinted from ref. 8 with permission from Springer.

The full Hill equation, or sigmoid $E_{\max}$ model, incorporates a curve-fitting parameter, γ, which describes the steepness of the concentration–effect relationship:

$$E = E_0 \pm \frac{E_{max} \times C_{\mathrm{p}}^{\gamma}}{\mathrm{EC}_{50}^{\gamma} + C_{\mathrm{p}}^{\gamma}}. \tag{4}$$

Initial estimates for this parameter can be determined using the linear slope of the effect versus log-concentration plot:

$$m = \frac{E_{\max} \times \gamma}{4}. \tag{5}$$

As the Hill coefficient increases from 1 to 5, the concentration–effect relationship becomes less graded, and values of 5 tend to result in quantal or all-or-none types of effects. In contrast, values less than 1 produce very shallow slopes.

Simple direct effect models have been utilized to characterize the adverse effects of a number of drugs. Arrhythmias may occur as a side effect of cardiac and noncardiac therapies, and an increasing number of studies are conducted with QTc intervals as the toxicodynamic endpoint. QTc prolongation in response to citalopram (7) and tacrolimus (8) has been modeled using a simple $E_{\max}$ function (Fig. 3). The simple $E_{\max}$ model incorporating baseline measurements of the dynamic endpoints was also used to model the cardiovascular toxicity of cocaine administration (9). The model

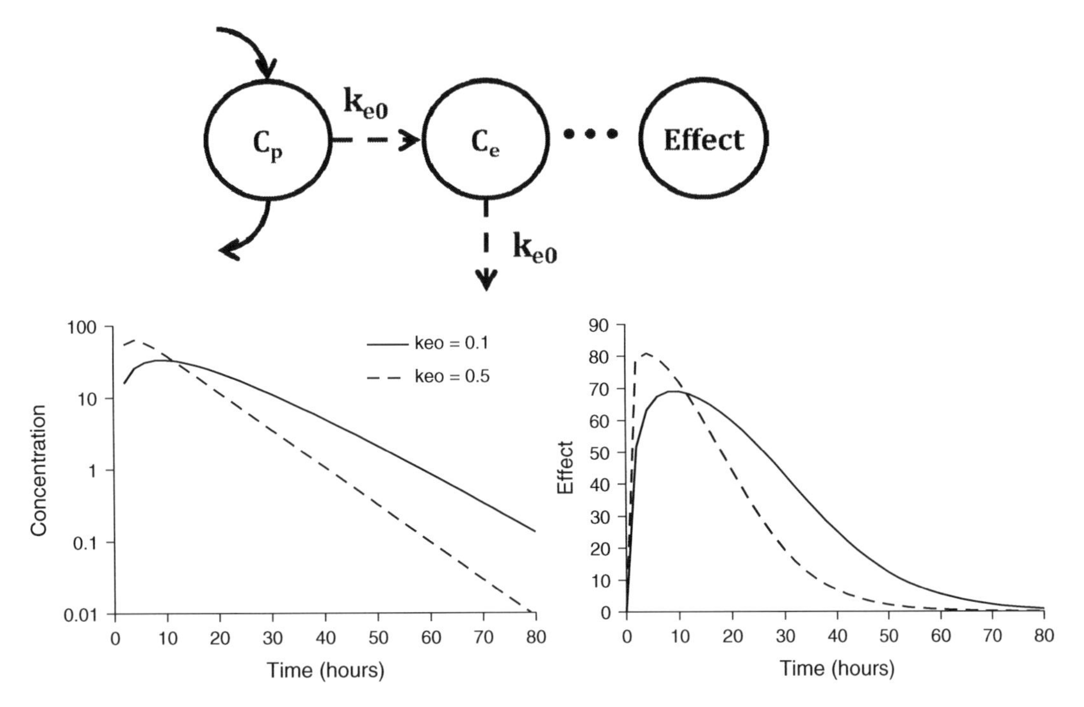

Fig. 4. Biophase model structure (*top panel*) and signature profiles for drug concentrations at the biophase (*left bottom panel*) and pharmacological effects (*right bottom panel*). Response curves were simulated using Eqs. 1 and 6 driven by drug concentrations following monoexponential disposition: $C_p = C^o e^{(-kt)}$. C^o was set to 100 units. Parameter values were $k = 0.12$/h, $k_{eo} = 0.1$ or 0.5/h, $E_{max} = 100$ units, and $EC_{50} = 15$ units.

reasonably described the effects of cocaine on multiple endpoints including heart rate and systolic and diastolic blood pressure. Both the E_{max} and sigmoid E_{max} models were evaluated for describing methemoglobin formation from dapsone metabolites (10); however, fitting criteria were not evaluated to select the best model.

3.2. Biophase Distribution

In many cases, the in vivo pharmacological effects will lag behind plasma drug concentrations. This results in the phenomenon of hysteresis, or a temporal disconnect in effect versus concentration plots. Distribution of drug to its site of action might represent a rate-limiting process that may account for the delay in drug effect. The term "biophase" was coined by Furchgott (11) to describe the drug site of action, and a mathematical approach to linking plasma concentrations and drug effect through a hypothetical effect compartment was popularized by Sheiner and colleagues (12) (Fig. 4, top panel). Plasma drug concentrations are described using an appropriate pharmacokinetic model, and the rate of change of drug concentrations at the biophase (C_e) is defined as

$$\frac{dC_e}{dt} = k_{eo} \times C_p - k_{eo} \times C_e, \qquad (6)$$

with k_{eo} as a first-order distribution rate constant. Although separate rate constants for production and loss were first proposed, they are often set as the same term (k_{eo}) for identifiability purposes. The amount of drug moving into and out of this compartment is assumed to be negligible, and therefore does not influence the pharmacokinetics of the drug. Biophase distribution is combined with Eq. 1 or 4, with C_e from Eq. 6 replacing C_p to drive the pharmacological effect. Figure 4 (bottom panels) illustrates the signature profile of the biophase model (i.e., biophase concentration and effect profiles) for a drug exhibiting monoexponential disposition. Peak drug effects are delayed relative to peak plasma concentrations; however, the time to peak effect is observed at the same time, independent of the dose level. The time to peak drug effect is related to k_{eo}, with smaller values resulting in later peak effects. Furthermore, for large dose levels, the slope of the decline of effect is linear and parallel between 20 and 80% of the maximum effect. Estimation of biophase model parameters can be done sequentially by fitting the pharmacokinetics and then fitting the biophase and pharmacodynamic parameters, or by simultaneously fitting all terms. The biophase model is only suitable for describing delayed responses due to drug distribution. As it was the first approach for describing such delayed drug responses, it has been commonly misapplied to describe systems in which the rate-limiting step is unrelated to drug distribution, resulting in poor fitting and/or unrealistic parameter values.

The biophase model was implemented for describing buprenorphine-induced respiratory depression in rats (13), and the clinical prediction of transient increases in blood pressure (14). Yassen and colleagues (13) utilized biophase distribution combined with a sigmoidal E_{max} model to characterize changes in ventilation following a range of dose levels of buprenorphine. In contrast, increases in blood pressure resulting from a drug in clinical development were described using the biophase model coupled with a more complex pharmacodynamic relationship incorporating changes from a blood pressure set point (14).

3.3. Indirect Response Models

Indirect response models represent a highly useful class of models wherein reversible drug–receptor interactions serve to alter the natural production or loss of biomarker response variables. A model reflecting inhibition of production was first utilized to characterize prothrombin activity in blood after oral warfarin administration (15). Dayneka and colleagues (16) were the first to formally propose four basic indirect response models whose structures are detailed in Fig. 5 (top panel). These models have been used to investigate the pharmacodynamics of a wide range of drug effects, and their mathematical properties have been well characterized (17, 18). The four basic models include inhibition of

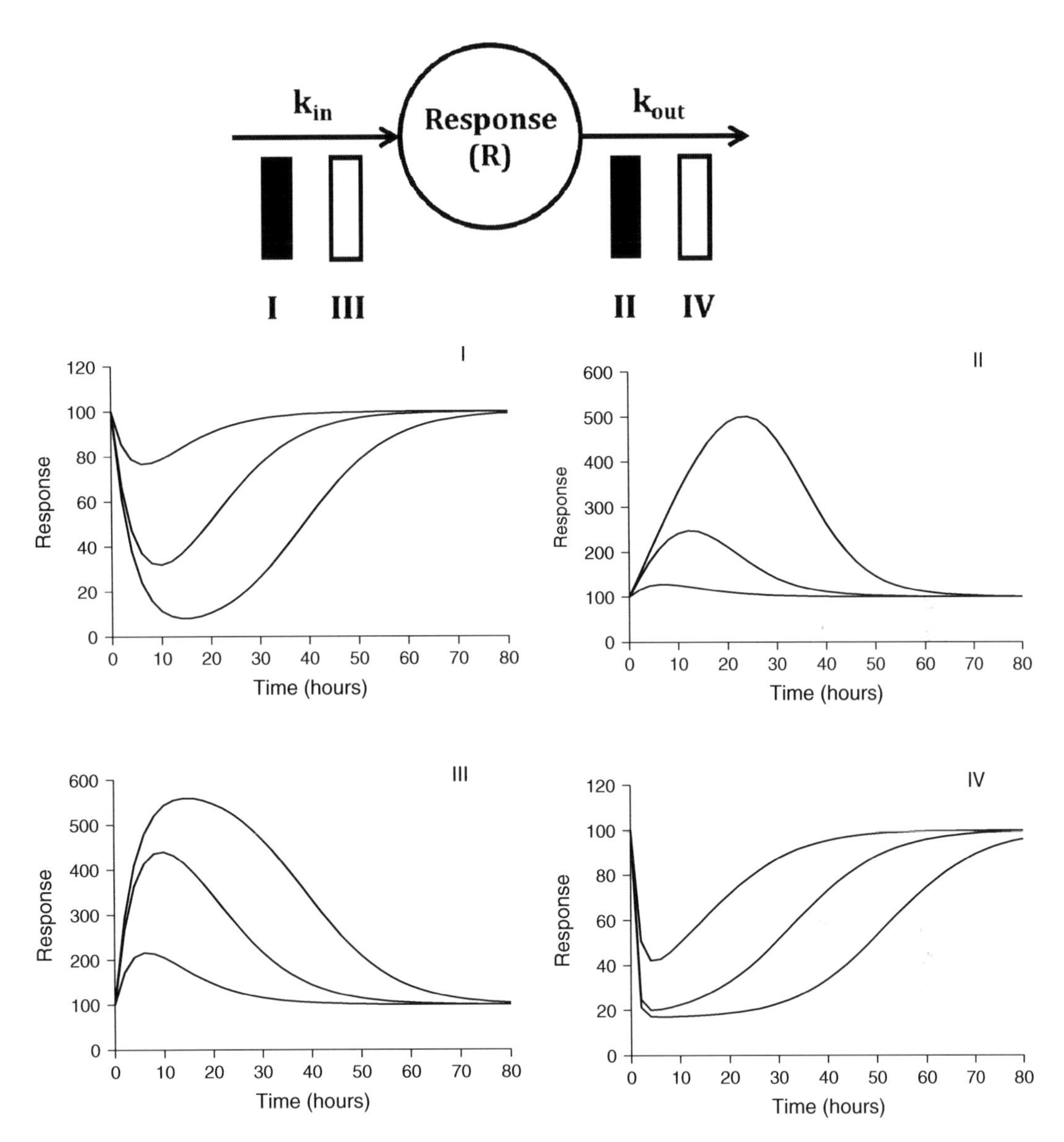

Fig. 5. Indirect response model structure (*top panel*) and signature profiles for the four basic indirect response models (*middle* and *bottom panels*). Response curves were simulated using Eqs. 7, 8, 9, and 10 driven by drug concentrations following monoexponential disposition: $C_p = C^0 e^{(-kt)}$. C^0 was set to 10, 100, or 1,000 units to achieve increasing doses. Parameter values were $k = 0.12$/h, $I_{max} = 1$ unit (Models I and II), $S_{max} = 10$ units (Models III and IV), $EC_{50} = 15$ units, $k_{out} = 0.25$/h, and $R^0 = 100$ units ($k_{in} = R^0 k_{out}$).

production (Model I) or dissipation (Model II) of response or stimulation of production (Model III) or dissipation of response (Model IV), and are defined by the following differential equations:

3.3.1. Model I

$$\frac{dR}{dt} = k_{in}\left(1 - \frac{I_{max} \times C_p}{IC_{50} + C_p}\right) - k_{out} \times R. \qquad (7)$$

3.3.2. Model II

$$\frac{dR}{dt} = k_{in} - k_{out}\left(1 - \frac{I_{max} \times C_p}{IC_{50} + C_p}\right)R. \tag{8}$$

3.3.3. Model III

$$\frac{dR}{dt} = k_{in}\left(1 + \frac{S_{max} \times C_p}{SC_{50} + C_p}\right) - k_{out} \times R. \tag{9}$$

3.3.4. Model IV

$$\frac{dR}{dt} = k_{in} - k_{out}\left(1 + \frac{S_{max} \times C_p}{SC_{50} + C_p}\right)R, \tag{10}$$

where k_{in} is a zero-order production rate constant, k_{out} is a first-order elimination rate constant, I_{max} and S_{max} are defined as the maximum fractional factors of inhibition ($0 < I_{max} \leq 1$) or stimulation ($S_{max} > 0$), and IC_{50} and SC_{50} are defined as the EC_{50}. Initial parameter estimates can be obtained from a graphical analysis of PK/PD data as previously described (17, 18). Signature profiles for these models in response to increasing dose levels are shown in Fig. 5 (middle and bottom panels). Interestingly, the time to peak responses are dose dependent, occurring at later times as the dose level is increased. This phenomenon is easily explained as the inhibition or stimulation effect will continue for larger doses, as drug remains above the EC_{50} for longer times. The initial condition for all models (R_0) is k_{in}/k_{out} which may be set constant or fitted as a parameter during model development. Ideally, a number of measurements should be obtained prior to drug administration to assess baseline conditions. Based on the determinants of R_0, typically the baseline and one of the turnover parameters are estimated, and the remaining rate constant is calculated as a function of the two estimated terms. This reduces the number of parameters to be estimated and maintains system stationarity.

The basic indirect response models can be extended to incorporate a precursor compartment (P). The following equations represent a general set of precursor-dependent indirect response models (Fig. 6, top panel) that were developed and characterized by Sharma and colleagues (19):

$$\frac{dP}{dt} = k_o\{1 \pm H_1(C_p)\} - (k_s + k_p\{1 \pm H_2(C_p)\})P, \tag{11}$$

$$\frac{dR}{dt} = k_p\{1 \pm H_2(C_p)\} \times P - k_{out} \times R, \tag{12}$$

where k_0 represents the zero-order rate constant for precursor production, k_p is a first-order rate constant for production of the response variable, and k_{out} is the first-order rate constant for dissipation of response. H_1 and H_2 represent the inhibition or stimulation of precursor production or production of response and are

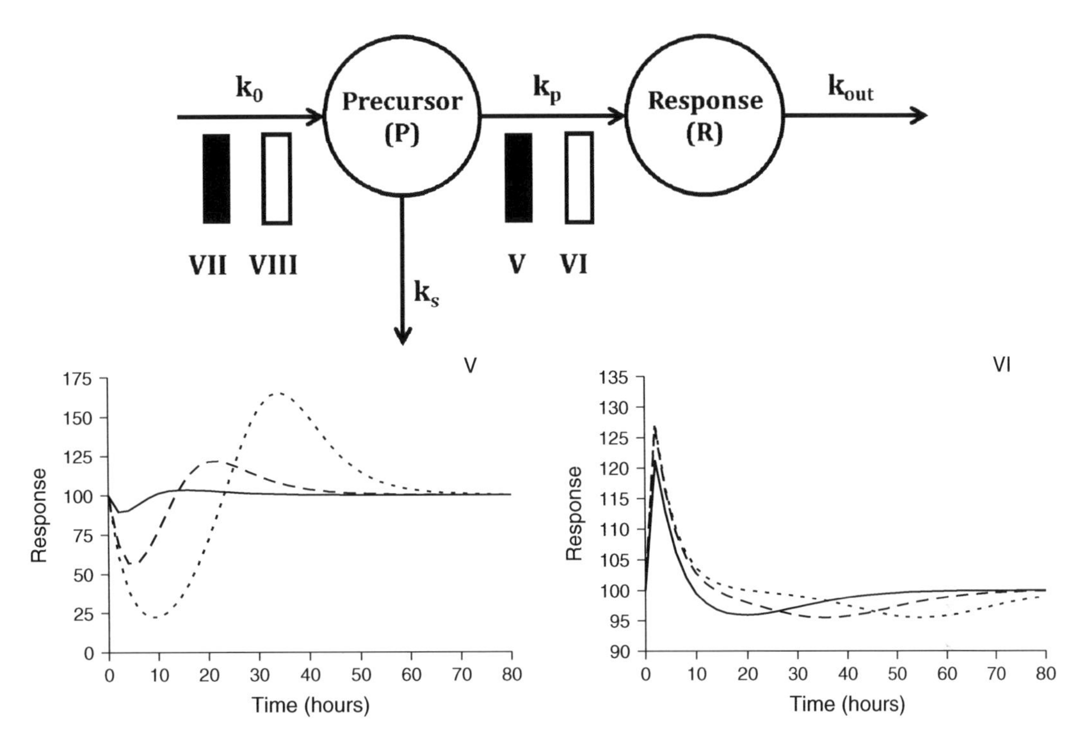

Fig. 6. Multiple compartment indirect response models (*top panel*) and signature profiles for Models V and VI (*bottom panel*). Response curves were simulated using Eqs. 11 and 12 driven by drug concentrations following monoexponential disposition: $C_p = C^0 e^{(-kt)}$. C^0 was set to 10, 100, or 1,000 units to achieve increasing doses. Parameter values were $k = 0.12$/h, $I_{max} = 1$ unit, $S_{max} = 10$ units, $EC_{50} = 15$ units, $k_0 = 25$ unit/h, $k_p = 0.5$/h, and $k_{out} = 0.25$/h.

analogous to the I_{max} and S_{max} functions presented in Eqs. 7 through 10. Stimulation or inhibition of k_p is more commonly observed than alterations in the production of precursor. The signature profiles for models V and VI are shown in Fig. 6 (bottom panels) and clearly demonstrate the rebound effect as drug washes out of the system. The data requirements for these models are similar to the basic indirect response models; however, sufficient data are needed to adequately capture baseline, maximum, and rebound effects, as well as the eventual gradual return to baseline conditions. Responses should be evaluated for two to three doses, with a sufficiently large dose to capture the maximum effect. The response measurements for the large dose should be used to determine initial parameter estimates followed by simultaneous fitting of all response data. Initial parameter estimates should be derived as previously described (19).

Indirect response models have been utilized to describe the pharmacodynamic effects of a wide range of compounds that alter the natural bioflux or turnover of endogenous substances or functions. A basic indirect response model for erythropoietin was extended to include multiple-compartments for describing the turnover of red blood cells and carboplatin-induced anemia (20).

This model nicely illustrates the development of a more complex model based on indirect mechanisms of drug action to simultaneously describe multiple in vivo processes.

3.4. Signal Transduction Models

Substantial time-delays in the observed pharmacodynamic response may result from multiple time-dependent steps occurring between drug–receptor binding and the ultimate pharmacological response. A transit compartment approach can be utilized to describe a lag between drug concentration and observed effects owing to time-dependent signal transduction (21, 22). Assuming rapid receptor binding, the following differential equation describes the rate of change of the initial transit compartment (M_1):

$$\frac{\mathrm{d}M_1}{\mathrm{d}t} = \frac{1}{\tau}\left(\frac{E_{max} \times C_{\mathrm{p}}}{\mathrm{EC}_{50} + C_{\mathrm{p}}} - M_1\right), \tag{13}$$

wherein the $E_{\max}$ model describes the drug–receptor interaction, and τ is the mean transit time through this compartment. Subsequent transit compartments may be added, and a general equation for the ith compartment can be defined as

$$\frac{\mathrm{d}M_i}{\mathrm{d}t} = \frac{1}{\tau}(M_{i-1} - M_i). \tag{14}$$

Later compartments will show a clear delay in the onset of response as well as substantial delays in achieving the maximum effect. Model development for signal transduction systems typically includes evaluating varied numbers of transit compartments and values for τ to determine the combination that best describes the data.

Chemotherapy-induced myelosuppression represents a classic example of the use of a transit compartment modeling to describe this adverse reaction to numerous chemotherapeutic agents (Fig. 7, top panel). The structural model was proposed by Friberg and colleagues (23) to describe myelosuppression induced by irinotecan, vinflunine (Fig. 7, bottom panel), and 2′-deoxy-2′-methylidenecytidine for a range of dose levels and various dosing regimens. This same structural model has been used to describe indisulam-induced myelosuppression (24), as well as the drug–drug interactions between indisulam and capecitabine (25), and pemetrexed and BI2536 (26).

3.5. Irreversible Effect Models

A wide range of compounds, including anticancer drugs, antimicrobial drugs, and enzyme inhibitors, elicit irreversible effects. A basic model for describing irreversible effects was developed by Jusko and includes simple cell killing (27):

$$\frac{\mathrm{d}R}{\mathrm{d}t} = -k \times C \times R, \tag{15}$$

where R represents cells or receptors, C is either C_{p} or C_{e}, and k is a second-order cell-kill rate constant. The initial condition for this

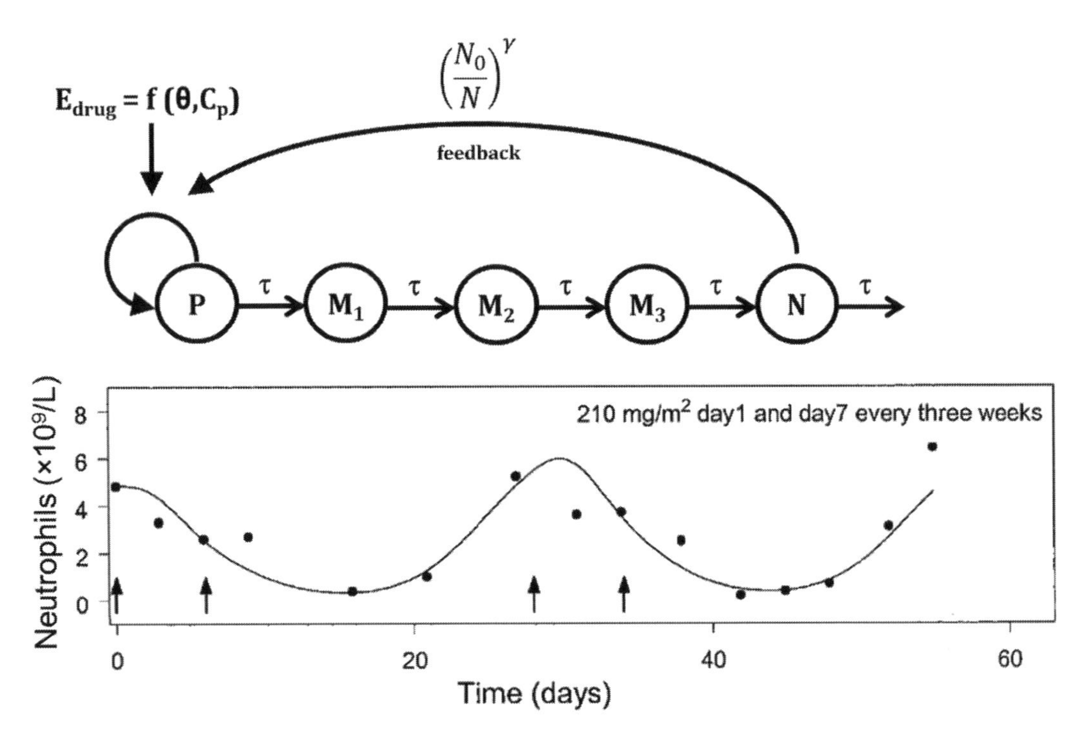

Fig. 7. Transit-compartment model of myelosuppression (*top panel*) including a proliferating progenitor pool (*P*), three transit compartments (M_i), and a plasma neutrophil compartment (*N*). Drug effect is driven by plasma drug concentration (C_p) and pharmacodynamic parameters (θ). An adaptive feedback function on the proliferation rate constant is governed by the ratio of initial neutrophils to current neutrophil count, raised to a power coefficient (γ). The time-course of neutrophils following vinflunine administration (*arrows* in *bottom panel*). Reprinted from ref. 23 with permission from the American Society of Clinical Oncology.

equation is the initial number of cells present within the system (R_0) often represented as a survival fraction. This approach is only applicable for non-proliferating cell populations, but may be extended to incorporate cell growth (27):

$$\frac{\mathrm{d}R}{\mathrm{d}t} = k_s \times R - k \times C \times R, \tag{16}$$

with k_s as an apparent first-order growth rate for proliferating cell populations, such as malignant cells or bacteria. This growth rate constant represents the net combination of natural growth and degradation of the cellular population, and its initial estimate can be determined from a control- or nondrug-treated cell population. The model diagram and corresponding signature profiles are shown in Fig. 8. The initial slope of the log survival fraction versus time curve out to time, t, and the plasma drug $AUC_{(0-t)}$ can be used to obtain an initial estimate for k ($k = -\ln S_{Ft}/AUC_{(0-t)}$), and the initial condition for Eq. 16 is the total cell population at time zero. In contrast to simple cell killing, the effect–time profiles are characterized by an initial cell kill phase, followed by an exponential growth phase, once drug concentrations are below an effective

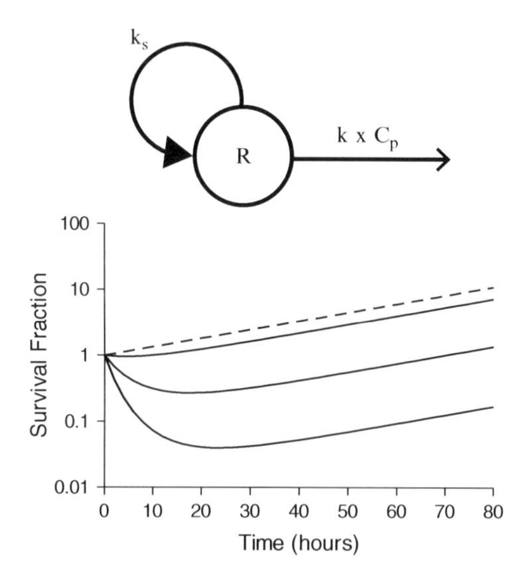

Fig. 8. Structural model for irreversible effects (*top panel*) and signature profiles for irreversible effect model with a proliferating cell population (*bottom panel*). Response curves were simulated using Eq. 16 driven by drug concentrations following monoexponential disposition: $C_p = C^o e^{(-kt)}$. C^o was set to 0, 10, 100, or 1,000 units to achieve a control population and increasing dose levels. Pharmacokinetic parameter was $k = 0.12$/h. Pharmacodynamic parameters were $k = 0.0005$ units/h, $k_s = 0.03$/h.

concentration (Fig. 8, bottom panel). Clearly the control group is needed to properly characterize the exponential growth rate constant in the untreated cell population.

The irreversible effect model can also be adapted to include the turnover or production and loss of a biomarker:

$$\frac{dR}{dt} = k_{in} - k_{out} R - k \times C \times R. \tag{17}$$

The initial condition for this model is the same as basic indirect responses or k_{in}/k_{out}. The signature profiles for this model are similar to the profiles for indirect response models I and IV (Fig. 5). It is important to understand the mechanism of action of the response that you are evaluating in order to determine which model should be utilized.

Irreversible effect models are commonly used to describe the cell killing action of chemotherapeutic agents and anti-infectives. This model was also applied to evaluate the formation of methemoglobin following the administration of a range of antimalarial agents (28). The final model characterized methemoglobin production resulting from the formation of an active drug metabolite.

3.6. More Complex Models

The main focus of this chapter was the introduction of commonly used mechanistic pharmacodynamic models that can be readily applied to toxicokinetics and dynamics. However, a number of mechanistic processes may be required to adequately describe the drug–system interactions under investigation. Slow receptor binding, tolerance phenomenon, drug interactions, opposing drug effects, and disease progression may add additional complexities to the analysis of toxicodynamic data. For example, Houze and colleagues (29) evaluated paraoxon-induced respiratory toxicity and its reversal with pralidoxime (PRX) administration in rats via the combination of multiple pharmacodynamic modeling components. Initially, the time-course of paraoxon inactivation of in vitro whole blood cholinesterase (WBChE) was modeled based on enzyme inactivation:

$$\frac{\mathrm{d}E_{\mathrm{A}}}{\mathrm{d}t} = -\left(\frac{kC_{\mathrm{PO}}}{\mathrm{EC}_{50,\mathrm{PO}} + C_{\mathrm{PO}}}\right)E_{\mathrm{A}} + k_{\mathrm{r}}E_{\mathrm{I}}, \quad (18)$$

where E_{A} is active enzyme, k is the maximal rate constant of enzyme inactivation, C_{PO} is paraoxon concentration, $\mathrm{EC}_{50,\mathrm{PO}}$ is the concentration of paraoxon that produces 50% of k, k_{r} is a first-order reactivation rate constant, and E_{I} is the inactive enzyme pool. The rate of change of the inactive enzyme (E_{I}) was defined as

$$\frac{\mathrm{d}E_{\mathrm{I}}}{\mathrm{d}t} = \left(\frac{kC_{\mathrm{PO}}}{\mathrm{EC}_{50,\mathrm{PO}} + C_{\mathrm{PO}}}\right)E_{A} - \left(k_{r} + k_{\mathrm{age}}\right)E_{\mathrm{I}}, \quad (19)$$

where k_{age} is a first-order rate constant of aging of inactive enzyme. The reactivation of this in vitro system by PRX was modeled as an indirect response, and Eqs. 18 and 19 were updated accordingly:

$$\frac{\mathrm{d}E_{\mathrm{A}}}{\mathrm{d}t} = -\left(\frac{kC_{\mathrm{PO}}}{\mathrm{EC}_{50,\mathrm{PO}} + C_{\mathrm{PO}}}\right)E_{\mathrm{A}} + k_{\mathrm{r}}\left(1 + \frac{E_{\max}C_{\mathrm{PRX}}^{h}}{\mathrm{EC}_{50,\mathrm{PRX}}^{h} + C_{\mathrm{PRX}}^{h}}\right)E_{\mathrm{I}} \quad (20)$$

$$\frac{\mathrm{d}E_{\mathrm{I}}}{\mathrm{d}t} = \left(\frac{kC_{\mathrm{PO}}}{\mathrm{EC}_{50,\mathrm{PO}} + C_{\mathrm{PO}}}\right)E_{\mathrm{A}} - k_{\mathrm{r}}\left(1 + \frac{E_{\max}C_{\mathrm{PRX}}^{h}}{\mathrm{EC}_{50,\mathrm{PRX}}^{h} + C_{\mathrm{PRX}}^{h}}\right)E_{\mathrm{I}} - \left(k_{\mathrm{age}}E_{\mathrm{I}}\right). \quad (21)$$

Interestingly, the estimated potency of PRX was in agreement with an empirical literature estimate. For the in vivo dynamics, a fixed pharmacokinetic function for PRX was introduced, and an empirical function was used to describe paraoxon-induced enzyme inactivation, as plasma concentrations were unavailable. The estimated parameters from the in vitro analysis were fixed (not identifiable from in vivo data only), and the toxicodynamic biomarker, expiratory time (T_{E}), was linked to apparent active enzyme (E_{A}) according to the following nonlinear transfer function:

$$T_E = T_E^0 + \frac{E_{\max,T_E}\left(\frac{E_0}{E_A} - 1\right)^n}{E_{50}^n + \left(\frac{E_0}{E_A} - 1\right)^n}, \quad (22)$$

where T_E^0 is the baseline expiratory time, $E_{\max,TE}$ is the maximal increase in T_E, E_0 is the baseline active enzyme (1 or 100%), E_{50} is the corrected enzyme ratio resulting in 50% of $E_{\max,TE}$, and n is a sigmoidicity coefficient. Expiratory profiles and the transient antidotal effect of PRX were described well, and this analysis highlights the integration of several basic modeling approaches described in this chapter. Further, the coupling of in vitro enzyme and in vivo toxicodynamic data demonstrates the versatility and multi-scale nature of the model.

An additional theoretical example of mechanism-based analysis of drug interactions was presented by Earp and colleagues (30), who examined drug interactions utilizing indirect response models. These more complex models typically consider multiple pharmacodynamic endpoints which require individual data sets and stepwise analysis for each endpoint. A corticosteroid model which considers mRNA dynamics of the glucocorticoid receptor and hepatic tyrosine aminotransferase mRNA and activity is an example of simultaneously characterizing multiple pharmacodynamic endpoints using an integration of basic modeling components (31).

The majority of mechanism-based pharmacodynamic models describe continuous physiological response variables. However, models are available for evaluating noncontinuous outcomes, such as the probability of a specific event occurring. Such responses are often more clinically relevant, and more research is needed to combine continuous mechanistic PK/PD models with clinical outcomes data. One example is the prediction of enoxaparin-induced bleeding events in patients undergoing various therapeutic dosing regimens (32). A population proportional-odds model was developed to predict the severity of bleeding event on an ordinal scale of 1–3 (32).

4. Prospectus

The future of mechanism-based pharmacodynamic modeling for both therapeutic and adverse drug responses is promising for model-based drug development and therapeutics, and many of the basic modeling concepts in this chapter will likely continue to represent key building components in more complex systems models. A diverse array of models is available with a minimal number of identifiable parameters to mimic mechanisms and the time-course of therapeutic and adverse drug effects. However, new methodologies will be needed to evolve these models further into translational platforms and prospectively predictive models of drug efficacy and safety. Network-based systems pharmacology models have shown utility for understanding drug-induced adverse events (1). Further research is needed to identify practical techniques for bridging systems pharmacology and in vivo PK/PD models to anticipate the clinical utility of new chemical entities from first principles.

Acknowledgments

The authors thank Dr. William J. Jusko (University at Buffalo, SUNY) for reviewing this chapter and providing insightful feedback. This work was supported by Grant No. GM57980 from the National Institutes of General Medicine, Grant No. DA023223 from the National Institute on Drug Abuse, and Hoffmann-La Roche Inc.

References

1. Berger SI, Iyengar R (2011) Role of systems pharmacology in understanding drug adverse events. Wiley Interdiscip Rev Syst Biol Med 3:129–135
2. Levy G (1964) Relationship between elimination rate of drugs and rate of decline of their pharmacologic effects. J Pharm Sci 53:342–343
3. Levy G (1966) Kinetics of pharmacologic effects. Clin Pharmacol Ther 7:362–372
4. Mager DE, Wyska E, Jusko WJ (2003) Diversity of mechanism-based pharmacodynamic models. Drug Metab Dispos 31:510–518
5. Yates FE (1975) On the mathematical modeling of biological systems: a qualified 'pro'. In: Vernberg FJ (ed) Physiological adaptation to the environment. Intext Educational Publishers, New York
6. Wagner JG (1968) Kinetics of pharmacologic response. I. Proposed relationships between response and drug concentration in the intact animal and man. J Theor Biol 20:173–201
7. Friberg LE, Isbister GK, Hackett LP, Duffull SB (2005) The population pharmacokinetics of citalopram after deliberate self-poisoning: a Bayesian approach. J Pharmacokinet Pharmacodyn 32:571–605
8. Minematsu T, Ohtani H, Yamada Y, Sawada Y, Sato H, Iga T (2001) Quantitative relationship between myocardial concentration of tacrolimus and QT prolongation in guinea pigs: pharmacokinetic/pharmacodynamic model incorporating a site of adverse effect. J Pharmacokinet Pharmacodyn 28:533–554
9. Laizure SC, Parker RB (2009) Pharmacodynamic evaluation of the cardiovascular effects after the coadministration of cocaine and ethanol. Drug Metab Dispos 37:310–314
10. Vage C, Saab N, Woster PM, Svensson CK (1994) Dapsone-induced hematologic toxicity: comparison of the methemoglobin-forming ability of hydroxylamine metabolites of dapsone in rat and human blood. Toxicol Appl Pharmacol 129:309–316
11. Furchgott RF (1955) The pharmacology of vascular smooth muscle. Pharmacol Rev 7:183–265
12. Sheiner LB, Stanski DR, Vozeh S, Miller RD, Ham J (1979) Simultaneous modeling of pharmacokinetics and pharmacodynamics: application to d-tubocurarine. Clin Pharmacol Ther 25:358–371
13. Yassen A, Kan J, Olofsen E, Suidgeest E, Dahan A, Danhof M (2007) Pharmacokinetic-pharmacodynamic modeling of the respiratory depressant effect of norbuprenorphine in rats. J Pharmacol Exp Ther 321:598–607
14. Stroh M, Addy C, Wu Y, Stoch SA, Pourkavoos N, Groff M, Xu Y, Wagner J, Gottesdiener K,

Shadle C, Wang H, Manser K, Winchell GA, Stone JA (2009) Model-based decision making in early clinical development: minimizing the impact of a blood pressure adverse event. AAPS J 11:99–108

15. Nagashima R, O'Reilly RA, Levy G (1969) Kinetics of pharmacologic effects in man: the anticoagulant action of warfarin. Clin Pharmacol Ther 10:22–35
16. Dayneka NL, Garg V, Jusko WJ (1993) Comparison of four basic models of indirect pharmacodynamic responses. J Pharmacokinet Biopharm 21:457–478
17. Jusko WJ, Ko HC (1994) Physiologic indirect response models characterize diverse types of pharmacodynamic effects. Clin Pharmacol Ther 56:406–419
18. Sharma A, Jusko WJ (1998) Characteristics of indirect pharmacodynamic models and applications to clinical drug responses. Br J Clin Pharmacol 45:229–239
19. Sharma A, Ebling WF, Jusko WJ (1998) Precursor-dependent indirect pharmacodynamic response model for tolerance and rebound phenomena. J Pharm Sci 87:1577–1584
20. Woo S, Krzyzanski W, Jusko WJ (2008) Pharmacodynamic model for chemotherapy-induced anemia in rats. Cancer Chemother Pharmacol 62:123–133
21. Sun YN, Jusko WJ (1998) Transit compartments versus gamma distribution function to model signal transduction processes in pharmacodynamics. J Pharm Sci 87:732–737
22. Mager DE, Jusko WJ (2001) Pharmacodynamic modeling of time-dependent transduction systems. Clin Pharmacol Ther 70:210–216
23. Friberg LE, Henningsson A, Maas H, Nguyen L, Karlsson MO (2002) Model of chemotherapy-induced myelosuppression with parameter consistency across drugs. J Clin Oncol 20:4713–4721
24. Zandvliet AS, Schellens JH, Copalu W, Beijnen JH, Huitema AD (2009) Covariate-based dose individualization of the cytotoxic drug indisulam to reduce the risk of severe myelosuppression. J Pharmacokinet Pharmacodyn 36:39–62
25. Zandvliet AS, Siegel-Lakhai WS, Beijnen JH, Copalu W, Etienne-Grimaldi MC, Milano G, Schellens JH, Huitema AD (2008) PK/PD model of indisulam and capecitabine: interaction causes excessive myelosuppression. Clin Pharmacol Ther 83:829–839
26. Soto E, Staab A, Freiwald M, Munzert G, Fritsch H, Doge C, Troconiz IF (2010) Prediction of neutropenia-related effects of a new combination therapy with the anticancer drugs BI 2536 (a Plk1 inhibitor) and pemetrexed. Clin Pharmacol Ther 88:660–667
27. Jusko WJ (1971) Pharmacodynamics of chemotherapeutic effects: dose-time-response relationships for phase-nonspecific agents. J Pharm Sci 60:892–895
28. Fasanmade AA, Jusko WJ (1995) An improved pharmacodynamic model for formation of methemoglobin by antimalarial drugs. Drug Metab Dispos 23:573–576
29. Houze P, Mager DE, Risede P, Baud FJ (2010) Pharmacokinetics and toxicodynamics of pralidoxime effects on paraoxon-induced respiratory toxicity. Toxicol Sci 116:660–672
30. Earp J, Krzyzanski W, Chakraborty A, Zamacona MK, Jusko WJ (2004) Assessment of drug interactions relevant to pharmacodynamic indirect response models. J Pharmacokinet Pharmacodyn 31:345–380
31. Hazra A, Pyszczynski N, DuBois DC, Almon RR, Jusko WJ (2007) Modeling receptor/gene-mediated effects of corticosteroids on hepatic tyrosine aminotransferase dynamics in rats: dual regulation by endogenous and exogenous corticosteroids. J Pharmacokinet Pharmacodyn 34:643–667
32. Barras MA, Duffull SB, Atherton JJ, Green B (2009) Modelling the occurrence and severity of enoxaparin-induced bleeding and bruising events. Br J Clin Pharmacol 68:700–711
33. Jusko WJ, Ko HC, Ebling WF (1995) Convergence of direct and indirect pharmacodynamic response models. J Pharmacokinet Biopharm 23:5–8

Index

A

B

Brad Reisfeld and Arthur N. Mayeno (eds.), *Computational Toxicology: Volume I*, Methods in Molecular Biology, vol. 929, DOI 10.1007/978-1-62703-050-2, © Springer Science+Business Media, LLC 2012

C

G

H

I

J

K

L

M

N

O

P

Q

R

S

V

W

X

Printed by Publishers' Graphics LLC
MO20121002.10.04.197